ELECTRONICS MATH

ELECTRONICS MATH

FOURTH EDITION

Bill Deem

PRENTICE HALL
Englewood Cliffs, New Jersey 07632

Library of Congress Cataloging-in-Publication Data

Deem, Bill R.
Electronics math / Bill Deem. – 4th ed.
p. cm.
Includes index.
ISBN 0-13-251299-8. — ISBN 0-13-252024-9
1. Electronics–Mathematics. I. Title.
TK7835.D36 1992
512'.1'0246213–dc20 92–16403
CIP

Production Editor: Adele Kupchik
Manufacturing Buyers: Ilene Levy and Ed O'Dougherty
Copy Editor: Bill Thomas
Editorial Assistant: Cathy Frank
Cover Photo: Paul C. Katz/The Image Bank
Cover Design: Laura Ierardi

A Simon & Schuster Company
Englewood Cliffs, New Jersey 07632

Printed in the United States of America
10 9 8 7 6 5 4 3

ISBN 0-13-251299-8
ISBN 0-13-252024-9 (Instructor's Examination Copy)

Prentice-Hall International (UK) Limited, *London*
Prentice-Hall of Australia Pty. Limited, *Sydney*
Prentice-Hall Canada Inc., *Toronto*
Prentice-Hall Hispanoamericana, S.A., *Mexico*
Prentice-Hall of India Private Limited, *New Delhi*
Prentice-Hall of Japan, Inc., *Tokyo*
Simon & Schuster Asia Pte. Ltd., *Singapore*
Editora Prentice-Hall do Brasil, Ltda., *Rio de Janeiro*

CONTENTS

CHAPTER 3
UNITS AND PREFIXES 64

CHAPTER 4
COMPUTER NUMBER SYSTEMS 88

CHAPTER 5
ALGEBRAIC TERMS: ROOTS AND POWERS 133

CHAPTER 6
FRACTIONS 153

CHAPTER 7
FRACTIONS AND LITERAL NUMBERS 193

CHAPTER 11 DC CIRCUIT ANALYSIS: KIRCHHOFF'S LAWS 278

CHAPTER 12 DC CIRCUIT ANALYSIS: OHM'S LAW 300

CHAPTER 13 DC CIRCUIT ANALYSIS: CIRCUIT THEOREMS 329

CHAPTER 14 GRAPHING 365

CHAPTER 15
SIMULTANEOUS LINEAR EQUATIONS 394

CHAPTER 16
COMPLEX NUMBERS 426

CHAPTER 17
THE RIGHT TRIANGLE 440

CHAPTER 18
TRIGONOMETRIC IDENTITIES 465

CHAPTER 22
FILTERS 559

CHAPTER 23
LOGARITHMS 574

CHAPTER 24
LOGARITHMIC EQUATIONS 594

CHAPTER 25
APPLICATIONS OF LOGARITHMS 607

PREFACE

The fourth edition of *Electronics Math* has been written to provide a thorough, complete, and practical study of electronics math and its relationship to the world of electronics.

In the fourth edition, we have added summaries at the end of chapters, called "Chapter at a Glance." Key points, rules, and equations are emphasized when first introduced. Additional examples, practice problems, and end-of-chapter problems have been added in areas where it was felt that they were needed.

In this book we have placed greater emphasis on certain areas of the discipline than does abstract math. For example, binary numbers are used in computers. Accordingly, binary notation and binary operations are developed along with decimal notation and operations. Also, because technicians utilize mathematical techniques that start and end with concrete numbers, concrete numbers are explained to the extent required by the beginning student.

There are no course prerequisites for this text. It is intended to be used as a separate text in electronics math or as a text that could be used as a reference throughout the study of electronics.

Because the calculator has become an integral part of all technical students' classroom tools, the use of the calculator in problem solving is introduced in the text as the need arises. Algorithms are presented when appropriate.

The text includes example problems, practice problems, self-tests, and end-of-chapter problems. There are over 300 examples to reinforce new concepts. There are over 1300 practice problems within chapters to ensure understanding. There are over 2600 end-of-chapter problems to test comprehension.

Chapter practice problems are directly related to end-of-chapter problems. For example, Practice Problems 12–3 and End-of-Chapter Problems 12–3 cover the same topic.

Self-tests are provided at the end of topics to help students determine whether or not they have mastered the topic material. Also, if the text is being used for self-study, the self-tests can be used to determine whether or not the student already possesses that skill.

DEVELOPMENT AND CLASS TESTING

Electronics Math has been used in several hundred classrooms over the last three editions. The fourth edition was developed and has been class tested over the last three years. To ensure accuracy, we have had technical reviewers work every example, practice problem, self-test, and end-of-chapter problem. In addition, each supplement has been technically checked by additional technical reviewers.

CHAPTER ORGANIZATION

Chapters 1, 2, and 3 deal with decimal numbers, powers of ten, and prefixes. These chapters introduce the student to the calculator and to problem solving involving electrical units.

Chapter 4 covers the various number systems (binary, octal, and hexadecimal) that are used in the study of computers. Conversions between the number systems and addition and subtraction in the various systems are covered.

Chapter 5 introduces the student to literal numbers, exponents, and radicals. Typical electrical problems involving literal numbers, squares, and square roots are solved using the calculator.

Chapters 6 and 7 are a review of addition, subtraction, multiplication, and division of fractions. Chapter 6 deals only with real numbers. Chapter 7 includes both real and literal numbers. Conversions between mixed numbers and decimal fractions and finding common denominators are covered here.

Chapters 8, 9, and 10 contain topics in algebra, including linear equations, second-degree equations, fractional equations, and factoring. Throughout these chapters, applications of problem solving using the calculator are presented.

In Chapters 11, 12, and 13, students utilize the skills they have developed by solving dc circuit problems using Kirchhoff's laws, Ohm's law, and various circuit theorems.

Graphical and algebraic solutions to circuit problems and linear equations are presented in Chapters 14 and 15. Practical applications are presented for each technique discussed.

Chapters 16 through 19 introduce those elements of algebra and trigonometry necessary for the solution of ac circuit problems. Angular velocity and the sine wave are introduced in these chapters. Problem solving using trigonometric functions and the calculator are presented here.

In Chapters 20, 21, and 22, ac series, parallel, and complex circuits are solved. In Chapter 20 the student learns how to express phasors in either polar or rectangular form. In Chapter 21, circuit theorems are again presented as an aid in solving complex circuit problems. These problem-solving techniques are used in Chapter 22 in determining the parameters that exist in different kinds of filter circuits. Chapters 23, 24, and 25 cover both common and natural logarithms and their applications. Logarithmic equations are covered in Chapter 24. Applications including the Bode plot are found in Chapter 25.

Chapter 26 discusses the basic logic functions inherent in all logic circuits and presents those theorems, laws, and postulates used in the simplification of logic expressions. Chapter 27, Karnaugh Maps, offers an alternative method of logic circuit simplification.

EXTENSIVE SUPPLEMENTS PACKAGE

This edition of *Electronics Math* comes with a wide variety of optimal supplements for both the instructor and student.

- A **Student Study Guide with Selected Solutions** contains chapter overviews and additional study questions for each chapter. It includes fully worked out solutions to selected end-of-chapter problems. It was authored by Wallace McIntyre, DeVry Institute of Technology, Chicago.
- An **Instructor's Solutions Manual with Transparency Masters** contains fully worked out solutions to end-of-chapter problems and provides the instructor with over 120 illustrations to use. It was developed by Bill Deem and John Deem.
- An **Instructor's Test Item File** contains over 750 additional test questions. It is also available on the IBM format diskette along with the Prentice Hall Test Manager. It was developed by Fred Kerr, DeVry Institute of Technology, Atlanta.

ACKNOWLEDGMENTS

The author wishes to acknowledge the assistance given him by the editorial and production staffs of Prentice Hall and the many students and teachers who aided and assisted him in the preparation of this edition. In particular, a special thanks to Barbara Snyder, who is always willing to read and critique the manuscript and who made many contributions and suggestions that were incorporated into this edition. Also, I wish to acknowledge the contribution of time and effort on the part of Pat Short for her help with the *Instructor's Solutions Manual* and my son John for his technical assistance in verifying answers to problems and in editing the *Instructor's Solutions Manual*. Finally, a great big thank you to my technical reviewers: Robert Arndt, Belleville Area College, Belleville, Illinois; Dennis Burden, ITT Technical Institute, Evansville, Indiana; Vrej Khachikian, DeVry Institute of Technology, Los Angeles, California; Wesley Johnstone, Okaloosa-Walton Community College, Niceville, Florida; Larry Moore, Midlands Technical College, Harbison Campus, Columbia, South Carolina; and Saeed Shaikh, Miami-Dade Community College, Miami, Florida.

ELECTRONICS MATH

THE DECIMAL NUMBER SYSTEM

CHAPTER 1

INTRODUCTION

The decimal number system is the number system that we have always used. We all grew up using this system for all computations. Many of the concepts we will discuss in this chapter we have applied for years without really thinking about them. We are going to spend some time discussing the decimal numbering system in order to help us better understand working with numbers and to pave the way for a quicker understanding of other number systems. These other number systems, such as the *binary number system* and the *hexadecimal number system,* are important number systems used in the computer world and will be discussed in Chapter 4.

The chapter is primarily a review of basic arithmetic. We will review the process of adding, subtracting, multiplying, and dividing mathematical terms and expressions. In this chapter we will convert between decimal numbers and decimal fractions, and we will learn how to round numbers. Rounding numbers is a common practice in electronics because we seldom need accuracies beyond three decimal digits. The concepts learned or relearned in this chapter are used throughout the text.

CHAPTER OBJECTIVES

In this chapter you will:

1. Develop an understanding of the decimal number system and the concept of place value.
2. Learn how to read whole numbers and decimal fractions.
3. Learn how to convert from decimal fractions to decimal numbers.
4. Learn the rules for rounding both whole and nonwhole numbers.
5. Learn how to work with signed numbers.
6. Learn how to add, subtract, multiply, and divide decimal numbers.
7. Learn how to recognize and simplify various mathematical expressions and terms.

1-1 DECIMAL NUMBER SYSTEM

Our number system is called the decimal number system. *Decimal* means ten. In the decimal number system there are ten symbols. These symbols are called *digits*. The ten digits are 0, 1, 2, 3, 4, 5, 6, 7, 8, 9. Zero is the digit having the least value. In counting, when the count reaches 9, a limit has been reached because 9 is the digit of greatest value. An additional count produces a *carry*. This carry is equal to 10 and is said to occupy the tens position; the 0 occupies the units position. This process continues each time 9 is reached in the units position. When a 9 appears in the tens position, the next carry to the tens position produces a carry to the hundreds position.

It is important to note that a 1 in the tens position possesses a value (*weight*) 10 times that of a 1 in the units position. The same is true for each position of greater weight. Because of this relationship, the decimal numbering system is called a *place value system* and its base is 10.

■ **KEY POINT** Significant digits, also called significant *figures,* are all the digits 1 through 9. Zero is also significant *except when in the leftmost position*. For example,

178	is correct to three significant digits
178.0	is correct to four significant digits
0.0178	is correct to three significant digits (zeros in the leftmost positions are not significant)

■ **KEY POINT** The most significant digit (MSD) is the leftmost nonzero digit. The least significant digit (LSD) is the right most digit.

Let's examine the number 687. The least significant digit (LSD) is 7. It occupies the units position and its value is 7. The next significant digit is 8. It occupies the 10's position and its value is 80 (8 × 10). The most significant digit (MSD) is 6. It occupies the hundreds position and its value is 600 (6 × 100).

Table 1-1 lists the names for places up to millions.

How do we treat zero when it is a digit? Consider the number 2073. We can see that there are 2 thousands, 0 hundreds, 7 tens, and 3 units. The zero is in the hundreds position and simply tells us how many hundreds there are. It does the same job as any other digit and is just as important. We certainly couldn't drop the zero. If we did we would no longer have the number 2073 (two thousand seventy-three) but would have the number 273 (two hundred seventy-three). We cannot drop the zero without changing the value of the number. We show this concept in two different ways in Table 1-2.

□ EXAMPLE 1-1

In the number 70,630 there are 7 ten-thousands, no thousands, 6 hundreds, 3 tens, and no units. The MSD (most significant digit) is 7 and the LSD (least significant digit) is 0.

TABLE 1-1

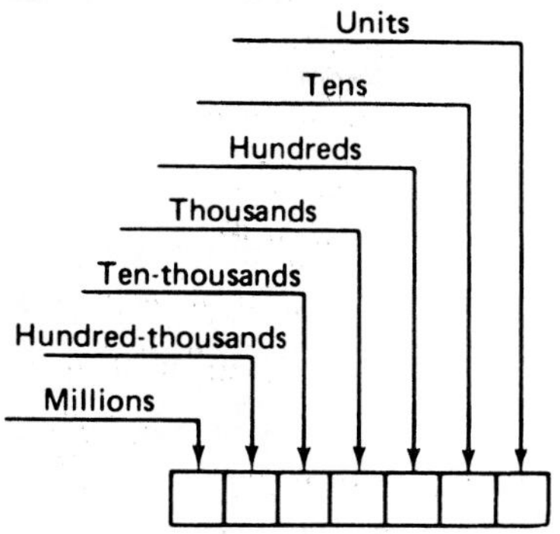

TABLE 1-2

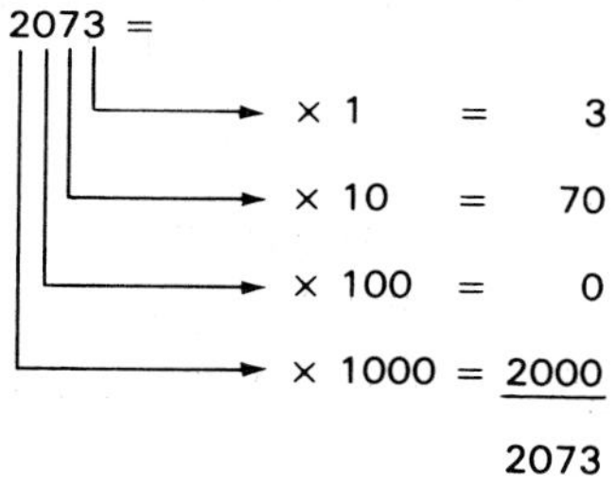

$(2 \times 1000) + (0 \times 100) + (7 \times 10) + (3 \times 1) = 2073$

☐ EXAMPLE 1-2

In the number 485,607 there are 4 hundred-thousands, 8 ten-thousands, 5 thousands, 6 hundreds, no tens, and 7 ones. The MSD is 4, and LSD is 7. ☐

PRACTICE PROBLEMS 1-1

1. In the number 4762:
(a) Which digit is the most significant digit (MSD)?
(b) Which digit is the least significant digit (LSD)?
(c) Which digit occupies the hundreds position?

2. In the number 7,623,418:
(a) How many ten-thousands are there?
(b) How many tens are there?
(c) How many hundred-thousands are there?
(d) How many units are there?

3. In the number 893,462:
(a) What weight does the 9 have?
(b) What weight does the 4 have?
(c) How many thousands are there?

4. How many significant digits are in the following numbers?
(a) 489.0 (b) 0.489 (c) 40,089 (d) 4809.0
(e) 0.00489

Solutions

1. (a) 4 (b) 2 (c) 7
2. (a) 2 (b) 1 (c) 6 (d) 8
3. (a) 9 ten-thousands or 90,000
   (b) 4 hundreds or 400 (c) 3
4. (a) 4 (b) 3 (c) 5 (d) 5 (e) 3

Additional practice problems are at the end of the chapter.

1-2 DECIMAL FRACTIONS

To discuss decimal digits, let's review some names used in dealing with fractions. In the fraction $\frac{7}{8}$ the number above the line (7) is called the *numerator*. The number below the line (8) is called the *denominator*. The line itself is called the *vinculum*.

A decimal fraction is a fraction whose denominator is 10 or a multiple of 10 (100, 1000, 10,000, and so on). If the denominator is 10, the fraction is read one-tenth if the number is $\frac{1}{10}$, two-tenths if the number is $\frac{2}{10}$, and so on.

Here are some decimal fractions and how they are read:

$$\frac{3}{100} \quad \text{3 hundred}\textit{ths}$$

$$\frac{1}{1000} \quad \text{1 thousand}\textit{th}$$

$$\frac{12}{10{,}000} \quad \text{12 ten-thousand}\textit{ths}$$

Notice that we read the denominator just as we read a whole number except that we add *th* or *ths*. Similarly, if we write "4 thousandths," we know this is a fraction (the *ths* at the end of thousand tells us this); so we know the fraction is $\frac{4}{1000}$.

□ EXAMPLE 1-3

(a) 7 tenths $= \dfrac{7}{10}$

(b) 23 hundredths $= \dfrac{23}{100}$

(c) 243 ten-thousandths $= \dfrac{243}{10{,}000}$

□

Table 1-3 gives the names of places from tenths to millionths. The tenths position is the first position to the right of the decimal point. The hundredths position is

TABLE 1-3

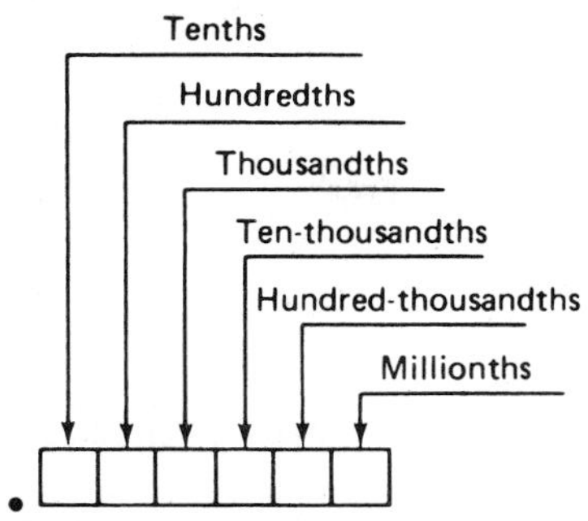

the next position, and so on, to the millionths position, which is the sixth position to the right of the decimal point.

Decimal fractions can be converted to decimal numbers by using the following rule:

■ **RULE 1-1** To convert decimal fractions to decimal numbers, determine the value of the denominator. Place the LSD of the numerator in this position.

□ EXAMPLE 1-4

Change the decimal fraction $\frac{17}{100}$ to a decimal number.

Solution The value of the denominator is one-hundredth. The LSD in the numerator is 7; therefore, the 7 is placed in the hundredths position. This puts the 1 in the tenths position and the number is 0.17.

□ EXAMPLE 1-5

Change the decimal fraction $\frac{17}{1000}$ to a decimal number.

Solution The denominator is one thousandth so the 7 must appear in the thousandths position. That puts the 1 in the hundredths position. As with whole numbers, zeros must be used as place holders when necessary. Therefore, a zero goes in the tenths position. The decimal number is written 0.017 (17 thousandths). □

Notice that in each case a zero was placed to the *left* of the decimal point. This is done to help us remember where the decimal point belongs.

■ **RULE 1-2** To convert decimal numbers to decimal fractions, determine the position of the rightmost digit. This tells us the value of the denominator. The numerator is the decimal number with the decimal point removed.

□ EXAMPLE 1-6

Convert 0.007 to a fraction.

Solution The rightmost digit, 7, appears in the thousandths position so the answer is $\frac{7}{1000}$ (seven-thousandths).

□ EXAMPLE 1-7

Convert 0.0023 to a fraction.

Solution The rightmost digit, 3, is in the ten-thousandths position so the answer is $\frac{23}{10,000}$ (twenty-three ten-thousandths). □

PRACTICE PROBLEMS 1-2

1. Convert the following fractions to decimal numbers:
   (a) $\frac{4}{100}$ (b) $\frac{23}{1000}$ (c) $\frac{203}{100,000}$
2. Write the following numbers as decimal fractions:
   (a) 0.05 (b) 0.00073 (c) 0.00009
3. Write each of the following first as a decimal fraction and then as a decimal number:
   (a) 7 tenths (b) 37 hundredths
   (c) 7 ten-thousandths
   (d) 417 hundred-thousandths (e) 6 millionths
4. In the number 0.00246, in which place does the 2 appear? The 6?
5. Write the names of the following numbers:
   (a) 0.03 (b) 0.0005 (c) 0.00073

Solutions

1. (a) 0.04 (b) 0.023 (c) 0.00203
2. (a) $\frac{5}{100}$ (b) $\frac{73}{100,000}$ (c) $\frac{9}{100,000}$
3. (a) $\frac{7}{10} = 0.7$ (b) $\frac{37}{100} = 0.37$
   (c) $\frac{7}{10,000} = 0.0007$ (d) $\frac{417}{100,000} = 0.00417$
   (e) $\frac{6}{1,000,000} = 0.000006$
4. The 2 appears in the thousandths position. The 6 appears in the hundred thousandths position.
5. (a) 3 hundredths (b) 5 ten-thousandths
   (c) 73 hundred-thousandths

Additional practice problems are at the end of the chapter.

Now let's look at numbers in which there is a whole-number part and a decimal-number part. To do this, let's prepare a table that is a combination of Tables 1-1 and 1-3. The result is shown in Table 1-4.

TABLE 1-4

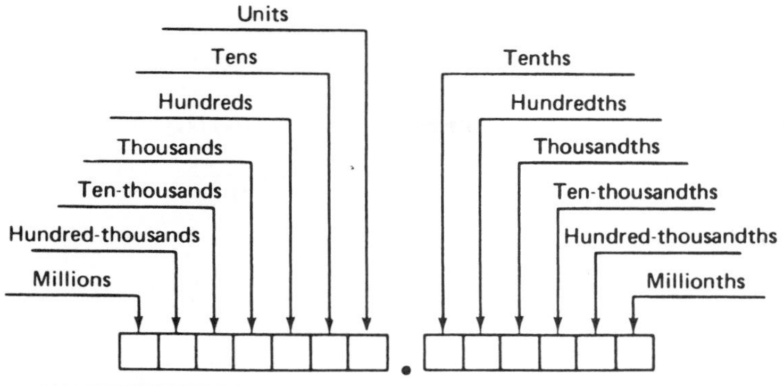

Consider the number 13.36. This would be read "thirteen *and* thirty-six hundredths." The word *and* is used in a number to define the position of the decimal point. We read the whole-number part, add the word *and*, and then read the decimal-number part. We could also write the number as a *mixed* number: $13\frac{36}{100}$. Mixed numbers indicate the addition of two numbers: a whole number and a fraction. $13.36 = 13\frac{36}{100} = 13 + \frac{36}{100}$. Typically, though, we wouldn't use the + sign but would simply write the number as a decimal number (13.36) or as a mixed number ($13\frac{36}{100}$).

Consider the number 432.243. This number is read "four hundred thirty-two *and* two hundred forty-three thousandths." Remember, the word *and* is used to connect the whole-number and decimal-number parts. As a mixed number we would write

$$432\frac{243}{1000}$$

■ **KEY POINT** Mixed numbers are numbers that are made up of a whole-number part and a fraction. When reading the number, the word *and* is used to connect the whole number and decimal number parts.

PRACTICE PROBLEMS 1-3

1. Convert each of the following decimal numbers to mixed numbers:
(a) 24.007 (b) 706.024 (c) 4.00017
(d) 7.400 (e) 370.006 (f) 47.113

2. Convert each of the following mixed numbers to decimal numbers:
(a) $1076\frac{7}{100}$ (b) $8\frac{23}{10{,}000}$ (c) $76\frac{14}{1000}$
(d) $35\frac{270}{100{,}000}$ (e) $713\frac{7}{1000}$ (f) $8\frac{87}{10{,}000}$

3. Express the following numbers as decimal fractions and as mixed numbers:
(a) Fifty-six and seventy-eight one-hundredths
(b) Fifteen and thirty-five thousandths
(c) One hundred six and eight ten-thousandths
(d) Seven hundred four and two hundred fourteen hundred-thousandths
(e) Four thousand seventy-five and fourteen ten-thousandths
(f) Eighty and forty-three thousandths

Solutions

1. (a) $24\frac{7}{1000}$ (b) $706\frac{24}{1000}$ (c) $4\frac{17}{100{,}000}$
   (d) $7\frac{400}{1000}$ (e) $370\frac{6}{1000}$ (f) $47\frac{113}{1000}$

2. (a) 1076.07 (b) 8.0023 (c) 76.014
   (d) 35.00270 (e) 713.007 (f) 8.0087

3. (a) $56.78 = 56\frac{78}{100}$ (b) $15.035 = 15\frac{35}{1000}$
   (c) $106.0008 = 106\frac{8}{10{,}000}$
   (d) $704.00214 = 704\frac{214}{100{,}000}$
   (e) $4075.0014 = 4075\frac{14}{10{,}000}$
   (f) $80.043 = 80\frac{43}{1000}$

Additional practice problems are at the end of the chapter.

1-3 ROUNDING WHOLE NUMBERS

In problem solving it is common practice to simplify multidigit numbers. This simplification is done by replacing nonzero digits with zeros. For example, the number 72,348 could be simplified to 72,350 to the nearest ten; 72,300 to the nearest hundred; 72,000 to the nearest thousand; or 70,000 to the nearest ten-thousand. This simplification is called *rounding*.

Rounding numbers is done to make problem solving easier. Some accuracy is lost as a result of the rounding. The more rounding we do, the less accurate the number. In electronics we work with components that typically have 5% or 10% tolerances. We normally make measurements with instruments whose tolerances are within the 5% or 10% range (or worse). For these reasons, great accuracies in problem solving are not necessary.

Let's look at the number 72,348 again. When we rounded to the nearest ten, we rounded to 72,350. We had the choice of rounding down to 72,340 or rounding up to 72,350. Since 48 is closer to 50 than to 40, we rounded up to 72,350. When we rounded to the nearest hundred, we rounded to 72,300. We had the choice of rounding down to 72,300 or rounding up to 72,400. Since 348 is closer to 300 than to 400, we rounded down to 72,300. The same logic was used in rounding to the nearest thousand. Since 2348 is closer to 2000 than to 3000, we rounded to 72,000. Finally, 72,000 was rounded to 70,000 because 72,000 is closer to 70,000 than to 80,000. The rule is:

RULE 1-3 When rounding numbers, round *up* if the digit is 5, 6, 7, 8, or 9. To round up, the digit is replaced with 0, and 1 is added to the next significant digit. Round *down* if the digit is 1, 2, 3, or 4. To round down, the digit is replaced with zero.

Although the digit 5 could indicate either up or down, it is typically rounded *up*.

☐ EXAMPLE 1-8

Round the number 2735 to the nearest ten, hundred, and thousand.

Solution To round to the nearest ten, we examine the digit in the units position. Since the digit is 5, we round *up* to 2740 by replacing 5 with 0 and adding 1 to the next significant digit. To round to the nearest hundred, we note that the digit in the tens position is 3; so we round *down* to 2700 by replacing 3 with 0. (The digit in the units position is also changed to 0 since we are rounding to the nearest hundred.) Since 7 is the digit in the hundreds position, we round *up* to the nearest thousand and the answer is 3000. ☐

PRACTICE PROBLEMS 1-4

Round the following numbers to the nearest ten, hundred, and thousand:

1. 2714
2. 6526
3. 43,258
4. 76,547
5. 82,803
6. 26,764
7. 78,226
8. 18,999
9. 4050
10. 243,671

Solutions

	To Nearest Tens	*To Nearest Hundreds*	*To Nearest Thousands*
1.	2710	2700	3000
2.	6530	6500	7000
3.	43,260	43,300	43,000
4.	76,550	76,500	77,000
5.	82,800	82,800	83,000
6.	26,760	26,800	27,000
7.	78,230	78,200	78,000
8.	19,000	19,000	19,000
9.	4050	4100	4000
10.	243,670	243,700	244,000

In problem 5, since a 0 already exists in the tens position, rounding to the nearest hundred is not necessary. In problem 8, when we round to the nearest ten, we round up because the units digit is 9. Remember that we round up by adding 1 in the next significant digit position (the tens position). Adding 1 to 9 in the tens position produces a 0 and a carry into the hundreds position. A carry into the hundreds position results in 0 and a carry into the thousands position, resulting in the answer 19,000.

In problem 9, since a 0 already exists in the units position, rounding to the nearest ten is not necessary. The 5 in the tens position causes a carry to the hundreds, resulting in the answer 4100 to the nearest hundred.

Additional practice problems are at the end of the chapter.

1-4 ROUNDING NONWHOLE NUMBERS

Nonwhole numbers are rounded by using the same rules as for whole numbers. As with whole numbers, we work from the least significant to the most significant digits. Consider the number 12.736. We could round to the nearest hundredth, tenth, unit, or ten. To round to the nearest hundredth, we would note the digit in the thousandths position. Since the digit is 6, we change the digit to 0 and add 1 to the next significant digit, making the number 12.74. We don't write the number as 12.740. The zero has no meaning (no place value) since we rounded to the nearest hundredth. Therefore, we just drop the zero. To round to the nearest tenth, we note the digit in the hundredths position. Since the number is 4, we simply drop it and the answer is 12.7. To round to the nearest unit, we note the digit in the tenths position. Because it is a 7, a number that is 5 or greater, we drop it and add 1 to the number in the units position. The answer is 13. To convert to the nearest ten, we change the 2 to a 0. The answer is 10. Remember that in rounding whole numbers the zero is a place holder and cannot be dropped. In a decimal fraction, if a zero exists in the LSD position as a result of rounding, the zero is dropped because it has no meaning and no value.

PRACTICE PROBLEMS 1-5

Round the following numbers to the nearest hundredth, tenth, unit, and ten:

1. 73.647
2. 26.401
3. 17.0419
4. 30.6908
5. 50.4736
6. 48.047
7. 33.781
8. 68.147
9. 78.6671
10. 15.5554

Solutions

	Hundredth	*Tenth*	*Unit*	*Ten*
1.	73.65	73.6	74	70
2.	26.40	26.4	26	30
3.	17.04	17.0	17	20
4.	30.69	30.7	31	30
5.	50.47	50.5	50	50
6.	48.05	48.0	48	50
7.	33.78	33.8	34	30
8.	68.15	68.1	68	70
9.	78.67	78.7	79	80
10.	15.56	15.6	16	20

Additional practice problems are at the end of the chapter.

SELF-TEST 1-1

1. In the number 20,378:
   (a) Which digit is the MSD?
   (b) Which digit is the LSD?
   (c) Which digit occupies the thousands position?
   (d) Which digit occupies the tens position?
   (e) What weight does the digit 3 have?
   (f) How many units are there?

2. Convert the following fractions to decimal numbers:
   (a) $\frac{41}{100}$ (b) $\frac{9}{1000}$ (c) $\frac{1783}{100{,}000}$

3. Write each of the following first as a decimal fraction and then as a decimal number:
   (a) Three-tenths
   (b) Eighty-five hundredths
   (c) Eighteen ten-thousandths

4. Write the names of the following numbers:
   (a) 0.33 (b) 0.004

5. Convert each of the following decimal numbers to mixed numbers:
   (a) 7.46 (b) 18.006

6. Convert each of the following mixed numbers to decimal numbers:
   (a) $76\frac{14}{100}$ (b) $6\frac{23}{1000}$

7. Express the following numbers as decimal fractions and as mixed numbers:
   (a) Three and seven-hundredths
   (b) Twenty-eight and sixty-three thousandths

8. Round to the nearest ten, hundred, and thousand:
   (a) 4765 (b) 9705 (c) 15,789
   (d) 37,046

9. Round to the nearest tenth, hundredth, and thousandth:
   (a) 0.0746 (b) 0.4605 (c) 0.4056
   (d) 0.3748

10. Round to the nearest ten, unit, tenth, and hundredth:
    (a) 17.486 (b) 23.462 (c) 36.547
    (d) 20.706

Answers to Self-test 1-1 are found at the end of the chapter.

1-5 SIGNED NUMBERS

In mathematics, numbers may be either positive (+) or negative (−). The plus sign is not usually written. If we saw the numbers 6, −4, 15, and −9 written, we would recognize that the numbers 6 and 15 are positive and the numbers 4 and 9 are negative.

In this section we introduce the symbols $<$ and $>$. The symbol $<$ means *less than* and the symbol $>$ means *greater than*.

In Figure 1-1 we have drawn a straight line that starts at zero and extends both right and left. Notice that numbers starting at zero and extending to the right are positive. Numbers starting at zero and extending to the left are negative. In comparing signed numbers, we see that the rightmost number is the greater and the leftmost number is the lesser. For example, compare 4 and −3. We could say that 4 is greater than −3 ($4 > -3$), or we could say that −3 is less than 4 ($-3 < 4$).

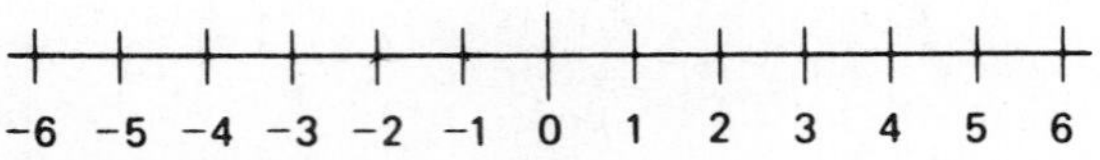

FIGURE 1-1

Consider the numbers −7, −2, and 6. We could say that $-7 < -2 < 6$. Or we could say that $6 > -2 > -7$. Notice that in comparing −2 and −7 we said that $-2 > -7$. Remember, on the number scale −2 is to the right of −7; therefore, it is greater. −2 is more positive than −7. −7 is more negative than −2. If we compare 3 and 11, we would say that $3 < 11$. We could also say that 3 is more negative than 11.

The *absolute* value of a number is that number without regard to sign. The absolute value of +5 and −5 is 5 and is symbolized as $|5|$. This tells us that no matter if 5 is − or +, the distance from the origin (0) is 5.

1-6 ADDITION AND SUBTRACTION

In addition to using the symbols + and − to denote the sign of a number, we also use these symbols to indicate the operations of addition and subtraction.

Anyone who has trouble visualizing adding signed numbers should refer to Figure 1-1. Adding two or more positive numbers does not present any problem with signs. $2 + 3 + 5 = 10$. Let's add two negative numbers: $(-2) + (-3)$. −2 and −3 are enclosed in parentheses to indicate that the − is the sign of the number and not a sign of subtraction. $(-2) + (-3) = -5$. $(-6) + 7 + (-9) = -8$. We could simplify the problem before finding the sum by adding the negative numbers together first:

$$-6 + 7 + (-9) = -15 + 7 = -8$$

In each case we move left on the number scale if the number is − and we move right if the number is +.

■ **RULE 1-4** To add numbers with like signs, add the numbers and affix the sign to the answer.

To add numbers with unlike signs, subtract the smaller from the larger and affix the sign of the larger to the answer.

In subtraction we call the first number the *minuend*. The second number is the *subtrahend*. The answer is the *difference*.

$$\begin{array}{rl} 6 & \text{minuend} \\ \underline{-3} & \text{subtrahend} \\ 3 & \text{difference} \end{array}$$

■ **RULE 1-5** To subtract, change the sign of the subtrahend and then add, following the rule for addition.

$$6 - (+3) = (-3) = 3$$

$$\begin{array}{r} 6 \\ \underline{-3} \\ 3 \end{array} = + \begin{array}{r} 6 \\ \underline{-3} \\ 3 \end{array}$$

When more than two numbers are subtracted, the same rule applies. Whenever subtracting is indicated, we change the sign of the following numbers and add. $17 - (+6) - (+3) = 17 + (-6) + (-3) = 17 + (-9) = 8$. When addition and subtraction are both indicated, we follow the rules for both.

$$25 - (+6) + 4 = 25 + (-6) + 4 = 23$$
$$-7 + 16 - (-4) = -7 + 16 + 4 = 13$$

PRACTICE PROBLEMS 1-6

Perform the indicated operations:

1. $6 - (+4)$
2. $-14 + (-6)$
3. $16 - (+4)$
4. $9 - (-3)$
5. $-10 - (-6)$
6. $-15 - (+4)$
7. $25 - (+3) - (+4)$
8. $4 - (-6) + (-15)$
9. $-6 - (-3) - (-10)$
10. $40 - (+30) - (-6)$

Solutions

1. 2
2. −20
3. 12
4. 12
5. −4
6. −19
7. 18
8. −5
9. 7
10. 16

Additional practice problems are at the end of the chapter.

1-7 MULTIPLICATION AND DIVISION

Both × and · are symbols used to indicate multiplication. The symbols ÷ and / are used to indicate division. Each part of a multiplication or division has a name as shown in Figure 1-2.

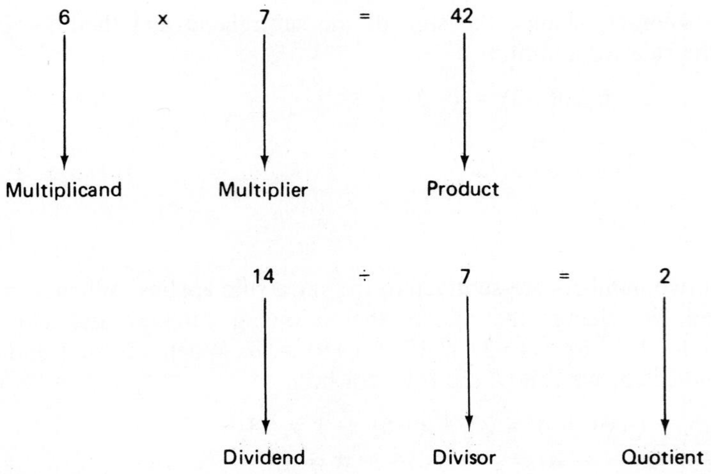

FIGURE 1-2

Signed numbers are multiplied and divided by using the same methods learned in arithmetic. The rules are:

■ **RULE 1-6** If all numbers are positive, the answer is positive. If there is an even number of negative signs, the answer is positive. If there is an odd number of negative signs, the answer is negative.

$$(+)\cdot(+) = +$$
$$(+)\cdot(-) = -$$
$$(-)\cdot(-) = +$$

$9 \times 8 = 72$	$(+)\cdot(+) = +$
$7 \times (-6) = -42$	$(+)\cdot(-) = -$
$-8 \times (-6) = 48$	$(-)\cdot(-) = +$
$16 \div 4 = 4$	$(+)\div(+) = +$
$12 \div (-4) = -3$	$(+)\div(-) = -$
$-20 \div 5 = -4$	$(-)\div(+) = -$
$-35 \div -7 = 5$	$(-)\div(-) = +$

PRACTICE PROBLEMS 1-7

Perform the indicated operations:

1. $17 \times (-5)$

2. $-14 \times (-16)$

3. -9×12

4. -12×14

5. $-14 \div 2$

6. $-42 \div 6$

7. $172 \div (-4)$

8. $98 \div (-14)$

9. $-121 \div (-11)$

10. $-144 \div (-12)$

Solutions

1. -85

2. 224

3. -108

4. -168

5. -7

6. -7

7. -43

8. -7

9. 11

10. 12

Additional practice problems are at the end of the chapter.

1-8 MATHEMATICAL EXPRESSIONS AND TERMS

Many problem-solving situations in electronics involve all four operations: addition, subtraction, multiplication, and division. Therefore, it is important that we understand the order in which these operations are performed.

Before we talk about these problem-solving situations, we must define some new words. Here and in future chapters we are going to be talking about mathematical *terms* and *expressions*.

- A mathematical *term* is a number preceded by a $+$ or $-$ sign.
- A mathematical *expression* is made up of one or more terms.

Thus, $4 + 7$ is an expression containing two terms. This expression can be reduced to *one* term (11).

$4 + (7 \times 3)$ is a two-term expression. 4 is one term and 7×3 is the other. We have used parentheses to help you identify the term. Remember, *terms* are separated by $+$ and $-$ signs. Signs of multiplication and division are signs within a term. In problem solving, all operations within a term are performed first and then the terms are combined. The rule is:

RULE 1-7 To simplify a mathematical expression, perform the multiplications and divisions first and then the additions and subtractions.

Here are some examples:

$$(7 \times 6) + (4 \times 2) = 42 + 8 = 50$$

$$(-14 \div 7) - (-6 \times 3) = -2 - (-18) = -2 + 18 = 16$$

$$4 + (-3) - (-4 \times 6) = 4 - 3 - (-24) = 4 - 3 + 24 = 25$$

My colleague, Barbara Snyder, uses the phrase "**M**y **D**ear **A**unt **S**ally" to help her students remember which function is performed first—**M**ultiply, **D**ivide, **A**dd, and **S**ubtract.

PRACTICE PROBLEMS 1-8

Simplify the following expressions to one term:

1. $8 + 6 \times 3$
2. $-3 + 12 \div (-2)$
3. $6 \times 3 - 2 \times 4$
4. $16 \div (-2) - (-4) + 6 \times (-6)$
5. $(-6) \times (-4) + 7 \times (-3)$
6. $6 \times (-4) \div 2 - 4 + (-6)$
7. $15 \times 3 \div 3 + 4$
8. $8 \times 4 - (-6) \times (-4)$

Solutions

1. 26
2. −9
3. 10
4. −40
5. 3
6. −22
7. 19
8. 8

Additional practice problems are at the end of the chapter.

Signs of grouping are another way of indicating the order in which problems are solved. These signs of grouping include parentheses ( ), brackets [ ], braces { }, and the vinculum (or bar) ———. Signs of grouping are also a way of indicating multiplication. For example, to multiply two numbers, we could symbolize the multiplication by using parentheses. (7)(6) or 7(6) means the same as 7×6 or $7 \cdot 6$. Thus $7 \times (6 + 2)$ could be written $7(6 + 2)$. In both cases we are finding the product of 7 and 8. $7(6 + 2) = 7(8) = 56$.

When signs of grouping are used, we first perform all the operations within the group and then we combine terms. Here are some examples:

$$7(6 + 3) = 7 \times 9 = 63$$

$$4 - (-6 + 3) \times 2 = 4 - (-3) \times 2 = 4 - (-6) = 4 + 6 = 10$$

When more than one sign of grouping is used, we work from the inside out.

$$3[4(3 + 2) - 4] = 3(4 \times 5 - 4) = 3(20 - 4) = 3(16) = 48$$

$$\begin{aligned} -3 - [2 + 4 \times (-6)] = -3 - [2 + (-24)] &= -3 - (-22) \\ &= -3 + 22 = 19 \end{aligned}$$

PRACTICE PROBLEMS 1-9

Perform the indicated operations:

1. $-6 - (-3) + 3(3 - 4)$
2. $6(-3) - [6(-3)]$
3. $14 - (-4 + 18 \div 3)$
4. $(4 + 3)(6 - 3)$
5. $4 - [6(7 - 3)]$
6. $6[18 \div (-6 + 3)] + 7 \times 4$
7. $10 - (6 + 3) \times 4$
8. $18 - 3[(3 + 4)(8 - 12)]$
9. $4[7 - (-3) + (2 - 6)(3 - 7)]$
10. $4 - 3\{-2[6(2 + 3)] - 6\}$

Solutions

1. −6
2. 0
3. 12
4. 21
5. −20
6. −8
7. −26
8. 102
9. 104
10. 202

SELF-TEST 1-2

Perform the indicated operations:

1. $14 - (+4) + (-5)$
2. $-16 - (-4) - (+6)$
3. -8×6
4. $-9 \times (-4)$
5. $4 \times (-3)$
6. $51 \div (-3)$
7. $-64 \div (-4)$
8. $-36 \div 4$
9. $5 \times (-3) + (-6) \times 7$
10. $-9 \div 3 + 4 \times (-6)$
11. $16 - (-6) + 27 \div (-3)$
12. $4 - (-6) + 4(5 - 2)$
13. $7 \times (-2) - [4(-6)]$
14. $7[15 \div (-3 + 6)] + 8 \div 4$
15. $6 - 5\{-4[8(4 + 2)] - 4\}$

Answers to Self-test 1-2 are at the end of the chapter.

CHAPTER 1 AT A GLANCE

Page	Key Point or Rule	Example
2	*Key point:* Significant digits or significant figures are all the digits 1 through 9. Zero is also significant except when in the leftmost position.	25 is correct to 2 significant digits. 205 is correct to 3 significant digits. 0.025 is correct to 2 significant digits.
2	*Key point:* The most significant digit (MSD) is the leftmost nonzero digit. The least significant digit (LSD) is the rightmost digit.	In the number 2740, the MSD is 2 and the LSD is 0.
5	*Rule 1-1:* To convert decimal fractions to decimal numbers, determine the value of the denominator. Place the LSD of the numerator in this position.	$\frac{37}{1000} = 0.037$ The digit 7 goes in thousandths position, which puts the 3 in the hundredths position. Zero is a place holder and goes in the tenths position.

Page	Key Point or Rule	Example
5	*Rule 1-2:* To convert decimal numbers to decimal fractions, determine the position of the rightmost digit. This tells us the value of the denominator. The numerator is the decimal number with the decimal point removed.	$0.0019 = \frac{19}{10{,}000}$ The 9 is in the ten-thousandths position, so the denominator is 10,000 and the numerator is 19.
7	*Key point:* Mixed numbers are numbers that are made up of a whole-number part and a fraction. When reading the number, the word *and* is used to connect the whole number and decimal number parts.	$6\frac{3}{10}$ $40\frac{11}{100}$
8	*Rule 1-3:* When rounding numbers, round *up* if the digit is 5, 6, 7, 8, 9. To round up, the digit is replaced with 0, and 1 is added to the next significant digit. Round *down* if the digit is 1, 2, 3, or 4. To round down, the digit is replaced with zero.	4747 rounded to the nearest ten is 4750. To the nearest hundred, it is 4700, and to the nearest 1000, it is 5000.
10	Apply Rule 1-3 to nonwhole numbers.	4.074 rounded to the nearest hundredth is 4.07. To the nearest tenth, it is 4.1, and to the nearest unit, it is 4.
12	*Rule 1-4:* To add numbers with like signs, add the numbers and affix the sign to the answer. To add numbers with unlike signs, subtract the smaller from the larger and affix the sign of the larger to the answer.	$-4 + (-3) = -7$ $-4 + 8 = 4$
13	*Rule 1-5:* To subtract, change the sign of the subtrahend and then add, following the rule for addition.	$-4 - (-3) = -1$
14	*Rule 1-6:* When multiplying or dividing signed numbers, if all numbers are positive, the answer is positive. If there is an even number of negative signs, the answer is positive. If there is an odd number of negative signs, the answer is negative.	$(+) \cdot (+) = +$ $(-) \cdot (-) = +$ $(+) \cdot (-) = -$
15	*Rule 1-7:* To simplify a mathematical expression, perform the multiplications and divisions first and then the additions and subtractions.	$(4 \times 3) - (9 \div 3) = 12 - 3 = 9$

End of Chapter Problems 1-1

1. In the number 8042:
(a) What is the MSD?
(b) What is the LSD?
(c) What digit occupies the hundreds position?
(d) What digit occupies the units position?
(e) What weight does the 8 have?
(f) How many tens are there?

2. In the number 4703:
(a) What is the MSD?
(b) What is the LSD?
(c) What digit occupies the hundreds position?
(d) What digit occupies the units position?
(e) What weight does the 4 have?
(f) How many tens are there?

3. In the number 67,403:
(a) What is the MSD?
(b) What is the LSD?
(c) What digit occupies the thousands position?
(d) What digit occupies the tens position?
(e) What weight does the 7 have?
(f) How many hundreds are there?

4. In the number 35,017:
(a) What is the MSD?
(b) What is the LSD?
(c) What digit occupies the thousands position?
(d) What digit occupies the tens position?
(e) What weight does the 5 have?
(f) How many hundreds are there?

5. In the number 894,307:
(a) What is the MSD?
(b) What is the LSD?
(c) What digit occupies the hundreds position?
(d) What digit occupies the ten-thousands position?
(e) What weight does the 8 have?
(f) How many thousands are there?

6. In the number 407,236:
(a) What is the MSD?
(b) What is the LSD?
(c) What digit occupies the hundreds position?
(d) What digit occupies the ten-thousands position?
(e) What weight does the 4 have?
(f) How many thousands are there?

End of Chapter Problems 1-2

1. Convert the following fractions to decimal numbers:
(a) $\frac{3}{10}$ (b) $\frac{16}{1000}$ (c) $\frac{278}{100{,}000}$
(d) $\frac{1763}{10{,}000}$ (e) $\frac{435}{1000}$ (f) $\frac{2060}{10{,}000}$

2. Convert the following fractions to decimal numbers:
(a) $\frac{495}{10{,}000}$ (b) $\frac{6}{100}$ (c) $\frac{75}{1000}$
(d) $\frac{6}{100{,}000}$ (e) $\frac{187}{1000}$ (f) $\frac{7}{10{,}000}$

3. Write the following numbers as decimal fractions:
(a) 0.007 (b) 0.0432 (c) 0.174
(d) 0.000065 (e) 0.00016 (f) 0.01234

4. Write the following numbers as decimal fractions:
(a) 0.173 (b) 0.0004 (c) 0.00473
(d) 0.000806 (e) 0.00008 (f) 0.00347

5. Write each of the following first as a decimal fraction and then as a decimal number:
(a) 17 thousandths (b) 4 hundredths
(c) 460 ten-thousandths (d) 27 millionths
(e) 1780 hundred-thousandths
(f) 65 thousandths

6. Write each of the following first as a decimal fraction and then as a decimal number:
(a) 6 tenths (b) 83 hundred-thousandths
(c) 780 millionths (d) 27 thousandths
(e) 284 ten-thousandths (f) 27 hundredths

7. In the number 0.001642, in which place does the
(a) one appear? (b) two appear?
(c) six appear?

8. In the number 0.008706, in which place does the
(a) seven appear? (b) six appear?
(c) eight appear?

9. In the number 0.508743, in which place does the
(a) seven appear? (b) three appear?
(c) zero appear?

10. In the number 0.34075, in which place does the
(a) four appear? (b) zero appear?
(c) five appear?

11. Write the names of the following:
(a) 0.006 (b) 0.147 (c) 0.00092
(d) 0.000007 (e) 0.0413 (f) 0.0101

12. Write the names of the following:
(a) 0.175 (b) 0.000035 (c) 0.00463
(d) 0.00056 (e) 0.076 (f) 0.000005

End of Chapter Problems 1-3

1. Convert each of the following decimal numbers to mixed numbers:
(a) 7.14 (b) 50.02 (c) 710.143 (d) 9.099
(e) 73.653 (f) 207.7834 (g) 28.00736
(h) 8.0706

2. Convert each of the following decimal numbers to mixed numbers:
(a) 47.3 (b) 8.75 (c) 56.075
(d) 9.475 (e) 450.00125 (f) 37.4073
(g) 48.00045 (h) 10.10738

3. Convert each of the mixed numbers to decimal numbers:
(a) $5\frac{68}{100}$ (b) $25\frac{7}{1000}$ (c) $7\frac{165}{10{,}000}$
(d) $70\frac{4}{10}$ (e) $473\frac{25}{1000}$ (f) $80\frac{743}{100{,}000}$
(g) $2475\frac{35}{1{,}000{,}000}$ (h) $307\frac{8}{100{,}000}$

4. Convert each of the following mixed numbers to decimal numbers:
(a) $45\frac{3}{10}$ (b) $365\frac{83}{100}$ (c) $3\frac{14}{10{,}000}$
(d) $17\frac{85}{1{,}000{,}000}$ (e) $819\frac{6}{10{,}000}$ (f) $83\frac{38}{1000}$
(g) $8\frac{17}{100}$ (h) $25\frac{275}{100{,}000}$

5. Express the following numbers as decimal numbers, and mixed numbers:
(a) Ninety-three and seven-tenths
(b) Thirty and four-hundredths
(c) Eleven and one ten-thousandths
(d) Nine hundred five and fifty-two thousandths

6. Express the following numbers as decimal numbers, and mixed numbers:
(a) Sixty-five and four-tenths
(b) Twelve and twenty-five hundredths
(c) Forty and thirty-seven thousandths
(d) Four hundred six and thirty-one thousandths

7. Express the following numbers as decimal numbers, and mixed numbers:
(a) Two hundred seventy-three and twenty-five hundred-thousandths
(b) Seven hundred four and seven hundred four-millionths
(c) Two thousand forty-four and five hundred four ten-thousands
(d) Ten thousand one hundred one and eighty-nine hundred-thousandths

8. Express the following numbers as decimal numbers, and mixed numbers:
(a) Eight hundred forty-five and sixty-five hundred-thousandths
(b) Seventy-four and thirty-one millionths
(c) Five thousand seventy-two and one hundred seven ten-thousandths
(d) Twelve thousand one hundred four and twenty-two hundred-thousandths

End of Chapter Problems 1-4

Round the following numbers to the nearest ten:

1. 17
2. 72
3. 45
4. 33
5. 64
6. 87
7. 127
8. 236
9. 874
10. 687

Round the following numbers to the nearest ten and hundred:

11. 273
12. 708
13. 356
14. 464
15. 1377
16. 8773
17. 1407
18. 2756
19. 8706
20. 7309

Round the following numbers to the nearest ten, hundred, and thousand:

21. 4817
22. 6706
23. 85,468
24. 27,046
25. 78,673
26. 84,615
27. 2784
28. 10,778
29. 35,486
30. 20,066
31. 68,448
32. 73,547
33. 73,654
34. 17,473
35. Round 465,432 to the nearest ten-thousand.
36. Round 756,700 to the nearest ten-thousand.

End of Chapter Problems 1-5

Round the following numbers to the nearest hundredth, tenth, unit, and ten:

1. 163.782
2. 89.076
3. 9.464
4. 273.926
5. 88.888
6. 25.546
7. 749.493
8. 20.506
9. 39.278
10. 76.567
11. 63.7478
12. 74.2636
13. 478.6706
14. 215.2535
15. 47.474
16. 88.077
17. 16.545
18. 44.574
19. 76.746
20. 613.657

End of Chapter Problems 1-6

Perform the indicated operations:

1. $12 - (+4) + (-3)$
2. $-3 - (+4) - (-6)$
3. $16 + 4 + (-6)$
4. $-15 - (-7) - (+10)$
5. $-20 - (-4) - (-10)$
6. $-30 - (-6) - (+10)$
7. $30 + (-3) + 7$
8. $40 + (-13) + 6$
9. $-10 - (+4) + (-14)$
10. $-14 - (+6) + (-18)$
11. $7 - (-7) - (+7)$
12. $14 + (-7) - (+7)$
13. $-5 - (-3) - (+12)$
14. $-6 - (-4) - (+16)$
15. $14 + (-3) - (-7)$
16. $20 + (-5) - (-8)$
17. $9 - (-4) - (-8)$
18. $7 - (-10) - (-12)$
19. $-20 - (+6) + (-7)$
20. $-25 - (+4) + (-8)$

End of Chapter Problems 1-7

Perform the indicated operations:

1. $14 \times (-3)$
2. -15×4
3. $-9 \times (-6)$
4. $25 \times (-6)$

5. -12×7
6. $-15 \times (-15)$
7. $-45 \times (-5)$
8. $210 \times (-4)$
9. $-310 \times (-3)$
10. -17×40
11. $28 \div (-4)$
12. $165 \div (-5)$
13. $-144 \div 6$
14. $-256 \div 16$
15. $-72 \div (-6)$
16. $-625 \div (-25)$
17. $512 \div (-16)$
18. $-128 \div (-8)$
19. $-700 \div (-25)$
20. $-600 \div 30$

End of Chapter Problems 1-8

Perform the indicated operations:

1. $(-7) \times 3 + 4$
2. $-6 - (-3) + 3 \times 3 - 4$
3. $6 \times (-3) - 6 \times (-3)$
4. $14 - (-4) + 18 \div 3$
5. $6 + 3 \times 5 - 4$
6. $21 + 4 \times 3 - 21 \div 7$
7. $6 \times 4 - 6 \div 2$
8. $(-5) + 6 \times (-4) - 3$
9. $(-7) - (-4) \times 6 + 1$
10. $(-18) \div 3 + 4 \times 6$
11. $-2 - (-2) \times 4 - 5 \times (-4)$
12. $-4 - (-5) \times 3 + 10 \div (-5)$
13. $16 \div (-4) - 7 \times 3 + 4$
14. $(-10) \times (-2) - 6 \times (-10) \div (-5) - 5$
15. $-4 \times (-5) - 3 + 15 \div (-3)$
16. $17 - 3 \times 4 - 24 \div 6$
17. $(-35) \div 5 + 4 + 3 \times 6$
18. $28 \div (-4) - 6 \times (-4) - 3$
19. $7 \times (-3) - 4 \times (-5) + 6 \times (-3)$
20. $(-3) \times (-8) - 4 + 27 \div (-3)$

End of Chapter Problems 1-9

Perform the indicated operations:

1. $-4 - (-2) + 6(6 - 3)$
2. $-9 - (-4) + 5(7 - 9)$
3. $7(-3) - [4(-7)]$
4. $5(-4) + [-3(-4)]$
5. $(5 + 4)(-6 + 2)$
6. $(-4 + 9)(7 - 3)$
7. $5 - 2 - 3[4(3 + 4)] - 2$
8. $-6 - 4 - [4(-3 + 1)] - 7$
9. $4[24 \div (-8 + 4)] + 7 \times 4$
10. $8[27 \div (-9 + 6)] + 6 \times 3$
11. $15 - 4[(-3 + 6)(9 - 12)]$
12. $12 - 5[(4 - 9)(-9 - 3)]$
13. $5[6 - (-4) + (3 - 7)(2 - 6)]$
14. $3[5 - (-5) + (2 - 7)(4 - 9)]$
15. $4[-6 - (-3)(4)] + 7 \times 3$
16. $(4 - 5)(-3 \times 4) - [-6(2 - 4)]$
17. $[9 + (3 \times -7)][18 - (2 \times 3)]$
18. $[24 - (4 \times 5)][14 - (7 \times 4)]$
19. $14 - [(-3) \times 4] + [3(4 - 5)(7 - 3) + 6]$
20. $[-8 - (-4 \times 5)][-15 - (14 \div -2)]$

Answers to Self-tests

SELF-TEST 1-1

1. (a) 2 (b) 8 (c) 0 (d) 7 (e) 300 (f) 8
2. (a) 0.41 (b) 0.009 (c) 0.01783

3. (a) $\frac{3}{10} = 0.3$ (b) $\frac{85}{100} = 0.85$

(c) $\frac{18}{10,000} = 0.0018$

4. (a) Thirty-three hundredths
(b) Four-thousandths

5. (a) $7\frac{46}{100}$ (b) $18\frac{6}{1000}$

6. (a) 76.14 (b) 6.023

7. (a) $3.07 = 3\frac{7}{100}$ (b) $28.063 = 28\frac{63}{1000}$

8.

	Ten	*Hundred*	*Thousand*
(a)	4770	4800	5000
(b)	9710	9700	10,000
(c)	15,790	15,800	16,000
(d)	37,050	37,000	37,000

9.

	Tenth	*Hundredth*	*Thousandth*
(a)	0.1	0.07	0.075
(b)	0.5	0.46	0.461
(c)	0.4	0.41	0.406
(d)	0.4	0.37	0.375

10.

	Ten	*Unit*	*Tenth*	*Hundredth*
(a)	20	17	17.5	17.49
(b)	20	23	23.5	23.46
(c)	40	37	36.5	36.55
(d)	20	21	20.7	20.71

SELF-TEST 1-2

1. 5
2. −18
3. −48
4. 36
5. −12
6. −17
7. 16
8. −9
9. −57
10. −27
11. 13
12. 22
13. 10
14. 37
15. 986

CHAPTER 2 POWERS OF TEN

INTRODUCTION

In our study of electronic circuits and components, we often work with very large numbers or very small numbers. Mathematical tools have been developed to make it easier to work with these numbers. These tools are a math shorthand called *powers of ten* or *scientific notation*. The use of powers of ten makes working with very large or very small numbers more manageable. In this chapter we will develop an understanding of how to use these tools to help in solving electronic problems.

The student will learn how to use a calculator in this chapter. See the Preface and Appendix A for information about calculators. Algorithms will be used where appropriate in the chapter to assist the student in the solution of more complex problems.

CHAPTER OBJECTIVES

In this chapter you will learn how to:

1. Convert numbers to powers of ten form.
2. Express numbers in scientific notation accurate to some given number of significant digits.
3. Add, subtract, multiply, and divide numbers in powers of ten form using a calculator.
4. Solve problems with complex denominators using a calculator.
5. Find the reciprocal of numbers using a calculator.
6. Find the square and square root of numbers using a calculator.

2-1 POSITIVE EXPONENTS

To simplify working with large numbers, we use a math shorthand called *powers of ten* or *scientific notation*. Let's see what these terms mean.

In Table 2-1 we have converted numbers that are 10 or multiples of 10 to powers of ten form. We could continue the table indefinitely, but this table should be long enough to understand the concept.

Let's look at $10 \times 10 \times 10 = 10^3$. In the expression 10^3, the number 10 is the *base* and 3 is the *exponent*. When a multiple of 10 is expressed in this way, it is said to be in the powers of ten form. 10^3 is read "ten to the third power."

The exponent tells us how many times 10 is used as a factor. We may also think of the exponent as telling us how many zeros are included to the right of the digit 1.

10^4 is read "ten to the fourth power." 4 is the exponent. Therefore, 10 is multiplied by itself four times: $10 \times 10 \times 10 \times 10 = 10{,}000$. Again, this is the same as saying that the answer is 1 followed by four zeros. For positive exponents the rule is:

■ **RULE 2-1** In converting from powers of ten form to a number, write the digit 1 and follow with as many zeros as indicated by the exponent.

PRACTICE PROBLEMS 2-1

Convert the following numbers from powers of ten form to regular numbers.

1. 10^2 **2.** 10^1
3. 10^6 **4.** 10^7
5. 10^9

Solutions

1. 100 **2.** 10
3. 1,000,000 **4.** 10,000,000
5. 1,000,000,000

Additional practice problems are at the end of the chapter.

TABLE 2-1

$1 \times 10 = 10 = 10^1$
$10 \times 10 = 100 = 10^2$
$10 \times 10 \times 10 = 1000 = 10^3$
$10 \times 10 \times 10 \times 10 = 10{,}000 = 10^4$
$10 \times 10 \times 10 \times 10 \times 10 = 100{,}000 = 10^5$
$10 \times 10 \times 10 \times 10 \times 10 \times 10 = 1{,}000{,}000 = 10^6$

Now let's consider a problem in which we wish to convert a number to powers of ten form. The power to which 10 is raised (the exponent) is determined by the number of zeros following the digit 1. $100 = 10^2$ because there are two zeros following the digit 1. $100{,}000 = 10^5$ because there are five zeros following the digit 1. Looking at it another way, $100 = 10 \times 10$. 100 equals 10 multiplied by itself two times; therefore, the exponent is 2. $100{,}000 = 10 \times 10 \times 10 \times 10 \times 10$. 100,000 equals 10 multiplied by itself five times; therefore, the exponent is 5. The rule is:

■ **RULE 2-2** In converting some multiple of 10 to powers of ten, count the number of zeros. This number is the exponent. The answer is 10 raised to that power.

PRACTICE PROBLEMS 2-2

Convert the following numbers to powers of ten form:

1. 100
2. 100,000
3. 10
4. 10,000,000
5. 1000
6. 10,000
7. 1,000,000
8. 100,000,000
9. 1,000,000,000

Solutions

1. 10^2
2. 10^5
3. 10^1
4. 10^7
5. 10^3
6. 10^4
7. 10^6
8. 10^8
9. 10^9

The answer to problem 3 could be written simply as 10 since $10^1 = 10$. In the future, when the number is raised to the first power, we will not write the exponent.

Additional practice problems are at the end of the chapter.

2-2 NEGATIVE EXPONENTS

When we dealt with numbers that were multiples of 10, we learned that the exponents were 1 or greater. If the number is 1, the exponent is zero. $10^0 = 1$. There are no zeros following the number 1; therefore, in powers of ten form, the exponent is zero.

Numbers whose values are less than 1 have negative exponents. Table 2-2 shows some of these numbers and their exponents. The exponent tells us how many places to the right of the decimal point the 1 is located.

The numbers in Table 2-2 could also be written as decimal fractions, as shown in Table 2-3.

Table 2-4 combines Table 2-2 and 2-3 and shows the relationship between negative powers of ten, decimal numbers, and decimal fractions.

TABLE 2-2

$1 = 10^0$
$0.1 = 10^{-1}$
$0.01 = 10^{-2}$
$0.001 = 10^{-3}$
$0.0001 = 10^{-4}$
$0.00001 = 10^{-5}$
$0.000001 = 10^{-6}$

TABLE 2-3

$0.1 = \frac{1}{10}$
$0.01 = \frac{1}{100}$
$0.001 = \frac{1}{1000}$
$0.0001 = \frac{1}{10{,}000}$
$0.00001 = \frac{1}{100{,}000}$
$0.000001 = \frac{1}{1{,}000{,}000}$

TABLE 2-4

$10^{-1} = 0.1 = \frac{1}{10} = \frac{1}{10^1}$
$10^{-2} = 0.01 = \frac{1}{100} = \frac{1}{10^2}$
$10^{-3} = 0.001 = \frac{1}{1000} = \frac{1}{10^3}$
$10^{-4} = 0.0001 = \frac{1}{10{,}000} = \frac{1}{10^4}$
$10^{-5} = 0.00001 = \frac{1}{100{,}000} = \frac{1}{10^5}$
$10^{-6} = 0.000001 = \frac{1}{1{,}000{,}000} = \frac{1}{10^6}$

From these relationships, we can develop the following rules:

RULE 2-3 In converting from powers of ten to decimal fractions, the negative exponent tells how many zeros follow the 1 in the denominator.

$$10^{-3} = \frac{1}{1000}$$

$$10^{-4} = \frac{1}{10{,}000}$$

RULE 2-4 In converting from powers of ten to decimal numbers, the negative exponent tells how far the 1 is located to the right of the decimal point.

$$10^{-3} = 0.001$$

$$10^{-4} = 0.0001$$

PRACTICE PROBLEMS 2-3

Convert the following powers of ten to (a) decimal numbers and (b) decimal fractions:

1. 10^{-4}
2. 10^{-2}
3. 10^{-1}
4. 10^{-8}

5. 10^{-3}

6. 10^{-7}

7. 10^{-6}

8. 10^{-5}

Solutions

1. (a) 0.0001 (b) $\frac{1}{10,000}$

2. (a) 0.01 (b) $\frac{1}{100}$

3. (a) 0.1 (b) $\frac{1}{10}$

4. (a) 0.00000001 (b) $\frac{1}{100,000,000}$

5. (a) 0.001 (b) $\frac{1}{1000}$

6. (a) 0.0000001 (b) $\frac{1}{10,000,000}$

7. (a) 0.000001 (b) $\frac{1}{1,000,000}$

8. (a) 0.00001 (b) $\frac{1}{100,000}$

Additional practice problems are at the end of the chapter.

Let's turn it around and consider problems in which we wish to convert decimal numbers and decimal fractions to powers of ten form.

$$0.01 = \frac{1}{100} = \frac{1}{10^2} = 10^{-2}$$

$$0.001 = \frac{1}{1000} = \frac{1}{10^3} = 10^{-3}$$

$$0.0001 = \frac{1}{10,000} = \frac{1}{10^4} = 10^{-4}$$

RULE 2-5 In converting from a decimal fraction to powers of ten form, count the number of zeros in the denominator. This number is the negative power to which 10 is raised.

$$\frac{1}{10,000} = 10^{-4}$$

$$\frac{1}{100} = 10^{-2}$$

RULE 2-6 In converting from a decimal number to powers of ten form, count how far the digit 1 is located to the right of the decimal point. This is the value of the negative exponent.

$$0.0001 = 10^{-4}$$

$$0.01 = 10^{-2}$$

PRACTICE PROBLEMS 2-4

Convert the following decimal numbers to powers of ten form:

1. 0.001
2. 0.1
3. 0.000001
4. 0.01
5. 0.00001
6. 0.0000001
7. 0.0001
8. 0.00000001

Convert the following decimal fractions to powers of ten form:

9. $\frac{1}{1000}$
10. $\frac{1}{100{,}000}$
11. $\frac{1}{10}$
12. $\frac{1}{1{,}000{,}000}$
13. $\frac{1}{100}$
14. $\frac{1}{10{,}000}$

Solutions

1. 10^{-3}
2. 10^{-1}
3. 10^{-6}
4. 10^{-2}
5. 10^{-5}
6. 10^{-7}
7. 10^{-4}
8. 10^{-8}
9. 10^{-3}
10. 10^{-5}
11. 10^{-1}
12. 10^{-6}
13. 10^{-2}
14. 10^{-4}

2-3 MULTIPLICATION IN POWERS OF TEN FORM

RULE 2-7 To multiply numbers in powers of ten form, *add the exponents*.

☐ EXAMPLE 2-1

Find $10^6 \times 10^{-2}$

Solution

$$10^6 \times 10^{-2} = 10^{6+(-2)} = 10^4$$

or

$$1{,}000{,}000 \times 0.01 = 10^4$$

or

$$1{,}000{,}000 \times \frac{1}{100} = 10{,}000 = 10^4$$

☐ EXAMPLE 2-2

Find $10^{-4} \times 10^{3}$.

Solution

$$10^{-4} \times 10^{3} = 10^{-4+3} = 10^{-1}$$

or

$$0.0001 \times 1000 = 0.1 = 10^{-1}$$

or

$$\frac{1}{10,000} \times 1000 = \frac{1}{10} = 0.01 = 10^{-1}$$

☐ EXAMPLE 2-3

Find $10^{-5} \times 10^{-2}$.

Solution

$$10^{-5} \times 10^{-2} = 10^{-7}$$

or

$$\frac{1}{100,000} \times \frac{1}{100} = \frac{1}{10,000,000} = 10^{-7}$$

or

$$0.00001 \times 0.01 = 0.0000001 = 10^{-7}$$

☐

PRACTICE PROBLEMS 2-5

Multiply the following numbers. Express the products in powers of ten form.

1. $10^{2} \times 10^{3}$
2. $10^{4} \times 10^{6}$
3. $10^{-1} \times 10^{-3}$
4. $10^{-3} \times 10^{-5}$
5. $10^{5} \times 10^{-2}$
6. $10^{6} \times 10^{-4}$
7. $10^{2} \times 10^{-3}$
8. $10^{3} \times 10^{-3}$
9. $10^{-4} \times 10^{7}$
10. $10^{-2} \times 10^{6}$

Solutions

1. 10^{5}
2. 10^{10}
3. 10^{-4}
4. 10^{-8}
5. 10^{3}
6. 10^{2}
7. 10^{-1}
8. 10^{0}
9. 10^{3}
10. 10^{4}

Additional practice problems are at the end of the chapter.

2-4 DIVISION IN POWERS OF TEN FORM

RULE 2-8 To divide numbers in powers of ten form, *subtract the exponents*.

☐ EXAMPLE 2-4

Find $\dfrac{10^5}{10^2}$.

Solution

$$\frac{10^5}{10^2} = 10^{5-2} = 10^3$$

or

$$\frac{100{,}000}{100} = 1000 = 10^3$$

☐ EXAMPLE 2-5

Find $\dfrac{10^{-5}}{10^2}$.

Solution

$$\frac{10^{-5}}{10^2} = 10^{-5-2} = 10^{-7}$$

or

$$\frac{0.00001}{100} = 0.0000001 = 10^{-7}$$

☐ EXAMPLE 2-6

Find $\dfrac{10^5}{10^{-2}}$.

Solution

$$\frac{10^5}{10^{-2}} = 10^{5-(-2)} = 10^7$$

or

$$\frac{100{,}000}{0.01} = 10{,}000{,}000 = 10^7$$

□ EXAMPLE 2-7

Find $\dfrac{10^{-5}}{10^{-2}}$.

Solution

$$\frac{10^{-5}}{10^{-2}} = 10^{-5-(-2)} = 10^{-3}$$

or

$$\frac{0.00001}{0.01} = 0.001 = 10^{-3}$$

□

PRACTICE PROBLEMS 2-6

Perform the indicated operations. Express your answers in powers of ten form.

1. $\dfrac{10^4}{10^2}$

2. $\dfrac{10^6}{10^3}$

3. $\dfrac{10^2}{10^5}$

4. $\dfrac{10^1}{10^6}$

5. $\dfrac{10^2}{10^{-2}}$

6. $\dfrac{10^3}{10^{-5}}$

7. $\dfrac{10^{-2}}{10^{-4}}$

8. $\dfrac{10^{-3}}{10^1}$

9. $\dfrac{10^{-4}}{10^{-6}}$

10. $\dfrac{10^{-1}}{10^{-4}}$

11. $\dfrac{10^{-5}}{10^{-3}}$

12. $\dfrac{10^{-9}}{10^{-6}}$

13. $\dfrac{1}{10^2}$

14. $\dfrac{1}{10^{-4}}$

15. $\dfrac{1}{10^6}$

Solutions

1. 10^2

2. 10^3

3. 10^{-3}

4. 10^{-5}

5. 10^4

6. 10^8

7. 10^2

8. 10^{-4}

9. 10^2

10. 10^3

11. 10^{-2}

12. 10^{-3}

13. 10^{-2}

14. 10^4

15. 10^{-6}

Additional practice problems are at the end of the chapter.

2-5 COMBINED MULTIPLICATION AND DIVISION IN POWERS OF TEN FORM

Multiplication and division may be performed in the same problem. We must take care to observe the signs of the exponents as we add or subtract.

$$\frac{10^4 \times 10^2}{10^3} = \frac{10^{4+2}}{10^3} = \frac{10^6}{10^3} = 10^{6-3} = 10^3$$

Notice that we performed the indicated multiplication by adding the exponents. Next we performed the indicated division by subtracting the exponent in the denominator from the exponent in the numerator.

$$\frac{10^{-5} \times 10^3}{10^{-2}} = \frac{10^{-5+3}}{10^{-2}} = \frac{10^{-2}}{10^{-2}} = 10^{-2+2} = 10^0 = 1$$

Remember $10^0 = 1$. In this case $\frac{0.01}{0.01} = 1$.

Let's try two more.

$$\frac{10^3 \times 10^4}{10^{-2} \times 10^6} = \frac{10^{3+4}}{10^{-2+6}} = \frac{10^7}{10^4} = 10^{7-4} = 10^3$$

$$\frac{1}{10^2 \times 10^4} = \frac{1}{10^{2+4}} = \frac{1}{10^6} = 10^{-6}$$

We arrive at the answer 10^{-6} because the numerator (1) can be written in powers of ten form as 10^0. Then we have

$$\frac{10^0}{10^6} = 10^{0-6} = 10^{-6}$$

A very useful rule to remember is:

■ **RULE 2-9** Whenever a power of ten is in the denominator, it can be moved to the numerator by changing the sign of the exponent.

Consider the example above in which we had

$$\frac{10^3 \times 10^4}{10^{-2} \times 10^6}$$

We could have solved by this method:

$$\frac{10^3 \times 10^4}{10^{-2} \times 10^6} = 10^3 \times 10^4 \times 10^2 \times 10^{-6} = 10^3$$

Notice that 10^{-2} in the denominator became 10^2 when it was moved to the numerator. 10^6 in the denominator became 10^{-6} in the numerator. When all powers of ten are in the numerator, we simply add the exponents.

$$10^{3+4+2-6} = 10^3$$

PRACTICE PROBLEMS 2-7

Perform the indicated operations. Express your answers in powers of ten form.

1. $\dfrac{10^2 \times 10^{-3}}{10^4}$
2. $\dfrac{10^{-1} \times 10^{-6}}{10^3}$
3. $\dfrac{10^1 \times 10^0 \times 10^6}{10^{-3}}$
4. $\dfrac{10^{-3} \times 10^4 \times 10^{-6}}{10^{-3}}$
5. $\dfrac{10^{-2} \times 10^4}{10^3 \times 10^2}$
6. $\dfrac{10^2 \times 10^6}{10^3 \times 10^{-1}}$
7. $\dfrac{1}{10^2 \times 10^4 \times 10^1}$
8. $\dfrac{1}{10^{-2} \times 10^6 \times 10^{-8}}$
9. $\dfrac{10^4 \times 10^{-1} \times 10^6}{10^2 \times 10^{-4} \times 10^5}$
10. $\dfrac{10^{-6} \times 10^{-3} \times 10^3}{10^{-9} \times 10^6 \times 10^{-3}}$

Solutions

1. 10^{-5}
2. 10^{-10}
3. 10^{10}
4. 10^{-2}
5. 10^{-3}
6. 10^6
7. 10^{-7}
8. 10^4
9. 10^6
10. 10^0

Additional practice problems are at the end of the chapter.

SELF-TEST 2-1

Convert the following numbers to powers of ten form:

1. 1000
2. 1,000,000
3. 0.001
4. 0.0000001
5. 0.0001
6. 0.00001

Determine the numbers represented by the following powers of ten:

7. 10^3
8. 10^5
9. 10^2
10. 10^0
11. 10^{-4}
12. 10^{-1}
13. 10^{-5}
14. 10^{-8}

Convert the following powers of ten to decimal fractions and to decimal numbers:

15. 10^{-2}
16. 10^{-3}
17. 10^{-5}
18. 10^{-1}

Perform the indicated operations. Express your answers in powers of ten form.

19. $10^2 \times 10^5$
20. $10^3 \times 10^{-4}$
21. $10^6 \times 10^{-2}$
22. $10^{-3} \times 10^{-6}$

23. $\dfrac{10^2}{10^5}$

24. $\dfrac{10^4}{10^{-6}}$

25. $\dfrac{10^{-3}}{10^6}$

26. $\dfrac{10^{-2}}{10^{-5}}$

27. $\dfrac{10^{-2} \times 10^4}{10^{-3}}$

28. $\dfrac{10^1 \times 10^{-5}}{10^4 \times 10^{-9}}$

29. $\dfrac{10^{-4} \times 10^{-1} \times 10^6}{10^6 \times 10^{-3} \times 10^1}$

30. $\dfrac{10^{-2} \times 10^{-3} \times 10^9}{10^2 \times 10^3 \times 10^4}$

31. $\dfrac{10^4 \times 10^{-4} \times 10^6}{10^3 \times 10^{-3} \times 10^{-2}}$

32. $\dfrac{10^1 \times 10^{-7} \times 10^{-2}}{10^{-3} \times 10^{-4} \times 10^2}$

33. $\dfrac{1}{10^2 \times 10^{-3} \times 10^4}$

Answers to Self-test 2-1 are at the end of the chapter.

2-6 CONVERTING BETWEEN REGULAR NUMBERS AND POWERS OF TEN NOTATION

Consider the number 470. Let's convert the number to powers of ten notation with positive exponents.

$$470 = 47.0 \times 10^1 = 4.70 \times 10^2 = 0.470 \times 10^3 = 0.0470 \times 10^4$$

and so on. This is the same as writing

$$470 = 47.0 \times 10 = 4.70 \times 100 = 0.470 \times 1000 = 0.0470 \times 10{,}000$$

and so on. As we move the decimal point *left*, the exponent becomes more positive.

$$470 = 47.0 \times 10^1$$

$$47.0 \times 10^1 = 4.70 \times 10^{1+1} = 4.70 \times 10^2$$

$$4.70 \times 10^2 = 0.470 \times 10^{2+1} = 0.470 \times 10^3$$

$$0.470 \times 10^3 = 0.0470 \times 10^{3+1} = 0.0470 \times 10^4$$

RULE 2-10 For each place we move the decimal point left in a number, the exponent increases by 1.

Said another way, as we make the exponent more positive, the decimal point moves left.

RULE 2-11 For each digit we make the exponent more positive, the decimal point moves 1 place left.

□ EXAMPLE 2-8

Change 275 to powers of ten form with a exponent of **(a)** 2, **(b)** 4.

Solution

(a) $275 = 2.75 \times 10^2$. The exponent changes by 2 in a positive direction, so the decimal point moves 2 places *left*.

(b) $275 = 0.0275 \times 10^4$. The exponent changes by 4 in a positive direction, so the decimal point moves 4 places *left*.

Or, in part (a) we changed the number so that the exponent is 2. If we start from there and change the exponent to 4, we just move the decimal point 2 places farther left.

$$2.75 \times 10^2 = 0.0275 \times 10^4.$$

□

Now let's convert the same number to powers of ten form with negative exponents.

$$470 = 4700 \times 10^{-1} = 47{,}000 \times 10^{-2} = 470{,}000 \times 10^{-3}$$

and so on. As we move the decimal point *right,* the exponent becomes more *negative*. For each place we move the point *right,* the exponent decreases by 1 (becomes more negative).

$$470 = 4700 \times 10^{-1}$$

$$4700 \times 10^{-1} = 47{,}000 \times 10^{-1+(-1)} = 47{,}000 \times 10^{-2}$$

$$47{,}000 \times 10^{-2} = 470{,}000 \times 10^{-2+(-1)} = 470{,}000 \times 10^{-3}$$

■ **RULE 2-12** For each place we move the decimal point *right* in a number, the exponent decreases by 1.

Said another way, as we make the exponent more negative, the decimal point moves *right*.

■ **RULE 2-13** For each digit we make the exponent more negative, the decimal point moves 1 place *right*.

□ EXAMPLE 2-9

Change 275 to powers of ten form with an exponent of **(a)** −2, **(b)** −4.

Solution

(a) $275 = 27{,}500 \times 10^{-2}$. The exponent changes by 2 in a negative direction, so the decimal point moves 2 places *right*.

(b) $275 = 2{,}750{,}000 \times 10^{-4}$. The exponent changes by 4 in a negative direction, so the decimal point moves 4 places *right*.

Or, in part (a) we changed the number so that the exponent is −2. To make the exponent −4, we just move the decimal point 2 places farther right.

$$27{,}500 \times 10^{-2} = 2{,}750{,}000 \times 10^{-4}$$

□

PRACTICE PROBLEMS 2-8

1. Change 38.6 to powers of ten form with an exponent of (a) 2, (b) 3, and (c) 4.
2. Change 5600 to powers of ten form with an exponent of (a) 2, (b) 3, and (c) 4.
3. Change 0.0000905 to powers of ten form with an exponent of (a) −2, (b) −6, and (c) −8.
4. Change 0.000000277 to powers of ten form with an exponent of (a) −5, (b) −7, and (c) −9.
5. Change 1.80 to powers of ten form with an exponent of (a) 1, (b) 3, (c) −1, and (d) −3.
6. Change 0.833 to powers of ten form with an exponent of (a) 6, (b) 3, (c) −3, and (d) −6.

Solutions

1. (a) 0.386×10^2, (b) 0.0386×10^3, (c) 0.00386×10^4
2. (a) 56.00×10^2, (b) 5.600×10^3, (c) 0.5600×10^4
3. (a) 0.00905×10^{-2}, (b) 90.5×10^{-6}, (c) 9050×10^{-8}
4. (a) 0.0277×10^{-5}, (b) 2.77×10^{-7}, (c) 277×10^{-9}
5. (a) 0.180×10, (b) 0.00180×10^3, (c) 18.0×10^{-1}, (d) 1800×10^{-3}
6. (a) 0.000000833×10^6, (b) 0.000833×10^3, (c) 833×10^{-3}, (d) $833{,}000 \times 10^{-6}$

Additional practice problems are at the end of the chapter.

2-7 SCIENTIFIC NOTATION

In science and engineering it is common practice to round answers to two, three, or four significant figures and then to change the number to a number between 1 and 10 times some power of ten. When a number is written in this manner, it is said to be expressed in *scientific notation*. The following examples are numbers expressed in scientific notation:

$$1.76 \times 10^4 \qquad 4.067 \times 10^{-2} \qquad 9.2 \times 10^6$$

As we mentioned in Chapter 1, great accuracy in problem solving is not realistic in electronics because we usually work with components whose values are given to two significant figures (color bands on resistors, for example), and they may be as much as ±10% different than that value. In addition, measuring instruments such as voltmeters and ohmmeters, if of the analog type like a multimeter, do not provide accuracy beyond two figures. Even digital meters at best provide 1% accuracies.

Therefore, throughout the text, with few exceptions, we will not require accuracies beyond three places, even though your calculator can provide eight or more.

Consider the number 764,832. Let's express this number in scientific notation accurate to three significant digits or simply "three places." Let's first round to three places. The number becomes 765,000. We rounded up because the fourth digit was 8. Now let's change the number to a number between 1 and 10. The answer is 7.65. To get 7.65, we had to move the decimal point five places *to the left*. To have an answer that equals 765,000, we must multiply 7.65 by 100,000 or 10^5.

765,000 = 7.65 × 10^5. For each place we move the decimal point left, we must multiply by 10. Since we moved the decimal point five places to the left, we multiplied by 10^5.

☐ EXAMPLE 2-10

Express the number 4703 in scientific notation rounded to three places.

Solution We must move the decimal point three places *left* in order to have a number between 1 and 10. That makes the exponent 3.

$$4703 = 4.703 \times 10^3$$

$$4.703 \times 10^3 = 4.70 \times 10^3, \quad \text{rounded to three places}$$

☐

RULE 2-14 To express numbers greater than 1 in scientific notation, count the number of places the decimal point must move left. This number is the exponent and is positive.

PRACTICE PROBLEMS 2-9

Round the following numbers to three significant figures and express your answers in scientific notation:

1. 1730
2. 26,745
3. 1,654,736
4. 42.69
5. 173,426
6. 47,394
7. 79.063
8. 678,091
9. 189.83
10. 6746

Solutions

1. 1.73×10^3
2. 2.67×10^4
3. 1.65×10^6
4. 4.27×10
5. 1.73×10^5
6. 4.74×10^4
7. 7.91×10^1
8. 6.78×10^5
9. 1.90×10^2
10. 6.75×10^3

Additional practice problems are at the end of the chapter.

Consider the number 0.00715. To change to scientific notation, we must move the decimal point three places to the *right* so that we have a number between 1 and 10. The number is now 7.15. We must multiply this new number by 10^{-3} so that the value of the number in scientific notation equals the value of the original number. For each place we move the decimal point *right*, we must multiply by 0.1 or 10^{-1}. We moved the point *right* three places, so the exponent is −3.

□ EXAMPLE 2-11

Express the number 0.00008706 in scientific notation rounded to three places.

Solution We must move the decimal point five places right to have a number between 1 and 10. That makes the exponent −5.

$$0.00008706 = 8.706 \times 10^{-5}$$

$$8.706 = 8.71 \times 10^{-5}, \quad \text{rounded to three places}$$

□

■ **RULE 2-15** To express numbers that are less than 1 in scientific notation, count the number of places the decimal point must move *right*. This number is the exponent and is negative.

PRACTICE PROBLEMS 2-10

Round the following numbers to three significant figures and express your answers in scientific notation:

1. 0.0167
2. 0.640732
3. 0.0003988
4. 0.00607
5. 0.1086
6. 0.00000706
7. 0.00001763
8. 0.009096
9. 0.07646
10. 0.000567

Solutions

1. 1.67×10^{-2}
2. 6.41×10^{-1}
3. 3.99×10^{-4}
4. 6.07×10^{-3}
5. 1.09×10^{-1}
6. 7.06×10^{-6}
7. 1.76×10^{-5}
8. 9.10×10^{-3}
9. 7.65×10^{-2}
10. 5.67×10^{-4}

Additional practice problems are at the end of the chapter.

2-7-1 Multiplication and Division

Let's work the following problems together.

□ EXAMPLE 2-12

Find the product of 76.3 and 446 using the calculator. Express the answer first in floating-point notation and then in scientific notation rounded to three places.

Solution Your calculator should be capable of displaying answers in various notations. The default notation (the notation it is in when first turned on or when cleared with the [AC] key) usually displays answers as 10-digit numbers. This notation is called *floating-decimal notation* or *floating-point* notation. Let's solve the problem using this notation.

[AC]	All Clear key clears everything and puts us in the floating-decimal notation.
76.3	Key in the first number.
[×]	The function is *multiply,* so we press the [×] key.
446	Enter the second number.
[=]	Press the [=] key to display the answer. If you didn't make any mistakes, the number 34029.8 is displayed. Because the last four digits in this answer are zero, only six digits are displayed.

We can also display answers in scientific notation on the calculator. We will call the keystrokes necessary to place your calculator into scientific notation mode [SCI].

We can choose how many digits we want to display to the right of the decimal point. We are rounding to three places, so we want two places displayed to the right of the decimal point. We will call the key combination necessary to display two digits to the right of the decimal point [FIX] 2.

Now, let's work the problem again. This time the answer will be displayed in scientific notation rounded to three places.

[SCI]	Display answer in *scientific notation.*
[FIX] 2	Display 2 digits right of the decimal point.
76.3	Enter the first number.
[×]	The function is *multiply.*
446	Enter the second number
[=]	Display the answer, which is 3.40×10^4.

If you make a mistake along the way, do not clear with the *all clear* key (usually labeled [AC]). Clear with the *clear entry* key, (usually labeled [CE]). Remember, the [AC] key clears everything and puts us in floating-decimal notation. The [CE] key clears the entry *only.* Use [AC] only when you want to clear everything.

If your calculator does not have the [FIX] feature, it will probably display answers to eight significant digits, or it will display answers in scientific notation accurate to five places. You must then make whatever conversions are necessary with pencil and paper.

□ EXAMPLE 2-13

Find $\frac{57{,}950 \times 10^3}{0.267}$.

Solution

[AC]	Clear the calculator
[SCI]	Change to scientific notation.
[FIX] 2	Display two digits to the right of the decimal point.
57950	Enter the first number.
[EE] 3	Enter the exponent.
[÷]	The function is *divide*.
.267	Enter the number.
[=]	The answer is 2.17×10^8.

□

Most problems containing three or more operations can be solved many different ways using the calculator. We will solve Example 2-14 three ways. These three ways will show the student some additional features of the calculator.

□ EXAMPLE 2-14

Find $\frac{0.00743 \times 10^2 \times 14}{5384 \times 0.04}$.

Solution

[AC]	Clear the calculator.
[SCI]	Change to scientific notation.
[FIX] 2	Display 2 digits to the right of the decimal point.
0.00743	Enter the first number.
[EE] 2	Enter the exponent.
[×]	The function is multiply.
14	Enter the next number.
[÷]	The function is divide.
5384	Enter the next number.
[÷]	The function is divide. The answer to

.04	this point is being *divided* by 0.04.
[=]	The answer is 4.83×10^{-2}.

Let's leave the calculator in this notation mode and solve the problem again.

.00743	Enter the first number.
[EE] 2	Enter the exponent.
[×]	The function is *multiply*.
14	Enter the second number.
[÷]	The function is divide.
[(]	Start a group.
5384	Enter one number of the denominator.
[×]	Multiply
.04	by the other number in the denominator.
[)]	End the group.
[=]	Again the answer is 4.83×10^{-2}.

The parenthesis is a sign of grouping. The calculator will perform all arithmetic operations within the parentheses and then, when the [=] key is pressed, division by this quantity is performed.

Let's try a third way. This time we will use the memory feature of the calculator. We will use [STO] to indicate memory storage and [RCL] to indicate memory recall.

5384	Enter one number of the denominator.
[×]	The function is *multiply*.
.04	Enter the other number in the denominator.
[=]	The product is 2.15×10^{2}.
[STO]	Store this number (the value of the denominator).
.00743	Enter one number in the numerator.
[EE] 2	Enter the exponent.
[×]	The function is *multiply*.
14	Enter the second number in the numerator.
[=]	The value of the numerator is 1.04×10^{1}.

÷	The function is *divide*.
RCL	Recall the value previously stored in memory (the value of the denominator).
=	Again, the answer is 4.83×10^{-2}.

Any of these solutions is valid. It is left up to the student to determine which method he or she prefers. Usually, one way is quicker (requires fewer steps) than another. Students should also refer to their calculator user's manual for additional help when needed.

□

Work the following problems by using your calculator. Because all the operations are multiplications and divisions, there is no preferred choice of the order in which they are performed.

PRACTICE PROBLEMS 2-11

Perform the indicated operations. Express your answers in scientific notation accurate to three places.

1. 47×106

2. 789×206

3. 24.6×0.00173

4. $473 \times 10^{-3} \times 26$

5. 2700×0.743

6. $33 \times 10^{3} \times 65 \times 10^{-4}$

7. $17 \times 43 \times 0.003$

8. $7800 \times 10^{-3} \times 27 \times 10^{-2}$

9. $680 \times 10^{3} \times 473 \times 10^{-7}$

10. $1.8 \times 10^{4} \times 2.3 \times 10^{-6}$

11. $9 \div 0.000742$

12. $25 \div 5600$

13. $15 \div 24 \times 10^{-3}$

14. $12 \div 680 \times 10^{2}$

15. $\dfrac{20 \times 22 \times 10^{3}}{720}$

16. $\dfrac{40 \times 213 \times 10^{-6}}{27 \times 10^{-4}}$

17. $\dfrac{18 \times 10^{3} \times 27 \times 10^{3}}{45 \times 10^{3}}$

18. $\dfrac{18 \times 0.00473}{673 \times 10^{-6}}$

19. $\dfrac{6700 \times 876}{914 \times 732}$

20. $\dfrac{743 \times 10^{2} \times 0.00365}{65 \times 10^{-1} \times 0.403}$

21. $\dfrac{0.000467 \times 10^{4} \times 7680}{2.6 \times 10^{-3} \times 526 \times 10^{2}}$

22. $\dfrac{14 \times 5600}{61.6 \times 10^{4}}$

23. $\dfrac{479 \times 10^{4} \times 5600 \times 10^{-2}}{48.7 \times 10^{-3} \times 9800 \times 10^{4}}$

24. $\dfrac{0.0406 \times 10^{-4} \times 5600 \times 10^{-2}}{7960 \times 10^{2} \times 0.00315 \times 10^{-5}}$

25. $\dfrac{37 \times 10^{3} \times 4.7 \times 10^{4}}{0.000183 \times 10^{2} \times 836 \times 10^{-3}}$

Solutions

1. 4.98×10^{3}

2. 1.63×10^{5}

3. 4.26×10^{-2}

4. 1.23×10^{1}

5. 2.01×10^3
6. 2.15×10^2
7. 2.19×10^0
8. 2.11×10^0
9. 3.22×10^1
10. 4.14×10^{-2}
11. 1.21×10^4
12. 4.46×10^{-3}
13. 6.25×10^2
14. 1.76×10^{-4}
15. 6.11×10^2
16. 3.16×10^0
17. 1.08×10^4
18. 1.27×10^2
19. 8.77×10^0
20. 1.04×10^2
21. 2.62×10^2
22. 1.27×10^{-1}
23. 5.62×10^1
24. 9.07×10^{-3}
25. 1.14×10^{11}

Often an answer like the answer to problem 7 (2.19×10^0) is simply written 2.19, since $10^0 = 1$ ($2.19 \times 10^0 = 2.19 \times 1 = 2.19$).

Additional practice problems are at the end of the chapter.

2-7-2 Addition and Subtraction

We can add or subtract numbers in powers of ten form or when using scientific notation *provided the exponents are the same*. For example, let's perform the following addition: $3 \times 10^2 + 4 \times 10^3$. The decimal numbers are 300 and 4000. If we add 300 and 4000, we get 4300, or 4.3×10^3. To perform the addition using powers of ten, we must first move the decimal point of one number (it doesn't matter which) so that the exponents are the same. Let's change 3×10^2 so that the exponent is 3. We are making the exponent one digit more positive, so the decimal point must move one place *left* (Rule 2-11).

$$3 \times 10^2 = 0.3 \times 10^3$$

Now we can perform the addition:

$$\begin{array}{r} 0.3 \times 10^3 \\ +4 \quad \times 10^3 \\ \hline 4.3 \times 10^3 \end{array}$$

Addition or subtraction is performed by adding the numbers together and multiplying by the indicated power of ten. The *exponents* are not added. The exponents merely tell us that we are adding 0.3 thousands to 4 thousands to get 4.3 *thousands*. We could have changed 4×10^3 to 40×10^2 and added.

$$\begin{array}{r} 3 \times 10^2 \\ 40 \times 10^2 \\ \hline 43 \times 10^2 \end{array}$$

$$43 \times 10^2 = 4.3 \times 10 \times 10^2 = 4.3 \times 10^3$$

The result is the same but more steps are involved.

□ EXAMPLE 2-15

Let's subtract 6×10^{-3} from 10^{-2}.

Solution

$$\begin{array}{r} 1 \times 10^{-2} \\ \underline{(-)6 \times 10^{-3}} \end{array}$$

10^{-2} is written in scientific notation form ($10^{-2} = 1 \times 10^{-2}$) because addition or subtraction of the decimal numbers is to be performed. In this problem we must again change one of the numbers so that both exponents are the same. Let's change 6×10^{-3} so that the exponent is -2. We are making the exponent one digit more positive, so the decimal point must move one place *right* (Rule 2-11).

$$6 \times 10^{-3} = 0.6 \times 10^{-2}$$

Now we can perform the subtraction:

$$\begin{array}{r} 1 \quad \times 10^{-2} \\ \underline{(-)0.6 \times 10^{-2}} \\ 0.4 \times 10^{-2} \end{array}$$

$$0.4 \times 10^{-2} = 4 \times 10^{-1} \times 10^{-2} = 4 \times 10^{-3}$$

We could have changed 1×10^{-2} to a number whose exponent is -3. $1 \times 10^{-2} = 0.01 = 10 \times 10^{-3}$. Because we wanted the exponent to equal -3, we had to move the decimal point one place to the right.

$$\begin{array}{r} 10 \times 10^{-3} \\ \underline{(-)6 \times 10^{-3}} \\ 4 \times 10^{-3} \end{array}$$

We can check our work by performing the subtraction in decimal.

$$\begin{array}{ll} 1 \times 10^{-2} = 0.01 & \quad\quad 0.01 \\ 6 \times 10^{-3} = 0.006 & \quad (-)\underline{0.006} \\ & \quad\quad 0.004 \end{array}$$

$$0.004 = 4 \times 10^{-3}$$

□

RULE 2-16 To add or subtract numbers in powers of ten form, the exponents must be the same.

PRACTICE PROBLEMS 2-12

Perform the indicated operations. Express your answers in scientific notation accurate to three places.

1. $10^2 + 10^3$

2. $10^1 + 10^{-1}$

3. $10^3 - 10^2$

4. $2 \times 10^2 + 3 \times 10^3$

5. $47 \times 10^{-2} + 560 \times 10^{-3}$
6. $40 \times 10^{3} - 300 \times 10^{2}$
7. $3 \times 10^{-2} - 45 \times 10^{-4}$
8. $270 \times 10^{-1} + 46.3 \times 10^{1}$
9. $6.8 \times 10^{3} + 6.8 \times 10^{4} + 56{,}000$
10. $213 \times 10^{-6} + 0.000043 + 51.3 \times 10^{-5}$

Solutions

1. 1.10×10^{3}
2. 1.01×10^{1}
3. 9.00×10^{2}
4. 3.20×10^{3}
5. 1.03×10^{0}
6. 1.00×10^{4}
7. 2.55×10^{-2}
8. 4.90×10^{2}
9. 1.31×10^{5}
10. 7.69×10^{-4}

Additional practice problems are at the end of the chapter.

2-8 PROBLEMS WITH COMPLEX DENOMINATORS

A typical Ohm's law problem might look like this:

$$\frac{25 \times 3.3 \times 10^{3}}{3.3 \times 10^{3} + 4.7 \times 10^{4}}$$

To solve this problem, we must first perform the indicated addition in the denominator.

$$\begin{aligned} 3.3 \times 10^{3} &= 0.33 \times 10^{4} \\ +4.7 \times 10^{4} &= \underline{4.7 \quad \times 10^{4}} \\ & 5.03 \times 10^{4} \end{aligned}$$

Now

$$\frac{25 \times 3.3 \times 10^{3}}{5.03 \times 10^{4}} = 1.64$$

All this math may be done by using the calculator:

25 [×] 3.3 [EE] 3 [÷] [(] 3.3 [EE] 3 [+] 4.7 [EE] 4 [)] [=]

Answer: 1.64

When using a calculator with the parentheses feature, enclosing the denominator is the easiest way to solve this kind of a problem.

☐ EXAMPLE 2-16

Perform the indicated operations. Express your answers in scientific notation rounded to three places.

(a) $\dfrac{9 \times 18 \times 10^{3}}{18 \times 10^{3} + 56 \times 10^{4}}$

(b) $\dfrac{150 \times 6800}{6800 + 39 \times 10^{3}}$

(c) $\dfrac{20 \times 1.2 \times 10^6}{1.2 \times 10^6 + 8.2 \times 10^6}$

(d) $\dfrac{14 \times 56 \times 10^3}{1.5 \times 10^3 + 5.6 \times 10^4}$

Solutions

(a) 2.80×10^{-1}

(b) 2.23×10^1

(c) 2.55×10^0

(d) 1.36×10^1

□

2-9 RECIPROCALS

■ KEY POINT The *reciprocal* of a number is that number divided into 1. $\frac{1}{3}$ is the reciprocal of 3, $\frac{1}{50}$ is the reciprocal of 50, and so on. In electronics it is often necessary to find the reciprocal of a number and then perform some other mathematical operation with that reciprocal.

□ EXAMPLE 2-17

Find the reciprocal of 27×10^3. Express the answer in scientific notation accurate to three places.

Solution Using the calculator, key in 27×10^3 and then press the reciprocal key [1/x]. The answer is 3.70×10^{-5}.

□ EXAMPLE 2-18

Find $\dfrac{1}{22 \times 10^3} + \dfrac{1}{47 \times 10^3}$.

Solution 22 [EE] 3 [1/x] [+] 47 [EE] 3 [1/x] [=]

Answer: 6.67×10^{-5}

□ EXAMPLE 2-19

Find $\dfrac{1}{\dfrac{1}{3.3 \times 10^3} + \dfrac{1}{6.8 \times 10^3}}$.

Solution 3.3 [EE] 3 [1/x] + 6.8 [EE] 3 [1/x] [=] [1/x]

Answer: 2.22×10^3

We took the reciprocal of the first number in the denominator and then added the reciprocal of the second number. At the end of this step ([=]), you should have displayed 4.50×10^{-4}. The reciprocal of this number is the answer, which is 2.22×10^3.

□

PRACTICE PROBLEMS 2-13

Perform the indicated operations. Express your answers in scientific notation accurate to three places.

1. $\dfrac{1}{470}$

2. $\dfrac{1}{33 \times 10^{4}}$

3. $\dfrac{1}{68 \times 10^{-9}}$

4. $\dfrac{1}{3.7 \times 10^{-3}}$

5. $\dfrac{1}{2760}$

6. $\dfrac{1}{5.6 \times 10^{3}}$

7. $\dfrac{1}{50 \times 10^{3}}$

8. $\dfrac{1}{370 \times 10^{-6}}$

9. $\dfrac{1}{4 \times 10^{-3}}$

10. $\dfrac{1}{10 \times 10^{3}}$

11. $\dfrac{1}{2700} + \dfrac{1}{1800}$

12. $\dfrac{1}{68 \times 10^{3}} + \dfrac{1}{47 \times 10^{3}}$

13. $\dfrac{1}{12 \times 10^{3}} + \dfrac{1}{7.5 \times 10^{3}}$

14. $\dfrac{25 \times 4700}{4700 + 47 \times 10^{3}}$

15. $\dfrac{20 \times 3.3 \times 10^{3}}{3.3 \times 10^{3} + 3.3 \times 10^{4}}$

16. $\dfrac{9 \times 10^{4}}{10^{3} + 10^{4}}$

17. $\dfrac{1}{\dfrac{1}{27 \times 10^{3}} + \dfrac{1}{18 \times 10^{3}}}$

18. $\dfrac{1}{\dfrac{1}{6800} + \dfrac{1}{5600}}$

19. $\dfrac{1}{27.6 \times 10^{3}} - \dfrac{1}{47 \times 10^{3}}$

20. $\dfrac{1}{\dfrac{1}{270} + \dfrac{1}{470} + \dfrac{1}{560}}$

Solutions

1. 2.13×10^{-3}

2. 3.03×10^{-6}

3. 1.47×10^{7}

4. 2.70×10^{2}

5. 3.62×10^{-4}

6. 1.79×10^{-4}

7. 2.00×10^{-5}

8. 2.70×10^{3}

9. 2.50×10^{2}

10. 1.00×10^{-4}

11. 9.26×10^{-4}

12. 3.60×10^{-5}

13. 2.17×10^{-4}

14. 2.27×10^{0}

15. 1.82×10^{0}

16. 8.18×10^{0}

17. 1.08×10^{4}

18. 3.07×10^{3}

19. 1.50×10^{-5}

20. 1.31×10^{2}

Additional practice problems are at the end of the chapter.

SELF-TEST 2-2

Perform the indicated operations. Express your answers in scientific notation accurate to three places.

1. 27×65
2. 0.00173×3900
3. $47{,}000 \times 0.0000423$
4. $1765 \times 24{,}673$
5. $33 \times 10^4 \times 27 \times 10^3$
6. $5.6 \times 10^3 + 4.7 \times 10^4$
7. $147 \times 10^{-7} + 0.833 \times 10^{-5} + 21.3 \times 10^{-6}$
8. $2 \times 3.14 \times 2500 \times 500 \times 10^{-9}$
9. $\dfrac{47.3 \times 10^2}{0.00846}$
10. $\dfrac{0.025 \times 10^{-6}}{483 \times 10^{-9}}$
11. $\dfrac{7800 \times 0.00273}{1.28 \times 10^{-4} \times 8000}$
12. $\dfrac{0.0000064 \times 84 \times 10^2}{6730 \times 0.045 \times 10^{-3}}$
13. $\dfrac{12 \times 3900}{56{,}000 + 3900}$
14. $\dfrac{1}{270} + \dfrac{1}{390}$
15. $\dfrac{25 \times 4700}{4700 + 47{,}000}$
16. $\dfrac{1}{3900} + \dfrac{1}{56 \times 10^3} + \dfrac{1}{20 \times 10^3}$
17. $\dfrac{1}{6.28 \times 12 \times 10^3 \times 470 \times 10^{-12}}$
18. $\dfrac{25 \times 2700}{2.7 \times 10^3 + 3.3 \times 10^4}$
19. $\dfrac{1}{\dfrac{1}{560 \times 10^2} + \dfrac{1}{47 \times 10^3}}$
20. $\dfrac{1}{\dfrac{1}{2700} + \dfrac{1}{3300} + \dfrac{1}{4700}}$

Answers to Self-test 2-2 are at the end of the chapter.

2-10 POWERS AND ROOTS IN BASE TEN

When we wrote 10^3, we learned that the exponent (3) told us to multiply 10 by itself three times.

$$10^3 = 10 \times 10 \times 10$$

If we wrote $(10^3)^2$, we would be raising 10^3 to the second power. That is,

$$(10^3)^2 = 10^3 \times 10^3 = 10^6$$

We need the parentheses to identify the quantity we are acting on. Consider the quantity $(10^{-2})^3$.

$$(10^{-2})^3 = 10^{-2} \times 10^{-2} \times 10^{-2} = 10^{-6}.$$

In each case the exponent can be thought of as a multiplier.

$$(10^3)^2 = 10^{3\times 2} = 10^6 \qquad \text{and} \qquad (10^{-2})^3 = 10^{-2\times 3} = 10^{-6}$$

☐ EXAMPLE 2-20

Perform the indicated operations.

(a) $(10^4)^{-3}$

(b) $\left(\frac{1}{10^4}\right)^2$

Solutions

(a) $(10^4)^{-3} = 10^{4\times(-3)} = 10^{-12}$

(b) $\left(\frac{1}{10^4}\right)^2 = \frac{1^2}{10^{4\times 2}} = \frac{1}{10^8} = 10^{-8}$

When the exponent is in the denominator, we usually move it to the numerator. ☐

■ **RULE 2-17** When raising a power to a power, multiply the exponents.

PRACTICE PROBLEMS 2-14

Perform the indicated operations. Express your answers in powers of ten form.

1. $(10^2)^3$
2. $(10^4)^2$
3. $(10^3)^3$
4. $(10^{-1})^2$
5. $(10^{-4})^3$
6. $(10^{-6})^2$
7. $(10^{-2})^{-3}$
8. $(10^{-3})^{-1}$
9. $(10^{-1})^{-4}$
10. $(10^3)^{-5}$
11. $(10^2)^{-1}$
12. $(10^6)^{-2}$
13. $\left(\frac{1}{10^2}\right)^4$
14. $\left(\frac{1}{10^{-3}}\right)^2$
15. $\left(\frac{1}{10^3}\right)^{-3}$

Solutions

1. 10^6
2. 10^8
3. 10^9
4. 10^{-2}
5. 10^{-12}
6. 10^{-12}
7. 10^6
8. 10^3
9. 10^4
10. 10^{-15}
11. 10^{-2}
12. 10^{-12}
13. 10^{-8}
14. 10^6
15. 10^9

Additional practice problems are at the end of the chapter.

To find the square root or cube root of powers of ten, we multiply the exponents. $(10^4)^{1/2} = 10^{4\times 1/2} = 10^2$. $(10^{12})^{1/3} = 10^{12\times 1/3} = 10^4$.

$$\left(\frac{1}{10^6}\right)^{1/2} = \frac{1}{10^{6\times 1/2}} = \frac{1}{10^3}$$

PRACTICE PROBLEMS 2-15

Perform the indicated operations. Express your answers in powers of ten form.

1. $(10^6)^{1/2}$
2. $(10^9)^{1/3}$
3. $(10^{-6})^{1/2}$
4. $(10^{-12})^{1/3}$
5. $\left(\frac{1}{10^{-3}}\right)^{1/3}$
6. $(10^{-6})^{1/3}$
7. $(10^3)^{1/3}$
8. $\left(\frac{1}{10^4}\right)^{1/2}$
9. $(10^{-15})^{1/3}$
10. $(10^{10})^{1/2}$

Solutions

1. 10^3
2. 10^3
3. 10^{-3}
4. 10^{-4}
5. 10^1
6. 10^{-2}
7. 10^1
8. 10^{-2}
9. 10^{-5}
10. 10^5

Additional practice problems are at the end of the chapter.

Let's square some numbers other than base 10 and express the answers in scientific notation accurate to three places.

☐ EXAMPLE 2-21

Find 14.7^2.

Solution This problem requires the use of another function key on the calculator. This key is usually labeled [x^2]. Pressing this key will square the contents of the x register (the value displayed). Use whatever keystrokes are necessary to perform this function. Using the calculator, 14.7 [x^2]. The answer, 2.16×10^2, is displayed.

☐ EXAMPLE 2-22

Find $(0.0236 \times 10^2)^2$.

Solution Using the calculator, 0.0236 [EE] 2 [x^2]. *Answer*: 5.57

☐ EXAMPLE 2-23

Find $\dfrac{278^2}{16}$.

Solution Using the calculator, 278 $\boxed{x^2}$ $\boxed{\div}$ 16 $\boxed{=}$. *Answer*: 4.83×10^3. Notice that only the number 278 is squared. The number 16 is not squared. ☐

PRACTICE PROBLEMS 2-16

Perform the indicated operations. Express your answers in scientific notation accurate to three places.

1. 6.3^2
2. 43^2
3. 0.0372^2
4. 0.00047^2
5. $(6.73 \times 10^2)^2$
6. $(1.04 \times 10^{-3})^2$
7. $(796 \times 10^{-2})^2$
8. 8764^2
9. 0.973^2
10. $(0.00567 \times 10^{-4})^2$
11. $(0.0000873 \times 10^4)^2$
12. $(1260 \times 10^{-3})^2$
13. $(0.0073)^2 \times 4700$
14. $(27 \times 10^{-6})^2 \times 68{,}000$
15. $20^2 \div 3300$
16. $9^2 \div 270$

Solutions

1. 3.97×10^1
2. 1.85×10^3
3. 1.38×10^{-3}
4. 2.21×10^{-7}
5. 4.53×10^5
6. 1.08×10^{-6}
7. 6.34×10^1
8. 7.68×10^7
9. 9.47×10^{-1}
10. 3.21×10^{-13}
11. 7.62×10^{-1}
12. 1.59×10^0
13. 2.50×10^{-1}
14. 4.96×10^{-5}
15. 1.21×10^{-1}
16. 3.00×10^{-1}

Additional practice problems are at the end of the chapter.

The following problems require using the square root key $\boxed{\sqrt{x}}$. We often use the notation $x^{1/2}$ instead of $\sqrt{x}$. Raising a number to the $\frac{1}{2}$ power is just another way of saying that the operation is "square root."

☐ EXAMPLE 2-24

Find the square root of 376.

Solution Using the calculator, 376 $\boxed{\sqrt{x}}$. The answer is 1.94×10.

☐ EXAMPLE 2-25

Find $\frac{1}{6.28(0.05 \times 5 \times 10^{-9})^{1/2}}$.

Solution Using the calculator:

0.05	We first find the square root of
[×]	the quantity inside the parentheses.
5	
[EE]	
[+/-]	
9	
[=]	$0.05 \times 5 \times 10^{-9} = 2.50 \times 10^{-10}$
[$\sqrt{x}$]	The square root of 2.50×10^{-10} is 1.58×10^{-5}.
[×]	
6.28	Multiply by 6.28.
[=]	9.93×10^{-5}.
[1/x]	The reciprocal of 9.93×10^{-5} is 1.01×10^{4}.

☐

PRACTICE PROBLEMS 2-17

Perform the indicated operations. Express your answer in scientific notation accurate to three places.

1. $36.5^{1/2}$
2. $0.925^{1/2}$
3. $(635 \times 10^{-2})^{1/2}$
4. $(1870 \times 10^{3})^{1/2}$
5. $(0.000675 \times 10^{2})^{1/2}$
6. $(27.3 \times 10^{2})^{1/2}$
7. $(5.46 \times 10^{-4})^{1/2}$
8. $\left(\frac{734 \times 10^{2}}{423}\right)^{1/2}$
9. $\left(\frac{0.00726 \times 10^{-3}}{77 \times 10^{-6}}\right)^{1/2}$
10. $\frac{1}{6.28(0.2 \times 3 \times 10^{-9})^{1/2}}$

Solutions

1. 6.04×10^{0}
2. 9.62×10^{-1}
3. 2.52×10^{0}
4. 1.37×10^{3}
5. 2.60×10^{-1}
6. 5.22×10

7. 2.34×10^{-2}
8. 1.32×10
9. 3.07×10^{-1}
10. 6.50×10^{3}

Additional practice problems are at the end of the chapter.

SELF-TEST 2-3

Perform the indicated operations. Express your answers in scientific notation accurate to three places.

1. $(10^3)^5$
2. $(10^2)^{-4}$
3. $\left(\frac{1}{10^{-2}}\right)^3$
4. $(10^{-6})^{1/2}$
5. $(10^6)^{1/3}$
6. $(10^{-3})^{1/3}$
7. 78.3^2
8. 0.000293^2
9. $(8796 \times 10^3)^2$
10. $(0.0173)^2 \times 1800$
11. $(39 \times 10^{-4})^2 \times 120$
12. $\frac{15^2}{1500}$
13. $273^{1/2}$
14. $(2.65 \times 10^{-4})^{1/2}$
15. $\left(\frac{0.0000467 \times 10^3}{63}\right)^{1/2}$
16. $\left(\frac{645 \times 10^2}{14}\right)^{1/2}$
17. $\frac{(24.5 \times 10^{-3})^{1/2}}{17}$
18. $(300 \times 10^{-3} \times 220)^{1/2}$
19. $\left(\frac{273 \times 10^4}{0.0012}\right)^{1/2}$
20. $\frac{1}{6.28(50 \times 10^{-3} \times 2 \times 10^{-6})^{1/2}}$

Answers to Self-test 2-3 are at the end of the chapter.

CHAPTER 2 AT A GLANCE

Page	Key Point or Rule	Example
25	*Rule 2-1:* In converting from powers of ten form to a number, write the digit 1 and follow with as many zeros as indicated by the exponent.	$10^3 = 1000$
26	*Rule 2-2:* In converting some multiple of 10 to powers of ten, count the number of zeros. This number is the exponent. The answer is 10 raised to that power.	$1000 = 10^3$
27	*Rule 2-3:* In converting from powers of ten to decimal fractions, the negative exponent tells	$10^{-2} = \frac{1}{100}$

Page	Key Point or Rule	Example
	how many zeros follow the 1 in the denominator.	
27	*Rule 2-4:* In converting from powers of ten to decimal numbers, the negative exponent tells how far the 1 is located to the right of the decimal point.	$10^{-2} = 0.01$
28	*Rule 2-5:* In converting from a decimal fraction to powers of ten form, count the number of zeros in the denominator. This number is the negative power to which 10 is raised.	$\frac{1}{100} = 10^{-2}$
28	*Rule 2-6:* In converting from a decimal number to powers of ten form, count how far the digit 1 is located to the right of the decimal point. This is the value of the negative exponent.	$0.01 = 10^{-2}$
29	*Rule 2-7:* To multiply numbers in powers of ten form, add the exponents.	$10^2 \times 10^3 = 10^{2+3} = 10^5$ $10^2 \times 10^{-3} = 10^{2+(-3)}$ $= 10^{-1}$
31	*Rule 2-8:* To divide numbers in powers of ten form, subtract the exponents.	$\frac{10^4}{10^2} = 10^{4-2} = 10^2$
33	*Rule 2-9:* Whenever a power of ten is in the denominator, it can be moved to the numerator by changing the sign of the exponent.	$\frac{10^2}{10^{-3}} = 10^2 \times 10^3$
35	*Rule 2-10:* For each place we move the decimal point *left,* the exponent increases by 1.	$1750 = 175 \times 10$ $= 17.5 \times 10^2$ $= 1.75 \times 10^3$
36	*Rule 2-12:* For each place we move the decimal point *right,* the exponent decreases by 1.	$0.00425 = 0.0425 \times 10^{-1}$ $= 0.425 \times 10^{-2}$ $= 4.25 \times 10^{-3}$
38	*Rule 2-14:* To express numbers greater than 1 in scientific notation, count the number of places the decimal point must move left. This number is the exponent and is positive.	$704 = 7.04 \times 10^2$ $37{,}654 = 3.77 \times 10^4$ (rounded to three places)
39	*Rule 2-15:* To express numbers that are less than 1 in scientific notation, count the number of places the decimal point must move right. This number is the exponent and is negative.	$0.00156 = 1.56 \times 10^{-3}$ $0.00027643 = 2.76 \times 10^{-4}$ (rounded to three places)

Page	Key Point or Rule	Example
45	*Rule 2-16:* To add or subtract numbers in powers of ten form, the exponents must be the same.	$3 \times 10^4 = 3 \times 10^4$ $+ 3 \times 10^3 = 0.3 \times 10^4$ $= 3.3 \times 10^4$
47	*Key point:* The reciprocal of a number is that number divided into 1.	Find the reciprocal of 150. $\frac{1}{150}$ $= 6.67 \times 10^{-3}$
50	*Rule 2-17:* When raising a power to a power, multiply the exponents.	$(10^2)^3 = 10^{2\times 3} = 10^6$ $(10^6)^{\frac{1}{2}} = 10^{6\times\frac{1}{2}} = 10^3$ $(3.6 \times 10^3)^2 = 1.30 \times 10^7$ $(746 \times 10^{-3})^{\frac{1}{2}} = 8.64 \times 10^{-1}$

End of Chapter Problems 2-1

Convert the following power of ten numbers to decimal numbers

1. 10^3
2. 10^5
3. 10^0
4. 10^8
5. 10^6
6. 10^4

End of Chapter Problems 2-2

Convert the following numbers to powers of ten form.

1. 10
2. 100,000
3. 1000
4. 10,000,000
5. 100
6. 1,000,000

End of Chapter Problems 2-3

Convert the following powers of ten numbers to (a) decimal fractions and (b) decimal numbers.

1. 10^{-2}
2. 10^{-5}
3. 10^{-3}
4. 10^{-6}
5. 10^{-1}
6. 10^{-4}

End of Chapter Problems 2-4

Convert the following decimal numbers to powers of ten form:

1. 0.0001
2. 0.01
3. 0.000001
4. 0.001

Convert the following decimal fractions to powers of ten form:

5. $\frac{1}{10,000}$

6. $\frac{1}{10,000,000}$

7. $\frac{1}{1000}$

8. $\frac{1}{1,000,000}$

End of Chapter Problems 2-5

Perform the indicated operations. Express your answers in powers of ten form.

1. $10^4 \times 10^6$
2. $10^3 \times 10^2$
3. $10^2 \times 10^4$
4. $10^3 \times 10^5$
5. $10^5 \times 10^{-1}$
6. $10^6 \times 10^{-3}$
7. $10^3 \times 10^{-7}$
8. $10^2 \times 10^{-6}$
9. $10^{-1} \times 10^4$
10. $10^{-2} \times 10^6$
11. $10^{-6} \times 10^{-4}$
12. $10^{-7} \times 10^{-3}$
13. $10^{-3} \times 10^{-9}$
14. $10^{-2} \times 10^{-5}$
15. $10^{-4} \times 10^{-5}$
16. $10^{-3} \times 10^{-6}$

End of Chapter Problems 2-6

Perform the indicated operations. Express your answers in powers of ten form.

1. $\frac{10^7}{10^3}$
2. $\frac{10^9}{10^4}$
3. $\frac{10^3}{10^8}$
4. $\frac{10^2}{10^5}$
5. $\frac{10^4}{10^{-4}}$
6. $\frac{10^7}{10^{-3}}$
7. $\frac{10^3}{10^{-9}}$
8. $\frac{10^4}{10^{-6}}$
9. $\frac{10^{-2}}{10^{-9}}$
10. $\frac{10^{-1}}{10^{-7}}$
11. $\frac{10^{-8}}{10^{-5}}$
12. $\frac{10^{-6}}{10^{-5}}$
13. $\frac{10^{-3}}{10^3}$
14. $\frac{10^{-6}}{10^1}$
15. $\frac{10^{-2}}{10^5}$
16. $\frac{10^{-4}}{10^7}$
17. $\frac{1}{10^{-3}}$
18. $\frac{1}{10^{-7}}$
19. $\frac{1}{10^4}$
20. $\frac{1}{10^6}$

End of Chapter Problems 2-7

Perform the indicated operations. Express your answers in powers of ten form.

1. $\dfrac{10^{-3} \times 10^{-6}}{10^{-2}}$

2. $\dfrac{10^{2} \times 10^{6}}{10^{4}}$

3. $\dfrac{10^{-6} \times 10^{4}}{10^{5}}$

4. $\dfrac{10^{-4} \times 10^{9}}{10^{3}}$

5. $\dfrac{1}{10^{2} \times 10^{-4} \times 10^{-9}}$

6. $\dfrac{1}{10^{-2} \times 10^{4} \times 10^{-7}}$

7. $\dfrac{1}{10^{-3} \times 10^{-6} \times 10^{-1}}$

8. $\dfrac{1}{10^{2} \times 10^{6} \times 10^{-3}}$

9. $\dfrac{10^{-4}}{10^{6} \times 10^{-2} \times 10^{4}}$

10. $\dfrac{10^{3}}{10^{4} \times 10^{6} \times 10^{-1}}$

11. $\dfrac{10^{-3} \times 10^{2} \times 10^{4}}{10^{2} \times 10^{-7}}$

12. $\dfrac{10^{-1} \times 10^{2} \times 10^{4}}{10^{4} \times 10^{6}}$

13. $\dfrac{10^{4} \times 10^{6} \times 10^{3}}{10^{2} \times 10^{-7} \times 10^{5}}$

14. $\dfrac{10^{-1} \times 10^{-7} \times 10^{6}}{10^{1} \times 10^{0} \times 10^{-6}}$

15. $\dfrac{10^{3} \times 10^{-5} \times 10^{6}}{10^{4} \times 10^{-3} \times 10^{7}}$

16. $\dfrac{10^{-2} \times 10^{-6} \times 10^{3}}{10^{3} \times 10^{6} \times 10^{-1}}$

17. $\dfrac{10^{-3} \times 10^{-6} \times 10^{-9}}{10^{3} \times 10^{6} \times 10^{9}}$

18. $\dfrac{10^{-9} \times 10^{3} \times 10^{-3}}{10^{4} \times 10^{-1} \times 10^{7}}$

End of Chapter Problems 2-8

Change the following numbers to powers of ten form with an exponent of (a) 2, (b) 4, and (c) 6.

1. 475

2. 593

3. 65.9

4. 10.7

5. 93,400

6. 27,000

7. 4780

8. 3300

Change the following numbers to powers of ten form with an exponent of (a) −2, (b) −4, and (c) −6.

9. 0.00465

10. 0.00758

11. 0.00000555

12. 0.00000906

13. 0.000673

14. 0.000825

15. 0.0000000108

16. 0.0000000745

Change the following numbers to powers of ten form with an exponent of (a) 3, and (b) −3.

17. 325

18. 722

19. 4.75

20. 8.04

End of Chapter Problems 2-9

Round the following numbers to three significant figures and express your answers in scientific notation.

1. 27,640
2. 17,896
3. 47.8
4. 276.7
5. 1,765,400
6. 3,706,400
7. 273.46
8. 45.986
9. 173,460
10. 407,600
11. 78.88
12. 15.706
13. 1,670,000
14. 15,673
15. 57,945
16. 83,636
17. 80,975
18. 40,783
19. 470,560
20. 399,999

End of Chapter Problems 2-10

Round the following numbers to three significant figures and express your answers in scientific notation.

1. 0.004783
2. 0.17600
3. 0.7474
4. 0.0002087
5. 0.01637
6. 0.01037
7. 0.000002778
8. 0.000005278
9. 0.0170983
10. 0.07096
11. 0.0008964
12. 0.0006794
13. 0.0045446
14. 0.0078449
15. 0.000050085
16. 0.000030067
17. 0.25465
18. 0.56374
19. 0.040045
20. 0.074037

End of Chapter Problems 2-11

Perform the indicated operations. Express your answers in scientific notation accurate to three places.

1. 273×78
2. 793×467
3. 0.0000273×4700
4. 0.00675×0.00147
5. $\dfrac{50}{0.00475}$
6. $\dfrac{12}{0.0000273}$
7. $\dfrac{468 \times 10^3}{76{,}300}$
8. $\dfrac{52{,}800 \times 10^{-2}}{9260}$
9. $\dfrac{176 \times 473}{27{,}800}$
10. $\dfrac{4630 \times 127}{4360}$
11. $\dfrac{0.00173 \times 287}{379}$
12. $\dfrac{0.000814 \times 706}{556}$

13. $\dfrac{76 \times 10^{-2} \times 906 \times 10^{4}}{0.000559}$

14. $\dfrac{283 \times 10^{-3} \times 990 \times 10^{4}}{0.00466}$

15. $\dfrac{47{,}600 \times 0.00316 \times 10^{-2}}{689 \times 10^{3}}$

16. $\dfrac{9240 \times 0.000643 \times 10^{4}}{0.00329 \times 10^{3}}$

17. $\dfrac{0.00790 \times 835 \times 10^{2}}{9840 \times 10^{-2}}$

18. $\dfrac{0.00238 \times 600 \times 10^{3}}{365 \times 10^{-3}}$

19. $\dfrac{651 \times 0.00179}{73.4 \times 0.874}$

20. $\dfrac{722 \times 0.000196}{0.673 \times 5800}$

21. $\dfrac{127 \times 10^{2} \times 0.024 \times 10^{-4}}{79 \times 10^{-6} \times 410 \times 10^{4}}$

22. $\dfrac{470 \times 10^{4} \times 0.00715 \times 10^{-2}}{27 \times 10^{-4} \times 330 \times 10^{-1}}$

23. $\dfrac{7630 \times 10^{-2} \times 875 \times 10^{3}}{43 \times 10^{3} \times 600 \times 10^{-6}}$

24. $\dfrac{2740 \times 10^{-3} \times 475 \times 10^{2}}{68 \times 10^{3} \times 400 \times 10^{-9}}$

End of Chapter Problems 2-12

Perform the indicated operations. Express your answers in scientific notation rounded to three places.

1. $10^{3} + 10^{4}$
2. $10^{2} + 10^{1}$
3. $10^{-1} + 10^{-1}$
4. $10^{-1} + 10^{-2}$
5. $33 \times 10^{2} + 56 \times 10^{1}$
6. $56 \times 10^{3} + 68 \times 10^{4}$
7. $470 \times 10^{3} + 1.2 \times 10^{4}$
8. $750 \times 10^{3} + 2.2 \times 10^{4}$
9. $27 \times 10^{-2} + 180 \times 10^{-4}$
10. $56 \times 10^{-2} + 270 \times 10^{-4}$
11. $45 \times 10^{-3} - 63 \times 10^{-4}$
12. $75 \times 10^{-3} - 24 \times 10^{-4}$
13. $8.2 \times 10^{3} + 8.2 \times 10^{4} + 9100$
14. $2.7 \times 10^{2} + 5.6 \times 10^{3} + 1200$
15. $43 \times 10^{4} - 3.9 \times 10^{3}$
16. $15 \times 10^{4} - 8.2 \times 10^{3}$

End of Chapter Problems 2-13

Perform the indicated operations. Express your answers in scientific notation accurate to three places.

1. $\dfrac{1}{6.28 \times 50 \times 10^{3} \times 30 \times 10^{-9}}$

2. $\dfrac{1}{6.28 \times 500 \times 10 \times 10^{-6}}$

3. $\dfrac{1}{6.28 \times 25 \times 10^{3} \times 50 \times 10^{-3}}$

4. $\dfrac{1}{6.28 \times 2 \times 10^{3} \times 5 \times 10^{-6}}$

5. $\dfrac{1}{680} + \dfrac{1}{560}$

6. $\dfrac{1}{470} + \dfrac{1}{330}$

7. $\dfrac{1}{3300} + \dfrac{1}{2700} + \dfrac{1}{1000}$

8. $\dfrac{1}{6800} + \dfrac{1}{5600} + \dfrac{1}{8200}$

9. $\dfrac{1}{56 \times 10^{3}} + \dfrac{1}{47 \times 10^{3}}$

10. $\dfrac{1}{27 \times 10^{4}} + \dfrac{1}{10 \times 10^{4}}$

11. $\dfrac{1}{470 \times 10^{3}} + \dfrac{1}{330 \times 10^{3}}$

12. $\dfrac{1}{6.8 \times 10^{3}} + \dfrac{1}{7.5 \times 10^{3}}$

13. $\dfrac{1}{27 \times 10^{3}} + \dfrac{1}{56 \times 10^{3}} + \dfrac{1}{18 \times 10^{3}}$

14. $\dfrac{1}{820 \times 10^{3}} + \dfrac{1}{430 \times 10^{3}} + \dfrac{1}{270 \times 10^{3}}$

15. $\dfrac{1}{\dfrac{1}{4700} + \dfrac{1}{5600}}$

16. $\dfrac{1}{\dfrac{1}{270} + \dfrac{1}{680}}$

17. $\dfrac{1}{\dfrac{1}{270} + \dfrac{1}{180} + \dfrac{1}{390}}$

18. $\dfrac{1}{\dfrac{1}{7500} + \dfrac{1}{5600} + \dfrac{1}{2200}}$

19. $\dfrac{1}{\dfrac{1}{56 \times 10^3} + \dfrac{1}{68 \times 10^3}}$

20. $\dfrac{1}{\dfrac{1}{2.2 \times 10^3} + \dfrac{1}{750}}$

21. $\dfrac{1}{\dfrac{1}{9.1 \times 10^3} + \dfrac{1}{8.2 \times 10^3} + \dfrac{1}{5.6 \times 10^3}}$

22. $\dfrac{1}{\dfrac{1}{18 \times 10^3} + \dfrac{1}{39 \times 10^3} + \dfrac{1}{47 \times 10^3}}$

23. $\dfrac{25 \times 3300}{3300 + 56{,}000}$

24. $\dfrac{20 \times 4700}{33{,}000 + 4700}$

25. $\dfrac{12 \times 5600}{5600 + 33{,}000}$

26. $\dfrac{5 \times 680}{680 + 1800}$

27. $\dfrac{12 \times 4.7 \times 10^3}{4.7 \times 10^4 + 4.7 \times 10^3}$

28. $\dfrac{15 \times 1.8 \times 10^3}{5.6 \times 10^4 + 1.8 \times 10^3}$

29. $\dfrac{20 \times 6.8 \times 10^3}{6.8 \times 10^3 + 4.7 \times 10^4}$

30. $\dfrac{5 \times 1.2 \times 10^3}{1.2 \times 10^3 + 1.5 \times 10^4}$

31. $\dfrac{25 \times 10^{-3} \times 213 \times 10^{-6}}{213 \times 10^{-6} + 303 \times 10^{-6}}$

32. $\dfrac{130 \times 10^{-6} \times 2 \times 10^{-3}}{2 \times 10^{-3} + 3 \times 10^{-3}}$

33. $\dfrac{15 \times 10^{-3} \times 303 \times 10^{-6}}{303 \times 10^{-6} + 370 \times 10^{-6}}$

34. $\dfrac{12 \times 10^{-3} \times 17.9 \times 10^{-6}}{17.9 \times 10^{-6} + 20 \times 10^{-6}}$

End of Chapter Problems 2-14

Perform the indicated operations.

1. $(10^2)^3$
2. $(10^2)^4$
3. $(10^5)^2$
4. $(10^{-2})^1$
5. $(10^3)^{-4}$
6. $(10^2)^{-6}$
7. $(10^{-4})^{-3}$
8. $\left(\dfrac{1}{10^3}\right)^2$
9. $\left(\dfrac{1}{10^{-4}}\right)^{-2}$
10. $\left(\dfrac{1}{10^{-3}}\right)^{-2}$

End of Chapter Problems 2-15

Perform the indicated operations.

1. $(10^6)^{1/2}$
2. $(10^8)^{1/2}$
3. $(10^{-4})^{1/2}$
4. $(10^{-2})^{1/2}$
5. $\left(\dfrac{1}{10^4}\right)^{1/2}$
6. $\left(\dfrac{1}{10^8}\right)^{1/2}$

7. $\left(\frac{1}{10^{-10}}\right)^{1/2}$

8. $\left(\frac{1}{10^{-6}}\right)^{1/2}$

9. $\left(\frac{10^2}{10^{-4}}\right)^{1/2}$

10. $\left(\frac{10^5}{10^{-3}}\right)^{1/2}$

End of Chapter Problems 2-16

Perform the indicated operations. Express your answers in scientific notation accurate to three places.

1. 79.3^2
2. 246^2
3. 1780^2
4. 5400^2
5. 0.04758^2
6. 0.06509^2
7. 0.00873^2
8. 0.00185^2
9. $(45 \times 10^3)^2$
10. $(364 \times 10^2)^2$
11. $(300 \times 10^3)^2$
12. $(500 \times 10^3)^2$
13. $(0.296 \times 10^{-3})^2$
14. $(0.555 \times 10^{-4})^2$
15. $(0.00903 \times 10^{-2})^2$
16. $(0.0033 \times 10^{-2})^2$
17. $(50 \times 10^{-3})^2 \times 2700$
18. $(375 \times 10^{-9})^2 \times 1200$
19. $(0.0765 \times 10^2)^2 \times 0.037$
20. $(0.00821 \times 10^3)^2 \times 0.976$

End of Chapter Problems 2-17

Perform the indicated operations. Express your answers in scientific notation accurate to three places.

1. $93^{\frac{1}{2}}$
2. $873^{\frac{1}{2}}$
3. $237^{\frac{1}{2}}$
4. $4756^{\frac{1}{2}}$
5. $0.00705^{\frac{1}{2}}$
6. $0.0065^{\frac{1}{2}}$
7. $0.000105^{\frac{1}{2}}$
8. $0.000387^{\frac{1}{2}}$
9. $(270 \times 10^5)^{\frac{1}{2}}$
10. $(43.2 \times 10^4)^{\frac{1}{2}}$
11. $(4500 \times 10^4)^{\frac{1}{2}}$
12. $(7400 \times 10^3)^{\frac{1}{2}}$
13. $(0.00923 \times 10^{-3})^{\frac{1}{2}}$
14. $(0.176 \times 10^4)^{\frac{1}{2}}$
15. $(0.000346 \times 10^5)^{\frac{1}{2}}$
16. $(0.0906 \times 10^{-6})^{\frac{1}{2}}$
17. $\left(\frac{0.289 \times 10^{-1}}{254}\right)^{\frac{1}{2}}$
18. $\left(\frac{0.00773 \times 10^{-4}}{330}\right)^{\frac{1}{2}}$
19. $\frac{1}{6.28(20 \times 10^{-3} \times 20 \times 10^{-9})^{1/2}}$
20. $\frac{1}{6.28(0.0047 \times 470 \times 10^{-12})^{1/2}}$
21. $\frac{1}{6.28(0.150 \times 680 \times 10^{-12})^{1/2}}$
22. $\frac{1}{6.28(100 \times 10^{-3} \times 30 \times 10^{-9})^{1/2}}$

Answers to Self-tests

SELF-TEST 2-1

1. 10^3
2. 10^6

3. 10^{-3}
4. 10^{-7}
5. 10^{-4}
6. 10^{-5}
7. 1000
8. 100,000
9. 100
10. 1
11. 0.0001
12. 0.1
13. 0.00001
14. 0.00000001
15. $\frac{1}{100} = 0.01$
16. $\frac{1}{1000} = 0.001$
17. $\frac{1}{100{,}000} = 0.00001$
18. $\frac{1}{10} = 0.1$
19. 10^{7}
20. 10^{-1}
21. 10^{4}
22. 10^{-9}
23. 10^{-3}
24. 10^{10}
25. 10^{-9}
26. 10^{3}
27. 10^{5}
28. 10
29. 10^{-3}
30. 10^{-5}
31. 10^{8}
32. 10^{-3}
33. 10^{-3}

SELF-TEST 2-2

1. 1.76×10^{3}
2. 6.75×10^{0}
3. 1.99×10^{0}
4. 4.35×10^{7}
5. 8.91×10^{9}
6. 5.26×10^{4}
7. 4.43×10^{-5}
8. 7.85×10^{-3}
9. 5.59×10^{5}
10. 5.18×10^{-2}
11. 2.08×10^{1}
12. 1.78×10^{-1}
13. 7.81×10^{-1}
14. 6.27×10^{-3}
15. 2.27×10^{0}
16. 3.24×10^{-4}
17. 2.82×10^{4}
18. 1.89×10^{0}
19. 2.56×10^{4}
20. 1.13×10^{3}

SELF-TEST 2-3

1. 10^{15}
2. 10^{-8}
3. 10^{6}
4. 10^{-3}
5. 10^{2}
6. 10^{-1}
7. 6.13×10^{3}
8. 8.58×10^{-8}
9. 7.74×10^{13}
10. 5.39×10^{-1}
11. 1.83×10^{-3}
12. 1.50×10^{-1}
13. 1.65×10^{1}
14. 1.63×10^{-2}
15. 2.72×10^{-2}
16. 6.79×10
17. 9.21×10^{-3}
18. 8.12×10^{0}
19. 4.77×10^{4}
20. 5.04×10^{2}

CHAPTER 3

UNITS AND PREFIXES

INTRODUCTION

In this chapter, we will continue our study of the calculator and number manipulating. The use of prefixes is introduced, and we show how the use of prefixes makes reading large and small quantities easy. In Chapter 2, we learned how to express answers in scientific notation. In this chapter, we will learn how to express answers in *engineering* notation. Engineering notation makes converting to prefixes easier, since both prefixes and engineering notation move the decimal point in increments of three. In the study of electronic principles and circuits, we will find that prefixes are used to express the values of current, resistance, capacitance, frequency, and many other quantities. Because prefixes are so widely used, it is essential that you understand how to use them.

The metric system is the international system of weights and measurements. Because of the need for a single, internationally accepted system, the metric system is gradually replacing the English system in this country.

In electronics, we already use the international system (SI) of measurements for most quantities, such as voltage, current, capacitance, and conductance, to name a few; but some measurements, such as distance, velocity, and weight, may be specified in either the English system or the metric system. For this reason, we must be able to convert from one system to the other.

CHAPTER OBJECTIVES

In this chapter you will learn how to:

1. Convert numbers to engineering notation.
2. Apply the various prefixes used in the electronics field.
3. Convert from one prefix to another.
4. Affix the correct prefix to electrical quantities.
5. Use the calculator in the engineering notation mode.
6. Apply the various prefixes used in the metric system.
7. Convert between the English and metric systems of measurements.

3-1 UNITS

Most of the units we deal with are either very large or very small. To deal with these quantities more easily, prefixes are used. The prefixes relate to powers of ten that are multiples of 3 or −3. *Engineering notation* is the process of moving the decimal point in a number so that the resulting power of ten is changed to a multiple of 3 or −3. Some examples are 10^3, 10^6, 10^{-3}, and 10^{-6}. Once we have changed a number into engineering notation, we can replace the power of ten with the prefix that represents that power of ten.

Before we begin to use engineering notation and these prefixes, let's review powers of ten by changing a number to powers of ten notation with positive exponents. The rule is:

> **RULE 3-1** If we make the exponent more positive, the decimal point moves to the left.

Consider the number 6700. Let's change the number using powers of ten notation with positive exponents.

$$6700 = 670 \times 10^1 = 67 \times 10^2 = 6.7 \times 10^3 = 0.67 \times 10^4$$

and so on. Notice that as we make the exponent more positive, the decimal point moves farther left.

Now let's convert some numbers to powers of ten form using engineering notation. We will use Rule 3-1 and work some example problems.

□ EXAMPLE 3-1

Change 4700 to a number times 10^3.

Solution We are changing the number to powers of ten form and we want the exponent to be 3, so the decimal point moves *left* three places.

$$4700 = 4.7 \times 10^3$$

□ EXAMPLE 3-2

Change 670×10^2 to a number times 10^3.

Solution We can solve this two ways. First, we could change 670×10^2 to a regular number. $670 \times 10^2 = 67{,}000$. Then, applying the same rule as before, $67{,}000 = 67 \times 10^3$. Another approach would be to note that the exponent is already 2. The difference between 10^2 and 10^3 is 10^1. Therefore, we need to move the decimal point one place to the left.

$$670 \times 10^2 = 67 \times 10^3$$

□ EXAMPLE 3-3

Change 330,000 to some number times 10^3 and then to some number times 10^6.

Solution Again, we move the decimal point three places to the left because the exponent is 3. Then we move the decimal point six places to the left because the exponent is 6.

$$330{,}000 = 330 \times 10^3$$

$$330{,}000 = 0.330 \times 10^6$$

Once we have 330×10^3, we can do this:

$$330 \times 10^3 = 0.330 \times 10^6$$

In this case, the exponent increases from 3 to 6, a change of 3 in a positive direction. Therefore, the decimal point moves three places to the left. □

Now let's change numbers to power of ten form with negative exponents. The rule is:

RULE 3-2 If the exponent is made more negative the decimal point moves to the right.

Consider the number 0.00275. Let's change to powers of ten notation using negative exponents and following Rule 3-2.

$$0.00275 = 0.0275 \times 10^{-1} = 0.275 \times 10^{-2} = 2.75 \times 10^{-3}$$

and so on. Notice that as we make the exponent more negative, the decimal point moves farther right.

Now, let's convert some numbers to powers of ten form using engineering notation. We will use Rule 3-2 and work some example problems.

□ EXAMPLE 3-4

Change 1.73 to some number times 10^{-3}.

Solution The exponent is −3. Therefore, we move the decimal point three places to the *right*.

$$1.73 = 1730 \times 10^{-3}$$

□ EXAMPLE 3-5

Change 27.7×10^{-4} to some number times 10^{-6}.

Solution We can solve the problem by changing 27.7×10^{-4} to a regular number and then moving the decimal point six places to the right.

$$27.7 \times 10^{-4} = 0.00277 = 2770 \times 10^{-6}$$

We could also solve the problem by noting that the exponent is already −4. Increasing its value to −6 is a change of 2 in a negative direction. Therefore, the decimal point moves two places to the right. Remember, if we make the exponent more negative, the decimal point moves to the *right*.

$$27.7 \times 10^{-4} = 2770 \times 10^{-6}$$

□

Now let's consider numbers for which we make the exponent more positive and then more negative.

□ EXAMPLE 3-6

Change 0.05×10^{-7} to some number times 10^{-9} and then to some number times 10^{-6}.

Solution When we change the power of ten from 10^{-7} to 10^{-9}, we are making the exponent more negative. The difference is -2. That is, $10^{-7-2} = 10^{-9}$. Therefore, the decimal point must move to the *right*.

$$0.05 \times 10^{-7} = 5 \times 10^{-9}$$

Next we need to change the exponent so that it is -6. We must add 1 to -7. $-7 + 1 = -6$. That is, $10^{-7+1} = 10^{-6}$. According to our first rule, if we *add* 1 to the exponent, we must move the decimal point to the *left*.

$$0.05 \times 10^{-7} = 0.005 \times 10^{-6}$$

□ EXAMPLE 3-7

Change 4.7 to some number times 10^3 and then to some number times 10^{-3}.

Solution

$$4.7 = 0.0047 \times 10^3$$

Since the exponent is 3, the decimal point moves to the *left*.

$$4.7 = 4700 \times 10^{-3}$$

Since the exponent is -3, the decimal point moves to the *right*.

□

KEY POINT Expressing numbers as numbers between 1 and 10 times the proper power of ten is called *scientific* notation. Expressing numbers as numbers times powers of ten that are multiples of 3 is called *engineering* notation.

PRACTICE PROBLEMS 3-1

Change the following numbers to numbers times 10^3 and to numbers times 10^6:

1. 27,000

2. 330,000

3. 5600

4. 390×10^2

5. 68×10^4

6. 1,200,000

7. 180×10^4

8. 1500×10^5

9. 1800×10^2

10. 51×10^5

Change the following numbers to numbers times 10^{-3} and to numbers times 10^{-6}:

11. 0.000423

12. 0.00716

13. 0.000014

14. 28.3×10^{-4}

15. 173×10^{-2}

16. 8.3×10^{-1}

17. 173×10^{-5}

18. 1.73×10^{-1}

19. 17.3×10^{-4}

20. 0.0706

Change the following numbers to numbers times 10^{-9} and to numbers times 10^{-12}:

21. 0.0046×10^{-4}

22. 0.413×10^{-7}

23. 6.73×10^{-10}

24. 6.73×10^{-8}

25. 67.3×10^{-11}

26. 0.0173×10^{-6}

27. 0.0563×10^{-8}

28. $76{,}500 \times 10^{-15}$

29. 6.43×10^{-10}

30. 906×10^{-10}

Change the following numbers to regular numbers, to numbers times 10^{-3}, and to numbers times 10^3:

31. 17×10^{-1}

32. 0.637×10^2

33. 716×10^{-2}

34. 47×10^1

35. 27.3×10^1

36. 3.93×10^2

37. 3.93×10^{-2}

38. 0.043×10^4

39. 0.706×10^{-1}

40. 89.1×10^2

Solutions

1. $27{,}000 = 27 \times 10^3 = 0.027 \times 10^6$

2. $330{,}000 = 330 \times 10^3 = 0.330 \times 10^6$

3. $5600 = 5.60 \times 10^3 = 0.0056 \times 10^6$

4. $390 \times 10^2 = 39 \times 10^3 = 0.039 \times 10^6$

5. $68 \times 10^4 = 680 \times 10^3 = 0.680 \times 10^6$

6. $1{,}200{,}000 = 1200 \times 10^3 = 1.20 \times 10^6$

7. $180 \times 10^4 = 1800 \times 10^3 = 1.8 \times 10^6$

8. $1500 \times 10^5 = 150{,}000 \times 10^3 = 150 \times 10^6$

9. $1800 \times 10^2 = 180 \times 10^3 = 0.180 \times 10^6$

10. $51 \times 10^5 = 5100 \times 10^3 = 5.1 \times 10^6$

11. $0.000423 = 0.423 \times 10^{-3} = 423 \times 10^{-6}$

12. $0.00716 = 7.16 \times 10^{-3} = 7160 \times 10^{-6}$

13. $0.000014 = 0.014 \times 10^{-3} = 14 \times 10^{-6}$

14. $28.3 \times 10^{-4} = 2.83 \times 10^{-3} = 2830 \times 10^{-6}$

15. $173 \times 10^{-2} = 1730 \times 10^{-3} = 1{,}730{,}000 \times 10^{-6}$

16. $8.3 \times 10^{-1} = 830 \times 10^{-3} = 830{,}000 \times 10^{-6}$

17. $173 \times 10^{-5} = 1.73 \times 10^{-3} = 1730 \times 10^{-6}$

18. $1.73 \times 10^{-1} = 173 \times 10^{-3} = 173{,}000 \times 10^{-6}$

19. $17.3 \times 10^{-4} = 1.73 \times 10^{-3} = 1730 \times 10^{-6}$

20. $0.0706 = 70.6 \times 10^{-3} = 70{,}600 \times 10^{-6}$

21. $0.0046 \times 10^{-4} = 460 \times 10^{-9} = 460{,}000 \times 10^{-12}$

22. $0.413 \times 10^{-7} = 41.3 \times 10^{-9} = 41{,}300 \times 10^{-12}$

23. $6.73 \times 10^{-10} = 0.673 \times 10^{-9} = 673 \times 10^{-12}$

24. $6.73 \times 10^{-8} = 67.3 \times 10^{-9} = 67{,}300 \times 10^{-12}$

25. $67.3 \times 10^{-11} = 0.673 \times 10^{-9} = 673 \times 10^{-12}$

26. $0.0173 \times 10^{-6} = 17.3 \times 10^{-9} = 17{,}300 \times 10^{-12}$

27. $0.0563 \times 10^{-8} = 0.563 \times 10^{-9} = 563 \times 10^{-12}$

28. $76{,}500 \times 10^{-15} = 0.0765 \times 10^{-9} = 76.5 \times 10^{-12}$

29. $6.43 \times 10^{-10} = 0.643 \times 10^{-9} = 643 \times 10^{-12}$

30. $906 \times 10^{-10} = 90.6 \times 10^{-9} = 90{,}600 \times 10^{-12}$

31. $17 \times 10^{-1} = 1.7 = 1700 \times 10^{-3} = 0.0017 \times 10^{3}$

32. $0.637 \times 10^{2} = 63.7 = 63{,}700 \times 10^{-3} = 0.0637 \times 10^{3}$

33. $716 \times 10^{-2} = 7.16 = 7160 \times 10^{-3} = 0.00716 \times 10^{3}$

34. $47 \times 10^{1} = 470 = 470{,}000 \times 10^{-3} = 0.47 \times 10^{3}$

35. $27.3 \times 10^{1} = 273 = 273{,}000 \times 10^{-3} = 0.273 \times 10^{3}$

36. $3.93 \times 10^{2} = 393 = 393{,}000 \times 10^{-3} = 0.393 \times 10^{3}$

37. $3.93 \times 10^{-2} = 0.0393 = 39.3 \times 10^{-3} = 0.0000393 \times 10^{3}$

38. $0.043 \times 10^{4} = 430 = 430{,}000 \times 10^{-3} = 0.43 \times 10^{3}$

39. $0.706 \times 10^{-1} = 0.0706 = 70.6 \times 10^{-3} = 0.0000706 \times 10^{3}$

40. $89.1 \times 10^{2} = 8910 = 8{,}910{,}000 \times 10^{-3} = 8.91 \times 10^{3}$

Additional practice problems are at the end of the chapter.

3-2 PREFIXES

Table 3-1 is a partial list of the International System of Units (SI). Only those units of interest to electronics students are included. Prefixes used with these electrical units are listed in Table 3-2. Prefixes are used to make it easier to communicate with each other. For example, suppose the value of a capacitance was 0.000000003 F. This is a very clumsy number to work with. If we express it in scientific notation, we get 3×10^{-9} F. That's better, but still awkward. Looking at our table of prefixes, we see that 10^{-9} is a *nano* unit and the symbol is n (lowercase n). We would write the value as 3 nF. "Three nanofarads" is the way we would say it.

TABLE 3-1

Quantity	***Unit of measure***	***Unit abbreviation***
Capacitance (*C*)	farad	F
Admittance (*Y*)	siemens	S
Conductance (G)	siemens	S
Current (*I*, *i*)	ampere	A
Energy (*w*)	joule	J
Frequency (*f*)	hertz	Hz
Impedance (*Z*)	ohm	Ω
Inductance (*L*)	henry	H
Power (*P*)	watt	W
Reactance (*X*)	ohm	Ω
Resistance (*R*)	ohm	Ω
Susceptance (*B*)	siemens	S
Time (*t*)	second	s
Wavelength (λ)	meter	m
Electromotive force (*E*)	volt	V
Difference in potential (*V*)	volt	V

TABLE 3-2

Prefix	*Symbol*	*Power of ten*
Exa	E	10^{18}
Peta	P	10^{15}
Tera	T	10^{12}
Giga	G	10^{9}
Mega	M	10^{6}
kilo	k	10^{3}
Units	none	10^{0}
milli	m	10^{-3}
micro	μ	10^{-6}
nano	n	10^{-9}
pico	p	10^{-12}
femto	f	10^{-15}
atto	a	10^{-18}

2,700,000 Ω would be simpler to express if we used the prefix equal to 10^6, which is *mega* and is symbolized as M.

$$2{,}700{,}000\ \Omega = 2.7 \times 10^6\ \Omega = 2.7\ \text{M}\Omega$$

Suppose we calculate a circuit current to be 5.6×10^{-4} A. We need to replace the power of ten with some prefix to make the quantity easier to read. Since our prefixes are multiples of 10^3 or 10^{-3}, we have to change 10^{-4} to either 10^{-3} or 10^{-6}. Let's do both. First change 5.6×10^{-4} to a number times 10^{-3}.

$$5.6 \times 10^{-4}\ \text{A} = 0.56 \times 10^{-3}\ \text{A}$$

To change the exponent from −4 to −3, we must add 1. $10^{-4+1} = 10^{-3}$. Because we made the exponent positive, the decimal point moved to the *left* (Rule 3-1). The prefix for 10^{-3} is *milli* (m). Then

$$5.6 \times 10^{-4}\ \text{A} = 0.56 \times 10^{-3}\ \text{A} = 0.56\ \text{mA}$$

Now let's change 5.6×10^{-4} to a number times 10^{-6}.

$$5.6 \times 10^{-4}\ \text{A} = 560 \times 10^{-6}\ \text{A}$$

To change the exponent from −4 to −6, we had to add −2. $10^{-4+(-2)} = 10^{-6}$. Because we made the exponent more negative, the decimal point moved to the *right* (Rule 3-2). The prefix for 10^{-6} is *micro* (the Greek letter μ is the symbol).

$$5.6 \times 10^{-4}\ \text{A} = 560 \times 10^{-6}\ \text{A} = 560\ \mu\text{A}$$

PRACTICE PROBLEMS 3-2

Using the indicated prefixes, change the following quantities:

1. 2700 Ω = ______________ kΩ = ______________ MΩ

2. 12,000 Hz = ______________ kHz = ______________ MHz

3. $0.00076\ A =$ __________ $mA =$ __________ μA
4. $0.000023\ A =$ __________ $mA =$ __________ μA
5. $0.0000002\ F =$ __________ $\mu F =$ __________ nF
6. $0.000004\ F =$ __________ $\mu F =$ __________ nF
7. $68{,}000\ \Omega =$ __________ $k\Omega =$ __________ $M\Omega$
8. $120{,}000\ \Omega =$ __________ $k\Omega =$ __________ $M\Omega$
9. $3{,}500{,}000\ Hz =$ __________ $kHz =$ __________ MHz
10. $0.00037\ S =$ __________ $mS =$ __________ μS
11. $5 \times 10^{-4}\ S =$ __________ $mS =$ __________ μS
12. $5.6 \times 10^{4}\ \Omega =$ __________ $k\Omega =$ __________ $M\Omega$
13. $21 \times 10^{-2}\ A =$ __________ $mA =$ __________ μA
14. $68 \times 10^{-10}\ F =$ __________ $nF =$ __________ pF
15. $45 \times 10^{-4}\ H =$ __________ $mH =$ __________ μH
16. $0.15 \times 10^{-2}\ V =$ __________ $mV =$ __________ μV

Solutions

1. $2700\ \Omega = 2.7\ k\Omega = 0.0027\ M\Omega$
2. $12{,}000\ Hz = 12\ kHz = 0.012\ MHz$
3. $0.00076\ A = 0.76\ mA = 760\ \mu A$
4. $0.000023\ A = 0.023\ mA = 23\ \mu A$
5. $0.0000002\ F = 0.2\ \mu F = 200\ nF$
6. $0.000004\ F = 4\ \mu F = 4000\ nF$
7. $68{,}000\ \Omega = 68\ k\Omega = 0.068\ M\Omega$
8. $120{,}000\ \Omega = 120\ k\Omega = 0.12\ M\Omega$
9. $3{,}500{,}000\ Hz = 3500\ kHz = 3.5\ MHz$
10. $0.00037\ S = 0.37\ mS = 370\ \mu S$
11. $5 \times 10^{-4}\ S = 0.5\ mS = 500\ \mu S$
12. $5.6 \times 10^{4}\ \Omega = 56\ k\Omega = 0.056\ M\Omega$
13. $21 \times 10^{-2}\ A = 210\ mA = 210{,}000\ \mu A$
14. $68 \times 10^{-10}\ F = 6.8\ nF = 6800\ pF$
15. $45 \times 10^{-4}\ H = 4.5\ mH = 4500\ \mu H$
16. $0.15 \times 10^{-2}\ V = 1.5\ mV = 1500\ \mu V$

Additional practice problems are at the end of the chapter.

Next, let's change quantities to basic units. The rule is:

RULE 3-3 To change numbers with prefixes to regular numbers, replace the prefix with the power of ten that it represents and then change to a regular number.

$$2200\ \mu S = 2200 \times 10^{-6}\ S = 0.0022\ S$$

$$68\ k\Omega = 68 \times 10^{3}\ \Omega = 68{,}000\ \Omega$$

$$30\ mA = 30 \times 10^{-3}\ A = 0.03\ A$$

$$25\ kHz = 25 \times 10^{3}\ Hz = 25{,}000\ Hz$$

PRACTICE PROBLEMS 3-3

Perform the indicated operations:

1. $46.7\ mA =$ __________ A
2. $407\ mA =$ __________ A
3. $68\ k\Omega =$ __________ Ω
4. $4.7\ k\Omega =$ __________ Ω

5. 2.73 kHz = ______________ Hz

6. 300 kHz = ______________ Hz

7. 670 μS = ______________ S

8. 37 mS = ______________ S

9. 55 mH = ______________ H

10. 465 μH = ______________ H

11. 120 kΩ = ______________ Ω

12. 2 MΩ = ______________ Ω

13. 150 mA = ______________ A

14. 0.43 mA = ______________ A

15. 0.68 kΩ = ______________ Ω

Solutions

1. 46.7 mA = 0.0467 A
2. 407 mA = 0.407 A
3. 68 kΩ = 68,000 Ω
4. 4.7 kΩ = 4700 Ω
5. 2.73 kHz = 2730 Hz
6. 300 kHz = 300,000 Hz
7. 670 μS = 0.00067 S
8. 37 mS = 0.037 S
9. 55 mH = 0.055 H
10. 465 μH = 0.000465 H
11. 120 kΩ = 120,000 Ω
12. 2 MΩ = 2,000,000 Ω
13. 150 mA = 0.15 A
14. 0.43 mA = 0.00043 A
15. 0.68 kΩ = 680 Ω

Additional practice problems are at the end of the chapter.

Now let's change from one prefix to another.

☐ EXAMPLE 3-8

20 mA = ______________ μA

Solution In changing from mA to μA, we are really changing the exponent from -3 to -6; so let's rewrite the problem substituting powers of ten for prefixes.

$$20 \times 10^{-3} = \underline{\hspace{3cm}} \times 10^{-6}$$

We are making the exponent more *negative*. Remember Rule 3-2: *If we make the exponent more negative, the decimal point moves to the right*. The change in the value of the exponent from 10^{-3} to 10^{-6} is a change of -3. Therefore, the decimal point moves *three places to the right*.

20 mA = 20,000 μA

☐ EXAMPLE 3-9

Change 450 μA to mA.

Solution In changing from μA to mA, we are really changing the exponent from -6 to -3; so let's rewrite the problem substituting powers of ten for prefixes.

$$450 \times 10^{-6}\text{ A} = \underline{\hspace{3cm}} \times 10^{-3}\text{ A}$$

We are making the exponent more *positive*. Remember Rule 3-1: *If we make the exponent more positive, the decimal point moves to the left*. The change in the value of the exponent from 10^{-6} to 10^{-3} is a change of +3. Therefore, the decimal point moves *three places to the left*.

$$450 \times 10^{-6} \text{ A} = 0.450 \times 10^{-3} \text{ A}$$

$$450\ \mu\text{A} = 0.450 \text{ mA}$$

□ EXAMPLE 3-10

Change 680 kΩ to MΩ.

Solution Substituting powers of ten for prefixes, we get

$$680 \times 10^{3}\ \Omega = \underline{\hspace{3cm}} \times 10^{6}\ \Omega$$

In going from kΩ to MΩ, 10^3 Ω to 10^6 Ω, we are making the exponent more positive. Therefore, the decimal point moves to the *left*. The change from 10^3 to 10^6 is a change in the exponent of +3. Using Rule 3-1, the decimal point moves three places to the left.

$$680 \text{ k}\Omega = 0.680 \text{ M}\Omega$$

□ EXAMPLE 3-11

Change 0.01 μF to pF.

Solution Substituting powers of ten for prefixes, we get

$$0.01 \times 10^{-6} \text{ F} = \underline{\hspace{3cm}} \times 10^{-12} \text{ F}$$

In going from μF to pF, 10^{-6} to 10^{-12}, we are making the exponent more negative. Therefore, the decimal point moves to the *right*. The change from 10^{-6} to 10^{-12} is a change in the exponent of −6. Using Rule 3-2, the decimal point moves six places to the right.

$$0.01\ \mu\text{F} = 10{,}000 \text{ pF}$$

□

In the following practice problems you will either change prefixes or change to basic units.

PRACTICE PROBLEMS 3-4

1. 20 mA = ______________ μA = ______________ A
2. 0.01 μF = ______________ nF = ______________ pF
3. 2 H = ______________ mH = ______________ μH
4. 2.5 kΩ = ______________ Ω = ______________ MΩ
5. 680 Ω = ______________ kΩ = ______________ MΩ
6. 1.8 V = ______________ mV = ______________ μV
7. 20 μA = ______________ mA = ______________ A
8. 1.5 MΩ = ______________ kΩ = ______________ Ω

9. 0.02 S = ______________ mS = ______________ μS
10. 50 nF = ______________ μF = ______________ pF
11. $4.7 \times 10^4\ \Omega$ = ______________ kΩ = ______________ MΩ
12. 16×10^{-7} F = ______________ μF = ______________ nF
13. 2.4×10^{-2} A = ______________ mA = ______________ μA
14. $8.4 \times 10^2\ \Omega$ = ______________ kΩ = ______________ MΩ
15. $56 \times 10^5\ \Omega$ = ______________ kΩ = ______________ MΩ
16. 27.3 mS = ______________ S = ______________ μS
17. 5600 μS = ______________ S = ______________ mS
18. 170 mW = ______________ W = ______________ μW
19. 0.000642 V = ______________ μV = ______________ mV
20. 30 kHz = ______________ MHz = ______________ Hz

Solutions

1. 20 mA = 20,000 μA = 0.02 A
2. 0.01 μF = 10 nF = 10,000 pF
3. 2 H = 2000 mH = 2,000,000 μH
4. 2.5 kΩ = 2500 Ω = 0.0025 MΩ
5. 680 Ω = 0.68 kΩ = 0.00068 MΩ
6. 1.8 V = 1800 mV = 1,800,000 μV
7. 20 μA = 0.02 mA = 0.00002 A
8. 1.5 MΩ = 1500 kΩ = 1,500,000 Ω
9. 0.02 S = 20 mS = 20,000 μS
10. 50 nF = 0.05 μF = 50,000 pF
11. $4.7 \times 10^4\ \Omega$ = 47 kΩ = 0.047 MΩ
12. 16×10^{-7} F = 1.6 μF = 1600 nF
13. 2.4×10^{-2} A = 24 mA = 24,000 μA
14. $8.4 \times 10^2\ \Omega$ = 0.84 kΩ = 0.00084 MΩ
15. $56 \times 10^5\ \Omega$ = 5600 kΩ = 5.6 MΩ
16. 27.3 mS = 0.0273 S = 27,300 μS
17. 5600 μS = 0.0056 S = 5.6 mS
18. 170 mW = 0.170 W = 170,000 μW
19. 0.000642 V = 642 μV = 0.642 mV
20. 30 kHz = 0.030 MHz = 30,000 Hz

Additional practice problems are at the end of the chapter.

SELF-TEST 3-1

Express your answers in scientific notation:

1. 4760 = ______________ $\times 10^3$ = ______________ $\times 10^6$
2. 32.4×10^4 = ______________ $\times 10^3$ = ______________ $\times 10^6$
3. 2.71×10^{-1} = ______________ $\times 10^{-3}$ = ______________ $\times 10^3$
4. 46.7×10^{-2} = ______________ $\times 10^{-3}$ = ______________ $\times 10^3$
5. 76.3 mA = ______________ A = ______________ μA
6. 0.0055 μF = ______________ nF = ______________ pF
7. 8340 μS = ______________ mS = ______________ S
8. 20 kΩ = ______________ Ω = ______________ MΩ
9. 75,000 Hz = ______________ MHz = ______________ kHz
10. 146 mS = ______________ S = ______________ μS

Answers to Self-test 3-1 are at the end of the chapter.

3-3 APPLICATIONS

In problem solving we perform many different operations and express answers in many different forms. In this section we will do some of these operations and express answers by using prefixes whenever practical.

For example, when resistors are connected in parallel, we usually convert from resistance to conductance for problem-solving purposes. (Refer to Table 3-1 for the symbols and units of resistance and conductance.)

☐ EXAMPLE 3-12

Two resistors are in parallel. $R_1 = 27\ \text{k}\Omega$ and $R_2 = 18\ \text{k}\Omega$. Find the total conductance (G_T) and the total resistance (R_T)

Solution To find the total resistance, we use the equation

$$R_T = \frac{1}{G_T}$$

where

$$G_T = G_1 + G_2$$

$$G_1 = \frac{1}{R_1}$$

$$G_2 = \frac{1}{R_2}$$

Since

$$G_1 = \frac{1}{R_1} \quad \text{and} \quad G_2 = \frac{1}{R_2}$$

we can write the equation like this:

$$R_T = \frac{1}{G_1 + G_2} = \frac{1}{\dfrac{1}{R_1} + \dfrac{1}{R_2}}$$

$$= \frac{1}{\dfrac{1}{27\ \text{k}\Omega} + \dfrac{1}{18\ \text{k}\Omega}} = \frac{1}{37\ \mu\text{S} + 55.6\ \mu\text{S}}$$

$$= \frac{1}{92.6\ \mu\text{S}} = 10.8\ \text{k}\Omega$$

The total conductance is 92.6 μS. The total resistance is 10.8 kΩ.

Using the calculator, first set the calculator to engineering notation with one digit to the right of the decimal point (if one digit to the right of the decimal point does not provide for a three-decimal-digit answer, then adjust [FIX]).

[AC] [ENG] [FIX] 1

Now work the problem:

27 [EE] 3 [1/x]	$\frac{1}{R_1} = 37.0 \times 10^{-6}$ S = 37.0 μS
[+]	
18 [EE] 3 [1/x]	$\frac{1}{R_2} = 55.6 \times 10^{-6}$ S = 55.6 μS
[=]	Total conductance. $G_T = 92.6 \times 10^{-6} = 92.6$ μS
[1/x]	Total resistance. $R_T = 10.8 \times 10^3\ \Omega = 10.8$ kΩ

□ EXAMPLE 3-13

Three resistors are connected in parallel. $R_1 = 470\ \Omega$, $R_2 = 220\ \Omega$, and $R_3 = 680\ \Omega$. Find the total conductance and the total resistance.

Solution

$$R_T = \frac{1}{G_T} = \frac{1}{G_1 + G_2 + G_3} = \frac{1}{\frac{1}{R_1} + \frac{1}{R_2} + \frac{1}{R_3}}$$

Since $R_1 = 470\ \Omega$, $R_2 = 220\ \Omega$, and $R_3 = 680\ \Omega$, then

$$R_T = \frac{1}{\frac{1}{470\ \Omega} + \frac{1}{220\ \Omega} + \frac{1}{680\ \Omega}} = \frac{1}{2.13\text{ mS} + 4.55\text{ mS} + 1.47\text{ mS}}$$

$$= \frac{1}{8.14\text{ mS}} = 123\ \Omega$$

The total conductance is 8.14 mS. The total resistance is 123 Ω.

With most calculators, all operations can be performed without having to store partial answers. Only the final answer need be written down:

470 [1/x] [+] 220 [1/x] [+] 680 [1/x] [=] *Answer*: $G_T = 8.14$ mS

[1/x] *Answer*: $R_T = 123\ \Omega$

□

To add units together when the prefixes are different, it is necessary to convert so that all prefixes are the same. For example, let's add 6300 μS and 15.3 mS. We cannot add milliunits and microunits together, just as we could not add 10^{-3} and 10^{-6} together. We must convert 6300 μS to milliunits or convert 15.3 mS to microunits.

$$6300\ \mu\text{S} + 15.3\text{ mS} =$$

$$6.30\text{ mS} + 15.3\text{ mS} = 21.6\text{ mS}$$

or $$6300\ \mu S + 15.3\ mS =$$

$$6300\ \mu S + 15{,}300\ \mu S = 21{,}600\ \mu S$$

In this case, converting to milliunits is better because it is easier to work with that number.

PRACTICE PROBLEMS 3-5

Perform the indicated operations:

1. 27 mA + 0.037 A = ______________ A = ______________ mA
2. 370 μS + 0.060 mS = ______________ mS = ______________ μS
3. $\dfrac{1}{68\ k\Omega} + \dfrac{1}{47\ k\Omega}$ = ______________ mS = ______________ μS
4. $\dfrac{1}{\dfrac{1}{4.7\ k\Omega} + \dfrac{1}{1.8\ k\Omega}}$ = ______________ Ω = ______________ kΩ
5. $\dfrac{20\ V}{27\ k\Omega}$ = ______________ A = ______________ mA = ______________ μA
6. $\dfrac{25\ V}{2.36\ mA}$ = ______________ Ω = ______________ kΩ = ______________ MΩ
7. $\dfrac{1}{6.28 \times 12\ kHz \times 50\ nF}$ = ______________ Ω = ______________ kΩ
8. $\dfrac{1}{2.7\ k\Omega} + \dfrac{1}{3.3\ k\Omega} + \dfrac{1}{4.7\ k\Omega}$ = ______________ mS = ______________ μS = ______________ S
9. $\dfrac{1}{\dfrac{1}{22\ k\Omega} + \dfrac{1}{33\ k\Omega} + \dfrac{1}{56\ k\Omega}}$ = ______________ Ω = ______________ kΩ = ______________ MΩ
10. $\dfrac{1}{6340\ \mu S}$ = ______________ Ω = ______________ kΩ
11. 2.73 kΩ $\times$ 0.43 mA = ______________ V = ______________ mV
12. 1.27 V + 48 mV + 5630 μV = ______________ V = ______________ mV = ______________ μV

Solutions

1. 27 mA + 0.037 A = 0.064 A = 64 mA
2. 370 μS + 0.060 mS = 0.430 mS = 430 μS
3. $\dfrac{1}{68\ k\Omega} + \dfrac{1}{47\ k\Omega} = 0.036\ mS = 36\ \mu S$
4. $\dfrac{1}{\dfrac{1}{4.7\ k\Omega} + \dfrac{1}{1.8\ k\Omega}} = 1300\ \Omega = 1.30\ k\Omega$
5. $\dfrac{20\ V}{27\ k\Omega} = 0.000741\ A = 0.741\ mA = 741\ \mu A$
6. $\dfrac{25\ V}{2.36\ mA} = 10{,}600\ \Omega = 10.6\ k\Omega = 0.0106\ M\Omega$

7. $\dfrac{1}{6.28 \times 12\text{ kHz} \times 50\text{ nF}} = 265\ \Omega = 0.265\text{ k}\Omega$

8. $\dfrac{1}{2.7\text{ k}\Omega} + \dfrac{1}{3.3\text{ k}\Omega} + \dfrac{1}{4.7\text{ k}\Omega} = 0.886\text{ mS} = 886\ \mu\text{S} = 0.000886\text{ S}$

9. $\dfrac{1}{\dfrac{1}{22\text{ k}\Omega} + \dfrac{1}{33\text{ k}\Omega} + \dfrac{1}{56\text{ k}\Omega}} = 10{,}700\ \Omega = 10.7\text{ k}\Omega = 0.0107\text{ M}\Omega$

10. $\dfrac{1}{6340\ \mu\text{S}} = 158\ \Omega = 0.158\text{ k}\Omega$

11. $2.73\text{ k}\Omega \times 0.43\text{ mA} = 1.17\text{ V} = 1170\text{ mV}$

12. $1.27\text{ V} + 48\text{ mV} + 5630\ \mu\text{V} = 1.32\text{ V} = 1320\text{ mV} = 1{,}320{,}000\ \mu\text{V}$

Additional practice problems are at the end of the chapter.

3-4 SYSTEMS OF MEASUREMENT

The systems of measurement with which we must become familiar are the metric system and the English system. The metric system uses the meter for linear measurement and the gram for mass or weight. The English system uses the yard for linear measurement and the pound for weight.

The metric system is the international system of weights and measurements. Because of the need for a single, internationally accepted system, the metric system is gradually replacing the English system in this country.

In electronics, we already use the international system (SI) of measurements for most quantities, such as voltage, current, capacitance, and conductance, to name a few, but some measurements, such as distance, velocity, and weight, may be specified in either the English system or the metric system. For this reason, we must be able to convert from one system to the other.

The most commonly used units of linear measurement in metric are the millimeter (mm), centimeter (cm), meter (m), and kilometer (km). The relationships between these units are

$$1\text{ millimeter} = 10^{-3}\text{ meter} = 0.001\text{ meter}$$

$$1\text{ centimeter} = 10^{-2}\text{ m} = 0.01\text{ m}$$

$$1\text{ kilometer} = 10^{3}\text{ m} = 1000\text{ m}$$

In the English system, we have

$$12\text{ inches (in)} = 1\text{ foot (ft)}$$

$$3\text{ ft} = 1\text{ yard (yd)}$$

$$5280\text{ ft} = 1760\text{ yd} = 1\text{ mile (mi)}$$

To convert from English to metric,

$$1 \text{ in} = 2.54 \text{ cm}$$

$$1 \text{ yd} = 0.9144 \text{ m}$$

$$1 \text{ mi} = 1.609 \text{ km}$$

To convert from metric to English,

$$1 \text{ cm} = 0.3937 \text{ in}$$

$$1 \text{ m} = 39.37 \text{ in}$$

$$1 \text{ m} = 1.094 \text{ yd}$$

$$1 \text{ km} = 0.6214 \text{ mi}$$

□ EXAMPLE 3-14

Convert the following: **(a)** 14 cm to inches; **(b)** 40 cm to inches; **(c)** 1.7 m to inches and yards; **(d)** 3 km to yards and miles.

Solutions

(a) 14 cm × 0.3937 in/cm = 5.512 in
(b) 40 cm × 0.3937 in/cm = 15.75 in
(c) 1.7 m × 39.37 in/m = 66.93 in
1.7 m × 1.094 yd/m = 1.860 yd
(d) 3000 m × 1.094 yd/m = 3282 yd
3000 m = 3 km; 3 km × 0.6214 mi/km = 1.864 mi

□ EXAMPLE 3-15

Convert the following: **(a)** 20 in to cm; **(b)** 60 in to cm and m; **(c)** 100 yd to m; **(d)** 2000 yd to m and km.

Solutions

(a) 20 in × 2.54 cm/in = 50.8 cm
(b) 60 in × 2.54 cm/in = 152.4 cm = 1.524 m
(c) 100 yd × 0.9144 m/yd = 91.44 m
(d) 2000 × 0.9144 m/yd = 1829 m = 1.829 km

□

The units of velocity are meters/second or kilometers/hour in metric and feet/second or miles/hour in English.

□ EXAMPLE 3-16

Convert the following: **(a)** 300 meters/second (m/s) to feet/second (ft/s); **(b)** 35 miles/hour (mi/h) to kilometers/hour (km/h); **(c)** 700 feet/second (ft/s) to meters/second (m/s); **(d)** 108 kilometers/hour (km/h) to miles/hour (mi/h).

Solutions

(a) 39.37 in = 1 m
39.37 in/m ÷ 12 in/ft = 3.281 ft/m
3.281 ft/m × 300 m/s = 984.3 ft/s
(b) 35 mi/h × 1.609 km/mi = 56.32 km/h
(c) 0.3048 ft = 1 m
700 ft/s × 0.3048 m/ft = 213.4 m/s
(d) 108 km/h × 0.6214 mi/km = 67.11 mi/h

□

The most commonly used units of mass or weight in metric are the milligram, gram, and kilogram. In the English system we have the ounce, pound, and ton.

16 ounces (oz) = 1 pound (lb)

2000 pounds = 1 ton (t)

To convert from metric to English,

1 gram (g) = 0.03527 ounces

1 kilogram (kg) = 35.27 ounces = 2.204 pounds

To convert from English to metric,

1 ounce = 28.35 grams

1 pound = 453.6 grams = 0.4536 kilogram

□ EXAMPLE 3-17

Convert the following: **(a)** 30 g to ounces; **(b)** 700 g to ounces and pounds; **(c)** 2.35 kg to ounces and pounds; **(d)** 0.255 oz to grams; **(e)** 7.5 oz to kilograms; **(f)** 1.5 lb to kilograms.

Solutions

(a) 30 g × 0.03527 g/oz = 1.058 oz
(b) 700 g × 0.03527 g/oz = 24.69 oz
24.69 oz ÷ 16 oz/lb = 1.543 lb
(c) 2.35 kg × 35.27 oz/kg = 82.88 oz
2.35 kg × 2.204 lb/kg = 5.180 lb
(d) 0.255 oz × 28.35 g/oz = 7.229 g
(e) 7.5 oz × 0.02835 kg/oz = 0.2126 kg (1 oz = 28.35 g = 0.02835 kg)
(f) 1.5 lb × 0.4536 lb/kg = 0.6804 kg

□

PRACTICE PROBLEMS 3-6

Make the following conversions:

1. 27 centimeters to inches

2. 3.4 meters to inches and yards

3. 5000 meters to yards and miles
4. 27 inches to centimeters
5. 440 yards to meters and kilometers
6. 2 miles to meters and kilometers
7. 400 meters/second to feet/second
8. 200 kilometers/hour to miles/hour
9. 300 feet/second to meters/second
10. 65 miles/hour to kilometers/hour
11. 70 grams to ounces
12. 1000 grams to ounces and pounds
13. 3.2 kilograms to ounces and pounds
14. 2 pounds to kilograms
15. 2.45 ounces to grams

Solutions

1. 10.63 in.
2. 133.9 in. = 3.720 yd
3. 5470 yd = 3.108 mi
4. 68.58 cm
5. 402.3 m = 0.4023 km
6. 3.218 km = 3218 m
7. 1312 ft/s
8. 124.3 mi/h
9. 91.44 m/s
10. 104.6 km/h
11. 2.469 oz
12. 35.27 oz = 2.204 lb
13. 112.9 oz = 7.054 lb
14. 0.9072 kg
15. 69.46 g

Additional practice problems are at the end of the chapter.

CHAPTER 3 AT A GLANCE

Page	Key Point or Rule	Example
65	*Rule 3-1:* If we make the exponent more positive, the decimal point moves to the *left*.	$6700 = 670.0 \times 10$ $= 67.00 \times 10^2$ $= 6.700 \times 10^3$
66	*Rule 3-2:* If the exponent is made more negative, the decimal point moves to the *right*.	$0.00275 = 0.0275 \times 10^{-1}$ $= 0.275 \times 10^{-2}$ $= 2.75 \times 10^{-3}$
67	*Key Point:* Expressing numbers as numbers between 1 and 10 times the proper power of ten is called *scientific* notation. Expressing numbers as numbers times powers of ten that are multiples of 3 is called *engineering* notation.	$47{,}000 = 4.70 \times 10^4$ in scientific notation to three places. $47{,}000 = 47.0 \times 10^3$ in engineering notation to three places.
71	*Rule 3-3:* To change numbers with prefixes to regular numbers, replace the prefix with the power of ten that it represents and then change to a regular number.	$30\text{ mA} = 30 \times 10^{-3}\text{ A}$ $= 0.030\text{ A}$ $33\text{ k}\Omega = 33 \times 10^3\ \Omega$ $= 33{,}000\ \Omega$

79	Some conversions of linear measure between English and metric are: 1 in = 2.54 cm 1 cm = 0.3937 in 1 yd = 0.9144 m 1 m = 1.094 yd 1 mi = 1.609 km 1 km = 0.6214 mi	20 cm = 7.874 in 12 in = 30.48 cm 100m = 109.4 yd 5 mi = 8.045 km
79	The units of velocity are meters/second or kilometers/hour in metric and feet/second or miles/hour in English.	15 m/s = 49.22 ft/s
80	Common units of weight in metric are the milligram, gram, and kilogram. In the English system, we have the ounce, pound, and ton. 16 oz = 1 lb 1 kg = 35.27 oz 2000 lb = 1 ton 1 lb = 453.6 gm	100 lb = 45.36 kg 100 kg = 220.5 lb

End of Chapter Problems 3-1

Change the following numbers to numbers times 10^3 and to numbers times 10^6:

1. 56,000
2. 1800
3. 220,000
4. 1,500,000
5. 39×10^4
6. 56×10^4
7. 1.8×10^5
8. 6.8×10^5
9. 4300×10^1
10. 3300×10^2

Change the following numbers to numbers times 10^{-3} and to numbers times 10^{-6}:

11. 0.00022
12. 0.000012
13. 0.00213
14. 0.0179
15. 0.00556×10^{-2}
16. 0.000133×10^{-1}
17. 1.22×10^{-4}
18. 40×10^{-2}
19. 25.6×10^{-5}
20. 11×10^{-4}

Change the following numbers to numbers times 10^{-9} and to numbers times 10^{-12}:

21. 0.00667×10^{-4}
22. 0.000233×10^{-4}
23. 0.179×10^{-7}
24. 0.041×10^{-8}
25. 67.4×10^{-10}
26. 27.3×10^{-10}
27. 1.77×10^{-11}
28. 3.75×10^{-8}
29. 700×10^{-10}
30. 0.0034×10^{-7}

Change the following numbers to regular numbers, to numbers times 10^{-3}, and to numbers times 10^3:

31. 73×10^{-1}
32. 0.042×10^2
33. 0.567×10
34. 17.4×10^{-1}
35. 1.78×10^{-2}
36. 9.25×10^{-1}

37. 783×10^{-1}

38. 0.00275×10^{2}

39. 0.0845×10^{4}

40. 3250×10^{-2}

End of Chapter Problems 3-2

1. 0.00026 A = ______________ mA = ______________ μA
2. 0.00736 A = ______________ mA = ______________ μA
3. 0.000632 S = ______________ mS = ______________ μS
4. 0.0024 S = ______________ mS = ______________ μS
5. 7630 Ω = ______________ kΩ = ______________ MΩ
6. 470,000 Ω = ______________ kΩ = ______________ MΩ
7. 17.3×10^{-3} A = ______________ mA = ______________ μA
8. 0.64×10^{-6} A = ______________ mA = ______________ μA
9. 71.3×10^{4} Ω = ______________ kΩ = ______________ MΩ
10. 6.8×10^{5} Ω = ______________ kΩ = ______________ MΩ
11. 5.63×10^{6} Hz = ______________ kHz = ______________ MHz
12. 487×10^{2} Hz = ______________ kHz = ______________ MHz
13. 2,000,000 Ω = ______________ kΩ = ______________ MΩ
14. 470,000 Ω = ______________ kΩ = ______________ MΩ
15. 23.7×10^{-5} S = ______________ mS = ______________ μS
16. 5.63×10^{-4} S = ______________ mS = ______________ μS
17. 30×10^{-8} F = ______________ μF = ______________ nF
18. 12×10^{-7} F = ______________ μF = ______________ nF
19. 0.000062 A = ______________ mA = ______________ μA
20. 0.00075 A = ______________ mA = ______________ μA

End of Chapter Problems 3-3

1. 800 mA = ______________ A
2. 70.3 mA = ______________ A
3. 2.5 mS = ______________ S
4. 600 μS = ______________ S
5. 33 kΩ = ______________ Ω
6. 3.3 kΩ = ______________ Ω
7. 0.47 kΩ = ______________ Ω
8. 0.56 MΩ = ______________ Ω
9. 12.5 kHz = ______________ Hz
10. 7.3 kHz = ______________ Hz
11. 100 μF = ______________ F
12. 50 μF = ______________ F
13. 0.25 mA = ______________ A
14. 500 mA = ______________ A
15. 900 μS = ______________ S
16. 25 mS = ______________ S
17. 750 kΩ = ______________ Ω
18. 1.2 kΩ = ______________ Ω
19. 30 mA = ______________ A
20. 300 mA = ______________ A

End of Chapter Problems 3-4

1. 0.00026 A = ______________ mA = ______________ μA
2. 0.00736 A = ______________ mA = ______________ μA

3. 0.00000632 S = ______________ mS = ______________ μS
4. 0.0000417 A = ______________ mA = ______________ μA
5. 56.2×10^{-3} S = ______________ mS = ______________ μS
6. 17.3×10^{-3} A = ______________ mA = ______________ μA
7. 613×10^{-6} S = ______________ mS = ______________ μS
8. 227×10^{-6} A = ______________ mA = ______________ μA
9. 7630 Ω = ______________ kΩ = ______________ MΩ
10. 12,500 Hz = ______________ kHz = ______________ MHz
11. 470,000 Ω = ______________ kΩ = ______________ MΩ
12. 1,200,000 Ω = ______________ kΩ = ______________ MΩ
13. 12.7×10^{4} Hz = ______________ kHz = ______________ MHz
14. 5.63×10^{6} Hz = ______________ kHz = ______________ MHz
15. 713×10^{3} Ω = ______________ kΩ = ______________ MΩ
16. 8.73×10^{5} Ω = ______________ kΩ = ______________ MΩ
17. 46.3 mA = ______________ μA = ______________ A
18. 213 mS = ______________ S = ______________ μS
19. 0.05 nF = ______________ μF = ______________ pF
20. 173 nF = ______________ pF = ______________ μF
21. 3.2 kΩ = ______________ Ω = ______________ MΩ
22. 17.8 kHz = ______________ Hz = ______________ MHz
23. 270 kΩ = ______________ MΩ = ______________ Ω
24. 47 kΩ = ______________ MΩ = ______________ Ω
25. 403 μS = ______________ S = ______________mS
26. 83.6 mS = ______________ S = ______________ μS
27. 1.03 MHz = ______________ Hz = ______________ kHz
28. 423 kHz = ______________ MHz = ______________ Hz
29. 1.43 nF = ______________ pF = ______________ μF
30. 0.002 μF = ______________ pF = ______________ nF
31. 55×10^{-4} S = ______________ mS = ______________ μS
32. 4.67×10^{-3} S = ______________ mS = ______________ μS
33. 106×10^{-2} nF = ______________ μF = ______________ pF
34. 7.36×10^{2} nF = ______________ μF = ______________ pF
35. 4.63×10^{2} kΩ = ______________ Ω = ______________ MΩ
36. 6.67×10^{2} kΩ = ______________ Ω = ______________ MΩ
37. 9.63×10^{4} Hz = ______________ kHz = ______________ MHz
38. 8.23×10^{2} Hz = ______________ kHz = ______________ MHz
39. 0.0078 A = ______________ mA = ______________ μA
40. 0.000104 A = ______________ mA = ______________ μA
41. 176 mV = ______________ V = ______________ μV
42. 0.00062 V = ______________ mV = ______________ μV
43. 1.73 W = ______________ mW = ______________ kW
44. 16.7 W = ______________ mW = ______________ kW
45. 0.025 H = ______________ mH = ______________ μH
46. 0.005 H = ______________ mH = ______________ μH

47. 173 mS = ______________ S = ______________ μS
48. 1460 μS = ______________ mS = ______________ S
49. 25×10^{-7} F = ______________ μF = ______________ nF
50. 2.5 mH = ______________ H = ______________ μH

End of Chapter Problems 3-5

1. 367 μA + 1.67 mA = ______________ mA = ______________ μA
2. 0.417 mA + 63 μA = ______________ mA = ______________ μA
3. 0.0173 mA + 78.4 μA = ______________ mA = ______________ μA
4. 650 μA + 2.35 mA = ______________ mA = ______________ μA
5. $\frac{1}{10\ \text{k}\Omega} + \frac{1}{15\ \text{k}\Omega}$ = ______________ mS = ______________ μS
6. $\frac{1}{6.8\ \text{k}\Omega} + \frac{1}{5.6\ \text{k}\Omega}$ = ______________ mS = ______________ μS
7. $\frac{1}{33\ \text{k}\Omega} + \frac{1}{56\ \text{k}\Omega} + \frac{1}{100\ \text{k}\Omega}$ = ______________ mS = ______________ μS
8. $\frac{1}{270\ \Omega} + \frac{1}{470\ \Omega} + \frac{1}{680\ \Omega}$ = ______________ mS = ______________ μS
9. $\frac{1}{1.2\ \text{k}\Omega} + \frac{1}{4.7\ \text{k}\Omega} + \frac{1}{2.7\ \text{k}\Omega}$ = ______________ mS = ______________ μS
10. $\frac{1}{33\ \text{k}\Omega} + \frac{1}{100\ \text{k}\Omega} + \frac{1}{68\ \text{k}\Omega}$ = ______________ mS = ______________ μS
11. $\frac{1}{\frac{1}{2.7\ \text{k}\Omega} + \frac{1}{1.5\ \text{k}\Omega}}$ = ______________ Ω = ______________ kΩ
12. $\frac{1}{\frac{1}{8.2\ \text{k}\Omega} + \frac{1}{7.5\ \text{k}\Omega}}$ = ______________ Ω = ______________ kΩ
13. $\frac{1}{\frac{1}{100\ \text{k}\Omega} + \frac{1}{220\ \text{k}\Omega}}$ = ______________ Ω = ______________ kΩ = ______________ MΩ
14. $\frac{1}{\frac{1}{470\ \text{k}\Omega} + \frac{1}{560\ \text{k}\Omega}}$ = ______________ Ω = ______________ kΩ = ______________ MΩ
15. $\frac{1}{\frac{1}{39\ \text{k}\Omega} + \frac{1}{12\ \text{k}\Omega} + \frac{1}{27\ \text{k}\Omega}}$ = ______________ Ω = ______________ kΩ = ______________ MΩ
16. $\frac{1}{\frac{1}{1.2\ \text{M}\Omega} + \frac{1}{2.2\ \text{M}\Omega} + \frac{1}{1\ \text{M}\Omega}}$ = ______________ Ω = ______________ kΩ = ______________ MΩ
17. $\frac{1}{\frac{1}{180\ \Omega} + \frac{1}{470\ \Omega} + \frac{1}{1\ \text{k}\Omega}}$ = ______________ Ω = ______________ kΩ

18. $\dfrac{1}{\dfrac{1}{330\ \Omega} + \dfrac{1}{560\ \Omega} + \dfrac{1}{910\ \Omega}}$ = ______________ Ω = ______________ kΩ

19. $\dfrac{1}{213\ \mu S}$ = ______________ Ω = ______________ kΩ

20. $\dfrac{1}{1.47\ mS}$ = ______________ Ω = ______________ kΩ

21. 500 μS + 1.76 mS + 0.000043 S = ______________ mS = ______________ μS

22. 670 μS + 2 mS + 0.0002 S = ______________ mS = ______________ μS

23. $\dfrac{9\ V}{33\ k\Omega}$ = ______________ mA = ______________ μA

24. $\dfrac{12\ V}{2.7\ k\Omega}$ = ______________ mA = ______________ μA

25. $\dfrac{14\ V}{200\ \mu A}$ = ______________ Ω = ______________ kΩ

26. $\dfrac{10\ V}{25\ mA}$ = ______________ Ω = ______________ kΩ

27. $\dfrac{50\ V}{1.2\ M\Omega}$ = ______________ mA = ______________ μA

28. $\dfrac{40\ V}{30\ \mu A}$ = ______________ kΩ = ______________ MΩ

29. 68 kΩ × 140 μA = ______________ V = ______________ mV

30. 330 kΩ × 0.17 mA = ______________ V = ______________ mV

31. 1.2 kΩ × 3.73 mA = ______________ V = ______________ mV

32. 4.7 kΩ × 63.4 μA = ______________ V = ______________ mV

33. $\dfrac{1}{6.28 \times 75\ kHz \times 25\ pF}$ = ______________ Ω = ______________ kΩ

34. $\dfrac{1}{6.28 \times 2.5\ kHz \times 30\ nF}$ = ______________ Ω = ______________ kΩ

35. $\dfrac{1}{6.28 \times 10\ kHz \times 0.05\ \mu F}$ = ______________ Ω = ______________ kΩ

36. $\dfrac{1}{6.28 \times 2.73\ kHz \times 200\ nF}$ = ______________ Ω = ______________ kΩ

37. $\dfrac{1}{6.28 \times 12\ kHz \times 2.7\ k\Omega}$ = ______________ μF = ______________ nF

38. $\dfrac{1}{6.28 \times 150\ Hz \times 600\ \Omega}$ = ______________ μF = ______________ nF

39. $\dfrac{1}{6.28(150\ mH \times 200\ nF)^{1/2}}$ = ______________ Hz = ______________ kHz

40. $\dfrac{1}{6.28(0.5\ mH \times 0.5\ nF)^{1/2}}$ = ______________ Hz = ______________ kHz

End of Chapter Problems 3-6

Make the following conversions:

1. 100 centimeters to inches
2. 80 centimeters to inches
3. 40 centimeters to inches
4. 120 centimeters to inches
5. 2.3 meters to inches and yards
6. 1.75 meters to inches and yards
7. 10 meters to yards
8. 20 meters to yards
9. 10,000 meters to yards and miles
10. 15,000 meters to yards and miles
11. 10 inches to centimeters
12. 12 inches to centimeters
13. 35.6 inches to centimeters
14. 40.3 inches to centimeters
15. 880 yards to meters and km
16. 1000 yards to meters and kilometers
17. 55 mi/h to km/h
18. 40 mi/h to km/h
19. 100 mi/h to km/h
20. 80 mi/h to km/h
21. 500 meters/s to feet/s
22. 200 meter/s to feet/s
23. 100 meters/s to feet/s
24. 80 meters/s to feet/s
25. 60 grams to ounces
26. 80 grams to ounces
27. 150 grams to ounces
28. 200 grams to ounces
29. 800 grams to ounces and pounds
30. 1200 grams to ounces and pounds
31. 5 kg to oz and lb
32. 6.5 kg to oz and lb
33. 1.35 lb to kg
34. 0.75 lb to kg
35. 4 oz to g
36. 45 oz to g
37. 0.275 oz to g
38. 0.75 oz to g

Answers to Self-tests

SELF-TEST 3-1

1. $4760 = 4.76 \times 10^3 = 0.00476 \times 10^6$
2. $32.4 \times 10^4 = 324 \times 10^3 = 0.324 \times 10^6$
3. $2.71 \times 10^{-1} = 271 \times 10^{-3} = 0.000271 \times 10^3$
4. $46.7 \times 10^{-2} = 467 \times 10^{-3} = 0.000467 \times 10^3$
5. 76.3 mA = 0.0763 A = 76,300 μA
6. 0.0055 μF = 5.5 nF = 5500 pF
7. 8340 μS = 8.34 mS = 0.00834 S
8. 20 kΩ = 20,000 Ω = 0.02 MΩ
9. 75,000 Hz = 0.075 MHz = 75 kHz
10. 146 mS = 0.146 S = 146,000 μS

CHAPTER 4

COMPUTER NUMBER SYSTEMS

INTRODUCTION

The basic tools for understanding computers are an understanding of the number systems used and an understanding of basic logic functions. Number systems are covered in this chapter and logic functions in Chapter 26.

We all know and understand the decimal number system, the number system we use daily, but the computer world makes use of the *binary* number system. All data are stored and manipulated inside the computer in binary. That is, within the computer all data are reduced to 1's and 0's. These 1's and 0's are stored in logic circuits called *registers* or they are stored in *memory*. The size of a register or memory location varies according to the kind of computer used. Microcomputers store 8 or 16 binary digits in each location. Large computers may store as many as 125 binary digits in each location.

We normally enter data into the computer in some number system other than binary because entering data in binary is too time consuming and too prone to error. There are too many 1's and 0's. Data are entered into the computer by means of the decimal, octal, or hexadecimal number system. Octal and hexadecimal are used most often because they are more closely related to binary than is the decimal system.

In this chapter we will work with numbers in the binary, octal, decimal, and hexadecimal number systems. We use subscripts to denote the base we are working in. This must be done when working with different bases to avoid errors. For example, the number 11 is a valid number in all the bases we study, but it has a different value in each. If it is a binary number, we write it this way: 11_2. If it is an octal number, it is written this way: 11_8. The importance of indicating the base is apparent when you consider that 11_2 has a decimal value of 3, whereas 11_8 has a decimal value of 9!

Because all these number systems are used in the computer world in one way or another, it is necessary that we understand how they are related to each other.

CHAPTER OBJECTIVES

In this chapter you will learn how to:

1. Read and write numbers in the binary, octal, and hexadecimal number systems.
2. Convert between the decimal, binary, octal, and hexadecimal number systems.
3. Add numbers in the various number systems.
4. Subtract numbers in the various number systems using both the direct and complement methods.

4-1 BINARY NUMBER SYSTEM

In Chapter 1, we talked about the decimal number system and how the value or *weight* of a digit depended on its position in the number. For example, in the number 7037, the 7 in the units position has a weight of 7, whereas the 7 in the thousands position has a weight of 7000. Furthermore, because our number system has ten digits (0 through 9), the next number after 9 causes a *carry* into the next position. In the decimal number system, we have no special symbol or character for numbers greater than 9, so we use combinations of these numbers in our *place value system* to indicate quantities greater than 9. Thus, the next number after 9 is 10. The one in the tens position has a weight of 10.

In powers of ten form in base 10, numbers in the units positions are raised to the zero power ($10^0 = 1$), numbers in the tens position are raised to the first power (10^1), numbers in the hundreds position are raised to the second power (10^2), and so on. Using the powers of ten form in the various positions, our number looks like this.

$$(7 \times 10^3) + (0 \times 10^2) + (3 \times 10^1) + 7 \times 10^0)$$
$$= 7000 + 0 + 30 + 7 = 7037$$

Those concepts, developed in Chapter 1 and reviewed here, apply to *all* number systems. The only difference is that other number systems have fewer (or more) characters.

In the binary number system, there are only two digits, 0 and 1. Digits in the units position can only have a value of 0 or 1 since that's all the digits there are. Numbers greater than 1 cause a carry into the next position. Just as in base 10, each position represents the base raised to a power. The units position has a power of 2^0, the next position is 2^1, and so on, as illustrated in Figure 4-1. In decimal, the weight or value of each position is a multiple of 10. In binary, the weight or value of each position is a multiple of 2.

Binary digits are called *bits* (a contraction of binary digits). Therefore, the digit in the units position is called the least significant *bit* (LSB), and so on, until the most significant bit (MSB) is reached.

MSB	4SB	3SB	2SB	LSB
2^4	2^3	2^2	2^1	2^0
16	8	4	2	1

FIGURE 4-1 Positional value of the first five positions for the binary number system.

4-1-1 Binary to Decimal Conversions

Converting from base 2 to base 10 is easy. If a 1 is present in a given position, the weight of that bit is added. If a 0 is present, the weight of that bit is not added. Consider the number 11010 in Example 4-1. The decimal equivalent is 26.

□ EXAMPLE 4-1

$11010_2 = (?)_{10}$.

Solution

$$(1 \times 2^4) + (1 \times 2^3) + (0 \times 2^2) + (1 \times 2^1) + (0 \times 2^0) =$$
$$16 + 8 + 0 + 2 + 0 = 26$$

□

Because there are only 1's and 0's in binary, each bit position either equals the weight of that digit position or it equals 0. In Example 4-1, the values of the bit positions are added to get the decimal equivalent. In the 2^4 or 16's position, there is a 1. Therefore, that bit position converts to 16. In the 2^3 or 8's position, there is a 1. That bit position converts to 8. In the 2^2 or 4's position, there is a 0. That bit position converts to 0. In the 2's position, there is a 1. That bit position converts to 2. There is a 0 in the units position. If we add the numbers together, we get decimal 26.

■ **RULE 4-1** In converting from binary to decimal, find the value or weight of the MSB. Work down to the LSB, adding the weight of that position if a 1 is present or a 0 if a 0 is present.

Table 4-1 is a conversion table for all the computer number systems.

□ EXAMPLE 4-2

Change the following binary numbers to decimal numbers: **(a)** 1100; **(b)** 10001; **(c)** 101011; **(d)** 111101.

Solutions

(a) $1100_2 = (1 \times 2^3) + (1 \times 2^2) + (0 \times 2^1) + (0 \times 2^0) =$
$$8 + 4 + 0 + 0 = 12_{10}$$

TABLE 4-1 Computer Numbering System

Numbering system		Decimal		Binary						Octal		Hexa-decimal	
Base		10		2						8		16	
Position weight	Power	10^1	10^0	2^5	2^4	2^3	2^2	2^1	2^0	8^1	8^0	16^1	16^0
	Value	10	1	32	16	8	4	2	1	8	1	16	1
											(LSD)		(LSD)
		0	0	0	0	0	0	0	0	0	0	0	0
			1	0	0	0	0	0	1		1		1
			2	0	0	0	0	1	0		2		2
			3	0	0	0	0	1	1		3		3
			4	0	0	0	1	0	0		4		4
			5	0	0	0	1	0	1		5		5
			6	0	0	0	1	1	0		6		6
			7	0	0	0	1	1	1		7		7
			8	0	0	1	0	0	0	1	0		8
			9	0	0	1	0	0	1	1	1		9
		1	0	0	0	1	0	1	0	1	2		A
		1	1	0	0	1	0	1	1	1	3		B
		1	2	0	0	1	1	0	0	1	4		C
		1	3	0	0	1	1	0	1	1	5		D
		1	4	0	0	1	1	1	0	1	6		E
		1	5	0	0	1	1	1	1	1	7		F
		1	6	0	1	0	0	0	0	2	0	1	0
		1	7	0	1	0	0	0	1	2	1	1	1
		1	8	0	1	0	0	1	0	2	2	1	2
		1	9	0	1	0	0	1	1	2	3	1	3
		2	0	0	1	0	1	0	0	2	4	1	4
		2	1	0	1	0	1	0	1	2	5	1	5
		2	2	0	1	0	1	1	0	2	6	1	6
		2	3	0	1	0	1	1	1	2	7	1	7
		2	4	0	1	1	0	0	0	3	0	1	8
		2	5	0	1	1	0	0	1	3	1	1	9
		2	6	0	1	1	0	1	0	3	2	1	A
		2	7	0	1	1	0	1	1	3	3	1	B
		2	8	0	1	1	1	0	0	3	4	1	C
		2	9	0	1	1	1	0	1	3	5	1	D
		3	0	0	1	1	1	1	0	3	6	1	E
		3	1	0	1	1	1	1	1	3	7	1	F
		3	2	1	0	0	0	0	0	4	0	2	0

Example: $1_{(10)} = 1_{(2)} = 1_{(8)} = 1_{(16)}$

Example: $10_{(10)} = 1010_{(2)} = 12_{(8)} = 0A_{(16)}$

Example: $25_{(10)} = 11001_{(2)} = 31_{(8)} = 19_{(16)}$

(b) $10001_2 = (1 \times 2^4) + (0 \times 2^3) + (0 \times 2^2) + (0 \times 2^1) + (1 \times 2^0)$

$16 + 0 + 0 + 0 + 1 = 17_{10}$

(c) $101011_2 = (1 \times 2^5) + (0 \times 2^4) + (1 \times 2^3) + (0 \times 2^2) + (1 \times 2^1) +$

$(1 \times 2^0) = 32 + 0 + 8 + 0 + 2 + 1$

$= 43_{10}$

(d) $111101_2 = (1 \times 2^5) + (1 \times 2^4) + (1 \times 2^3) + (1 \times 2^2) + (0 \times 2^1) +$

$(1 \times 2^0) = 32 + 16 + 8 + 4 + 0 + 1$

$= 61_{10}$

□

4-1-2 Decimal To Binary Conversions

Let's turn the process around and find the binary equivalent of a decimal number.

□ EXAMPLE 4-3

Convert 11_{10} to binary:

Solution

Step 1. $\frac{11}{2} = 5 + 1$ (LSB)

Step 2. $\frac{5}{2} = 2 + 1$

Step 3. $\frac{2}{2} = 1 + 0$

Step 4. $\frac{1}{2} = 0 + 1$ — MSB → 1 0 1 1

Since we are converting to base 2, we first divide the decimal number by 2 (step 1). The answer is 5 with a remainder of 1. This remainder is the LSB in the answer. The whole number left after this initial division (5) is again divided by 2 (step 2). The answer is 2 with a remainder of 1. This 1 is the bit in the 2^1 position. The whole number left (2) is divided by 2 (step 3). The result is 1 and a remainder of 0. This 0 is the bit in the 2^2 position. Step 4 is the final step because no further divisions are possible. The remainder in the final step is the MSB.

□

A second method of converting from decimal to binary is shown in the following example.

□ EXAMPLE 4-4

Convert 167 to binary.

Solution The highest multiple of 2 contained in 167 is 128 (2^7). Therefore, we need a 128. So we start writing our binary number with a 1 in the MSB position, remembering that this "1" has a value of 128. Next we determine if we need a 64. If we do, we put a 1 in the next bit position. If we don't need a 64, we put a 0 in this position. Since 128 + 64 is greater than 167, we must put a zero in this position. At this point we have 10, which is 128 + 0.

This procedure is followed for each bit position. Can we use 32? Yes. We now have 101. Remember, we started at the MSB position and are working down to the LSB position. Can we use a 16? No; a 16 is too much. Put a zero in the 16 (2^4) position. We now have 1010. Can we use an 8? No; so put a zero in the 2^3 position. Now we have 10100. Can we use a 4? Yes. Our number is now 101001. Can we use a 2? Yes. We now have 1010011. Do we need a 1 or a 0 in the LSB position? We need a 1. Our binary number is 10100111. Any decimal number can be converted to binary in this manner. With practice, this method is faster than the method shown previously. The process is detailed in Figure 4-2.

MSB ↓							LSB ↓
2^7	2^6	2^5	2^4	2^3	2^2	2^1	2^0
128	64	32	16	8	4	2	1
1	0	1	0	0	1	1	1

$128 + 0 + 32 + 0 + 0 + 4 + 2 + 1 = 167_{10}$

Binary No.	167	Original decimal number
1	−128	2^7 is the largest binary number that we can subtract from 167, so the MSB is in the 2^7 position.
	39	Remainder
0	−0	The 2^6 position must equal 0 (64 is greater than 39), so we must put a 0 in the 2^6 position.
1	−32	We can subtract 32 from 39, so there is a 1 in the 2^5 position.
	7	Remainder
0	−0	The 2^4 position must equal 0 (can't subtract 16 from 7).
0	−0	The 2^3 position must equal 0 (can't subtract 8 from 7).
1	−4	We can subtract 4 from 7, so there is a 1 in the 2^4 position.
	3	Remainder
1	−2	We can subtract 2 from 3, so there is a 1 in the 2^1 position.
1	1	Remainder: there is a 1 in the LSB position. (If there had been no remainder, the LSB would equal 0.)

FIGURE 4-2 Conversion of 167_{10} to binary.

□ EXAMPLE 4-5

Convert the following decimal numbers to binary numbers: **(a)** 7; **(b)** 34; **(c)** 89; **(d)** 203.

Solutions

(a)

$$\frac{7}{2} = 3 + 1$$
$$\frac{3}{2} = 1 + 1$$
$$\frac{1}{2} = 0 + 1 \rightarrow 1 \quad 1 \quad 1$$

$7_{10} = 111_2$

(b)

$$\frac{34}{2} = 17 + 0$$
$$\frac{17}{2} = 8 + 1$$
$$\frac{8}{2} = 4 + 0$$
$$\frac{4}{2} = 2 + 0$$
$$\frac{2}{2} = 1 + 0$$
$$\frac{1}{2} = 0 + 1 \rightarrow 1 \quad 0 \quad 0 \quad 0 \quad 1 \quad 0$$

$34_{10} = 100010_2$

(c)

$$\frac{89}{2} = 44 + 1$$
$$\frac{44}{2} = 22 + 0$$
$$\frac{22}{2} = 11 + 0$$
$$\frac{11}{2} = 5 + 1$$
$$\frac{5}{2} = 2 + 1$$
$$\frac{2}{2} = 1 + 0$$
$$\frac{1}{2} = 0 + 1 \rightarrow 1 \quad 0 \quad 1 \quad 1 \quad 0 \quad 0 \quad 1$$

$89_{10} = 1011001_2$

(d) $$\frac{203}{2} = 101 + 1$$

$$\frac{101}{2} = 50 + 1$$

$$\frac{50}{2} = 25 + 0$$

$$\frac{25}{2} = 12 + 1$$

$$\frac{12}{2} = 6 + 0$$

$$\frac{6}{2} = 3 + 0$$

$$\frac{3}{2} = 1 + 1$$

$$\frac{1}{2} = 0 + 1 \rightarrow 1\ 1\ 0\ 0\ 1\ 0\ 1\ 1$$

$$203_{10} = 11001011_2$$

□

PRACTICE PROBLEMS 4-1

Convert the following binary numbers to decimal numbers:

1. 101_2
2. 111_2
3. 1001_2
4. 1110_2
5. 10110_2
6. 11001_2
7. 100101_2
8. 111011_2
9. 1010100_2
10. 1111001_2

Convert the following decimal numbers to binary numbers:

11. 4
12. 8
13. 10
14. 27
15. 39
16. 57
17. 78
18. 100
19. 150
20. 200

Solutions

1. 5
2. 7
3. 9
4. 14
5. 22
6. 25
7. 37
8. 59
9. 84
10. 121

11. 100_2
12. 1000_2
13. 1010_2
14. 11011_2
15. 100111_2
16. 111001_2
17. 1001110_2
18. 1100100_2
19. 10010110_2
20. 11001000_2

Additional practice problems are at the end of the chapter.

4-2 OCTAL NUMBER SYSTEM

Octal, or base 8, is an important computer number system. Since there are 8 digits in octal (0,1,2,3,4,5,6,7), the digit of greatest value is 7, which is 1 less than the base. In counting, the sequence is 0 to 7. On the next count, we return to 0 in the units position (just as we do when we reach 9 in decimal) and a carry is generated. The resulting number is 10 and is read "one-zero." This number is equal to decimal 8 because the carry is equal to 8 (base 8). This process continues; the sequence goes from 0 to 7 and back to 0 again. Each time a count greater than 7 is reached, a carry is generated.

In the octal number system, digits in the units position have the indicated value. Digits in the next position have *place value*. That is, digits in this position indicate the number of 8's present. This, then, is the 8's position. The next position indicates the number of 64's present, and so on, as shown in Figure 4-3.

Again, the concept of *place value* must be understood so that we can work in this (or any) base. Consider the number 567_{10}. The five has a value of 500 because it is in the hundreds or 10^2 position ($5 \times 10^2 = 500$). The six has a value of 60 because it is in the tens position ($6 \times 10 = 60$). If the number were 567_8, the five would have a value of 320 in decimal because it is in the 8^2 or 64s position ($5 \times 8^2 = 320$). The six has a value of 48, in decimal, because it is in the eights (8^1) position. $6 \times 8 = 48$. In both bases the seven is in the units position and has a value of 7.

$$567_8 = (5 \times 8^2) + (6 \times 8) + 7 = 375_{10}$$

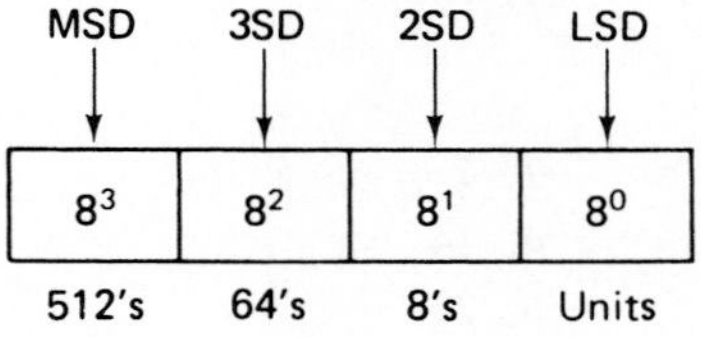

FIGURE 4-3 Positional value of the first four positions.

4-2-1 Octal to Decimal Conversions

Let's examine the number 1025 in octal. We might be tempted to read this as "one thousand twenty-five." Such language implies base 10. The number should be read "one-zero-two-five," which does not imply any base. We normally think in decimal, and decimal is the number system we use most; therefore, let's convert the number to base 10. There is a 1 in the 512's position + a 0 in the 64's position + a 2 in the 8's position + a 5 in the units position.

$$1025_8 = (1 \times 8^3) + (0 \times 8^2) + (2 \times 8^1) + (5 \times 8^0) = 533_{10}$$

$$1025_8 = (512) + (0) + (16) + (5) = 533_{10}$$

RULE 4-2 In converting from octal to decimal, find the weight of the digit in each position. Add the values of the digits in each position to determine the decimal equivalent.

☐ EXAMPLE 4-6

Change the following octal numbers to decimal: **(a)** 73_8; **(b)** 432_8; **(c)** 600_8; **(d)** 1234_8.

Solutions

(a) $73_8 = (7 \times 8^1) + (3 \times 8^0) = (56 + 3) = 59_{10}$
(b) $432_8 = (4 \times 8^2) + (3 \times 8^1) + (2 \times 8^0) = 256 + 24 + 2 = 282_{10}$
(c) $600_8 = (6 \times 8^2) + (0 \times 8^1) + (0 \times 8^0) = 384 + 0 + 0 = 384_{10}$
(d) $1234_8 = (1 \times 8^3) + (2 \times 8^2) + (3 \times 8^1) + (4 \times 8^0) = 512 + 128$
$+ 24 + 4 = 668_{10}$

☐

4-2-2 Decimal to Octal Conversions

Let's reverse the process. In Example 4-7, the number 189 is converted from decimal to octal.

☐ EXAMPLE 4-7

$189_{10} = (?)_8$.

Solution

Step 1. $\frac{189}{8} = 23 + 5$

Step 2. $\frac{23}{8} = 2 + 7$

Step 3. $\frac{2}{8} = 0 + 2$ — MSD → 2 7 5 $\quad 189_{10} = 275_8$

Since we are converting to base 8, we first divide the decimal number by 8 (step 1). The answer is 23 with a remainder of 5. This remainder is the LSD (least significant digit) of our base 8 number. The whole number left after this initial division (23) is again divided by 8 to determine the value of the digit in the next (8^1) position. This division leaves a remainder of 7, which becomes the digit in the 8^1 position. The whole number resulting from this division (2) is again divided by 8, and the remainder (2) is the digit in the 8^2 position. This digit is the MSD (most significant digit) since no further division is possible.

☐ EXAMPLE 4-8

Convert the following decimal numbers to octal numbers: **(a)** 78; **(b)** 376; **(c)** 1463.

Solutions

(a)

$$\frac{78}{8} = 9 + 6$$

$$\frac{9}{8} = 1 + 1$$

$$\frac{1}{8} = 0 + 1 \rightarrow 1 \quad 1 \quad 6 \qquad 78_{10} = 116_8$$

(b)

$$\frac{376}{8} = 47 + 0$$

$$\frac{47}{8} = 5 + 7$$

$$\frac{5}{8} = 0 + 5 \rightarrow 5 \quad 7 \quad 0 \qquad 376_{10} = 570_8$$

(c)

$$\frac{1463}{8} = 182 + 7$$

$$\frac{182}{8} = 22 + 6$$

$$\frac{22}{8} = 2 + 6$$

$$\frac{2}{8} = 0 + 2 \rightarrow 2 \quad 6 \quad 6 \quad 7 \qquad 1463_{10} = 2667_8$$

☐

PRACTICE PROBLEMS 4-2

1. Convert the following numbers from base 8 to base 10:
   (a) 46_8 (b) 111_8 (c) 204_8 (d) 400_8 (e) 777_8
2. Convert the following numbers from base 10 to base 8:
   (a) 27_{10} (b) 89_{10} (c) 100_{10} (d) 256_{10} (e) 400_{10}

Solutions

1. (a) 38_{10} (b) 73_{10} (c) 132_{10} (d) 256_{10} (e) 511_{10}
2. (a) 33_8 (b) 131_8 (c) 144_8 (d) 400_8 (e) 620_8

Additional practice problems are at the end of the chapter.

4-3 HEXADECIMAL NUMBER SYSTEM

Another important computer number system is the hexadecimal system (base 16). How many digits exist in hexadecimal? Sixteen, because the base is 16. What is the value of the largest digit in hexadecimal? Fifteen, one less than the base. However, 15 is a decimal number and requires *two digits* (15). In the hexadecimal number system, 16 *different* symbols must be used. Zero through 9 are used for the first ten, and A through F are used for the remaining six, as shown in Table 4-2. (Also refer to Table 4-1 on page 91 for the complete conversion table.)

Let's look at the number 567 again. Recall from the previous section that, in the number 567_{10}, the 5 is in the 10^2 position and so has a value of 500 ($5 \times 10^2 = 500$). Six is in the 10s position and so has a value of 60 ($6 \times 10 = 60$). In the number 567_{16}, the 5 is in the 16^2 position, so it has a value of 1280 in decimal ($5 \times 16^2 = 1280$). The 6 is in the 16 position and so has a value of 80 ($6 \times 16 = 96$). Again, the 7 is in the units position and so has a value of 7.

$$567_{16} = (5 \times 16^2) + (6 \times 16) + 7 = 1383$$

TABLE 4-2

Decimal	*Hexadecimal*	*Decimal*	*Hexadecimal*
0	0	9	9
1	1	10	A
2	2	11	B
3	3	12	C
4	4	13	D
5	5	14	E
6	6	15	F
7	7	16	10
8	8		

4-3-1 Hexadecimal to Decimal Conversions

The hexadecimal number 3AC2 may be converted to base 10 as follows (see Figure 4-4):

Three 4096's, which equal	12,288
+ten 256's, which equal	2,560
+twelve 16's, which equal	192
+two 1's, which equal	2
SUM	$15{,}042_{10}$

or

$$(3 \times 16^3) + (10 \times 16^2) + (12 \times 16^1) + (2 \times 16^0) = 15{,}042_{10}$$

RULE 4-3 In converting from hexadecimal to decimal, find the weight of the digit in each position. Add the values of the digits in each position to determine the decimal equivalent.

□ EXAMPLE 4-9

Convert the following hexadecimal numbers to decimal numbers: **(a)** 72; **(b)** C29; **(c)** 12AB; **(d)** 1A2A.

Solutions

(a) $72_{16} = (7 \times 16^1) + (2 \times 16^0) = 112 + 2 = 114_{10}$
(b) $C29_{16} = (12 \times 16^2) + (2 \times 16^1) + (9 \times 16^0) = 3113_{10}$
(c) $12AB_{16} = (1 \times 16^3) + (2 \times 16^2) + (10 \times 16^1) + (11 \times 16^0) = 4779_{10}$
(d) $1A2A_{16} = (1 \times 16^3) + (10 \times 16^2) + (2 \times 16^1) + (10 \times 16^0) = 6698_{10}$

□

4-3-2 Decimal to Hexadecimal Conversions

The conversion of decimal numbers to hexadecimal is identical to conversion using other bases. In Example 4-10, the number 1324 is converted from decimal to hexadecimal. Since the hexadecimal number system has a base of 16, division is by 16. In working with remainders greater than 9, we substitute the characters A, B, C, D, E, and F as required.

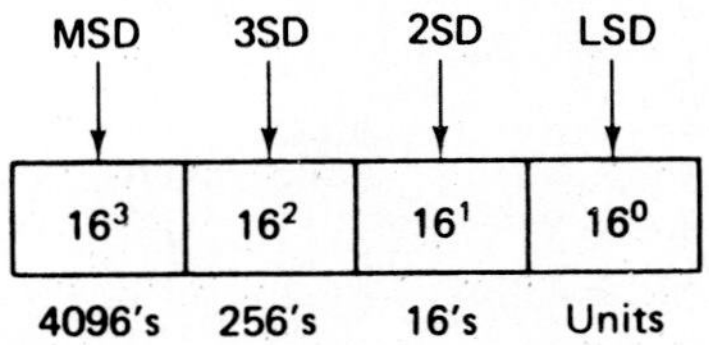

FIGURE 4-4 Positional value of a four-digit hexadecimal number.

EXAMPLE 4-10

$1324_{10} = (?)_{16}$.

Solution

Step 1. $\frac{1324}{16} = 82 + 12$ → LSD

Step 2. $\frac{82}{16} = 5 + 2$

Step 3. $\frac{5}{16} = 0 + 5$ — MSD → 5 2 C $\quad 1324_{10} = 52C_{16}$

EXAMPLE 4-11

Convert the following decimal numbers to hexadecimal numbers:
(a) 672; **(b)** 1763; **(c)** 12760.

Solutions

(a) $\frac{672}{16} = 42 + 0$

$\frac{42}{16} = 2 + 10$

$\frac{2}{16} = 0 + 2 \rightarrow 2 \quad A \quad 0 \quad 672_{10} = 2A0_{16}$

(b) $\frac{1763}{16} = 110 + 3$

$\frac{110}{16} = 6 + 14$

$\frac{6}{16} = 0 + 6 \rightarrow 6 \quad E \quad 3 \quad 1763_{10} = 6E3_{16}$

(c) $\frac{12760}{16} = 797 + 8$

$\frac{797}{16} = 49 + 13$

$\frac{49}{16} = 3 + 1$

$\frac{3}{16} = 0 + 3 \rightarrow 3 \quad 1 \quad D \quad 8 \quad 12760_{10} = 31D8_{16}$

PRACTICE PROBLEMS 4-3

1. Convert the following hexadecimal numbers to decimal numbers:
   (a) D_{16} (b) $3F_{16}$ (c) $A4_{16}$ (d) $1CD_{16}$ (e) $10BE_{16}$
2. Convert the following decimal numbers to hexadecimal numbers:
   (a) 27_{10} (b) 85_{10} (c) 100_{10} (d) 256_{10} (e) 1500_{10}

Solutions

1. (a) 13_{10} (b) 63_{10} (c) 164_{10}
   (d) 461_{10} (e) 4286_{10}
2. (a) $1B_{16}$ (b) 55_{16} (c) 64_{16}
   (d) 100_{16} (e) $5DC_{16}$

Additional practice problems are at the end of the chapter.

4-4 BINARY TO OCTAL TO HEXADECIMAL CONVERSIONS

4-4-1 Binary to Octal Conversions

Many 1's and 0's are required when we use binary numbers to represent large quantities. If we convert from binary to octal, the number of digits required is reduced by a factor of 3 because one octal digit equals 3 bits. That is, all octal digits, 0 through 7, can be represented by 3 bits.

0 = 000	4 = 100
1 = 001	5 = 101
2 = 010	6 = 110
3 = 011	7 = 111

■ **KEY POINT** To convert from binary to octal, separate the bits into 3-bit groups, starting with the LSB and moving left to the MSB.

☐ EXAMPLE 4-12

Convert 110111100001_2 to octal:

Solution

110	111	100	001	(binary number divided into groups of three)
6	7	4	1	(octal substitution for each 3-bit group)

☐ EXAMPLE 4-13

Convert 11101001_2 to octal:

Solution 011 101 001
3 5 1 $11101001_2 = 351_8$

Notice that a zero was added to the leftmost group of bits to remind us that 3 bits represent one octal digit.

☐

The procedure may be reversed to convert from octal to binary.

☐ EXAMPLE 4-14

Convert 417_8 to binary:

Solution 4 1 7 $417_8 = 100001111_2$
100 001 111

☐ EXAMPLE 4-15

Convert 362_8 to binary:

Solution 3 6 2 $362_8 = 11110010_2$
011 110 010

☐

In each example, 3 bits are substituted for each octal digit. A zero in the leftmost bit position has no meaning and need not be written in the answer.

4-4-2 Binary to Hexadecimal Conversions

If we convert from binary to hexadecimal, the number of digits required is reduced by a factor of 4 because one hexadecimal digit equals 4 bits. That is, all hexadecimal digits (0 through F) can be represented by 4 bits.

0 = 0000	8 = 1000
1 = 0001	9 = 1001
2 = 0010	A = 1010
3 = 0011	B = 1011
4 = 0100	C = 1100
5 = 0101	D = 1101
6 = 0110	E = 1110
7 = 0111	F = 1111

■ **KEY POINT** To convert from binary to hexadecimal, separate the bits into 4-bit groups, starting with the LSB and moving left to the MSB.

☐ EXAMPLE 4-16

Convert 110110011011_2 to hexadecimal:

Solution

1101	1001	1011	(binary number divided into 4-bit groups)
D	9	B	(hexadecimal substitution for each 4-bit group)

☐

PRACTICE PROBLEMS 4-4

Convert the following binary numbers to (a) octal numbers and then (b) hexadecimal numbers:

1. 1101001101_2

2. 1000011001_2

3. 1100110001_2

4. 111001110101_2

5. 111101011000110_2

Convert the following numbers to binary numbers:

6. 73_8

7. 127_8

8. 617_8

9. 506_8

10. 1207_8

11. $A6_{16}$

12. $4C_{16}$

13. $C0D_{16}$

14. $BA1_{16}$

15. $F23_{16}$

Solutions

1. (a) 1515_8 (b) $34D_{16}$

2. (a) 1031_8 (b) 219_{16}

3. (a) 1461_8 (b) 331_{16}

4. (a) 7165_8 (b) $E75_{16}$

5. (a) 75306_8 (b) $7AC6_{16}$

6. 111011_2

7. 1010111_2

8. 110001111_2

9. 101000110_2

10. 1010000111_2

11. 10100110_2

12. 1001100_2

13. 110000001101_2

14. 101110100001_2

15. 111100100011_2

Additional practice problems are at the end of the chapter.

4-5 BINARY TO OCTAL TO DECIMAL TO HEXADECIMAL CONVERSIONS

4-5-1 The Calculator

Conversion between number systems is possible with many calculators. However, use the calculator only after you have a thorough understanding of the concepts.

4-5-2 Converting between the Various Number Systems

The simplest octal to hexadecimal conversion is to first convert the number to binary and then make a conversion from binary to the other number system. We have converted from decimal to the other systems, and we have converted between binary, octal, and hexadecimal. Now let's practice some conversions among all the number systems we have studied. For example, let's convert 200_{10} to binary, octal, and hexadecimal.

☐ EXAMPLE 4-17

$$200_{10} = \underline{\qquad\qquad}_2 = \underline{\qquad\qquad}_8 = \underline{\qquad\qquad}_{16}$$

Solution There are many ways to solve this problem. We either could convert from 200_{10} to each of the other systems, or we could convert to base 2 and then convert to base 8 and base 16 from base 2. Another way, and probably the one requiring fewest steps, is to convert from base 10 to base 16, then from base 16 to base 2, and then from base 2 to base 8. Remember, fewer steps mean fewer chances for error.

$$\frac{200}{16} = 12 + 8 \longrightarrow 8$$

$$\frac{12}{16} = 0 + 12 \longrightarrow C \quad 8$$

$$200_{10} = C8_{16}$$

Convert to binary:

C	8	
1100	1000	$C8_{16} = 11001000_2$

Convert to octal:

011	001	000	$11001000_2 = 310_8$
3	1	0	

$$200_{10} = 11001000_2 = 310_8 = C8_{16}$$

☐ EXAMPLE 4-18

$C27_{16} = \_\_\_\_\_\_\_\__2 = \_\_\_\_\_\_\_\__8 = \_\_\_\_\_\_\_\__{10}$

Solution In this example, we have a straight conversion to binary and then to octal. Convert to binary:

C	2	7
1100	0010	0111

$C27_{16} = 110000100111_2$

Convert to octal:

110	000	100	111
6	0	4	7

$110000100111_2 = 6047_8$

The conversion to decimal can be made from any of the other systems, but conversion from base 16 requires fewer steps.

$$\begin{aligned} C27_{16} &= (12 \times 16^2) + (2 \times 16) + (7 \times 16^0) \\ &= 3072 + 32 + 7 \\ &= 3111_{10} \\ C27_{16} &= 110000100111_2 = 6047_8 = 3111_{10} \end{aligned}$$

☐

You will find that some conversions in the following problems will exceed the limits of the calculator. Consult the users manual and become familiar with the maximum numbers that can be stored and displayed in systems other than decimal.

PRACTICE PROBLEMS 4-5

1. $73_8 = \_\_\_\_\_\_\_\__{16} = \_\_\_\_\_\_\_\__2 = \_\_\_\_\_\_\_\__{10}$
2. $273_8 = \_\_\_\_\_\_\_\__{16} = \_\_\_\_\_\_\_\__2 = \_\_\_\_\_\_\_\__{10}$
3. $A3_{16} = \_\_\_\_\_\_\_\__8 = \_\_\_\_\_\_\_\__2 = \_\_\_\_\_\_\_\__{10}$
4. $12E_{16} = \_\_\_\_\_\_\_\__8 = \_\_\_\_\_\_\_\__2 = \_\_\_\_\_\_\_\__{10}$
5. $1101001_2 = \_\_\_\_\_\_\_\__8 = \_\_\_\_\_\_\_\__{16} = \_\_\_\_\_\_\_\__{10}$
6. $10111100_2 = \_\_\_\_\_\_\_\__8 = \_\_\_\_\_\_\_\__{16} = \_\_\_\_\_\_\_\__{10}$

Solutions

1. $73_8 = 3B_{16} = 111011_2 = 59_{10}$
2. $273_8 = BB_{16} = 10111011_2 = 187_{10}$
3. $A3_{16} = 243_8 = 10100011_2 = 163_{10}$
4. $12E_{16} = 456_8 = 100101110_2 = 302_{10}$
5. $1101001_2 = 151_8 = 69_{16} = 105_{10}$
6. $10111100_2 = 274_8 = BC_{16} = 188_{10}$

Additional practice problems are at the end of the chapter.

SELF-TEST 4-1

Convert the following numbers to decimal numbers:

1. 46_8
2. 276_8
3. $F6_{16}$
4. $C3A_{16}$
5. 1011011_2
6. 11001010_2

Convert the following decimal numbers to (a) binary, (b) octal, and (c) hexadecimal:

7. 10
8. 28
9. 187
10. 625

Convert the following to the bases indicated.

11. 93_{10} = ______$_2$ = ______$_8$ = ______$_{16}$
12. $E1A_{16}$ = ______$_2$ = ______$_8$ = ______$_{10}$
13. 11010010110_2 = ______$_8$ = ______$_{10}$ = ______$_{16}$
14. 573_8 = ______$_2$ = ______$_{10}$ = ______$_{16}$

Answers to Self-test 4-1 are at the end of the chapter.

4-6 ADDITION

4-6-1 Adding Decimal Numbers

Add the numbers 23 and 45. When numbers are added, they are added in columns; one number is placed below the other. Either number may be placed first. In Example 4-19, the numbers 23 and 45 are placed in the position for addition. The digits are separated to emphasize positional differences. Eight is in the units position (3 + 5), and 6 is in the tens position (2 + 4). The answer is 68. (Six 10's and eight 1's.)

☐ **EXAMPLE 4-19**

23 + 45 = ?

```
2   3
4   5
-----
6   8
```

☐ **EXAMPLE 4-20**

Add the numbers 64 and 87 in base 10:

Solution

Carry	1	1	
		6	4
		8	7
	1	(15)	(11)
(subtract value of carry)		10	10
	1	5	1

(remainders taken as decimal sum)

□

In Example 4-20, there is an 11 in the units position. This number is greater than 9, which tells us there must be a carry (10). The digit remaining in the units position is the difference of 10 and 11, or 1. The tens position now contains 6 + 8 + a carry of 1, or 15. Again, since the number is greater than 9, a carry is indicated. Subtracting 10 from 15, we get 5 with a carry of 1. The third position contains only the carry. Since we have been adding for years in decimal, our experience allows us to perform these operations automatically. We have taken time to discuss the arithmetic steps because this is the addition process in all bases.

□ EXAMPLE 4-21

Add the following decimal numbers using the procedure given in Example 4-20: **(a)** 738 + 417; **(b)** 9706 + 463; **(c)** 1774 + 7268.

Solutions

(a)

1		1		
	7	3	8	
	4	1	7	
1	(11)	5	(15)	
	10		10	738 + 417 = 1155
1	1	5	5	

(b)

1	1				
	9	7	0	6	
		4	6	3	
1	(10)	(11)	6	9	9706 + 463 = 10169
	10	10			
1	0	1	6	9	

(c)

1	1	1		
1	7	7	4	
7	2	6	8	
9	(10)	(14)	(12)	1774 + 7268 = 9042
	10	10	10	
9	0	4	2	

□

4-6-2 Adding Octal Numbers

☐ EXAMPLE 4-22

Add the octal numbers 736 and 215.

Solution

(carry)	1		1	
		7	3	6
		2	1	5
(decimal sum)	1	(9)	5	(11)
(subtract value of carry)		8		8
	1	1	5	3

$736_8 + 215_8 = 1153_8$

Again, the LSD or units position is added first. The sum of 6 and 5 is 11. The number is greater than 7 (7 is the largest digit in base 8), which tells us there must be a carry. Subtracting 8 (the value of the carry) leaves 3 with a carry of 1. The 8^1 position now contains 3 + 1 + a carry of 1, or 5. Since 5 is less than 7, no carry exists. In the 8^2 position, 7 + 2 is 9. Since the number is greater than 7, a carry is generated. Subtracting the carry (8) yields a remainder of 1 with a carry of 1. Since the 8^3 position contains only the carry, the answer is 1153_8.

☐ EXAMPLE 4-23

Add the following octal numbers:

(a)	**(b)**	**(c)**
243	764	604
172	414	777

Solutions

(a)

1			
2	4	3	
1	7	2	
4	(11)	5	
	8		
4	3	5	$= 435_8$

(b)

1	1	1		
7	6	4		
	4	1	4	
1	(12)	(8)	(8)	
	8	8	8	
1	4	0	0	$= 1400_8$

(c)

1	1	1		
	6	0	4	
	7	7	7	
1	(14)	(8)	(11)	
	8	8	8	
1	6	0	3	$= 1603_8$

☐

PRACTICE PROBLEMS 4-6

Add the following octal numbers on page 110.

1. 204
   316

2. 447
   173

3. 605
   726

4. 517
   567

5. 176
   761

Solutions

1. 522_8
2. 642_8
3. 1533_8
4. 1306_8
5. 1157_8

Additional practice problems are at the end of the chapter.

4-6-3 Adding Hexadecimal Numbers

☐ EXAMPLE 4-24

Add the hexadecimal numbers 7AF and 579.

Solution

(carry)	1	1		
	7	A	F	
	5	7	9	
(decimal sum)	(13)	(18)	(24)	
(subtract value of carry)		16	16	
	D	2	8	$= D28_{16}$

Adding the units position yields 24. Twenty-four is greater than the largest digit (F), which is equal to decimal 15. This indicates that a carry should exist. Subtracting the value of a carry (16) leaves a remainder of 8 with a carry of 1. In the 16^1 position, the sum of A (decimal 10) + 7 + a carry of 1 is 18. Again, a carry is generated, leaving a remainder of 2 with a carry of 1. In the 16^2 position, the carry +7 + 5 equals decimal 13. In hexadecimal, 13 is D; therefore, the answer is $D28_{16}$.

☐ EXAMPLE 4-25

Add the following hexadecimal numbers:

(a) 4D3
818

(b) 789
C47

(c) F347
E006

Solutions

(a)

4	D	3	
8	1	8	
(12)	(14)	(11)	
C	E	B	$= CEB_{16}$

(b)

1		1		
	7	8	9	
	C	4	7	
1	(19)	(13)	(16)	
	16		16	
1	3	D	0	$= 13D0_{16}$

(c)

1 (carry)					
	F	3	4	7	
	E	0	0	6	
1	(29)	3	4	(13)	
	16				
1	D	3	4	D	= $1D34D_{16}$

□

PRACTICE PROBLEMS 4-7

Add the following hexadecimal numbers:

1. C4
   1B
2. A4
   B7
3. A28
   6F9
4. 9B5
   4C8
5. FACE
   1701

Solutions

1. DF_{16}
2. $15B_{16}$
3. 1121_{16}
4. $E7D_{16}$
5. $111CF_{16}$

Additional practice problems are found at the end of the chapter.

4-6-4 Adding Binary Numbers

□ EXAMPLE 4-26

Add the binary numbers 1101 and 1001.

Solution

	1			1		(carry)
		1	1	0	1	
		1	0	0	1	
(decimal sum)		(2)	1	1	(2)	
(subtract value of carry)		2			2	
	1	0	1	1	0	= 10110_2

The method of addition is the same as that used with other bases. Notice that 1 + 1 equals decimal 2. Since 2 is invalid in base 2, a carry is generated into the next position. We see from the example that a carry is generated whenever a sum greater than 1 results from the addition of numbers in a column.

□ EXAMPLE 4-27

Add the following binary numbers on page 112.

(a) 10	**(b)** 111	**(c)** 1001	**(d)** 11100	
11	101	1010	01010	

Solutions

(a)

1			
	1	0	
	1	1	
1	(2)	1	
	2		
1	0	1	$= 101_2$

(b)

1	1	1		
	1	1	1	
	1	0	1	
1	(3)	(2)	(2)	
	2	2	2	
1	1	0	0	$= 1100_2$

(c)

1					
	1	0	0	1	
	1	0	1	0	
1	(2)	0	1	1	
	2				
1	0	0	1	1	$= 10011_2$

(d)

1	1					
	1	1	1	0	0	
	0	1	0	1	0	
1	(2)	(2)	1	1	0	
	2	2				
1	0	0	1	1	0	$= 100110_2$

□

From working with these and other examples of binary addition, we see that:

1. 0 + 0 always equals 0.
2. 0 + 1 always equals 1.
3. 1 + 1 always equals 0 and a carry.
4. 1 + 1 + a carry always equals 1 and a carry.

PRACTICE PROBLEMS 4-8

Add the following binary numbers:

1. 1011
1100

2. 1001
1111

3. 10111
10001

4. 11001
11110

5. 101110
111100

Solutions

1. 10111_2

2. 11000_2

3. 101000_2
4. 110111_2
5. 1101010_2

Additional practice problems are at the end of the chapter.

SELF-TEST 4-2

Add the following numbers:

1. AF_{16}
   $\underline{36_{16}}$
2. $C4A_{16}$
   $\underline{17E_{16}}$
3. $F4A9_{16}$
   $\underline{FE7A_{16}}$
4. 73_8
   $\underline{66_8}$
5. 63_8
   $\underline{41_8}$
6. 73_8
   $\underline{46_8}$
7. 101_2
   $\underline{111_2}$
8. 1101_2
   $\underline{1011_2}$
9. 10110_2
   $\underline{11110_2}$

Answers to Self-test 4-2 are at the end of the chapter.

4-7 SUBTRACTION

4-7-1 Subtracting Decimal Numbers

Subtraction in base 10 is so automatic for us that we may do it without thinking of the rules involved. Since these rules in base 10 are identical to those for other bases, and since we are more familiar with base 10 than with other bases, let's review the process of subtraction in base 10.

□ EXAMPLE 4-28

Subtract 25 from 43:

Solution

	3		borrow 1
(minuend)	~~4~~ ⇆	(1) 3	
(subtrahend)	2	5	
(difference)	1	8	

As you know, we start with the least significant position and subtract. Since we cannot subtract 5 from 3, we *borrow* from the next significant position of the minuend. We may now interpret the 3 as 13 and find the difference, which is 8. Moving to the next position, we must reduce the minuend by 1 because of the borrow. We now subtract the number in the subtrahend from the new minuend and the difference is 1. Remember, the borrow carries with it a weight equal to that of the base.

□ EXAMPLE 4-29

Subtract the following decimal numbers. Show the process of borrowing.

(a) $\begin{array}{r} 624 \\ -276 \\ \hline \end{array}$ **(b)** $\begin{array}{r} 854 \\ -235 \\ \hline \end{array}$ **(c)** $\begin{array}{r} 8534 \\ -6748 \\ \hline \end{array}$ **(d)** $\begin{array}{r} 705 \\ -528 \\ \hline \end{array}$

Solutions

(a)

$$\begin{array}{rrr} 5 & (1)1 & \\ \not{6} & \not{2} & (1)4 \\ -2 & 7 & 6 \\ \hline 3 & 4 & 8 \end{array}$$

(b)

$$\begin{array}{rrr} & 4 & \\ 8 & \not{5} & (1)4 \\ -2 & 3 & 5 \\ \hline 6 & 1 & 9 \end{array}$$

(c)

$$\begin{array}{rrrr} 7 & (1)4 & (1)2 & \\ \not{8} & \not{5} & \not{3} & (1)4 \\ -6 & 7 & 4 & 8 \\ \hline 1 & 7 & 8 & 6 \end{array}$$

(d)

$$\begin{array}{rrr} 6 & 9 & \\ \not{7} & (\not{1})\not{0} & (1)5 \\ -5 & 2 & 8 \\ \hline 1 & 7 & 7 \end{array}$$

□

Notice in Example 4-29(d) that, since 8 is greater than 5, a borrow is indicated. However, the next significant digit is 0 and no borrow can take place (we can't borrow something from nothing). Therefore, we must borrow from the next significant digit (7), giving the second position a value of 10. Now we can borrow from that position to yield 15 in the least significant position.

4-7-2 Subtracting Octal Numbers

In Example 4-30, the number 267_8 is subtracted from 512_8. Notice that a borrow is necessary in the first position since 7 is greater than 2. The borrow creates a new number in the minuend, 12. (This number is read "one-two," *not* "twelve," because twelve is a decimal number. One-two in base 8 means 8 + 2 in base 10.)

□ EXAMPLE 4-30

Subtract 267_8 from 512_8:

Solution

$$\begin{array}{lrrr} & 4 & \nearrow 10 & \\ \text{(minuend)} & \not{5} & \not{1} \rightarrow & (1)2 \\ \text{(subtrahend)} & 2 & 6 & 7 \\ \hline \text{(difference)} & 2 & 2 & 3 = 223_8 \end{array}$$

One-two minus 7 equals 3. If you think in base 10, then (8 + 2) − 7 equals 3. In the next position, 1 was borrowed from the number in the minuend, making its value 0. A borrow is again required; it makes the value of the minuend 10 (or 8 + 0 in base 10). Subtracting 6 from 10 yields a difference of 2. In the next position, 1 was borrowed from the minuend, making its value 4. The difference in this position is 2.

EXAMPLE 4-31

Solve the following subtraction problems in base 8:

(a)
```
    43
(−)16
```

(b)
```
   161
(−)72
```

(c)
```
  2172
(−)1717
```

Solutions

(a)

	3	
	~~4~~	(1)3
(−)	1	6
	2	5

$= 25_8$

(b)

	0	(1)5	
	~~1~~	~~6~~	(1)1
(−)		7	2
		6	7

$= 67_8$

(c)

	1		6		
	~~2~~	(1)1	~~7~~	(1)2	
(−)	1	7	1	7	
		2	5	3	

$= 253_8$

PRACTICE PROBLEMS 4-9

Subtract the following octal numbers:

1.
```
    73
(−)56
```

2.
```
   376
(−)277
```

3.
```
   673
(−)176
```

4.
```
  2133
(−)1574
```

5.
```
  6014
(−)3247
```

Solutions

1. 15_8

2. 77_8

3. 475_8

4. 337_8

5. 2545_8

Additional practice problems are at the end of the chapter.

4-7-3 Subtracting Hexadecimal Numbers

Applying the same rules as in base 8, let's subtract the hexadecimal number 85E from the number C37.

☐ EXAMPLE 4-32

C37 − 85E = $(?)_{16}$.

Solution

	B	↗(1)2		
(minuend)	~~C~~	~~3~~ ⟶	(1)7	
(subtrahend)	8	5	E	
(difference)	3	D	9	$= 3D9_{16}$

Notice that a borrow is necessary in the first position since E is greater than 7. One-seven (*not* seventeen) minus E equals 9 or, in base 10, (16 + 7) − 14 equals 9. (Remember, the borrow has a weight of 16.) In the next position, a borrow is required, so 5 is subtracted from 12 (one-two), which equals D. (In base 10, it would be 16 + 2 − 5 = 13.) Finally, in the next position, subtracting 8 from B (remember that we borrowed 1, making C a B) yields 3 or, in base 10, 11 − 8 = 3.

☐ EXAMPLE 4-33

Subtract the following hexadecimal numbers:

(a)	A29	**(b)**	3A12	**(c)**	F1B2
	−7BF		−29A0		−ABC7

Solutions

(a)

9	↗(1)1		
~~A~~	~~2~~→(1)9		
7	B	F	
2	6	A	$= 26A_{16}$

(b)

	9			
3	~~A~~→(1)1		2	
2	9	A	0	
1	0	7	2	$= 1072_{16}$

(c)

E	↗(1)0	↗(1)A		
~~F~~	~~1~~	~~B~~→(1)2		
A	B	C	7	
4	5	E	B	$= 45EB_{16}$

☐

PRACTICE PROBLEMS 4-10

Subtract the following hexadecimal numbers:

1. A17
(−)26B

2. B27
(−)6A4

3. 7A29
(−)2BC9

4. 4A9B
(−)100F

5. $\begin{array}{r} \text{FADE} \\ (-)\underline{\text{2C3F}} \end{array}$

Solutions

1. $7AC_{16}$

2. 483_{16}

3. $4E60_{16}$

4. $3A8C_{16}$

5. $CE9F_{16}$

Additional practice problems are at the end of the chapter.

4-7-4 Subtracting Binary Numbers

The process of subtraction in base 2 is similar to that in the other bases, as was discussed earlier in this chapter.

☐ EXAMPLE 4-34

Subtract 0101 from 1110:

Solution

borrow			0		
(minuend)	1	1	1̸→(1)0		
(subtrahend)	0	1	0	1	
(difference)	1	0	0	1	$= 1001_2$

☐ EXAMPLE 4-35

Subtract 01101 from 10011:

Solution

borrow	0	1				
(minuend)	1→(1̸)0̸→(1)0			1	1	
(subtrahend)	0	1	1	0	1	
	0	0	1	1	0	$= 00110_2$

☐

In Example 4-34, a borrow is necessary in the least significant position since the minuend is 0. The borrow has a weight of 2 because we are working in base 2. Note that in Example 4-35 a borrow is required in the 2^2 position. However, since the 2^3 position contains a 0, the borrow must come from the 2^4 position to the 2^3 position. A borrow is now possible from the 2^3 position, and the minuend in the 2^2 position is 10 (one-zero), resulting in a difference of 1. The new minuend in the next position (2^3) is now 1, and the difference is 0. Since our original borrow came from the next position (2^4), the minuend is 0 and the difference is 0.

PRACTICE PROBLEMS 4-11

Subtract the following binary numbers:

1. $\begin{array}{r} 1010 \\ (-)\underline{0101} \end{array}$

2. $\begin{array}{r} 10110 \\ (-)\underline{01101} \end{array}$

3. $\begin{array}{r} 110010 \\ (-)\underline{100111} \end{array}$

4. $\begin{array}{r} 100111 \\ (-)\underline{001101} \end{array}$

5. $\begin{array}{r} 1100110 \\ (-)\underline{1011001} \end{array}$

Solutions

1. 101_2
2. 1001_2
3. 1011_2
4. 11010_2
5. 1101_2

Additional practice problems are at the end of the chapter.

4-7-5 Complement Method of Subtraction

In the modern computer, subtraction is usually performed by addition of the complement of the subtrahend. This is done to simplify the design of the arithmetic section of the computer. To help us understand the complement method of subtraction in binary, we will first use the complement method of subtraction in decimal.

To subtract numbers in decimal, we must first find the nine's complement of the subtrahend. The nine's complement of a digit is the difference between 9 and that digit and can be found by a simple inspection procedure. Let's find the nine's complement of 235:

$$\begin{array}{rl} 999 & \\ -\ 235 & \\ \hline 764 & \text{(nine's complement of 235)} \end{array}$$

The nine's complement may be converted to the ten's complement by adding 1. Find the ten's complement of 5789:

$$\begin{array}{rl} 9999 & \\ -\ 5789 & \\ \hline 4210 & \text{(nine's complement)} \\ +\quad 1 & \\ \hline 4211 & \text{(ten's complement)} \end{array}$$

□ EXAMPLE 4-36

Subtract 676 from 835 using the complement method.

Solution Find the ten's complement of the subtrahend; add this number to the minuend and drop the carry.

999			835	(minuend)
− 676			324	(ten's complement of 676)
323	(nine's complement)	~~1~~	159	(difference = 159)
+ 1				
324	(ten's complement)			

☐ EXAMPLE 4-37

Subtract 4678 from 6392 using the complement method.

Solution

9999		6392	(minuend)
− 4678		+ 5322	(ten's complement of 4678)
5321	~~1~~	1714	(difference = 1714)
+ 1			
5322			

☐

4-7-6 One's Complement

The one's complement of a binary number is similar to the nine's complement of a decimal number. The binary number to be complemented is subtracted from a binary number made up of all 1's. In the following example the one's complement of 10110010 is found.

☐ EXAMPLE 4-38

```
   1 1 1 1 1 1 1 1
 − 1 0 1 1 0 0 1 0
   ---------------
   0 1 0 0 1 1 0 1   (one's complement of 10110010)
```

☐

Careful inspection reveals that the one's complement of a binary number may be produced by changing all the ones in the number to zeros and changing all of the zeros to ones.

```
1 0 1 1 0 0 1 0   (invert this number to get the one's complement)
0 1 0 0 1 1 0 1
```

The computer is able to perform this operation easily by means of a circuit called an *inverter*. Notice that carries or borrows are not involved in this operation.

To subtract by using the one's complement method, the one's complement of the subtrahend is added to the minuend. The carry out of the highest bit position is then added to the *lowest bit position* (LSB). This is called *end-around carry*.

☐ EXAMPLE 4-39

Find the difference between 10110011 and 01101101:

Solution

```
                                       1
      (minuend)   10110011              10110011
   (subtrahend)   01101101              10010010   (one's complement)
                                       101000101
                                               1   (end-around carry)
                                        --------
                                        01000110   (difference)
```

□ EXAMPLE 4-40

Find the difference between 11011000 and 10110011:

Solution

```
                                       1
      (minuend)   11011000              11011000
   (subtrahend)   10110011              01001100   (one's complement)
                                       100100100
                                               1   (end-around carry)
                                        --------
                                        00100101   (difference)
```

□

4-7-7 Two's Complement

■ **KEY POINT** Most computers subtract by using the two's complement method.

The two's complement of a number may be found by adding 1 to the one's complement. Let's find the two's complement of 10011101. First, we change all 0's to 1's and all 1's to 0's (one's complement). Then we add 1 as follows:

```
10011101        0 1 1 0 0 0 1 0   (one's complement)
          (+)                 1
                ---------------
                0 1 1 0 0 0 1 1   (two's complement)
```

To subtract, we obtain the two's complement of the subtrahend and then add to the minuend. A quick way to find the two's complement of a number is to copy all bits starting at the LSB and continuing through the first 1 bit. Then complement the remaining bits. This is the two's complement of the original binary number. Consider the number 10010100. To find the two's complement, copy all the bits starting at the LSB and continuing through the first 1 bit. This gives us the three least significant bits (100). Next, complement the remaining bits. The answer is 01101100.

□ EXAMPLE 4-41

Let's subtract 01001010 from 01100111:

Solution

```
      (minuend)   01100111        01100111
   (subtrahend)   01001010     (+)10110110   (two's complement)
                                1 00011101   (difference)
        (this carry is ignored)
```

□

The carry resulting from the most significant bit position is ignored. There is no end-around carry as there is with the one's complement method. The computer obtains the one's complement by using an inverter. The two's complement is then obtained by starting the addition process with a carry, as shown in Example 4-42.

☐ EXAMPLE 4-42

Subtract 00110110 from 01011011.

Solution

(minuend)	01011011		01011011	
(subtrahend)	00110110		11001001	(one's complement)
			+ 1	(add 1 to get the two's complement)
(carry is dropped) →		1	00100101	(difference)

☐

The number of bit positions in the subtrahend must agree with the number of bit positions in the minuend. This is usually accomplished by providing a fixed number of storage cells in the computer. In the preceding examples, eight bit positions were used; therefore, eight storage cells were used to store the numbers. Eight cells were also used to store the difference, causing the carry out of the most significant bit position to be dropped because there was no place to store it. Each of the above eight-cell combinations is called a *register*. Example 4-43 illustrates the concept of registers with fixed numbers of bit positions.

☐ EXAMPLE 4-43

Subtract 00001100 from 01001010 (Figure 4-5).

Solution The minuend is placed in the Y register; the one's complement of the subtrahend is placed in the X register. The LSB is in the rightmost bit position. A 1 is placed in the LSB position of the carry register. All the other positions of the carry register store a zero at the start. If a carry is produced with the addition of a column, a 1 is stored in the *next* bit position of the carry register. Since the output

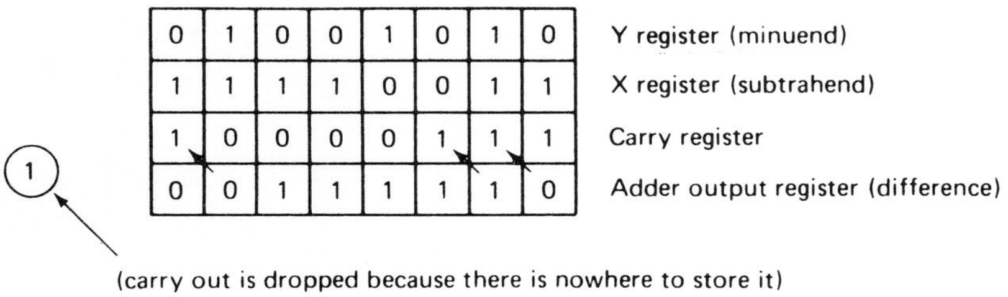

FIGURE 4-5

register is limited to eight bit positions, the carry out of the MSB position is dropped. □

PRACTICE PROBLEMS 4-12

For the following decimal numbers, find (a) the nine's complement and (b) the ten's complement:

1. 17
2. 36
3. 88
4. 276
5. 8976

For the following binary numbers, find (a) the one's complement and (b) the two's complement:

6. 1101
7. 1001
8. 100110
9. 110001
10. 11000011

Subtract the following binary numbers by using (a) the one's complement method and (b) the two's complement method:

11.
```
   11001
(−)01101
```

12.
```
   101101
(−)001011
```

13.
```
   1100011
(−)0011001
```

14.
```
   10110010
(−)00011101
```

15.
```
   11100100
(−)00011011
```

Solutions

1. (a) 82 (b) 83
2. (a) 63 (b) 64
3. (a) 11 (b) 12
4. (a) 723 (b) 724
5. (a) 1023 (b) 1024
6. (a) 0010 (b) 0011
7. (a) 0110 (b) 0111
8. (a) 011001 (b) 011010
9. (a) 001110 (b) 001111
10. (a) 00111100 (b) 00111101

11. (a)
```
  1
   11001
   10010   (one's complement)
  -------
1  01011
└──────→1
  -------
   01100
```
(b)
```
  1
   11001
   10011   (two's complement)
  -------
1̸  01100
```

12. (a)
```
  1
   101101
   110100   (one's complement)
  --------
1  100001
└───────→1
  --------
   100010
```
(b)
```
  1
   101101
   110101   (two's complement)
  --------
1̸  100010
```

13. (a)
```
1       11
        1100011
        1100110   (one's complement)
      1 1001001
      └────────→1
        1001010
```
(b)
```
1       111
        1100011
        1100111   (two's complement)
      1̸ 1001010
```

14. (a)
```
1
        10110010
        11100010   (one's complement)
      1 10010100
      └─────────→1
        10010101
```
(b)
```
1       11   1
        10110010
        11100011   (two's complement)
      1̸ 10010101
```

15. (a)
```
1       11   1
        11100100
        11100100   (one's complement)
      1 11001000
      └─────────→1
        11001001
```
(b)
```
1       11   1
        11100100
        11100101   (two's complement)
      1̸ 11001001
```

Additional practice problems are at the end of the chapter.

4-7-8 Sixteen's and Eight's Complement

When large signed binary numbers are added or subtracted, it is convenient to use the sixteen's complement method. The sixteen's complement is obtained by taking the fifteen's complement and adding one. The fifteen's complement is obtained by substituting each digit with the difference between its value and 15. For example, the sixteen's complement of 3B8 is

```
   FFF     C47   (fifteen's complement)
(−)3B8       1   (plus 1)
   C47     C48   (sixteen's complement)
```

In many systems, octal is used to represent large binary numbers. Therefore, it is convenient to use the eight's complement method for performing signed binary number additions and subtractions. The seven's complement is first obtained by substituting each octal digit with the difference between its value and 7. Then 1 is added to produce the eight's complement. For example, the eight's complement of 136_8 is

```
   777     641   (seven's complement)
(−)136       1   (plus 1)
   641     642   (eight's complement)
```

The following subtraction is performed by using the two's complement method:

```
   0011 1100 1011 ──→ 0011 1100 1011   (minuend)
(−)0010 0110 0111 ──→ 1101 1001 1001   (two's complement)
(carry is dropped) ──→ 1̸ 0001 0110 0100   (difference)
```

The same problem may be solved by using the sixteen's complement method:

$$
\begin{array}{lrrrl}
 & \text{3CB} & & \text{3CB} & \text{(minuend)}\\
 & \underline{(-)\,\text{267}} & & \underline{\text{D99}} & \text{(subtrahend in sixteen's complement form)}\\
 & & \not{1} & \text{164} & \text{(difference)}\\
\text{(carry is dropped)} & \longrightarrow & \uparrow & &
\end{array}
$$

If the above binary numbers are expressed in octal, the eight's complement method may be used:

$$
\begin{array}{lrrrl}
 & 1713_8 & & 1713 & \text{(minuend)}\\
 & \underline{(-)\;1147_8} & & \underline{6631} & \text{(subtrahend in eight's complement form)}\\
 & & \not{1} & 0544 & \text{(difference)}\\
\text{(carry is dropped)} & \longrightarrow & \uparrow & &
\end{array}
$$

PRACTICE PROBLEMS 4-13

For the following octal numbers, find (a) the seven's complement and (b) the eight's complement:

1. 6

2. 12

3. 56

4. 173

5. 726

For the following hexadecimal numbers, find (a) the fifteen's complement and (b) the sixteen's complement.

6. A

7. 2C

8. 87

9. 4A3

10. F209

Perform the following subtractions: (a) in binary by using the two's complement method, (b) in octal by using the eight's complement method, and (c) in hexadecimal by using the sixteen's complement method.

11. $\begin{array}{r} 11010011 \\ \underline{(-)00111101} \end{array}$

12. $\begin{array}{r} 10011101 \\ \underline{(-)00110011} \end{array}$

13. $\begin{array}{r} 10111100 \\ \underline{(-)01001110} \end{array}$

14. $\begin{array}{r} 11110000 \\ \underline{(-)00111100} \end{array}$

15. $\begin{array}{r} 11001111 \\ \underline{(-)10000011} \end{array}$

Solutions

1. (a) 1 (b) 2

2. (a) 65 (b) 66

3. (a) 21 (b) 22

4. (a) 604 (b) 605

5. (a) 051 (b) 052

6. (a) 5 (b) 6

7. (a) D3 (b) D4

8. (a) 78 (b) 79

9. (a) B5C (b) B5D

10. (a) 0DF6 (b) 0DF7

11. (a) 11010011
11000011 (two's complement)
~~1~~ 10010110_2

(b) 1101 0011 = D3
1100 0011 = C3 (sixteen's complement)
~~1~~ 96_{16}

(c) 011 010 011 = 323
111 000 011 = 703 (eight's complement)
~~1~~ 226_8

The subtrahend is 00111101. The two's complement is 11000011. Since we have already found the two's complement of 00111101 to be 11000011, we separate the two's complement into two 4-bit groups. Then 1100 0011 = C3 which is the sixteen's complement. To find the eight's complement, we must separate the subtrahend into three 3-bit groups. Since we have only 8 bits, we add a zero to the subtrahend in the original problem to get 000 111 101. The two's complement is 111 000 011. The eight's complement is 703_8.

12. (a) 10011101
11001101 (two's complement)
~~1~~ 01101010_2

(b) 1001 1101 = 9D
1100 1101 = CD (sixteen's complement)
~~1~~ $6A_{16}$

(c) 010 011 101 = 235
111 001 101 = 715 (eight's complement)
~~1~~ 152_8

13. (a) 10111100
10110010 (two's complement)
~~1~~ 01101110_2

(b) 1011 1100 = BC
1011 0010 = B2 (sixteen's complement)
~~1~~ $6E_{16}$

(c) 010 111 100 = 274
110 110 010 = 662 (eight's complement)
~~1~~ 156_8

14. (a) 11110000
11000100 (two's complement)
~~1~~ 10110100_2

(b) 1111 0000 = F0
1100 0100 = C4 (sixteen's complement)
~~1~~ $B4_{16}$

(c) 011 110 000 = 360
111 000 100 = 704 (eight's complement)
~~1~~ 264_8

15. (a) 11001111
01111101 (two's complement)
~~1~~ 01001100

(b) 1100 1111 = CF
0111 1101 = 7D (sixteen's complement)
~~1~~ $4C_{16}$

(c) 011 001 111 = 317
101 111 101 = 575 (eight's complement)

$\cancel{1}\ 114_8$

Additional practice problems are at the end of the chapter.

SELF-TEST 4-3

Work the following problems by using (a) the straight subtraction method and (b) the complement method:

1. 11101011_2 (−)$\underline{01001101_2}$

2. 11000110_2 (−)$\underline{10011011_2}$

3. 73_8 (−)$\underline{46_8}$

4. 603_8 (−)$\underline{364_8}$

5. $F6_{16}$ (−)$\underline{3A_{16}}$

6. $B03_{16}$ (−)$\underline{8AE_{16}}$

Answers to Self-Test 4-3 are at the end of the chapter.

CHAPTER 4 AT A GLANCE

Page	Key Point or Rule	Example
90	*Rule 4-1:* In converting from binary to decimal, find the value of the MSB. Work down to the LSB, adding the weight of that position if a 1 is present or a 0 if a 0 is present.	$101_2 = 1 \times 2^2 + 0 \times 2^1 + 1 \times 2^0 = 5_{10}$
97	*Rule 4-2:* In converting from octal to decimal, find the weight of the digit in each position. Add the values of the digits in each position to determine the decimal equivalent.	$473_8 = 4 \times 8^2 + 7 \times 8^1 + 3 \times 8^0 = 315_{10}$
100	*Rule 4-3:* In converting from hexadecimal to decimal, find the weight of the digit in each position. Add the values of the digits in each position to determine the decimal equivalent.	$2BC_{16} = 2 \times 16^2 + 11 \times 16^1 + 2 \times 16^0 = 700_{10}$
102	*Key Point:* To convert from binary to octal, separate the bits into 3-bit groups, starting with the LSB and moving left to the MSB.	$10110110_2 = 266_8$
104	*Key Point:* To convert from binary to hexadecimal, separate the bits into 4-bit groups, starting with the LSB and moving left to the MSB.	$10011011_2 = 9B_{16}$

120	*Key Point:* Computers subtract by using the two's complement method. The two's complement of the subtrahend is added to the minuend. The carry out of the MSB is ignored.	01100111 − 01001010 Solution: 01100111 (+)10110110 (two's complement) ~~1~~ 00011101 (difference)

End of Chapter Problems 4-1

Convert the following binary numbers to decimal numbers:

1. 111_2
2. 110_2
3. 1010_2
4. 1011_2
5. 1101_2
6. 1000_2
7. 10111_2
8. 11100_2
9. 100110_2
10. 110011_2
11. 111000_2
12. 101011_2
13. 1100101_2
14. 1000111_2
15. 1110111_2
16. 1111111_2
17. 11110000_2
18. 10101010_2
19. 11000011_2
20. 10011001_2

Convert the following decimal numbers to binary numbers:

21. 5
22. 6
23. 12
24. 15
25. 21
26. 31
27. 45
28. 53
29. 59
30. 62
31. 68
32. 80
33. 96
34. 115
35. 135
36. 160
37. 210
38. 245

End of Chapter Problems 4-2

Convert the following numbers from octal to decimal:

1. 14_8
2. 27_8
3. 77_8
4. 100_8
5. 276_8
6. 667_8
7. 4176_8
8. 6437_8
9. $11{,}714_8$
10. $23{,}476_8$

Convert the following numbers from decimal to octal:

11. 20_{10}
12. 42_{10}
13. 80_{10}
14. 98_{10}
15. 360_{10}
16. 517_{10}
17. 1417_{10}
18. 5716_{10}
19. 3789_{10}
20. 7063_{10}

End of Chapter Problems 4-3

Convert the following numbers from hexadecimal to decimal:

1. $1A_{16}$
2. $A1_{16}$
3. $4C_{16}$
4. $1BC_{16}$
5. 200_{16}
6. 400_{16}
7. $11AA_{16}$
8. $FACE_{16}$
9. $A02B_{16}$
10. $94AB_{16}$

Convert the following numbers from decimal to hexadecimal:

11. 22_{10}
12. 40_{10}
13. 97_{10}
14. 127_{10}
15. 512_{10}
16. 783_{10}
17. 2700_{10}
18. 4906_{10}
19. 6075_{10}
20. 8880_{10}

End of Chapter Problems 4-4

Convert the following binary numbers to (a) octal numbers and then (b) hexadecimal numbers.

1. 11001010_2
2. 10010110_2
3. 10011011_2
4. 11110011_2
5. 10000001_2
6. 10001111_2
7. 110011001001_2
8. 111100001111_2
9. 101011000011_2
10. 110100111010_2

Convert the following numbers to binary numbers.

11. 17_8
12. 23_8
13. 44_8
14. 53_8
15. 200_8
16. 400_8
17. 560_8
18. 777_8
19. 1000_8
20. 1002_8
21. C_{16}
22. F_{16}
23. 24_{16}
24. 30_{16}
25. $4F_{16}$
26. $D4_{16}$

27. $E3_{16}$

28. $F7_{16}$

29. $19A_{16}$

30. $2FF_{16}$

End of Chapter Problems 4-5

1. 75_8 = ________$_2$ = ________$_{10}$ = ________$_{16}$
2. 103_8 = ________$_2$ = ________$_{10}$ = ________$_{16}$
3. 140_8 = ________$_2$ = ________$_{10}$ = ________$_{16}$
4. 175_8 = ________$_2$ = ________$_{10}$ = ________$_{16}$
5. $2A_{16}$ = ________$_2$ = ________$_8$ = ________$_{10}$
6. $6C_{16}$ = ________$_2$ = ________$_8$ = ________$_{10}$
7. $1A7_{16}$ = ________$_2$ = ________$_8$ = ________$_{10}$
8. $2FB_{16}$ = ________$_2$ = ________$_8$ = ________$_{10}$
9. 100110_2 = ________$_8$ = ________$_{10}$ = ________$_{16}$
10. 110101_2 = ________$_8$ = ________$_{10}$ = ________$_{16}$
11. 111000101_2 = ________$_8$ = ________$_{10}$ = ________$_{16}$
12. 101111011111_2 = ________$_8$ = ________$_{10}$ = ________$_{16}$
13. 10_{10} = ________$_2$ = ________$_8$ = ________$_{16}$
14. 100_{10} = ________$_2$ = ________$_8$ = ________$_{16}$
15. 290_{10} = ________$_2$ = ________$_8$ = ________$_{16}$
16. 413_{10} = ________$_2$ = ________$_8$ = ________$_{16}$

End of Chapter Problems 4-6

Add the following octal numbers:

1. 123
   234

2. 451
   116

3. 456
   317

4. 517
   116

5. 667
   107

6. 273
   606

7. 604
   617

8. 176
   412

9. 726
   161

10. 471
    707

End of Chapter Problems 4-7

Add the following hexadecimal numbers:

1. 4A
   A3

2. B2
   1C

3. A07
29E

4. B18
398

5. 4AB
8AB

6. C17
8B9

7. ABF
CDE

8. FED
CBA

9. F0C6
E9A6

10. BA09
E6E4

End of Chapter Problems 4-8

Add the following binary numbers:

1. 1100
1001

2. 1001
1111

3. 11100
10111

4. 10111
11101

5. 10011
10110

6. 10101
11101

7. 110011
101111

8. 111000
011100

9. 101110
111011

10. 111101
100111

End of Chapter Problems 4-9

Subtract the following octal numbers:

1. 73
(−)26

2. 64
(−)40

3. 616
(−)554

4. 416
(−)267

5. 540
(−)273

6. 405
(−)167

7. 2004
(−)1445

8. 3001
(−)2563

9. 6334
(−)4617

10. 5503
(−)3473

Subtract the following hexadecimal numbers:

11. A6
(−)8C

12. CA
(−)8B

13. E06
(−)A25

14. C0A
(−)149

15. $\begin{array}{r} 9A0 \\ (-)\underline{3AF} \end{array}$

16. $\begin{array}{r} 8B4 \\ (-)\underline{3C6} \end{array}$

17. $\begin{array}{r} C00 \\ (-)\underline{103} \end{array}$

18. $\begin{array}{r} B00 \\ (-)\underline{608} \end{array}$

19. $\begin{array}{r} B029 \\ (-)\underline{A04D} \end{array}$

20. $\begin{array}{r} C24B \\ (-)\underline{174C} \end{array}$

Subtract the following binary numbers:

21. $\begin{array}{r} 1011 \\ (-)\underline{0110} \end{array}$

22. $\begin{array}{r} 1101 \\ (-)\underline{0110} \end{array}$

23. $\begin{array}{r} 1101 \\ (-)\underline{1011} \end{array}$

24. $\begin{array}{r} 1001 \\ (-)\underline{0110} \end{array}$

25. $\begin{array}{r} 11001 \\ (-)\underline{10111} \end{array}$

26. $\begin{array}{r} 111011 \\ (-)\underline{011101} \end{array}$

27. $\begin{array}{r} 10010 \\ (-)\underline{01111} \end{array}$

28. $\begin{array}{r} 110010 \\ (-)\underline{100111} \end{array}$

29. $\begin{array}{r} 110011 \\ (-)\underline{011101} \end{array}$

30. $\begin{array}{r} 111001 \\ (-)\underline{010110} \end{array}$

End of Chapter Problems 4-10

Find the two's complement of:

1. 1010_2
2. 11001_2
3. 101101_2
4. 11000010_2

Find the eight's complement of:

5. 64_8
6. 526_8
7. 635_8
8. 2723_8

Find the sixteen's complement of:

9. $A3_{16}$
10. $3B4_{16}$
11. $C1E_{16}$
12. $3F03_{16}$

Use the one's complement method to find:

13. $\begin{array}{r} 11001_2 \\ (-)\underline{01100_2} \end{array}$

14. $\begin{array}{r} 11100_2 \\ (-)\underline{10011_2} \end{array}$

15. $\begin{array}{r} 101101_2 \\ (-)\underline{001010_2} \end{array}$

16. $\begin{array}{r} 1101001_2 \\ (-)\underline{0010110_2} \end{array}$

Use the two's complement method to find:

17. $\begin{array}{r} 10110_2 \\ (-)\underline{01011_2} \end{array}$

18. $\begin{array}{r} 10011_2 \\ (-)\underline{00100_2} \end{array}$

19. $\begin{array}{r} 1011001_2 \\ (-)\underline{0010110_2} \end{array}$

20. $\begin{array}{r} 11100101_2 \\ (-)\underline{10011010_2} \end{array}$

Use the eight's complement method to find:

21. $\begin{array}{r} 63_8 \\ (-)\underline{25_8} \end{array}$

22. $\begin{array}{r} 503_8 \\ (-)\underline{127_8} \end{array}$

23. $\begin{array}{r} 626_8 \\ (-)\underline{377_8} \end{array}$

24. $\begin{array}{r} 6026_8 \\ (-)\underline{4562_8} \end{array}$

Use the sixteen's complement method to find:

25. $\begin{array}{r} A2_{16} \\ (-)\underline{8C_{16}} \end{array}$

26. $\begin{array}{r} E4D_{16} \\ (-)\underline{BCA_{16}} \end{array}$

27. $\begin{array}{r} 6A2_{16} \\ (-)\underline{4FF_{16}} \end{array}$

28. $\begin{array}{r} F40B_{16} \\ (-)\underline{2AEC_{16}} \end{array}$

Answers to Self-tests

SELF-TEST 4-1

1. 38
2. 190
3. 246
4. 3130
5. 91
6. 202
7. (a) 1010_2 (b) 12_8 (c) A_{16}
8. (a) 11100_2 (b) 34_8 (c) $1C_{16}$
9. (a) 10111011_2 (b) 273_8 (c) BB_{16}
10. (a) 1001110001_2 (b) 1161_8 (c) 271_{16}
11. (a) 1011101_2 (b) 135_8 (c) $5D_{16}$
12. (a) 111000011010_2 (b) 7032_8 (c) 3610_{10}
13. (a) 3226_8 (b) 1686_{10} (c) 696_{16}
14. (a) 101111011_2 (b) 379_{10} (c) $17B_{16}$

SELF-TEST 4-2

1. $E5_{16}$
2. $DC8_{16}$
3. $1F323_{16}$
4. 161_8
5. 124_8
6. 141_8
7. 1100_2
8. 11000_2
9. 110100_2

SELF-TEST 4-3

1. 10011110
2. 101011_2
3. 25_8
4. 217_8
5. BC_{16}
6. 255_{16}

ALGEBRAIC TERMS: ROOTS AND POWERS

CHAPTER 5

INTRODUCTION

When we move into areas new to us, whether that area is mathematics or electronics or some other technology, there are always new words to learn. To communicate with other people in the field, the same words in our vocabularies must have the same meaning if we are to understand each other. The words that we will define, such as *algebraic expression, term,* and *literal number,* are probably words you have heard and used before.

We will learn how *variables* are used, and we will assign values to variables and solve some simple electronics problems. We will also solve problems that include squares and square roots.

The skills we develop in this chapter and the concepts we learn will be used throughout the text.

CHAPTER OBJECTIVES

In this chapter, you will learn how to:

1. Identify algebraic expressions and terms.
2. Solve problems that contain both real and literal numbers.
3. Find the square and square root of both real and literal numbers.
4. Solve problems using various forms of Ohm's law.
5. Develop problem-solving techniques.

5-1 LITERAL AND REAL NUMBERS

We may divide numbers into two groups. The first group we can define as *real* numbers. These are the numbers we use all the time: 7, 468, 3.25, −6, and $2\frac{1}{2}$ are all examples of real numbers. Real numbers always have the same value. The number 6 always has a value of 6. It can never mean some other amount. Real numbers are also referred to as *constants*.

Often, in electronics, we have quantities that are not constant. For example, circuit current will depend on the values we assign to circuit voltage and resistance. Voltages, currents, resistances, and other circuit parameters are called *circuit variables*. Letters, called *literal numbers* are used to represent these quantities. $E = IR$ is Ohm's law written as an equation, where E represents the voltage source, I represents circuit current, and R is the circuit resistance. These letters E, I, and R, represent the variables voltage, current, and resistance. These literal numbers can represent different values. The letter E can stand for one value in one circuit and an entirely different value in another circuit. E might equal 10 volts in one circuit, 25 volts in a second circuit, and so on. Literal numbers can also be used to represent constants, but we usually use them to represent variables. Later in this chapter, and in other chapters, we will solve many equations that contain variables.

5-2 ALGEBRAIC EXPRESSIONS AND TERMS

An *algebraic expression* is a number or a group of numbers. The numbers can be real numbers, literal numbers, or both. $6A$, $3 + 2x$, $7(x - 2)$, $1.6 + 7x$, and $3ab + 4c - 5ac$ are all examples of algebraic expressions. If we knew the value of the literal numbers, then we could determine the numerical value of the expression.

When numbers are separated into groups with + or − signs, each group in the expression is called a *term*. The algebraic expression $3ab + 4c - 5ac$ contains three terms: $3ab$, $4c$, and $5ac$. $6A$ is one term. The expression $1.6 + 7x$ contains two terms: 1.6 and $7x$.

When an algebraic expression contains one term, it is called a *monomial*. Where an expression contains more than one term, it is called a *polynomial*. Two-term expressions are also called *binomials*. Three-term expressions are also called *trinomials*. These words, binomial and trinomial, are more descriptive than polynomial.

So, an algebraic expression can be a monomial or a polynomial. Furthermore, monomials and polynomials can contain either real or literal numbers or both.

5-3 NUMERICAL COEFFICIENTS

The *numerical coefficient* of an algebraic term is the real-number part of the term. If no numerical value is given, it is understood to be 1. In the term $6xy$, 6 is the numerical coefficient. In the term ax, 1 is the coefficient.

PRACTICE PROBLEMS 5-1

List the number of terms in the following expressions. Also indicate the kind of expression (monomial, binomial, and so on).

1. $2a + 3y$

2. $2x + y + 1$

3. $4(x + 1) - 3$

4. $\frac{3a}{8} + 26$

5. $\frac{4x}{5} + 1$

6. $3(a - 1) + 2b$

7. $\frac{3a - 1}{4} + 6$

8. $\frac{4c + 3}{2d} - b$

9. $\frac{7y}{8} + 6 - 4(x - 3)$

10. $3a + 6b - 4c - 3$

11. $4c - 3a + 4$

12. $\frac{4x - 3y}{4} - 6x + 3$

In the following expressions, list the numerical coefficients and the literal numbers in each term.

13. $a + 2b + 4c$

14. $\frac{3x}{4} + y$

Solutions

1. 2, binomial

2. 3, trinomial

3. 2, binomial

4. 2, binomial

5. 2, binomial

6. 2, binomial

7. 2, binomial

8. 2, binomial

9. 3, trinomial

10. 4, polynomial

11. 3, trinomial

12. 3, trinomial

The binomials and trinomials may also be called polynomials.

13. 1, 2, and 4 are the numerical coefficients. a, b, and c are the literal numbers.

14. $\frac{3}{4}$ and 1 are the numerical coefficients. x and y are the literal numbers.

Additional practice problems are at the end of the chapter.

5-4 EXPONENTS

In Chapter 2, we discussed the exponent whose base was 10. The exponent told us how many times 10, the base, was multiplied by itself. This is the function of an exponent in any base. The base may be any real number. For example,

$$5^3 = 5 \cdot 5 \cdot 5 = 125$$

The base can also be a literal number:

$$x^3 = x \cdot x \cdot x = ?$$

We were able to find the value of 5^3 because 5 is a real number and we can deal with real numbers directly. We can find x^3 only if we are given the value for x. Suppose we assign a value of 4 to x.
Then

$$x^3 = x \cdot x \cdot x = 4 \cdot 4 \cdot 4 = 64$$

■ **KEY POINT** We can solve a problem that contains literal numbers only if we know their value.

PRACTICE PROBLEMS 5-2

Solve the following problems. Round your answers to three significant figures.

1. 7^2
2. 6^3
3. 4.5^2
4. 8.73^2
5. 24^2

Solve the following problems where $x = 3$, $y = 4$:

6. x^3
7. y^2
8. y^3
9. x^2
10. x^4

Solutions

1. 49.0
2. 216
3. 20.3
4. 76.2
5. 576
6. 27.0
7. 16.0
8. 64.0
9. 9.00
10. 81.0

Additional practice problems are at the end of the chapter.

When real numbers and literal numbers appear in the same term of an algebraic expression, we can determine the value of the term if we know the value of the literal numbers. For example, in the term $6x^2$, if $x = 3$, then $6x^2 = 6 \cdot x^2 = 6 \cdot 3^2 = 6 \cdot 9 = 54$. When one or more of the numbers in a term is raised to a power, only that number is affected.

When 3 is substituted for x in our example, we get $6 \cdot 3^2$. Our first step is to find 3^2. Then we complete the multiplication. We can find the value of any term by multiplying the real numbers times the known values of the literal numbers.

□ EXAMPLE 5-1

Find **(a)** 5^3; **(b)** a^2, where $a = 7.3$; **(c)** b^3, where $b = 14.5$. Round your answers to three significant figures.

Solutions

(a) $5^3 = 5 \times 5 \times 5 = 125$.

Using the calculator,

5 [×] 5 [×] 5 [=] *Answer:* 125

or 5 [y^x] 3 [=] *Answer:* 125

When using the [y^x] key, the number is keyed in *first*, and then the exponent is keyed in. (Some calculators use [x^y] instead of [y^x].

(b) $a^2 = 7.3^2 = 7.3 \times 7.3 = 53.3$.

Using the calculator, 7.3 [x^2]. The answer is displayed and is 53.3. Using the [x^2] key saves key strokes, which decreases the chance for errors.

(c) $b^3 = 14.5^3 = 14.5 \times 14.5 \times 14.5 = 3050$.

Using the calculator,

14.5 [×] 14.5 [×] 14.5 [=] *Answer:* 3050

It's better to use the [y^x] key:

14.5 [y^x] 3 [=] *Answer:* 3050

Notice in (b) and (c) that we had to be given a value for a and a value for b in order to solve the problems.

□ EXAMPLE 5-2

Let $x = 3$ and $y = 4$. Find **(a)** xy; **(b)** x^2y; **(c)** x^2y^2; **(d)** $4x^3y^2$.

Solutions

(a) $xy = 3 \cdot 4 = 12$

(b) $x^2y = 3^2 \cdot 4 = 9 \cdot 4 = 36$

(c) $x^2y^2 = 3^2 \cdot 4^2 = 9 \cdot 16 = 144$

(d) $4x^3y^2 = 4 \cdot 3^3 \cdot 4^2 = 4 \cdot 27 \cdot 16 = 1728$

The algorithm for (d) looks like this:

4 [×] 3 [y^x] 3 [×] 4 [x^2] [=]   *Answer:* 1728

□

■ **KEY POINT** If a sign of grouping is used, then all numbers within that group are acted on by the exponent.

For example, in the term $(ab)^2$, both *a and b* are raised to the second power: $(ab)^2 = a^2 \cdot b^2$.

□ EXAMPLE 5-3

Find $(ab)^2$, where $a = 3$ and $b = 4$.

Solution

$$(ab)^2 = a^2 \cdot b^2 = 3^2 \cdot 4^2 = 9 \cdot 16 = 144$$

It would also be correct to find the product of *a* and *b* and then raise that value to the second power.

$$(ab)^2 = (3 \cdot 4)^2 = 12^2 = 144$$

Either method is correct and either method may be used in problem solving.

Using the calculator,

3 [×] 4 [=] [x^2]   *Answer:* 144.

Or

3 [×] 4 = [y^x] 2 [=]   *Answer:* 144.

Whenever the exponent is 2, the first method is better because it requires fewer steps.

□

If there is more than one term in an algebraic expression, we find the value of each term and then perform the indicated addition or subtraction.

□ EXAMPLE 5-4

Find $3a^2b - ab^2$, where $a = 5$ and $b = 6$.

Solution

$$\begin{aligned} 3a^2b - ab^2 &= 3 \cdot 5^2 \cdot 6 - 5 \cdot 6^2 \\ &= 3 \cdot 25 \cdot 6 - 5 \cdot 36 \\ &= 450 - 180 = 270 \end{aligned}$$

Using the calculator,

3 [×] 5 [x^2] [×] 6 [−] 5 [×] 6 [x^2] [=]   *Answer:* 270

□

PRACTICE PROBLEMS 5-3

Find the values of the following algebraic expressions where $x = 3$, $y = 4$, and $z = 5$:

1. xy
2. x^2y
3. $(xy)^2$
4. $2y^2z$
5. $3(xz)^2$
6. $x^3y - z$
7. $y^3 - z^2$
8. $(x^2y - z^2)^2$

Solutions

1. 12
2. 36
3. 144
4. 160
5. 675
6. 103
7. 39
8. 121

Additional practice problems are at the end of the chapter.

5-5 ROOTS

The *square root* of a number is one of its two equal factors. $6 \cdot 6 = 36$. The two equal factors are 6; therefore, 6 is the square root of 36. Actually, there are *two* square roots of 36 because $(-6)(-6)$ also equals 36. All positive numbers have two square roots: one negative and one positive. Usually, in solving electrical problems, the negative root is not considered.

The symbol used to denote the square root of a number is $\sqrt{}$ and is called the *radical*. $\sqrt{36} = 6$. Raising a number to the one-half power is another way of denoting square root. $36^{1/2} = 6$. Both notations will be used in this text.

We can extract from memory the square root of numbers like 25, 36, 64, and 81. If we go much beyond $\sqrt{100}$, most of us have to use the calculator. To find the square root of a number by using the calculator, we punch in the number and then the $\boxed{\sqrt{x}}$ key. We can check our answer by squaring it.

□ EXAMPLE 5-5

Find $\sqrt{40}$.

Solution

$$\sqrt{40} = 6.32 \quad \text{(rounded to three places)}$$

Using the calculator, $40\,\boxed{\sqrt{x}}$. *Answer:* 6.32.

Check: $6.32 \cdot 6.32 \simeq 40$

Because we rounded to three places, we don't get exactly 40 when we check. □

The cube root of a number is one of its three equal parts. The symbol is $\sqrt[3]{\ }$. Raising a number to the one-third power is another way of indicating cube root. $\sqrt[3]{x} = x^{1/3}$. Because we seldom work with cube roots or higher roots in electronics, we will limit problem solving in this chapter to square roots.

PRACTICE PROBLEMS 5-4

Solve the following problems accurate to three places:

1. $5^{1/2}$
2. $7^{1/2}$
3. $27^{1/2}$
4. $256^{1/2}$
5. $512^{1/2}$

Solutions

1. 2.24
2. 2.65
3. 5.20
4. 16.0
5. 22.6

Additional practice problems are at the end of the chapter.

KEY POINT When we extract the square root of an expression, we perform all mathematical functions under the radical *first*. Then we find the square root.

☐ EXAMPLE 5-6

Find the value of the following algebraic expressions, where $x = 4$, $y = 5$, and $z = 6$. Express answers in scientific notation rounded to three places.

(a) $\sqrt{2x + 3y}$ **(b)** $\sqrt{3xy^2}$ **(c)** $\sqrt{x^2y^2} + 4$

(d) $\sqrt{x^2y^3z^3 - x^3y^4z} - 4y$

Solutions

(a) $\sqrt{2x + 3y} = \sqrt{8 + 15} = \sqrt{23} = 4.80$

2 [×] 4 [+] 3 [×] 5 = [√x] *Answer:* 4.80

(b) $\sqrt{3xy^2} = \sqrt{3 \cdot 4 \cdot 5^2} = \sqrt{12 \cdot 25} = \sqrt{300} = 1.73 \times 10$

3 [×] 4 [×] 5 [x^2] = [√x] *Answer:* 1.73 × 10

(c) $\sqrt{x^2y^2} + 4 = \sqrt{4^2 \cdot 5^2} + 4 = \sqrt{16 \cdot 25} + 4$

$= \sqrt{400} + 4 = 20 + 4 = 2.40 \times 10$

4 [x^2] [×] 5 [x^2] [=] [√x] + 4 [=] *Answer:* 2.40 × 10

(d) $\sqrt{x^2y^3z^3 - x^3y^4z} - 4y = \sqrt{4^2 \cdot 5^3 \cdot 6^3 - 4^3 \cdot 5^4 \cdot 6} - 4 \cdot 5$
$= \sqrt{16 \cdot 125 \cdot 216 - 64 \cdot 625 \cdot 6} - 20 = \sqrt{4.32 \times 10^5 - 2.40 \times 10^5} - 20$
$= 438 - 20 = 4.18 \times 10^2$

4 [x^2] [×] 5 [y^x] 3 [×] 6 [y^x] 3 [−] [(] 4 [y^x] 3 [×] 5 [y^x] 4 [×] 6 [)] [=]
[$\sqrt{x}$] [−] 4 [×] 5 [=]   *Answer:* 4.18×10^2

□

PRACTICE PROBLEMS 5-5

Find the value of the following algebraic expressions, where $x = 3$, $y = 4$, and $z = 5$. Round answers to three places.

1. $\sqrt{xy}$
2. $\sqrt{z^2}$
3. $\sqrt{y^2z^2}$
4. $\sqrt{5x^2z}$
5. $\sqrt{x^2y^2}$
6. $\sqrt{4y^2 + z^2}$
7. $\sqrt{x^4 - 8y}$
8. $\sqrt{x^2yz^2 + 10}$
9. $\sqrt{3x^2y^3 - x^3z^2} + 3y$
10. $\sqrt{2x^4yz^2 - 3xy^3z} - 4xz$

Solutions

1. 3.46
2. 5.00
3. 20.0
4. 15.0
5. 12.0
6. 9.43
7. 7.00
8. 30.2
9. 44.4
10. 55.4

Additional practice problems are at the end of the chapter.

SELF-TEST 5-1

Find the value of the following expressions, where $x = 3$, $y = 5$, and $z = 4$. Consider only the positive root. Round answers to three places.

1. x^2y
2. $\sqrt{xz}$
3. $(2xy)^2$
4. $\sqrt{3xy^2}$
5. $xy + 2z$
6. $3y^2 - x^2z$
7. $\sqrt{3x^3 + 2yz}$
8. $\sqrt{3xy + yz^2}$
9. $\sqrt{2xz^2 + 3y^2z}$
10. $\sqrt{4y^2z - 2xz^2}$

Answers to Self-test 5-1 are at the end of the chapter.

5-6 PRACTICAL APPLICATIONS

Literal numbers are used in equations to represent the relationship between variables in electronics circuits. In this section we will use some of these equations and solve for circuit variables. Where square roots are found, only the positive root is considered.

One form of Ohm's law is $E = IR$. E is the applied voltage and is measured in volts; I is the circuit current and is measured in amperes; R is the circuit resistance and is measured in ohms.

To solve the problems in the following examples with your calculator, set the calculator up in engineering notation if you have that feature; otherwise, set it up in scientific notation rounded to three significant figures.

□ EXAMPLE 5-7

Find E when $I = 5$ mA and $R = 2.7$ kΩ.

Solution Recall from Chapter 3 that the prefix milli (m) is 10^{-3} and kilo (k) is 10^3. Then

$$E = IR = 5 \text{ mA} \times 2.7 \text{ k}\Omega = 5 \times 10^{-3} \times 2.7 \times 10^3 = 13.5 \text{ V}$$

5 [EE] [+/-] 3 [×] 2.7 [EE] 3 [=] *Answer*: 13.5 V

□

The power dissipated in a circuit can be found by using the equation $P = I^2R$. Power is measured in watts.

□ EXAMPLE 5-8

Find P when $I = 50$ mA and $R = 220$ Ω.

Solution

$$P = I^2R = (50 \text{ mA})^2 \times 220\ \Omega = (50 \times 10^{-3})^2 \times 220 = 0.55 \text{ W}$$

50 [EE] [+/-] 3 [x^2] [×] 220 = *Answer:* $550 \times 10^{-3} = 0.550$ W

□

When the power dissipated and the resistance are known, circuit current may be found by using the equation $I = \sqrt{\dfrac{P}{R}}$.

□ EXAMPLE 5-9

Find I when $P = 700$ mW and $R = 3.3$ kΩ.

Solution

$$I = \sqrt{\frac{P}{R}} = \sqrt{\frac{700 \text{ mW}}{3.3 \text{ k}\Omega}} = \sqrt{\frac{700 \times 10^{-3}}{3.3 \times 10^{3}}} = 1.46 \times 10^{-2} \text{ A}$$

$$= 14.6 \times 10^{-3} \text{ A} = 14.6 \text{ mA}$$

700 [EE] [+/-] 3 [÷] 3.3 [EE] 3 [=] [$\sqrt{x}$]

Answer: $14.6 \times 10^{-3} = 14.6$ mA

□

■ **KEY POINT** Prefixes are changed to their powers of ten values to solve problems, but it is standard practice to use prefixes instead of powers of ten in answers.

The applied voltage may be found when P and R are known by using the equation $E = \sqrt{PR}$.

□ EXAMPLE 5-10

Find E when $P = 100$ mW and $R = 10$ kΩ.

Solution

$$E = \sqrt{PR} = \sqrt{100 \text{ mW} \times 10 \text{ k}\Omega}$$

$$= \sqrt{100 \times 10^{-3} \times 10 \times 10^{3}} = 31.6 \text{ V}$$

100 [EE] [+/-] 3 [×] 10 [EE] 3 [=] [$\sqrt{x}$] *Answer:* 31.6 = 31.6 V

□

P can be found when E and R are known by using the equation $P = \dfrac{E^2}{R}$.

□ EXAMPLE 5-11

Find P when $E = 15$ V and $R = 2$ kΩ.

Solution

$$P = \frac{E^2}{R} = \frac{15^2}{2 \text{ k}\Omega} = \frac{15^2}{2 \times 10^3} = 0.113 \text{ W}$$

15 [x^2] [÷] 2 [EE] 3 [=] *Answer:* $113 \times 10^{-3} = 0.113$ W

□

PRACTICE PROBLEMS 5-6

In the following problems, round answers to three significant figures. Use prefixes where appropriate.

$E = IR$. Find E when:

1. $I = 2$ mA, $R = 2.7$ kΩ
2. $I = 4.3$ mA, $R = 3.3$ kΩ
3. $I = 400$ μA, $R = 22$ kΩ
4. $I = 40$ mA, $R = 680$ Ω
5. $I = 5.6$ mA, $R = 4.7$ kΩ

$P = I^2R$. Find P when:

6. $I = 40$ mA, $R = 1$ kΩ
7. $I = 170$ mA, $R = 470$ Ω
8. $I = 2.3$ A, $R = 33$ Ω
9. $I = 1.76$ mA, $R = 27$ kΩ
10. $I = 400$ μA, $R = 56$ kΩ

$I = \sqrt{\frac{P}{R}}$. Find I when:

11. $P = 250$ mW, $R = 120$ Ω
12. $P = 1.2$ W, $R = 22$ kΩ
13. $P = 37.3$ mW, $R = 680$ Ω
14. $P = 4.73$ mW, $R = 12$ kΩ
15. $P = 680$ mW, $R = 1.8$ kΩ

$E = \sqrt{PR}$. Find E when:

16. $P = 56$ mW, $R = 2$ kΩ
17. $P = 475$ mW, $R = 100$ Ω
18. $P = 2.7$ W, $R = 150$ Ω
19. $P = 780$ mW, $R = 2.7$ kΩ
20. $P = 30$ mW, $R = 3.9$ kΩ

$P = \frac{E^2}{R}$. Find P when:

21. $E = 12$ V, $R = 470$ Ω
22. $E = 9$ V, $R = 1.8$ kΩ
23. $E = 20$ V, $R = 4.7$ kΩ
24. $E = 30$ V, $R = 120$ Ω
25. $E = 15$ V, $R = 12$ kΩ

Solutions

1. 5.40 V
2. 14.2 V
3. 8.80 V
4. 27.2 V
5. 26.3 V
6. 1.60 W
7. 13.6 W
8. 175 W
9. 83.6 mW
10. 8.96 mW
11. 45.6 mA
12. 7.39 mA
13. 7.41 mA
14. 628 μA
15. 19.4 mA
16. 10.6 V
17. 6.89 V
18. 20.1 V
19. 45.9 V
20. 10.8 V
21. 306 mW
22. 45 mW
23. 85.1 mW
24. 7.50 W
25. 18.8 mW

Additional practice problems are at the end of the chapter.

Capacitive reactance in an ac circuit is measured in ohms. The equation is

$$X_C = \frac{1}{2\pi f C}$$

where π (the Greek letter pi) is a constant and equals 3.14 accurate to three places, f is the frequency in hertz (Hz), and C is the capacitance in farads (F).

□ EXAMPLE 5-12

Find the capacitive reactance (X_C) when $f = 7.5$ kHz and $C = 0.5\ \mu$F.

Solution

$$X_C = \frac{1}{2\pi f C} = \frac{1}{2 \times \pi \times 7.5\text{ kHz} \times 0.5\ \mu\text{F}} = 42.4\ \Omega$$

2 [×] [π] [×] 7.5 [EE] 3 [×] .5 [EE] [+/-] 6 [=] [1/x] *Answer:* 42.4 Ω

□

An important parameter in ac circuits is the resonant frequency (f_r). This frequency is found by using the following equation:

$$f_r = \frac{1}{2\pi\sqrt{LC}}$$

where L is the circuit inductance in henries (H) and C is the circuit capacitance in farads (F).

□ EXAMPLE 5-13

Find the resonant frequency when $L = 200$ mH and $C = 300$ pF.

Solution

$$f_r = \frac{1}{2\pi\sqrt{LC}} = \frac{1}{2 \times \pi \times \sqrt{200\text{ mH} \times 300\text{ pF}}}$$

$$= \frac{1}{6.28 \times \sqrt{6 \times 10^{-11}}} = \frac{1}{6.28 \times 7.75 \times 10^{-6}}$$

$$= \frac{1}{4.87 \times 10^{-5}} = 2.05 \times 10^4\text{ Hz} = 20.5\text{ kHz}$$

2 [×] [π] [×] [(] 200 [EE] [+/-] 3 [×] 300 [EE] [+/-] 12 [)] [√x] [=] [1/x]

Answer: $20.5 \times 10^3 = 20.5$ kHz

If you were going to solve a number of problems where you had to find $2 \times \pi$, you might save time by storing the product:

2 [×] [π] [=] [STO]

Now, with 2π stored, we could do this:

200 [EE] [+/-] 3 [×] 300 [EE] [+/-] 12 [=] [$\sqrt{x}$] [×] [RCL] [=] [1/x]

Answer: 20.5×10^3

Again the answer is 20.5×10^3. Notice that if you key in *RCL first* you will have to use parentheses, as we did in the original solution.

Always use prefixes instead of powers of ten form in your final answers. □

Another important parameter in ac circuits is *impedance*. Impedance is measured in ohms. One method used to find impedance when both resistance and reactance are known is

$$Z = \sqrt{R^2 + X^2}$$

where R is the circuit resistance and X is the circuit reactance.

□ EXAMPLE 5-14

Find Z when $R = 5.6\ \text{k}\Omega$ and $X = 4\ \text{k}\Omega$.

Solution

$$Z = \sqrt{R^2 + X^2} = \sqrt{(5.6\ \text{k}\Omega)^2 + (4\ \text{k}\Omega)^2} = \sqrt{(5.6 \times 10^3)^2 + (4 \times 10^3)^2}$$
$$= \sqrt{3.14 \times 10^7 + 1.6 \times 10^7} = \sqrt{4.74 \times 10^7}$$
$$= 6.88 \times 10^3\ \Omega = 6.88\ \text{k}\Omega$$

5.6 [EE] 3 [x^2] [+] 4 [EE] 3 [x^2] [=] [$\sqrt{x}$]

Answer: $6.88 \times 10^3 = 6.88\ \text{k}\Omega$

Remember, when adding numbers, the exponents to the base 10 must be the same. (When using the calculator to perform the addition, no changes need to be made.) □

PRACTICE PROBLEMS 5-7

In the following problems, round answers to three significant figures. Use prefixes where appropriate.

$X_C = \dfrac{1}{2\pi f C}$. Find X_C when:

1. $f = 120$ Hz, $C = 2\ \mu$F
2. $f = 2.5$ kHz, $C = 300$ nF
3. $f = 420$ Hz, $C = 0.5\ \mu$F
4. $f = 60$ Hz, $C = 10\ \mu$F
5. $f = 1.2$ kHz, $C = 50$ nF

$f_r = \dfrac{1}{2\pi\sqrt{LC}}$. Find f_r when:

6. $L = 300$ mH, $C = 100$ nF
7. $L = 500$ mH, $C = 470$ pF
8. $L = 200$ mH, $C = 2\ \mu$F
9. $L = 1.7$ H, $C = 20$ nF
10. $L = 37$ mH, $C = 75$ nF

$Z = \sqrt{R^2 + X^2}$. Find Z when:

11. $R = 1.8$ kΩ, $X = 3$ kΩ
12. $R = 27$ kΩ, $X = 50$ kΩ
13. $R = 1$ kΩ, $X = 600$ Ω
14. $R = 4.7$ kΩ, $X = 3.1$ kΩ
15. $R = 12$ kΩ, $X = 16$ kΩ

Solutions

1. 663 Ω
2. 212 Ω
3. 758 Ω
4. 265 Ω
5. 2.65 kΩ
6. 919 Hz
7. 10.4 kHz
8. 252 Hz
9. 863 Hz
10. 3.02 kHz
11. 3.50 kΩ
12. 56.8 kΩ
13. 1.17 kΩ
14. 5.63 kΩ
15. 20.0 kΩ

SELF-TEST 5-2

1. $E = IR$. Find E if $I = 7.5$ mA and $R = 2.7$ kΩ.
2. $P = I^2R$. Find P if $I = 27$ mA and $R = 150$ Ω.
3. $I = \sqrt{\frac{P}{R}}$. Find I when $P = 750$ mW and $R = 3.3$ kΩ.
4. $E = \sqrt{PR}$. Find E when $P = 60$ mW and $R = 6.8$ kΩ.
5. $P = \frac{E^2}{R}$. Find P when $E = 15$ V and $R = 1.8$ kΩ.
6. $X_C = \frac{1}{2\pi fC}$. Find X_C when $f = 3.2$ kHz and $C = 150$ nF.
7. $f_r = \frac{1}{2\pi\sqrt{LC}}$. Find f_r when $L = 700$ mH and $C = 470$ pF.
8. $Z = \sqrt{R^2 + X^2}$. Find Z when $R = 5.6$ kΩ and $X = 8$ kΩ.

Answers to Self-test 5-2 are at the end of the chapter.

CHAPTER 5 AT A GLANCE

Page	Key Point or Rule	Example
136	*Key point:* We can solve problems that contain literal numbers only if we know their value.	Find x^3 if $x = 5$. $x^3 = 5^3 = 125$

138	*Key point:* If a sign of grouping is used, then all numbers within that group are acted on by the exponent.	Find $(ab)^2$ if $a = 3$ and $b = 4$. $(3 \times 4)^2 = 12^2 = 144$
140	*Key point:* When we extract the square root of an expression, we perform all mathematical functions under the radical *first*.	Let $a = 4$, $b = 5$. $\sqrt{a + b^2} = 5.39$
143	*Key point:* Prefixes are changed to their powers of ten values to solve problems, but use prefixes in answers.	Find I when $P = 700$ mW and $R = 3.3$ kΩ. $I = \sqrt{\frac{P}{R}} = \sqrt{\frac{700 \text{ mW}}{3.3 \text{ k}\Omega}} = \sqrt{\frac{700 \times 10^{-3}}{3.3 \times 10^{3}}} = 1.46 \times 10^{-2}$ A $= 14.6 \times 10^{-3}$ A $= 14.6$ mA

End of Chapter Problems 5-1

List the number of terms in the following expressions. Also indicate the kind of expression (monomial, binomial, and so on).

1. $\frac{a}{2} + 2b$
2. $4a - \frac{3b}{4}$
3. $4 + 6b + c$
4. $6 + 2a - 3b$
5. $\frac{6a}{7b} + \frac{4c}{2}$
6. $2x + \frac{7z}{3}$
7. $\frac{4(3 - x)}{a} + 7b$
8. $\frac{6(2 - y)}{z} + 4b$
9. $\frac{2a + 3b}{4} + 3 - \frac{c}{7}$
10. $\frac{3x - 4y}{5} + 6 + \frac{z}{2}$

In the following expressions, list the numerical coefficients and the literal numbers in each term:

11. $3a - 4b + 2c$
12. $5x + 2y - 7z$
13. $\frac{2x}{3} + 7y$
14. $\frac{2a}{3} - 3b$
15. $\frac{5}{2a} - \frac{4}{b}$
16. $\frac{3}{x} + \frac{7y}{8}$

End of Chapter Problems 5-2

Solve the following problems. Round to three significant figures.

1. 2.3^2
2. 3.6^2
3. 7.4^2
4. 9.2^2
5. 16^2
6. 18^2
7. 8^3
8. 9^3
9. 3.6^3
10. 4.7^3
11. 5.5^3
12. 10.4^3

Solve the following problems where $a = 4$ and $b = 7$:

13. a^2
14. b^2
15. a^3
16. b^3
17. a^4
18. b^4

End of Chapter Problems 5-3

Solve the following problems, where $a = 5$, $b = 3$, and $c = 6$. Round answers to three significant figures.

1. ac
2. b^2c
3. a^2c
4. $(bc)^2$
5. $(ac)^2$
6. $a^3 - b^3$
7. $2(ab)^2$
8. $3(bc)^2$
9. $a^2b^2 - 3c^2$
10. $abc^2 - b^2c^2$
11. $ac^2 - 2abc$
12. $b^2c + a^2$
13. $(a^2b - 2c^2)^2$
14. $(2a^2b - 4c^2)^2$
15. $3a^2b^3c - 2a^3b^2c^2$
16. $4a^3b^3c^4 - 6a^2b^4c^3$
17. $3ab^3c + 4a^2bc^3 - 2a^4bc^2$
18. $2a^3b^3c - 4a^2b^4c^2 + 3b^3c^3$
19. $(3a^2c^4 - 2a^3b^2c)^3$
20. $(2a^3b^4c^3 - 4ab^2c^3)^3$

End of Chapter Problems 5-4

Solve the following problems. Round to three significant figures.

1. $30^{\frac{1}{2}}$
2. $18^{\frac{1}{2}}$
3. $40^{\frac{1}{2}}$
4. $80^{\frac{1}{2}}$
5. $200^{\frac{1}{2}}$
6. $625^{\frac{1}{2}}$
7. $1040^{\frac{1}{2}}$
8. $1440^{\frac{1}{2}}$
9. $2000^{\frac{1}{2}}$
10. $2500^{\frac{1}{2}}$

End of Chapter Problems 5-5

Solve the following problems, where $x = 3$, $y = 4$, and $z = 5$. Round answers to three significant figures.

1. $\sqrt{yz}$
2. $\sqrt{xz}$
3. $\sqrt{y^2}$
4. $\sqrt{3x^2}$
5. $\sqrt{6xz^2}$
6. $\sqrt{5y^2z}$
7. $\sqrt{x^2 + 3y^2}$
8. $\sqrt{6y^2 + x^2}$
9. $\sqrt{6.4y^3 - z^2}$
10. $\sqrt{3.7y^3 - z^2}$
11. $\sqrt{y^2z^2} + x$
12. $\sqrt{x^2z^2} + y$
13. $\sqrt{x^3} + 2z$
14. $\sqrt{y^3} + 3x$
15. $\sqrt{x^3z^2 + 3x^4y^2}$
16. $\sqrt{x^3y^3 + 4x^4y^3}$
17. $\sqrt{3y^3z^3 - 5x^3z^3}$
18. $\sqrt{6x^3z^4 - 3y^3z^3}$
19. $\sqrt{4x^3y^4z^2 - 2x^2y^3z} - z$
20. $\sqrt{3x^4y^4z^3 - 4x^3y^4z^2} - y^3$
21. $\sqrt{5.6x^4y^3z^4 - 3x^2y^4z^3} - 4z^3$
22. $\sqrt{2.7x^3y^3z^4 - 5x^3y^4z^2} - 8y^3$

End of Chapter Problems 5-6

In the following problems, round answers to three significant figures. Use prefixes where appropriate.

$E = IR$. Find E when:

1. $I = 660\ \mu A$, $R = 6.8\ k\Omega$
2. $I = 1.73$ mA, $R = 12\ k\Omega$
3. $I = 3.7$ mA, $R = 18\ k\Omega$
4. $I = 3.7$ mA, $R = 1.8\ k\Omega$
5. $I = 50$ mA, $R = 2.7\ k\Omega$
6. $I = 125\ \mu A$, $R = 75\ k\Omega$
7. $I = 430$ mA, $R = 560\ \Omega$
8. $I = 84.5$ mA, $R = 68\ \Omega$

$P = I^2R$. Find P when:

9. $I = 7.5$ mA, $R = 680\ \Omega$
10. $I = 760\ \mu A$, $R = 1.8\ k\Omega$
11. $I = 36.5$ mA, $R = 3.3\ k\Omega$
12. $I = 2$ mA, $R = 120\ \Omega$
13. $I = 1.2$ A, $R = 5\ \Omega$
14. $I = 620\ \mu A$, $R = 1.5\ k\Omega$
15. $I = 23.5$ mA, $R = 810\ \Omega$
16. $I = 65.5$ mA, $R = 680\ \Omega$

$I = \sqrt{\dfrac{P}{R}}$. Find I when:

17. $P = 65.4$ mW, $R = 5.6\ k\Omega$
18. $P = 100$ mW, $R = 1\ k\Omega$
19. $P = 50$ mW, $R = 18\ k\Omega$
20. $P = 500$ mW, $R = 330\ \Omega$
21. $P = 4.6$ W, $R = 10\ \Omega$
22. $P = 25$ W, $R = 8\ \Omega$
23. $P = 650\ \mu W$, $R = 56\ \Omega$
24. $P = 275\ \mu W$, $R = 100\ \Omega$

$E = \sqrt{PR}$. Find E when:

25. $P = 8.3$ mW, $R = 27\ k\Omega$
26. $P = 78.3$ mW, $R = 5.6\ k\Omega$
27. $P = 130$ mW, $R = 750\ \Omega$
28. $P = 500$ mW, $R = 1.8\ k\Omega$
29. $P = 18$ mW, $R = 6.8\ k\Omega$
30. $P = 1.2$ W, $R = 220\ \Omega$
31. $P = 100$ mW, $R = 100\ \Omega$
32. $P = 700\ \mu W$, $R = 180\ k\Omega$

$P = \dfrac{E^2}{R}$. Find P when:

33. $E = 16$ V, $R = 6.8$ kΩ

34. $E = 4.7$ V, $R = 560$ Ω

35. $E = 13.3$ V, $R = 33$ kΩ

36. $E = 8.73$ V, $R = 2.7$ kΩ

37. $E = 18.2$ V, $R = 10$ kΩ

38. $E = 25$ V, $R = 1.5$ kΩ

39. $E = 150$ V, $R = 4.7$ kΩ

40. $E = 360$ mV, $R = 1.2$ kΩ

End of Chapter Problems 5-7

In the following problems, round answers to three significant figures. Use prefixes where appropriate.

$X_C = \dfrac{1}{2\pi fC}$. Find X_C when:

1. $f = 5$ kHz, $C = 100$ nF

2. $f = 10$ kHz, $C = 1$ μF

3. $f = 200$ Hz, $C = 200$ nF

4. $f = 15$ kHz, $C = 470$ pF

5. $f = 6.3$ kHz, $C = 2$ μF

6. $f = 500$ Hz, $C = 500$ nF

7. $f = 65.4$ kHz, $C = 250$ pF

8. $f = 57.5$ kHz, $C = 500$ pF

9. $f = 130$ kHz, $C = 750$ pF

10. $f = 250$ kHz, $C = 500$ pF

$f_r = \dfrac{1}{2\pi\sqrt{LC}}$. Find f_r when:

11. $L = 80$ mH, $C = 150$ nF

12. $L = 1$ mH, $C = 10$ μF

13. $L = 150$ μH, $C = 50$ nF

14. $L = 6.5$ mH, $C = 600$ nF

15. $L = 50$ mH, $C = 250$ nF

16. $L = 300$ mH, $C = 50$ nF

17. $L = 400$ mH, $C = 5$ nF

18. $L = 550$ mH, $C = 7.5$ nF

19. $L = 0.25$ H, $C = 25$ μF

20. $L = 0.75$ H, $C = 50$ μF

$Z = \sqrt{R^2 + X^2}$. Find Z when:

21. $R = 120$ kΩ, $X = 80$ kΩ

22. $R = 100$ Ω, $X = 100$ Ω

23. $R = 200$ Ω, $X = 100$ Ω

24. $R = 3$ kΩ, $X = 4$ kΩ

25. $R = 33$ kΩ, $X = 20$ kΩ

26. $R = 27$ kΩ, $X = 20$ kΩ

27. $R = 4.7$ kΩ, $X = 7.35$ kΩ

28. $R = 8.2$ kΩ, $X = 10$ kΩ

29. $R = 2.2$ MΩ, $X = 6.75$ MΩ

30. $R = 910$ kΩ, $X = 1.34$ MΩ

Answers to Self-tests

SELF-TEST 5-1

1. 45.0

2. 3.46

3. 900

4. 15.0

5. 23.0

6. 39.0

7. 11.0

8. 11.2

9. 19.9

10. 17.4

SELF-TEST 5-2

1. 20.3 V
2. 109 mW
3. 15.1 mA
4. 20.2 V
5. 125 mW
6. 332 Ω
7. 8.77 kHz
8. 9.77 kΩ

CHAPTER 6

FRACTIONS

INTRODUCTION

Any study of mathematics involves the use of fractions. In electronics, we constantly use equations that contain fractions or deal with numbers that are nonwhole numbers. In this chapter we will review how to add, subtract, multiply, and divide fractions. We will also learn how to express fractions as decimal fractions or as proper fractions and how to convert from one to the other.

Only fractions that are real numbers are considered in this chapter. Fractions containing literal numbers are discussed in Chapter 7.

CHAPTER OBJECTIVES

In this chapter you will learn how to:

1. Identify prime numbers.
2. Find the prime factors of real numbers.
3. Find the lowest common multiple of numbers.
4. Multiply and divide whole and fractional numbers.
5. Add and subtract fractions with common denominators and with unlike denominators.
6. Express numbers as either mixed numbers or as improper fractions.
7. Multiply, divide, add, and subtract improper fractions and mixed numbers.

6-1 PRIME NUMBERS AND PRIME FACTORS

A number that has no whole-number factors except 1 and itself is called a *prime number*. Some examples of prime numbers are 2, 3, 5, 7, and 11. A number that is not prime is divisible by more than one prime number. For example, 4, 6, 8, 9, and 10 are not prime numbers because they are all divisible by more than one prime number. 4 is divisible by 2, 6 is divisible by 2 or 3, and so on. The prime numbers that make up a nonprime number are called *prime factors*. In determining prime factors, we ignore 1 as a factor.

Let's look at some nonprime numbers and their prime factors.

☐ EXAMPLE 6-1

Find the prime factors of 10, 15, 21, and 26.

Solutions

$$10 = 2 \cdot 5$$

$$15 = 3 \cdot 5$$

$$21 = 3 \cdot 7$$

$$26 = 2 \cdot 13$$

☐

Often a prime factor appears more than once as a factor, as shown in the following examples.

☐ EXAMPLE 6-2

Find the prime factors of 8, 9, 36, and 150.

Solutions

$$8 = 2 \cdot 2 \cdot 2 = 2^3$$

$$9 = 3 \cdot 3 = 3^2$$

$$36 = 2 \cdot 2 \cdot 3 \cdot 3 = 2^2 \cdot 3^2$$

$$150 = 2 \cdot 3 \cdot 5 \cdot 5 = 2 \cdot 3 \cdot 5^2$$

Notice that we expressed the prime factors of 8 as $2 \cdot 2 \cdot 2$ or 2^3 and the prime factors of 9 as $3 \cdot 3$ or 3^2. Either form is acceptable and both forms are used.

☐

PRACTICE PROBLEMS 6-1

Find the prime factors of:

1. 12

2. 24

3. 39

4. 45

5. 70

6. 72

7. 100

8. 120

9. 210

10. 300

Solutions

1. $2 \cdot 2 \cdot 3$

2. $2 \cdot 2 \cdot 2 \cdot 3$

3. $3 \cdot 13$

4. $3 \cdot 3 \cdot 5$

5. $2 \cdot 5 \cdot 7$

6. $2^3 \cdot 3^2$

7. $2^2 \cdot 5^2$

8. $2^3 \cdot 3 \cdot 5$

9. $2 \cdot 3 \cdot 5 \cdot 7$

10. $2^2 \cdot 3 \cdot 5^2$

Additional practice problems are at the end of the chapter.

6-2 REDUCING FRACTIONS TO LOWEST TERMS

Fractions have been reduced to lowest terms when there are no common factors in the numerator and denominator. To reduce fractions to lowest terms, we first find the prime factors of the numerator and denominator and then cancel the common factors. The product of the remaining factors is the fraction reduced to lowest terms. Let's look at some fractions and reduce them to lowest terms.

□ EXAMPLE 6-3

Examine the following fractions and reduce to lowest terms when necessary:

(a) $\frac{4}{5}$ **(b)** $\frac{11}{130}$ **(c)** $\frac{12}{21}$ **(d)** $\frac{39}{45}$ **(e)** $\frac{54}{96}$

Solutions

(a) $\frac{4}{5} = \frac{2 \cdot 2}{5}$

There are no common factors, so the fraction is already at lowest terms.

(b) $\frac{11}{130} = \frac{11}{2 \cdot 5 \cdot 13}$

There are no common factors, so the fraction is already at lowest terms.

(c) $\frac{12}{21} = \frac{2 \cdot 2 \cdot 3}{3 \cdot 7}$

Three is a factor common to both numerator and denominator, so cancel these factors:

$$\frac{2 \cdot 2 \cdot \not{3}}{\not{3} \cdot 7} = \frac{4}{7}$$

The fraction is now reduced to lowest terms.

(d) $\frac{39}{45} = \frac{3 \cdot 13}{3 \cdot 3 \cdot 5}$

Cancel like terms:

$$\frac{\not{3} \cdot 13}{\not{3} \cdot 3 \cdot 5} = \frac{13}{15}$$

The fraction is now reduced to lowest terms.

(e) $\frac{54}{96} = \frac{2 \cdot 3 \cdot 3 \cdot 3}{2 \cdot 2 \cdot 2 \cdot 2 \cdot 2 \cdot 3}$

Cancel like terms:

$$\frac{\not{2} \cdot \not{3} \cdot 3 \cdot 3}{\not{2} \cdot 2 \cdot 2 \cdot 2 \cdot 2 \cdot \not{3}} = \frac{9}{16}$$

The fraction is now reduced to lowest terms. □

PRACTICE PROBLEMS 6-2

Examine the following fractions and reduce to lowest terms when necessary:

1. $\frac{18}{39}$

2. $\frac{3}{8}$

3. $\frac{20}{32}$

4. $\frac{15}{45}$

5. $\frac{9}{16}$

6. $\frac{12}{33}$

7. $\frac{15}{34}$

8. $\frac{7}{35}$

9. $\frac{18}{32}$

10. $\frac{24}{64}$

11. $\frac{48}{108}$

12. $\frac{125}{625}$

13. $\frac{156}{195}$

14. $\frac{495}{3300}$

Solutions

1. $\frac{18}{39} = \frac{2 \cdot \not{3} \cdot 3}{\not{3} \cdot 13} = \frac{6}{13}$

2. $\frac{3}{8} = \frac{3}{2 \cdot 2 \cdot 2} = \frac{3}{8}$

Fraction is already in lowest terms.

3. $\frac{20}{32} = \frac{\not{2} \cdot \not{2} \cdot 5}{\not{2} \cdot \not{2} \cdot 2 \cdot 2 \cdot 2} = \frac{5}{8}$

4. $\frac{15}{45} = \frac{\not{3} \cdot \not{5}}{\not{3} \cdot 3 \cdot \not{5}} = \frac{1}{3}$

5. $\frac{9}{16} = \frac{3 \cdot 3}{2 \cdot 2 \cdot 2 \cdot 2} = \frac{9}{16}$

6. $\frac{12}{33} = \frac{2 \cdot 2 \cdot \not{3}}{\not{3} \cdot 11} = \frac{4}{11}$

Fraction is already in lowest terms.

7. $\dfrac{15}{34} = \dfrac{3 \cdot 5}{2 \cdot 17} = \dfrac{15}{34}$

Fraction is already in lowest terms.

8. $\dfrac{7}{35} = \dfrac{\cancel{7}}{5 \cdot \cancel{7}} = \dfrac{1}{5}$

9. $\dfrac{18}{32} = \dfrac{\cancel{2} \cdot 3 \cdot 3}{\cancel{2} \cdot 2 \cdot 2 \cdot 2 \cdot 2} = \dfrac{9}{16}$

10. $\dfrac{24}{64} = \dfrac{\cancel{2} \cdot \cancel{2} \cdot \cancel{2} \cdot 3}{\cancel{2} \cdot \cancel{2} \cdot \cancel{2} \cdot 2 \cdot 2 \cdot 2} = \dfrac{3}{8}$

11. $\dfrac{48}{108} = \dfrac{\cancel{2} \cdot \cancel{2} \cdot 2 \cdot 2 \cdot \cancel{3}}{\cancel{2} \cdot \cancel{2} \cdot \cancel{3} \cdot 3 \cdot 3} = \dfrac{4}{9}$

12. $\dfrac{125}{625} = \dfrac{\cancel{5} \cdot \cancel{5} \cdot \cancel{5}}{\cancel{5} \cdot \cancel{5} \cdot \cancel{5} \cdot 5} = \dfrac{1}{5}$

13. $\dfrac{156}{195} = \dfrac{2 \cdot 2 \cdot \cancel{3} \cdot \cancel{13}}{\cancel{3} \cdot 5 \cdot \cancel{13}} = \dfrac{4}{5}$

14. $\dfrac{495}{3300} = \dfrac{\cancel{3} \cdot 3 \cdot \cancel{5} \cdot \cancel{11}}{2 \cdot 2 \cdot \cancel{3} \cdot \cancel{5} \cdot 5 \cdot \cancel{11}} = \dfrac{3}{20}$

SELF-TEST 6-1

Find the prime factors of the following numbers:

1. 40
2. 48
3. 56
4. 84
5. 180
6. 220

Examine the following fractions and reduce to lowest terms when necessary:

7. $\dfrac{16}{25}$
8. $\dfrac{14}{21}$
9. $\dfrac{24}{56}$
10. $\dfrac{70}{150}$

Answers to Self-test 6-1 are at the end of the chapter.

6-3 MULTIPLICATION AND DIVISION OF FRACTIONS

6-3-1 Multiplication of Fractions

We learned how to multiply and divide fractions in arithmetic. At that time we learned the following rule:

■ **RULE 6-1** To multiply fractions, we (1) multiply the numerators (this product is the numerator in the answer), (2) multiply the denominators (this product is the denominator in the answer), and (3) cancel like factors to reduce to lowest terms.

Let's try a few together:

☐ EXAMPLE 6-4

Find **(a)** $\frac{2}{3} \times \frac{4}{5}$ **(b)** $\frac{1}{4} \times \frac{3}{4} \times \frac{3}{8}$ **(c)** $\frac{3}{8} \times \frac{4}{5}$ **(d)** $\frac{3}{4} \times \frac{2}{3} \times \frac{5}{6}$

Express answers in lowest terms.

Solutions

(a) $\frac{2}{3} \times \frac{4}{5} = \frac{8}{15} = \frac{2 \cdot 2 \cdot 2}{3 \cdot 5}$

There are no like factors in the numerator and denominator, so $\frac{8}{15}$ is the answer reduced to lowest terms.

(b) $\frac{1}{4} \times \frac{3}{4} \times \frac{3}{8} = \frac{9}{128} = \frac{3 \cdot 3}{2^7}$

There are no like factors, so the answer is $\frac{9}{128}$.

(c) $\frac{3}{8} \times \frac{4}{5} = \frac{12}{40} = \frac{\cancel{2} \cdot \cancel{2} \cdot 3}{\cancel{2} \cdot \cancel{2} \cdot 2 \cdot 5} = \frac{3}{10}$

Like terms are canceled, and the result, $\frac{3}{10}$, is the answer reduced to lowest terms.

(d) $\frac{3}{4} \times \frac{2}{3} \times \frac{5}{6} = \frac{30}{72} = \frac{\cancel{2} \cdot \cancel{3} \cdot 5}{\cancel{2} \cdot 2 \cdot 2 \cdot \cancel{3} \cdot 3} = \frac{5}{12}$

Like terms are canceled, and the result, $\frac{5}{12}$, is the answer reduced to lowest terms.

☐

PRACTICE PROBLEMS 6-3

Find the product of the following fractions and reduce to lowest terms:

1. $\frac{1}{8} \times \frac{1}{2}$

2. $\frac{2}{5} \times \frac{1}{6}$

3. $\frac{2}{3} \times \frac{3}{10}$

4. $\frac{3}{16} \times \frac{1}{3}$

5. $\frac{4}{15} \times \frac{3}{5} \times \frac{1}{2}$

6. $\frac{4}{21} \times \frac{3}{5} \times \frac{5}{8}$

7. $\frac{1}{2} \times \frac{2}{3} \times \frac{3}{5}$

8. $\frac{7}{8} \times \frac{1}{3} \times \frac{1}{2}$

9. $\frac{5}{12} \times \frac{4}{25} \times \frac{9}{16}$

10. $\frac{7}{30} \times \frac{3}{14} \times \frac{10}{21}$

Solutions

1. $\frac{1}{8} \times \frac{1}{2} = \frac{1}{16}$

Fraction is already at lowest terms.

2. $\frac{2}{5} \times \frac{1}{6} = \frac{2}{30} = \frac{\cancel{2}}{\cancel{2} \cdot 3 \cdot 5} = \frac{1}{15}$

3. $\frac{2}{3} \times \frac{3}{10} = \frac{6}{30} = \frac{\cancel{2} \cdot \cancel{3}}{\cancel{2} \cdot \cancel{3} \cdot 5} = \frac{1}{5}$

4. $\frac{3}{16} \times \frac{1}{3} = \frac{3}{48} = \frac{\cancel{3}}{2 \cdot 2 \cdot 2 \cdot 2 \cdot \cancel{3}} = \frac{1}{16}$

5. $\frac{4}{15} \times \frac{3}{5} \times \frac{1}{2} = \frac{12}{150} = \frac{\cancel{2} \cdot 2 \cdot \cancel{3}}{\cancel{2} \cdot \cancel{3} \cdot 5 \cdot 5} = \frac{2}{25}$

6. $\frac{4}{21} \times \frac{3}{5} \times \frac{5}{8} = \frac{60}{840} = \frac{\cancel{2} \cdot \cancel{2} \cdot \cancel{3} \cdot \cancel{5}}{\cancel{2} \cdot \cancel{2} \cdot 2 \cdot \cancel{3} \cdot \cancel{5} \cdot 7} = \frac{1}{14}$

7. $\frac{1}{2} \times \frac{2}{3} \times \frac{3}{5} = \frac{6}{30} = \frac{\cancel{2} \cdot \cancel{3}}{\cancel{2} \cdot \cancel{3} \cdot 5} = \frac{1}{5}$

8. $\frac{7}{8} \times \frac{1}{3} \times \frac{1}{2} = \frac{7}{48} = \frac{7}{2 \cdot 2 \cdot 2 \cdot 2 \cdot 3} = \frac{7}{48}$

Fraction is already at lowest terms.

9. $\frac{5}{12} \times \frac{4}{25} \times \frac{9}{16} = \frac{180}{3600} = \frac{\cancel{2} \cdot \cancel{2} \cdot \cancel{3} \cdot 3 \cdot \cancel{5}}{\cancel{2} \cdot \cancel{2} \cdot 2 \cdot 2 \cdot 2 \cdot 2 \cdot \cancel{3} \cdot \cancel{5} \cdot 5} = \frac{3}{80}$

10. $\frac{7}{30} \times \frac{3}{14} \times \frac{10}{21} = \frac{210}{8820} = \frac{\cancel{2} \cdot \cancel{3} \cdot \cancel{5} \cdot \cancel{7}}{\cancel{2} \cdot 2 \cdot \cancel{3} \cdot 3 \cdot \cancel{5} \cdot \cancel{7} \cdot 7} = \frac{1}{42}$

Additional practice problems are at the end of the chapter.

6-3-2 Division of Fractions

In arithmetic we learned that when we divide fractions we simply invert the divisor and then multiply. Here are some examples.

□ EXAMPLE 6-5

Perform the indicated divisions and reduce to lowest terms:

(a) $\frac{1}{5} \div \frac{1}{3}$ **(b)** $\frac{5}{8} \div 3$ **(c)** $\frac{3}{16} \div 2$ **(d)** $\frac{3}{8} \div \frac{3}{4}$

Solutions

(a) $\frac{1}{5} \div \frac{1}{3} = \frac{1}{5} \times 3 = \frac{3}{5}$

(b) $\frac{5}{8} \div 3 = \frac{5}{8} \times \frac{1}{3} = \frac{5}{24}$

(c) $\frac{3}{16} \div 2 = \frac{3}{16} \times \frac{1}{2} = \frac{3}{32}$

(d) $\frac{3}{8} \div \frac{3}{4} = \frac{3}{8} \times \frac{4}{3} = \frac{12}{24} = \frac{1}{2}$

Notice that in each example we inverted the divisor and then multiplied. Like factors in the numerator and denominator of the answers were canceled, as before. □

PRACTICE PROBLEMS 6-4

Perform the indicated divisions. Reduce answers to lowest terms.

1. $\frac{7}{8} \div \frac{3}{5}$

2. $\frac{5}{7} \div \frac{5}{3}$

3. $\frac{2}{15} \div \frac{4}{5}$

4. $\frac{4}{9} \div \frac{2}{3}$

5. $\frac{7}{30} \div \frac{5}{6}$

6. $\frac{1}{3} \div \frac{2}{5}$

7. $\frac{2}{5} \div \frac{4}{5}$

8. $\frac{2}{9} \div 3$

9. $\frac{3}{7} \div 4$

10. $\frac{7}{8} \div 2$

Solutions

1. $\frac{7}{8} \div \frac{3}{5} = \frac{7}{8} \times \frac{5}{3} = \frac{35}{24}$

2. $\frac{5}{7} \div \frac{5}{3} = \frac{5}{7} \times \frac{3}{5} = \frac{15}{35} = \frac{3}{7}$

3. $\frac{2}{15} \div \frac{4}{5} = \frac{2}{15} \times \frac{5}{4} = \frac{10}{60} = \frac{1}{6}$

4. $\frac{4}{9} \div \frac{2}{3} = \frac{4}{9} \times \frac{3}{2} = \frac{12}{18} = \frac{2}{3}$

5. $\frac{7}{30} \div \frac{5}{6} = \frac{7}{30} \times \frac{6}{5} = \frac{42}{150} = \frac{7}{25}$

6. $\frac{1}{3} \div \frac{2}{5} = \frac{1}{3} \times \frac{5}{2} = \frac{5}{6}$

7. $\frac{2}{5} \div \frac{4}{5} = \frac{2}{5} \times \frac{5}{4} = \frac{10}{20} = \frac{1}{2}$

8. $\frac{2}{9} \div 3 = \frac{2}{9} \times \frac{1}{3} = \frac{2}{27}$

9. $\frac{3}{7} \div 4 = \frac{3}{7} \times \frac{1}{4} = \frac{3}{28}$

10. $\frac{7}{8} \div 2 = \frac{7}{8} \times \frac{1}{2} = \frac{7}{16}$

Additional practice problems are at the end of the chapter.

SELF-TEST 6-2

Perform the indicated operations:

1. $\frac{2}{5} \times \frac{7}{8}$

2. $\frac{3}{8} \times \frac{2}{3}$

3. $\frac{2}{3} \times \frac{1}{5} \times \frac{1}{4}$

4. $\frac{3}{4} \times \frac{2}{5} \times \frac{4}{15}$

5. $\frac{1}{5} \div \frac{2}{3}$

6. $\frac{7}{9} \div \frac{1}{6}$

7. $\frac{4}{15} \div \frac{2}{3}$

8. $\frac{3}{5} \div 3$

Answers to Self-test 6-2 are at the end of the chapter.

6-4 ADDITION AND SUBTRACTION OF FRACTIONS

6-4-1 Adding Fractions with Common Denominators

Two or more fractions can be combined into one when the denominators are identical. If fractions to be added have the same denominator, the denominator in the answer is that number. The numerator in the answer is found by adding all the numerators together.

☐ EXAMPLE 6-6

Perform the indicated additions. Reduce answers to lowest terms:

(a) $\frac{1}{5} + \frac{2}{5}$ **(b)** $\frac{3}{8} + \frac{1}{8}$ **(c)** $\frac{4}{11} + \frac{3}{11} + \frac{2}{11}$ **(d)** $\frac{2}{15} + \frac{4}{15} + \frac{4}{15}$

Solutions

(a) $\frac{1}{5} + \frac{2}{5} = \frac{1+2}{5} = \frac{3}{5}$ **(b)** $\frac{3}{8} + \frac{1}{8} = \frac{3+1}{8} = \frac{4}{8} = \frac{1}{2}$

(c) $\frac{4}{11} + \frac{3}{11} + \frac{2}{11} = \frac{4+3+2}{11} = \frac{9}{11}$

(d) $\frac{2}{15} + \frac{4}{15} + \frac{4}{15} = \frac{2+4+4}{15} = \frac{10}{15} = \frac{2}{3}$

☐

In each of the examples above, the denominators are the same. That is, each fraction had a denominator of 5 in (a), a denominator of 8 in (b), and so on. Thus, 5 is the *common denominator* in (a) and 8 is the common denominator in (b). Said another way:

KEY POINT When the denominators of fractions are identical, that number is the *common denominator*.

When adding fractions, then, the first step is to determine if the denominators are identical. If they are, the next step is to add the numerators, as in the examples above, and reduce to lowest terms.

6-4-2 Subtracting Fractions with Common Denominators

Up to this point we have been working with positive fractions. Let's consider a fraction in which a part or all of the fraction is negative. Some examples are shown next.

☐ EXAMPLE 6-7

(a) $-\frac{3}{8}$ **(b)** $\frac{-3}{8}$ **(c)** $\frac{3}{-8}$ **(d)** $\frac{-3}{-8}$ **(e)** $-\frac{-3}{8}$ **(f)** $-\frac{-3}{-8}$

☐

Consider that there are three signs associated with any fraction: the sign of the numerator, the sign of the denominator, and the sign of the entire fraction. In the fraction $\frac{3}{8}$, we know the signs are positive because no sign is given. In Example 6-7(a), the sign of the fraction is negative but the numerator and denominator are positive. In Example 6-7(b), the numerator is negative and the denominator and the fraction are positive, and so on, for each fraction.

We don't leave fractions with negative numerators or denominators if we can avoid it. We change the signs so that the fraction in the answer is either positive or negative and both numerator and denominator are positive. To change the signs, the rule is:

■ **RULE 6-2** Any two signs of a fraction may be changed without changing the value of the fraction.

Consider Example 6-7(a), applying the above rule:

$$-\frac{+3}{+8} = +\frac{-3}{+8} = -\frac{-3}{-8} = +\frac{+3}{-8}$$

(The plus signs were included here to help us change signs.) Notice that in each case *two* signs were changed.

Suppose we want to subtract one fraction from another as in Example 6-8.

□ EXAMPLE 6-8

Perform the indicated operation:

$$\frac{7}{8} - \frac{3}{8}$$

Solution The subtraction is accomplished by changing the sign of the fraction and adding. However, if we change the sign of the fraction, we must also change *one more sign*. Remember, we must change *two* signs. The other sign we will change is the sign of the numerator. We don't want to change the sign of the denominator because, to add the fractions, the denominators must be identical. Eight and −8 are not identical.

To solve the problem in Example 6-8 then, we must first change the sign of the second fraction and add (remembering to also change the sign of the numerator).

$$\frac{7}{8} - \frac{3}{8} = \frac{7}{8} + \frac{-3}{8} = \frac{7-3}{8} = \frac{4}{8} = \frac{1}{2}$$

We perform the addition as before. When we add the numerators, we are adding +7 and −3. The result is 4. The resultant fraction, $\frac{4}{8}$, is reduced to lowest terms, or $\frac{1}{2}$.

□

Here are some more examples:

☐ EXAMPLE 6-9

Perform the indicated operations:

(a) $\frac{7}{9} - \frac{3}{9}$ **(b)** $\frac{4}{15} + \frac{3}{15} - \frac{2}{15}$ **(c)** $\frac{3}{8} - \frac{7}{8} + \frac{1}{8}$ **(d)** $\frac{11}{12} + \frac{5}{12} - \frac{7}{12}$

Solutions

(a) $\frac{7}{9} - \frac{3}{9} = \frac{7}{9} + \frac{-3}{9} = \frac{7 - 3}{9} = \frac{4}{9}$

(b) $\frac{4}{15} + \frac{3}{15} - \frac{2}{15} = \frac{4}{15} + \frac{3}{15} + \frac{-2}{15} = \frac{4 + 3 - 2}{15} = \frac{5}{15} = \frac{1}{3}$

Of course, we soon recognize that whenever a subtraction is indicated, we merely subtract the numerator instead of adding, so when we see a problem like (b) above, we can skip a step.

(b) $\frac{4}{15} + \frac{3}{15} - \frac{2}{15} = \frac{4 + 3 - 2}{15} = \frac{5}{15} = \frac{1}{3}$

(c) $\frac{3}{8} - \frac{7}{8} + \frac{1}{8} = \frac{3 - 7 + 1}{8} = \frac{-3}{8} = -\frac{3}{8}$

Instead of leaving the answer as a positive fraction with a negative numerator, it is standard practice to change the signs so that the numerator is positive and the fraction is negative.

(d) $\frac{11}{12} + \frac{5}{12} - \frac{7}{12} = \frac{11 + 5 - 7}{12} = \frac{9}{12} = \frac{3}{4}$

☐

PRACTICE PROBLEMS 6-5

Perform the indicated operations.

1. $\frac{1}{3} + \frac{2}{3}$

2. $\frac{1}{5} + \frac{2}{5}$

3. $\frac{5}{7} - \frac{2}{7}$

4. $\frac{9}{16} - \frac{5}{16}$

5. $\frac{1}{24} + \frac{3}{24} + \frac{11}{24}$

6. $\frac{2}{15} + \frac{3}{15} - \frac{4}{15}$

7. $\frac{9}{32} - \frac{4}{32} + \frac{5}{32}$

8. $\frac{13}{16} - \frac{3}{16} - \frac{5}{16}$

9. $\frac{4}{5} - \frac{3}{5} - \frac{2}{5}$

10. $\frac{7}{32} + \frac{4}{32} - \frac{15}{32}$

Solutions

1. $\dfrac{1+2}{3} = \dfrac{3}{3} = 1$

2. $\dfrac{1+2}{5} = \dfrac{3}{5}$

3. $\dfrac{5-2}{7} = \dfrac{3}{7}$

4. $\dfrac{9-5}{16} = \dfrac{4}{16} = \dfrac{1}{4}$

5. $\dfrac{1+3+11}{24} = \dfrac{15}{24} = \dfrac{5}{8}$

6. $\dfrac{2+3-4}{15} = \dfrac{1}{15}$

7. $\dfrac{9-4+5}{32} = \dfrac{10}{32} = \dfrac{5}{16}$

8. $\dfrac{13-3-5}{16} = \dfrac{5}{16}$

9. $\dfrac{4-3-2}{5} = -\dfrac{1}{5}$

10. $\dfrac{7+4-15}{32} = -\dfrac{4}{32} = -\dfrac{1}{8}$

Additional practice problems are at the end of the chapter.

6-4-3 Lowest Common Multiple

When adding or subtracting fractions, it is often necessary to find the *lowest common multiple* of the numbers that make up the denominators. This lowest common multiple will be the common denominator of the fractions.

■ **KEY POINT** The lowest common multiple of two or more numbers is the lowest multiple that is common to each of them.

For example, consider the number 6 and 8. Some multiples of 6 are 12, 18, 24, 36, 42, and 48. Some multiples of 8 are 16, 24, 32, 40, and 48. Of these groups, the only *common* multiples are 24 and 48. The *lowest* common multiple of 6 and 8 is 24.

■ **RULE 6-3** To find the lowest common multiple (LCM) of two or more numbers:

1. Factor each term.
2. Determine the *maximum* number of times a prime factor appears in any one term.
3. Multiply the resulting prime factors found in step 2. This number is the LCM.

■ **KEY POINT** The lowest common multiple is the *lowest* multiple of each number that is common to all numbers.

□ EXAMPLE 6-10

Find the LCM of 6 and 8.

Solution Find the prime factors of each number:

$$6 = 2 \cdot 3$$
$$8 = 2 \cdot 2 \cdot 2$$

The maximum number of times 2 appears as a factor is three times; therefore, $2 \cdot 2 \cdot 2$ is part of the LCM. Three appears a maximum of one time; therefore, 3 is part of the LCM. This takes care of all the prime factors. Our next step is to multiply these factors together to get the LCM.

$$2 \cdot 2 \cdot 2 \cdot 3 = 24$$

24 is the lowest number that contains both 6 and 8 as factors. That is, 6 is a factor of 24 ($6 \times 4 = 24$), and 8 is a factor of 24 ($8 \times 3 = 24$).

□

There are other common multiples of 6 and 8: 48, 120, and 240 to name a few. But 24 is the *lowest*.

□ EXAMPLE 6-11

Find the LCM of 30 and 36.

Solution

$$30 = 2 \cdot 3 \cdot 5$$
$$36 = 2 \cdot 2 \cdot 3 \cdot 3$$

Two appears as a factor a maximum of two times, 3 appears as a factor a maximum of two times, and 5 appears as a factor once; therefore,

$$\text{LCM} = 2 \cdot 2 \cdot 3 \cdot 3 \cdot 5 = 180$$

180 is the lowest number that has both 30 and 36 as factors.

□ EXAMPLE 6-12

Find the LCM of 50, 60, and 80.

Solution

$$50 = 2 \cdot 5 \cdot 5$$
$$60 = 2 \cdot 2 \cdot 3 \cdot 5$$
$$80 = 2 \cdot 2 \cdot 2 \cdot 2 \cdot 5$$

Two appears as a factor a maximum of four times, 3 appears as a factor a maximum of one time, and 5 appears as a factor a maximum of two times. Therefore,

$$\text{LCM} = 2 \cdot 2 \cdot 2 \cdot 2 \cdot 3 \cdot 5 \cdot 5 = 1200$$

1200 is the lowest number that has 50, 60, and 80 as factors.

□

PRACTICE PROBLEMS 6-6

Find the lowest common multiple (LCM) of the following numbers:

1. 3, 8	**2.** 9, 15
3. 3, 4, 8	**4.** 4, 8, 12
5. 12, 15, 25	**6.** 12, 33, 132
7. 10, 15, 18	**8.** 10, 26, 65
9. 14, 70, 210	**10.** 52, 65, 78

Solutions

1. 24	**2.** 45
3. 24	**4.** 24
5. 300	**6.** 132
7. 90	**8.** 130
9. 210	**10.** 780

Additional practice problems are at the end of the chapter.

6-4-4 Adding and Subtracting Fractions with Unlike Denominators

When the fractions we are adding or subtracting have unlike denominators, we cannot add them directly as before. We must first find a common denominator. We do this by finding the least common multiple (LCM) of all the denominators. Then we change each fraction so that each has this LCM as its denominator. *The LCM is the common denominator*. Once we have a common denominator (CD), we can add and subtract as before.

RULE 6-4 To add fractions with unlike denominators: (1) find the CD, (2) change the denominator of all fractions to this denominator, (3) divide the CD by the old denominator of each fraction then multiply this number by the old numerator to find the new numerator of the fraction, and (4) add the fractions, combining all like terms in the numerator. Reduce the answer to lowest terms.

☐ EXAMPLE 6-13

Perform the indicated operation:

$$\frac{1}{4} + \frac{1}{6}$$

Solution Since the denominators in this example are different, we find the CD:

$$4 = 2 \cdot 2$$
$$6 = 2 \cdot 3$$
$$\text{CD} = 2 \cdot 2 \cdot 3 = 12$$

Next we change each fraction so that its denominator is 12 (the CD).

$$\frac{1}{4} = \frac{?}{12}$$

If we change the denominator of a fraction, we must also change the numerator; otherwise, the fractions are not equal.

$$\frac{1}{4} \neq \frac{1}{12}$$

If we divide the new denominator (12) by the old denominator (4), we get 3. Multiplying this number by the old numerator (1) we get $1 \times 3 = 3$.

$$\frac{1}{4} = \frac{1 \times 3}{12} = \frac{3}{12}$$

If we reduce $\frac{3}{12}$, we get the original fraction $\frac{1}{4}$. The fractions are equal.

The second fraction is changed in the same manner.

$$\frac{1}{6} = \frac{?}{12}$$
$$\frac{1}{6} = \frac{1 \times 2}{12} = \frac{2}{12}$$
$$\frac{1}{4} + \frac{1}{6} = \frac{3}{12} + \frac{2}{12} = \frac{3 + 2}{12} = \frac{5}{12}$$

□ EXAMPLE 6-14

Perform the indicated operation:

$$\frac{1}{6} + \frac{3}{8}$$

Solution

Step 1. Find the common denominator (remember, the LCM is the CD):

$$6 = 2 \cdot 3$$
$$8 = 2 \cdot 2 \cdot 2$$
$$\text{CD} = 2 \cdot 2 \cdot 2 \cdot 3 = 24$$

Step 2. Change each fraction so that its denominator is the CD.

$$\frac{1}{6} = \frac{1 \times 4}{24} = \frac{4}{24}$$
$$\frac{3}{8} = \frac{3 \times 3}{24} = \frac{9}{24}$$

Step 3. Add the new fractions together. Reduce to lowest terms where possible.

$$\frac{4}{24} + \frac{9}{24} = \frac{4 + 9}{24} = \frac{13}{24}$$

□ EXAMPLE 6-15

Perform the indicated operation:

$$\frac{1}{4} + \frac{6}{18} - \frac{9}{24}$$

Solution

Step 1. Find the common denominator:

$$4 = 2 \cdot 2$$
$$18 = 2 \cdot 3 \cdot 3$$
$$24 = 2 \cdot 2 \cdot 2 \cdot 3$$
$$\text{CD} = 2^3 \cdot 3^2 = 72$$

Step 2. Change each fraction so that its denominator is the CD.

$$\frac{1}{4} = \frac{1 \times 18}{72} = \frac{18}{72}$$
$$\frac{6}{18} = \frac{6 \times 4}{72} = \frac{24}{72}$$
$$\frac{9}{24} = \frac{9 \times 3}{72} = \frac{27}{72}$$

Step 3. Perform the indicated operation. Reduce to lowest terms where possible.

$$\frac{1}{4} + \frac{5}{18} - \frac{7}{24} = \frac{18 + 24 - 27}{72} = \frac{15}{72} = \frac{5}{24}$$

□

PRACTICE PROBLEMS 6-7

Perform the indicated operations. Reduce to lowest terms.

1. $\frac{7}{20} + \frac{5}{24}$

2. $\frac{7}{12} - \frac{5}{28}$

3. $\dfrac{6}{35} - \dfrac{3}{14}$

4. $\dfrac{3}{16} + \dfrac{2}{5}$

5. $\dfrac{1}{3} + \dfrac{1}{4} + \dfrac{1}{5}$

6. $\dfrac{1}{6} + \dfrac{1}{9} + \dfrac{1}{10}$

7. $\dfrac{5}{12} + \dfrac{5}{8} - \dfrac{1}{3}$

8. $\dfrac{5}{16} + \dfrac{5}{12} - \dfrac{1}{6}$

9. $\dfrac{7}{12} + \dfrac{5}{8} - \dfrac{5}{6}$

10. $\dfrac{5}{12} + \dfrac{5}{18} - \dfrac{5}{36}$

11. $\dfrac{4}{15} - \dfrac{5}{6} - \dfrac{1}{4}$

12. $\dfrac{7}{20} - \dfrac{7}{30} - \dfrac{3}{10}$

Solutions

1. CD = 120. $\dfrac{42}{120} + \dfrac{25}{120} = \dfrac{42 + 25}{120} = \dfrac{67}{120}$

2. CD = 84. $\dfrac{49}{84} - \dfrac{15}{84} = \dfrac{49 - 15}{84} = \dfrac{34}{84} = \dfrac{17}{42}$

3. CD = 70. $\dfrac{12}{70} - \dfrac{15}{70} = \dfrac{12 - 15}{70} = -\dfrac{3}{70}$

4. CD = 80. $\dfrac{15}{80} + \dfrac{32}{80} = \dfrac{15 + 32}{80} = \dfrac{47}{80}$

5. CD = 60. $\dfrac{20}{60} + \dfrac{15}{60} + \dfrac{12}{60} = \dfrac{20 + 15 + 12}{60} = \dfrac{47}{60}$

6. CD = 90. $\dfrac{15}{90} + \dfrac{10}{90} + \dfrac{9}{90} = \dfrac{15 + 10 + 9}{90} = \dfrac{34}{90} = \dfrac{17}{45}$

7. CD = 24. $\dfrac{10}{24} + \dfrac{15}{24} - \dfrac{8}{24} = \dfrac{10 + 15 - 8}{24} = \dfrac{17}{24}$

8. CD = 48. $\dfrac{15}{48} + \dfrac{20}{48} - \dfrac{8}{48} = \dfrac{15 + 20 - 8}{48} = \dfrac{27}{48} = \dfrac{9}{16}$

9. CD = 24. $\dfrac{14}{24} + \dfrac{15}{24} - \dfrac{20}{24} = \dfrac{14 + 15 - 20}{24} = \dfrac{9}{24} = \dfrac{3}{8}$

10. CD = 36. $\dfrac{15}{36} + \dfrac{10}{36} - \dfrac{5}{36} = \dfrac{15 + 10 - 5}{36} = \dfrac{20}{36} = \dfrac{5}{9}$

11. CD = 60. $\dfrac{16}{60} - \dfrac{50}{60} - \dfrac{15}{60} = \dfrac{16 - 50 - 15}{60} = -\dfrac{49}{60}$

12. CD = 60. $\dfrac{21}{60} - \dfrac{14}{60} - \dfrac{18}{60} = \dfrac{21 - 14 - 18}{60} = -\dfrac{11}{60}$

Additional practice problems are at the end of the chapter.

SELF-TEST 6-3

Find the LCM of the following numbers:

1. 12, 15
2. 20, 24
3. 35, 42
4. 45, 75
5. 10, 21, 30
6. 14, 35, 105
7. 40, 60, 80
8. 45, 65, 91

Perform the indicated operations. Reduce to lowest terms.

9. $\frac{1}{8} + \frac{3}{8}$

10. $\frac{5}{24} + \frac{7}{24} + \frac{1}{24}$

11. $\frac{5}{12} - \frac{7}{12}$

12. $\frac{4}{15} + \frac{7}{15} - \frac{2}{15}$

13. $\frac{5}{12} - \frac{7}{15}$

14. $\frac{3}{8} + \frac{1}{6}$

15. $\frac{3}{10} + \frac{5}{12} - \frac{13}{16}$

16. $\frac{5}{24} - \frac{11}{18} - \frac{11}{36}$

Answers to Self-test 6-3 are at the end of the chapter.

6-5 IMPROPER FRACTIONS AND MIXED NUMBERS

KEY POINT When the numerator is greater than the denominator, the fraction is called an *improper* fraction. Some examples are $\frac{4}{3}$, $\frac{6}{5}$, and $\frac{17}{12}$. Improper fractions always have a value greater than 1.

KEY POINT Numbers that have a whole number part and a fractional part are called *mixed numbers*. Some examples of mixed numbers are $2\frac{2}{3}$ and $5\frac{1}{4}$. When we write the number $2\frac{2}{3}$, we are really saying that we have 2 plus $\frac{2}{3}$: $2 + \frac{2}{3}$. Even though we don't write the "+", we know it is implied.

We can express nonwhole numbers as either mixed numbers or as improper fractions. In problem solving, it is easier to work with improper fractions. Answers, however, are usually written as mixed numbers. The conversion from one form to the other is shown in the following examples.

☐ EXAMPLE 6-16

Change $\frac{4}{3}$ to a mixed number.

Solution

Step 1. Perform the indicated division. Write the remainder as a fraction.

$$\begin{array}{r} 1 + \frac{1}{3} \\ 3\overline{)4} \\ \underline{3} \\ 1 \end{array}$$

Step 2. Write the answer as a mixed number.

$$1 + \frac{1}{3} = 1\frac{1}{3}$$

Remember, $1\frac{1}{3}$ means $1 + \frac{1}{3}$.

☐ EXAMPLE 6-17

Change $\frac{17}{12}$ to a mixed number.

Solution

$$\begin{array}{r} 1 + \frac{5}{12} \\ 12\overline{)17} \\ \underline{12} \\ 5 \end{array} \qquad 1 + \frac{5}{12} = 1\frac{5}{12}$$

☐ EXAMPLE 6-18

Change $\frac{16}{3}$ to a mixed number.

Solution

$$\begin{array}{r} 5 + \frac{1}{3} \\ 3\overline{)16} \\ \underline{15} \\ 1 \end{array} \qquad 5 + \frac{1}{3} = 5\frac{1}{3}$$

☐ EXAMPLE 6-19

Change $2\frac{2}{3}$ to an improper fraction.

Solution

Step 1. Separate the whole number and fractional parts.

$$2\frac{2}{3} = 2 + \frac{2}{3}$$

Step 2. Find the CD and add. The denominators are $1\left(2 = \frac{2}{1}\right)$ and 3.

$$2 = \frac{6}{3}$$

Therefore, $\frac{6}{3} + \frac{2}{3} = \frac{6 + 2}{3} = \frac{8}{3}$.

$$2\frac{2}{3} = \frac{8}{3}$$

A shortcut method is: Multiply the whole number and the denominator of the fraction.

$$2 \times 3 = 6$$

Add the numerator of the fraction to this number.

$$6 + 2 = 8$$

This number is the numerator in the answer. The denominator is unchanged.

$$2\frac{2}{3} = \frac{6 + 2}{3} = \frac{8}{3}$$

☐ EXAMPLE 6-20

Using the shortcut method, change $5\frac{1}{4}$ to an improper fraction.

Solution $5\frac{1}{4} = \frac{5 \times 4 + 1}{4} = \frac{20 + 1}{4} = \frac{21}{4}$

☐ EXAMPLE 6-21

Change $8\frac{7}{8}$ to an improper fraction.

Solution $8\frac{7}{8} = \frac{8 \times 8 + 7}{8} = \frac{64 + 7}{8} = \frac{71}{8}$ ☐

PRACTICE PROBLEMS 6-8

Change the following improper fractions to mixed numbers:

1. $\frac{27}{4}$

2. $\frac{23}{8}$

3. $\frac{27}{5}$

4. $\frac{17}{3}$

5. $\frac{17}{6}$

6. $\frac{21}{4}$

7. $\frac{28}{3}$

8. $\frac{13}{4}$

Change the following mixed numbers to improper fractions:

9. $3\frac{5}{8}$

10. $5\frac{3}{4}$

11. $7\frac{1}{3}$

12. $2\frac{1}{2}$

13. $4\frac{1}{8}$

14. $6\frac{1}{3}$

15. $3\frac{1}{12}$

16. $5\frac{3}{8}$

Solutions

1. $6\frac{3}{4}$

2. $2\frac{7}{8}$

3. $5\frac{2}{5}$

4. $5\frac{2}{3}$

5. $2\frac{5}{6}$

6. $5\frac{1}{4}$

7. $9\frac{1}{3}$

8. $3\frac{1}{4}$

9. $\frac{29}{8}$

10. $\frac{23}{4}$

11. $\frac{22}{3}$

12. $\frac{5}{2}$

13. $\frac{33}{8}$

14. $\frac{19}{3}$

15. $\frac{37}{12}$

16. $\frac{43}{8}$

Additional practice problems are at the end of the chapter.

6-5-1 Multiplication and Division

Multiplying and dividing improper fractions is no different from multiplying and dividing proper fractions. When the answer is an improper fraction, we change it to a mixed number.

□ EXAMPLE 6-22

(a) $\frac{4}{3} \times \frac{8}{7} = \frac{32}{21} = 1\frac{11}{21}$

(b) $\frac{5}{3} \times 2 = \frac{10}{3} = 3\frac{1}{3}$

(c) $\frac{5}{2} \div \frac{6}{5} = \frac{5}{2} \times \frac{5}{6} = \frac{25}{12} = 2\frac{1}{12}$

(d) $\frac{8}{3} \div \frac{1}{4} = \frac{8}{3} \times 4 = \frac{32}{3} = 10\frac{2}{3}$

□

KEY POINT When we multiply or divide mixed numbers, we change the mixed numbers to improper fractions and perform the multiplication or division as before.

EXAMPLE 6-23

(a) $2\frac{1}{3} \times 3\frac{7}{8} = \frac{7}{3} \times \frac{31}{8} = \frac{217}{24} = 9\frac{1}{24}$

(b) $3\frac{1}{4} \times 4\frac{1}{3} = \frac{13}{4} \times \frac{13}{3} = \frac{169}{12} = 14\frac{1}{12}$

(c) $5\frac{1}{5} \div 3\frac{2}{3} = \frac{26}{5} \div \frac{11}{3} = \frac{26}{5} \times \frac{3}{11} = \frac{78}{55} = 1\frac{23}{55}$

(d) $2\frac{1}{3} \div 6\frac{1}{4} = \frac{7}{3} \div \frac{25}{4} = \frac{7}{3} \times \frac{4}{25} = \frac{28}{75}$

No change is necessary here since the answer is a proper fraction.

PRACTICE PROBLEMS 6-9

Perform the indicated operation. If the answer is an improper fraction, change it to a mixed number.

1. $1\frac{7}{8} \times 3\frac{3}{4}$
2. $\frac{7}{8} \times \frac{16}{9}$
3. $3 \times \frac{8}{3}$
4. $\frac{5}{3} \times 3\frac{1}{2}$
5. $\frac{9}{5} \times 2\frac{2}{5}$
6. $\frac{9}{7} \div \frac{4}{3}$
7. $2\frac{3}{4} \div \frac{4}{3}$
8. $4 \div 1\frac{2}{3}$
9. $5\frac{1}{6} \div 4\frac{5}{12}$
10. $5\frac{1}{2} \div \frac{3}{4}$

Solutions

1. $7\frac{1}{32}$
2. $1\frac{5}{9}$
3. 8
4. $5\frac{5}{6}$
5. $4\frac{8}{25}$
6. $\frac{27}{28}$
7. $2\frac{1}{16}$
8. $2\frac{2}{5}$

9. $1\frac{9}{53}$

10. $7\frac{1}{3}$

Additional practice problems are at the end of the chapter.

6-5-2 Addition and Subtraction

■ **KEY POINT** To add and subtract, we change all mixed numbers to improper fractions and perform the additions and subtractions as we did with proper fractions. Let's work some together.

□ EXAMPLE 6-24

(a) $\frac{9}{8} + \frac{5}{4}$

Find the common denominator and add. The CD is 8.

$$\frac{9}{8} + \frac{5}{4} = \frac{9}{8} + \frac{10}{8} = \frac{9 + 10}{8} = \frac{19}{8} = 2\frac{3}{8}$$

(b) $\frac{7}{3} + 2\frac{5}{6}$

Change $2\frac{5}{6}$ to an improper fraction:

$$2\frac{5}{6} = \frac{17}{6}$$

Find the CD. The CD is 6.

$$\frac{7}{3} + \frac{17}{6} = \frac{14}{6} + \frac{17}{6} = \frac{14 + 17}{6} = \frac{31}{6} = 5\frac{1}{6}$$

(c) $3\frac{3}{4} - 1\frac{7}{16}$

Change to improper fractions:

$$3\frac{3}{4} = \frac{15}{4} \text{ and } 1\frac{7}{16} = \frac{23}{16}$$

Find the CD. The CD is 16.

$$\frac{15}{4} - \frac{23}{16} = \frac{60}{16} - \frac{23}{16} = \frac{60 - 23}{16} = \frac{37}{16} = 2\frac{5}{16}$$

(d) $\frac{9}{4} - 7\frac{2}{3}$

Change $7\frac{2}{3}$ to an improper fraction:

$$7\frac{2}{3} = \frac{23}{3}$$

Find the CD. The CD is 12.

$$\frac{9}{4} - \frac{23}{3} = \frac{27}{12} - \frac{92}{12} = \frac{27 - 92}{12} = \frac{-65}{12} = -\frac{65}{12} = -5\frac{5}{12}$$

Remember, when the numerator or denominator is negative, we change *two* signs. In this case we change the sign of the numerator and the sign of the fraction.

PRACTICE PROBLEMS 6-10

Perform the indicated operations. Express your answers as mixed numbers or as proper fractions.

1. $\frac{16}{9} + \frac{4}{3}$

2. $3\frac{1}{3} + \frac{8}{15}$

3. $4\frac{7}{8} + \frac{9}{4}$

4. $5\frac{1}{3} + 6\frac{5}{8}$

5. $\frac{9}{8} - \frac{1}{2}$

6. $2\frac{1}{7} - \frac{8}{3}$

7. $4\frac{2}{5} - 2\frac{1}{10}$

8. $2\frac{9}{10} - 4\frac{1}{15}$

9. $\frac{9}{7} + \frac{8}{3} - 1\frac{2}{3}$

10. $5\frac{5}{6} - 1\frac{2}{3} + 3\frac{3}{4}$

Solutions

1. $\frac{28}{9} = 3\frac{1}{9}$

2. $\frac{58}{15} = 3\frac{13}{15}$

3. $\frac{57}{8} = 7\frac{1}{8}$

4. $\frac{287}{24} = 11\frac{23}{24}$

5. $\frac{5}{8}$

6. $-\frac{11}{21}$

7. $\frac{23}{10} = 2\frac{3}{10}$

8. $-\frac{35}{30} = -\frac{7}{6} = -1\frac{1}{6}$

9. $\frac{48}{21} = 2\frac{6}{21} = 2\frac{2}{7}$

10. $\frac{190}{24} = \frac{95}{12} = 7\frac{11}{12}$

Additional practice problems are at the end of the chapter.

SELF-TEST 6-4

Change to mixed numbers:

1. $\frac{29}{8}$

2. $\frac{33}{5}$

Change to improper fractions:

3. $3\frac{3}{16}$

4. $8\frac{2}{3}$

Perform the indicated operation:

5. $2\frac{5}{8} \times \frac{9}{4}$

6. $3\frac{1}{3} \times 4\frac{1}{2}$

7. $\frac{9}{8} \div 1\frac{7}{8}$

8. $5\frac{1}{3} \div 2\frac{5}{6}$

9. $\frac{8}{3} + 1\frac{1}{2}$

10. $2\frac{7}{8} + 5\frac{1}{4}$

11. $6\frac{1}{2} - 1\frac{3}{32}$

12. $8\frac{1}{3} - 4\frac{5}{12}$

13. $1\frac{1}{4} + \frac{9}{8} - \frac{10}{3}$

14. $6\frac{1}{2} + 4\frac{1}{3} - 5\frac{1}{5}$

Answers to Self-test 6-4 are at the end of the chapter.

6-6 DECIMAL FRACTIONS

When working with fractions, it is often desirable or necessary to express answers as decimal fractions rather than as mixed numbers. We worked with decimal fractions in Chapter 1. Let's perform all four functions on two fractions and give the answers as decimal fractions and then as proper fractions or as mixed numbers.

☐ EXAMPLE 6-25

Add, subtract, multiply, and divide the following two fractions: $\frac{3}{8}$ and $\frac{2}{3}$. Express answers as decimal numbers accurate to three places and then as proper fractions or as mixed numbers.

Solutions This problem is easily solved as decimal fractions using your calculator, but don't forget the hierarchy: Multiply, divide, add, and subtract.

$\frac{3}{8} + \frac{2}{3}$	3 [÷] 8 [+] 2 [÷] 3 [=]	*Answer:* 1.04
$\frac{3}{8} - \frac{2}{3}$	3 [÷] 8 [−] 2 [÷] 3 [=]	*Answer:* −0.292
$\frac{3}{8} \times \frac{2}{3}$	3 [÷] 8 [×] 2 [÷] 3 [=]	*Answer:* 0.250
$\frac{3}{8} \div \frac{2}{3}$	3 [÷] 8 [÷] [(] 2 [÷] 3 [)] [=]	*Answer:* 0.563

(We had to use parentheses here to show that 3 ÷ 8 was to be divided by the fraction $\frac{2}{3}$.)

As proper fractions or mixed numbers,

$$\frac{3}{8} + \frac{2}{3} = \frac{9 + 16}{24} = \frac{25}{24} = 1\frac{1}{24}$$

$$\frac{3}{8} - \frac{2}{3} = \frac{9 - 16}{24} = \frac{-7}{24} = -\frac{7}{24}$$

$$\frac{3}{8} \times \frac{2}{3} = \frac{6}{24} = \frac{1}{4}$$

$$\frac{3}{8} \div \frac{2}{3} = \frac{3}{8} \times \frac{3}{2} = \frac{9}{16}$$

□

□ EXAMPLE 6-26

Add, subtract, multiply, and divide the following two mixed numbers: $3\frac{1}{4}$ and $2\frac{1}{6}$. Express answers first as decimal numbers accurate to three places and then as proper fractions or as mixed numbers.

Solutions As decimal fractions,

$$3\frac{1}{4} + 2\frac{1}{6} = 5.24$$

3 [+] 1 [÷] 4 [+] 2 [+] 1 [÷] 6 [=] *Answer:* 5.42

When there are a number of functions to perform, the use of parentheses is often necessary to let the calculator know how to handle the numbers.

$$3\frac{1}{4} - 2\frac{1}{6} = 1.08$$

3 [+] 1 [÷] 4 [−] [(] 2 [+] 1 [÷] 6 [)] [=] *Answer:* 1.08

The parentheses tells the calculator that some function is to be performed on a *group*. In this case the group is the mixed number $2\frac{1}{6}$. Without the parentheses, the calculator would follow the algebraic logic and subtract 2 and then *add* $\frac{1}{6}$. To subtract the entire number $2\frac{1}{6}$, we put the number inside the parentheses. Once we perform the operations within the parentheses $(2 + \frac{1}{6})$, then that answer is subtracted from $3\frac{1}{4}$. When in doubt as to whether or not you need the parentheses, it is a good idea to go ahead and use them.

$$3\frac{1}{4} \times 2\frac{1}{6} = 7.04$$

[(] 3 [+] 1 [÷] 4 [)] [×] [(] 2 [+] 1 [÷] 6 [)] [=] *Answer:* 7.04

and, finally,

$$3\frac{1}{4} \div 2\frac{1}{6} = 1.50$$

[(] 3 [+] 1 [÷] 4 [)] [÷] [(] 2 [+] 1 [÷] 6 [)] [=] *Answer:* 1.50

As proper fractions or as mixed numbers,

$$3\frac{1}{4} + 2\frac{1}{6} = \frac{13}{4} + \frac{13}{6} = \frac{39 + 26}{12} = \frac{65}{12} = 5\frac{5}{12}$$

$$3\frac{1}{4} - 2\frac{1}{6} = \frac{13}{4} - \frac{13}{6} = \frac{39 - 26}{12} = \frac{13}{12} = 1\frac{1}{12}$$

$$3\frac{1}{4} \times 2\frac{1}{6} = \frac{13}{4} \times \frac{13}{6} = \frac{169}{24} = 7\frac{1}{24}$$

$$3\frac{1}{4} \div 2\frac{1}{6} = \frac{13}{4} \div \frac{13}{6} = \frac{13}{4} \times \frac{6}{13} = \frac{39}{26} = 1\frac{1}{2}$$

□

PRACTICE PROBLEMS 6-11

Add, subtract, multiply, and divide the following numbers. Express answers as decimal numbers and then as proper fractions or as mixed numbers.

1. $\frac{1}{3}$ and $\frac{5}{8}$
2. $\frac{6}{7}$ and $\frac{3}{5}$
3. $\frac{9}{16}$ and $\frac{3}{32}$
4. $\frac{7}{16}$ and $\frac{2}{3}$
5. $1\frac{3}{5}$ and $7\frac{1}{8}$
6. $9\frac{3}{16}$ and $2\frac{3}{4}$
7. $3\frac{1}{3}$ and $1\frac{5}{32}$
8. $4\frac{2}{9}$ and $2\frac{1}{3}$
9. $2\frac{11}{16}$ and $8\frac{1}{8}$
10. $6\frac{1}{2}$ and $4\frac{7}{8}$

Solutions

As decimal fractions:

1. $\frac{1}{3} + \frac{5}{8} = 0.958$

$\frac{1}{3} - \frac{5}{8} = -0.292$

$\frac{1}{3} \times \frac{5}{8} = 0.208$

$\frac{1}{3} \div \frac{5}{8} = 0.533$

2. $\frac{6}{7} + \frac{3}{5} = 1.46$

$\frac{6}{7} - \frac{3}{5} = 0.257$

$\frac{6}{7} \times \frac{3}{5} = 0.514$

$\frac{6}{7} \div \frac{3}{5} = 1.43$

3. $\dfrac{9}{12} + \dfrac{3}{32} = 0.656$

$\dfrac{9}{16} - \dfrac{3}{32} = 0.469$

$\dfrac{9}{16} \times \dfrac{3}{32} = 0.0527$

$\dfrac{9}{16} \div \dfrac{3}{32} = 6.00$

4. $\dfrac{7}{16} + \dfrac{2}{3} = 1.10$

$\dfrac{7}{16} - \dfrac{2}{3} = -0.229$

$\dfrac{7}{16} \times \dfrac{2}{3} = 0.292$

$\dfrac{7}{16} \div \dfrac{2}{3} = 0.656$

5. $1\dfrac{3}{5} + 7\dfrac{1}{8} = 8.73$

$1\dfrac{3}{5} - 7\dfrac{1}{8} = -5.53$

$1\dfrac{3}{5} \times 7\dfrac{1}{8} = 11.4$

$1\dfrac{3}{5} \div 7\dfrac{1}{8} = 0.225$

6. $9\dfrac{3}{16} + 2\dfrac{3}{4} = 11.9$

$9\dfrac{3}{16} - 2\dfrac{3}{4} = 6.43$

$9\dfrac{3}{16} \times 2\dfrac{3}{4} = 25.3$

$9\dfrac{3}{16} \div 2\dfrac{3}{4} = 3.34$

7. $3\dfrac{1}{3} + 1\dfrac{5}{32} = 4.49$

$3\dfrac{1}{3} - 1\dfrac{5}{32} = 2.18$

$3\dfrac{1}{3} \times 1\dfrac{5}{32} = 3.85$

$3\dfrac{1}{3} \div 1\dfrac{5}{32} = 2.88$

8. $4\dfrac{2}{9} + 2\dfrac{1}{3} = 6.56$

$4\dfrac{2}{9} - 2\dfrac{1}{3} = 1.89$

$4\dfrac{2}{9} \times 2\dfrac{1}{3} = 9.85$

$4\dfrac{2}{9} \div 2\dfrac{1}{3} = 1.81$

9. $2\dfrac{11}{16} + 8\dfrac{1}{8} = 10.8$

$2\dfrac{11}{16} - 8\dfrac{1}{8} = -5.44$

$2\dfrac{11}{16} \times 8\dfrac{1}{8} = 21.8$

$2\dfrac{11}{16} \div 8\dfrac{1}{8} = 0.331$

10. $6\dfrac{1}{2} + 4\dfrac{7}{8} = 11.4$

$6\dfrac{1}{2} - 4\dfrac{7}{8} = 1.63$

$6\dfrac{1}{2} \times 4\dfrac{7}{8} = 31.7$

$6\dfrac{1}{2} \div 4\dfrac{7}{8} = 1.33$

As proper fractions or as mixed numbers:

1. $\dfrac{1}{3} + \dfrac{5}{8} = \dfrac{8 + 15}{24} = \dfrac{23}{24}$

$\dfrac{1}{3} - \dfrac{5}{8} = \dfrac{8 - 15}{24} = -\dfrac{7}{24}$

$\dfrac{1}{3} \times \dfrac{5}{8} = \dfrac{5}{24}$

$\dfrac{1}{3} \div \dfrac{5}{8} = \dfrac{1}{3} \times \dfrac{8}{5} = \dfrac{8}{15}$

2. $\dfrac{6}{7} + \dfrac{3}{5} = \dfrac{30 + 21}{35} = \dfrac{51}{35} = 1\dfrac{16}{35}$

$\dfrac{6}{7} - \dfrac{3}{5} = \dfrac{30 - 21}{35} = \dfrac{9}{35}$

$\dfrac{6}{7} \times \dfrac{3}{5} = \dfrac{18}{35}$

$\dfrac{6}{7} \div \dfrac{3}{5} = \dfrac{6}{7} \times \dfrac{5}{3} = \dfrac{30}{21} = \dfrac{10}{7} = 1\dfrac{3}{7}$

3. $\frac{9}{16} + \frac{3}{32} = \frac{18 + 3}{32} = \frac{21}{32}$

$\frac{9}{16} - \frac{3}{32} = \frac{18 - 3}{32} = \frac{15}{32}$

$\frac{9}{16} \times \frac{3}{32} = \frac{27}{512}$

$\frac{5}{16} \div \frac{3}{32} = \frac{5}{16} \times \frac{32}{3} = \frac{10}{3} = 3\frac{1}{3}$

4. $\frac{7}{16} + \frac{2}{3} = \frac{21 + 32}{48} = \frac{53}{48} = 1\frac{5}{48}$

$\frac{7}{16} - \frac{2}{3} = \frac{21 - 32}{48} = -\frac{11}{48}$

$\frac{7}{16} \times \frac{2}{3} = \frac{14}{48} = \frac{7}{24}$

$\frac{7}{16} \div \frac{2}{3} = \frac{7}{16} \times \frac{3}{2} = \frac{21}{32}$

5. $1\frac{3}{5} + 7\frac{1}{8} = \frac{8}{5} + \frac{57}{8} = \frac{64 + 285}{40} = \frac{349}{40} = 8\frac{29}{40}$

$\frac{8}{5} - \frac{57}{8} = \frac{64 - 285}{40} = -\frac{221}{40} = -5\frac{21}{40}$

$\frac{8}{5} \times \frac{57}{8} = \frac{57}{5} = 11\frac{2}{5}$

$\frac{8}{5} \div \frac{57}{8} = \frac{8}{5} \times \frac{8}{57} = \frac{64}{285}$

6. $9\frac{3}{16} + 2\frac{3}{4} = \frac{147}{16} + \frac{11}{4} = \frac{147 + 44}{16} = \frac{191}{16}$

$= 11\frac{15}{16}$

$\frac{147}{16} - \frac{11}{4} = \frac{147 - 44}{16} = \frac{103}{16} = 6\frac{7}{16}$

$\frac{147}{16} \times \frac{11}{4} = \frac{1617}{64} = 25\frac{17}{64}$

$\frac{147}{16} \div \frac{11}{4} = \frac{147}{16} \times \frac{4}{1} = \frac{147}{44} = 3\frac{15}{44}$

7. $3\frac{1}{3} + 1\frac{5}{32} = \frac{10}{3} + \frac{37}{32} = \frac{320 + 111}{96} = \frac{431}{96}$

$= 4\frac{47}{96}$

$\frac{10}{3} - \frac{37}{32} = \frac{320 - 111}{96} = \frac{209}{96} = 2\frac{17}{96}$

$\frac{10}{3} \times \frac{37}{32} = \frac{5}{3} \times \frac{37}{16} = \frac{185}{48} = 3\frac{41}{48}$

$\frac{10}{3} \div \frac{37}{32} = \frac{10}{3} \times \frac{32}{37} = \frac{320}{111} = 2\frac{98}{111}$

8. $4\frac{2}{9} + 2\frac{1}{3} = \frac{38}{9} + \frac{7}{3} = \frac{38 + 21}{9} = \frac{59}{9} = 6\frac{5}{9}$

$\frac{38}{9} - \frac{7}{3} = \frac{38 - 21}{9} = \frac{17}{9} = 1\frac{8}{9}$

$\frac{38}{9} \times \frac{7}{3} = \frac{266}{27} = 9\frac{23}{27}$

$\frac{38}{9} \div \frac{7}{3} = \frac{38}{9} \times \frac{3}{7} = \frac{38}{21} = 1\frac{17}{21}$

9. $2\frac{11}{16} + 8\frac{1}{8} = \frac{43}{16} + \frac{65}{8} = \frac{43 + 130}{16} = \frac{173}{16}$

$= 10\frac{13}{16}$

$\frac{43}{16} - \frac{65}{8} = \frac{43 - 130}{16} = -\frac{87}{16} = -5\frac{7}{16}$

$\frac{43}{16} \times \frac{65}{8} = \frac{2795}{128} = 21\frac{107}{128}$

$\frac{43}{16} \div \frac{65}{8} = \frac{43}{16} \times \frac{8}{65} = \frac{43}{130}$

10. $6\frac{1}{2} + 4\frac{7}{8} = \frac{13}{2} + \frac{39}{8} = \frac{52 + 39}{8} = \frac{91}{8} = 11\frac{3}{8}$

$\frac{13}{2} - \frac{39}{8} = \frac{52 - 39}{8} = \frac{13}{8} = 1\frac{5}{8}$

$\frac{13}{2} \times \frac{39}{8} = \frac{507}{16} = 31\frac{11}{16}$

$\frac{13}{2} \div \frac{39}{8} = \frac{13}{2} \times \frac{8}{39} = \frac{4}{3} = 1\frac{1}{3}$

CHAPTER 6 AT A GLANCE

Page	Key Point or Rule	Example
161	*Key point:* When the denominators of fractions are identical, that number is the *common denominator* (CD).	In the fractions $\frac{1}{8}, \frac{3}{8}$, and $\frac{7}{8}$, 8 is the CD
162	*Rule 6-2:* Any two signs of a fraction may be changed without changing the value of the fraction.	$-\frac{+3}{+8} = +\frac{-3}{+8} = -\frac{-3}{-8} = +\frac{+3}{-8}$
164	*Key point:* The *lowest common multiple* (LCM) of two or more numbers is the lowest multiple that is common to each of them.	Some multiples of 4 and 6 are 12, 24, 36, and 48. The LCM is 12.
164	*Rule 6-3:* To find the LCM of two or more numbers: (1) Factor each term; (2) determine the *maximum* number of times a prime factor appears in any one term; (3) multiply the resulting prime factors found in step 2. This number is the low LCM.	Find the LCM of 30 and 36. $30 = 2 \cdot 3 \cdot 5$ $36 = 2 \cdot 2 \cdot 3 \cdot 3$ $\text{LCM} = 2 \cdot 2 \cdot 3 \cdot 3 \cdot 5$ $= 180$
166	*Rule 6-4:* To add or subtract fractions with unlike denominators: (1) find the CD (2) change the denominator of all fractions to this denominator, add the fractions, combining all like terms in the numerator. Reduce to lowest terms.	$\frac{1}{3} + \frac{1}{4} = \frac{4}{12} + \frac{3}{12}$ $= \frac{4+3}{12} = \frac{7}{12}$
170	*Key point:* When the numerator is greater than the denominator, the fraction is called an improper fraction.	$\frac{9}{8}, \frac{25}{16}$, and $\frac{17}{3}$ are examples.
170	*Key point:* Numbers that have a whole number part and a fractional part are called *mixed numbers*.	$4\frac{3}{4}$, and $6\frac{7}{8}$ are examples.
174	*Key point:* When we multiply or divide mixed numbers, we change the mixed numbers to improper fractions and perform the function as before.	$2\frac{2}{3} \times 3\frac{1}{4} = \frac{8}{3} \times \frac{13}{4}$ $= \frac{104}{12} = 8\frac{2}{3}$
175	*Key point:* To add and subtract, we change all mixed numbers to improper fractions and perform the function as with proper fractions.	$2\frac{2}{3} + 3\frac{1}{4} = \frac{8}{3} + \frac{13}{4}$ $= \frac{32+39}{12} = \frac{71}{12} = 5\frac{11}{12}$

SELF-TEST 6-5

Add, subtract, multiply, and divide the following numbers. Express answers as decimal numbers accurate to three places and then as proper fractions or as mixed numbers.

1. $\frac{9}{16}$ and $\frac{3}{8}$

2. $\frac{2}{3}$ and $\frac{1}{6}$

3. $5\frac{7}{8}$ and $2\frac{1}{4}$

4. $3\frac{3}{16}$ and $6\frac{1}{3}$

Answers to Self-test 6-5 are at the end of the chapter.

End of Chapter Problems 6-1

Find the prime factors of the following:

1. 18

2. 27

3. 44

4. 50

5. 63

6. 45

7. 92

8. 105

9. 231

10. 195

11. 56

12. 72

13. 147

14. 66

15. 84

16. 182

17. 210

18. 225

19. 455

20. 495

End of Chapter Problems 6-2

Examine the following fractions and reduce to lowest terms:

1. $\frac{3}{21}$

2. $\frac{5}{30}$

3. $\frac{5}{8}$

4. $\frac{12}{36}$

5. $\frac{22}{32}$

6. $\frac{16}{64}$

7. $\frac{18}{45}$

8. $\frac{24}{54}$

9. $\frac{21}{40}$

10. $\frac{20}{32}$

11. $\frac{66}{110}$

12. $\frac{195}{260}$

13. $\frac{182}{195}$

14. $\frac{330}{462}$

15. $\frac{273}{441}$

16. $\frac{220}{330}$

17. $\frac{294}{330}$

18. $\frac{385}{5005}$

19. $\frac{1190}{3570}$

20. $\frac{2450}{3850}$

End of Chapter Problems 6-3

Find the product of the following fractions and reduce to lowest terms:

1. $\frac{1}{2} \times \frac{1}{3}$

2. $\frac{3}{4} \times \frac{1}{5}$

3. $\frac{3}{8} \times \frac{2}{9}$

4. $\frac{3}{4} \times \frac{2}{3}$

5. $\frac{9}{16} \times \frac{8}{15}$

6. $\frac{5}{8} \times \frac{3}{16}$

7. $\frac{22}{33} \times \frac{15}{28}$

8. $\frac{21}{40} \times \frac{16}{35}$

9. $\frac{3}{4} \times \frac{7}{8}$

10. $\frac{3}{8} \times \frac{2}{3}$

11. $\frac{3}{8} \times \frac{4}{9} \times \frac{2}{3}$

12. $\frac{3}{5} \times \frac{2}{9} \times \frac{5}{6}$

13. $\frac{1}{3} \times \frac{1}{4} \times \frac{2}{5}$

14. $\frac{2}{9} \times \frac{3}{3} \times \frac{1}{5}$

15. $\frac{1}{5} \times \frac{4}{7} \times \frac{1}{2}$

16. $\frac{2}{3} \times \frac{3}{4} \times \frac{4}{5}$

17. $\frac{6}{35} \times \frac{14}{15} \times \frac{5}{12}$

18. $\frac{10}{21} \times \frac{7}{12} \times \frac{9}{25}$

19. $\frac{8}{15} \times \frac{5}{16} \times \frac{22}{25}$

20. $\frac{15}{39} \times \frac{14}{30} \times \frac{3}{4}$

End of Chapter Problems 6-4

Perform the following divisions. Reduce your answers to lowest terms.

1. $\frac{3}{4} \div \frac{9}{16}$

2. $\frac{4}{5} \div \frac{10}{3}$

3. $\frac{3}{16} \div \frac{7}{8}$

4. $\frac{20}{3} \div \frac{5}{6}$

5. $\frac{7}{9} \div \frac{2}{3}$

6. $\frac{2}{15} \div \frac{2}{5}$

7. $\frac{9}{16} \div \frac{21}{32}$

8. $\frac{9}{20} \div \frac{8}{15}$

9. $\frac{8}{25} \div \frac{16}{35}$

10. $\frac{3}{10} \div \frac{9}{16}$

11. $\frac{5}{36} \div \frac{15}{54}$

12. $\frac{25}{48} \div \frac{15}{28}$

13. $\frac{7}{16} \div \frac{11}{32}$

14. $\frac{9}{32} \div \frac{13}{64}$

15. $\frac{35}{64} \div \frac{15}{32}$

16. $\frac{25}{64} \div \frac{5}{32}$

17. $\frac{92}{225} \div \frac{23}{175}$

18. $\frac{76}{325} \div \frac{19}{425}$

19. $\frac{105}{121} \div \frac{320}{363}$

20. $\frac{120}{169} \div \frac{252}{845}$

End of Chapter Problems 6-5

Perform the indicated operations. Reduce to lowest terms where possible.

1. $\frac{1}{8} + \frac{3}{8}$

2. $\frac{5}{16} + \frac{7}{16}$

3. $\frac{2}{7} + \frac{3}{7}$

4. $\frac{9}{32} + \frac{11}{32}$

5. $\frac{11}{12} - \frac{5}{12}$

6. $\frac{5}{8} - \frac{3}{8}$

7. $\frac{9}{14} - \frac{5}{14}$

8. $\frac{7}{9} - \frac{3}{9}$

9. $\frac{5}{12} + \frac{7}{12} + \frac{11}{12}$

10. $\frac{3}{18} + \frac{5}{18} + \frac{1}{18}$

11. $\frac{3}{16} - \frac{5}{16} + \frac{7}{16}$

12. $\frac{5}{12} + \frac{7}{12} - \frac{1}{12}$

13. $\frac{7}{24} + \frac{11}{24} - \frac{5}{24}$

14. $\frac{23}{32} - \frac{17}{32} + \frac{5}{32}$

15. $\frac{7}{9} - \frac{5}{9} - \frac{4}{9}$

16. $\frac{5}{11} - \frac{9}{11} + \frac{2}{11}$

17. $\frac{11}{16} + \frac{5}{16} - \frac{7}{16}$

18. $\frac{3}{10} - \frac{7}{10} + \frac{9}{10}$

19. $\frac{17}{18} - \frac{11}{18} + \frac{5}{18}$

20. $\frac{7}{30} + \frac{11}{30} - \frac{13}{30}$

End of Chapter Problems 6-6

Find the least common multiple (LCM) of the following numbers:

1. 6, 15

2. 24, 36

3. 55, 65

4. 45, 63

5. 84, 90

6. 50, 60

7. 22, 26

8. 34, 51

9. 8, 24, 36

10. 15, 18, 24

11. 63, 75, 105

12. 16, 24, 54

13. 18, 36, 72

14. 14, 56, 196

15. 20, 52, 65

16. 21, 35, 105

17. 50, 70, 175

18. 54, 66, 99

19. 45, 75, 225

20. 30, 51, 85

End of Chapter Problems 6-7

Perform the indicated operations. Reduce to lowest terms.

1. $\frac{1}{4} + \frac{1}{6}$

2. $\frac{1}{6} + \frac{1}{8}$

3. $\frac{2}{5} + \frac{4}{15}$

4. $\frac{3}{7} + \frac{5}{21}$

5. $\frac{2}{3} - \frac{1}{6}$

6. $\frac{5}{12} - \frac{3}{8}$

7. $\frac{7}{8} - \frac{3}{16}$

8. $\frac{11}{16} - \frac{1}{3}$

9. $\frac{17}{24} - \frac{5}{6}$

10. $\frac{5}{18} - \frac{5}{6}$

11. $\frac{5}{12} - \frac{13}{16}$

12. $\frac{7}{30} - \frac{8}{15}$

13. $\frac{2}{3} + \frac{1}{6} + \frac{1}{12}$

14. $\frac{2}{5} + \frac{2}{15} + \frac{7}{30}$

15. $\frac{4}{15} + \frac{3}{25} + \frac{1}{3}$

16. $\frac{7}{30} + \frac{4}{35} + \frac{1}{5}$

17. $\frac{11}{42} + \frac{2}{3} - \frac{5}{28}$

18. $\frac{7}{36} + \frac{1}{2} - \frac{7}{48}$

19. $\frac{9}{56} - \frac{13}{42} - \frac{1}{21}$

20. $\frac{5}{44} - \frac{11}{66} - \frac{5}{12}$

21. $\frac{4}{55} - \frac{3}{25} + \frac{3}{5}$

22. $\frac{4}{39} + \frac{15}{78} - \frac{7}{26}$

23. $\frac{3}{65} + \frac{2}{15} - \frac{4}{39}$

24. $\frac{1}{2} + \frac{5}{26} - \frac{13}{22}$

End of Chapter Problems 6-8

Change the following improper fractions to mixed numbers:

1. $\frac{17}{2}$

2. $\frac{9}{4}$

3. $\frac{37}{4}$

4. $\frac{24}{5}$

5. $\frac{19}{5}$

6. $\frac{17}{5}$

7. $\frac{43}{7}$

8. $\frac{36}{8}$

9. $\frac{27}{4}$

10. $\frac{19}{6}$

11. $\frac{67}{7}$

12. $\frac{53}{7}$

13. $\frac{87}{16}$

14. $\frac{55}{16}$

15. $\frac{89}{12}$

16. $\frac{100}{12}$

17. $\frac{60}{9}$

18. $\frac{100}{9}$

19. $\frac{37}{15}$

20. $\frac{49}{15}$

Change the following mixed numbers to improper fractions:

21. $10\frac{1}{2}$

22. $8\frac{3}{4}$

23. $4\frac{3}{16}$

24. $3\frac{2}{5}$

25. $3\frac{5}{8}$

26. $4\frac{1}{3}$

27. $9\frac{1}{3}$

28. $8\frac{1}{4}$

29. $5\frac{1}{6}$

30. $6\frac{2}{3}$

31. $8\frac{7}{8}$

32. $5\frac{3}{4}$

33. $12\frac{1}{3}$

34. $12\frac{3}{4}$

35. $7\frac{1}{3}$

36. $8\frac{3}{4}$

37. $10\frac{3}{5}$

38. $10\frac{1}{4}$

39. $4\frac{3}{16}$

40. $5\frac{7}{16}$

End of Chapter Problems 6-9

Perform the indicated operations. If the answer is an improper fraction, change to a mixed number.

1. $\frac{3}{8} \times \frac{10}{3}$

2. $\frac{8}{5} \times \frac{5}{3}$

3. $\frac{3}{4} \times \frac{8}{5}$

4. $\frac{2}{3} \times \frac{16}{7}$

5. $\frac{3}{8} \times \frac{5}{3}$

6. $\frac{5}{16} \times \frac{32}{11}$

7. $\frac{9}{4} \times 2\frac{1}{5}$

8. $\frac{8}{5} \times 2\frac{5}{8}$

9. $3 \times 3\frac{1}{4}$

10. $2 \times 5\frac{5}{6}$

11. $2\frac{1}{3} \times 3\frac{1}{4}$

12. $1\frac{5}{6} \times 2\frac{7}{8}$

13. $3\frac{1}{3} \times 4\frac{1}{5}$

14. $2\frac{7}{8} \times 5\frac{3}{4}$

15. $3\frac{3}{8} \times 6\frac{1}{3}$

16. $4\frac{3}{8} \times 2\frac{3}{5}$

17. $2\frac{5}{6} \times 3\frac{2}{3}$

18. $4\frac{1}{3} \times 3\frac{1}{4}$

19. $5\frac{5}{6} \times 4\frac{2}{5}$

20. $6\frac{2}{5} \times 5\frac{5}{8}$

21. $\frac{7}{2} \div \frac{4}{3}$

22. $\frac{9}{5} \div \frac{5}{3}$

23. $\frac{7}{16} \div \frac{1}{4}$

24. $\frac{7}{9} \div \frac{2}{3}$

25. $3 \div \frac{7}{5}$

26. $4 \div \frac{8}{3}$

27. $6 \div \frac{5}{8}$

28. $5 \div \frac{2}{3}$

29. $\frac{9}{4} \div 1\frac{2}{3}$

30. $\frac{8}{7} \div 2\frac{1}{4}$

31. $\frac{7}{4} \div 2\frac{4}{5}$

32. $\frac{11}{9} \div 3\frac{1}{3}$

33. $3\frac{1}{2} \div 4\frac{1}{3}$

34. $5\frac{2}{3} \div 3\frac{2}{3}$

35. $2\frac{1}{3} \div 6\frac{1}{2}$

36. $1\frac{7}{16} \div 3\frac{3}{8}$

37. $5\frac{3}{5} \div 3\frac{1}{3}$

38. $3\frac{2}{3} \div 1\frac{15}{18}$

39. $4\frac{2}{5} \div 1\frac{8}{25}$

40. $6\frac{3}{8} \div 1\frac{1}{16}$

End of Chapter Problems 6-10

Perform the indicated operations. If the answer is an improper fraction, change to a mixed number:

1. $\frac{16}{5} + \frac{10}{3}$

2. $\frac{15}{2} + \frac{9}{7}$

3. $4\frac{5}{6} + \frac{5}{7}$

4. $6\frac{1}{3} + \frac{2}{5}$

5. $2\frac{1}{3} + \frac{9}{5}$

6. $3\frac{3}{8} + \frac{10}{9}$

7. $3\frac{7}{8} + 4\frac{1}{2}$

8. $4\frac{2}{5} + 3\frac{3}{10}$

9. $\frac{7}{8} - \frac{7}{32}$

10. $\frac{23}{32} - \frac{3}{16}$

11. $\frac{10}{3} - \frac{1}{4}$

12. $\frac{12}{5} - \frac{1}{3}$

13. $3\frac{5}{6} - \frac{9}{4}$

14. $7\frac{1}{2} - \frac{9}{8}$

15. $3\frac{1}{8} - 1\frac{1}{4}$

16. $3\frac{7}{12} - 5\frac{3}{5}$

17. $4\frac{5}{6} - 2\frac{2}{3}$

18. $6\frac{9}{16} - 3\frac{5}{64}$

19. $\frac{9}{4} + \frac{7}{2} - 2\frac{1}{3}$

20. $\frac{8}{3} + \frac{11}{5} - 1\frac{1}{2}$

21. $7\frac{1}{2} - 2\frac{1}{3} + 3\frac{1}{4}$

22. $6\frac{2}{3} + 3\frac{1}{5} - 2\frac{1}{4}$

23. $2\frac{7}{8} + 4\frac{3}{4} - 3\frac{11}{16}$

24. $7\frac{1}{3} - 4\frac{5}{6} + 2\frac{2}{3}$

25. $3\frac{7}{16} - \frac{15}{32} + 2\frac{3}{8}$

26. $6\frac{7}{8} - 2\frac{3}{16} + 3\frac{7}{16}$

27. $7\frac{2}{3} - 3\frac{2}{5} + 2\frac{4}{15}$

28. $2\frac{2}{7} + 3\frac{2}{3} - 2\frac{3}{7}$

29. $8\frac{3}{8} + 4\frac{2}{3} - 3\frac{1}{12}$

30. $6\frac{1}{6} + 3\frac{1}{3} - 2\frac{1}{9}$

End of Chapter Problems 6-11

Add, subtract, multiply, and divide the following numbers. Express answers as decimal numbers rounded to three places.

1. $\frac{3}{8}$ and $\frac{3}{32}$

2. $\frac{3}{5}$ and $\frac{1}{3}$

3. $\frac{7}{16}$ and $\frac{2}{7}$

4. $\frac{13}{32}$ and $\frac{5}{64}$

5. $\frac{5}{8}$ and $\frac{37}{64}$

6. $\frac{5}{16}$ and $\frac{13}{32}$

7. $\frac{3}{5}$ and $\frac{3}{10}$

8. $\frac{5}{12}$ and $\frac{7}{10}$

9. $3\frac{3}{8}$ and $2\frac{1}{3}$

10. $7\frac{5}{32}$ and $4\frac{1}{12}$

11. $4\frac{2}{5}$ and $2\frac{15}{64}$

12. $9\frac{7}{16}$ and $4\frac{3}{32}$

13. $7\frac{1}{2}$ and $3\frac{5}{9}$

14. $8\frac{5}{16}$ and $5\frac{23}{64}$

15. $4\frac{7}{8}$ and $8\frac{7}{16}$

16. $5\frac{4}{7}$ and $9\frac{1}{12}$

17. $3\frac{2}{3}$ and $6\frac{2}{3}$

18. $6\frac{3}{64}$ and $9\frac{1}{2}$

19. $1\frac{7}{12}$ and $3\frac{2}{3}$

20. $2\frac{9}{16}$ and $4\frac{9}{32}$

Answers to Self-tests

SELF-TEST 6-1

1. $2 \cdot 2 \cdot 2 \cdot 5 = 2^3 \cdot 5$

2. $2 \cdot 2 \cdot 2 \cdot 2 \cdot 3 = 2^4 \cdot 3$

3. $2 \cdot 2 \cdot 2 \cdot 7 = 2^3 \cdot 7$

4. $2 \cdot 2 \cdot 3 \cdot 7 = 2^2 \cdot 3 \cdot 7$

5. $2 \cdot 2 \cdot 3 \cdot 3 \cdot 5 = 2^2 \cdot 3^2 \cdot 5$

6. $2 \cdot 2 \cdot 5 \cdot 11 = 2^2 \cdot 5 \cdot 11$

7. $\frac{16}{25}$

8. $\frac{2}{3}$

9. $\frac{3}{7}$

10. $\frac{7}{15}$

SELF-TEST 6-2

1. $\frac{7}{20}$

2. $\frac{1}{4}$

3. $\frac{1}{30}$

4. $\frac{2}{25}$

5. $\frac{3}{10}$

6. $\frac{14}{3}$

7. $\frac{2}{5}$

8. $\frac{1}{5}$

SELF-TEST 6-3

1. 60

2. 120

3. 210

4. 225

5. 210

6. 210

7. 240

8. 4095

9. $\frac{1}{2}$

10. $\frac{13}{24}$

11. $-\frac{1}{6}$

12. $\frac{3}{5}$

13. $-\frac{1}{20}$

14. $\frac{13}{24}$

15. $-\frac{23}{240}$

16. $-\frac{17}{24}$

SELF-TEST 6-4

1. $3\frac{5}{8}$
2. $6\frac{3}{5}$
3. $\frac{51}{16}$
4. $\frac{26}{3}$
5. $5\frac{29}{32}$
6. 15
7. $\frac{3}{5}$
8. $1\frac{15}{17}$
9. $4\frac{1}{6}$
10. $8\frac{1}{8}$
11. $5\frac{13}{32}$
12. $3\frac{11}{12}$
13. $-\frac{23}{24}$
14. $5\frac{19}{30}$

SELF-TEST 6-5

Answers as decimal fractions:

1. $\frac{9}{16} + \frac{3}{8} = 0.938$
   $\frac{9}{16} - \frac{3}{8} = 0.188$
   $\frac{9}{16} \times \frac{3}{8} = 0.211$
   $\frac{9}{16} \div \frac{3}{8} = 1.50$
2. $\frac{2}{3} + \frac{1}{6} = 0.833$
   $\frac{2}{3} - \frac{1}{6} = 0.500$
   $\frac{2}{3} \times \frac{1}{6} = 0.111$
   $\frac{2}{3} \div \frac{1}{6} = 4.00$
3. $5\frac{7}{8} + 2\frac{1}{4} = 8.13$
   $5\frac{7}{8} - 2\frac{1}{4} = 3.63$
   $5\frac{7}{8} \times 2\frac{1}{4} = 13.2$
   $5\frac{7}{8} \div 2\frac{1}{4} = 2.61$
4. $3\frac{3}{16} + 6\frac{1}{3} = 9.52$
   $3\frac{3}{16} - 6\frac{1}{3} = -3.15$
   $3\frac{3}{16} \times 6\frac{1}{3} = 20.2$
   $3\frac{3}{16} \div 6\frac{1}{3} = 0.503$

Answers as proper fractions or as mixed numbers:

1. $\frac{9}{16} + \frac{3}{8} = \frac{15}{16}$
   $\frac{9}{16} - \frac{3}{8} = \frac{3}{16}$
   $\frac{9}{16} \times \frac{3}{8} = \frac{27}{128}$
   $\frac{9}{16} \div \frac{3}{8} = 1\frac{1}{2}$
2. $\frac{2}{3} + \frac{1}{6} = \frac{5}{6}$
   $\frac{2}{3} - \frac{1}{6} = \frac{1}{2}$
   $\frac{2}{3} \times \frac{1}{6} = \frac{1}{9}$
   $\frac{2}{3} \div \frac{1}{6} = 4$

3. $5\frac{7}{8} + 2\frac{1}{4} = 8\frac{1}{8}$

$5\frac{7}{8} - 2\frac{1}{4} = 3\frac{5}{8}$

$5\frac{7}{8} \times 2\frac{1}{4} = 13\frac{7}{32}$

$5\frac{7}{8} \div 2\frac{1}{4} = 2\frac{11}{18}$

4. $3\frac{3}{16} + 6\frac{1}{3} = 9\frac{25}{48}$

$3\frac{3}{16} - 6\frac{1}{3} = -3\frac{7}{48}$

$3\frac{3}{16} \times 6\frac{1}{3} = 20\frac{3}{16}$

$3\frac{3}{16} \div 6\frac{1}{3} = \frac{153}{304}$

FRACTIONS AND LITERAL NUMBERS

CHAPTER 7

INTRODUCTION

In Chapter 6, we worked with fractions that contained only real numbers. In this chapter, we will continue our study of fractions and will include literal numbers. The study of fractions that contain literal numbers is extremely important to the electronics student because almost all fractions we encounter contain literal numbers. To deal with the various equations we work with in problem solving, a basic understanding of fractions and literal numbers is necessary.

CHAPTER OBJECTIVES

In this chapter you will learn how to:

1. Find the prime factors of literal numbers.
2. Find the lowest common multiple of literal numbers.
3. Multiply and divide whole and fractional numbers.
4. Add and subtract fractions with common denominators.
5. Add and subtract fractions with unlike denominators.

7-1 PRIME NUMBERS

In Chapter 6, we learned how to define and recognize prime numbers. Literal numbers are prime when they are raised to the first power. x is prime. x^2 is not prime because $x^2 = x \cdot x$. The prime factors of x^3y^2 are $x \cdot x \cdot x \cdot y \cdot y$.

Real numbers and literal numbers often appear in the same term. Such terms can be factored by finding the prime factors of the real numbers as before, and then finding the prime factors of the literal numbers. Let's factor some of these numbers.

☐ EXAMPLE 7-1

Factor $40a^2b^3$.

Solution

$$40a^2b^3 = 2 \cdot 2 \cdot 2 \cdot 5 \cdot a \cdot a \cdot b \cdot b \cdot b$$

☐ EXAMPLE 7-2

Factor $33x^2y^2z$.

Solution

$$33x^2y^2z = 3 \cdot 11 \cdot x \cdot x \cdot y \cdot y \cdot z$$

☐

PRACTICE PROBLEMS 7-1

Find the prime factors of the following:

1. $6ab^2$
2. $8a^2bc^2$
3. $22x^3$
4. $38x^2y^2$
5. $30x^2y^3$
6. $115z^2$
7. $136x^2z^3$
8. $168a^2c^2$
9. $175b^2c^3$
10. $200y^3$

Solutions

1. $2 \cdot 3 \cdot a \cdot b \cdot b$
2. $2 \cdot 2 \cdot 2 \cdot a \cdot a \cdot b \cdot c \cdot c$
3. $2 \cdot 11 \cdot x \cdot x \cdot x$
4. $2 \cdot 19 \cdot x \cdot x \cdot y \cdot y$
5. $2 \cdot 3 \cdot 5 \cdot x \cdot x \cdot y \cdot y \cdot y$
6. $5 \cdot 23 \cdot z \cdot z$
7. $2 \cdot 2 \cdot 2 \cdot 17 \cdot x \cdot x \cdot z \cdot z \cdot z$
8. $2 \cdot 2 \cdot 2 \cdot 3 \cdot 7 \cdot a \cdot a \cdot c \cdot c$
9. $5 \cdot 5 \cdot 7 \cdot b \cdot b \cdot c \cdot c \cdot c$
10. $2 \cdot 2 \cdot 2 \cdot 5 \cdot 5 \cdot y \cdot y \cdot y$

Additional practice problems are at the end of the chapter.

7-2 LOWEST COMMON MULTIPLE

Finding the lowest common multiple of numbers containing literal numbers is similar to finding the LCM of real numbers. The rule is:

RULE 7-1 To find the lowest common multiple of numbers containing literal numbers:

1. Factor each term.
2. Determine the *maximum number* of times a prime factor appears in any one term.
3. Multiply the resulting prime factors found in step 2. This number is the LCM.

The lowest common multiple is the *lowest* multiple of each term that is common to all terms.

□ EXAMPLE 7-3

Find the LCM of $6a$ and $18a^2$.

Solution Find the prime factors of each term:

$$6a = 2 \cdot 3 \cdot a$$

$$18a^2 = 2 \cdot 3 \cdot 3 \cdot a \cdot a$$

The maximum number of times 2 appears as a factor is once; therefore, 2 is part of the LCM. Three appears a maximum of two times; therefore, $3 \cdot 3$ or 3^2 is part of the LCM. Since a appears a maximum of two times, $a \cdot a$ or a^2 is part of the LCM. This takes care of all the prime factors. Our next step is to multiply these factors together to get the LCM.

$$2 \cdot 3 \cdot 3 \cdot a \cdot a = 18a^2$$

$18a^2$ is the lowest number that contains both $6a$ and $18a^2$ as factors. That is, $6a$ is a factor of $18a^2$ ($6a \cdot 3a = 18a^2$) and $18a^2$ is a factor ($18a^2 \cdot 1 = 18a^2$). There are other common multiples: $36a^2$, $36a^3$, and $54a^3$ to name a few. But $18a^2$ is the *lowest*.

□ EXAMPLE 7-4

Find the LCM of $10xy^2$ and $15x^3y$.

Solution

$$10xy^2 = 2 \cdot 5 \cdot x \cdot y \cdot y$$

$$15x^3y = 3 \cdot 5 \cdot x \cdot x \cdot x \cdot y$$

Two, three, and five each appear as factors once. Therefore, 2 and 3 and 5 are part of the LCM. Since x appears a maximum of three times, x^3 is part of the LCM. Since y appears a maximum of two times, y^2 is part of the LCM.

$$\begin{aligned} \text{LCM} &= 2 \cdot 3 \cdot 5 \cdot x^3 \cdot y^2 \\ &= 30x^3y^2 \end{aligned}$$

$30x^3y^2$ is the lowest number that has both $10xy^2$ and $15x^3y$ as factors.

□ EXAMPLE 7-5

Find the LCM of $14a^2b$, $42abc^2$, and $12b$.

Solution

$$\begin{aligned} 14a^2b &= 2 \cdot 7 \cdot a \cdot a \cdot b \\ 42abc^2 &= 2 \cdot 3 \cdot 7 \cdot a \cdot b \cdot c \cdot c \\ 12b &= 2 \cdot 2 \cdot 3 \cdot b \\ \text{LCM} &= 2 \cdot 2 \cdot 3 \cdot 7 \cdot a \cdot a \cdot b \cdot c \cdot c = 84a^2bc^2 \end{aligned}$$

$84a^2bc^2$ is the smallest number that has $14a^2b$, $42abc^2$, and $12b$ as factors. That is, $84a^2bc^2$ is the smallest number that is divisible by all three factors. Just as with real numbers, literal numbers do not always appear as factors in each term. Remember the rule: *determine the maximum number of times a prime factor appears in* ***any*** *one term.* □

PRACTICE PROBLEMS 7-2

Find the LCM of the following numbers:

1. a^2b^3, a^3b^2
2. a^2b^4, a^2b^3
3. a^2b^2, a^3b, ab
4. a^3b^2, a^2b^3, a^3b^3
5. $a^3b^3c^3$, abc, $a^2b^3c^2$
6. $12a^2b$, $18ab^2$
7. $3ab^2$, $8a^2b^2$
8. $12xyz^2$, $30x^2yz^2$, $45xyz$
9. $11xz^2$, $44x^2y$, y^2z
10. $108xy^4$, $72y^4$, x^2z
11. $81a^3b^2$, $99a^3b$, $77\,ab$
12. $64ab^2$, $96a^2b^2$, $160a^3b$

Solutions

1. a^3b^3
2. a^2b^4
3. a^3b^2
4. a^3b^3
5. $a^3b^3c^3$
6. $36a^2b^2$
7. $24a^2b^2$
8. $180x^2yz^2$
9. $44x^2y^2z^2$
10. $216x^2y^4z$
11. $6237a^3b^2$
12. $960a^3b^2$

Additional practice problems are at the end of the chapter.

SELF-TEST 7-1

Find the prime factors of the following numbers:

1. $28xy^2$
2. $54a^3b^2$
3. $76xy^3$
4. $120x^2z^2$
5. $130a^3b^2$
6. $315a^2b^2$

Find the LCM of the following numbers.

7. a^2b, ab^3
8. x^2yz^3, x^3yz^2
9. $8xz^2, 9xz$
10. $14ab^2, 35a^2b^3$
11. $18a^2b, 24ac^2, 36b^2c$
12. $32a^3b^2, 64b^2c^2, 128a^2c^2$

Answers to Self-test 7-1 are at the end of the chapter.

7-3 MULTIPLICATION AND DIVISION OF MONOMIALS

7-3-1 Multiplication

To multiply monomials, we multiply the real numbers and then multiply the literal numbers by using the laws of exponents learned in previous chapters. Recall from arithmetic that each part of a multiplication problem has a name.

$$\begin{array}{rl} 6a & \text{(multiplicand)} \\ \times 7a & \text{(multiplier)} \\ \hline 42a^2 & \text{(product)} \end{array}$$

The product cannot be simplified further because we have not been given a value for a. Remember that we can indicate the operation to be performed, in this case multiplication, in a number of ways. We indicate multiplication with $\times$ or $\cdot$ or ( ) or other signs of grouping.

In the problem $4a^2b \cdot 3ab^3$, we multiply 4×3 and get 12. $a^2 \cdot a = a^{2+1} = a^3$ and $b \times b^3 = b^{1+3} = b^4$. The answer is $12a^3b^4$. With practice, we can do the addition of exponents of like bases in our heads.

■ **KEY POINT** In multiplication, all the literal numbers contained in the multiplicand and the multiplier must appear in the product unless the exponent is zero.

□ EXAMPLE 7-6

$3x^3y^{-2}z \times 5x^2y^2z^2 = 15x^5z^3$. If we put like bases together, we can more easily see the result:

$$3 \cdot 5 \cdot x^3 \cdot x^2 \cdot y^{-2} \cdot y^2 \cdot z \cdot z^2 = 15 \cdot x^{3+2} \cdot y^{-2+2} \cdot z^{1+2}$$
$$= 15x^5y^0z^3$$
$$= 15x^5z^3$$

Remember, any number raised to the zero power equals 1.

☐ EXAMPLE 7-7

$$4x^{-3}yz^{-2} \cdot 2xyz^2 \cdot 3x^{-1}y^{-1}z^2 = 24x^{-3}yz^2$$
$$4 \times 2 \times 3 = 24$$
$$x^{-3+1-1} = x^{-3}$$
$$y^{1+1-1} = y^1 = y$$
$$z^{-2+2+2} = z^2$$

Multiplying the resulting numbers together gives the product $24x^{-3}yz^2$. Separating the numbers and adding the exponents of like bases may help you see the solution. Don't forget to multiply the parts together to get the final product!

☐

PRACTICE PROBLEMS 7-3

Perform the indicated operations:

1. $2ab \cdot 3a^2b$
2. $4a^3b \cdot 3ab^2$
3. $3a^4b \cdot 3a^{-3}b^{-4}$
4. $6x^{-2}y^4 \cdot 8x^4y^{-1}$
5. $2x^3y^2 \cdot 3x^{-2}y^{-1} \cdot 3x^4y^3$
6. $5x^6y^2z^2 \cdot x^{-2}y^{-4}z^{-3} \cdot x^{-1}y^3$
7. $3x^3y^{-3}z^{-2} \cdot 3x^{-3}y^3 \cdot 3xyz^4$
8. $7x^5y^{-2}z^2 \cdot 3x^{-1}y^5z^4 \cdot 2x^{-1}y^{-1}z^{-1}$

Solutions

1. $6a^3b^2$
2. $12a^4b^3$
3. $9ab^{-3}$
4. $48x^2y^3$
5. $18x^5y^4$
6. $5x^3yz^{-1}$
7. $27xyz^2$
8. $42x^3y^2z^5$

Additional practice problems are at the end of the chapter.

7-3-2 Division

Recall from arithmetic that the parts of a division problem are as follows:

(a) $\dfrac{18a}{6a} = 3$

- $18a$ ← dividend
- 3 ← quotient
- $6a$ ← divisor

$$\text{(b)}\quad 6a\overline{)18a}^{\;3} \qquad \begin{array}{l} 3 \leftarrow \text{quotient} \\ 18a \leftarrow \text{dividend} \\ 6a \leftarrow \text{divisor} \end{array}$$

When written as a fraction as in (a), the dividend is the numerator and the divisor is the denominator.

In arithmetic we learned how to do "long division." Here, our long division is done by means of a calculator, but we must do the division of literal numbers ourselves because calculators can only deal with real numbers.

In dividing monomials, we find the prime factors of the dividend and the divisor. Then we cancel each factor that appears in both the dividend and the divisor. The resultant number is the quotient.

☐ EXAMPLE 7-8

$$\frac{10a^3}{2a^2}.$$

Solution Find the prime factors and cancel those appearing in both numerator and denominator.

$$\frac{\not{2}\cdot 5\cdot \not{a}\cdot \not{a}\cdot a}{\not{2}\cdot \not{a}\cdot \not{a}} = \frac{5a}{1} = 5a$$

Another way to solve the problem would be to move all literal numbers into the numerator and multiply. Recall that we can move numbers from the numerator to the denominator or from the denominator to the numerator by changing the sign of the exponent. Using this method in Example 7-8, we get

$$\frac{10a^3}{2a^2} = \frac{10a^3\cdot a^{-2}}{2} = 5a^{3-2} = 5a$$

☐ EXAMPLE 7-9

$$\frac{6a^2b^3}{18a^4b}.$$

Solution

$$\frac{\not{2}\cdot \not{3}\cdot \not{a}\cdot \not{a}\cdot \not{b}\cdot b\cdot b}{\not{2}\cdot \not{3}\cdot 3\cdot \not{a}\cdot \not{a}\cdot a\cdot a\cdot \not{b}} = \frac{b^2}{3a^2} = \frac{a^{-2}b^2}{3}$$

or

$$\frac{6a^2a^{-4}b^3b^{-1}}{18} = \frac{6a^{-2}b^2}{18} = \frac{a^{-2}b^2}{3}$$

☐ EXAMPLE 7-10

$$\frac{40x^2y^{-1}}{10x^{-1}y^{-3}}.$$

Solution

$$\frac{40x \cdot x \cdot \cancel{y^{-1}}}{10x^{-1} \cdot \cancel{y^{-1}} \cdot y^{-1} \cdot y^{-1}} = \frac{4x^2}{x^{-1}y^{-2}}$$

$$\frac{4x^2 \cdot x}{y^{-2}} = \frac{4x^3}{y^{-2}} = 4x^3y^2$$

and also $$\frac{40x^2y^{-1}}{10x^{-1}y^{-3}} = \frac{40x^2 \cdot x \cdot y^{-1} \cdot y^3}{10} = 4x^3y^2$$

□

The second method is usually quicker. Both methods of solution are important in that they help develop important math concepts and ideas. The first method helps us develop an understanding of factoring so that we may deal more easily with fractions. The second method helps us strengthen our understanding of exponents and bases.

PRACTICE PROBLEMS 7-4

Perform the following divisions by (a) canceling like factors and (b) moving all literal numbers to the numerator and multiplying:

1. $\dfrac{15a^4}{5a}$
2. $\dfrac{36a^2}{6a^3}$
3. $\dfrac{42a^3b^2}{7ab}$
4. $\dfrac{56ab^3}{7a^3b}$
5. $\dfrac{20a^4b^2c}{3a^2b^2c^2}$
6. $\dfrac{39a^4bc^4}{3a^2b^3c}$
7. $\dfrac{7x^{-2}y}{63xy^2}$
8. $\dfrac{72xy^2}{9x^{-3}y^{-4}}$
9. $\dfrac{84x^{-1}y^2z^{-3}}{6x^3y^{-1}z^{-2}}$
10. $\dfrac{46x^{-1}y^2z^{-3}}{23xy^{-1}z^2}$

Solutions

1. (a) $\dfrac{3 \cdot \cancel{5} \cdot \cancel{a} \cdot a \cdot a \cdot a}{\cancel{5}\cancel{a}} = 3a^3$

(b) $\dfrac{15a^4 \cdot a^{-1}}{5} = 3a^3$

2. (a) $\dfrac{\cancel{2} \cdot 2 \cdot \cancel{3} \cdot 3 \cdot \cancel{a} \cdot \cancel{a}}{\cancel{2} \cdot \cancel{3} \cdot \cancel{a} \cdot \cancel{a} \cdot a} = \dfrac{6}{a} = 6a^{-1}$

(b) $\dfrac{36a^2 \cdot a^{-3}}{6} = 6a^{-1}$

3. (a) $\dfrac{2 \cdot 3 \cdot \cancel{7} \cdot \cancel{a} \cdot a \cdot a \cdot \cancel{b} \cdot b}{\cancel{7} \cdot \cancel{a} \cdot \cancel{b}} = 6a^2b$

(b) $\dfrac{42a^3 \cdot a^{-1} \cdot b^2 \cdot b^{-1}}{7} = 6a^2b$

4. (a) $\dfrac{2 \cdot 2 \cdot 2 \cdot \cancel{7} \cdot \cancel{a} \cdot \cancel{b} \cdot b \cdot b}{\cancel{7} \cdot \cancel{a} \cdot a \cdot a \cdot \cancel{b}} = \dfrac{8b^2}{a^2} = 8a^{-2}b^2$

(b) $\dfrac{56a \cdot a^{-3} \cdot b^3 \cdot b^{-1}}{7} = 8a^{-2}b^2$

5. (a) $\dfrac{2 \cdot 2 \cdot 5 \cdot \cancel{a} \cdot \cancel{a} \cdot a \cdot a \cdot \cancel{b} \cdot \cancel{b} \cdot \cancel{c}}{3 \cdot \cancel{a} \cdot \cancel{a} \cdot \cancel{b} \cdot \cancel{b} \cdot \cancel{c} \cdot c} = \dfrac{20a^2}{3c} = \dfrac{20a^2c^{-1}}{3}$

(b) $\dfrac{20a^4 \cdot a^{-2} \cdot b^2 \cdot b^{-2} \cdot c \cdot c^{-2}}{3} = \dfrac{20a^2b^0c^{-1}}{3} = \dfrac{20a^2c^{-1}}{3}$

6. (a) $\dfrac{\cancel{3}\cdot 13\cdot \cancel{a}\cdot \cancel{a}\cdot a\cdot a\cdot \cancel{b}\cdot \cancel{c}\cdot c\cdot c\cdot c}{\cancel{3}\cdot \cancel{a}\cdot \cancel{a}\cdot \cancel{b}\cdot b\cdot b\cdot \cancel{c}} = \dfrac{13a^2c^3}{b^2} = 13a^2b^{-2}c^3$

(b) $\dfrac{39a^4\cdot a^{-2}\cdot b\cdot b^{-3}\cdot c^4\cdot c^{-1}}{3} = 13a^2b^{-2}c^3$

7. (a) $\dfrac{\cancel{7}\cdot x^{-1}\cdot x^{-1}\cdot \cancel{y}}{3\cdot 3\cdot \cancel{7}\cdot x\cdot \cancel{y}\cdot y} = \dfrac{x^{-2}}{9xy} = \dfrac{x^{-2}\cdot x^{-1}\cdot y^{-1}}{9} = \dfrac{x^{-3}y^{-1}}{9}$

(b) $\dfrac{7x^{-2}\cdot x^{-1}\cdot y\cdot y^{-2}}{63} = \dfrac{x^{-3}y^{-1}}{9}$

8. (a) $\dfrac{2\cdot 2\cdot 2\cdot \cancel{3}\cdot \cancel{3}\cdot x\cdot y\cdot y}{\cancel{3}\cdot \cancel{3}\cdot x^{-1}\cdot x^{-1}\cdot x^{-1}\cdot y^{-1}\cdot y^{-1}\cdot y^{-1}\cdot y^{-1}} = \dfrac{8xy^2}{x^{-3}\cdot y^{-4}} = 8x\cdot x^3\cdot y^2\cdot y^4 = 8x^4y^6$

(b) $\dfrac{72x\cdot x^3\cdot y^2\cdot y^4}{9} = 8x^4y^6$

9. (a) $\dfrac{\cancel{2}\cdot 2\cdot \cancel{3}\cdot 7\cdot x^{-1}\cdot y\cdot y\cdot \cancel{z^{-1}}\cdot \cancel{z^{-1}}\cdot z^{-1}}{\cancel{2}\cdot \cancel{3}\cdot x\cdot x\cdot x\cdot y^{-1}\cdot \cancel{z^{-1}}\cdot \cancel{z^{-1}}} = \dfrac{14x^{-1}y^2z^{-1}}{x^3y^{-1}} = 14x^{-1}x^{-3}y^2y^1z^{-1} = 14x^{-4}y^3z^{-1}$

(b) $\dfrac{84x^{-1}\cdot x^{-3}\cdot y^2\cdot y\cdot z^{-3}\cdot z^2}{6} = 14x^{-4}y^3z^{-1}$

10. (a) $\dfrac{2\cdot 23\cdot x^{-1}\cdot y\cdot y\cdot z^{-1}\cdot z^{-1}\cdot z^{-1}}{23x\cdot y^{-1}\cdot z\cdot z} = 2x^{-1}\cdot x^{-1}\cdot y^2\cdot y\cdot z^{-3}\cdot z^{-2} = 2x^{-2}y^3z^{-5}$

(b) $\dfrac{46x^{-1}\cdot x^{-1}\cdot y^2\cdot y\cdot z^{-3}\cdot z^{-2}}{23} = 2x^{-2}y^3z^{-5}$

Additional practice problems are at the end of the chapter.

SELF-TEST 7-2

Perform the indicated operations:

1. $3x\cdot 5y$
2. $3x^2y\cdot 4x^2y^3$
3. $5a^{-1}b^2\cdot 6a^3b^{-1}$
4. $3a^2b^{-1}\cdot 5ab^{-2}\cdot 2a^{-4}b$
5. $6ab^{-3}c^2\cdot 7a^{-4}b^2c\cdot a^{-1}$
6. $\dfrac{48x^2y}{6y^2}$
7. $\dfrac{24x^{-2}y}{3xy^2}$
8. $\dfrac{56a^2bc^{-3}}{8b}$
9. $\dfrac{63a^{-1}b^2c^{-4}}{7a^3c^{-3}}$
10. $\dfrac{48xyz^{-3}}{3x^{-1}y^3z}$

Answers to Self-test 7-2 are at the end of the chapter.

7-4 MULTIPLICATION AND DIVISION OF FRACTIONS

7-4-1 Multiplication

■ **KEY POINT** To multiply fractions, we (1) multiply the numerators (this product is the numerator in the answer), (2) multiply the denominators (this product is the denominator in the answer), and (3) cancel like factors to reduce to lowest terms.

□ **EXAMPLE 7-11**

Multiply the following fractions:

(a) $\frac{3a}{4} \times \frac{6a^2}{10}$ **(b)** $\frac{7x^2y}{2x} \times \frac{3x^3y^3}{14x^2}$ **(c)** $\frac{3a^{-1}b^2}{4a^2b^3} \times \frac{4a^3b^{-3}}{9a^{-1}b^{-1}} \times \frac{2a^2b^4}{a^3b^2}$

Solutions

(a) $\frac{3a}{4} \times \frac{6a^2}{10} = \frac{3a \cdot 6a^2}{4 \cdot 10} = \frac{18a^3}{40} = \frac{\not{2} \cdot 3 \cdot 3 \cdot a^3}{\not{2} \cdot 2 \cdot 2 \cdot 5} = \frac{9a^3}{20}$

(b) $\frac{7x^2y}{2x} \times \frac{3x^3y^3}{14x^2} = \frac{7x^2y \cdot 3x^3y^3}{2x \cdot 14x^2} = \frac{21x^5y^4}{28x^3} = \frac{3 \cdot 7 \cdot x^5 \cdot y^4}{2 \cdot 2 \cdot 7 \cdot x^3}$

$= \frac{3 \cdot \not{7} \cdot x^5 \cdot x^{-3} \cdot y^4}{2 \cdot 2 \cdot \not{7}} = \frac{3x^2y^4}{4}$

(c) $\frac{3a^{-1}b^2}{4a^2b^3} \times \frac{4a^3b^{-3}}{9a^{-1}b^{-1}} \times \frac{2a^2b^4}{a^3b^2} = \frac{3a^{-1}b^2 \cdot 4a^3b^{-3} \cdot 2a^2b^4}{4a^2b^3 \cdot 9a^{-1}b^{-1} \cdot a^3b^2}$

At this point we can save some time by examining both numerator and denominator for like factors. The factors don't have to be prime; they only have to be *like* factors. 4, a^2, a^3, b^2, a^{-1} are found in both numerator and denominator and can be canceled at this time. We are now left with

$$\frac{3\not{a^{-1}}\not{b^2} \cdot \not{4}\not{a^3}b^{-3} \cdot 2\not{a^2}b^4}{\not{4}\not{a^2}b^3 \cdot 9\not{a^{-1}}b^{-1} \cdot \not{a^3}\not{b^2}} = \frac{3 \cdot b^{-3} \cdot 2b^4}{b^3 \cdot 9b^{-1}} = \frac{6b}{9b^2}$$

$$= \frac{2 \cdot \not{3} \cdot \not{b}}{\not{3} \cdot 3 \cdot \not{b} \cdot b} = \frac{2}{3b} = \frac{2b^{-1}}{3}$$

In each case, we reduce to lowest terms by canceling like terms. In part (c), the literal number, b, was moved to the numerator in the final answer. Although it is mathematically correct to leave literal numbers in the denominator, we normally give answers with all literal numbers in the numerator.

□

PRACTICE PROBLEMS 7-5

Find the product of the following numbers. Reduce answers to lowest terms.

1. $\dfrac{3x}{5} \times \dfrac{2x}{4}$
2. $\dfrac{4x^2}{7} \times \dfrac{3x}{4}$
3. $\dfrac{5x^2}{8y} \times \dfrac{6xy}{5}$
4. $\dfrac{2x^3}{3y^2} \times \dfrac{3y^3}{4x}$
5. $\dfrac{3xy^{-1}}{5z} \times \dfrac{5y^3z^2}{9x^{-2}}$
6. $\dfrac{4a^{-1}b^2}{5c^2} \times \dfrac{b^{-3}c}{3a^2} \times \dfrac{3a}{4b^{-1}c^3}$
7. $\dfrac{2x^2y^{-1}}{5z^2} \times \dfrac{20y^3z^3}{30xz^{-1}} \times \dfrac{xz^2}{y^3}$

Solutions

1. $\dfrac{6x^2}{20} = \dfrac{3x^2}{10}$
2. $\dfrac{12x^3}{28} = \dfrac{3x^3}{7}$
3. $\dfrac{30x^3y}{40y} = \dfrac{3x^3}{4}$
4. $\dfrac{6x^3y^3}{12xy^2} = \dfrac{x^2y}{2}$
5. $\dfrac{15xy^2z^2}{45x^{-2}z} = \dfrac{x^3y^2z}{3}$
6. $\dfrac{12b^{-1}c}{60a^2b^{-1}c^5} = \dfrac{a^{-2}c^{-4}}{5}$
7. $\dfrac{40x^3y^2z^5}{150xy^3z} = \dfrac{4x^2y^{-1}z^4}{15}$

Additional practice problems are at the end of the chapter.

7-4-2 Division

As with fractions that include only real numbers, division is performed by inverting the divisor and then multiplying.

☐ EXAMPLE 7-12

Divide the following fractions:

(a) $\dfrac{3a}{4} \div \dfrac{4a}{5}$ **(b)** $\dfrac{15x^2y}{8z^2} \div \dfrac{12x^2z}{21y^3}$ **(c)** $\dfrac{28a^2c}{42b^3} \div \dfrac{12a^3b^2}{35c^2}$

Solutions

(a) $\dfrac{3a}{4} \div \dfrac{4a}{5} = \dfrac{3a}{4} \times \dfrac{5}{4a} = \dfrac{15a}{16a} = \dfrac{15}{16}$

(b) $\dfrac{15x^2y}{8z^2} \div \dfrac{12x^2y}{21y^3} = \dfrac{15x^2y}{8z^2} \times \dfrac{21y^3}{12x^2z} = \dfrac{315x^2y^4}{96x^2z^3}$

$= \dfrac{\not{3} \cdot 3 \cdot 5 \cdot 7\not{x^2}y^4}{\not{3} \cdot 2 \cdot 2 \cdot 2 \cdot 2 \cdot 2 \cdot \not{x^2} \cdot z^3} = \dfrac{105y^4z^{-3}}{32}$

(c) $\dfrac{28a^2c}{42b^3} \div \dfrac{12a^3b^2}{35c^2} = \dfrac{28a^2c}{42b^3} \times \dfrac{35c^2}{12a^3b^2} = \dfrac{\not{2} \cdot \not{2} \cdot \not{7} \cdot 5 \cdot 7 \cdot a^2 \cdot c^3}{\not{2} \cdot 3 \cdot \not{7} \cdot \not{2} \cdot 2 \cdot 3 \cdot a^3 \cdot b^5}$

An alternate method of solution is offered here. Rather than multiply the numerators and denominators and then find the factors, the prime factors are found. Then like factors in both numerator and denominator are canceled.

$$\frac{5 \cdot 7 \cdot a^2 \cdot a^{-3} \cdot b^{-5} \cdot c^3}{2 \cdot 3 \cdot 3} = \frac{35a^{-1}b^{-5}c^3}{18}$$

□

PRACTICE PROBLEMS 7-6

Perform the following divisions. Reduce answers to lowest terms.

1. $\dfrac{4a}{5} \div \dfrac{6a^2}{7}$
2. $\dfrac{5a^2}{6} \div \dfrac{15a}{18}$
3. $\dfrac{7xy^2}{9z} \div \dfrac{14x^2y}{3z^2}$
4. $\dfrac{42x^{-2}y}{15y^{-1}z^2} \div \dfrac{7xz}{3y}$
5. $\dfrac{20a^2b^{-1}c}{9} \div \dfrac{5ab^3c^{-2}}{27}$
6. $\dfrac{10a^{-2}b^3}{7c^2} \div \dfrac{5b^{-1}}{28ac^4}$

Solutions

1. $\dfrac{28a}{30a^2} = \dfrac{14a^{-1}}{15}$
2. $\dfrac{90a^2}{90a} = a$
3. $\dfrac{21xy^2z^2}{126x^2yz} = \dfrac{x^{-1}yz}{6}$
4. $\dfrac{126x^{-2}y^2}{105xy^{-1}z^3} = \dfrac{6x^{-3}y^3z^{-3}}{5}$
5. $\dfrac{540a^2b^{-1}c}{45ab^3c^{-2}} = 12ab^{-4}c^3$
6. $\dfrac{280a^{-1}b^3c^4}{35b^{-1}c^2} = 8a^{-1}b^4c^2$

Additional practice problems are at the end of the chapter.

SELF-TEST 7-3

Perform the indicated operations:

1. $\dfrac{3x^2}{7} \times \dfrac{2x^2}{5}$
2. $\dfrac{3ab^2}{5c} \times \dfrac{2bc^{-1}}{3a^2}$
3. $\dfrac{2a^2bc^{-1}}{3b^{-1}} \times \dfrac{4b^3}{2ac^2} \times \dfrac{7a^{-3}}{8}$
4. $\dfrac{5a}{6} \div \dfrac{3a^2}{4}$

5. $\frac{36a^2b}{42c^{-3}} \div \frac{9b^3}{21a^3c}$

6. $\frac{12b}{21a^2c} \div \frac{3b^2}{7a^{-1}c^2}$

7. $\frac{15x^{-1}y^{-1}}{27z^3} \div \frac{5x^{-1}y}{9z}$

Answers to Self-test 7-3 are at the end of the chapter.

7-5 ADDITION AND SUBTRACTION OF FRACTIONS

7-5-1 Adding and Subtracting Fractions with Common Denominators

Recall that when adding or subtracting fractions we simply add or subtract the numerators if the denominators are identical. The denominator in the answer is the common denominator.

☐ EXAMPLE 7-13

Perform the indicated operations. Reduce answers to lowest terms:

(a) $\frac{2a}{7} + \frac{3a}{7}$ **(b)** $\frac{4}{5a} + \frac{3}{5a}$ **(c)** $\frac{a}{12} + \frac{5a}{12} + \frac{7a}{12}$

(d) $\frac{7x}{9} - \frac{3x}{9}$ **(e)** $\frac{11a}{12} + \frac{5a}{12} - \frac{7a}{12}$

Solutions

(a) $\frac{2a}{7} + \frac{3a}{7} = \frac{2a + 3a}{7} = \frac{5a}{7}$

(b) $\frac{4}{5a} + \frac{3}{5a} = \frac{4 + 3}{5a} = \frac{7}{5a}$

(c) $\frac{a}{12} + \frac{5a}{12} + \frac{7a}{12} = \frac{a + 5a + 7a}{12} = \frac{13a}{12}$

(d) $\frac{7x}{9} - \frac{3x}{9} = \frac{7x}{9} + \frac{-3x}{9} = \frac{7x - 3x}{9} = \frac{4x}{9}$

(e) $\frac{11a}{12} + \frac{5a}{12} - \frac{7a}{12} = \frac{11a + 5a - 7a}{12} = \frac{16a - 7a}{12} = \frac{9a}{12} = \frac{3a}{4}$

In each of these examples, the denominators were identical (common denominators), so we added or subtracted the numerators to find the answer and then reduced to lowest terms.

When the numerators contain unlike terms, they cannot be added. In these fractions, all the unlike terms are written over the common denominator. In the following examples, the numerators contain unlike terms.

☐ EXAMPLE 7-14

Perform the indicated operations. Reduce to lowest terms:

(a) $\frac{2a}{5} + \frac{3b}{5}$ **(b)** $\frac{3xy}{7} + \frac{2x}{7} + \frac{xy}{7}$ **(c)** $\frac{3}{8x} + \frac{5a}{8x} + \frac{2}{8x}$

(d) $\frac{7a}{12} - \frac{4b}{12}$ **(e)** $\frac{6ab}{3} + \frac{4b}{3} - \frac{3ac}{3}$ **(f)** $\frac{4a^2c}{7} + \frac{3ac}{7} - \frac{2a^2c}{7}$

Solutions

(a) $\frac{2a}{5} + \frac{3b}{5} = \frac{2a + 3b}{5}$

The answer cannot be simplified further because $2a$ and $3b$ are unlike terms.

(b) $\frac{3xy}{7} + \frac{2x}{7} + \frac{xy}{7} = \frac{3xy + 2x + xy}{7} = \frac{4xy + 2x}{7}$

We can combine the like terms. $3xy + xy = 4xy$.

(c) $\frac{3}{8x} + \frac{5a}{8x} + \frac{2}{8x} = \frac{3 + 5a + 2}{8x} = \frac{5 + 5a}{8x}$

We can combine the like terms. $3 + 2 = 5$.

(d) $\frac{7a}{12} - \frac{4b}{12} = \frac{7a - 4b}{12}$

(e) $\frac{6ab}{3} + \frac{4b}{3} - \frac{3ac}{3} = \frac{6ab + 4b - 3ac}{3}$

(f) $\frac{4a^2c}{7} + \frac{3ac}{7} - \frac{2a^2c}{7} = \frac{4a^2c + 3ac - 2a^2c}{7} = \frac{2a^2c + 3ac}{7}$

We can combine the like terms. $4a^2c - 2a^2c = 2a^2c$.

☐

If the denominators are common, we add or subtract the numerators and combine like terms. Remember, we can only combine *like* terms. In Example 7-14(f), we were able to combine $4a^2c$ and $-2a^2c$ because they were *like* terms. That is, the literal numbers were the same (a^2c).

PRACTICE PROBLEMS 7-7

Perform the indicated operations. Reduce to lowest terms.

1. $\frac{3}{5a} + \frac{1}{5a}$

2. $\frac{3}{8ab} + \frac{1}{8ab} + \frac{5}{8ab}$

3. $\frac{2}{9x} + \frac{5}{9x} - \frac{2}{9x}$

4. $\frac{5x}{11} + \frac{2x}{11}$

5. $\frac{7x}{20} - \frac{3x}{20}$

6. $\frac{3xy}{16} + \frac{xy}{16} + \frac{5xy}{16}$

7. $\frac{9ab}{20} - \frac{5ab}{20} - \frac{3ab}{20}$

8. $\frac{3a}{8b} + \frac{5}{8b} + \frac{2a}{8b}$

9. $\frac{2xy}{7} - \frac{3y}{7} + \frac{2y}{7}$

10. $\frac{3}{4y} - \frac{5}{4y}$

11. $\frac{8a}{21} - \frac{14a}{21} + \frac{2a}{21}$

Solutions

1. $\frac{3 + 1}{5a} = \frac{4}{5a}$

2. $\frac{3 + 1 + 5}{8ab} = \frac{9}{8ab}$

3. $\frac{2 + 5 - 2}{9x} = \frac{5}{9x}$

4. $\frac{5x + 2x}{11} = \frac{7x}{11}$

5. $\frac{7x - 3x}{20} = \frac{4x}{20} = \frac{x}{5}$

6. $\frac{3xy + xy + 5xy}{16} = \frac{9xy}{16}$

7. $\frac{9ab - 5ab - 3ab}{20} = \frac{ab}{20}$

8. $\frac{3a + 5 + 2a}{8b} = \frac{5a + 5}{8b}$

9. $\frac{2xy - 3y + 2y}{7} = \frac{2xy - y}{7}$

10. $\frac{3 - 5}{4y} = \frac{-2}{4y} = \frac{-1}{2y} = -\frac{1}{2y}$

11. $\frac{8a - 14a + 2a}{21} = \frac{10a - 14a}{21} = \frac{-4a}{21} = -\frac{4a}{21}$

Additional practice problems are at the end of the chapter.

7-5-2 Adding and Subtracting Fractions with Unlike Denominators

As with real numbers, when working with literal numbers and unlike denominators, we must first find the common denominator (the LCM) and then change each fraction so that all fractions have this common denominator. If we change the denominator, we must also change the numerator. We find the new numerator by using the rules given in Chapter 6.

RULE 7-2 To add fractions with unlike denominators: (1) find the CD; (2) change the denominator of all fractions to this denominator; (3) divide the CD by the old denominator of each fraction and then multiply this number by the old numerator to find the new numerator of the fraction; (4) add the fractions, combining all like terms in the numerator. Reduce the answer to the lowest terms.

The following examples illustrate the process.

□ EXAMPLE 7-15

$$\frac{3a}{20} + \frac{2a}{15} + \frac{3a}{10}.$$

Solution

Step 1. Find the common denominator:

$$20 = 2 \cdot 2 \cdot 5$$
$$15 = 3 \cdot 5$$
$$10 = 2 \cdot 5$$
$$\text{CD} = 2 \cdot 2 \cdot 3 \cdot 5 = 60$$

Step 2. Change each fraction so that its denominator is the CD.

$$\frac{3a}{20} = \frac{3a \times 3}{60} = \frac{9a}{60}$$
$$\frac{2a}{15} = \frac{2a \times 4}{60} = \frac{8a}{60}$$
$$\frac{3a}{10} = \frac{3a \times 6}{60} = \frac{18a}{60}$$

Step 3. Add the new fractions together. Reduce to lowest terms where possible.

$$\frac{9a}{60} + \frac{8a}{60} + \frac{18a}{60} = \frac{9a + 8a + 18a}{60} = \frac{35a}{60} = \frac{7a}{12}$$

□ EXAMPLE 7-16

$$\frac{3}{10a^2} + \frac{4}{15a} + \frac{b}{6a^2}.$$

Solution

$$10a^2 = 2 \cdot 5 \cdot a \cdot a$$
$$15a = 3 \cdot 5 \cdot a$$
$$6a^2 = 2 \cdot 3 \cdot a \cdot a$$
$$\text{CD} = 2 \cdot 3 \cdot 5 \cdot a \cdot a = 30a^2$$
$$\frac{3}{10a^2} = \frac{3 \times 3}{30a^2} = \frac{9}{30a^2}$$
$$\frac{4}{15a} = \frac{4 \times 2a}{30a^2} = \frac{8a}{30a^2}$$

$$\frac{b}{6a^2} = \frac{b \times 5}{30a^2} = \frac{5b}{30a^2}$$

$$\frac{9}{30a^2} + \frac{8a}{30a^2} + \frac{5b}{30a^2} = \frac{9 + 8a + 5b}{30a^2}$$

No reduction of the numerator is possible here. We have no like terms. □

PRACTICE PROBLEMS 7-8

Perform the indicated operations. Reduce to lowest terms where possible.

1. $\frac{3a}{4} + \frac{2b}{3}$
2. $\frac{2a}{3} + \frac{3a}{2}$
3. $\frac{b}{3a} + \frac{2b}{5}$
4. $\frac{x}{5y} + \frac{3x}{4}$
5. $\frac{x}{2y} + \frac{y}{5x}$
6. $\frac{3x}{y} + \frac{2y}{x}$
7. $\frac{2xy}{15z} + \frac{3xz}{5y}$
8. $\frac{5}{6a} + \frac{4}{15a^3} + \frac{1}{4a^2}$
9. $\frac{7a}{8} - \frac{2a}{5} + \frac{3a}{20}$
10. $\frac{3bc}{16a} - \frac{5ab}{8c}$
11. $\frac{11a}{15} - \frac{3a}{5} + \frac{2a}{3}$
12. $\frac{2a^2c}{3b} - \frac{4a^2c}{15b} + \frac{2a^2c}{5b}$

Solutions

1. $\frac{9a}{12} + \frac{8b}{12} = \frac{9a + 8b}{12}$
2. $\frac{4a}{6} + \frac{9a}{6} = \frac{4a + 9a}{6} = \frac{13a}{6}$
3. $\frac{5b}{15a} + \frac{6ab}{15a} = \frac{5b + 6ab}{15a}$
4. $\frac{4x}{20y} + \frac{15xy}{20y} = \frac{4x + 15xy}{20y}$
5. $\frac{5x^2}{10xy} + \frac{2y^2}{10xy} = \frac{5x^2 + 2y^2}{10xy}$
6. $\frac{3x^2}{xy} + \frac{2y^2}{xy} = \frac{3x^2 + 2y^2}{xy}$
7. $\frac{2xy^2}{15yz} + \frac{9xz^2}{15yz} = \frac{2xy^2 + 9xz^2}{15yz}$
8. $\frac{50a^2}{60a^3} + \frac{16}{60a^3} + \frac{15a}{60a^3} = \frac{50a^2 + 16 + 15a}{60a^3}$
9. $\frac{35a}{40} + \frac{-16a}{40} + \frac{6a}{40} = \frac{35a - 16a + 6a}{40} = \frac{25a}{40} = \frac{5a}{8}$
10. $\frac{3bc^2}{16ac} + \frac{-10a^2b}{16ac} = \frac{3bc^2 - 10a^2b}{16ac}$
11. $\frac{11a}{15} + \frac{-9a}{15} + \frac{10a}{15} = \frac{11a - 9a + 10a}{15} = \frac{12a}{15} = \frac{4a}{5}$
12. $\frac{10a^2c}{15b} + \frac{-4a^2c}{15b} + \frac{6a^2c}{15b} = \frac{10a^2c - 4a^2c + 6a^2c}{15b} = \frac{12a^2c}{15b} = \frac{4a^2c}{5b}$

Additional practice problems are at the end of the chapter.

SELF-TEST 7-4

Perform the indicated operations:

1. $\frac{8}{9x} - \frac{2}{9x}$
2. $\frac{2}{7a} + \frac{3}{7a}$
3. $\frac{2a}{9c} + \frac{4a}{9c} - \frac{a}{9c}$
4. $\frac{15a}{12} - \frac{a}{12}$
5. $\frac{3}{5x} + \frac{2}{7x^2} + \frac{3}{10x^3}$
6. $\frac{4a}{5b} + \frac{2}{3}$
7. $\frac{4}{15b} - \frac{5b}{18a^2c} + \frac{3}{10}$

Answers to Self-test 7-4 are at the end of the chapter.

CHAPTER 7 AT A GLANCE

Page	Key Point or Rule	Example
195	*Rule 7-1:* To find the lowest common multiple of numbers containing literal numbers: (1) factor each term; (2) determine the maximum number of times a prime factor appears in any one term; (3) multiply the resulting prime factors found in step 2. This number is the LCM.	Find the LCM of $10xy^2$ and $15x^3y$. $10xy^2 = 2 \cdot 5 \cdot x \cdot y \cdot y$ $15x^3y = 3 \cdot 5 \cdot x \cdot x \cdot x \cdot y$ $\text{LCM} = 3 \cdot 5 \cdot x^3 \cdot y^2$
197	*Key point:* In multiplication, all the literal numbers contained in the multiplicand and the multiplier must appear in the product unless the exponent is zero.	$3x^3y^{-2}z \times 5x^2y^2z^2$ $= 15x^2y^0z^3 = 15x^5z^3$
202	*Key point:* To multiply fractions, we (1) multiply the numerators, (2) multiply the denominators, and (3) cancel like factors to reduce to lowest terms.	$\frac{3a}{4b} \times \frac{6a^2b}{10} = \frac{3a \cdot 6a^2b}{4b \cdot 10}$ $= \frac{\not{2} \cdot 3 \cdot 3 \cdot a^3 \cdot \not{b}}{\not{2} \cdot 2 \cdot 2 \cdot 5 \cdot \not{b}} = \frac{9a^3}{20}$
207	*Rule 7-2:* To add, change fractions to fractions with a CD. Divide the CD by the old denominator and multiply this number by the old numerator to find the new numerator.	Add $\frac{3a}{2b}$ and $\frac{2b}{3a}$. Change to fractions with a CD; then add. $\frac{3a}{2b} = \frac{9a^2}{6ab}$ $\frac{2b}{3a} = \frac{4b^2}{6ab}$ $\frac{3a}{2b} + \frac{2b}{3a} = \frac{9a^2 + 4b^2}{6ab}$

End of Chapter Problems 7-1

Find the prime factors of the following:

1. $8a^2b^2$
2. $12x^2$
3. $24x^2y$
4. $32ab^2$
5. $42c^3$
6. $56yz^3$
7. $88y^2z$
8. $132a^3b$
9. $135a^2c^2$
10. $162ab^2$

End of Chapter Problems 7-2

Find the LCM of the following numbers:

1. $ab^3,\ a^3b$
2. $a^3b^3,\ a^4b$
3. $a^3bc^3,\ ab^3c$
4. $a^3b^2c,\ a^2b^2c^2$
5. $x^2y^3z^4,\ xy^2z^2$
6. $x^2yz^3,\ xy^4z^2$
7. $ab^2,\ a^3b^3,\ a^3b^2$
8. $a^4b^2,\ a^4bc^3,\ ab^3$
9. $a^2bc^3,\ a^4b^2c^2$
10. $ab^2c^2,\ a^3bc^2,\ a^2b^2c^2$
11. $a^4b^2c,\ a^4b^3c,\ abc^2$
12. $15x^2y^2,\ 27x^3y$
13. $26xy^3,\ 39x^2y^2$
14. $8x^4y^2z,\ 32xy$
15. $16x^2y^2z,\ 32xy,\ 8x^3z^2$
16. $64x^2y,\ 16x^3y^3,\ 128xy^2$
17. $30a^2b,\ 75ab^2,\ 50abc$
18. $15xz^2,\ 20x^2y^2,\ 24x^2y^2z^2$
19. $18x^2z^3,\ 27y^4z,\ 30x^2y^2z$
20. $20x^3z^2,\ 28y^2z,\ 35x^2yz^2$

End of Chapter Problems 7-3

Perform the indicated operations:

1. $3a^2b \cdot 7a^2b^2$
2. $4a^3b \cdot 4ab$
3. $5a^{-1}b^{-3} \cdot 4a^{-3}b^{-2}$
4. $6a^{-2}b^{-2} \cdot 5a^{-3}b^{-4}$
5. $10x^2yz \cdot x^3y^2z \cdot 2xz^2$
6. $8x^3y^2z \cdot x^2z^2 \cdot 3xyz$
7. $a^{-1}bc^2 \cdot 3a^2b^{-2}c^{-4} \cdot 2a^2b^{-1}c$
8. $x^{-2}y^{-1}z^2 \cdot 4x^2y^{-1}z \cdot 3x^4y^{-2}z^3$
9. $3x^2y^{-1}z \cdot 4x^4y^{-2}z \cdot 5x^{-1}y^{-1}z^2$
10. $4x^{-3}yz^2 \cdot 5x^{-3}y^{-3}z \cdot x^2y^2z^{-3}$

End of Chapter Problems 7-4

Reduce to lowest terms:

1. $\dfrac{18a^3}{2a}$
2. $\dfrac{27a^5}{3a^2}$
3. $\dfrac{48a^2b^3}{6ab}$
4. $\dfrac{54a^4b^5}{9a^3b^2}$
5. $\dfrac{25a^4b^2c}{5abc^2}$
6. $\dfrac{8ab^3c^2}{16a^3bc^3}$
7. $\dfrac{9x^{-2}y^2}{45x^{-3}y}$
8. $\dfrac{63x^3y^{-2}}{7x^2y^{-4}}$

9. $\dfrac{44x^{-3}yz^{-2}}{4x^2y^2z^2}$

10. $\dfrac{96x^2y^{-3}z}{12x^{-3}y^2z^{-2}}$

End of Chapter Problems 7-5

Perform the indicated multiplications. Reduce to lowest terms where possible.

1. $\dfrac{4b}{15} \times \dfrac{3b}{8}$

2. $\dfrac{6b}{7} \times \dfrac{14b}{3}$

3. $\dfrac{3}{2b} \times \dfrac{5}{7b}$

4. $\dfrac{5}{6b} \times \dfrac{7}{3b}$

5. $\dfrac{5a}{9} \times \dfrac{6a^2}{7}$

6. $\dfrac{7a^2}{8} \times \dfrac{6a^3}{7}$

7. $\dfrac{3x^2}{10y} \times \dfrac{5x}{8y}$

8. $\dfrac{10a}{21b^2} \times \dfrac{3a^2}{5b^2}$

9. $\dfrac{3}{4y^2} \times \dfrac{8xy}{9}$

10. $\dfrac{10x^2y}{12} \times \dfrac{5}{x}$

11. $\dfrac{8x^{-1}y}{5} \times \dfrac{10x^2y^{-3}}{3z^{-1}}$

12. $\dfrac{6xy^{-2}}{5} \times \dfrac{12x^{-3}y^3}{6z^{-2}}$

13. $\dfrac{11a^2}{32b^3} \times \dfrac{4a^2b}{33c}$

14. $\dfrac{7x^2y^{-2}}{8z^2} \times \dfrac{3z^3}{4x^2y^{-3}}$

15. $\dfrac{7x^2}{24y^2} \times \dfrac{4xy^{-2}}{21x}$

16. $\dfrac{9a^3}{16b^2} \times \dfrac{8ab^4}{3c}$

17. $\dfrac{4a^3c^{-2}}{5b^2} \times \dfrac{7b^{-1}c^3}{16a^2} \times \dfrac{20a^3b^{-1}}{21c}$

18. $\dfrac{3x^4}{5y^2} \times \dfrac{2y^3z^3}{9x^2} \times \dfrac{45yz}{3x^{-1}}$

19. $\dfrac{6xy^{-2}}{15z^2} \times \dfrac{25x^{-3}z^2}{18y} \times \dfrac{12y^2}{5x^{-1}z^{-2}}$

20. $\dfrac{5x^{-3}z^2}{12y^2} \times \dfrac{24yz^3}{35x^2} \times \dfrac{14y}{3x^{-1}z^{-1}}$

End of Chapter Problems 7-6

1. $\dfrac{3x}{8} \div \dfrac{7x^3}{3}$

2. $\dfrac{7x^3}{9} \div \dfrac{28x}{21}$

3. $\dfrac{7x}{16} \div \dfrac{21x^2}{32}$

4. $\dfrac{5a^2}{16} \div \dfrac{25a^4}{24}$

5. $\dfrac{16b}{3} \div \dfrac{8b^4}{27}$

6. $\dfrac{27a^4}{5} \div \dfrac{9a^2}{35}$

7. $\dfrac{14ab^2}{9c^3} \div \dfrac{35b}{c^2}$

8. $\dfrac{35a^3b^{-1}}{24c^2} \div \dfrac{7ab^2}{12c^3}$

9. $\dfrac{3xy^3}{10x^{-1}} \div \dfrac{3xy^3}{10x^{-1}}$

10. $\dfrac{10a}{21b^2c^3} \div \dfrac{20a^{-2}}{28b^{-1}c^2}$

11. $\dfrac{32a}{45b^2} \div \dfrac{16a^2}{25bc}$

12. $\dfrac{12xy^3}{21z} \div \dfrac{32x^3y}{35z^4}$

13. $\dfrac{15a^4b}{64c^3} \div \dfrac{75ab}{32c^2}$

14. $\dfrac{16a^5c^2}{35b^3} \div \dfrac{18a^2c}{25}$

15. $\dfrac{20x^3y^2}{21z^2} \div \dfrac{30xy}{7z^2}$

16. $\dfrac{18x^2y^3}{35z^3} \div \dfrac{24xy^2}{7z^3}$

17. $\dfrac{21a^3b^{-2}}{4c^2} \div \dfrac{14a^{-1}b^2}{3c^{-4}}$

18. $\dfrac{16a^4b^{-1}}{15c^3} \div \dfrac{24a^{-2}b^2}{21c}$

19. $\dfrac{42yz^{-2}}{55x} \div \dfrac{21x^2y^2}{22z}$

20. $\dfrac{56y^2z^{-3}}{25x^2} \div \dfrac{21x^3y^2}{20z^{-1}}$

End of Chapter Problems 7-7

Perform the following operations. Reduce to lowest terms.

1. $\dfrac{2}{7x} + \dfrac{3}{7x}$

2. $\dfrac{2}{9a} + \dfrac{4}{9a}$

3. $\dfrac{6}{7x} - \dfrac{2}{7x}$

4. $\dfrac{5}{3x} - \dfrac{4}{3x}$

5. $\dfrac{1}{12ab} + \dfrac{5}{12ab} + \dfrac{3}{12ab}$

6. $\dfrac{1}{14ab} + \dfrac{3}{14ab} + \dfrac{5}{14ab}$

7. $\dfrac{3}{14a} + \dfrac{9}{14a} - \dfrac{5}{14a}$

8. $\dfrac{9}{16x} - \dfrac{5}{16x} + \dfrac{3}{16x}$

9. $\dfrac{5x}{24} + \dfrac{7x}{24}$

10. $\dfrac{3a}{20} + \dfrac{5a}{20}$

11. $\dfrac{8x}{15} - \dfrac{2x}{15}$

12. $\dfrac{7ab}{9} - \dfrac{4ab}{9}$

13. $\dfrac{7ab}{25} + \dfrac{6ab}{25} + \dfrac{2ab}{25}$

14. $\dfrac{2xy}{15} + \dfrac{4xy}{15} + \dfrac{6xy}{15}$

15. $\dfrac{9ab}{25} - \dfrac{2ab}{25} + \dfrac{8ab}{25}$

16. $\dfrac{3xy}{20} + \dfrac{7xy}{20} - \dfrac{9xy}{20}$

17. $\dfrac{2x}{9} + \dfrac{4x}{9} + \dfrac{4x}{9}$

18. $\dfrac{3ab}{14} + \dfrac{3ab}{14} + \dfrac{5ab}{14}$

19. $\dfrac{3x}{7} - \dfrac{5x}{7}$

20. $\dfrac{7}{11a} - \dfrac{9}{11a}$

21. $\dfrac{9x}{24} - \dfrac{15x}{24} + \dfrac{x}{24}$

22. $\dfrac{6a}{15} - \dfrac{7a}{15} - \dfrac{2a}{15}$

23. $\dfrac{4ab}{8} + \dfrac{3b}{8} - \dfrac{2ab}{8}$

24. $\dfrac{6ab}{4} + \dfrac{2a}{4} - \dfrac{3ab}{4}$

25. $\dfrac{6abc}{4d} - \dfrac{3ac}{4d} + \dfrac{abc}{4d}$

26. $\dfrac{14xyz}{6b} - \dfrac{6xz}{6b} + \dfrac{xyz}{6b}$

27. $\dfrac{3abc}{4x} - \dfrac{5abc}{4x} - \dfrac{3ab}{4x}$

28. $\dfrac{4abc}{9} - \dfrac{7abc}{9} - \dfrac{2ab}{9}$

29. $\frac{6xy}{5} - \frac{4xz}{5} + \frac{3yz}{5}$

30. $\frac{6xz}{7} - \frac{5xy}{7} - \frac{6yz}{7}$

End of Chapter Problems 7-8

Perform the indicated operations. Reduce to lowest terms where possible.

1. $\frac{3x}{5} + \frac{2x}{15}$
2. $\frac{3x}{8} + \frac{3x}{16}$
3. $\frac{2a}{3} + \frac{3a}{2}$
4. $\frac{b}{3} + \frac{2b}{5}$
5. $\frac{x}{2} - \frac{2x}{3}$
6. $\frac{2x}{3} - \frac{3x}{7}$
7. $\frac{4x}{5} - \frac{3x}{10}$
8. $\frac{5a}{12} - \frac{7a}{8}$
9. $\frac{y}{2x} + \frac{5y}{8}$
10. $\frac{2b}{3a} + \frac{3a}{2b}$
11. $\frac{3x}{4y} + \frac{2y}{3x}$
12. $\frac{13ab}{16c} - \frac{ac}{4b}$
13. $\frac{2b}{5x} + \frac{3}{10y}$
14. $\frac{a}{4b} + \frac{3b}{8c}$
15. $\frac{5ab^2}{16c} - \frac{15ab}{32c^2}$
16. $\frac{5x^2z}{8y^2} - \frac{3z^2}{4y}$
17. $\frac{3b}{4} + \frac{3b}{8} - \frac{15b}{16}$
18. $\frac{3a}{5} + \frac{3a}{8} - \frac{3a}{4}$
19. $\frac{2a}{7} + \frac{7a}{15} - \frac{a}{3}$
20. $\frac{5x}{12} - \frac{3x}{8} + \frac{5x}{16}$
21. $\frac{x}{10y^2} - \frac{2x}{5y^2} + \frac{x}{3y^2}$
22. $\frac{2x}{5y^2} - \frac{3x}{10y^2} + \frac{7x}{20y^2}$
23. $\frac{4x}{3yz} + \frac{2y}{5x} - \frac{z}{6y}$
24. $\frac{7y}{9xz} - \frac{5x}{8y} + \frac{z}{36x}$
25. $\frac{4a}{33b} + \frac{5b}{55c} - \frac{3c}{11a}$
26. $\frac{3b}{8c} - \frac{4b}{15c} + \frac{5b}{24a}$
27. $\frac{5y^2}{6x^2y} + \frac{4xy}{15y^2} - \frac{2x^2}{9y}$
28. $\frac{5}{12x^2y} - \frac{z}{6xy^2} + \frac{2y}{15x}$
29. $\frac{3b}{20a^2} + \frac{7c}{10ab} - \frac{5a}{12b^3}$
30. $\frac{4a}{7b^2} + \frac{2c}{9ab^2} - \frac{4ab^2}{21}$

Answers To Self-tests

SELF-TEST 7-1

1. $2^2 \cdot 7 \cdot x \cdot y^2$
2. $2 \cdot 3^3 \cdot a^3 \cdot b^2$

3. $2^2 \cdot 19 \cdot x \cdot y^3$
4. $2^3 \cdot 3 \cdot 5 \cdot x^2 \cdot z^2$
5. $2 \cdot 5 \cdot 13 \cdot a^3 \cdot b^2$
6. $3^2 \cdot 5 \cdot 7 \cdot a^2 \cdot b^2$
7. a^2b^3
8. x^3yz^3
9. $72xz^2$
10. $70a^2b^3$
11. $72a^2b^2c^2$
12. $128a^3b^2c^2$

SELF-TEST 7-2

1. $15xy$
2. $12x^4y^4$
3. $30a^2b$
4. $30a^{-1}b^{-2}$
5. $42a^{-4}b^{-1}c^3$
6. $8x^2y^{-1}$
7. $8x^{-3}y^{-1}$
8. $7a^2c^{-3}$
9. $9a^{-4}b^2c^{-1}$
10. $16x^2y^{-2}z^{-4}$

SELF-TEST 7-3

1. $\dfrac{6x^4}{35}$
2. $\dfrac{2a^{-1}b^3c^{-2}}{5}$
3. $\dfrac{7a^{-2}b^5c^{-3}}{6}$
4. $\dfrac{10a^{-1}}{9}$
5. $2a^5b^{-2}c^4$
6. $\dfrac{4a^{-3}b^{-1}c}{3}$
7. $y^{-2}z^{-2}$

SELF-TEST 7-4

1. $\dfrac{2}{3x}$
2. $\dfrac{5}{7a}$
3. $\dfrac{5a}{9c}$
4. $\dfrac{7a}{6}$
5. $\dfrac{42x^2 + 20x + 21}{70x^3}$
6. $\dfrac{12a + 10b}{15b}$
7. $\dfrac{24a^2c - 25b^2 + 27a^2bc}{90a^2bc}$

CHAPTER 8 LINEAR EQUATIONS

INTRODUCTION

When algebraic terms or expressions are separated by an equals sign, the resulting algebraic statement is called an *equation*. In simplest equations, the literal numbers are raised to the first power. These equations are called *linear* equations. If the literal numbers are raised to the second power, we have *second-degree* equations. We will be solving both linear and second-degree equations in this chapter.

All linear equations can be solved provided there is only one unknown—one literal number. Ohm's law equations may be either linear or second-degree equations, depending on the variables used. $E = IR$ is a linear equation. If two of the variables are known, we can solve for the third. $P = I^2R$ is a second-degree equation because I is raised to the second power. These equations can be solved, again provided that there is only one unknown variable.

The use of equations such as these are part of any study of electronics. The solution of any electronics problem will involve the use of some sort of equation. It may be a simple linear equation or a second-degree equation, like the ones we will study in this chapter, or it might be a more complex equation of the type we will study in future chapters. Any course of study in problem solving must include a thorough understanding of how to solve linear equations and second-degree equations. In this chapter we will also show how the calculator is used in solving equations.

CHAPTER OBJECTIVES

In this chapter, you will learn how to:

1. Recognize and identify the parts of an equation.

2. Solve various kinds of linear equations containing one unknown variable.

3. Solve various second-degree equations containing one unknown variable.
4. Use the calculator to solve various kinds of electrical problems that involve the use of linear or second-degree equations.
5. Apply the rules that have been developed to help in solving equations.
6. Verify the solutions of equations that have been solved.

8-1 IDENTIFYING EQUATIONS

Algebraic expressions separated by an equals sign (=) are called equations. The equals sign implies that the expressions are equal. Let's call the expression to the left of the equals sign the *left side* and the expression to the right of the equals sign the *right side*. All the following algebraic statements are equations.

☐ EXAMPLE 8-1

(a) $16 + 4 = 20$ **(b)** $16 + x = 20$ **(c)** $x + 4 = 20$
(d) $x^3 + 4 = 20$ **(e)** $x^2 + x = 20$ **(f)** $x^2 + 4 = 20$
(g) $4x = 20$ **(h)** $\frac{x}{4} = 20$ **(i)** $x - 4 = 20$

☐

In Example 8.1(a) all numbers are real numbers. Both sides have a value of 20. In all the other examples, a part of the left side is unknown. The value of the unknown (x) must be of such a value as to make the left side equal to 20. Sometimes the value of x can be found by inspection. This should be the case in Example 8-1(b) and (c). In (b), $x = 4$:

$$16 + x = 20$$
$$16 + 4 = 20$$
$$20 = 20$$

and in (c), x must equal 16.

$$x + 4 = 20$$
$$16 + 4 = 20$$
$$20 = 20$$

In other examples we must work a little harder. Notice in Examples 8-1(d) and (e) that we have x^3 and x^2 terms. In all other equations, x is raised to the first power. Equations in which the unknown is raised to the first power are called *linear* equations. Equations containing x^2 terms are called *second-degree* equations. Equations containing x^3 terms are called *third-degree* equations.

8-2 LINEAR EQUATIONS

Here is the basic rule to use when solving equations.

RULE 8-1 We can perform any mathematical operation on one side of an equation, *provided we perform the same operation on the other side*.

This rule means that we can make any change we want in the value of one side as long as we make the same change in the other side. We can add, subtract, multiply, divide, square, and so on. We can perform any of these operations as long as we maintain the equality of sides.

In solving equations, we typically solve for some unknown value. In simple linear equations, there is only one unknown and only one value for that unknown.

In this and following chapters, we will use a five-step method of solution for both linear and second-degree equations.

RULE 8-2 To solve both linear and second-degree equations:

Step 1. Clear all fractions (if any).

Step 2. Move all terms containing the unknown to one side and combine them into one term. It is usually preferable to move unknowns to the left side.

Step 3. Move all constants to the other side.

Step 4. Perform the operation or operations necessary to make the coefficient of the unknown equal to 1 (don't forget to perform this operation on *both* sides).

Step 5. Substitute the value of the unknown back into the original equation to verify solution.

These five steps do not have to be done in the order stated, except that step 5 would always be the last step. Also, not all equations would require all five steps.

Consider Example 8-1(c). We know from inspection that $x = 16$, but let's work it mathematically. In solving equations we must somehow get the unknown on one side by itself. That is, we have to move all terms that are real numbers to one side and leave the term containing the unknown on the other side. It is standard practice to put the term containing the unknown on the left side. We would solve the equation in the following manner.

☐ EXAMPLE 8-1(c)

$$x + 4 = 20.$$

Solution We want the unknown variable, x, on the left side by itself. This can only be done by subtracting 4. $x + 4 - 4 = x$. But if we subtract 4 from the left side, we must subtract 4 from the right side to maintain the equality of sides.

$$x + 4 - 4 = 20 - 4$$

$$x = 16$$

Then we replace the unknown with its value and check for equality:

$$x + 4 = 20$$

$$16 + 4 = 20$$

$$20 = 20$$

□

When we want to move a term from one side to the other, we do it by adding or subtracting. In the preceding example, we subtracted. In Example 8-1(i) we will add.

□ EXAMPLE 8-1(i)

$$x - 4 = 20.$$

Solution

$$x - 4 + 4 = 20 + 4 \qquad \text{(add 4 to both sides)}$$

$$x = 24$$

Check

$$24 - 4 = 20$$

$$20 = 20$$

□

From these examples we can make up the following rule:

■ **RULE 8-3** Terms may be moved from one side of an equation to the other side by changing their signs.

Notice in Example 8-1(c) that 4 was positive when it was on the left side, and it became negative when it was moved to the right side. In Example 8-1(i), 4 was negative when it was on the left side, and it became positive when it was moved to the right side.

$$x + 4 = 20$$

$$x = 20 - 4 \qquad \text{(move 4 to right side and change its sign)}$$

$$x = 16$$

$$x - 4 = 20$$

$$x = 20 + 4 \qquad \text{(move } -4 \text{ to right side and change its sign)}$$

$$x = 24$$

☐ EXAMPLE 8-1(g)

$$4x = 20.$$

Solution

$$4x = 20$$

$$\frac{4x}{4} = \frac{20}{4} \quad \text{(divide } \textit{both sides} \text{ by 4)}$$

$$x = 5$$

Check

$$4(5) = 20$$

$$20 = 20$$

In this example, the left-side term is $4x$. Since we have to find the value of x, we must somehow get rid of the 4. That is, we want the left side to equal x. We can make the left side equal to x by dividing by 4.

☐ EXAMPLE 8-1(h)

$$\frac{x}{4} = 20$$

Solution How can we make the left side equal to x? Multiplication by 4 is the only way it can be done.

$$\frac{x}{4} \times 4 = \frac{4x}{4} = x$$

Solving the equation for x, we get

$$\frac{x}{4} = 20$$

$$\frac{x}{4}(4) = 20(4)$$

$$x = 80$$

Check

$$\frac{80}{4} = 20$$

$$20 = 20$$

☐

These two examples (g and h) show how we can move parts of terms from one side to another by multiplying or dividing.

A shortcut method of multiplying or dividing terms in order to simplify the terms is called *cross-multiplication*. When we cross-multiply, we multiply the numerator of one term and the denominator of the other term. Consider Example 8-1(h) again. (We have shown the 1 in the denominator of the right side just to help you see the process.)

$$\frac{x}{4} = \frac{20}{1} \qquad \text{If we cross-multiply, we get}$$

$$x \cdot 1 = 20 \cdot 4$$

$$x = 80$$

Cross-multiplication in this example is really the multiplication of both sides by 4 and then by 1. Often the process can be done by inspection. It's a quick way of getting rid of a fraction when it is part of the equation.

☐ EXAMPLE 8-2

$$\frac{20}{x} = 5. \text{ Find } x.$$

Solution We must get x out of the denominator. There are two ways we can do this. The first way is to take the reciprocal of both sides:

$$\frac{20}{x} = 5$$

$$\frac{x}{20} = \frac{1}{5}$$

Next we multiply both sides by 20. This is the only way we can move 20 to the right side.

$$\frac{x}{20}(20) = \frac{1}{5}(20)$$

$$\frac{20x}{20} = \frac{20}{5}$$

$$x = 4$$

Check

$$\frac{20}{4} = 5$$

$$5 = 5$$

The other way to solve the problem is to multiply both sides by x.

$$\frac{20}{x}(x) = 5x$$

$$\frac{20x}{x} = 5x$$

$$20 = 5x$$

Next we divide both sides by 5.

$$\frac{20}{5} = \frac{5x}{5}$$

$$4 = x$$

or

$$x = 4$$

Of course, we could have cross-multiplied here also, but we wanted to go through the actual mathematical operations.

□ EXAMPLE 8-3

$$\frac{x}{3} = -5. \text{ Find } x.$$

Solution

$$\frac{x}{3}(3) = -5(3) \qquad \text{(multiply both sides by 3)}$$

$$x = -15$$

Check

$$\frac{-15}{3} = -5$$

$$-5 = -5$$

□ EXAMPLE 8-4

$$\frac{24}{-x} = -2. \text{ Find } x.$$

Solution

$$\frac{-x}{24} = -\frac{1}{2} \qquad \text{(take the reciprocal of both sides)}$$

$$\frac{-x}{24}(24) = -\frac{1}{2}(24)$$

$$-x = -12$$

$$x = 12 \qquad \text{(change the sign of both sides)}$$

Check

$$\frac{24}{-12} = -2$$

$$-2 = -2$$

□ **EXAMPLE 8-5**

$$\frac{2x}{5} = \frac{3}{4} \qquad \text{(Find } x\text{)}$$

Solution

$$2x \cdot 4 = 3 \cdot 5 \qquad \text{(cross-multiply)}$$

$$8x = 15$$

$$x = \frac{15}{8} = 1\frac{7}{8} \qquad \text{(divide both sides by 8 and reduce to lowest terms)}$$

□

PRACTICE PROBLEMS 8-1

Solve for the unknown in the following equations. Check your answers by plugging the answer back into the original problem.

1. $x - 3 = 14$
2. $a + 5 = 6$
3. $4 - a = 6$
4. $5 + y = 6$
5. $3x = 12$
6. $5a = 30$
7. $6a = -42$
8. $-3a = 18$
9. $\frac{x}{3} = 4$
10. $\frac{a}{7} = 3$
11. $\frac{a}{5} = -3$
12. $\frac{-x}{6} = 7$
13. $\frac{24}{x} = 6$
14. $\frac{28}{a} = 7$
15. $\frac{-32}{b} = 4$
16. $\frac{-48}{y} = -6$
17. $\frac{2x}{3} = 4$
18. $\frac{4}{3x} = 8$
19. $5 + 6x = 23$
20. $3a + 4 = 25$
21. $\frac{7x}{4} = \frac{5}{8}$
22. $\frac{3}{7x} = \frac{3}{11}$

Solutions

1. $x = 17$
2. $a = 1$
3. $a = -2$
4. $y = 1$
5. $x = 4$
6. $a = 6$
7. $a = -7$
8. $a = -6$
9. $x = 12$
10. $a = 21$
11. $a = -15$
12. $x = -42$

13. $x = 4$

14. $a = 4$

15. $b = -8$

16. $y = 8$

17. $x = 6$

18. $x = \frac{1}{6}$

19. $x = 3$

20. $a = 7$

21. $x = \frac{5}{14}$

22. $x = 1\frac{4}{7}$

Additional practice problems are at the end of the chapter.

8-3 SECOND-DEGREE EQUATIONS

Up to this point we have solved equations by means of addition, subtraction, multiplication, or division. Some equations we use in electronics require additional operations. Let's consider some general case equations.

☐ EXAMPLE 8-6

$x^2 = 81$. Find x.

Solution Whenever a number is squared, we can find that number by finding its square root. $\sqrt{6^2} = 6$, $\sqrt{7^2} = 7$, and $\sqrt{x^2} = x$. If we take the square root of the left side, we must also take the square root of the right side.

$$x^2 = 81$$
$$\sqrt{x^2} = \sqrt{81}$$
$$x = 9$$

As in previous chapters, we are not considering the negative root (−9 in this example).

☐

The following mathematical manipulations are very useful in problem solving:

KEY POINT The square root of the square of a number is that number.

We can turn this rule around and say:

KEY POINT The square of the square root of a number is that number.

An example of how to apply this rule follows.

□ EXAMPLE 8-7

$$\sqrt{x} = 5. \text{ Find } x.$$

Solution We need x where we have $\sqrt{x}$. We must square the term to get rid of the square root.

$$(\sqrt{x})^2 = x \quad \text{or} \quad (x^{1/2})^2 = x^1 = x$$

We must also square the right side.

$$\sqrt{x} = 5$$
$$(\sqrt{x})^2 = 5^2$$
$$x = 25$$

Check

$$\sqrt{25} = 5, \qquad 5 = 5$$

□

Let's look at some more examples of problems involving square and square roots.

□ EXAMPLE 8-8

$$\frac{x^2}{4} = 6. \text{ Find } x.$$

Solution

$$\frac{x^2}{4}(4) = 6(4) \qquad \text{(multiply both sides by 4)}$$

$$x^2 = 24$$

$$\sqrt{x^2} = \sqrt{24} \qquad \text{(take the square root of both sides)}$$

$$x = 4.90$$

As we get into more complex problems, we need to use the calculator when we verify our answers.

Check

$$\frac{4.90^2}{4} = 6$$

4.90 [x^2] [÷] 4 [=]. The answer is 6.0025. The answer is not *exactly* 6 because x did not equal exactly 4.90. We rounded to 4.90. However, if you had stored $\sqrt{24}$ when you found it,

$\sqrt{24}$ = 24 [$\sqrt{x}$] [STO]

and if you recall the stored value when you check it,

[RCL] [x^2] [÷] 4 [=]

you will get *exactly* 6.0000. So, get into the habit of storing your answers to these equations to make it easier and quicker to verify your answers.

□ EXAMPLE 8-9

$$\frac{\sqrt{y}}{2} = 4. \text{ Find } y.$$

Solution

$$\frac{\sqrt{y}}{2}(2) = 4(2) \qquad \text{(multiply both sides by 2)}$$

$$\sqrt{y} = 8$$

$$y = 64 \qquad \text{(square both sides)}$$

Check

$$\frac{\sqrt{64}}{2} = 4$$

$$\frac{8}{2} = 4$$

$$4 = 4$$

□ EXAMPLE 8-10

$$15 = 3xy. \text{ Find } x.$$

Solution

$$\frac{15}{3} = \frac{\cancel{3}xy}{\cancel{3}} \qquad \text{(divide by 3)}$$

$$5 = xy$$

$$\frac{5}{y} = \frac{x\cancel{y}}{\cancel{y}} \qquad \text{(divide by } y\text{)}$$

$$\frac{5}{y} = x$$

or

$$x = \frac{5}{y}$$

Check

$$15 = 3\left(\frac{5}{y}\right)y$$

$$15 = \frac{3 \cdot 5 \cdot \cancel{y}}{\cancel{y}}$$

$$15 = 15$$

□ EXAMPLE 8-11

$$4 = \frac{1}{3a\sqrt{b}}. \text{ Find } a \text{ and then } b.$$

Solution

$$4(a) = \frac{1}{3a\sqrt{b}}(a) \qquad \text{(multiply by } a\text{)}$$

$$4a = \frac{1}{3\sqrt{b}}$$

$$\frac{4a}{4} = \frac{1}{3\sqrt{b}}\left(\frac{1}{4}\right) \qquad \text{(divide by 4)}$$

$$a = \frac{1}{12\sqrt{b}}$$

Check

$$4 = \frac{1}{3\left(\frac{1}{12\sqrt{b}}\right)(\sqrt{b})} = \frac{1}{3\left(\frac{1}{12}\right)} = \frac{1}{\frac{3}{12}} = \frac{12}{3} = 4$$

$$4 = \frac{1}{3a\sqrt{b}}. \text{ Find } b.$$

$$4(\sqrt{b}) = \frac{1}{3a\sqrt{b}}(\sqrt{b}) \qquad \text{(multiply by } \sqrt{b}\text{)}$$

$$4\sqrt{b} = \frac{1}{3a}$$

$$\frac{4\sqrt{b}}{4} = \frac{1}{3a}\left(\frac{1}{4}\right) \qquad \text{(divide by 4)}$$

$$\sqrt{b} = \frac{1}{12a}$$

$$(\sqrt{b})^2 = \left(\frac{1}{12a}\right)^2 \qquad \text{(square both sides)}$$

$$b = \frac{1}{144a^2}$$

or

$$b = \frac{a^{-2}}{144}$$

Check

$$4 = \frac{1}{3a\sqrt{\frac{1}{144a^2}}}$$

$$4 = \frac{1}{3a\left(\frac{1}{12}a\right)} = \frac{1}{\frac{3a}{12a}} = \frac{12}{3} = 4$$

□

We can't cover all the kinds of equations that might be encountered, but the preceding examples should provide enough problem-solving situations so that any similar equations can be solved.

PRACTICE PROBLEMS 8-2

Solve for the unknown in the following equations. Express real-number answers accurate to *four* places:

1. $x^2 = 36$

2. $\frac{x^2}{5} = 7$

3. $25 = 5x^2$

4. $3x^2 = 48$

5. $\frac{a^2}{3} = 15$

6. $\frac{3x^2}{4} = 60$

7. $\frac{14}{5b^2} = 6$

8. $\frac{21}{5} = 4x^2$

9. $\sqrt{x} = 6$

10. $\frac{\sqrt{b}}{6} = 2$

11. $4 = \frac{1}{3\sqrt{x}}$

12. $\sqrt{a} = 9$

13. $\frac{\sqrt{b}}{5} = 5$

14. $4 = \frac{1}{2\sqrt{x}}$

15. $5 = \frac{5}{2\sqrt{x}}$

16. $3\sqrt{b} = \frac{3}{5}$

17. $6 = 3ab$

18. $2a = 3b$

19. $4\ \text{k}\Omega = R_1 + R_2$

20. $I^2 = \frac{P}{R}$

21. $X_L = 2\pi fL$ (Do not solve for π. π is a constant, not a variable. π always equals 3.142, to four places.)

22. $f_r = \frac{1}{2\pi\sqrt{LC}}$

23. $4 = \frac{1}{3a\sqrt{b}}$

24. $6 = \frac{2}{3\sqrt{ab}}$

25. $X_C = \frac{1}{2\pi fC}$

Solutions

1. $x = 6.000$

2. $x = 5.916$

3. $x = 2.236$

4. $x = 4.000$

5. $a = 6.708$

6. $x = 8.944$

7. $b = 0.6831$

8. $x = 1.025$

9. $x = 36.00$

10. $b = 144.0$

11. $x = 0.006944$

12. $a = 81.00$

13. $b = 625.0$

14. $x = 0.01563$

15. $x = 0.2500$

16. $b = 0.04000$

17. $a = \frac{2}{b}, b = \frac{2}{a}$

18. $a = \frac{3b}{2}, b = \frac{2a}{3}$

19. $R_1 = 4\ \text{k}\Omega - R_2, R_2 = 4\ \text{k}\Omega - R_1$

20. $I = \sqrt{\frac{P}{R}}, R = \frac{P}{I^2}, P = I^2R$

21. $f = \frac{X_L}{2\pi L}, L = \frac{X_L}{2\pi f}$

22. $L = \frac{1}{(2\pi f_r)^2 C}, C = \frac{1}{(2\pi f_r)^2 L}$

23. $a = \frac{1}{12\sqrt{b}}, b = \frac{1}{144a^2}$

24. $a = \frac{1}{81b}, b = \frac{1}{81a}$

25. $f = \frac{1}{2\pi C X_C}, C = \frac{1}{2\pi f X_C}$

SELF-TEST 8-1

Solve for the unknown in the following equations:

1. $y - 6 = 4$

2. $4x = 20$

3. $\frac{-a}{4} = 2$

4. $\frac{5}{2y} = 2$

5. $a^2 = 49$

6. $\frac{a^2}{3} = 3$

7. $\sqrt{a} = 5$

8. $\frac{\sqrt{y}}{5} = 6$

9. $2 = \frac{1}{4\sqrt{x}}$

Solve for each literal number in the following equations:

10. $3xy = 7$

11. $R = \frac{E^2}{P}$

12. $3 = \frac{1}{4x\sqrt{y}}$

Answers to Self-test 8-1 are at the end of the chapter.

8-4 APPLICATIONS

Often, in electronics problem solving, we are solving an equation—solving for the unknown. In this section we will change equations so that the unknown is on one side and all known values are on the other. The following are examples of problems that have to be solved when working with electrical circuits.

☐ EXAMPLE 8-12

$E = IR$. Find I when $E = 12$ V and $R = 12$ kΩ.

Solution First solve the equation for I.

$$E = IR$$

$$\frac{E}{R} = \frac{IR}{R} \qquad \text{(divide by } R\text{)}$$

$$\frac{E}{R} = I$$

$$I = \frac{E}{R} \qquad \text{(change sides so that the unknown is on the left side)}$$

Now plug in the values of E and R and solve.

$$I = \frac{12\text{ V}}{12\text{ k}\Omega} = 1\text{ mA}$$

☐ EXAMPLE 8-13

$P = \frac{E^2}{R}$. Find E when $P = 300$ mW and $R = 1.5$ kΩ.

Solution Solve for E.

$$PR = \frac{E^2}{R}(R) \qquad \text{(multiply by } R\text{)}$$

$$PR = E^2$$

$$\sqrt{PR} = \sqrt{E^2} \qquad \text{(take the square root of both sides)}$$

$$\sqrt{PR} = E$$

$$E = \sqrt{PR}$$

Plug in the known values.

$$E = \sqrt{300\text{ mW} \times 1.5\text{ k}\Omega} = \sqrt{450}$$

$$E = 21.2\text{ V}$$

☐ EXAMPLE 8-14

$\frac{N_P}{N_S} = \sqrt{\frac{Z_P}{Z_S}}$. Find Z_P when $Z_S = 8\ \Omega$, $N_P = 500$, and $N_S = 25$.

Solution Solve for Z_p.

$$\left(\frac{N_P}{N_S}\right)^2 = \sqrt{\left(\frac{Z_P}{Z_S}\right)}^2 \qquad \text{(square both sides)}$$

$$\left(\frac{N_P}{N_S}\right)^2 = \frac{Z_P}{Z_S}$$

$$\left(\frac{N_P}{N_S}\right)^2 \times Z_S = \frac{Z_P}{Z_S}(Z_S) \qquad \text{(multiply by } Z_S\text{)}$$

$$\left(\frac{N_P}{N_S}\right)^2 Z_S = Z_P$$

$$Z_P = \left(\frac{N_P}{N_S}\right)^2 Z_S$$

Plug in the known values.

$$Z_P = \left(\frac{500}{25}\right)^2 \times 8\ \Omega$$

$$= 3200\ \Omega = 3.2\ \text{k}\Omega$$

Often the student has a choice when attempting a solution using the calculator. This is a good example of two different ways to go:

[(] 500 [÷] 25 [)] [x^2] [×] 8 [=]

yields 3.2 kΩ, as does

500 [÷] 25 [=] [x^2] [×] 8 [=]

☐ EXAMPLE 8-15

$f_r = \frac{1}{2\pi\sqrt{LC}}$. Find L if $C = 400$ pF and $f_r = 15$ kHz.

Solution Solve for L.

$$f_r\sqrt{LC} = \frac{1}{2\pi} \qquad \text{(multiply by } \sqrt{LC}\text{)}$$

$$\sqrt{LC} = \frac{1}{2\pi f_r} \qquad \text{(divide by } f_r\text{)}$$

$$LC = \left(\frac{1}{2\pi f_r}\right)^2 \qquad \text{(square both sides)}$$

$$LC = \frac{1}{(2\pi f_r)^2} \qquad (1^2 = 1)$$

$$L = \frac{1}{(2\pi f_r)^2 C} \qquad \text{(divide by } C\text{)}$$

Plug in the known values.

$$L = \frac{1}{(2\pi \times 15 \text{ kHz})^2 \times 400 \text{ pF}} = \frac{1}{3.55}$$

$$= 2.81 \times 10^{-1} \text{ H} = 281 \text{ mH}$$

[(] 2 [×] [π] [×] 15 [EE] 3 [)] [x^2] [×] 400 [EE] [+/-] 12 [=] [1/x]

Answer: 2.81×10^{-1}

□

PRACTICE PROBLEMS 8-3

1. $R_T = R_1 + R_2$. Find R_2 when $R_T = 678\ \Omega$ and $R_1 = 238\ \Omega$.
2. $G_T = G_1 + G_2$. Find G_2 when $G_T = 600\ \mu S$ and $G_1 = 170\ \mu S$.
3. $A_V = \frac{V_o}{V_{in}}$. Find V_o if $A_V = 48$ and $V_{in} = 25$ mV.
4. $A_V = \frac{V_o}{V_{in}}$. Find V_{in} if $A_V = 37$ and $V_o = 3.83$ V.
5. $P = \frac{E^2}{R}$. Find R if $E = 30$ V and $P = 780$ mW.
6. $P = \frac{E^2}{R}$. Find E if $P = 720$ mW and $R = 1.2$ kΩ.
7. $P = I^2R$. Find I if $P = 170$ mW and $R = 1.2$ kΩ.
8. $P = I^2R$. Find R if $P = 1.2$ W and $I = 500$ mA.
9. $\frac{N_P^2}{N_S^2} = \frac{Z_P}{Z_S}$. Find N_S if $N_P = 200$, $Z_P = 2$ kΩ, and $Z_S = 16\ \Omega$.
10. $\frac{N_P^2}{N_S^2} = \frac{Z_P}{Z_S}$. Find N_P if $N_S = 50$, $Z_P = 800\ \Omega$, and $Z_S = 25\ \Omega$.
11. $X_C = \frac{1}{2\pi fC}$. Find C if $f = 75$ Hz and $X_C = 13.7$ kΩ.
12. $X_C = \frac{1}{2\pi fC}$. Find f if $C = 20$ nF and $X_C = 3.7$ kΩ.
13. $X_L = 2\pi fL$. Find L if $f = 10$ kHz and $X_L = 2.73$ kΩ
14. $X_L = 2\pi fL$. Find f if $L = 2.3$ H and $X_L = 4.55$ kΩ.
15. $B_C = 2\pi fC$. Find C if $f = 2.7$ kHz and $B_C = 200\ \mu S$.
16. $B_C = 2\pi fC$. Find f if $C = 400$ nF and $B_C = 4.35$ mS.
17. $f_r = \frac{1}{2\pi\sqrt{LC}}$. Find L if $f_r = 7.85$ kHz and $C = 20$ nF.
18. $f_r = \frac{1}{2\pi\sqrt{LC}}$. Find C if $f_r = 12$ kHz and $L = 400$ mH.

Solutions

1. $R_2 = 440\ \Omega$
2. $G_2 = 430\ \mu S$
3. $V_o = 1.2$ V
4. $V_{in} = 0.104$ V
5. $R = 1.15$ kΩ
6. $E = 29.4$ V

7. $I = 11.9$ mA
8. $R = 4.80\ \Omega$
9. $N_S = 17.9$
10. $N_P = 283$
11. $C = 155$ nF
12. $f = 2.15$ kHz
13. $L = 43.4$ mH
14. $f = 315$ Hz
15. $C = 11.8$ nF
16. $f = 1.73$ kHz
17. $L = 20.6$ mH
18. $C = 440$ pF

SELF-TEST 8-2

Solve for the unknown in the following problems. Express answers accurate to three places.

1. $P = \dfrac{E^2}{R}$. Find E if $P = 12.6$ W and $R = 470\ \Omega$.
2. $P = I^2R$. Find I if $P = 780$ mW and $R = 18$ kΩ.
3. $\dfrac{N_P^2}{N_S^2} = \dfrac{Z_P}{Z_S}$. Find N_P if $N_S = 2000$, $Z_P = 50\ \Omega$, and $Z_S = 4$ kΩ.
4. $X_L = 2\pi fL$. Find f if $L = 600$ mH and $X_L = 5.35$ kΩ.
5. $f_r = \dfrac{1}{2\pi\sqrt{LC}}$. Find L if $f_r = 10$ kHz and $C = 200$ pF.

Answers to Self-test 8-2 are at the end of the chapter.

CHAPTER 8 AT A GLANCE

Page	Key Point or Rule	Example
218	*Rule 8-1:* We can perform any mathematical operation on one side of an equation, *provided we perform the same operation on the other side*.	If $x = 5$, then $\frac{x}{2} = \frac{5}{2}$, $x + 3 = 5 + 3$, $x^2 = 5^2$
218	*Rule 8-2:* To solve both linear and second-degree equations: (1) Clear all fractions; (2) move all terms containing the unknown to one side; (3) move all constants to the other side; (4) perform the operations necessary to make the coefficient of the unknown equal to 1; (5) verify the solution.	$2x + 3x = 20$ $5x = 20$ $x = 4$ Verify: $2(4) + 3(4) = 20$ $8 + 12 = 20$
219	*Rule 8-3:* Terms may be moved from one side of an equation to the other side by changing their signs.	$x + 4 = 20$ $x = 20 - 4$

224	*Key point:* The square root of the square of a number is that number.	$\sqrt{a^2} = a^{2 \cdot \frac{1}{2}} = a$
224	*Key point:* The square of the square root of a number is that number.	$\left(\sqrt{a}\right)^2 = a^{\frac{1}{2} \cdot 2} = a$

End of Chapter Problems 8-1

Express answers that are nonwhole numbers as fractions or mixed numbers.

1. $a - 7 = 3$

2. $a - 5 = 3$

3. $5 - x = 3$

4. $6 - x = -2$

5. $8 - x = 5$

6. $b + 3 = 10$

7. $7 - x = -2$

8. $a - 4 = 5$

9. $b + 2 = 6$

10. $2 - x = 7$

11. $3 - x = 7$

12. $4 = x - 3$

13. $5y = 20$

14. $6y = 24$

15. $2b = -18$

16. $6a = -24$

17. $\frac{a}{4} = 3$

18. $\frac{x}{5} = 10$

19. $-6x = 42$

20. $7b = -28$

21. $\frac{c}{3} = 4$

22. $-3x = 21$

23. $\frac{a}{5} = -2$

24. $\frac{c}{6} = 2$

25. $\frac{x}{-2} = -8$

26. $\frac{a}{4} = -6$

27. $\frac{x}{-4} = -3$

28. $\frac{4a}{7} = 4$

29. $\frac{3}{5y} = 6$

30. $\frac{4}{3y} = 1$

31. $\frac{3x}{4} = 6$

32. $\frac{2x}{5} = 4$

33. $\frac{7}{2y} = 14$

34. $\frac{5}{3a} = 4$

35. $\frac{5y}{-3} = 10$

36. $\frac{6y}{-5} = 8$

37. $\frac{-6}{7y} = 3$

38. $\frac{-3}{2a} = 9$

39. $\frac{-8}{3b} = 2$

40. $\frac{-3}{2b} = 15$

41. $\frac{4}{x} = \frac{2}{3}$

42. $\frac{5}{x} = \frac{3}{4}$

43. $\dfrac{2a}{5} = \dfrac{3}{7}$

44. $\dfrac{5a}{9} = \dfrac{2}{3}$

45. $\dfrac{14x}{5} = \dfrac{3}{10}$

46. $\dfrac{9x}{4} = \dfrac{7}{2}$

47. $\dfrac{7}{2a} = \dfrac{21}{4}$

48. $\dfrac{6}{5a} = \dfrac{3}{7}$

49. $\dfrac{4}{9} = \dfrac{3}{5x}$

50. $\dfrac{3}{14} = \dfrac{5}{4x}$

End of Chapter Problems 8-2

Solve for the unknown in the following equations:

1. $x^2 = 25$
2. $x^2 = 16$
3. $\dfrac{a^2}{3} = 18$
4. $\dfrac{x^2}{6} = 5$
5. $\dfrac{a^2}{4} = 14$
6. $\dfrac{x^2}{5} = 3$
7. $\dfrac{6}{b^2} = 4$
8. $\dfrac{3}{a^2} = 6$
9. $12 = \dfrac{72}{a^2}$
10. $3 = \dfrac{27}{x^2}$
11. $\sqrt{a} = 7$
12. $\sqrt{a} = 4$
13. $\dfrac{\sqrt{x}}{3} = 2$
14. $\dfrac{\sqrt{y}}{4} = 3$
15. $\dfrac{\sqrt{a}}{6} = 2.3$
16. $\dfrac{\sqrt{b}}{1.7} = 4.6$
17. $2 = \dfrac{1}{2\sqrt{x}}$
18. $3 = \dfrac{1}{3\sqrt{x}}$
19. $3 = \dfrac{1}{4\sqrt{x}}$
20. $2 = \dfrac{1}{3\sqrt{a}}$
21. $\dfrac{3}{4\sqrt{a}} = 6$
22. $\dfrac{2}{3\sqrt{a}} = 2$
23. $\dfrac{2\sqrt{x}}{3} = 6$
24. $\dfrac{3.7\sqrt{x}}{4.2} = 7.3$

Solve for each literal number in the following equations:

25. $2ax = 12$
26. $6bc = 18$
27. $8bc = 14$
28. $12xy = 15$
29. $P = \dfrac{E^2}{R}$
30. $B_L = \dfrac{1}{2\pi fL}$
31. $B_C = 2\pi fC$
32. $6 = \dfrac{1}{2a\sqrt{x}}$

33. $3 = \dfrac{1}{2a\sqrt{b}}$

34. $3 = \dfrac{4}{\sqrt{bc}}$

35. $5 = \dfrac{6}{\sqrt{xy}}$

36. $P = I^2R$

37. $X_L = 2\pi fL$

38. $X_C = \dfrac{1}{2\pi fC}$

39. $f_r = \dfrac{1}{2\pi\sqrt{LC}}$

40. $Z_P = \left(\dfrac{N_P}{N_S}\right)^2 Z_S$

End of Chapter Problems 8-3

1. $R_T = R_1 + R_2$. Find R_1 if $R_T = 12.4$ kΩ and $R_2 = 5.6$ kΩ.
2. $R_T = R_1 + R_2$. Find R_2 if $R_T = 27$ kΩ and $R_1 = 12$ kΩ.
3. $G_T = G_1 + G_2$. Find G_1 if $G_T = 54.9$ μS and $G_2 = 37$ μS.
4. $G_T = G_1 + G_2$. Find G_2 if $G_T = 516$ μS and $G_1 = 303$ μS.
5. $A_V = \dfrac{V_O}{V_{in}}$. Find V_O if $A_V = 150$ and $V_{in} = 55$ mV.
6. $A_V = \dfrac{V_O}{V_{in}}$. Find V_{in} if $A_V = 63$ and $V_O = 3.15$ mV.

If $P = \dfrac{E^2}{R}$:

7. Find R if $P = 2$ W and $E = 15$ V.
8. Find R if $P = 17.3$ W and $E = 35$ V.
9. Find R if $P = 150$ mW and $E = 450$ mV.
10. Find R if $P = 27.3$ W and $E = 7.32$ V.
11. Find R if $P = 30$ mW and $E = 32$ V.
12. Find R if $P = 132$ mW and $E = 17.3$ V.
13. Find E if $P = 6.73$ W and $R = 27$ Ω.
14. Find E if $P = 680$ mW and $R = 68$ kΩ.
15. Find E if $P = 10.3$ W and $R = 1.8$ kΩ.
16. Find E if $P = 300$ mW and $R = 270$ Ω.
17. Find E if $P = 1.73$ W and $R = 47$ kΩ.
18. Find E if $P = 635$ mW and $R = 3.3$ kΩ.

If $P = I^2R$:

19. Find I if $P = 600$ mW and $R = 680$ Ω.
20. Find I if $P = 3.35$ W and $R = 750$ Ω.
21. Find I if $P = 500$ mW and $R = 2.7$ kΩ.
22. Find I if $P = 3.83$ W and $R = 810$ Ω.
23. Find I if $P = 14.7$ W and $R = 100$ Ω.
24. Find I if $P = 100$ W and $R = 120$ Ω.
25. Find R if $P = 500$ mW and $I = 10$ mA.
26. Find R if $P = 2$ W and $I = 610$ mA.
27. Find R if $P = 600$ mW and $I = 47.3$ mA.
28. Find R if $P = 420$ mW and $I = 540$ μA.
29. Find R if $P = 104$ W and $I = 3.77$ A.
30. Find R if $P = 40$ W and $I = 250$ mA.

If $\dfrac{N_P^2}{N_S^2} = \dfrac{Z_P}{Z_S}$:

31. Find N_P if $N_S = 100$, $Z_P = 2.5$ kΩ, and $Z_S = 8$ Ω.
32. Find N_P if $N_S = 200$, $Z_P = 600$ Ω, and $Z_S = 10$ Ω.
33. Find N_P if $N_S = 50$, $Z_P = 40$ Ω, and $Z_S = 500$ Ω.
34. Find N_P if $N_S = 100$, $Z_P = 65$ Ω, and $Z_S = 2.35$ kΩ.
35. Find N_P if $N_S = 25$, $Z_P = 750$ Ω, and $Z_S = 75$ Ω.
36. Find N_P if $N_S = 1000$, $Z_P = 40$ Ω, and $Z_S = 1.2$ kΩ.
37. Find Z_S if $N_P = 1200$, $N_S = 100$, and $Z_P = 1$ kΩ.
38. Find Z_S if $N_P = 2000$, $N_S = 50$, and $Z_P = 500$ Ω.
39. Find Z_S if $N_P = 100$, $N_S = 1800$, and $Z_P = 2.3$ Ω.
40. Find Z_S if $N_P = 250$, $N_S = 2500$, and $Z_P = 4$ Ω.
41. Find Z_S if $N_P = 800$, $N_S = 60$, and $Z_P = 325$ Ω.
42. Find Z_S if $N_P = 3000$, $N_S = 300$, and $Z_P = 1$ kΩ.

43. Find N_S if $N_P = 3000$, $Z_S = 12\ \Omega$, and $Z_P = 1.5\ \text{k}\Omega$.

44. Find N_S if $N_P = 250$, $Z_S = 100\ \text{k}\Omega$, and $Z_P = 1.25\ \text{k}\Omega$.

45. Find N_S if $N_P = 2000$, $Z_S = 4\ \Omega$, and $Z_P = 800\ \Omega$.

46. Find N_S if $N_P = 200$, $Z_S = 870\ \Omega$, and $Z_P = 2.4\ \Omega$.

47. Find Z_P if $N_P = 100$, $N_S = 2500$, and $Z_S = 2\ \text{k}\Omega$.

48. Find Z_P if $N_P = 250$, $N_S = 10{,}000$, and $Z_S = 4.5\ \text{k}\Omega$.

49. Find Z_P if $N_P = 5000$, $N_S = 1000$, and $Z_S = 40\ \Omega$.

50. Find Z_P if $N_P = 7500$, $N_S = 500$, and $Z_S = 12\ \Omega$.

If $X_C = \dfrac{1}{2\pi fC}$:

51. Find C if $f = 2.7$ kHz and $X_C = 700\ \Omega$.

52. Find C if $f = 10.7$ kHz and $X_C = 370\ \Omega$.

53. Find f if $C = 10$ nF and $X_C = 5\ \text{k}\Omega$.

54. Find f if $C = 400$ nF and $X_C = 2\ \text{k}\Omega$.

If $X_L = 2\pi fL$:

55. Find L if $f = 4.8$ kHz and $X_L = 3.85\ \text{k}\Omega$.

56. Find L if $f = 1.2$ kHz and $X_L = 25\ \text{k}\Omega$.

57. Find f if $L = 54.1$ mH and $X_L = 170\ \Omega$.

58. Find f if $L = 150$ mH and $X_L = 3.7\ \text{k}\Omega$.

If $f_r = \dfrac{1}{2\pi\sqrt{LC}}$.

59. Find L if $f_r = 30$ kHz and $C = 100$ pF.

60. Find L if $f_r = 3.5$ kHz and $C = 250$ nF.

61. Find C if $f_r = 100$ Hz and $L = 100$ mH.

62. Find C if $f_r = 20$ kHz and $L = 100$ mH.

Answers to Self-tests

SELF-TEST 8-1

1. $y = 10$

2. $x = 5$

3. $a = -8$

4. $y = \dfrac{5}{4}$

5. $a = 7$

6. $a = 3$

7. $a = 25$

8. $y = 900$

9. $x = \dfrac{1}{64}$

10. $x = \dfrac{7}{3y}$, $y = \dfrac{7}{3x}$

11. $P = \dfrac{E^2}{R}$, $E = \sqrt{PR}$

12. $x = \dfrac{1}{12\sqrt{y}}$, $y = \dfrac{1}{144x^2}$

SELF-TEST 8-2

1. $E = 77.0$ V

2. $I = 6.58$ mA

3. $N_P = 224$

4. $f = 1.42$ kHz

5. $L = 1.27$ H

CHAPTER 9 FACTORING ALGEBRAIC EXPRESSIONS

INTRODUCTION

In previous chapters we multiplied or divided monomials. Many of the algebraic expressions we work with are polynomials, that is, algebraic expressions that contain more than one term.

As we encounter more complex equations, we must develop ways to simplify these equations so that we can solve them. Many times, equations can be simplified by factoring. Factoring is the process of finding the common factors of polynomials and then extracting these factors from each term. In this chapter you will learn how to factor polynomials.

CHAPTER OBJECTIVES

In this chapter you will learn how to:

1. Multiply polynomials by monomials.
2. Multiply binomials by binomials.
3. Divide polynomials by monomials and by binomials.
4. Determine common factors of polynomials.
5. Determine the factors of trinomials.

9-1 MULTIPLICATION OF POLYNOMIALS BY MONOMIALS

Let's multiply the polynomial $(4x + 2)$ by the monomial $3x$. There are two ways to multiply a polynomial by a monomial. We will show both ways in the following examples.

□ EXAMPLE 9-1

$$3x(4x + 2) = 3x \cdot 4x + 3x \cdot 2 = 12x^2 + 6x$$

The sign of grouping (the parentheses) tells us that each term within the group must be multiplied by $3x$. Since there are two terms, and each term is multiplied by $3x$, there will be two terms in the answer.

□ EXAMPLE 9-2

$$\begin{array}{r} 4x + 2 \\ \times \quad\quad 3x \\ \hline 12x^2 + 6x \end{array}$$

In this example, we set up the problem the way we set up a multiplication problem in arithmetic. Each term in the multiplicand $(4x + 2)$ is multiplied by the multiplier $(3x)$.

□

Most problems are given in the form shown in Example 9-1. Most students prefer this method of solution, so this is the method we will use in the rest of the examples.

Always keep in mind the following rule when working problems of this kind.

RULE 9-1 When multiplying a polynomial by a monomial, there must be as many terms in the answer as there are in the polynomial.

□ EXAMPLE 9-3

Multiply $4(2x^2 + 3x - 4)$.

Solution Multiply each term in the group by 4. This yields

$$4 \cdot 2x^2 + 4 \cdot 3x - 4 \cdot 4 = 8x^2 + 12x - 16$$

Notice that we put a minus sign in front of $4 \cdot 4$ because the sign of the term was negative and the sign of the multiplier was positive. Remember $(+) \cdot (-) = (-)$.

□ EXAMPLE 9-4

Perform the indicated multiplication and combine like terms.

$$4a(a + 2b) - 3a(a + 3b)$$

Solution We have two different multiplications to perform:

$$4a(a + 2b) \quad \text{and} \quad -3a(a + 3b)$$

$$4a(a + 2b) = 4a^2 + 8ab$$

$$-3a(a + 3b) = -3a^2 - 9ab$$

Putting them together, we get

$$4a(a + 2b) - 3a(a + 3b) = 4a^2 + 8ab - 3a^2 - 9ab$$

When we combine like terms, we get

$$4a^2 - 3a^2 + 8ab - 9ab = a^2 - ab$$

□ EXAMPLE 9-5

Perform the indicated multiplication and combine like terms:

$$3x^2 + 6x(2x - 3y - 2) - 2x(3x - 4y - 3)$$

Solution

$$6x(2x - 3y - 2) = 12x^2 - 18xy - 12x$$

$$-2x(3x - 4y - 3) = -6x^2 + 8xy + 6x$$

Notice that in each part of the problem where there are three terms in the polynomial there are three terms in that part of the answer (Rule 9-1).

Putting it all together and combining like terms, we get

$$3x^2 + 6x(2x - 3y - 2) - 2x(3x - 4y - 3)$$
$$= 3x^2 + 12x^2 - 18xy - 12x - 6x^2 + 8xy + 6x$$
$$= 9x^2 - 10xy - 6x$$

□

PRACTICE PROBLEMS 9-1

Perform the indicated operations. Combine like terms.

1. $4(x + 2)$
2. $3(a - 4)$
3. $4a - 3(4a - 3)$
4. $5x + 2(3x - 4)$
5. $2x(3 + 4y)$
6. $3a(2 - 4b)$
7. $2x(3x + 2y)$
8. $3a(2a - 6b)$
9. $3x(4x - 2y + 4)$
10. $4(2x^2 + 3x - 4)$
11. $5(3a^2 - a - 3)$
12. $3a(a + b) - 2a(a + 4b)$
13. $2x(x - 2y) - 3x(x + 4y)$
14. $x(xy + 4y) - 2(x^2y - 3xy)$
15. $x(3xy - 3y) + 4(x^2y - 4xy)$
16. $2a(a^2 + 9a + 20) + 3a(a^2 - a - 12)$

17. $3x(2x^2 - 7x + 1) - 2x(2x^2 - 3x + 4)$

18. $4x^2 + 2x(x + y - 3) - 2x(2x - y)$

19. $4a(a - 2b + 3) - 2a^2 + 3a(2a - 2b - 2)$

20. $3a(a^2 - 2ab + 2) - 6(a^2 - a^2b + 2)$

Solutions

1. $4x + 8$
2. $3a - 12$
3. $-8a + 9$
4. $11x - 8$
5. $6x + 8xy$
6. $6a - 12ab$
7. $6x^2 + 4xy$
8. $6a^2 - 18ab$
9. $12x^2 - 6xy + 12x$
10. $8x^2 + 12x - 16$
11. $15a^2 - 5a - 15$
12. $a^2 - 5ab$
13. $-x^2 - 16xy$
14. $-x^2y + 10xy$
15. $7x^2y - 19xy$
16. $5a^3 + 15a^2 + 4a$
17. $2x^3 - 15x^2 - 5x$
18. $2x^2 + 4xy - 6x$
19. $8a^2 - 14ab + 6a$
20. $3a^3 - 6a^2 + 6a - 12$

Additional practice problems are at the end of the chapter.

The following three examples show how we can multiply polynomials that contain fractions. Again, each term in the polynomial is multiplied by the monomial.

□ EXAMPLE 9-6

$$3a\left(2a + 3 - \frac{4a^2}{3a}\right) = 6a^2 + 9a - \frac{12a^3}{3a}$$

$$= 6a^2 + 9a - 4a^2 = 2a^2 + 9a$$

The third term, $-\dfrac{12a^3}{3a}$, reduces to $-4a^2$.

□ EXAMPLE 9-7

$$4x\left(\frac{1}{x} + \frac{3x}{2} + 4\right) = \frac{4x}{x} + \frac{12x^2}{2} + 16x = 4 + 6x^2 + 16x$$

$$= 6x^2 + 16x + 4 \qquad \text{(in descending order)}$$

□ EXAMPLE 9-8

$$3a\left(\frac{1}{3} + \frac{2a^2}{4a} - \frac{6a}{5}\right) = \frac{3a}{3} + \frac{6a^3}{4a} - \frac{18a^2}{5} = a + \frac{3a^2}{2} - \frac{18a^2}{5}$$

$$= \frac{10a + 15a^2 - 36a^2}{10} = \frac{10a - 21a^2}{10}$$

$$= -\frac{21a^2 - 10a}{10} \qquad \text{(in descending order)}$$

□

PRACTICE PROBLEMS 9-2

Perform the indicated operations:

1. $5a\left(4a - \frac{3a^2}{5a}\right)$

2. $3a\left(2a + \frac{5a^2}{3a}\right)$

3. $5x\left(2x + 2 - \frac{3x^2}{5x}\right)$

4. $3x\left(\frac{1}{x} + \frac{2x}{3} + 1\right)$

5. $2y\left(\frac{2}{y} + \frac{4y}{2} - 3\right)$

6. $3y\left(\frac{2}{3} - \frac{3y^2}{2y} + \frac{6y}{4}\right)$

7. $2b\left(\frac{3}{5} + \frac{4b^2}{3b} - \frac{5b}{4}\right)$

8. $5b\left(\frac{3}{10} - \frac{5b^2}{15b} + \frac{2b}{5}\right)$

Solutions

1. $17a^2$

2. $11a^2$

3. $7x^2 + 10x$

4. $2x^2 + 3x + 3$

5. $4y^2 - 6y + 4$

6. $2y$

7. $\frac{b^2}{6} + \frac{6b}{5}$

8. $\frac{b^2}{3} + \frac{3b}{2}$

Additional practice problems are at the end of the chapter.

9-2 MULTIPLICATION OF BINOMIALS BY BINOMIALS

■ **KEY POINT** When multiplying a binomial by a binomial, multiply each term in the multiplicand by each term in the multiplier.

□ EXAMPLE 9-9

$$\begin{array}{r} a + 4 \\ (\times)\ a - 2 \\ \hline \end{array}$$

Solution We must multiply each term in the multiplicand by each term in the multiplier.

$$\begin{array}{rrl} & a + 4 & \\ & a - 2 & \\ \hline a^2 + 4a & & \text{result of } a(a + 4) \\ - 2a & - 8 & \text{result of } -2(a + 4) \\ \hline a^2 + 2a & - 8 & \end{array}$$

Notice that like terms are placed in the same column. This lets us combine like terms when the two rows are added together. No matter how many terms are in the multiplicand and multiplier, this method can be used.

□

Let's look at it again.

□ EXAMPLE 9-10

$$(a + 4)(a - 2)$$

Solution This time let's do it mentally; the procedure is to multiply each term in the second binomial by each term in the first binomial and combine like terms:

$$a \cdot a = a^2$$

$$a \cdot -2 = -2a$$

$$(a + 4)(a - 2)$$

$$4 \cdot a = 4a$$

$$4 \cdot -2 = -8$$

The four resulting terms are $a^2 - 2a + 4a - 8$. Combining like terms, we get $a^2 + 2a - 8$.

□

In problems such as this, we may say that the middle term equals *the product of the means plus the product of the extremes*. The *means* refers to the product of the two inter terms (4 and a in this problem). The *extremes* refers to the product of the two outer terms (a and -2 in this problem). The result, $-2a + 4a$ yields the middle term, $2a$. This is often a quick way to find the value of the middle term.

Let's try two more.

□ EXAMPLE 9-11

$$(x - 3y)(x + 7y)$$

Solution Multiplying each term in the second binomial by each term in the first yields

$$x \cdot x = x^2$$

$$x \cdot 7y = 7xy$$

$$-3y \cdot x = -3xy$$

$$-3y \cdot 7y = -21y^2$$

$$(x - 3y)(x + 7y) = x^2 + 7xy - 3xy - 21y^2 = x^2 + 4xy - 21y^2$$

□ EXAMPLE 9-12

$$(3a - 2b)(4a - 4b)$$

Solution Multiplying each term in the second binomial by each term in the first binomial yields

$$3a \cdot 4a = 12a^2$$

$$3a \cdot -4b = -12ab$$

$$-2b \cdot 4a = -8ab$$

$$-2b \cdot -4b = 8b^2$$

$$(3a - 2b)(4a - 4b) = 12a^2 - 12ab - 8ab + 8b^2$$

$$= 12a^2 - 20ab + 8b^2$$

□ EXAMPLE 9-13

$$(a + 3)^2$$

Solution Identical binomials are often expressed in this manner. Remember that the exponent 2 means that the binomial is multiplied by itself two times.

$$(a + 3)^2 = (a + 3)(a + 3)$$

Now we can solve the problem just as we did in the previous examples by multiplying each term in the second binomial by each term in the first:

$$a \cdot a = a^2$$

$$3 \cdot a = 3a$$

$$a \cdot 3 = 3a$$

$$3 \cdot 3 = 9$$

$$(a + 3)^2 = a^2 + 3a + 3a + 9 = a^2 + 6a + 9$$

The middle terms will always be the same when multiplying identical binomials, so we can shorten the process by simply finding the product of one middle term and then multiplying by 2:

$$(a + 3)^2 = a^2 + 2(3a) + 9 = a^2 + 6a + 9$$

□

PRACTICE PROBLEMS 9-3

Perform the indicated operations:

1. $(a + 2)^2$

2. $(a - 4)^2$

3. $(2a - 3)^2$

4. $(3a - 2)^2$

5. $(x + 2y)^2$

6. $(2x - 3y)^2$

7. $(a + 6)(a + 2)$
8. $(a + 2)(a - 5)$
9. $(a - 3)(a + 4)$
10. $(a - 4)(a - 5)$
11. $(x + 3)(x - 3)$
12. $(x - 5)(x + 5)$
13. $(a + b)(a + 2b)$
14. $(a + 3b)(a - 4b)$
15. $(2a + 3b)^2$
16. $(4x - 3y)^2$

Solutions

1. $a^2 + 4a + 4$
2. $a^2 - 8a + 16$
3. $4a^2 - 12a + 9$
4. $9a^2 - 12a + 4$
5. $x^2 + 4xy + 4y^2$
6. $4x^2 - 12xy + 9y^2$
7. $a^2 + 8a + 12$
8. $a^2 - 3a - 10$
9. $a^2 + a - 12$
10. $a^2 - 9a + 20$
11. $x^2 - 9$
12. $x^2 - 25$
13. $a^2 + 3ab + 2b^2$
14. $a^2 - ab - 12b^2$
15. $4a^2 + 12ab + 9b^2$
16. $16x^2 - 24xy + 9y^2$

Additional practice problems are at the end of the chapter.

9-3 DIVISION OF POLYNOMIALS

KEY POINT When dividing a polynomial by a monomial, divide each term in the polynomial by the monomial.

The following examples illustrate the process.

☐ EXAMPLE 9-14

Perform the indicated division:

$$(6x^3 + 12x^2 - 3x) \div 3x$$

Solution Each term in the dividend is divided by $3x$, the divisor. We could set up the problem as we would a long-division problem and solve like this:

$$\begin{array}{r l}
2x^2 + 4x - 1 & \\
3x\overline{)6x^3 + 12x^2 - 3x} & \\
\underline{6x^3 \qquad\qquad\qquad} & 3x \cdot 2x^2 = 6x^3 \\
0 + 12x^2 \qquad & \\
\underline{12x^2 \qquad} & 3x \cdot 4x = 12x^2 \\
0 - 3x & \\
\underline{-3x} & 3x \cdot -1 = -3x \\
0 & \text{no remainder}
\end{array}$$

$3x$ is divided into the first term. The result of this division is $2x^2$. $3x \cdot 2x^2 = 6x^3$. The process continues until all terms have been divided by $3x$. Usually, a problem like this can be done by inspection. If there is a remainder, it is written as a fraction. The following example illustrates this.

☐ EXAMPLE 9-15

$$(7x^2 + 4x) \div 4x$$

Solution

$$\begin{array}{r} x + 1 + \dfrac{3x^2}{4x} \\ 4x\overline{)7x^2 + 4x} \\ \underline{4x^2 \quad\quad} \\ 3x^2 + 4x \\ \underline{4x} \\ 3x^2 + 0 \end{array}$$

$3x^2$ is left over; therefore, the remainder is $\dfrac{3x^2}{4x}$.

Check the answer by multiplying the divisor and the quotient. The product should equal the dividend:

$$4x\left(x + 1 + \frac{3x^2}{4x}\right) = 4x^2 + 4x + \frac{12x^3}{4x}$$
$$= 4x^2 + 4x + 3x^2 = 7x^2 + 4x$$

☐

Let's try dividing a polynomial by a binomial.

KEY POINT When dividing a polynomial by a binomial, set it up as a long division problem with the literal numbers in descending order; then perform the division.

The following examples illustrate the process.

☐ EXAMPLE 9-16

$$(a^2 + 3a - 54) \div (a + 9)$$

Solution First, we set it up as a long-division problem, and then we make sure that the literal numbers in both the dividend and divisor are in descending order, as shown:

$$a + 9\overline{)a^2 + 3a - 54}$$

Then we determine how many times the first term in the divisor goes into the first term of the dividend. a goes into a^2 a times. Multiplying this a times each term in the divisor results in $a^2 + 9a$. Subtracting this result from the dividend leaves a remainder of $-6a$. After we bring the next term down, we have the partial solution shown:

$$\begin{array}{r} a \quad\quad\quad\quad \\ a + 9\overline{)a^2 + 3a - 54} \\ \underline{a^2 + 9a \quad\quad} \\ 0 - 6a - 54 \end{array}$$

Next we determine how many times a goes into $-6a$. The answer is -6. Now we multiply -6 by $(a + 9)$. There is no remainder and the solution is complete.

$$
\begin{array}{r}
a - 6 \\
a + 9\overline{)a^2 + 3a - 54} \\
\underline{a^2 + 9a} \\
-6a - 54 \\
\underline{-6a - 54} \\
0
\end{array}
$$

We can check our answer:

$$(a + 9)(a - 6) = a^2 + 3a - 54$$

Let's try another one.

EXAMPLE 9-17

$$(6a^2 - 23ab + 20b^2) \div (3a - 4b)$$

Solution

$$
\begin{array}{r}
2a - 5b \\
3a - 4b\overline{)6a^2 - 23ab + 20b^2} \\
\underline{6a^2 - 8ab} \\
-15ab + 20b^2 \\
\underline{-15ab + 20b^2} \\
0
\end{array}
$$

Check

$$(3a - 4b)(2a - 5b) = 6a^2 - 23ab + 20b^2$$

Let's do one more.

EXAMPLE 9-18

$$(x^2 - 49) \div (x + 7)$$

Solution

$$
\begin{array}{r}
x \qquad - 7 \\
x + 7\overline{)x^2 \qquad - 49} \\
\underline{x^2 + 7x} \\
-7x - 49 \\
\underline{-7x - 49} \\
0
\end{array}
$$

Check

$$(x + 7)(x - 7) = x^2 - 49$$

□ EXAMPLE 9-19

Here's one with a remainder.

$$(x^2 + 4x + 7) \div (x - 3)$$

Solution

$$\begin{array}{r} x \quad\quad + 7 \\ x - 3\overline{)x^2 + 4x + 7} \\ \underline{x^2 - 3x \quad\quad} \\ 7x + 7 \\ \underline{7x - 21} \\ 28 \end{array}$$

The remainder is $\frac{28}{x-3}$, so the answer is

$$x + 7 + \frac{28}{x - 3}$$

Check the answer:

$$(x - 3)\left(x + 7 + \frac{28}{x - 3}\right) = (x - 3)\left(\frac{x^2 - 3x + 7x - 21 + 28}{x - 3}\right)$$

$$= x^2 - 3x + 7x - 21 + 28 = x^2 + 4x + 7$$

□

PRACTICE PROBLEMS 9-4

Perform the indicated operations:

1. $(a^3 + 2a^2 - a) \div a$
2. $(4a^4 - 6a^3 + 8a^2) \div 2a$
3. $(x^2 + 13x + 42) \div (x + 6)$
4. $(x^2 - 36) \div (x + 6)$
5. $(9x^2 - 12x + 4) \div (3x - 2)$
6. $(4a^2 + 12ab + 9b^2) \div (2a + 3b)$
7. $(a^2 - ab - 12b^2) \div (a - 4b)$
8. $(x^2 - 9x + 20) \div (x - 5)$
9. $(3ab + a^2 + 2b^2) \div (b + a)$
10. $(9 + 4x^2 + 12x) \div (3 + 2x)$
11. $(2x^2 + 3x + 14) \div (x + 1)$
12. $(6x^2 + 2x - 7) \div (2x + 4)$
13. $(x^2 + 2xy - y^2) \div (x + y)$
14. $(2x^2 - 4xy + 2y^2) \div (2x + 2y)$

Solutions

1. $a^2 + 2a - 1$
2. $2a^3 - 3a^2 + 4a$
3. $x + 7$
4. $x - 6$
5. $3x - 2$
6. $2a + 3b$
7. $a + 3b$
8. $x - 4$
9. $a + 2b$
10. $2x + 3$
11. $2x + 1 + \frac{13}{x + 1}$
12. $3x - 5 + \frac{13}{2x + 4}$

13. $x + y - \dfrac{2y^2}{x + y}$

14. $x - 3y + \dfrac{8y^2}{2x + 2y}$

Additional practice problems are at the end of the chapter.

SELF-TEST 9-1

Perform the indicated operations:

1. $3a(a - 6)$
2. $2xy(3x + 2y)$
3. $(a - 9)^2$
4. $(4a + 2)^2$
5. $(a + 6)(a - 1)$
6. $(x + 2y)(x - 6y)$
7. $(2x + 3y)(4x + 4y)$
8. $3(a - 2b)(2a - 2b)$
9. $(6a^3 - 12a^2) \div 3a$
10. $(x^2 + 2x - 48) \div (x + 8)$
11. $(x^2 + 13x + 36) \div (x + 9)$
12. $(a^2 + 12ab + 20b^2) \div (a + 2b)$
13. $(13x + 30 + x^2) \div (x + 9)$
14. $(10ab + a^2 + 20b^2) \div (a + 2b)$

Answers to Self-test 9-1 are at the end of the chapter.

9-4 FACTORING POLYNOMIALS

Earlier in this chapter we multiplied monomials times polynomials. Here we are going to reverse the process. We are going to determine what the common factors are of various algebraic expressions. This is simply a process of determining what factors are common to each term in the expression.

■ **RULE 9-2** To factor a polynomial:

1. Find the prime factors of each term.
2. Determine the factors common to all terms.
3. Factor the polynomial by dividing the polynomial by the common factors.

Consider the expression $6a^2 + 12a - 18a^3$. If we factor each term, we get

$$6a^2 = 2 \cdot 3 \cdot a^2$$

$$12a = 2 \cdot 2 \cdot 3 \cdot a$$

$$18a^3 = 2 \cdot 3 \cdot 3 \cdot a^3$$

The common factors are the factors that appear in all terms. These common factors are 2, 3, and a or $6a$. Since $6a$ is common to each term, we can divide the expression by $6a$. This yields $a + 2 - 3a^2$.

$$\frac{6a^2 + 12a - 18a^3}{6a} = a + 2 - 3a^2$$

The factors of $6a^2 + 12a - 18a^3$ are $6a$ and $a + 2 - 3a^2$.

$$6a^2 + 12a - 18a^3 = 6a(a + 2 - 3a^2)$$

Since we normally write algebraic expressions either in descending order or in ascending order, let's rearrange the answer so that it looks like this:

$$6a(-3a^2 + a + 2)$$

☐ EXAMPLE 9-20

Factor $28a^3 - 35a^2$.

Solution The prime factors are

$$28a^3 = 2 \cdot 2 \cdot 7 \cdot a^3$$

$$35a^2 = 5 \cdot 7 \cdot a^2$$

The common factors are 7 and a^2. Therefore, we divide each term by $7a^2$ to find the factors of the original problem.

$$\frac{28a^3 - 35a^2}{7a^2} = 4a - 5$$

The factors of $28a^3 - 35a^2$ are $7a^2$ and $4a - 5$.

$$28a^3 - 35a^2 = 7a^2(4a - 5)$$

☐

PRACTICE PROBLEMS 9-5

Factor the following problems:

1. $4a + 2$
2. $6a + 8$
3. $16 - 2y$
4. $21 - 3y$
5. $x^3 + 2x^2$
6. $4x^4 + 2x^2$
7. $18a^2 - 9a$
8. $15a^2 - 3a$
9. $14x^2y - 14x^2z + 20xyz$
10. $15a^2b - 21a^3b^2 + 9a^2b^2$

Solutions

1. $2(2a + 1)$
2. $2(3a + 4)$
3. $2(8 - y)$
4. $3(7 - y)$
5. $x^2(x + 2)$
6. $2x^2(2x^2 + 1)$
7. $9a(2a - 1)$
8. $3a(5a - 1)$
9. $2x(7xy - 7xz + 10yz)$
10. $3a^2b(5 - 7ab + 3b)$

Additional practice problems are at the end of the chapter.

9-5 FACTORS OF TRINOMIALS

A trinomial of the form $ax^2 + bx + c$ can be factored if we can find two numbers whose sum is the coefficient of the second term and whose product is the third term.

For example, consider the expression $x^2 + 13x + 42$. From the work we did earlier in the chapter, we know that the answer will be in the form $(x + n)(x + m)$. In this expression the coefficient of the second term is 13, and the third term is 42. We must find two numbers whose sum is 13 and whose product is 42. We find these numbers by trial and error. That is, we consider all the combination of numbers whose sum is 13 and select the one combination whose product is 42. In this example, the combinations are

$$\begin{array}{rr} 12 + 1 & 12 \times 1 = 12 \\ 11 + 2 & 11 \times 2 = 22 \\ 10 + 3 & 10 \times 3 = 30 \\ 9 + 4 & 9 \times 4 = 36 \\ 8 + 5 & 8 \times 5 = 40 \\ 7 + 6 & 7 \times 6 = 42 \end{array}$$

The two numbers must be 7 and 6:

$$x^2 + 13x + 42 = (x + 7)(x + 6)$$

It would also be correct to write the answer as $(x + 6)(x + 7)$. With practice, we can usually determine the correct combination by inspection.

□ EXAMPLE 9-21

Factor $x^2 - 6x - 27$.

Solution The coefficient of the middle term is -6 and the product is -27. Any time the third term is negative, the factors must be of this form: $(x + n)(x - m)$. The unknown number m must be greater than n because the middle term is negative. There are so many combinations of numbers that equal -6 that it is simpler in this case to consider numbers whose product is -27.

$$\begin{array}{rr} 1 \cdot -27 = -27 & 1 + (-27) = -26 \\ -1 \cdot 27 = -27 & -1 + 27 = 26 \\ 3 \cdot -9 = -27 & 3 + (-9) = -6 \\ -3 \cdot 9 = -27 & -3 + 9 = 6 \end{array}$$

Of the four possibilities, only 3 and -9 equal -6 when added together:

$$x^2 - 6x - 27 = (x + 3)(x - 9)$$

☐ EXAMPLE 9-22

Factor $x^2 + 2x - 8$.

Solution The third term is again negative; therefore, the answer is again of the form $(x + m)(x - n)$. Let's again consider all the possible combinations that yield a product of -8:

$$2 \cdot -4 = -8 \qquad 2 + (-4) = -2$$
$$-2 \cdot 4 = -8 \qquad -2 + 4 = 2$$
$$1 \cdot -8 = -8 \qquad 1 + (-8) = -7$$
$$-1 \cdot 8 = -8 \qquad -1 + 8 = 7$$

Of the four possibilities, only -2 and 4 equal 2 when added together:

$$x^2 + 2x - 8 = (x + 4)(x - 2)$$

☐ EXAMPLE 9-23

Factor $x^2 + 12x + 36$.

Solution Inspection shows that the factors are $(x + 6)$ and $(x + 6)$ or $(x + 6)^2$:

$$x^2 + 12x + 36 = (x + 6)^2$$

☐ EXAMPLE 9-24

Factor $x^2 - 36$.

Solution Note that there is no middle term and that the last term is negative. The only way the middle term can drop out is for the two numbers to be equal. The only two equal numbers whose product is -36 are 6 and -6. Then

$$x^2 - 36 = (x + 6)(x - 6)$$

☐ EXAMPLE 9-25

Factor $x^2 - 12x + 36$.

Solution The middle term is negative and the last term is positive. This combination can only result if both numbers are negative. The numbers must be -6 because $-6 + (-6) = -12$ and $-6 \cdot -6 = 36$.

$$x^2 - 12x + 36 = (x - 6)^2$$

☐

In Example 9-26 we have added another squared term, y^2. When trinomials are of the form $ax^2 + bxy + cy^2$, the factors must be $(px + my)(qx + ny)$. Otherwise, the method of solution is the same.

☐ EXAMPLE 9-26

Factor $x^2 + 2xy + y^2$.

Solution By inspection we see that the factors must be $(x + y)$ and $(x + y)$ or $(x + y)^2$.

☐ EXAMPLE 9-27

Factor $x^2 - 4xy - 32y^2$.

Solution The coefficient of $y^2 = -32$ and the coefficient of the middle term is -4. The possible solutions are

$$-32 \cdot 1 \qquad -32 + 1 = -31$$

$$-16 \cdot 2 \qquad -16 + 2 = -14$$

$$-8 \cdot 4 \qquad -8 + 4 = -4$$

The correct combination is -8 and $+4$. Because the middle term is negative, the larger of the two digits must be the negative digit. Therefore, the three combinations in which the smaller of the two digits is negative were not considered.

$$x^2 - 4xy - 32y^2 = (x + 4y)(x - 8y)$$

☐ EXAMPLE 9-28

Factor $4x^2 + 13xy + 10y^2$.

Solution This example shows a coefficient of x^2 that is some number other than 1, in this case 4. The coefficient of y^2 is 10. We are looking for two numbers whose product is 4 (because the first term is $4x^2$) and two numbers whose product is 10 (the last term is $10y^2$). The product of the means plus the product of the extremes must equal 13, the coefficient of the middle term. The possible combinations are

$$(2 + 2)(2 + 5) \qquad 4 + 10 = 14$$

$$(2 + 5)(2 + 2) \qquad 10 + 4 = 14$$

$$(1 + 5)(4 + 2) \qquad 20 + 2 = 22$$

$$(1 + 2)(4 + 5) \qquad 8 + 5 = 13$$

In the right column, we added the product of the means and the product of the extremes. We tried different combinations until we finally got a sum equal to 13, the coefficient of the middle term. Of course, there are still other combinations we didn't consider, but we got the right answer on the fourth try, so there was no need to try any more.

$$4x^2 + 13xy + 10y^2 = (x + 2y)(4x + 5y)$$

☐

In this chapter we have presented trinomials that were factorable. In reality, not all trinomials are factorable using the methods we have developed in this chapter. In such cases, we must resort to other methods of solution; the quadratic is the most common. The use of the quadratic in finding the factors of trinomials is covered in Chapter 10.

PRACTICE PROBLEMS 9-6

Factor the following:

1. $x^2 + 3x + 2$

2. $x^2 + 7x + 12$

3. $a^2 + a - 6$

4. $a^2 - a - 12$

5. $a^2 + 15a + 56$

6. $a^2 + 2a - 15$

7. $x^2 + 12x + 36$

8. $x^2 - 64$

9. $x^2 + 12xy + 35y^2$

10. $x^2 + 16xy + 63y^2$

11. $2a^2 - 12ab - 32b^2$

12. $2a^2 + 4ab - 48b^2$

Solutions

1. $(x + 2)(x + 1)$

2. $(x + 3)(x + 4)$

3. $(a + 3)(a - 2)$

4. $(a - 4)(a + 3)$

5. $(a + 7)(a + 8)$

6. $(a + 5)(a - 3)$

7. $(x + 6)^2$

8. $(x + 8)(x - 8)$

9. $(x + 5y)(x + 7y)$

10. $(x + 7y)(x + 9y)$

11. $2(a + 2b)(a - 8b)$

12. $2(a + 6b)(a - 4b)$

Additional practice problems are at the end of the chapter.

SELF-TEST 9-2

Factor the following:

1. $9x - 36$

2. $56y^2 + 8y$

3. $a^2 + 15a + 36$

4. $a^2 - 7a - 30$

5. $x^2 + 2xy - 35y^2$

6. $x^2 + 12xy + 36y^2$

7. $2x^2 + 16x + 32$

8. $3a^2 - 3b^2$

9. $a^2 - 3ab - 54b^2$

10. $a^2 - 6ab + 9b^2$

Answers to Self-test 9-2 are at the end of the chapter.

CHAPTER 9 AT A GLANCE

Page	Key Point or Rule	Example
239	*Rule 9-1:* When multiplying a polynomial by a monomial, there must be as many terms in the answer as there are in the polynomial.	$3x(4x - 2y + 4)$ $= 12x^2 - 6xy + 12x$
242	*Key point:* When multiplying a binomial by a binomial, multiply each term in the multiplicand by each term in the multiplier.	$(a + 4)(a - 2)$ $= a^2 + 4a - 2a - 8$ $= a^2 + 2a - 8$
245	*Key point:* When dividing a polynomial by a monomial, divide each term in the polynomial by the monomial.	$(6x^3 + 12x^2 - 3x) \div 3x$ $= 2x^2 + 4x - 1$
246	*Key point:* When dividing a polynomial by a binomial, set it up as a long division problem with the literal numbers in descending order; then perform the division.	$(a^2 + 3a - 54) \div (a + 9)$ $\begin{array}{r} a - 6 \\ a + 9\overline{)a^2 + 3a - 54} \\ \underline{a^2 + 9a} \\ -6a - 54 \\ \underline{-6a - 54} \\ 0 \end{array}$
249	*Rule 9-2:* To factor a polynomial: 1. Find the prime factors of each term. 2. Determine the factors common to all terms. 3. Factor the polynomial by dividing the polynomial by the common factors.	Factor: $6a^2 + 12a - 18a^3$ $6a^2 = 2 \cdot 3 \cdot a^1$ $12a = 2 \cdot 2 \cdot 3 \cdot a$ $18a^3 = 2 \cdot 3 \cdot 3 \cdot a^3$ The common factors are $2 \cdot 3 \cdot a = 6a$ $(6a^2 + 12a - 18a^3) \div 6a$ $= a + 2 - 3a^2$ $6a^2 + 12a - 18a^3$ $= 6a(a + 2 - 3a^2)$ $= 6a(-3a^2 + a + 2)$
251	*Key point:* Trinomials can be factored if we can find two numbers whose sum is the coefficient of the second term and whose product is the third term.	Factor $x^2 + 13x + 42$. $(x + 7)(x + 6)$

End of Chapter Problems 9-1

Perform the indicated operations. Combine like terms.

1. $3(a + 3)$

2. $5(x + 2)$

3. $4(a - 1)$

4. $6(a - 3)$

5. $7(4 - b)$

6. $8(3 - b)$

7. $3a - 2(3a - 1)$
8. $4a - 3(2a - 3)$
9. $6a - 3(4a + 2)$
10. $5a - 5(2a + 2)$
11. $4(2x - 3) - 2x$
12. $5(3x + 4) - 8x$
13. $3x(2 + 2y)$
14. $6x(3 - 2y)$
15. $3a(2a + 3b)$
16. $4a(3a + 2b)$
17. $3(3x^2 + 6x - 12)$
18. $4(2x^2 + 3x - 6)$
19. $2(4a^2 - 3a + 2)$
20. $4(3a^2 + 3a - 3)$
21. $4a(a + b) - 3a(a + 5b)$
22. $5a(a + b) - 5a(a + 2b)$
23. $a(2ab + 2b) - 3(a^2b - 2ab)$
24. $2a(ab + 2b) - 4(2a^2b - ab)$
25. $2x(4xy - 3y) - 3(x^2y + xy)$
26. $3x(2xy + 4y) - 4(2x^2y - 2xy)$
27. $2x(3xy - 4y) - 4(x^2y + 3xy)$
28. $3x(xy - 2y) - 5(x^2y + xy)$
29. $x(x^2 + 8x + 15) + 2x(x^2 - x - 6)$
30. $2x(x^2 + 7x + 12) + x(x^2 - 6x + 9)$
31. $4a(a^2 - 3a + 6) - 3a(a^2 + 5a - 6)$
32. $5a(a^2 - 2a - 3) + 2a(a^2 - 3a + 4)$
33. $3a^2 + 2a(a - b + 2) - 2a(2a + b)$
34. $2x^2 + 3x(x - 2y + 3) - 3x(2x + 2y)$
35. $4x(2x - 3y + 2) - 2x^2 - 3x(x + 2y - 3)$
36. $3a(4a - 2 + 3b) + 4a^2 - 4a(2a - 3 + b)$

End of Chapter Problems 9-2

1. $4x\left(3x + \dfrac{2x^2}{3x}\right)$
2. $5x\left(2x + \dfrac{3x^2}{10x}\right)$
3. $2y\left(3y - \dfrac{3y^2}{6y}\right)$
4. $3y\left(4y - \dfrac{6y^2}{9y}\right)$
5. $3a\left(2a + 3 + \dfrac{4a^2}{6a}\right)$
6. $4a\left(2a + 4 + \dfrac{3a^2}{8a}\right)$
7. $2b\left(4b - 3 - \dfrac{6b^2}{4b}\right)$
8. $3b\left(2b - 4 - \dfrac{4b^2}{6b}\right)$
9. $4a\left(\dfrac{2}{a} + \dfrac{3a}{2} - 4\right)$
10. $2a\left(\dfrac{3}{a} + \dfrac{6a}{4} - 3\right)$
11. $3a\left(\dfrac{3}{2a} + \dfrac{2a}{3} - 2\right)$
12. $5a\left(\dfrac{2}{3a} + \dfrac{3a}{5} - 3\right)$
13. $3x\left(\dfrac{1}{3} - \dfrac{2x^2}{3x} + 2\right)$
14. $2x\left(\dfrac{1}{8} - \dfrac{3x^2}{2x} + 3\right)$
15. $6x\left(\dfrac{1}{2} + \dfrac{4x^2}{3x} - 3\right)$
16. $10x\left(\dfrac{1}{5} + \dfrac{3x^2}{2x} - 2\right)$
17. $8b\left(\dfrac{1}{4} + \dfrac{5b^2}{8b} - 1\right)$
18. $7b\left(\dfrac{3}{7} + \dfrac{4b^2}{7b} - 2\right)$
19. $3b\left(\dfrac{3}{4} + \dfrac{2x^2}{5b} + 1\right)$
20. $5b\left(\dfrac{5}{6} + \dfrac{3b^2}{4b} + 2\right)$

End of Chapter Problems 9-3

Perform the indicated operations:

1. $(a + 3)^2$
2. $(a + 5)^2$
3. $(a - 6)^2$
4. $(a - 2)^2$

5. $(2x - 4)^2$
6. $(3x - 2)^2$
7. $(4x + 3)^2$
8. $(3x + 3)^2$
9. $(2a + 3b)^2$
10. $(a^4 + 2b)^2$
11. $(x^3 - 4y)^2$
12. $(x^4 - 2y)^2$
13. $(2x - 3y)^2$
14. $(3x - 4y)^2$
15. $(a + 1)(a + 3)$
16. $(y + 2)(y + 7)$
17. $(y + 3)(y - 6)$
18. $(a + 5)(a - 1)$
19. $(a - 2)(a + 6)$
20. $(a - 8)(a + 3)$
21. $(x - 4)(x - 1)$
22. $(x - 6)(x - 5)$
23. $(a + 4)(a - 4)$
24. $(a + 8)(a - 8)$
25. $(2x + 3)(2x - 3)$
26. $(3x - 2)(3x + 2)$
27. $(x + y)(x + 3y)$
28. $(x + 2y)(x + 3y)$
29. $(x - 5y)(x + 4y)$
30. $(x - 6y)(x + 2y)$
31. $(a + 2b)(4a - 3b)$
32. $(a + 3b)(3a - 3b)$
33. $(a - 4b)(2a - 2b)$
34. $(a - 5b)(a - 2b)$
35. $(2a + 3b)(3a - 4b)$
36. $(4a + 2b)(3a - 3b)$
37. $(3x + 4y)(2x + 5y)$
38. $(2x + 3y)(3x + 3y)$
39. $(2x - 5y)(2x - 3y)$
40. $(3x - 2y)(4x - 5y)$

End of Chapter Problems 9-4

Perform the indicated operations:

1. $(x^3 + 3x^2 + 2x) \div x$
2. $(x^4 + 2x^3 + x^2) \div x^2$
3. $(6x^3 + 2x^2 + 4x) \div 2x$
4. $(12x^4 - 6x^3 + 9x^2 - 3x) \div 3x$
5. $(16x^4 - 12x^3 + 8x^2) \div 4x^2$
6. $(3x^4 - 3x^3 - 6x) \div 3x$
7. $(x^2 + 5xy + 6y^2) \div (x + 2y)$
8. $(x^2 + 4xy + 3y^2) \div (x + y)$
9. $(x^2 + 4x - 21) \div (x + 7)$
10. $(x^2 + 2x - 8) \div (x - 2)$
11. $(x^2 - 49) \div (x + 7)$
12. $(x^2 - 9) \div (x - 3)$
13. $(9x^2 - 12x + 4) \div (3x - 2)$
14. $(9x^2 + 18x + 9) \div (3x + 3)$
15. $(a^2 - 25) \div (a + 5)$
16. $(b^2 - 11b + 30) \div (b - 6)$
17. $(6x^2 + xy - 12y^2) \div (3x - 4y)$
18. $(12x^2 - 6xy - 6y^2) \div (4x + 2y)$
19. $(9a^2 - 24ab + 16b^2) \div (3a - 4b)$
20. $(24a + 16a^2 + 9) \div (3 + 4a)$
21. $(x^2 - 6xy - 27y^2) \div (x + 3y)$
22. $(8b^2 + 15a^2 - 26ab) \div (5a - 2b)$
23. $(16 - 12x^2 - 16x) \div (2 - 3x)$
24. $(12 - 14x - 10x^2) \div (4 + 2x)$
25. $(12x^2 - 4x - 5) \div (3x + 2)$
26. $(20 + 24a^2 - 50a) \div (3a - 4)$
27. $(8x^2 + 20xy + 15y^2) \div (4x + 5y)$
28. $(10a^2 + 4b^2 - 17ab) \div (5a - 4b)$
29. $(4a^2 + 2a + 2) \div (a + 1)$
30. $(a^2 + 3a - 4) \div (a - 2)$
31. $(6a^2 - 3a - 2) \div (a - 1)$
32. $(4a^2 - 4a + 4) \div (2a + 1)$
33. $(a^2 + 4ab + b^2) \div (a + b)$
34. $(a^2 - 3ab - 2b^2) \div (a - b)$
35. $(6x^3 + 3x^2 + 4x + 4) \div (x + 1)$
36. $(4a^3 - 6a^2 + 4a - 3) \div (2a - 1)$

End of Chapter Problems 9-5

Factor the following:

1. $3a + 9$
2. $12b - 18$
3. $7b - 21$
4. $6x + 12$
5. $4x^2y + 6xy$
6. $6x^2y - 3xy$
7. $16a^2 + 2a$
8. $18a^2 + 3a$
9. $24a^2bc^3 - 16ab^2c^2$
10. $140a^3bc^2 - 28a^2b^2c^3$
11. $12a^3b^2 + 9a^2b^4 - 6a^4b^4$
12. $21a^4b^3 + 7ab^3 - 14a^2b^2$
13. $10x^3y^2 - 20x^4y^4 + 15xy^3$
14. $18xy^2 - 9x^2y^3 + 18x^2y$
15. $15a^3b^2c^4 - 3a^2b^3c^4 - 12b^2c$
16. $16a^2bc^3 - 4a^2b^2c^4 - 8a^3b^2c^2$
17. $35x^4y^2z^3 + 25x^2y^4z^4 - 30x^2y^2z^2$
18. $42x^2y^3z^4 - 21x^3yz^4 + 63x^4y^3z^3$
19. $24x^3y^2z^4 - 42x^2y^3z^2 - 72x^2yz^2$
20. $48x^2y^2z + 96x^3y^3z - 144x^3y^2z^2$
21. $\frac{7x^2z^3}{12} - \frac{14x^3y^2z^2}{9}$
22. $\frac{15x^3z^2}{28} - \frac{45x^2y^2z^3}{14}$
23. $\frac{12x^4y^2z}{35} + \frac{24x^2y^3z^3}{55} - \frac{18x^3y^3z^4}{25}$
24. $\frac{21x^2y^3z^4}{32} - \frac{63x^3y^3z^3}{64} + \frac{35x^4y^2z^4}{16}$
25. $\frac{21a^3c^2}{22b^2} + \frac{28a^2c^2}{33b^3} - \frac{42a^3c^3}{44b^2}$
26. $\frac{52a^3b^2}{55c} - \frac{26a^2b^3}{100c^2} + \frac{104a^2b^2}{165c}$

End of Chapter Problems 9-6

Factor the following:

1. $x^2 + 5x + 6$
2. $x^2 + 9x + 14$
3. $x^2 + 4x - 32$
4. $a^2 - 3a - 28$
5. $a^2 + 6a + 9$
6. $a^2 + 18a + 81$
7. $a^2 - 16$
8. $x^2 - 64$
9. $x^2 - 49$
10. $x^2 - 25$
11. $x^2 - 8x + 15$
12. $y^2 - 13y + 42$
13. $a^2 - 12a + 35$
14. $a^2 - 14a + 45$
15. $a^2 - 11a + 28$
16. $a^2 - 9a + 18$
17. $y^2 + 5y - 36$
18. $y^2 + 3y - 40$
19. $x^2 - 2x - 35$
20. $x^2 + 7x - 18$
21. $x^2 - 2x - 24$
22. $x^2 - 5x - 14$
23. $6x^2 + 5x - 6$
24. $3x^2 + 10x + 8$
25. $8a^2 - 20a + 8$
26. $4a^2 - 19a + 12$
27. $9x^2 + 6x - 8$
28. $8x^2 - 10x - 12$
29. $6b^2 - 18b + 12$
30. $15b^2 - 29b + 12$
31. $10a^2 + 31a + 15$
32. $18a^2 + 6a - 4$
33. $2x^2 + 7xy + 6y^2$
34. $20a^2 + 5ab - 15b^2$
35. $20a^2 - 7ab - 6b^2$
36. $12x^2 - 20xy - 8y^2$
37. $8x^2 + 20xy - 12y^2$
38. $20x^2 + 18xy - 14y^2$
39. $12x^2 + 18xy + 6y^2$
40. $15x^2 + 26xy + 8y^2$

Answers to Self-tests

SELF-TEST 9-1

1. $3a^2 - 18a$
2. $6x^2y + 4xy^2$
3. $a^2 - 18a + 81$
4. $16a^2 + 16a + 4$
5. $a^2 + 5a - 6$
6. $x^2 - 4xy - 12y^2$
7. $8x^2 + 20xy + 12y^2$
8. $6a^2 - 18ab + 12b^2$
9. $2a^2 - 4a$
10. $x - 6$
11. $x + 4$
12. $a + 10b$
13. $x + 4 - \dfrac{6}{x + 9}$
14. $a + 8b + \dfrac{4b^2}{a + 2b}$

SELF-TEST 9-2

1. $9(x - 4)$
2. $8y(7y + 1)$
3. $(a + 12)(a + 3)$
4. $(a + 3)(a - 10)$
5. $(x + 7y)(x - 5y)$
6. $(x + 6y)^2$
7. $2(x + 4)^2$
8. $3(a + b)(a - b)$
9. $(a + 6b)(a - 9b)$
10. $(a - 3b)^2$

CHAPTER 10

FRACTIONAL EQUATIONS

INTRODUCTION

Many of the equations we work with in electronics are more complex than the ones we worked with in Chapter 7. Sometimes the unknown is part of a polynomial that can be in either the numerator or the denominator of a term.

In Chapters 8 and 9 we laid the groundwork for the solution of most complex equations. In this chapter we will be solving problems similar to most of the problem types you will encounter in your electronics studies. You will be given practical problems to solve and will be given various methods to solve them.

In solving second-degree equations, there are always two possible solutions or *roots*. Such equations are called *quadratic equations*. The roots may or may not be equal and may or may not have the same sign. We will learn how to use the quadratic to help in solving problems that we cannot factor.

CHAPTER OBJECTIVES

In this chapter you will learn how to:

1. Solve fractional linear equations.
2. Solve linear equations where the unknown is part of a polynomial.
3. Solve electrical problems where one of the quantities is unknown.
4. Use the quadratic equation to solve second-degree equations.

10-1 GENERAL EQUATIONS

When we have an equation to solve that involves fractions or mixed numbers, we usually have to get rid of the fraction in order to simplify the solution.

An easy rule to remember is:

RULE 10-1 To solve an equation that includes fractions or mixed numbers: (1) Get rid of the fraction (or mixed number); (2) solve for the unknown; (3) verify your answer.

Consider the equation in Example 10-1.

☐ EXAMPLE 10-1

Solve for a.

$$\frac{1}{3} + \frac{a}{4} = 2$$

Solution The first step in the solution is to get rid of the fraction. We get rid of the fraction by first finding the CD (common denominator) and then multiplying each term by it. In this example, the CD is 12. Now we multiply each term by 12.

$$\frac{1}{3}(12) + \frac{a}{4}(12) = 2(12)$$

$$4 + 3a = 24$$

Remember, whatever we do to one term, we must do to *all* terms on both sides of the equation. The last step is to solve for a.

$$4 + 3a = 24$$

$$3a = 20$$

$$a = \frac{20}{3} = 6\frac{2}{3}$$

(When we have an improper fraction, we change it to a mixed number.)

Check

$$\frac{1}{3} + \frac{\frac{20}{3}}{4} = 2$$

$$\frac{1}{3} + \frac{5}{3} = 2$$

$$2 = 2$$

Note: $$\frac{\frac{20}{3}}{4} = \frac{20}{3} \div 4 = \frac{20}{3} \times \frac{1}{4} = \frac{20}{12} = \frac{5}{3}$$

□

Let's try one in which the unknown is in the denominator.

□ EXAMPLE 10-2

Solve for x.

$$\frac{5}{x-1} - \frac{1}{2} = 2$$

Solution In this example the denominators are 2 and $x - 1$. This makes the CD $2(x - 1)$. To get rid of the fractions, we multiply both sides by this CD. Remember, we must multiply *each* term in each member by this CD.

$$\frac{5 \cdot 2(x-1)}{x-1} - \frac{2(x-1)}{2} = 2 \cdot 2(x-1)$$

Cancel like terms:

$$\frac{5 \cdot 2\cancel{(x-1)}}{\cancel{x-1}} - \frac{\cancel{2}(x-1)}{\cancel{2}} = 2 \cdot 2(x-1)$$

$$10 - x + 1 = 4x - 4$$

$$11 - x = 4x - 4$$

$$11 = 5x - 4$$ (move $-x$ to the right side and it becomes $+x$; that is, $4x + x = 5x$)

$$15 = 5x$$ (move -4 to the left side and it becomes $+4$, that is, $11 + 4 = 15$)

$$3 = x$$

or $$x = 3$$

Check

$$\frac{5}{3-1} - \frac{1}{2} = 2$$

$$\frac{5}{2} - \frac{1}{2} = 2$$

$$2 = 2$$

□

Let's consider an equation that includes more than one variable and solve for each of them.

☐ EXAMPLE 10-3

$\frac{2}{x} - \frac{3}{y} = 4$. Solve for x and y.

Solution In this example the CD is xy. Multiplying each term by xy, we get

$$\frac{2xy}{x} - \frac{3xy}{y} = 4xy$$

$$2y - 3x = 4xy \qquad (10\text{-}1)$$

This cleared the fractions and we now have a simple equation. Let's first solve for x. To do so, we first put all terms containing x on one side of the equation.

$$2y = 4xy + 3x \quad \text{(move } -3x \text{ to the right side)}$$

or

$$4xy + 3x = 2y$$

Now we factor out the x.

$$x(4y + 3) = 2y$$

Next we move the $4y + 3$ to the right side.

$$\frac{x(4y + 3)}{4y + 3} = \frac{2y}{4y + 3} \quad \text{(divide both sides by } 4y + 3\text{)}$$

$$\frac{x\cancel{(4y + 3)}}{\cancel{4y + 3}} = \frac{2y}{4y + 3} \quad \text{(cancel like terms)}$$

$$x = \frac{2y}{4y + 3}$$

Now let's solve for y. After multiplying all terms of the original equation by xy, we had

$$2y - 3x = 4xy \qquad (10.1)$$

Now we put all terms containing y on one side (because we are solving for y) and move all other terms to the other side.

$$2y - 3x - 4xy = 0 \quad \text{(move } 4xy \text{ to the left side)}$$

$$2y - 4xy = 3x \quad \text{(move } -3x \text{ to the right side)}$$

Next we factor out y.

$$y(2 - 4x) = 3x$$

Now we move $2 - 4x$ to the right side.

$$\frac{y(2-4x)}{2-4x} = \frac{3x}{2-4x} \qquad \text{(divide each side by } 2 - 4x\text{)}$$

$$y = \frac{3x}{2-4x}$$

□ EXAMPLE 10-4

$\frac{2}{x+2} + \frac{3}{x-3} = \frac{4}{x+2}$. Solve for x.

Solution In this example, the CD is $(x + 2)(x - 3)$. These terms, $(x + 2)$ and $(x - 3)$, are prime factors, so we leave them as they are. That is, we don't find their product. Multiply each term by the CD.

$$\frac{2(x+2)(x-3)}{x+2} + \frac{3(x+2)(x-3)}{x-3} = \frac{4(x+2)(x-3)}{x+2}$$

Cancel like terms in numerators and denominators.

$$\frac{2\cancel{(x+2)}(x-3)}{\cancel{x+2}} + \frac{3(x+2)\cancel{(x-3)}}{\cancel{x-3}} = \frac{4\cancel{(x+2)}(x-3)}{\cancel{x+2}}$$

$$2(x-3) + 3(x+2) = 4(x-3)$$

Combine like terms and solve for x.

$$2x - 6 + 3x + 6 = 4x - 12$$

$$2x + 3x - 4x = 6 - 6 - 12$$

$$x = -12$$

Check

$$\frac{2}{-12+2} + \frac{3}{-12-3} = \frac{4}{-12+2}$$

$$\frac{2}{-10} + \frac{3}{-15} = \frac{4}{-10}$$

$$-\frac{1}{5} - \frac{1}{5} = -\frac{2}{5}$$

$$-\frac{2}{5} = -\frac{2}{5}$$

□

PRACTICE PROBLEMS 10-1

Solve for the unknown in the following problems. Express your answers as fractions or mixed numbers.

1. $\frac{8x}{2} - 4 = 10$
2. $\frac{R}{2} + \frac{R}{4} = 3$
3. $\frac{Z + 1}{3} - Z = 4 - 2Z$
4. $\frac{4}{I + 2} + 2 = 5$
5. $\frac{3I}{2} - 2 = \frac{I}{2} + 4$
6. $\frac{4x}{2} - \frac{3 + x}{5} = \frac{-2x + 3}{4}$
7. $\frac{6 - x}{x} - \frac{4}{x} = \frac{3}{x}$
8. $\frac{4}{x + 3} - \frac{1}{3} = 4$
9. $\frac{x}{a} + \frac{x}{b} = 2$. Solve for a, b, and x.
10. $3(2x + 1) = 2y + x - 3$. Solve for x and y.
11. $\frac{1}{a + 4} - \frac{2}{a - 2} = \frac{3}{a + 4}$
12. $\frac{2}{3a + 2} - \frac{3}{2a + 4} = \frac{5}{3a + 2}$

Solutions

1. $x = 3\frac{1}{2}$
2. $R = 4$
3. $Z = 2\frac{3}{4}$
4. $I = -\frac{2}{3}$
5. $I = 6$
6. $x = \frac{27}{46}$
7. $x = -1$
8. $x = -2\frac{1}{13}$
9. $a = \frac{bx}{2b - x}, b = \frac{ax}{2a - x}, x = \frac{2ab}{a + b}$
10. $x = \frac{2y - 6}{5} \quad y = \frac{5x + 6}{2}$
11. $a = -1$
12. $a = -1\frac{1}{5}$

Additional practice problems are at the end of the chapter.

10-2 SOME REAL EQUATIONS

In this section we will deal with equations that are used in circuit analysis. We have already used some of the equations in previous chapters. For instance, we have solved this equation for V_1, where we were given the values for the other variables.

$$V_1 = \frac{ER_1}{R_1 + R_2}$$

■ **KEY POINT** When unknown variables appear in more than one term, we must factor out the unknown to arrive at a solution.

In the following example such a case is presented.

☐ EXAMPLE 10-5

Given the equation $V_1 = \dfrac{ER_1}{R_1 + R_2}$, solve for R_1.

Solution Notice that R_1 appears in both the numerator and denominator of the right side. The R_1's can't cancel because there are two terms in the denominator, and R_1 appears in only one of them. R_1 would have to be common to *all* terms in the fraction to be canceled. Our first step will be to get rid of the fraction. We can do that by multiplying each term by $R_1 + R_2$.

$$V_1(R_1 + R_2) = \frac{ER_1(R_1 + R_2)}{R_1 + R_2} \qquad \text{(multiplying both sides by } R_1 + R_2 \text{ gets rid of the fraction)}$$

$$V_1(R_1 + R_2) = ER_1$$

Next, we need to remove the parentheses on the left side because the variable we are solving for, R_1, is inside. We could remove the need for the parentheses by dividing each side by V_1, but we would still have R_1 in the right side as part of a fraction. A better step is to perform the indicated multiplication.

$$V_1R_1 + V_1R_2 = ER_1$$

Next we put all terms containing R_1 on the left side and move all other terms to the right side.

$$V_1R_1 + V_1R_2 - ER_1 = 0 \qquad \text{(move } ER_1 \text{ to the left side)}$$

$$V_1R_1 - ER_1 = -V_1R_2 \qquad \text{(move } V_1R_2 \text{ to the right side)}$$

We're solving for R_1, so we factor out R_1:

$$R_1(V_1 - E) = -V_1R_2$$

And finally we divide by $V_1 - E$.

$$\frac{R_1(V_1 - E)}{V_1 - E} = \frac{-V_1R_2}{V_1 - E}$$

$$R_1 = \frac{-V_1R_2}{V_1 - E} \qquad \text{(divide both sides by } V_1 - E\text{)}$$

We can get rid of the minus sign in the numerator by changing that sign and the sign of the denominator or the sign of the fraction. Remember, there are three signs associated with a fraction: the sign of the numerator, the sign of the denominator, and the sign of the fraction. We can change any two signs without changing the value of the fraction. Don't forget, we must change the sign of *each* term when changing the sign of the numerator or denominator.

We could have avoided the problem with the signs if we had moved V_1R_2 to the right side in our second step. Then

$$V_1R_1 + V_1R_2 = ER_1$$
$$V_1R_2 = ER_1 - V_1R_1$$
$$V_1R_2 = R_1(E - V_1)$$
$$\frac{V_1R_2}{E - V_1} = R_1$$

or

$$R_1 = \frac{V_1R_2}{E - V_1}$$

□ EXAMPLE 10-6

Given the equation $r_i = \beta\left(r_e + \frac{r_b}{\beta}\right)$. Solve for r_e.

Solution In this example, as in many others, we find that there is more than one approach to the solution. The trick is to find the easiest one—the one that requires the fewest steps. In this case, let's divide both sides by β. This step isolates r_e in one term.

$$\frac{r_i}{\beta} = r_e + \frac{r_b}{\beta} \qquad \text{(divide both sides by } \beta\text{)}$$

$$r_e = \frac{r_i}{\beta} - \frac{r_b}{\beta} \qquad \text{(subtract } \frac{r_b}{\beta} \text{ from both sides)}$$

□

PRACTICE PROBLEMS 10-2

1. $I_2 = \frac{I_TG_2}{G_1 + G_2}$. Solve for I_T, G_1, and G_2.

2. $V_1 = \frac{ER_1}{R_1 + R_2}$. Solve for E, R_1, and R_2.

3. $F = \frac{9}{5}C + 32$. Solve for C.

4. $A_i = \frac{h_{fe}}{h_{oe}R_L + 1}$. Solve for R_L.

5. $S = \frac{R_E + R_B}{R_E + R_B(1 - \alpha)}$. Solve for R_B.

6. $R_S = \frac{R_iR_o}{R_i + R_o}$. Solve for R_o.

7. $r_i = \beta\left(r_e + \frac{r_b}{\beta}\right)$. Solve for r_b.

8. $A_V = \frac{r_e}{r_e + r_e'}$. Solve for r_e.

Solutions

1. $I_T = \frac{I_2(G_1 + G_2)}{G_2}$

$G_1 = \frac{I_TG_2 - I_2G_2}{I_2}$

$G_2 = \frac{I_2G_1}{I_T - I_2}$

2. $E = \frac{V_1(R_1 + R_2)}{R_1}$

$R_1 = \frac{V_1R_2}{E - V_1}$

$R_2 = \frac{ER_1 - V_1R_1}{V_1}$

3. $C = \dfrac{5F - 160}{9}$

4. $R_L = \dfrac{h_{fe} - A_i}{A_i h_{oe}}$

5. $R_B = \dfrac{R_E - SR_E}{S - S\alpha - 1}$

6. $R_o = \dfrac{R_S R_i}{R_i - R_S}$

7. $r_b = r_i - \beta r_e$

8. $r_e = \dfrac{A_V r_e'}{1 - A_V}$

Additional practice problems are at the end of the chapter.

10-3 APPLICATIONS

Now let's carry the work in the previous section one step further. Let's determine actual values of unknowns just as we would do in circuit analysis.

☐ EXAMPLE 10-7

Given the equation $I_1 = \dfrac{I_T G_1}{G_1 + G_2}$, where $I_1 = 2.73$ mA, $I_T = 6.75$ mA, and $G_1 = 370\ \mu S$, find G_2.

Solution Let's first multiply both sides by $G_1 + G_2$ to get rid of the fraction. This results in

$$I_1(G_1 + G_2) = \frac{I_T G_1(G_1 + G_2)}{G_1 + G_2}$$

$$I_1(G_1 + G_2) = I_T G_1 \qquad (10\text{-}2)$$

Now we have two possible solutions. Let's examine both.

First solution:

$$G_1 + G_2 = \frac{I_T G_1}{I_1} \qquad \text{(divide both sides by } I_1\text{)}$$

$$G_2 = \frac{I_T G_1}{I_1} - G_1 \qquad \text{(move } G_1 \text{ to the right side)}$$

Then we plug in known values.

$$G_2 = \frac{6.75 \text{ mA} \times 370\ \mu\text{S}}{2.73 \text{ mA}} - 370\ \mu\text{S}$$

Using the calculator,

6.75 [EE] [+/-] 3 [×] 370 [EE] [+/-] 6 [÷]
2.73 [EE] [+/-] 3 [−] 370 [EE] [+/-] 6 [=]

Answer: 545×10^{-6}

$G_2 = 545\ \mu S$

Second solution:

$$I_1(G_1 + G_2) = I_T G_1 \qquad \text{(10-2)}$$

$$I_1 G_1 + I_1 G_2 = I_T G_1 \qquad \text{(remove parentheses in left side)}$$

$$I_1 G_2 = I_T G_1 - I_1 G_1 \qquad \text{(move } I_1 G_1 \text{ term to right side)}$$

$$G_2 = \frac{I_T G_1 - I_1 G_1}{I_1} \qquad \text{(divide by } I_1\text{)}$$

Then we plug in known values.

$$G_2 = \frac{6.75 \text{ mA} \times 370\ \mu\text{S} - 2.73 \text{ mA} \times 370\ \mu\text{S}}{2.73 \text{ mA}}$$

Using the calculator,

6.75 [EE] [+/-] 3 [×] 370 [EE] [+/-] 6 [−] 2.73 [EE] [+/-] 3 [×] 370 [EE] [+/-] 6 [=] [÷] 2.73 [EE] [+/-] 3 [=]

Answer: 545×10^{-6}

$$G_2 = 545\ \mu\text{S}$$

The first solution appears to be easier, but both resulted in the same answer.

□ EXAMPLE 10-8

Given the equation $R_{\text{TH}} = \dfrac{V_{\text{OC}} - V_L}{I_L}$, solve for V_{OC} when $R_{\text{TH}} = 1.5\ \text{k}\Omega$, $V_L = 3.75$ V, and $I_L = 5$ mA.

Solution We need to rearrange the equation and solve for V_{OC}; therefore, let's do it this way:

$$R_{\text{TH}} I_L = V_{\text{OC}} - V_L \qquad \text{(multiply by } I_L\text{)}$$

$$R_{\text{TH}} I_L + V_L = V_{\text{OC}} \qquad \text{(move } V_L \text{ to the left side)}$$

We then plug in known values.

$$V_{\text{OC}} = 1.5\ \text{k}\Omega \times 5 \text{ mA} + 3.75 \text{ V}$$

$$= 7.5 \text{ V} + 3.75 \text{ V} = 11.25 \text{ V}$$

□

PRACTICE PROBLEMS 10-3

1. $V_1 = \dfrac{ER_1}{R_1 + R_2}$

(a) Solve for E when $V_1 = 34$ V, $R_1 = 270\ \Omega$, and $R_2 = 330\ \Omega$.

(b) Solve for R_1 when $E = 25$ V, $V_1 = 14.7$ V, and $R_2 = 3.3\ \text{k}\Omega$.

2. $I_L = \dfrac{I_{\text{SC}} G_L}{G_L + G_N}$

(a) Solve for G_L when $I_{\text{SC}} = 10$ mA, $I_L = 3$ mA, and $G_N = 300\ \mu$S.

(b) Solve for G_N when $I_{\text{SC}} = 600\ \mu$A, $I_L = 200\ \mu$A, and $G_L = 2.7$ mS.

3. $A_i = \dfrac{h_{fe}}{1 + h_{oe} R_L}$
Solve for h_{fe} when $A_i = 65$, $h_{oe} = 10\ \mu S$, and $R_L = 5\ k\Omega$.

4. $\beta = \dfrac{\alpha}{1 - \alpha}$
Solve for α when $\beta = 90$.

5. $R_{TH} = \dfrac{V_{OC} - V_L}{I_L}$
Solve for V_L when $R_{TH} = 1.7\ k\Omega$, $V_{OC} = 12$ V, and $I_L = 750\ \mu A$.

Solutions

1. (a) $E = 75.6$; (b) $R_1 = 4.71\ k\Omega$
2. (a) $G_L = 129\ \mu S$; (b) $G_N = 5.40$ mS
3. $h_{fe} = 68.3$
4. 0.989
5. 10.7 V

Additional practice problems are at the end of the chapter.

SELF-TEST 10-1

1. $\dfrac{R + 2}{3} - R = 2 - R$
Solve for R.

2. $\dfrac{2}{I - 2} + 3 = 5$
Solve for I.

3. $\dfrac{a}{2} + \dfrac{3}{b} = \dfrac{2}{x}$
Solve for a, b, and x.

4. $I_L = \dfrac{I_{SC} G_L}{G_L + G_N}$
Solve for G_L when $I_{SC} = 750\ \mu A$, $I_L = 200\ \mu A$, and $G_N = 300\ \mu S$.

5. $R_T = \dfrac{R_1 R_2}{R_1 + R_2}$
Solve for R_2 when $R_T = 38.7\ k\Omega$ and $R_1 = 68\ k\Omega$.

Answers to Self-test 10-1 are at the end of the chapter.

10-4 QUADRATIC EQUATIONS

■ **KEY POINT** Quadratic equations are second-degree equations and have two solutions or roots.

In other words, the unknown is raised to the second power. Some examples of these quadratic equations are $x^2 = 25$, $2x^2 + x + 2 = 0$, and $x^2 + x = 4$. A quadratic equation in standard form looks like this:

$$ax^2 + bx + c = 0 \qquad (10\text{-}3)$$

where a and b are the coefficients of x^2 and x and where c is a constant. If the equation is solved for x, the result is

$$x = \frac{-b \pm \sqrt{b^2 - 4ac}}{2a} \tag{10-4}$$

When we solve for the unknown in a second-degree equation, there will always be two possible solutions called *roots*. Sometimes we can factor the quadratic to find the two roots. If the quadratic is not factorable, then we must resort to the quadratic equation. In the following examples we will first solve for the unknown by factoring, and then we will solve by using the quadratic equation.

□ EXAMPLE 10-9

Solve the equation $x^2 - 25 = 0$.

Solution Quadratic equations that have no x term are called *pure quadratics*. Such equations are readily solved by moving the constant to the right member and finding the square root of both members.

$$x^2 = 25$$
$$x = \sqrt{25} = 5$$

The equation is satisfied when $x = 5$ or -5.

$$5^2 = 25 \qquad (-5)^2 = 25$$

□

10-4-1 Solving by Factoring

In Chapter 9 we determined that *trinomials can be factored if we can find two numbers whose sum is the coefficient of the second term and whose product is the third term*. In the following examples, we show how to solve for the unknown when the equation is factorable.

□ EXAMPLE 10-10

Find x in the equation $x^2 + 7x + 12 = 0$.

Solution

$$x^2 + 7x + 12 = 0$$
$$(x + 3)(x + 4) = 0 \qquad \text{(factor the left side)}$$

Let each factor equal zero in turn, and solve for x.

$$x + 3 = 0$$
$$x = -3$$
$$x + 4 = 0$$
$$x = -4$$

The equation is satisfied when $x = -3$ or $x = -4$.

☐ EXAMPLE 10-11

$x^2 - 3x - 6 = 4$. Solve for x.

Solution Put the equation in standard form and then factor the left member.

$$x^2 - 3x - 10 = 0$$
$$(x + 2)(x - 5) = 0$$

Equating each factor to zero, we get

$$x + 2 = 0$$
$$x = -2$$
$$x - 5 = 0$$
$$x = 5$$

The equation is satisfied when $x = -2$ or $x = 5$. ☐

PRACTICE PROBLEMS 10-4

Solve the following equations for x.

1. $x^2 = 49$
2. $x^2 - x - 30 = 0$
3. $x^2 + 5x + 6 = 0$
4. $x^2 - 4x - 10 = 11$
5. $x^2 + 10x + 12 = -12$

Solutions

1. $x = 7, x = -7$
2. $x = -5, x = 6$
3. $x = -2, x = -3$
4. $x = 7, x = -3$
5. $x = -4, x = -6$

Additional practice problems are at the end of the chapter.

10-4-2 Solving by the Quadratic Equation

When we have a second-degree equation to solve, we first put it into the form $ax^2 + bx + c = 0$. If the equation is factorable, we merely factor as before. If the equation is not factorable, then we resort to using the quadratic equation.

☐ EXAMPLE 10-12

Given the equation $x^2 + 4x + 5 = 13$, solve for x.

Solution First, we put the equation in standard form:

$$x^2 + 4x - 8 = 0$$

Next we determine if the equation can be factored. Since this one does not factor, we will solve it by using the quadratic equation:

$$x = \frac{-b \pm \sqrt{b^2 - 4ac}}{2a} \qquad (10\text{-}4)$$

An examination of the equation shows that a (the coefficient of x^2) = 1; b (the coefficient of x) = 4; and c (the constant) = −8. Substituting these values in our equation results in the following:

$$x = \frac{-4 \pm \sqrt{4^2 - (4)(1)(-8)}}{2(1)} = \frac{-4 \pm \sqrt{16 + 32}}{2}$$

$$= \frac{-4 \pm \sqrt{48}}{2} = \frac{-4 \pm 6.93}{2}$$

At this point we see that there are two solutions:

1. $x = \dfrac{-4 + 6.93}{2} = \dfrac{2.93}{2} = 1.46$

2. $x = \dfrac{-4 - 6.93}{2} = \dfrac{-10.93}{2} = -5.46$

□

PRACTICE PROBLEMS 10-5

Solve the following equations for x:

1. $2x^2 + 3x - 9 = 0$

2. $x^2 - 5x + 2 = 10$

3. $3x^2 - 6x = 7$

4. $x^2 - 3x - 12 = -5$

Solutions

1. $x = 1.5$, $x = -3$

2. $x = 6.28$, $x = -1.28$

3. $x = 2.83$, $x = -0.83$

4. $x = -1.54$, $x = 4.54$

Additional practice problems are at the end of the chapter.

SELF-TEST 10-2

Solve for x in the following problems:

1. $x^2 - 121 = 0$

2. $x^2 - 6x - 27 = 0$

3. $x^2 - 10x + 30 = 3x - 12$

4. $2x^2 + 4x - 5 = 0$

5. $2x^2 - 5x - 3 = 0$

Answers to Self-test 10-2 are at the end of the chapter.

CHAPTER 10 AT A GLANCE

Page	Key Point or Rule	Example
261	*Rule 10-1:* To solve an equation that includes fractions or mixed numbers: (1) get rid of the fraction (or mixed number); (2) solve for the unknown; (3) verify your answer.	$\frac{1}{3} + \frac{a}{4} = 2$ $\frac{1}{3}(12) + \frac{a}{4}(12) = 2(12)$ $4 + 3a = 24$ $a = \frac{20}{3} = 6\frac{2}{3}$ See page 261.
266	*Key point:* When unknown variables appear in more than one term, we must factor out the unknown to arrive at a solution.	$V_1R_1 + V_2 + R_2 = ER_1$ Solve for R_1. $R_1(V_1 - E) = V_2R_2$ $R_1 = \frac{V_2R_2}{V_1 - E}$
270	*Key point:* Quadratic equations are second-degree equations and have two solutions or *roots*.	$x^2 + 7x + 12 = 0$ $(x + 3)(x + 4) = 0$ $x = -3, x = -4$

End of Chapter Problems 10-1

Solve for the unknown in the following problems. Express your answers as proper fractions or as mixed numbers.

1. $\frac{5x}{2} + 3 = 6$

2. $\frac{6x}{4} - 2 = 5$

3. $\frac{4a}{3} - 5 = 2$

4. $\frac{2b}{5} - 4 = 12$

5. $\frac{2I}{4} + \frac{I}{6} = 2$

6. $\frac{3R}{2} + \frac{2R}{3} = 6$

7. $\frac{2x}{3} - \frac{3x}{4} = 1$

8. $\frac{3x}{4} - \frac{7x}{2} = 3$

9. $\frac{4R}{2} = 3 + \frac{R}{3}$

10. $\frac{3R}{4} - 2 = \frac{2R}{3}$

11. $\frac{R + 3}{4} - 2 = \frac{3R + 1}{2} + 4$

12. $\frac{2R - 3}{3} - 4 = \frac{2R + 3}{5} - 3$

13. $\frac{R - 2}{5} + 3 = \frac{2R + 2}{3} + 4$

14. $\frac{a + 3}{2} + 4 = \frac{3a - 2}{3} + 1$

15. $\frac{c - 3}{4} - \frac{c + 1}{3} = 2 + 3c$

16. $\frac{R + 2}{3} - \frac{R - 1}{2} = 3 - 2R$

17. $\frac{3}{I - 3} + 3 = 4$

18. $\frac{4}{I - 5} + 2 = 5$

19. $\dfrac{4}{R-2} + 2 = 3$

20. $\dfrac{5}{R+3} - 4 = 2$

21. $\dfrac{5x}{2} + \dfrac{x-2}{6} = \dfrac{3x-6}{4}$

22. $\dfrac{2x}{4} - \dfrac{3-x}{2} = \dfrac{2x-1}{6}$

23. $\dfrac{4a}{3} + \dfrac{a+2}{5} = \dfrac{2a-3}{5}$

24. $\dfrac{3c}{6} - \dfrac{c-3}{3} = \dfrac{3c-2}{2}$

25. $\dfrac{4-x}{2x} - \dfrac{3}{x} = \dfrac{1}{x}$

26. $\dfrac{3x-1}{2x} + \dfrac{1}{x} = \dfrac{2}{x}$

27. $\dfrac{3-y}{2y} - \dfrac{2}{y} = \dfrac{3}{y}$

28. $\dfrac{4+y}{3y} + \dfrac{3}{y} = \dfrac{1}{y}$

29. $\dfrac{3}{G-3} - \dfrac{1}{4} = 5$

30. $\dfrac{2}{R+3} - \dfrac{1}{3} = 1$

31. $\dfrac{2}{R+3} - \dfrac{1}{3} = 4$

32. $\dfrac{3}{R-5} - \dfrac{2}{3} = 6$

33. $\dfrac{1}{a} + \dfrac{1}{b} = 2$. Solve for a and b.

34. $\dfrac{2}{a} + \dfrac{3}{b} = 4$. Solve for a and b.

35. $\dfrac{2}{a} - 3 = \dfrac{1}{b}$. Solve for a and b.

36. $\dfrac{3}{a} - 4 = \dfrac{2}{b}$. Solve for a and b.

37. $\dfrac{x}{2a} + \dfrac{x}{3b} = 2$. Solve for x, a, and b.

38. $\dfrac{2x}{3a} - \dfrac{x}{b} = 3$. Solve for x, a, and b.

39. $\dfrac{3x}{2a} - \dfrac{x}{3b} = 4$. Solve for x, a, and b.

40. $\dfrac{x}{6a} - \dfrac{4x}{5b} = 1$. Solve for x, a, and b.

41. $\dfrac{3}{a-2} + \dfrac{2}{a+2} = \dfrac{b}{a+2}$. Solve for a and b.

42. $\dfrac{1}{x+3} - \dfrac{2}{x-4} = \dfrac{2y}{x-4}$. Solve for x and y.

43. $\dfrac{2}{2a-1} + \dfrac{3}{3a+2} = \dfrac{2b}{2a-1}$. Solve for a and b.

44. $\dfrac{3}{3a-3} + \dfrac{2}{2a+3} = \dfrac{3b}{2a+3}$. Solve for a and b.

45. $\dfrac{5}{3b-2} - \dfrac{4}{3b-1} = \dfrac{3}{3b-1}$. Solve for b.

46. $\dfrac{3}{2R+2} + \dfrac{5}{3R-3} = \dfrac{2}{3R-3}$. Solve for R.

End of Chapter Problems 10-2

1. $V_L = \dfrac{V_{OC} R_L}{R_{TH} + R_L}$. Solve for V_{OC}, R_L, and R_{TH}.

2. $V_2 = \dfrac{ER_2}{R_1 + R_2}$. Solve for E, R_1, and R_2.

3. $I_L = \dfrac{I_{SC} G_L}{G_L + G_N}$. Solve for I_{SC}, G_L, and G_N.

4. $I_1 = \dfrac{I_T G_1}{G_1 + G_2}$. Solve for I_T, G_1, and G_2.

5. $R_S = \dfrac{R_i R_o}{R_i + R_o}$. Solve for R_i and R_o.

6. $S = \dfrac{R_E + (R/\beta)}{R_E + I(1-\alpha)}$. Solve for R and R_E.

7. $r_i = \beta\left(r_e + \dfrac{r_b}{\beta}\right)$. Solve for r_e and r_b.

8. $A_V = \dfrac{r_e}{r_e + r_i}$. Solve for r_e and r_i.

9. $R_T = \dfrac{R_1 R_2}{R_1 + R_2}$. Solve for R_1 and R_2.

10. $Z_T = \dfrac{Z_1 Z_2}{Z_1 + Z_2}$. Solve for Z_1 and Z_2.

11. $A = \dfrac{B}{1 + (1/A)}$. Solve for B.

12. $B = \dfrac{A}{2 + (1/B)}$. Solve for A.

13. $f = \dfrac{1}{2\pi C(R_1 + R_2)}$. Solve for R_1.

14. $I_C = \dfrac{V_{CC}}{R_E + (R_B/\beta)}$. Solve for β, R_E, and R_b.

15. $I_B = \dfrac{V_{CC} - V_B}{R_1 + R_2}$. Solve for V_B and R_1.

16. $K = \dfrac{1}{1 + (\beta R_E/R_B)}$. Solve for R_B and R_E.

End of Chapter Problems 10-3

1. $V_2 = \dfrac{ER_2}{R_1 + R_2}$
   (a) Solve for R_2 when $E = 35$ V, $V_2 = 10$ V, and $R_1 = 12$ kΩ.
   (b) Solve for R_1 when $E = 15$ V, $V_2 = 6.3$ V, and $R_2 = 560$ Ω.

2. $V_2 = \dfrac{ER_2}{R_1 + R_2}$
   (a) Solve for R_2 when $E = 20$ V, $V_2 = 11.1$ V, and $R_1 = 12$ kΩ.
   (b) Solve for R_1 when $E = 10$ V, $V_2 = 4.13$ V, and $R_2 = 330$ Ω.

3. $I_1 = \dfrac{I_T G_1}{G_1 + G_2}$
   (a) Solve for I_T when $I_1 = 4.95$ mA, $G_1 = 21.3$ μS, and $G_2 = 30.3$ μS.
   (b) Solve for G_2 when $I_T = 100$ μA, $I_1 = 45.1$ μA, and $G_1 = 1.47$ mS.

4. $I_1 = \dfrac{I_T G_1}{G_1 + G_2}$
   (a) Solve for I_T when $I_1 = 6.35$ mA, $G_1 = 3$ mS, and $G_2 = 5.83$ mS.
   (b) Solve for G_2 when $I_T = 275$ μA, $I_1 = 120$ μA, and $G_1 = 50$ μS.

5. $A_i = \dfrac{h_{fe}}{1 + h_{oe} R_L}$
   (a) Solve for h_{oe} when $A_i = 37$, $h_{fe} = 70$, and $R_L = 12$ kΩ (h_{oe} is measured in siemens).
   (b) Solve for R_L when $A_i = 80$, $h_{fe} = 100$, and $h_{oe} = 12$ μS.

6. $a = \dfrac{\beta}{1 + \beta}$
   Solve for β when $a = 0.993$.

7. $R_T = \dfrac{R_1 R_2}{R_1 + R_2}$
   Solve for R_1 when $R_T = 1.83$ kΩ and $R_2 = 4.7$ kΩ.

8. $R_T = \dfrac{R_1 R_2}{R_1 + R_2}$
   Solve for R_1 when $R_T = 9.64$ kΩ and $R_2 = 27$ kΩ.

9. $Z_T = \dfrac{Z_1 Z_2}{Z_1 + Z_2}$
   Solve for Z_2 when $Z_T = 64.4$ kΩ and $Z_1 = 91$ kΩ.

10. $Z_T = \dfrac{Z_1 Z_2}{Z_1 + Z_2}$
    Solve for Z_2 when $Z_T = 27.6$ kΩ and $Z_1 = 56$ kΩ.

11. $I_B = \dfrac{V_{CC} - V_B}{R_1 + R_2}$
    Solve for R_2 when $V_{CC} = 20$ V, $V_B = 1.3$ V, $I_B = 312$ μA, and $R_1 = 56$ kΩ.

12. $I_B = \dfrac{V_{CC} - V_B}{R_1 + R_2}$
    Solve for R_1 when $V_{CC} = 9$ V, $V_B = 1.54$ V, $I_B = 187$ μA, and $R_2 = 33$ kΩ.

13. $I_C = \dfrac{V_{CC}}{R_E + (R_B/\beta)}$
    Find R_B when $V_{CC} = 15$ V, $I_C = 1.2$ mA, $R_E = 1.5$ kΩ, and $\beta = 50$.

14. $I_C = \dfrac{V_{CC}}{R_E + (R_B/\beta)}$
    Find R_E when $V_{CC} = 10$ V, $I_C = 500$ μA, $R_B = 560$ kΩ, and $\beta = 70$.

15. $f = \dfrac{1}{2\pi C(R_1 + R_2)}$
    Solve for R_1 when $f = 8.76$ kHz, $C = 500$ pF, and $R_2 = 27$ kΩ.

16. $f = \dfrac{1}{2\pi C(R_1 + R_2)}$
    Solve for R_1 when $f = 1.70$ kHz, $C = 20$ nF, and $R_2 = 1$ kΩ.

End of Chapter Problems 10-4

Solve the following quadratics by factoring:

1. $x^2 - 16 = 0$
2. $x^2 = 81$
3. $x^2 + 6x + 8 = 0$
4. $x^2 + 12x + 35 = 0$
5. $x^2 + 3x - 18 = 0$
6. $x^2 - x - 20 = 0$
7. $x^2 - 6x + 8 = 0$
8. $x^2 - 14x + 48 = 0$
9. $x^2 - x - 56 = 0$
10. $x^2 - 4x - 32 = 0$
11. $x^2 - 15 = 3x + 25$
12. $x^2 - 2 = 10 - 4x$
13. $x^2 - 3x + 16 = 10x - 20$
14. $x^2 - 3x = 4x - 12$
15. $x^2 + 2x = 5x + 18$
16. $x^2 - 5x + 10 = 4x - 10$
17. $x^2 - 2x + 25 = 4 - 12x$
18. $x^2 + 40 = 10 - 11x$
19. $x^2 = 3x - 2$
20. $x^2 + 10x - 8 = 16 + 5x$

End of Chapter Problems 10-5

Solve for x in the following problems:

1. $x^2 + 8x + 12 = 0$
2. $x^2 + 7x + 3 = 0$
3. $x^2 + 6x + 6 = 0$
4. $x^2 + 14x + 9 = 0$
5. $x^2 + 2x - 14 = 0$
6. $x^2 + 5x - 10 = 0$
7. $x^2 - 15x - 15 = 0$
8. $x^2 - 20x - 24 = 0$
9. $x^2 - 3x = 4x + 4$
10. $3x^2 - 8x + 4 = 0$
11. $3x^2 + 2x - 1 = 0$
12. $x^2 - 6x = 2x + 5$
13. $2x^2 - 5x - 5 = 0$
14. $2x^2 - 7x + 5 = 0$
15. $4x^2 + 12x + 8 = 0$
16. $3x^2 - 9x = 8$
17. $3x^2 + 4x = -3x - 3$
18. $2x^2 + 7x = 25 - 5x$
19. $3x^2 + 6x = 20x + 20$
20. $5x^2 - 7x + 5 = 13x - 13$

Answers to Self-tests

SELF-TEST 10-1

1. $R = 4$
2. $I = 3$
3. $x = \dfrac{4b}{ab + 6}, a = \dfrac{4b - 6x}{bx}, b = \dfrac{6x}{4 - ax}$
4. $G_L = 109\ \mu S$
5. $R_2 = 89.8\ k\Omega$

SELF-TEST 10-2

1. $x = \pm 11$
2. $x = 9, x = -3$
3. $x = 6, x = 7$
4. $x = 0.870, x = -2.87$
5. $x = 3, x = -\frac{1}{2}$

CHAPTER 11

DC CIRCUIT ANALYSIS: KIRCHHOFF'S LAWS

INTRODUCTION

Three basic laws deal with circuit analysis. These are Kirchhoff's current and voltage laws and Ohm's law. Theorems have been developed from these basic laws to help in solving the more complex problems. These theorems will be discussed in Chapter 13. Since Kirchhoff's laws and Ohm's law are basic to the understanding of all electrical and electronics circuits, these laws will be discussed in detail in this and the following chapter.

All electrical circuits consist of three basic parts: an *electrical energy source,* a *transfer network,* and a *load*. Some typical energy loads are lamps, motors, and radios. If a load is an energy user, it is said to have the electrical property of resistance. In this kind of load, electrical energy is converted to some other form, usually heat and light. If the load is an energy storer, it has the property of capacitance or inductance. In this chapter we will limit ourselves to loads that are energy users, that is, those loads that are resistive.

The transfer network in an electrical circuit is usually a copper conductor that provides a path for energy to get from the source to the load. An example of such a simple circuit is shown in Figure 11-1.

CHAPTER OBJECTIVES

In this chapter you will learn how to:

1. Use Kirchhoff's current law to find the currents in series, parallel, and series–parallel circuits.
2. Use Kirchhoff's voltage law to find the voltage drops in series, parallel, and series–parallel circuits.
3. Find the polarity of voltage drops in dc circuits.

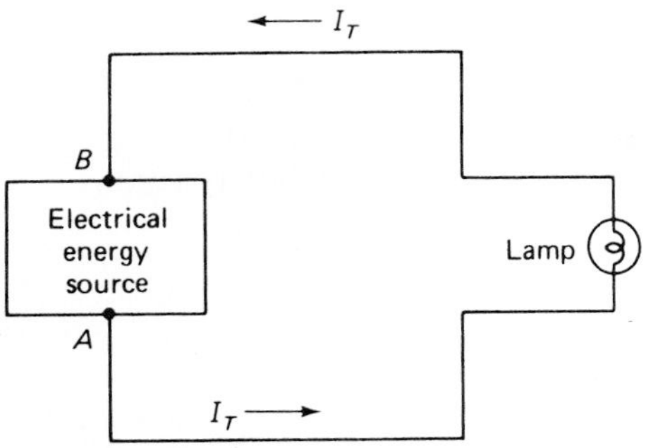

FIGURE 11-1

11-1 KIRCHHOFF'S CURRENT LAW

The physical duality of Kirchhoff's current law can be visualized by using the following analogy. Consider water flowing in a pipe. If we were to choose a point anywhere along the length of the pipe, the rate of water flow (say gallons per minute) *into* that point would have to equal the rate of water flow *leaving* it, as illustrated in Figure 11-2.

Charges moving in a given direction in an electrical conductor are analogous to water flowing in a pipe. When a charge in motion approaches a point already containing a charge, there is a repelling force between them. If there is a continuous supply of charges, incoming charges cause other charges to move because like charges repel each other. If charges, then, are removed from a point, other charges take their place. This flow of charges in an electrical circuit is called *current*.

Kirchhoff's current law essentially expresses the law of conservation of charges and further defines their behavior at a point in an electrical circuit. It can be stated as follows:

■ **KEY POINT** The number of charges per second (current) flowing *to* a point in a conductor must equal the number of charges flowing *from* that point.

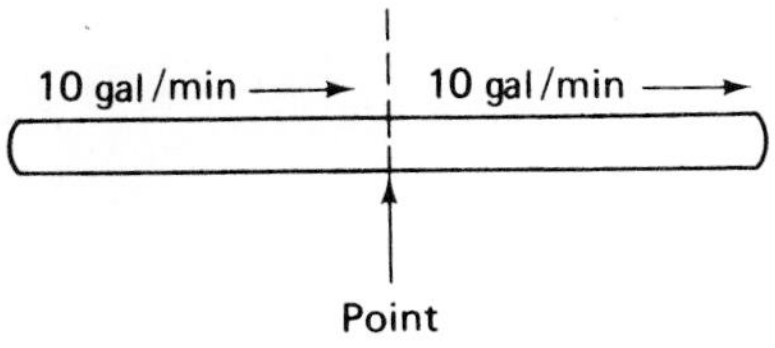

FIGURE 11-2

Stated another way:

■ **KEY POINT** The algebraic sum of currents *into* and *out* of a point must equal zero if currents toward the point are given an arbitrary plus sign (+) and if currents leaving the point are given an arbitrary negative sign (−) (Figure 11-3).

11-1-1 Current in Series Circuits

Practical applications of Kirchhoff's current law may be further visualized by the following: Note that the current (I in Figure 11-1) has but one path. Therefore, according to Kirchhoff's current law, the magnitude of the current must be the same at any point in the closed circuit. If, for example, 1 A leaves point A in the diagram, 1 A must return to point B at that same instant. The current through the energy source, the lamp, and the conductor must be the same. Such an electrical circuit is called a *series circuit*. A series circuit is a circuit in which there is only one path for current.

The diagram in Figure 11-4 consists of two lamps tied end to end. Again, note that the current has but one path and its magnitude must be the same at any point in that path.

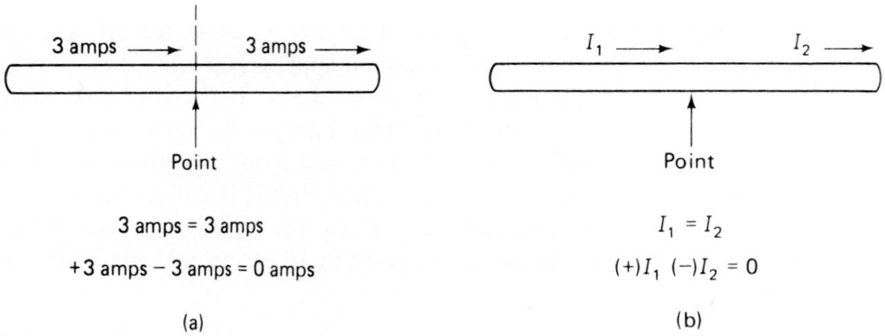

FIGURE 11-3

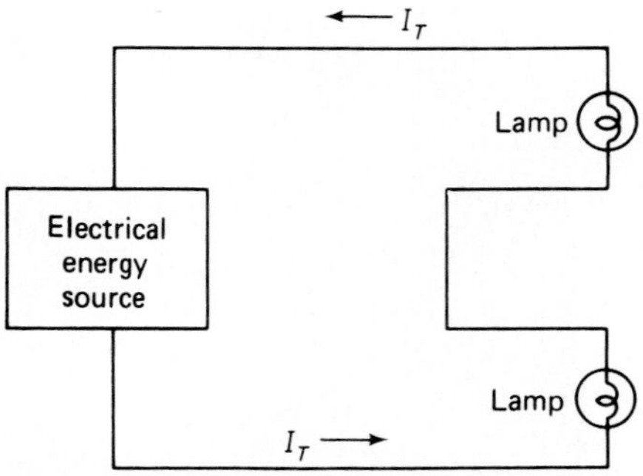

FIGURE 11-4

11-1-2 Current in Parallel Circuits

Let us now go back to our water pipe analogy. This time we will add a T to our pipe. At a given instant, if 10 gal/min enter the T, then 10 gal/min must leave at that same instant, as shown in Figure 11-5. The rate of water flowing *into* the T is equal to the rate of water flowing *out* of the T.

The electrical analogy of the T is called a *junction*. A junction is a point in an electrical circuit where three or more conductors meet. Kirchhoff's current law for a junction can be stated as follows:

■ **KEY POINT** The sum of currents toward a junction must equal the sum of currents leaving that junction (as illustrated in Figure 11-6).

Stated another way:

■ **KEY POINT** The algebraic sum of currents into and out of a junction must equal zero if currents toward the junction are given an arbitrary (+) sign and if currents leaving the point are given an arbitrary (−) sign.

In Figure 11-7 the current into the junction is

$$I_1 + I_2 = 2\text{ A} + 3\text{ A} = 5\text{ A}$$

Then, according to Kirchhoff's current law, the current out of the junction (I_3) must equal 5 A. Algebraically,

$$I_1 + I_2 = I_3$$

or

$$I_1 + I_2 - I_3 = 0$$

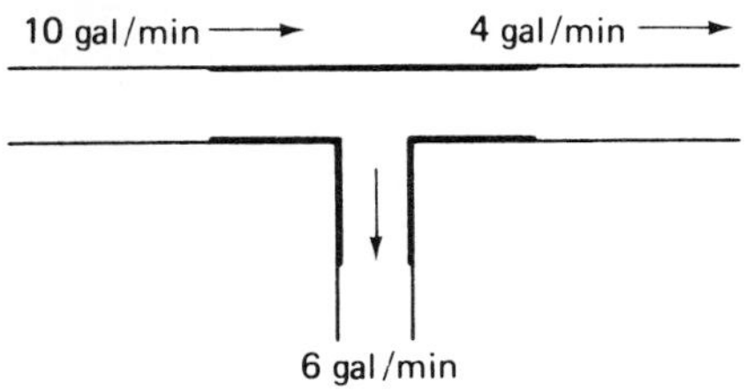

FIGURE 11-5

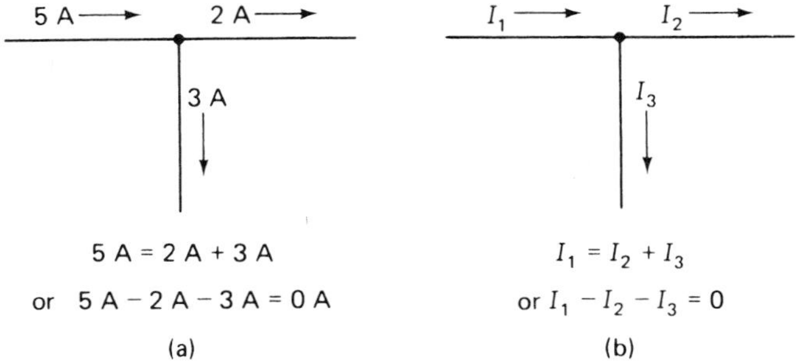

FIGURE 11-6

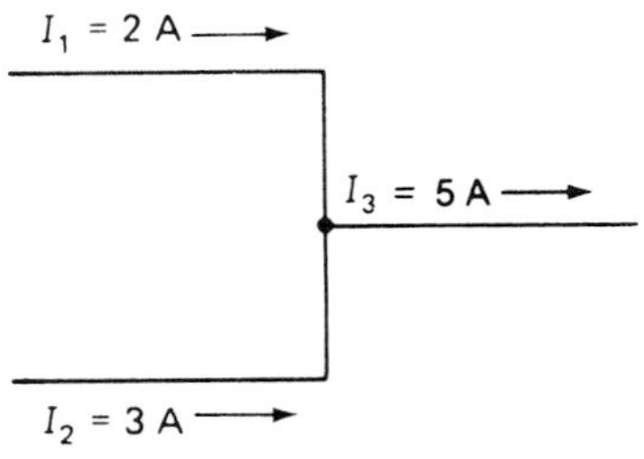

FIGURE 11-7

In Figure 11-8 the current into the junction, I_1, equals 8 A. Therefore, the current out of the junction must equal 8 A, or

$$I_2 + I_3 + I_4 = 8 \text{ A}$$
$$1 \text{ A} + 3 \text{ A} + I_4 = 8 \text{ A}$$

Therefore,

$$I_4 = 4 \text{ A}$$

Algebraically,

$$I_1 = I_2 + I_3 + I_4$$

or

$$I_1 - I_2 - I_3 - I_4 = 0$$

Solve for I_1 and I_3 in Figure 11-9.

$$I_1 + I_2 + I_3 = I_4 + I_5$$

Then

$$I_1 = I_4 + I_5 - I_3 - I_2$$

and

$$I_3 = I_4 + I_5 - I_2 - I_1$$

The diagram in Figure 11-10 consists of two lamps connected *across* an energy source. Note that the current, I_T, into junction A must equal the currents out of that junction ($I_1 + I_2$). Therefore, from Kirchhoff's current law,

$$I_T = I_1 + I_2$$

If the current, I_T, in Figure 11-10 is 5 A and I_1 is measured to be 2 A, then I_2 must equal 3 A.

$$I_T = I_1 + I_2$$
$$5 \text{ A} = 2 \text{ A} + I_2$$
$$5 \text{ A} - 2 \text{ A} = I_2$$
$$3 \text{ A} = I_2$$

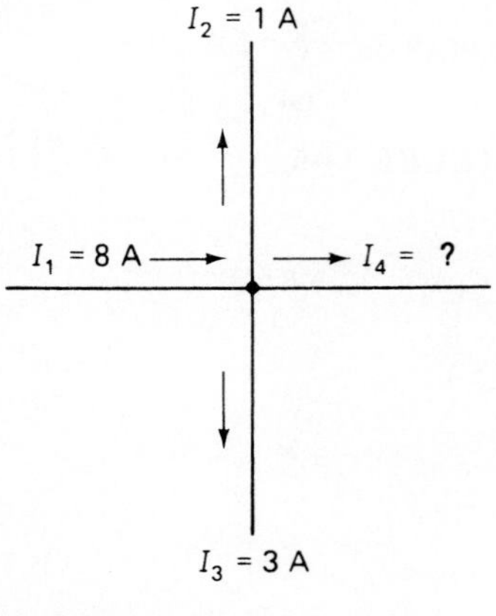

FIGURE 11-8

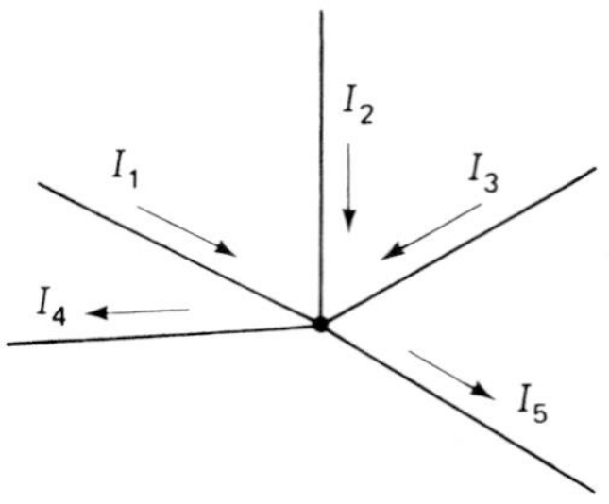

FIGURE 11-9

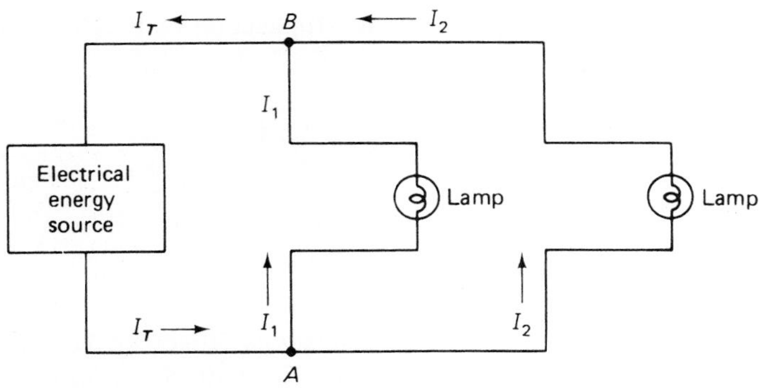

FIGURE 11-10

The branch currents into and out of junction B must also be equal to 5 A. Circuits that have multiple current paths are called *parallel circuits*.

11-1-3 Series–Parallel Circuits

An electrical circuit that has both series- and parallel-circuit characteristics is called a *series–parallel circuit*. Any number of energy sources and loads may make up a series–parallel circuit. For example, Figure 11-11 shows three lamps and a battery connected in a series–parallel arrangement. The part of the transfer network between points A and B carries the total current. At point B the current divides (parallel circuit): one part flows through lamp L_2 and the remainder flows through lamp L_3. The current recombines at point C, and total current flows in the transfer network between point C and lamp L_1 and on to point D (series circuit). So we have one circuit component (L_1) in which all the current flows (series portion) and two circuit components in which part of the total current flows (parallel portion).

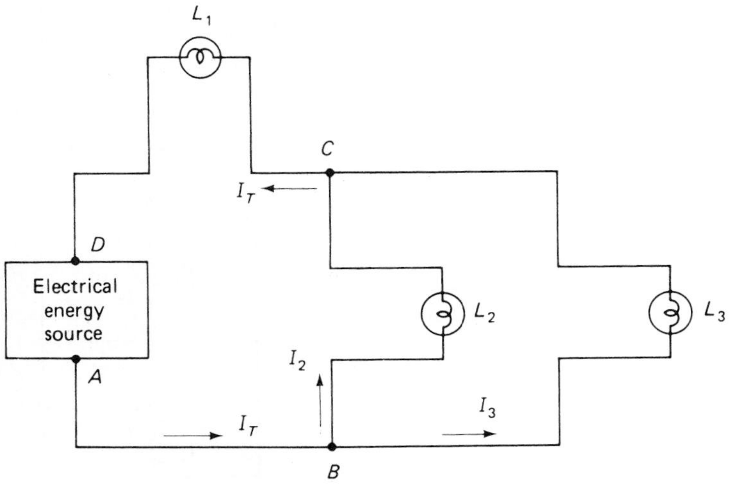

FIGURE 11-11

Figure 11-12 shows another circuit configuration. Application of Kirchhoff's current law yields these current relationships:

$$I_T = I_1$$
$$I_T = I_4 + I_3$$
$$I_T = I_4 + I_2$$
$$I_2 = I_3$$
$$I_1 = I_2 + I_4$$

Inspection of the circuit shows that R_2 and R_3 are in series. Together they are in parallel with R_4. Finally, this whole combination is in series with R_1. In the circuit, then, total current flows through R_1, and a division takes place at point B, resulting in two currents, I_4 and $I_2(I_T = I_2 + I_4)$. Since R_2 and R_3 are in series, the current through them is the same ($I_2 = I_3$). A recombination of currents takes place at the junction of R_3 and R_4 (point C). Then total current flows from point C to the electrical energy source (point D).

□ EXAMPLE 11-1

In Figure 11-13, if $I_T = 1$ A, $I_3 = 0.4$ A, and $I_4 = 0.5$ A, find I_1 and I_2.

Solution Using Kirchhoff's current law, we can write equations for various currents:

$$I_T = I_1 + I_4$$
$$I_4 = I_2 + I_3$$
$$I_T = I_1 + I_2 + I_3$$

Choosing an equation containing only one unknown, we get

$$I_T = I_1 + I_4$$

or $$I_1 = I_T - I_4 = 1\text{ A} - 0.5\text{ A} = 0.5\text{ A}$$

and $$I_4 = I_2 + I_3$$

or $$I_2 = I_4 - I_3 = 0.5\text{ A} - 0.4\text{ A} = 0.1\text{ A}$$

□

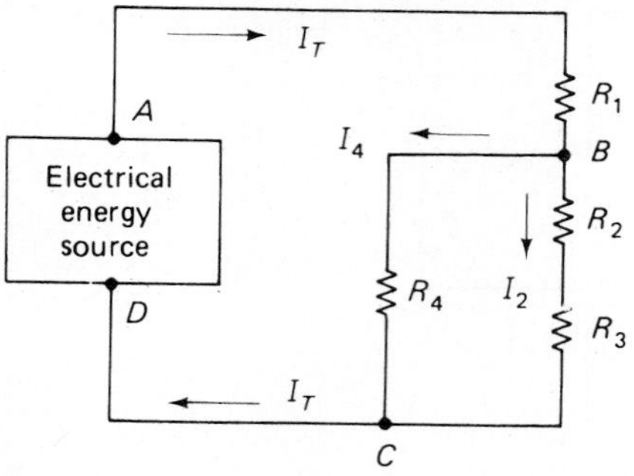

FIGURE 11-12

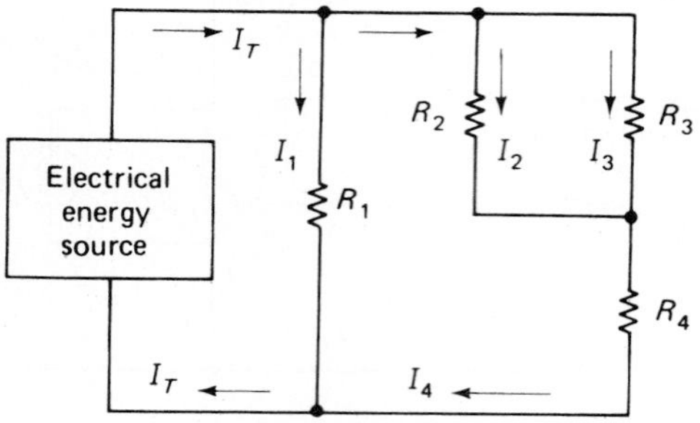

FIGURE 11-13

PRACTICE PROBLEMS 11-1

Refer to Figure 11-14.

1. Find I_T if $I_1 = 1$ A, $I_2 = 2$ A, and $I_3 = 3$ A.

2. Find I_3 if $I_T = 12$ mA, $I_1 = 1$ mA, and $I_2 = 7$ mA.

3. Find I_2 if $I_T = 10\ \mu$A, $I_1 = 4\ \mu$A, and $I_3 = 2\ \mu$A

4. Find I_1 if $I_T = 2$ A, $I_2 = 1.2$ A, and $I_3 = 300$ mA.

Refer to Figure 11-15.

5. $I_1 = 4$ mA and $I_2 = 3$ mA. Find I_T and I_3.

6. $I_T = 2.3$ mA and $I_3 = 600\ \mu$A. Find I_1 and I_2.

7. $I_T = 400$ mA and $I_1 = 300$ mA. Find I_2 and I_3.

Refer to Figure 11-12.

8. $I_1 = 400$ mA, $I_2 = 150$ mA. Find I_T, I_3, and I_4.

9. $I_3 = 600\ \mu$A, $I_4 = 800\ \mu$A. Find I_T, I_1, and I_2.

Refer to Figure 11-13.

10. $I_T = 17$ mA, $I_2 = 6$ mA, $I_4 = 13$ mA. Find I_1 and I_3.

11. $I_1 = 500\ \mu$A, $I_2 = 1.3$ mA, $I_3 = 850\ \mu$A. Find I_T and I_4.

12. Refer to Figure 11-16. Does current flow through R_5 from A to B or from B to A? What is the magnitude of this current?

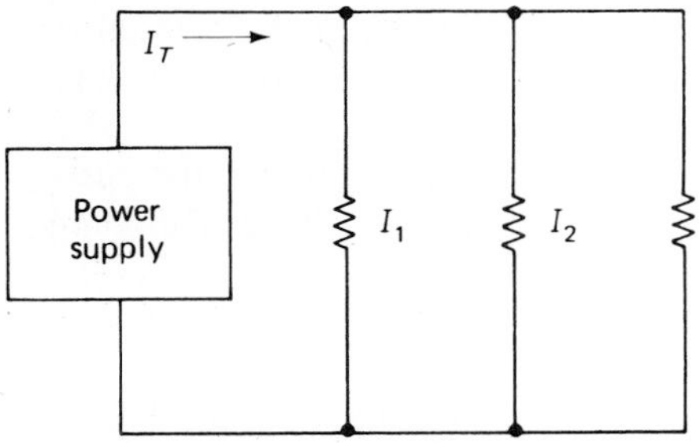

FIGURE 11-14

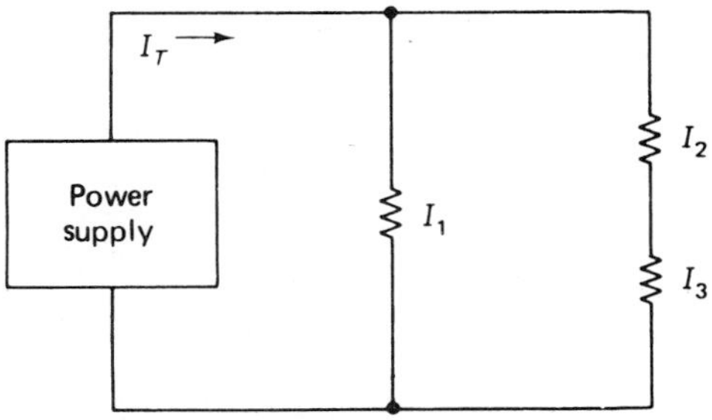

FIGURE 11-15

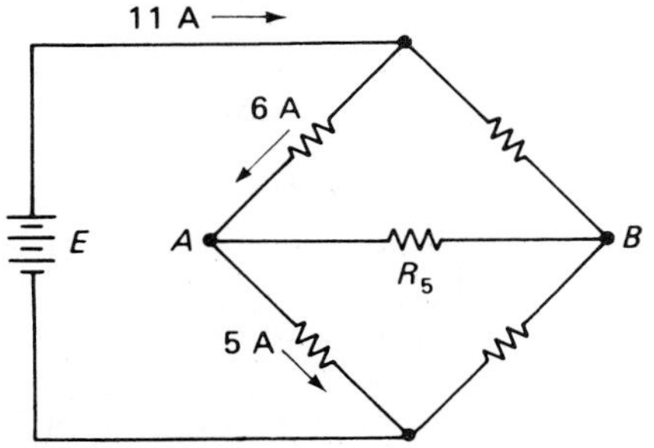

FIGURE 11-16

Solutions

1. 6 A

2. 4 mA

3. 4 μA

4. 500 mA

5. $I_T = 7$ mA, $I_3 = 3$ mA
6. $I_1 = 1.7$ mA, $I_2 = 600\ \mu$A
7. $I_2 = 100$ mA, $I_3 = 100$ mA
8. $I_T = 400$ mA, $I_3 = 150$ mA, $I_4 = 250$ mA
9. $I_T = 1.4$ mA, $I_1 = 1.4$ mA, $I_2 = 600\ \mu$A
10. $I_1 = 4$ mA, $I_3 = 7$ mA
11. $I_T = 2.65$ mA, $I_4 = 2.15$ mA
12. From A to B. $I_5 = 1$ A.

Additional practice problems are at the end of the chapter.

11-2 KIRCHHOFF'S VOLTAGE LAW

If a charge travels around any closed path in an electrical network and returns to its starting point, its net potential energy change must be zero. That is, it must go through equal potential energy gains and potential energy losses.

Kirchhoff's voltage law expresses the law of conservation of energy and can be stated as follows:

KEY POINT Around any closed path the sum of potential rises must equal the sum of potential drops ($\Sigma E = \Sigma V$).

Potential *rises* are the electrical energy sources and could be the 110 volts supplied by the local electric company, a battery, or some other source. A potential *rise* is also called an electromotive force, (emf), source voltage, and applied voltage. The symbol is E and the unit is the volt.

Potential *drops* refer to the energy used in causing current through the energy users, like the lamps in Figure 11-17. These potential drops are also called *voltage drops*. The symbol is V, and the unit is the volt.

Stated another way:

KEY POINT The algebraic sum of emf's (E) and voltage drops (V) around a closed path is equal to zero ($\Sigma E - \Sigma V = 0$).

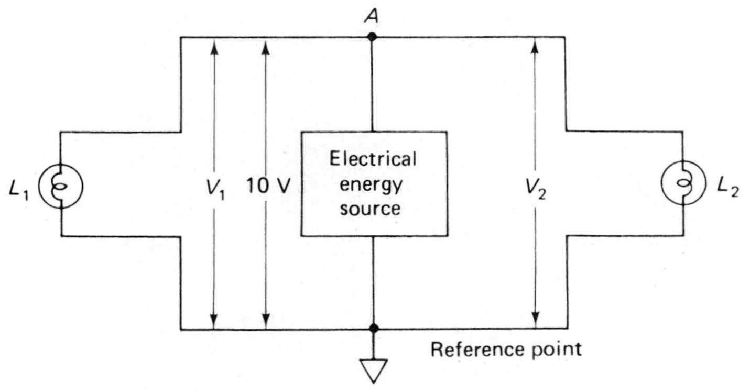

FIGURE 11-17

In the circuit of Figure 11-17, both L_1 and L_2 are connected *across* the electrical energy source. Kirchhoff's voltage law tells us that in any closed path the emf's and voltage drops must be equal. For loop 1, $E = V_1$, (the voltage drop across L_1), and for loop 2, $E = V_2$ (the voltage drop across L_2). If $E = V_1 = V_2$, then $V_1 = V_2$.

Consider a single charge at the reference point in the circuit. The electrical energy source raises the potential energy of the charge to 10 V at point A. For the charge to return to the reference point, it must travel through lamp 1 or lamp 2. No matter which path it takes, it must give up the potential energy that the electrical energy source provided it. Therefore, in parallel circuits the voltage drop is the same for each branch and equals the emf.

From this and previous discussions we may conclude the following:

■ **KEY POINT** In a parallel circuit; (1) the voltage drop is the same across each branch: $V_1 = V_2 = V_N$; (2) The total current into and out of the branches equals the sum of the individual branch currents: $I_T = I_1 + I_2 + \cdots + I_N$.

In Figure 11-18 we have an electrical energy source connected in series with two lamps. Once again, let's consider a single charge at the reference point. The electrical energy source raises the potential energy of the charge to 10 V (point A). In this circuit there is only one path the charge may take. It must pass through lamp 1 and then through lamp 2. When it has passed through lamp 1, it has converted some, but not all, of the energy provided by the electrical energy source. The remainder of its energy is converted when it travels through lamp 2. The charge drops across a potential difference of 6 V for lamp 1. Since it drops across a total potential difference of 10 V, there remains 4 V to drop across lamp 2. When the charge finally passes through lamp 2, all the potential energy provided to it by the electrical energy source is converted, and the charge has made a complete trip around the circuit and returned to the reference point.

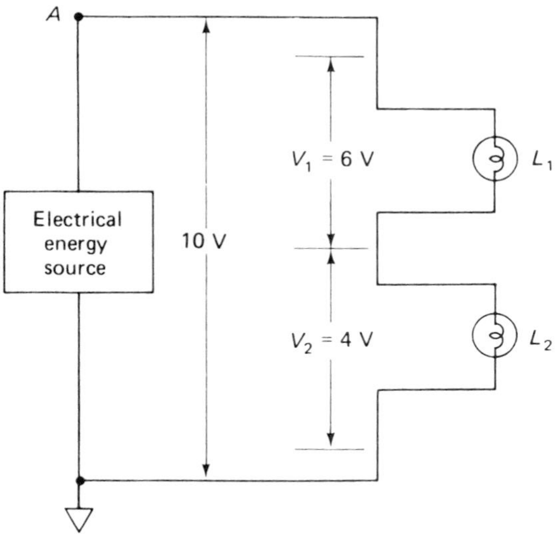

FIGURE 11-18

From this and previous discussions we may conclude the following:

KEY POINT In a series circuit, (1) the current is the same throughout the circuit: $I_T = I_1 = I_2 = I_N$; (2) the sum of the voltage drops must equal the emf: $E = V_1 + V_2 + \cdots + V_N$

According to Kirchhoff's voltage law, the total rise in potential (E) in Figure 11-18 must equal the total drop in potential:

$$E = V_1 + V_2$$

If, for the above example, the potential rise is equal to 10 V, then the sum of potential drops ($V_1 + V_2$) must also equal 10 V.

☐ EXAMPLE 11-2

In Figure 11-18, find the magnitude of V_2 if $E = 30$ V and V_1 equals 10 V.

Solution

$$E = V_1 + V_2$$

Then
$$V_2 = E - V_1$$
$$V_2 = 30 \text{ V} - 10 \text{ V} = 20 \text{ V}$$

☐ EXAMPLE 11-3

In Figure 11-19, determine V_2.

Solution

$$E = V_1 + V_2 + V_3$$

Then
$$V_2 = E - V_1 - V_3 = 20 \text{ V} - 5 \text{ V} - 7 \text{ V} = 8 \text{ V}$$

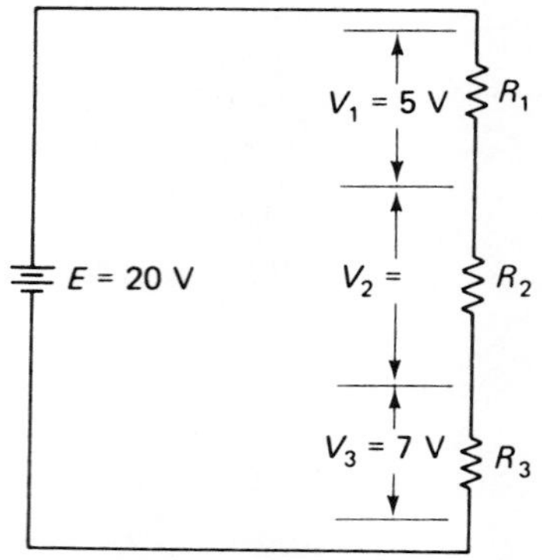

FIGURE 11-19

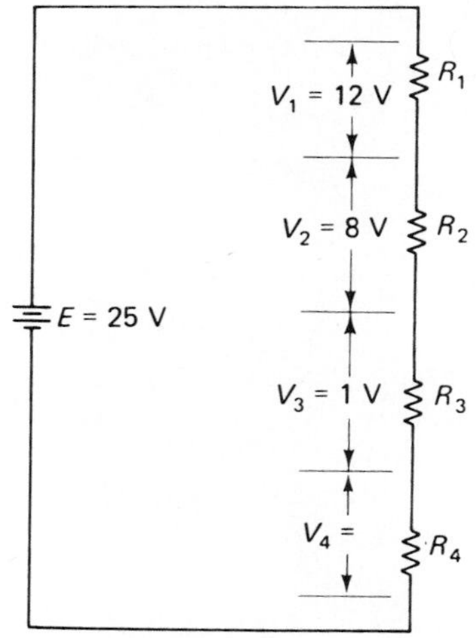

FIGURE 11-20

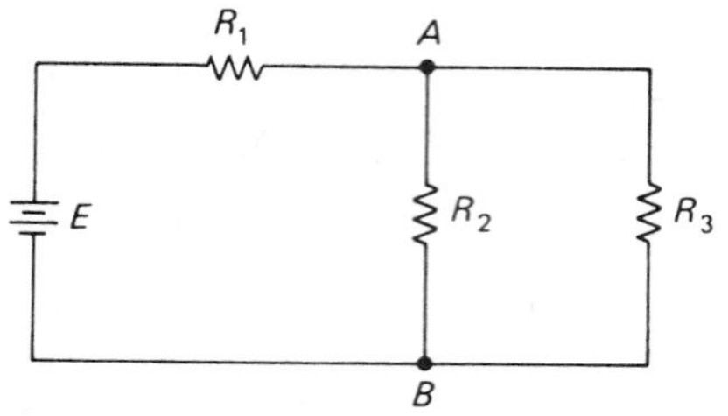

FIGURE 11-21

□ EXAMPLE 11-4

In Figure 11-20, determine V_4.

Solution

$$E = V_1 + V_2 + V_3 + V_4$$

Then

$$V_4 = E - V_1 - V_2 - V_3 = 25\text{ V} - 12\text{ V} - 8\text{ V} - 1\text{ V} = 4\text{ V}$$

□

In Figure 11-21, Kirchhoff's voltage law is satisfied because part of the source voltage is dropped across the circuit between points A and B (across R_2 and R_3) and part is dropped across R_1.

$$E = V_1 + V_2$$

and

$$E = V_1 + V_3$$

Since there are two paths, either may be chosen in writing the equation. In Figure 11-22, application of Kirchhoff's voltage law yields

$$E = V_1 + V_2 + V_3$$

or

$$E = V_1 + V_4$$

We can write it either way since R_4 is in parallel with the series circuit consisting of R_2 and R_3. Because of this relationship between R_2, R_3, and R_4, we can also say that

$$V_4 = V_2 + V_3$$

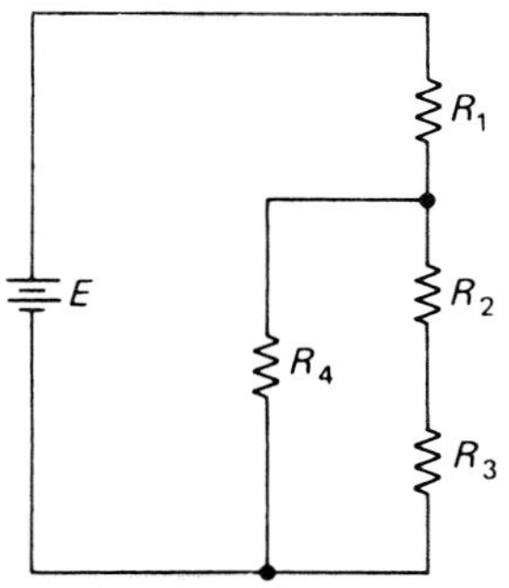

FIGURE 11-22

Recalling that *the algebraic sum of emf's and voltage drops around a closed path is equal to zero,* we could also write the equations like this:

$$E - V_1 - V_2 - V_3 = 0$$

and

$$E - V_1 - V_4 = 0$$

also

$$V_4 - V_2 - V_3 = 0$$

That is, we may choose *any* closed path (or loop), and the sum of the drops and emf's in that loop must equal zero.

□ EXAMPLE 11-5

In Figure 11-22, let $E = 10$ V, $V_2 = 5$ V, and $V_3 = 2$ V. Find V_1 and V_4.

Solution

$$V_4 = V_2 + V_3$$

$$V_4 = 5\text{ V} + 2\text{ V} = 7\text{ V}$$

$$E = V_1 + V_2 + V_3$$

$$V_1 = E - V_2 - V_3 = 10\text{ V} - 5\text{ V} - 2\text{ V} = 3\text{ V}$$

□

PRACTICE PROBLEMS 11-2

1. In Figure 11-23, if $E = 25$ V and $V_1 = 10$ V, find V_2.
2. In Figure 11-19, if $E = 20$ V, $V_1 = 5$ V, and $V_2 = 7$ V, find V_3.
3. In Figure 11-19, if $V_1 = 3$ V, $V_2 = 30$ V, and $V_3 = 10$ V, find E.

Refer to Figure 11-21.

4. $E = 50$ V, $V_3 = 20$ V. Find V_1 and V_2.
5. $E = 50$ V, $V_1 = 15$ V. Find V_2 and V_3.
6. $V_1 = 7$ V, $V_3 = 14$ V. Find E and V_2.

Refer to Figure 11-22.

7. $E = 10$ V, $V_2 = 5$ V, $V_3 = 2$ V. Find V_1 and V_4.
8. $E = 150$ mV, $V_2 = 30$ mV, $V_4 = 80$ mV. Find V_1 and V_3.
9. $V_1 = 7$ V, $V_2 = 5$ V, $V_4 = 8$ V. Find E and V_3.

Refer to Figure 11-24.

10. $E = 35$ V, $V_1 = 12$ V, $V_3 = 10$ V, $V_5 = 5$ V. Find V_2, V_4, and V_6.

Solutions

1. 15 V
2. 8 V
3. 43 V
4. $V_1 = 30$ V, $V_2 = 20$ V
5. $V_2 = 35$ V, $V_3 = 35$ V
6. $E = 21$ V, $V_2 = 14$ V
7. $V_1 = 3$ V, $V_4 = 7$ V
8. $V_1 = 70$ mV, $V_3 = 50$ mV
9. $E = 15$ V, $V_3 = 3$ V
10. $V_2 = 18$ V, $V_4 = 8$ V, $V_6 = 35$ V

11-3 POLARITY

■ **KEY POINT** In an electrical circuit, the polarity of the voltage drops across the individual resistances depends on the direction of current through them.

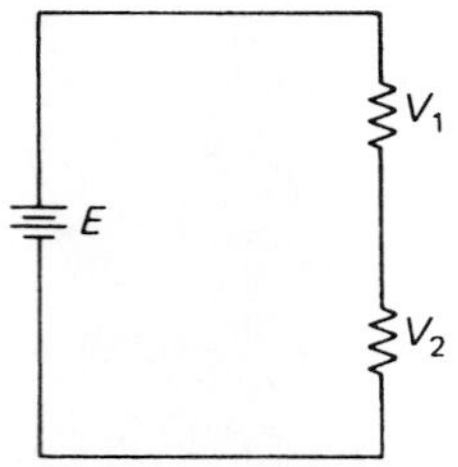

FIGURE 11-23

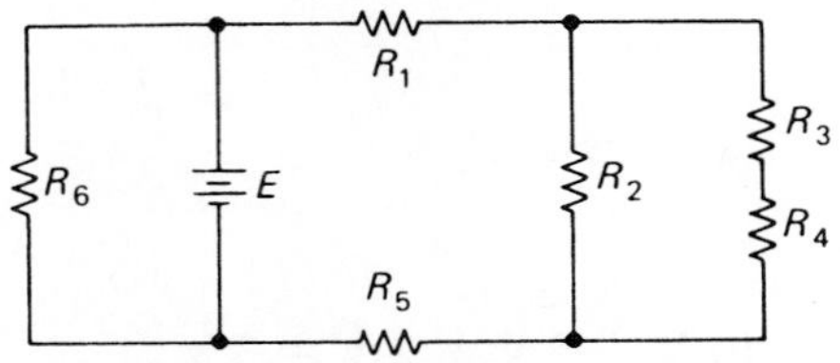

FIGURE 11-24

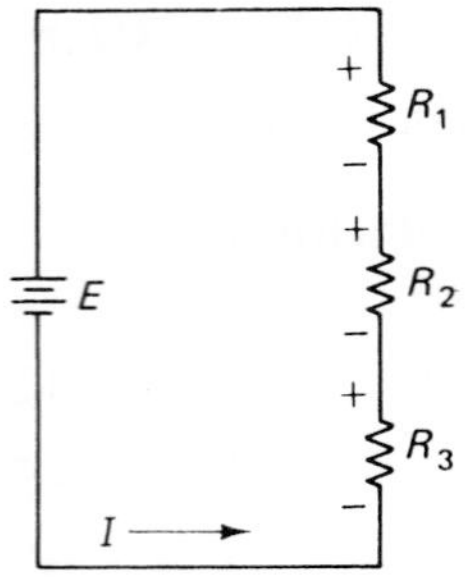

FIGURE 11-25

If we use electron current, as in Figure 11-25, the polarities are as shown. The end of the resistor where current enters is assigned a negative (−) sign. This makes the other end positive (+). If we use conventional current, as in Figure 11-26, the end where current enters is assigned a positive (+) sign. This makes the other end negative (−). We see in Figures 11-25 and 11-26 that the resultant polarities are the same regardless of whether we use electron current or conventional current. In our circuits polarities are assigned, but usually we choose the direction of current with which we are most comfortable.

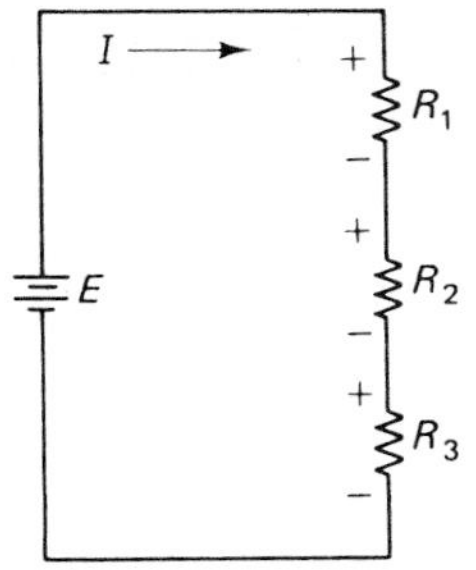

FIGURE 11-26

In Figure 11-27, polarities are shown as dictated by the direction of current. Suppose that $V_1 = 5$ V, $V_2 = 7$ V, and $V_3 = 8$ V. The general equation states that

$$E + V_1 + V_2 + V_3 = 0$$

Observing polarity, let's start at point A and go around the loop in a counterclockwise direction. Using the sign we encounter first as we go through each source and resistance, we get

$$+20 \text{ V} - 8 \text{ V} - 7 \text{ V} - 5 \text{ V} = 0$$

$$20 \text{ V} - 20 \text{ V} = 0$$

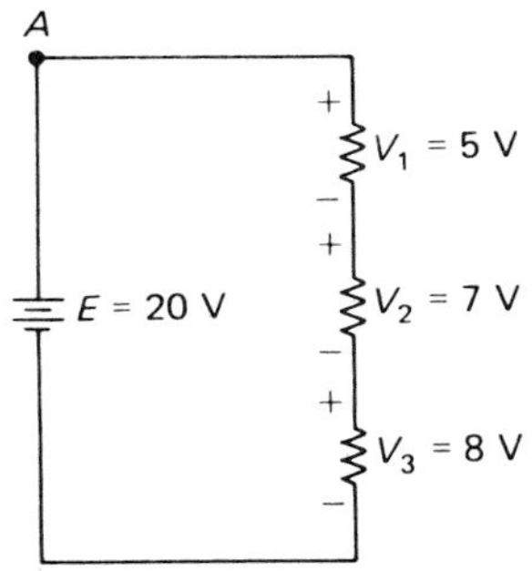

FIGURE 11-27

In Figure 11-28, R_2 and R_3 are in parallel and they are in series with R_1. Because R_2 and R_3 are in parallel, $V_2 = V_3$. One loop includes E, R_1, and R_2. Therefore, $E = V_1 + V_2$. A second loop includes E, R_1, and R_3. Therefore, $E = V_1 + V_3$. In each case, 25 V = 15 V + 10 V.

Let's find V_2 and V_4 in Figure 11-29. One loop includes E, R_1, R_2 and R_4. Since we have two unknown voltage drops in this loop, we can't use it. Another loop includes E, R_3, and R_4. In this loop the only unknown is V_4.

$$E = V_3 + V_4$$

$$V_4 = E - V_3 = 30 \text{ V} - 10 \text{ V} = 20 \text{ V}$$

Now that we know V_4, we can use the loop that includes E, R_1, and R_4 to find V_2.

$$E = V_1 + V_2 + V_4$$

$$V_2 = E - V_1 - V_4$$

$$= 30 \text{ V} - 3 \text{ V} - 20 \text{ V} = 30 \text{ V} - 23 \text{ V} = 7 \text{ V}$$

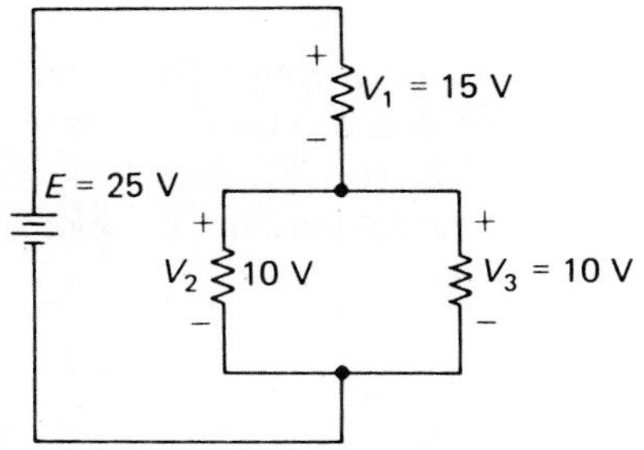

FIGURE 11-28

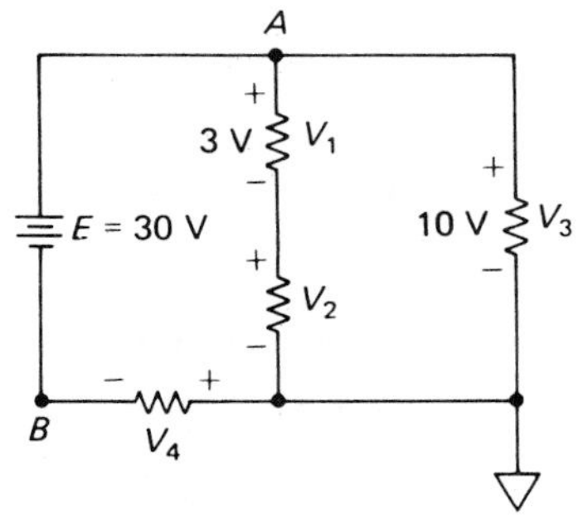

FIGURE 11-29

We could also use the loop consisting of R_1, R_2, and R_3 to find V_2. Observing polarity and moving around the loop clockwise from R_2, we get

$$-V_2 - V_1 + V_3 = 0$$

$$V_2 = V_3 - V_1$$

$$= 10\text{ V} - 3\text{ V} = 7\text{ V}$$

The loop we use may or may not include a source.

Suppose we were interested in the potential at one point in a circuit with reference to another point. For example, in Figure 11-29 we might ask the question, "What is the potential at point A with respect to common?" "With respect to" means "with reference to." Said another way, "If common is our reference point, what is the potential at point A?" The answer is 10 V. Point A is 10 V more positive than common. This can be determined three different ways:

1. The circuit from point A to common through R_3 shows that point A is the positive side of R_3 and common is the negative side. The drop across R_3 is 10 V; so point A is 10 V positive with respect to common.
2. Starting at point A and going to common through the path consisting of R_1 and R_2 yields $+3\text{ V} + 7\text{ V} = 10\text{ V}$.
3. Starting at point A and going to common through the path consisting of E and R_4, we get $+30\text{ V} - 20\text{ V} = 10\text{ V}$.

In each case we observed polarity and added (algebraically) the emf's and voltage drops.

What is the potential at point B with respect to common? The answer is -20 V. The easiest way to determine this is to observe the polarity of the drop across R_4. We could arrive at the same answer by using the path consisting of E and R_3 or by using the path consisting of E, R_1, and R_2. Using E and R_3, we get $-30\text{ V} + 10\text{ V} = -20\text{ V}$. Using E, R_1, and R_2, we get $-30\text{ V} + 3\text{ V} + 7\text{ V} = -20\text{ V}$.

□ EXAMPLE 11-6

In Figure 11-30,

(a) What is the drop across R_2 and what is its polarity?
(b) What is the drop across R_3 and what is its polarity?
(c) What is the potential at point A with respect to common?

Solutions

(a) The drop across $R_2(V_2)$ can be found by going around the outer loop. The other choice is the loop consisting of E_2, R_2, and R_3, but we don't know V_3; therefore, we can't use that loop. The outer loop consists of E_1, E_2, R_1, and R_2.

Let's start at the left side of R_2, which is point B, and go around the loop in a clockwise direction. The general equation for the loop is

$$E_1 + V_1 + E_2 + V_2 = 0$$

Observing polarity, we get

$$-50\text{ V} + 5\text{ V} + 30\text{ V} + V_2 = 0$$

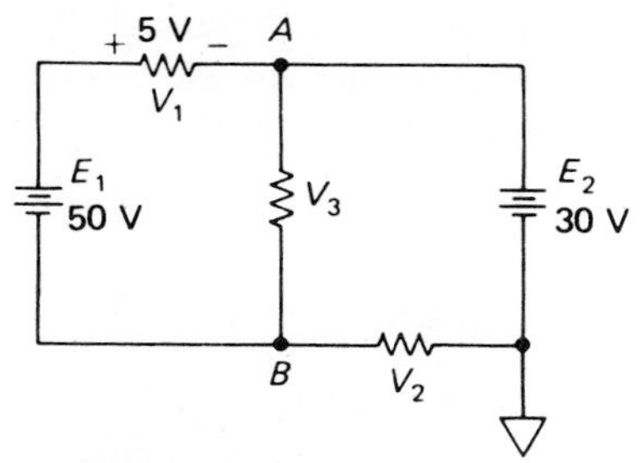

FIGURE 11-30

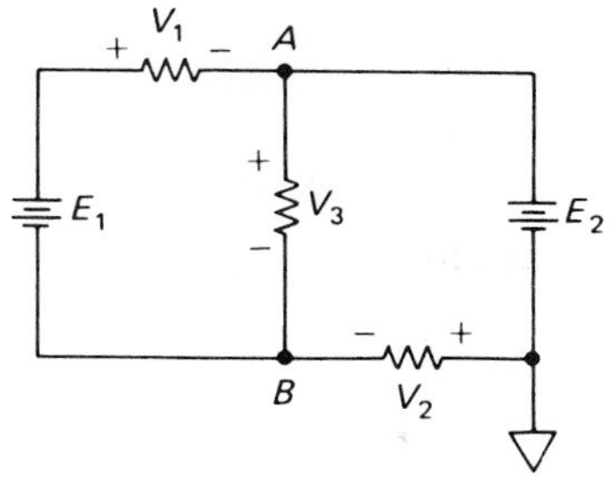

FIGURE 11-31

Notice that a positive polarity was assigned to V_2. Because we don't know the polarity of V_2 yet, we give it a + sign until we solve the equation. Solving, we get

$$-15\ \text{V} + V_2 = 0$$

$$V_2 = 15\ \text{V}$$

Notice that V_2 is positive. This tells us that the first sign we encounter when we get to R_2 is a + sign. We went in a clockwise direction, so the + sign must be assigned to the right side of R_2 (which makes the left side negative), as shown in Figure 11-31. If we had gone in a counterclockwise direction from point B, we would have this equation:

$$V_2 + E_2 + V_1 + E_1 = 0$$

Assigning polarities and values, we get

$$V_2 - 30\ \text{V} - 5\ \text{V} + 50\ \text{V} = 0$$

$$V_2 + 15\ \text{V} = 0$$

$$V_2 = -15\ \text{V}$$

Going counterclockwise we see that V_2 is *negative* 15 V. That is, the left side of R_2 is negative (which makes the right side positive) and agrees with our first method.

(b) Knowing both V_1 and V_2 allows us to choose either the loop that includes R_3, E_1, and R_1 or the loop that includes R_3, R_2, and E_2. Let's use the loop that includes R_3, E_1, and R_1. Starting at point A and moving clockwise (remember we can go in either direction), we get

$$V_3 + E_1 + V_1 = 0 \qquad \text{(general equation)}$$

$$V_3 - 50\ \text{V} + 5\ \text{V} = 0 \qquad \text{(plug in known values and observe polarity)}$$

$$V_3 - 45\ \text{V} = 0$$

$$V_3 = 45\ \text{V}$$

We started at point A. Since the first sign we encountered was a + sign ($V_3 = 45$ V), the polarity is as shown in Figure 11-31.

(c) Notice that there are three paths from point A to common. One is through E_2. Another is through R_2 and R_3. A third is through R_1, E_1, and R_2. No matter which is chosen, the result should show that point A is 30 V positive with respect to common.

$$P_t A \text{ to common} = E_2 = 30\text{ V}$$
$$= V_3 + V_2 = 45\text{ V} - 15\text{ V} = 30\text{ V}$$
$$= V_1 + E_1 + V_2 = -5\text{ V} + 50\text{ V} - 15\text{ V}$$
$$= 30\text{ V}$$

□

PRACTICE PROBLEMS 11-3

1. Refer to Figure 11-29. Let $V_1 = 7$ V and $V_3 = 15$ V. Find (a) V_2, (b) V_4, (c) potential at point A with respect to common, (d) potential at point A with respect to point B, and (e) potential at point B with respect to common.
2. Refer to Figure 11-32. Find (a) V_1, (b) V_3, (c) potential at point A with respect to common, (d) potential at point B with respect to common, and (e) potential at point A with respect to point B.
3. Refer to Figure 11-33. Find (a) V_3, (b) V_4, (c) potential at point A with respect to common, and (d) potential at point B with respect to point A.

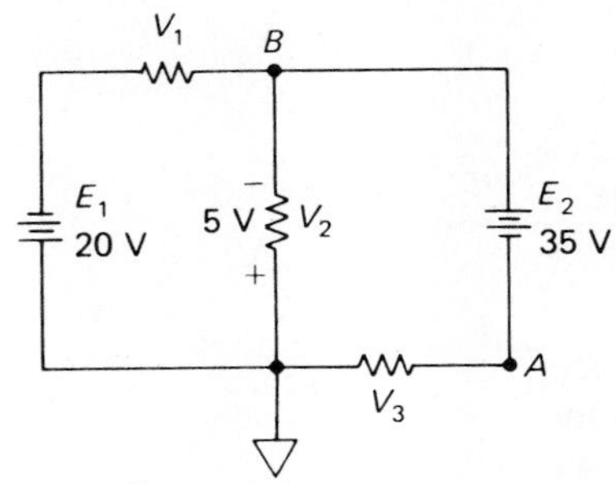

FIGURE 11-32

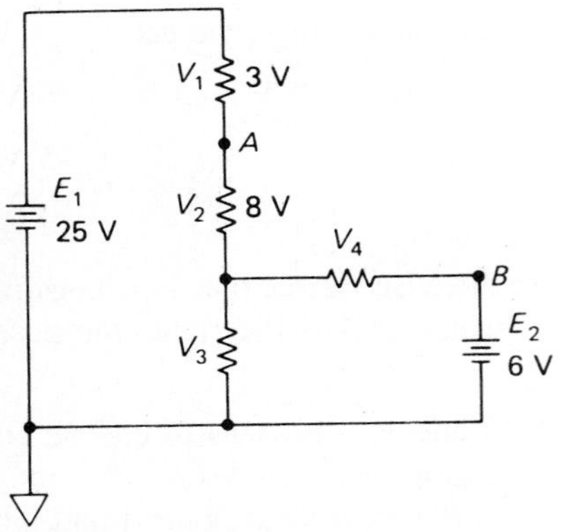

FIGURE 11-33

Solutions

1. (a) 8 V (b) 15 V (c) 15 V (d) 30 V (e) −15 V
2. (a) 15 V (b) 40 V (c) −40 V (d) −5 V (e) −35 V
3. (a) 14 V (b) 8 V (c) 22 V (d) −16 V

Additional practice problems are at the end of the chapter.

SELF-TEST 11-1

1. Refer to Figure 11-14. Find I_3 if $I_T = 16$ mA, $I_1 = 4$ mA, and $I_2 = 2$ mA.
2. Refer to Figure 11-15. $I_1 = 40\ \mu$A and $I_2 = 60\ \mu$A. Find I_T and I_3.
3. Refer to Figure 11-15. $I_T = 3$ mA and $I_2 = 2.6$ mA. Find I_1 and I_3.
4. Refer to Figure 11-12. $I_T = 800\ \mu$A, $I_2 = 308\ \mu$A. Find I_1, I_3, and I_4.
5. Refer to Figure 11-13. $I_1 = 15$ mA, $I_2 = 5$ mA, $I_3 = 7$ mA. Find I_T and I_4.
6. Refer to Figure 11-19. $E = 30$ V, $V_1 = 7$ V, $V_2 = 10$ V. Find V_3.
7. Refer to Figure 11-19. $V_1 = 200$ mV, $V_2 = 1.3$ V, $V_3 = 400$ mV. Find E.
8. Refer to Figure 11-21. $E = 20$ V, $V_1 = 7$ V. Find V_2 and V_3.
9. Refer to Figure 11-22. $V_1 = 3$ V, $V_2 = 10$ V, $V_4 = 12$ V. Find E and V_3.
10. Refer to Figure 11-34. Find V_2 and V_3.
11. Refer to Figure 11-29. Let $V_1 = 13$ V and $V_3 = 21$ V. Find (a) V_2, (b) V_4, (c) potential at point B with respect to point A, and (d) potential at point A with respect to common.
12. Refer to Figure 11-33. Let $V_1 = 4$ V and $V_2 = 12$ V. Find (a) V_3, (b) V_4, (c) potential at common with respect to point A, and (d) potential at point B with respect to point A.

Answers to Self-test 11-1 are at the end of the chapter.

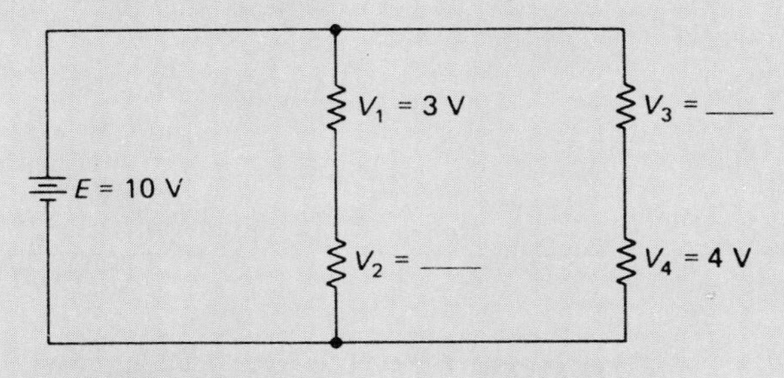

FIGURE 11-34

CHAPTER 11 AT A GLANCE

Page	Key Point	Example
279	*Key point:* The current flowing to a point in a conductor must equal the current flowing from that point.	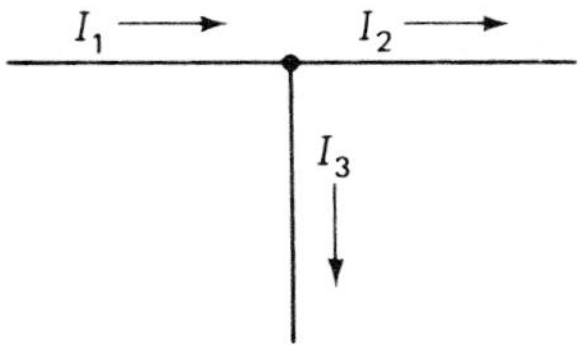 $I_1 = I_2 + I_3$ or $I_1 - I_2 - I_3 = 0$ **FIGURE 11-35**
286	*Key point:* Around any closed path, the sum of potential rises must equal the sum of potential drops.	$E = V_1 + V_2$
287	*Key point:* In a parallel circuit: (1) the voltage drop is the same across each branch; (2) the	$V_1 = V_2 = V_N$ $I_T = I_1 + I_2 + \cdots + I_N$

total current into and out of the branches equals the sum of the individual branch currents.

288 *Key point:* In a series circuit: (1) the current is the same throughout the circuit; (2) the sum of the voltage drops must equal the emf.

$$I_T = I_1 = I_2 = I_N$$
$$E = V_1 + V_2 + \cdots + V_N$$

290 *Key point:* In an electrical circuit the polarity of the voltage drops across the individual resistances depends on the direction of current through them.

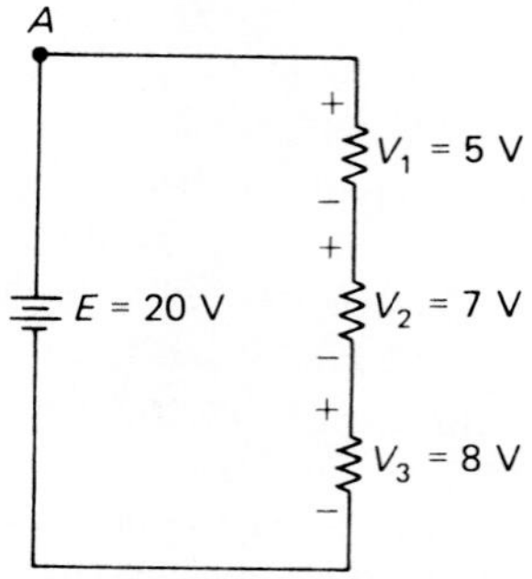

FIGURE 11-36

End of Chapter Problems 11-1

Refer to Figure 11-37.

1. Find I_T if $I_1 = 3$ A, $I_2 = 2$ A, and $I_3 = 4$ A.
2. Find I_T if $I_1 = 10$ mA, $I_2 = 15$ mA, and $I_3 = 2$ mA.
3. Find I_3 if $I_T = 20$ mA, $I_1 = 6$ mA, and $I_2 = 11$ mA.
4. Find I_3 if $I_T = 400$ mA, $I_1 = 100$ mA, and $I_2 = 150$ mA.
5. Find I_2 if $I_T = 65$ μA, $I_1 = 10$ μA, and $I_3 = 10$ μA.
6. Find I_2 if $I_T = 7$ mA, $I_1 = 500$ μA, and $I_3 = 3.2$ mA.
7. Find I_1 if $I_T = 1.3$ A, $I_2 = 300$ mA, and $I_3 = 40$ mA.
8. Find I_1 if $I_T = 400$ mA, $I_2 = 120$ mA, $I_3 = 150$ mA.

Refer to Figure 11-38.

9. $I_1 = 10$ mA and $I_2 = 15$ mA. Find I_T and I_3.
10. $I_1 = 600$ μA and $I_2 = 1.2$ mA. Find I_T and I_3.

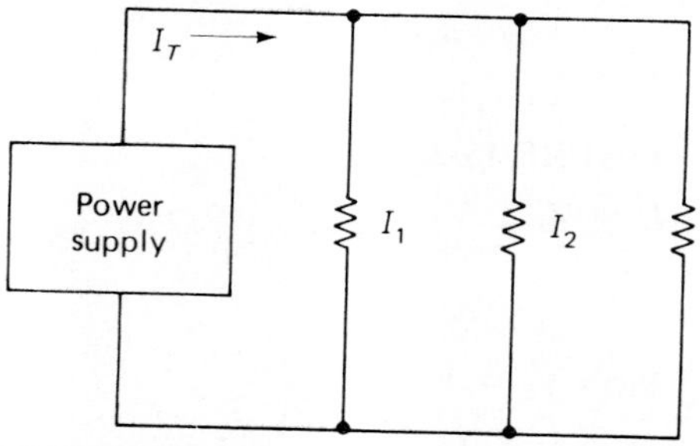

FIGURE 11-37

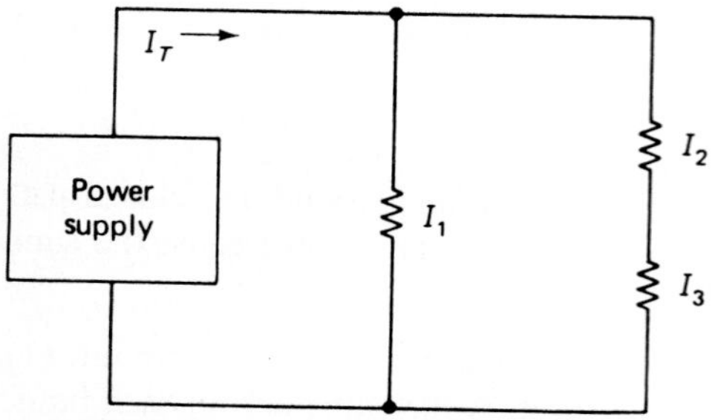

FIGURE 11-38

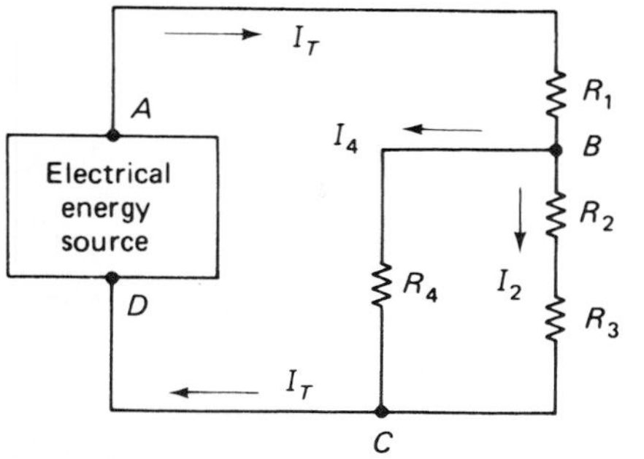

FIGURE 11-39

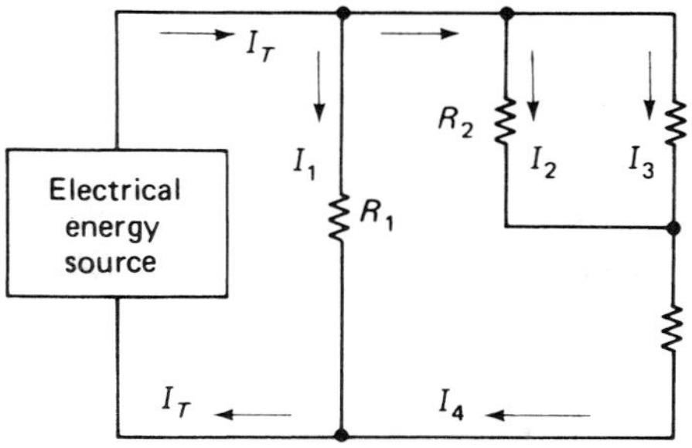

FIGURE 11-40

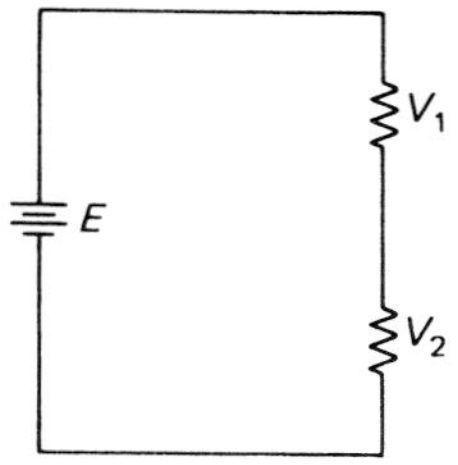

FIGURE 11-41

11. $I_T = 1.6$ A and $I_3 = 800$ mA. Find I_1 and I_2.

12. $I_T = 350$ mA and $I_3 = 150$ mA. Find I_1 and I_2.

13. $I_T = 600$ μA and $I_1 = 400$ μA. Find I_2 and I_3.

14. $I_T = 2.1$ mA and $I_1 = 700$ μA. Find I_2 and I_3.

Refer to Figure 11-39.

15. $I_1 = 1.6$ A, $I_2 = 800$ mA. Find I_T, I_3, and I_4.

16. $I_1 = 750$ mA, $I_2 = 300$ mA. Find I_T, I_3, and I_4.

17. $I_3 = 500$ μA, $I_4 = 2.7$ mA. Find I_T, I_1, and I_2.

18. $I_3 = 3$ mA, $I_4 = 3.6$ mA. Find I_T, I_1, and I_2.

Refer to Figure 11-40.

19. $I_T = 6.7$ mA, $I_2 = 1.2$ mA, $I_4 = 5$ mA. Find I_1 and I_3.

20. $I_T = 206$ mA, $I_2 = 65$ mA, $I_4 = 160$ mA. Find I_1 and I_3.

21. $I_1 = 6$ mA, $I_3 = 3.4$ mA, $I_4 = 8.2$ mA. Find I_T and I_2.

22. $I_1 = 760$ mA, $I_3 = 300$ mA, $I_4 = 465$ mA. Find I_T and I_2.

23. Refer to Figure 11-24. $I_T = 100$ mA, $I_2 = 20$ mA, and $I_5 = 30$ mA. Find I_1, I_3, I_4, and I_6.

End of Chapter Problems 11-2

Refer to Figure 11-41.

1. $E = 30$ V and $V_1 = 12$ V. Find V_2.

2. $E = 17$ V and $V_1 = 4$ V. Find V_2.

3. $V_1 = 7$ V and $V_2 = 6$ V. Find E.

4. $V_1 = 500$ mV and $V_2 = 1.2$ V. Find E.

Refer to Figure 11-42.

5. $E = 70$ V, $V_1 = 22$ V, $V_2 = 37$ V. Find V_3.

6. $E = 12$ V, $V_1 = 2$ V, $V_2 = 6$ V. Find V_3.

7. $V_1 = 3$ V, $V_2 = 8$ V, $V_3 = 7$ V. Find E.

8. $V_1 = 40$ V, $V_2 = 22$ V, $V_3 = 15$ V. Find E.

9. $E = 37$ V, $V_2 = 20$ V, $V_3 = 3$ V. Find V_1.

10. $E = 45$ V, $V_2 = 15$ V, $V_3 = 20$ V. Find V_1.

Refer to Figure 11-43.

11. $E = 20$ V, $V_3 = 8$ V. Find V_1 and V_2.

12. $E = 7.8$ V, $V_3 = 3.6$ V. Find V_1 and V_2.

13. $E = 18$ V, $V_1 = 5$ V. Find V_2 and V_3.

14. $E = 40$ V, $V_1 = 18$ V. Find V_2 and V_3.

15. $V_1 = 12$ V, $V_3 = 5$ V. Find E and V_2.

16. $V_1 = 300$ mV, $V_3 = 1.1$ V. Find E and V_2.

Refer to Figure 11-44.

17. $E = 12$ V, $V_2 = 3$ V, $V_3 = 6$ V. Find V_1 and V_4.

18. $E = 25$ V, $V_2 = 8$ V, $V_3 = 7$ V. Find V_1 and V_4.

19. $E = 3$ V, $V_2 = 500$ mV, $V_4 = 1.3$ V. Find V_1 and V_3.

20. $E = 35$ V, $V_2 = 10$ V, $V_4 = 12$ V. Find V_1 and V_3.

21. $V_1 = 20$ V, $V_2 = 30$ V, $V_4 = 40$ V. Find E and V_3.

22. $V_1 = 3.8$ V, $V_2 = 1.6$ V, $V_4 = 4$ V. Find E and V_3.

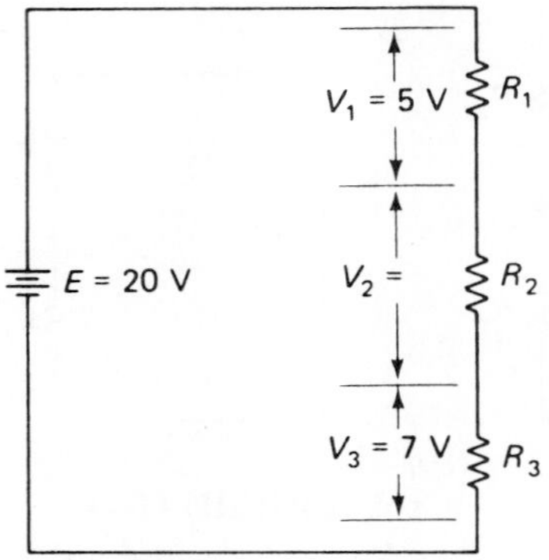

FIGURE 11-42

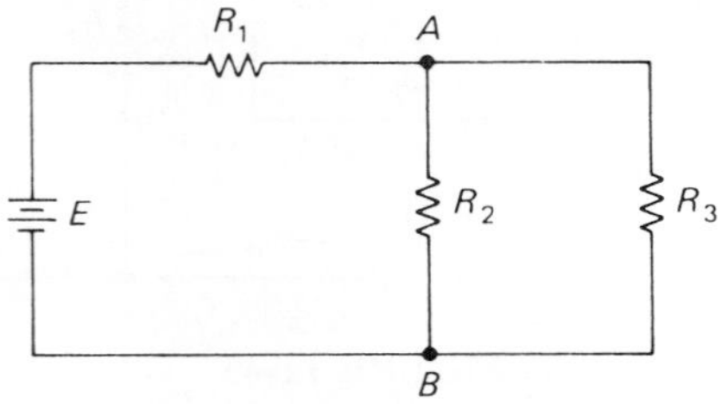

FIGURE 11-43

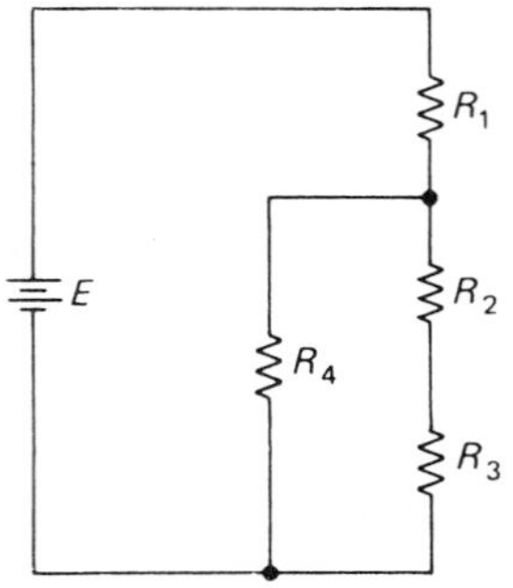

FIGURE 11-44

Refer to Figure 11-45.

23. $E = 50$ V, $V_1 = 10$ V, $V_3 = 8$ V, $V_5 = 15$ V. Find V_2, V_4, and V_6.

24. $E = 20$ V, $V_1 = 6.7$ V, $V_3 = 3.2$ V, $V_5 = 4.7$ V. Find V_2, V_4, and V_6.

25. $V_1 = 9$ V, $V_4 = 6.5$ V, $V_5 = 4$ V, $V_6 = 25$ V. Find E, V_2, and V_3.

26. $V_1 = 12$ V, $V_4 = 12$ V, $V_5 = 12$ V, $V_6 = 40$ V. Find E, V_2, and V_3.

End of Chapter Problems 11-3

1. Refer to Figure 11-46. Let $V_1 = 15$ V and $V_3 = 25$ V. Find (a) V_2, (b) V_4, (c) potential at point A with respect to common, (d) potential at point A with respect to point B, and (e) potential at point B with respect to common.

2. Refer to Figure 11-46. Let $V_1 = 6$ V and $V_3 = 15$ V. Find (a) V_2, (b) V_4, (c) potential at point A with respect to common, (d) potential at point A with respect to point B, and (e) potential at point B with respect to common.

3. Refer to Figure 11-32. Let $V_2 = 8$ V. Find (a) V_1, (b) V_3, (c) potential at point A with respect to common, (d) potential at point A with respect to point B, and (e) potential at point B with respect to common.

4. Refer to Figure 11-32. Let $V_2 = 14$ V. Find (a) V_1, (b) V_3, (c) potential at point A with respect to common, (d) potential at point A with respect to point B, and (e) potential at point B with respect to common.

5. Refer to Figure 11-33. Let $V_1 = 6$ V and $V_2 = 8$ V. Find (a) V_3, (b) V_4, (c) potential at point A with respect to common, and (d) potential at point B with respect to point A.

6. Refer to Figure 11-33. Let $V_1 = 3$ V and $V_2 = 3$ V. Find (a) V_3, (b) V_4, (c) potential at point A with respect to common, and (d) potential at point B with respect to point A.

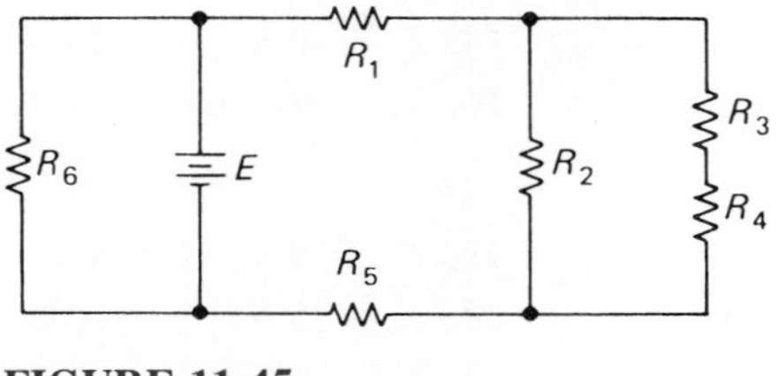

FIGURE 11-45

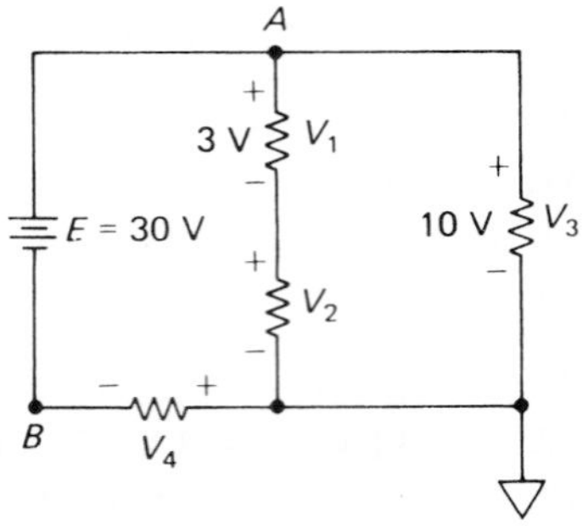

FIGURE 11-46

7. Refer to Figure 11-47. Let $E_1 = 40$ V, $E_2 = 10$ V, and $V_3 = 6$ V with polarity as shown. Find (a) V_1, (b) V_2, (c) potential at point A with respect to common, and (d) potential at point C with respect to point B.

8. Refer to Figure 11-47. Let $E_1 = 30$ V, $E_2 = 5$ V, and $V_3 = 4$ V with polarity as shown. Find (a) V_1, (b) V_2, (c) potential at point A with respect to common, and (d) potential at point C with respect to point B.

Answers to Self-test

SELF-TEST 11-1

1. 10 mA
2. $I_T = 100\ \mu$A, $I_3 = 60\ \mu$A
3. $I_1 = 400\ \mu$A, $I_3 = 2.6$ mA
4. $I_1 = 800\ \mu$A, $I_3 = 308\ \mu$A, $I_4 = 492\ \mu$A
5. $I_T = 27$ mA, $I_4 = 12$ mA
6. 13 V
7. 1.9 V
8. $V_2 = 13$ V, $V_3 = 13$ V
9. $E = 15$ V, $V_3 = 2$ V
10. $V_2 = 7$ V, $V_3 = 6$ V
11. (a) 8 V (b) 9 V (c) −30 V (d) 21 V
12. (a) 9 V (b) 3 V (c) −21 V (d) −15 V

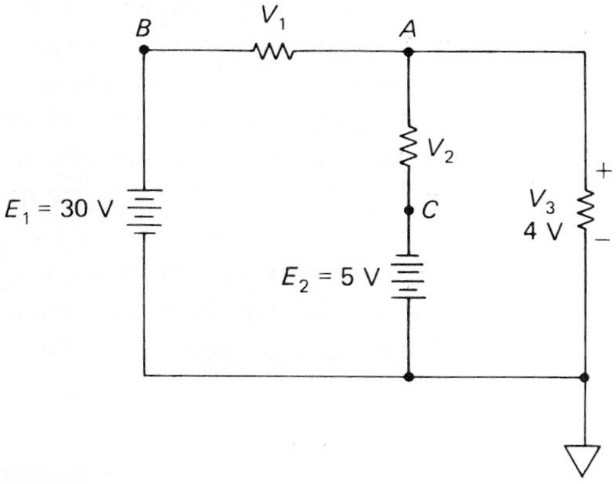

FIGURE 11-47

CHAPTER 12

DC CIRCUIT ANALYSIS: OHM'S LAW

INTRODUCTION

In Chapter 11 we learned about the relationship between voltage drops and the source voltage in series, parallel, and series–parallel circuits. These relationships were developed using Kirchhoff's voltage law. Kirchhoff's current law explained to us the relationship between the source current and the individual branch currents in these same circuits. Now that we understand these relationships, we are ready to learn how to compute these voltage drops and currents and how voltage, current, and resistance relate one to the other.

Ohm's law expresses the relationship between voltage, current, and resistance in an electrical circuit. Ohm's law states that circuit current is directly proportional to voltage and inversely proportional to resistance. Expressed as an equation, Ohm's law is

$$I = \frac{E}{R}$$

where E is the source voltage, I is the circuit current, and R is the circuit resistance. If we know the source voltage and the circuit resistance, we can calculate the circuit current. If E and I are known, we can rearrange the equation to solve for R.

$$R = \frac{E}{I}$$

If I and R are known, we can solve for E.

$$E = IR$$

We must always know two of the variables in order to solve for the third.

These are the three basic forms of Ohm's law. In this chapter we will learn how to apply these equations in the solution of various kinds of electrical circuit problems. Ohm's law and Kirchhoff's laws are the basic laws that express the relationship between voltage, current, and resistance in electrical circuits. A thorough understanding of these laws and how to use them is essential to the study of electronics.

CHAPTER OBJECTIVES

In this chapter you will learn how to:

1. Compute total resistance in a series circuit.
2. Compute total conductance in a parallel circuit.
3. Relate between conductance and resistance.
4. Compute total resistance in a series–parallel circuit.
5. Use Ohm's law and Kirchhoff's laws to solve series, parallel, and series–parallel circuits.
6. Use the calculator in solving circuit problems.

12-1 CIRCUIT RESISTANCE

Before we get into solving circuit problems in terms of voltage drops and currents, let's learn how to calculate circuit resistance. If there is only one resistance in an electrical circuit, the total circuit resistance is equal to that resistance. Such a circuit is shown in Figure 12-1(a). $R_T = R$. In Figure 12-1(b), there are two resistances in series. To identify which resistance we are talking about, they are labeled R_1 and R_2.

KEY POINT In a series circuit the total resistance equals the sum of the individual resistances.

$$R_T = R_1 + R_2 + R_3 + \cdots + R_N$$

Therefore, in Figure 12-1(b), $R_T = R_1 + R_2$

☐ EXAMPLE 12-1

In Figure 12-2(a), let $R_1 = 2.7\ \text{k}\Omega$ and $R_2 = 4.7\ \text{k}\Omega$. Find R_T.

Solution

$$R_T = R_1 + R_2$$
$$= 2.7\ \text{k}\Omega + 4.7\ \text{k}\Omega = 7.4\ \text{k}\Omega$$

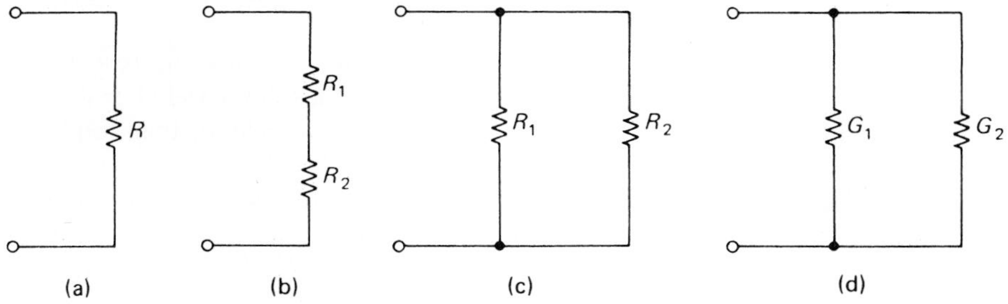

FIGURE 12-1

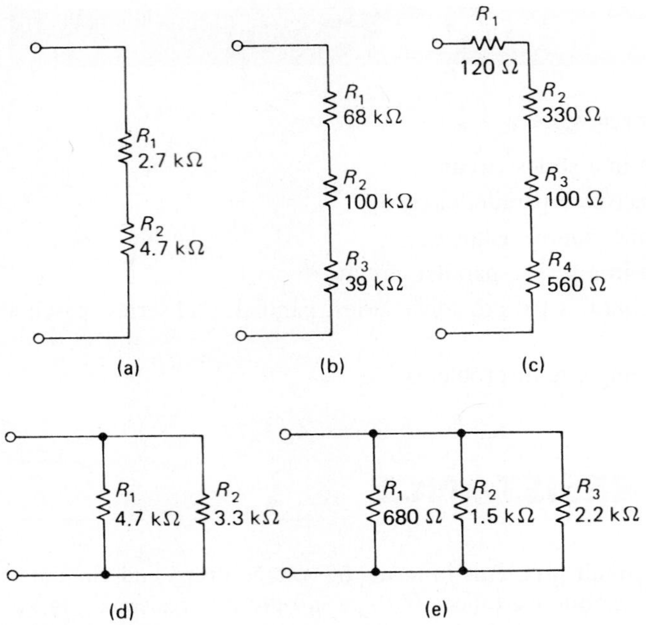

FIGURE 12-2

☐ EXAMPLE 12-2

In Figure 12-2(b), let $R_1 = 68\ \text{k}\Omega$, $R_2 = 100\ \text{k}\Omega$, and $R_3 = 39\ \text{k}\Omega$. Find R_T.

Solution

$$R_T = R_1 + R_2 + R_3$$
$$= 68\ \text{k}\Omega + 100\ \text{k}\Omega + 39\ \text{k}\Omega = 207\ \text{k}\Omega$$

☐ EXAMPLE 12-3

In Figure 12-2(c), let $R_1 = 120\ \Omega$, $R_2 = 330\ \Omega$, $R_3 = 100\ \Omega$, and $R_4 = 560\ \Omega$. Find R_T.

Solution

$$R_T = R_1 + R_2 + R_3 + R_4$$
$$= 120\ \Omega + 330\ \Omega + 100\ \Omega + 560\ \Omega = 1.11\ \text{k}\Omega$$

☐

In Figure 12-1(c), R_1 and R_2 are connected in parallel. Connecting resistances in parallel *reduces* total circuit resistance. This is so because as parallel paths are added, more circuit current flows. If total circuit current is increased, then total circuit resistance must decrease.

It is more logical to think of the parallel circuit in terms of circuit conductances. Conductance and resistance are reciprocals.

$$G = \frac{1}{R}, \qquad R = \frac{1}{G}$$

Figure 12-1(d) is the same as Figure 12-1(c) except that we are using the symbol for conductance (G) to show that we will work with that unit in the parallel circuit.

■ **KEY POINT** In a parallel circuit the total conductance equals the sum of the individual branch conductances.

$$G_T = G_1 + G_2 + G_3 + \cdots + G_N$$

□ EXAMPLE 12-4

In Figure 12-2(d), let $R_1 = 4.7\ \text{k}\Omega$ and $R_2 = 3.3\ \text{k}\Omega$. Find R_T.

Solution Because we are dealing with a parallel circuit, we must first find G_1 and G_2.

$$G_1 = \frac{1}{R_1} = \frac{1}{4.7\ \text{k}\Omega} = 213\ \mu\text{S}$$

$$G_2 = \frac{1}{R_2} = \frac{1}{3.3\ \text{k}\Omega} = 303\ \mu\text{S}$$

Now we can find G_T:

$$G_T = G_1 + G_2 = 213\ \mu\text{S} + 303\ \mu\text{S} = 516\ \mu\text{S}$$

R_T is the reciprocal of G_T.

$$R_T = \frac{1}{G_T} = \frac{1}{516\ \mu\text{S}} = 1.94\ \text{k}\Omega$$

We can put the whole solution into one equation:

$$R_T = \frac{1}{\dfrac{1}{R_1} + \dfrac{1}{R_2}}$$

or

$$R_T = \frac{1}{G_1 + G_2}$$

Using the calculator,

4.7 [EE] 3 [1/x] [+] 3.3 [EE] 3 [1/x] [=] [1/x] *Answer*: 1.94×10^3

Notice that when the [=] sign was pressed G_T (516 μS) was displayed.

□ EXAMPLE 12-5

In Figure 12-2(e), let $R_1 = 680\ \Omega$, $R_2 = 1.5\ \text{k}\Omega$, and $R_3 = 2.2\ \text{k}\Omega$. Find R_T.

Solution

$$G_T = G_1 + G_2 + G_3$$
$$= 1.47\ \text{mS} + 667\ \mu\text{S} + 445\ \mu\text{S} = 2.59\ \text{mS}$$

$$R_T = \frac{1}{G_T} = \frac{1}{2.59\ \text{mS}} = 386\ \Omega$$

■ **KEY POINT** It is always true that *in a parallel circuit, R_T is always less than the smallest branch resistance.*

Using the calculator;

680 [1/x] [+] 1.5 [EE] 3 [1/x] [+] 2.2 [EE] 3 [1/x] [=] [1/x]

Answer: 386

$R_T = 386\ \Omega$

□

PRACTICE PROBLEMS 12-1

1. Refer to Figure 12-2(a). Find R_T when (a) $R_1 = 82\ \text{k}\Omega$, $R_2 = 100\ \text{k}\Omega$, and (b) $R_1 = 33\ \text{k}\Omega$, $R_2 = 15\ \text{k}\Omega$.

2. Refer to Figure 12-2(b). Find R_T when (a) $R_1 = 120\ \Omega$, $R_2 = 220\ \Omega$, $R_3 = 1\ \text{k}\Omega$, and (b) $R_1 = 68\ \text{k}\Omega$, $R_2 = 120\ \text{k}\Omega$, $R_3 = 39\ \text{k}\Omega$.

3. Refer to Figure 12-2(d). Find G_T and R_T when (a) $R_1 = 6.8\ \text{k}\Omega$, $R_2 = 2.7\ \text{k}\Omega$, (b) $R_1 = 7.5\ \text{k}\Omega$, $R_2 = 15\ \text{k}\Omega$, (c) $R_1 = 47\ \text{k}\Omega$, $R_2 = 120\ \text{k}\Omega$, and (d) $R_1 = 68\ \Omega$, $R_2 = 220\ \Omega$.

4. Refer to Figure 12-2(e). Find G_T and R_T when (a) $R_1 = 1.5\ \text{k}\Omega$, $R_2 = 4.7\ \text{k}\Omega$, $R_3 = 3.3\ \text{k}\Omega$, (b) $R_1 = 680\ \Omega$, $R_2 = 1.8\ \text{k}\Omega$, $R_3 = 470\ \Omega$, (c) $R_1 = 120\ \text{k}\Omega$, $R_2 = 120\ \text{k}\Omega$, $R_3 = 180\ \text{k}\Omega$, and (d) $R_1 = 22\ \text{k}\Omega$, $R_2 = 100\ \text{k}\Omega$, $R_3 = 10\ \text{k}\Omega$.

Solutions

1. (a) $R_T = R_1 + R_2 = 82\ \text{k}\Omega + 100\ \text{k}\Omega = 182\ \text{k}\Omega$

(b) $R_T = R_1 + R_2 = 33\ \text{k}\Omega + 15\ \text{k}\Omega = 48\ \text{k}\Omega$

2. (a) $R_T = R_1 + R_2 + R_3 = 120\ \Omega + 220\ \Omega + 1\ \text{k}\Omega = 1.34\ \text{k}\Omega$

(b) $R_T = R_1 + R_2 + R_3 = 68\ \text{k}\Omega + 120\ \text{k}\Omega + 39\ \text{k}\Omega = 227\ \text{k}\Omega$

3. (a) $G_T = G_1 + G_2 = 147\ \mu\text{S} + 370\ \mu\text{S} = 517\ \mu\text{S}$

$$R_T = \frac{1}{G_T} = \frac{1}{517\ \mu\text{S}} = 1.93\ \text{k}\Omega$$

(b) $G_T = G_1 + G_2 = 133\ \mu\text{S} + 66.7\ \mu\text{S} = 200\ \mu\text{S}$

$$R_T = \frac{1}{G_T} = \frac{1}{200\ \mu\text{S}} = 5\ \text{k}\Omega$$

(c) $G_T = G_1 + G_2 = 21.3\ \mu\text{S} + 8.33\ \mu\text{S} = 29.6\ \mu\text{S}$

$$R_T = \frac{1}{G_T} = \frac{1}{29.6\ \mu\text{S}} = 33.8\ \text{k}\Omega$$

(d) $G_T = G_1 + G_2 = 14.7\ \text{mS} + 4.55\ \text{mS} = 19.3\ \text{mS}$

$$R_T = \frac{1}{G_T} = \frac{1}{19.3\ \text{mS}} = 51.9\ \Omega$$

4. (a) $G_T = G_1 + G_2 + G_3 = 667\ \mu\text{S} + 213\ \mu\text{S} + 303\ \mu\text{S} = 1.18\ \text{mS}$

$$R_T = \frac{1}{G_T} = \frac{1}{1.18\ \text{mS}} = 846\ \Omega$$

(b) $G_T = G_1 + G_2 + G_3 = 1.47\ \text{mS} + 556\ \mu\text{S} + 2.13\ \text{mS} = 4.15\ \text{mS}$

$$R_T = \frac{1}{G_T} = \frac{1}{4.15\ \text{mS}} = 241\ \Omega$$

(c) $G_T = G_1 + G_2 + G_3 = 8.33\ \mu\text{S} + 8.33\ \mu\text{S} + 5.56\ \mu\text{S} = 22.2\ \mu\text{S}$

$$R_T = \frac{1}{G_T} = \frac{1}{22.2\ \mu\text{S}} = 45\ \text{k}\Omega$$

(d) $G_T = G_1 + G_2 + G_3 = 45.5\ \mu S + 10\ \mu S + 100\ \mu S = 155\ \mu S$

$$R_T = \frac{1}{G_T} = \frac{1}{155\ \mu S} = 6.43\ k\Omega$$

Additional practice problems are at the end of the chapter.

Now let's look at some series–parallel circuits.

RULE 12-1 To find the total resistance in a series–parallel circuit: (1) determine which resistors are in parallel and reduce that part of the circuit to an equivalent resistance; (2) add series resistances when possible; (3) continue steps 1 and 2 until the circuit is reduced to one resistance.

Let's find the total resistance of the circuit in Figure 12-3(a), where $R_1 = 4.7\ k\Omega$, $R_2 = 6.8\ k\Omega$, and $R_3 = 12\ k\Omega$. Notice that R_2 and R_3 are in parallel and they are in series with R_1. The equation is

$$R_T = R_1 + R_2 \parallel R_3$$

(The symbol $\parallel$ means "in parallel with "). Let's first reduce R_2 and R_3 to an equivalent series resistance, which we will call R_X. We will solve for R_X the same way we solved for R_T in Figure 12-2(d).

$$G_X = G_2 + G_3 = 147\ \mu S + 83.3\ \mu S = 230\ \mu S$$

$$R_X = \frac{1}{G_X} = \frac{1}{230\ \mu S} = 4.34\ k\Omega$$

We now have a circuit that looks like Figure 12-3(b). Then

$$R_T = R_1 + R_X = 4.7\ k\Omega + 4.34\ k\Omega = 9.04\ k\Omega$$

We have reduced a complex circuit to a single resistance, as shown in Figure 12-3(c).

$$R_T = R_1 + \frac{1}{\dfrac{1}{R_2} + \dfrac{1}{R_3}}$$

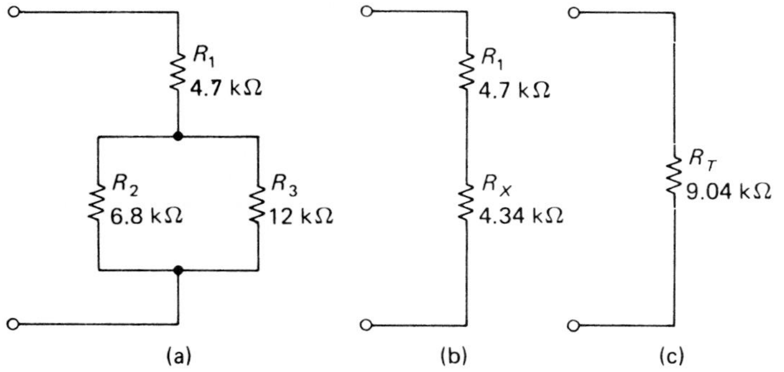

FIGURE 12-3

Using the calculator,

4.7 [EE] 3 [+] [(] 6.8 [EE] 3 [1/x] [+] 12 [EE] 3 [1/x] [)] [1/x] [=]

Answer: 9.04×10^3

$$R_T = 9.04 \text{ k}\Omega$$

Let's try another one. In Figure 12-4(a), let $R_1 = 20 \text{ k}\Omega$, $R_2 = 30 \text{ k}\Omega$, and $R_3 = 68 \text{ k}\Omega$. Notice that R_1 and R_2 are in series. The equivalent resistance (R_X) is equal to $R_1 + R_2$.

$$R_X = R_1 + R_2 = 20 \text{ k}\Omega + 30 \text{ k}\Omega = 50 \text{ k}\Omega$$

The circuit now looks like the one in Figure 12-4(b). The final step is to reduce the circuit to a single resistance, R_T.

$$G_T = G_X + G_3 = 20 \ \mu\text{S} + 14.7 \ \mu\text{S} = 34.7 \ \mu\text{S}$$

$$R_T = \frac{1}{G_T} = \frac{1}{34.7 \ \mu\text{S}} = 28.8 \text{ k}\Omega$$

Let's try one more. In Figure 12-5(a), let $R_1 = 1.2 \text{ k}\Omega$, $R_2 = 6.8 \text{ k}\Omega$, $R_3 = 10 \text{ k}\Omega$, and $R_4 = 3.3 \text{ k}\Omega$. Notice that R_2, R_3, and R_4 are connected in parallel. To find R_X, we must first find G_X.

$$G_X = G_2 + G_3 + G_4 = 147 \ \mu\text{S} + 100 \ \mu\text{S} + 303 \ \mu\text{S} = 550 \ \mu\text{S}$$

$$R_X = \frac{1}{G_X} = \frac{1}{550 \ \mu\text{S}} = 1.82 \text{ k}\Omega$$

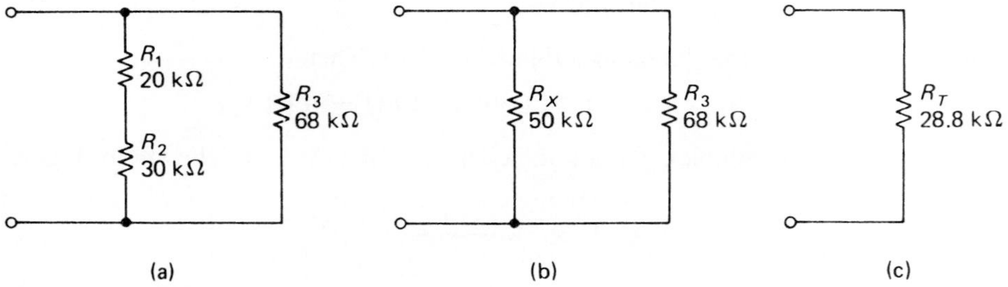

FIGURE 12-4

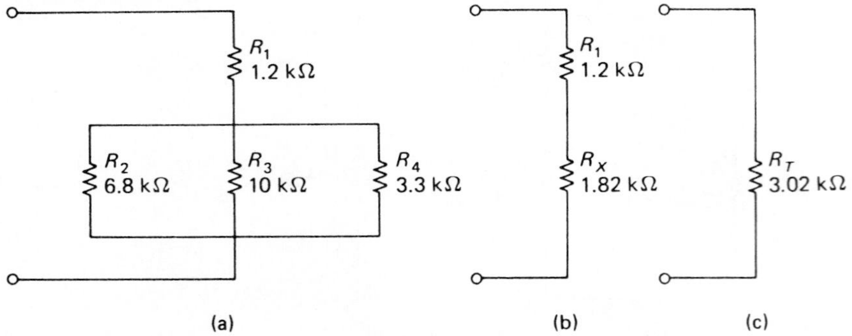

FIGURE 12-5

This equivalent resistance, R_X, is in series with R_1, as shown in Figure 12-5(b). R_T then is equal to R_1 and R_X in series.

$$R_T = R_X + R_1 = 1.82\ \text{k}\Omega + 1.2\ \text{k}\Omega = 3.02\ \text{k}\Omega$$

The circuit reduces to the equivalent resistance shown in Figure 12-5(c).

PRACTICE PROBLEMS 12-2

1. Refer to Figure 12-3(a) and find R_T when (a) $R_1 = 330\ \Omega$, $R_2 = 680\ \Omega$, $R_3 = 910\ \Omega$, and (b) $R_1 = 82\ \text{k}\Omega$, $R_2 = 270\ \text{k}\Omega$, $R_3 = 180\ \text{k}\Omega$.
2. Refer to Figure 12-4(a) and find R_T when (a) $R_1 = 150\ \Omega$, $R_2 = 560\ \Omega$, $R_3 = 1.5\ \text{k}\Omega$, and (b) $R_1 = 22\ \text{k}\Omega$, $R_2 = 56\ \text{k}\Omega$, $R_3 = 39\ \text{k}\Omega$.
3. Refer to Figure 12-5(a) and find R_T when (a) $R_1 = 750\ \text{k}\Omega$, $R_2 = 1.2\ \text{M}\Omega$, $R_3 = 2.2\ \text{M}\Omega$, $R_4 = 910\ \text{k}\Omega$, and (b) $R_1 = 2.7\ \text{k}\Omega$, $R_2 = 1\ \text{k}\Omega$, $R_3 = 10\ \text{k}\Omega$, $R_4 = 8.2\ \text{k}\Omega$.

Solutions

1. (a) $R_T = R_1 + R_2 \parallel R_3 = 330\ \Omega + 389\ \Omega = 719\ \Omega$
   (b) $R_T = R_1 + R_2 \parallel R_3 = 82\ \text{k}\Omega + 108\ \text{k}\Omega = 190\ \text{k}\Omega$
2. (a) $R_T = (R_1 + R_2) \parallel R_3 = 710\ \Omega \parallel 1.5\ \text{k}\Omega = 482\ \Omega$
   (b) $R_T = (R_1 + R_2) \parallel R_3 = 78\ \text{k}\Omega \parallel 39\ \text{k}\Omega = 26.0\ \text{k}\Omega$
3. (a) $R_1 + R_2 \parallel R_3 \parallel R_4 = 750\ \text{k}\Omega + 419\ \text{k}\Omega = 1.17\ \text{M}\Omega$
   (b) $R_1 + R_2 \parallel R_3 \parallel R_4 = 2.7\ \text{k}\Omega + 818\ \Omega = 3.52\ \text{k}\Omega$

Additional practice problems are at the end of the chapter.

There are many kinds of series–parallel circuits. We have examined only a few. In all problem solving we must recognize which resistors are in parallel and reduce that part of the circuit to an equivalent resistance. We then add series resistances and continue both operations until we have reduced the circuit to one resistance (R_T).

SELF-TEST 12-1

1. Refer to Figure 12-2(b). Find R_T when $R_1 = 39\ \text{k}\Omega$, $R_2 = 56\ \text{k}\Omega$, and $R_3 = 27\ \text{k}\Omega$.
2. Refer to Figure 12-2(d). Find R_T when $R_1 = 820\ \Omega$ and $R_2 = 1.2\ \text{k}\Omega$.
3. Refer to Figure 12-2(e). Find R_T when $R_1 = 33\ \text{k}\Omega$, $R_2 = 100\ \text{k}\Omega$, and $R_3 = 82\ \text{k}\Omega$.
4. Refer to Figure 12-3(a). Find R_T when $R_1 = 12\ \text{k}\Omega$, $R_2 = 27\ \text{k}\Omega$, and $R_3 = 47\ \text{k}\Omega$.
5. Refer to Figure 12-4(a). Find R_T when $R_1 = 330\ \Omega$, $R_2 = 1.2\ \text{k}\Omega$, and $R_3 = 910\ \Omega$.
6. Refer to Figure 12-5(a). Find R_T when $R_1 = 270\ \text{k}\Omega$, $R_2 = 1.2\ \text{M}\Omega$, $R_3 = 750\ \text{k}\Omega$, and $R_4 = 470\ \text{k}\Omega$.

Answers to Self-test 12-1 are at the end of the chapter.

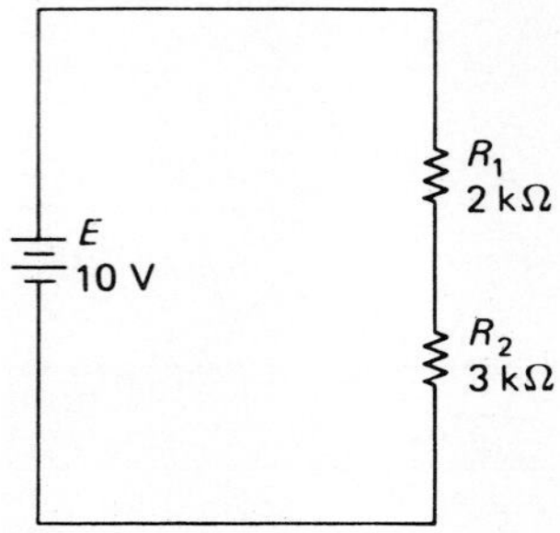

FIGURE 12-6

12-2 SERIES CIRCUITS

Consider the circuit in Figure 12-6. Let's find the current, I, and the voltage drops across R_1 and R_2. The general Ohm's law equation is

$$I = \frac{E}{R}$$

When dealing with a specific situation, we must be more specific in labeling the variables in our equation. For example, in a series circuit we know from Kirchhoff's law that current is constant. $I_T = I_1 = I_2$. Therefore, we can label the current I and not use a subscript. There are three resistances though, R_1, R_2, and R_T, and there are three voltages, E, V_1, and V_2. If we use R_T in the equation, we must use E. If we use R_1, we must use V_1, and so on. To find I, we could write the equation three ways then, depending on known values:

$$I = \frac{E}{R_T}$$

$$I = \frac{V_1}{R_1}$$

$$I = \frac{V_2}{R_2}$$

In our circuit we know E and can find R_T:

$$R_T = R_1 + R_2 = 5\ \text{k}\Omega$$

Then

$$I = \frac{E}{R_T} = \frac{10\ \text{V}}{5\ \text{k}\Omega} = 2\ \text{mA}$$

Knowing I, we can solve for V_1 and V_2 by rearranging the equation and solving for the unknown:

$$V_1 = IR_1 = 2\ \text{mA} \times 2\ \text{k}\Omega = 4\ \text{V}$$

$$V_2 = IR_2 = 2\ \text{mA} \times 3\ \text{k}\Omega = 6\ \text{V}$$

Using Kirchhoff's voltage law to check our answers, we get

$$E = V_1 + V_2$$

$$10\ \text{V} = 4\ \text{V} + 6\ \text{V}$$

Let's look at the problem from another angle. Because the current is constant and $IR = V$, by inspection we can see that the greater voltage drops across R_2.

■ **KEY POINT** In any series circuit the greater voltage drops across the greater resistance.

Furthermore, the voltage drops are always in proportion to the resistances. In this problem the resistances are in the ratio of 2 to 3. Expressed as a proportion,

$$\frac{R_1}{R_2} = \frac{V_1}{V_2} \qquad (12\text{-}1)$$

We could also say that

$$\frac{R_1}{R_T} = \frac{V_1}{E} \qquad (12\text{-}2)$$

$$\frac{R_2}{R_T} = \frac{V_2}{E} \qquad (12\text{-}3)$$

R_1 is 0.4 or 40% of total resistance. Therefore, 40% of the total voltage

$$\frac{R_1}{R_T} = \frac{2\ \text{k}\Omega}{5\ \text{k}\Omega} = 0.4$$

must drop across it. We can rearrange our proportion (Equation 12-2):

$$\frac{R_1}{R_T} = \frac{V_1}{E} \qquad (12\text{-}2)$$

$$V_1 = E\left(\frac{R_1}{R_T}\right)$$

or

$$V_1 = \frac{ER_1}{R_T} \qquad (12\text{-}4)$$

Then it follows that

$$V_2 = \frac{ER_2}{R_T} \qquad (12\text{-}5)$$

We will use these equations often in problem solving. Looking at Equation 12-4 again, we see

$$V_1 = \left(\frac{E}{R_T}\right)R_1$$

or

$$V_1 = IR_1 \qquad \left(I = \frac{E}{R_T}\right)$$

☐ EXAMPLE 12-6

In Figure 12-6, let $E = 12$ V, $R_2 = 1.5$ kΩ, and $V_1 = 4$ V. Find R_1, V_2, I, and R_T.

Solution In this problem, V_2 is the only unknown variable that can be found from the information given. We can't find R_1 because we don't know I. We can't find R_T because we don't know I or R_1. We can find V_2 by using Kirchhoff's voltage law.

$$E = V_1 + V_2$$
$$12\text{ V} = 4\text{ V} + V_2$$
$$8\text{ V} = V_2$$

Once we have determined the value of V_2 we can find I.

$$I = \frac{V_2}{R_2} = \frac{8\text{ V}}{1.5\text{ k}\Omega} = 5.33\text{ mA}$$

Now that we know the value of I we can find R_1 and R_T.

$$R_1 = \frac{V_1}{I} = \frac{4\text{ V}}{5.33\text{ mA}} = 750\ \Omega$$
$$R_T = R_1 + R_2 = 750\ \Omega + 1.5\text{ k}\Omega = 2.25\text{ k}\Omega$$

Also, $$R_T = \frac{E}{I} = \frac{12\text{ V}}{5.33\text{ mA}} = 2.25\text{ k}\Omega$$

□ EXAMPLE 12-7

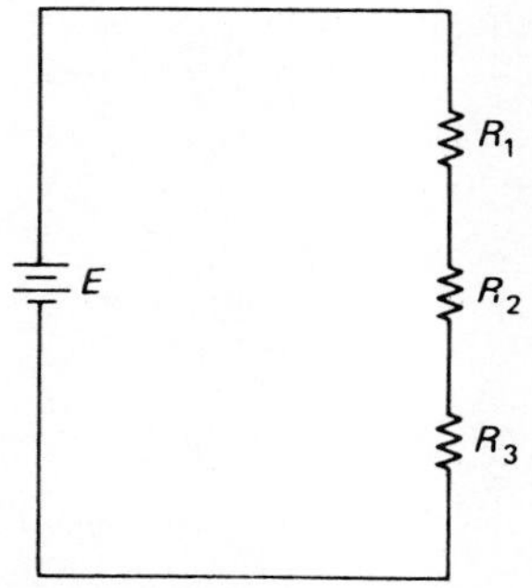

FIGURE 12-7

In Figure 12-7, let $E = 12$ V, $V_1 = 2.73$ V, $I = 38.7$ mA, and $R_2 = 100\ \Omega$. Find V_2, V_3, R_1, R_3 and R_T.

Solution A check of given values shows that with the information given we can find R_T because we know E and I.

$$R_T = \frac{E}{I} = \frac{12\text{ V}}{38.7\text{ mA}} = 310\ \Omega$$

We can also find R_1 because we know I and V_1.

$$R_1 = \frac{V_1}{I} = \frac{2.73\text{ V}}{38.7\text{ mA}} = 70.5\ \Omega$$

We can find V_2 because we know I and R_2.

$$V_2 = IR_2 = 38.7\text{ mA} \times 100\ \Omega = 3.87\text{ V}$$

The given values plus the unknowns we have found provide us with enough information to solve for the rest of the unknowns, which are V_3 and R_3. Although other solutions may be possible, we will solve for these variables as follows:

$$V_3 = E - V_1 - V_2 = 12\text{ V} - 2.73\text{ V} - 3.87\text{ V} = 5.40\text{ V}$$

Then

$$R_3 = \frac{V_3}{I} = \frac{5.40\text{ V}}{38.7\text{ mA}} = 140\ \Omega$$

Check R_T:

$$R_T = R_1 + R_2 + R_3 = 70.5\ \Omega + 100\ \Omega + 140\ \Omega = 311\ \Omega$$

(rounding causes a difference of 1 in the LSD position)

Check E:

$$E = V_1 + V_2 + V_3 = 2.73\text{ V} + 3.87\text{ V} + 5.40\text{ V} = 12\text{ V}$$

□

PRACTICE PROBLEMS 12-3

1. Refer to Figure 12-6. Find I, R_T, V_1, and V_2 when (a) $E = 25$ V, $R_1 = 2.7$ kΩ, and $R_2 = 4.7$ kΩ, (b) $E = 30$ V, $R_1 = 10$ kΩ, and $R_2 = 15$ kΩ, and (c) $E = 12$ V, $R_1 = 470$ Ω, and $R_2 = 1.2$ kΩ.
2. Refer to Figure 12-6. Find R_T, R_1, V_1, and V_2 when $E = 10$ V, $I = 150$ μA, and $R_2 = 20$ kΩ.
3. Refer to Figure 12-6. Find E, R_T, V_2, and R_1 when $I = 12.6$ mA, $V_1 = 6.7$ V, and $R_2 = 810$ Ω.
4. Refer to Figure 12-7. Find I, R_T, V_1, V_2, and V_3 when (a) $E = 10$ V, $R_1 = 3.3$ kΩ, $R_2 = 6.8$ kΩ, and $R_3 = 2.2$ kΩ, (b) $E = 20$ V, $R_1 = 4.7$ kΩ, $R_2 = 10$ kΩ, and $R_3 = 12$ kΩ, and (c) $E = 15$ V, $R_1 = 12$ kΩ, and $R_2 = 33$ kΩ, and $R_3 = 18$ kΩ.
5. Refer to Figure 12-7. Find V_2, V_3, R_1, R_3, and R_T when $E = 18$ V, $V_1 = 8.2$ V, $I = 40$ μA, and $R_2 = 180$ kΩ.
6. Refer to Figure 12-7. Find E, R_2, R_3, V_1, and V_3 when $I = 37.6$ mA, $R_1 = 47$ Ω, $V_2 = 1.92$ V, and $R_T = 137$ Ω.

Solutions

1. (a) $R_T = R_1 + R_2 = 2.7\text{ k}\Omega + 4.7\text{ k}\Omega = 7.4\text{ k}\Omega$

   $I = \dfrac{E}{R_T} = \dfrac{25\text{ V}}{7.4\text{ k}\Omega} = 3.38\text{ mA}$

   $V_1 = IR_1 = 3.38\text{ mA} \times 2.7\text{ k}\Omega = 9.12\text{ V}$

   $V_2 = IR_2 = 3.38\text{ mA} \times 4.7\text{ k}\Omega = 15.9\text{ V}$

   (b) $R_T = 25$ kΩ, $I = 1.2$ mA, $V_1 = 12$ V, $V_2 = 18$ V

   (c) $R_T = 1.67$ kΩ, $I = 7.19$ mA, $V_1 = 3.38$ V, $V_2 = 8.62$ V
2. $R_T = 66.7$ kΩ, $R_1 = 46.7$ kΩ, $V_1 = 7$ V, $V_2 = 3$ V
3. $E = 16.9$ V, $V_2 = 10.2$ V, $R_1 = 532$ Ω, $R_T = 1.34$ kΩ
4. (a) $R_T = R_1 + R_2 + R_3 = 3.3\text{ k}\Omega + 6.8\text{ k}\Omega + 2.2\text{ k}\Omega = 12.3\text{ k}\Omega$

   $I = \dfrac{E}{R_T} = \dfrac{10\text{ V}}{12.3\text{ k}\Omega} = 813\ \mu\text{A}$

   $V_1 = IR_1 = 813\ \mu\text{A} \times 3.3\text{ k}\Omega = 2.68\text{ V}$

   $V_2 = IR_2 = 813\ \mu\text{A} \times 6.8\text{ k}\Omega = 5.53\text{ V}$

   $V_3 = IR_3 = 813\ \mu\text{A} \times 2.2\text{ k}\Omega = 1.79\text{ V}$

   (b) $R_T = 26.7$ kΩ, $I = 749$ μA, $V_1 = 3.52$ V, $V_2 = 7.49$ V, $V_3 = 8.99$ V

   (c) $R_T = 63$ kΩ, $I = 238$ μA, $V_1 = 2.86$ V, $V_2 = 7.86$ V, $V_3 = 4.29$ V
5. $R_T = 450$ kΩ, $V_2 = 7.2$ V, $V_3 = 2.6$ V, $R_3 = 65$ kΩ, $R_1 = 205$ kΩ
6. $E = 5.15$ V, $V_1 = 1.77$ V, $V_3 = 1.46$ V, $R_2 = 51.1$ Ω, $R_3 = 38.8$ Ω

Additional practice problems are at the end of the chapter.

12-3 PARALLEL CIRCUITS

Consider the circuit in Figure 12-8.
Let's find V, R_T, I_T, and I_2. The general equation is

$$V = IR = I\left(\frac{1}{G}\right) = \frac{I}{G}$$

Specifically,

$$V = \frac{I_T}{G_T} = \frac{I_1}{G_1} = \frac{I_2}{G_2}$$

Since we are given I_T, we will find V by using

$$V = \frac{I_T}{G_T}$$

$$= \frac{10 \text{ mA}}{2 \text{ mS} + 3 \text{ mS}} = \frac{10 \text{ mA}}{5 \text{ mS}} = 2 \text{ V}$$

Now we can find I_1 and I_2:

$$I = \frac{V}{R} = VG$$

$$I_1 = VG_1 = 2 \text{ V} \times 2 \text{ mS} = 4 \text{ mA}$$

$$I_2 = VG_2 = 2 \text{ V} \times 3 \text{ mS} = 6 \text{ mA}$$

We must keep in mind that V is the constant in a parallel circuit. Since we know G_T,

$$R_T = \frac{1}{G_T} = \frac{1}{5 \text{ mS}} = 200\ \Omega$$

We see that I_2 was the greater current. That was because G_2 was the greater conductance.

■ **KEY POINT** In a parallel circuit the greater current flows through the greater conductance.

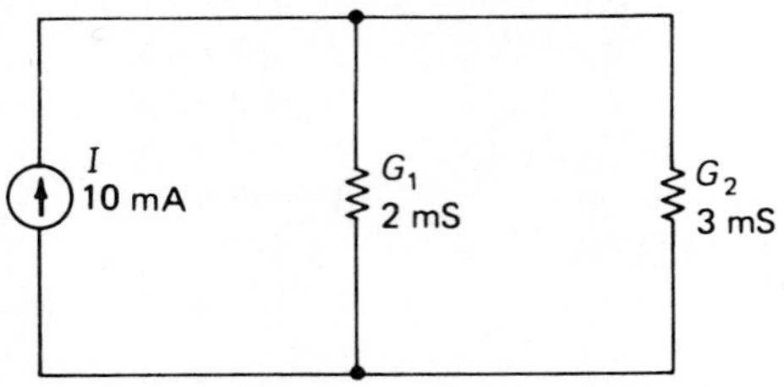

FIGURE 12-8

The currents are always in proportion to the conductance. In this problem the conductances are in the ratio of 2 to 3. Expressed as a proportion,

$$\frac{G_1}{G_2} = \frac{I_1}{I_2} \tag{12-6}$$

We could also say that

$$\frac{G_1}{G_T} = \frac{I_1}{I_T} \tag{12-7}$$

and

$$\frac{G_2}{G_T} = \frac{I_2}{I_T} \tag{12-8}$$

G_1 is 0.4 or 40% of the total conductance. Therefore, 40% of the total current flows through it. We can rearrange our proportion (Equation 12-7):

$$\frac{G_1}{G_T} = \frac{I_1}{I_T} \tag{12-7}$$

$$I_1 = I_T\left(\frac{G_1}{G_T}\right)$$

or

$$I_1 = \frac{I_T G_1}{G_T} \tag{12-9}$$

Then it follows that

$$I_2 = \frac{I_T G_2}{G_T} \tag{12-10}$$

We will use these equations very often in problem solving.

Notice the duality or similarity between Equations 12-4 and 12-9. The relationship between voltage drops and resistance in a series circuit is the same as the relationship between current and conductance in a parallel circuit.

□ EXAMPLE 12-8

In Figure 12-8, let $I_T = 400\ \mu\text{A}$, $R_1 = 8.1\ \text{k}\Omega$, and $I_2 = 150\ \mu\text{A}$. Solve for V, I_1, R_2, G_T, and R_T.

Solution An examination of known variables shows that we have one unknown current, I_1. We can find this current by applying Kirchhoff's current law. We don't have enough information to find any of the other unknowns at this time.

$$I_T = I_1 + I_2$$

$$400\ \mu\text{A} = I_1 + 150\ \mu\text{A}$$

$$250\ \mu\text{A} = I_1$$

Now that we know I_1 we can find V.

$$V = I_1 R_1 = 250\ \mu\text{A} \times 8.1\ \text{k}\Omega = 2.03\ \text{V}$$

With V we can find R_T, G_T, and R_2.

$$R_2 = \frac{V}{I_2} = \frac{2.03\ \text{V}}{150\ \mu\text{A}} = 13.5\ \text{k}\Omega$$

$$G_T = \frac{I_T}{V} = \frac{400\ \mu\text{A}}{2.03\ \text{V}} = 198\ \mu\text{S}$$

$$R_T = \frac{1}{G_T} = \frac{1}{198\ \mu\text{S}} = 5.06\ \text{k}\Omega$$

There may be alternate solutions that result in the same answers. Typically, more than one valid approach can be used to solve problems like these.

☐ EXAMPLE 12-9

Refer to Figure 12-9. Let $V = 50$ V, $G_1 = 90\ \mu$S, $I_2 = 400\ \mu$A, and $G_T = 250\ \mu$S. Find R_1, R_2, R_3, R_T, I_1, I_3, and I_T.

Solution From the given data we can find I_T.

$$I_T = VG_T = 50\ \text{V} \times 250\ \mu\text{S} = 12.5\ \text{mA}$$

We can also find R_2:

$$R_2 = \frac{V}{I_2} = \frac{50\ \text{V}}{400\ \mu\text{A}} = 125\ \text{k}\Omega$$

Since we know G_1, we can find R_1:

$$R_1 = \frac{1}{G_1} = \frac{1}{90\ \mu\text{S}} = 11.1\ \text{k}\Omega$$

Since we know G_1 and V, we can find I_1:

$$I_1 = VG_1 = 50\ \text{V} \times 90\ \mu\text{S} = 4.5\ \text{mA}$$

Since we know G_T, we can find R_T:

$$R_T = \frac{1}{G_T} = \frac{1}{250\ \mu\text{S}} = 4\ \text{k}\Omega$$

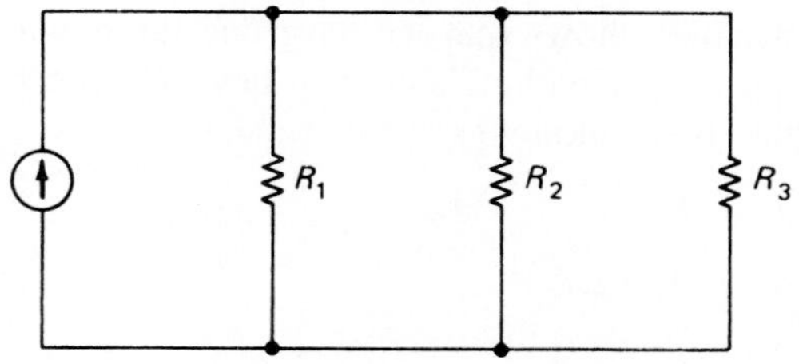

FIGURE 12-9

The rest of the unknowns can now be found:

$$I_3 = I_T - I_1 - I_2 = 12.5 \text{ mA} - 4.5 \text{ mA} - 400\ \mu\text{A} = 7.6 \text{ mA}$$

$$R_3 = \frac{V}{I_3} = \frac{50 \text{ V}}{7.6 \text{ mA}} = 6.58 \text{ k}\Omega$$

□

PRACTICE PROBLEMS 12-4

1. Refer to Figure 12-8. Find G_T, R_T, V, I_1, and I_2 when (a) $I_T = 20$ mA, $R_1 = 2.2$ kΩ, and $R_2 = 4.7$ kΩ, and (b) $I_T = 200\ \mu$A, $R_1 = 15$ kΩ, $R_2 = 27$ kΩ.
2. Refer to Figure 12-8. Find I_1, V, R_2, G_T, and R_T when $I_T = 1.5$ mA, $R_1 = 10$ kΩ, and $I_2 = 800\ \mu$A.
3. Refer to Figure 12-8. Find I_1, I_2, R_2, G_T, and R_T when $I_T = 1$ mA, $V = 2.7$ V, and $R_1 = 5.6$ kΩ.
4. Refer to Figure 12-9. Find G_T, R_T, V, I_1, I_2, and I_3 when (a) $I_T = 10$ mA, $R_1 = 330\ \Omega$, $R_2 = 470\ \Omega$, and $R_3 = 680\ \Omega$, and (b) $I_T = 250\ \mu$A, $R_1 = 5.6$ kΩ, $R_2 = 2.2$ kΩ, and $R_3 = 3.9$ kΩ.
5. Refer to Figure 12-9. Given $I_T = 50$ mA, $I_1 = 20$ mA, $V = 9$ V, and $R_2 = 1$ kΩ, find R_1, R_3, I_2, I_3, R_T, and G_T.
6. Refer to Figure 12-9. Given $V = 12$ V, $G_1 = 17.9\ \mu$S, $I_2 = 255\ \mu$A, and $G_T = 69.4\ \mu$S, find R_1, R_2, R_3, I_1, I_3, and I_T.

Solutions

1. (a) $G_T = G_1 + G_2 = 455\ \mu\text{S} + 213\ \mu\text{S} = 667\ \mu\text{S}$

$$R_T = \frac{1}{G_T} = \frac{1}{667\ \mu\text{S}} = 1.5 \text{ k}\Omega$$

$$V = \frac{I_T}{G_T} = I_T R_T = 30 \text{ V} \qquad \text{(either expression yields 30 V)}$$

$$I_1 = VG_1 = 30 \text{ V} \times 455\ \mu\text{S} = 13.6 \text{ mA} \qquad \left(\text{we could have used } \frac{V}{R_1}\right)$$

$$I_2 = VG_2 = 30 \text{ V} \times 213\ \mu\text{S} = 6.38 \text{ mA}$$

(b) $G_T = 104\ \mu$S
$R_T = 9.64$ kΩ
$V = 1.93$ V
$I_1 = 129\ \mu$A
$I_2 = 71.4\ \mu$A

2. $I_1 = 700\ \mu$A, $V = 7$ V, $R_2 = 8.75$ kΩ, $R_T = 4.67$ kΩ, $G_T = 214\ \mu$S
3. $I_1 = 482\ \mu$A, $I_2 = 518\ \mu$A, $R_2 = 5.21$ kΩ, $G_T = 370\ \mu$S, $R_T = 2.7$ kΩ
4. (a) $G_T = G_1 + G_2 + G_3 = 3.03 \text{ mS} + 2.13 \text{ mS} + 1.47 \text{ mS} = 6.63 \text{ mS}$

$$R_T = \frac{1}{G_T} = \frac{1}{6.63 \text{ mS}} = 151\ \Omega$$

$$V = I_T R_T = 10 \text{ mA} \times 151\ \Omega = 1.51 \text{ V}$$
$$I_1 = VG_1 = 1.51 \text{ V} \times 3.03 \text{ mS} = 4.57 \text{ mA}$$
$$I_2 = VG_2 = 1.51 \text{ V} \times 2.13 \text{ mS} = 3.21 \text{ mS}$$
$$I_3 = VG_3 = 1.51 \text{ V} \times 1.47 \text{ mS} = 2.22 \text{ mS}$$

(b) $G_T = 890\ \mu$S
$R_T = 1.12$ kΩ
$V = 281$ mV
$I_1 = 50.2\ \mu$A
$I_2 = 128\ \mu$A
$I_3 = 72.1\ \mu$A

5. $I_2 = 9$ mA, $I_3 = 21$ mA, $R_1 = 450\ \Omega$, $R_3 = 429\ \Omega$, $R_T = 180\ \Omega$, $G_T = 5.56$ mS
6. $R_T = 14.4$ kΩ, $R_1 = 55.9$ kΩ, $R_2 = 47.1$ kΩ, $R_3 = 33.1$ kΩ, $I_1 = 215\ \mu$A, $I_3 = 363\ \mu$A, $I_T = 833\ \mu$A

Additional practice problems are at the end of the chapter.

SELF-TEST 12-2

1. Refer to Figure 12-6. Let $E = 25$ V, $R_1 = 18$ kΩ, and $V_2 = 6.5$ V. Find I, R_T, R_2, and V_1.
2. Refer to Figure 12-7. Let $I = 800$ μA, $R_1 = 68$ kΩ, $V_2 = 50$ V, and $R_T = 200$ kΩ. Find E, R_2, R_3, V_1, and V_3.
3. Refer to Figure 12-8. Let $I_T = 30$ mA, $V = 1$ V, and $R_1 = 100$ Ω. Find I_1, I_2, R_2, G_T, and R_T.
4. Refer to Figure 12-9. Let $I_T = 1$ mA, $I_1 = 200$ μA, $V = 10$ V, and $R_2 = 27$ kΩ. Find R_1, R_3, I_2, I_3, G_T, and R_T.

Answers to Self-test 12-2 are at the end of the chapter.

12-4 SERIES–PARALLEL CIRCUITS

Consider the circuit in Figure 12-10. Let's find the unknown variables, which are I_T, I_1, I_2, I_3, R_T, V_1, V_2, and V_3. Since we know all values of R and we know E, we will first find R_T. Then we can find I_T.

$$R_T = R_1 + R_2 \parallel R_3$$

Let

$$R_X = R_2 \parallel R_3$$

Then

$$G_X = G_2 + G_3 = 333\ \mu\text{S} + 179\ \mu\text{S} = 512\ \mu\text{S}$$

$$R_X = \frac{1}{G_X} = \frac{1}{512\ \mu\text{S}} = 1.95\ \text{k}\Omega$$

$$R_T = R_1 + R_X = 2\ \text{k}\Omega + 1.95\ \text{k}\Omega = 3.95\ \text{k}\Omega$$

$$I_T = \frac{E}{R_T} = \frac{20\ \text{V}}{3.95\ \text{k}\Omega} = 5.06\ \text{mA}$$

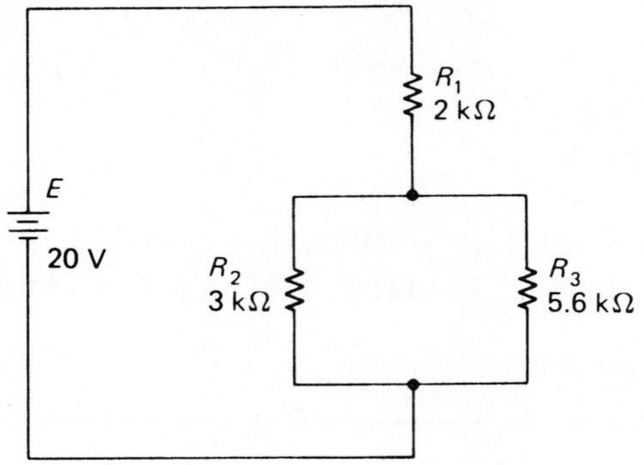

FIGURE 12-10

Since R_1 is a series resistor, $I_T = I_1$.

$$I_1 = I_T = 5.06 \text{ mA}$$

Now we can find V_1.

$$V_1 = I_1 R_1 = 5.06 \text{ mA} \times 2 \text{ k}\Omega = 10.1 \text{ V}$$

Applying Kirchhoff's voltage law, we get

$$V_2 = V_3 = E - V_1 = 20 \text{ V} - 10.1 \text{ V} = 9.9 \text{ V}$$

We could also find V_2 and V_3 by using

$$V_2 = V_3 = \frac{ER_X}{R_T} = \frac{20 \text{ V} \times 1.95 \text{ k}\Omega}{3.95 \text{ k}\Omega} = 9.87 \text{ V}$$

[The difference in the two methods (9.9 V versus 9.87 V) resulted from rounding V_1. V_1 was actually 10.12 V. Either answer is acceptable.] Now that we know V_2 and V_3, we can find I_2 and I_3.

$$I_2 = \frac{V_2}{R_2} = \frac{9.9 \text{ V}}{3 \text{ k}\Omega} = 3.3 \text{ mA}$$

$$I_3 = \frac{V_3}{R_3} = \frac{9.9 \text{ V}}{5.6 \text{ k}\Omega} = 1.77 \text{ mA}$$

□ EXAMPLE 12-10

Solve for R_T, I_T, I_3, I_1, V_1, and V_2 in Figure 12-11.

Solution R_1 and R_2 are in series. This makes the branch resistance equal to $R_1 + R_2$. This branch is in parallel with R_3.

$$R_T = (R_1 + R_2) \parallel R_3$$

Let

$$R_X = R_1 + R_2$$

Then

$$R_X = 15 \text{ k}\Omega + 27 \text{ k}\Omega = 42 \text{ k}\Omega$$

$$G_T = G_X + G_3 = 23.8\ \mu\text{S} + 100\ \mu\text{S} = 12.4\ \mu\text{S}$$

$$R_T = \frac{1}{G_T} = \frac{1}{124\ \mu\text{S}} = 8.08 \text{ k}\Omega$$

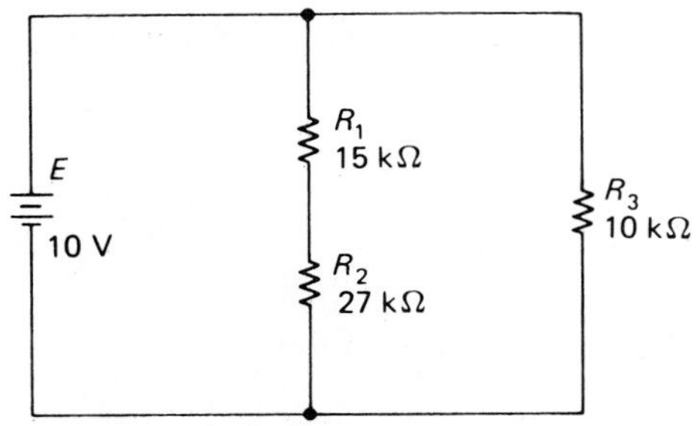

FIGURE 12-11

Using the calculator,

15 [EE] 3 [+] 27 [EE] 3 [=] [1/x] [+] 10 [EE] 3 [1/x] [=] [1/x]

Answer: 8.08×10^3

$$R_T = 8.08\ \text{k}\Omega$$

Now we can find I_T:

$$I_T = \frac{E}{R_T} = \frac{10\ \text{V}}{8.08\ \text{k}\Omega} = 1.24\ \text{mA}$$

Knowing I_T, we can find I_3.

$$I_3 = \frac{I_T G_3}{G_T} = \frac{1.24\ \text{mA} \times 100\ \mu\text{S}}{124\ \mu\text{S}} = 1\ \text{mA}$$

Using the calculator,

1.24 [EE] [+/-] 3 [×] 100 [EE] [+/-] 6 [÷] 124 [EE] [+/-] 6 [=]

Answer: 1×10^{-3}

$$I_3 = 1\ \text{mA}$$

or

$$I_3 = \frac{V_3}{R_3} = \frac{10\ \text{V}}{10\ \text{k}\Omega} = 1\ \text{mA}$$

(We knew $V_3 = 10$ V because $V_3 = E$. The second method was easier. We must be aware that very often there is more than one path to take in finding the unknowns in complex circuits.) Since $I_3 = 1$ mA,

$$I_1 = I_2 = I_T - I_3 = 1.24\ \text{mA} - 1\ \text{mA} = 240\ \mu\text{A}$$

$$V_1 = \frac{ER_1}{R_X} = \frac{10\ \text{V} \times 15\ \text{k}\Omega}{42\ \text{k}\Omega} = 3.57\ \text{V}$$

$$V_2 = E - V_1 = 10\ \text{V} - 3.57\ \text{V} = 6.43\ \text{V}$$

☐ EXAMPLE 12-11

Consider the circuit in Figure 12-12. Let's find all the currents and voltage drops and the total resistance.

Solution Since we are given all the circuit resistances, we can find R_T. A look at the circuit shows that R_2 and R_3 are in parallel. Let

$$R_A = R_2 \parallel R_3$$

$$R_A = \frac{1}{G_2 + G_3} = \frac{1}{435\ \mu\text{S}} = 2.3\ \text{k}\Omega$$

This resistance, R_A, is in series with R_1. The total branch resistance is $R_A + R_1$. Let's call this resistance R_B.

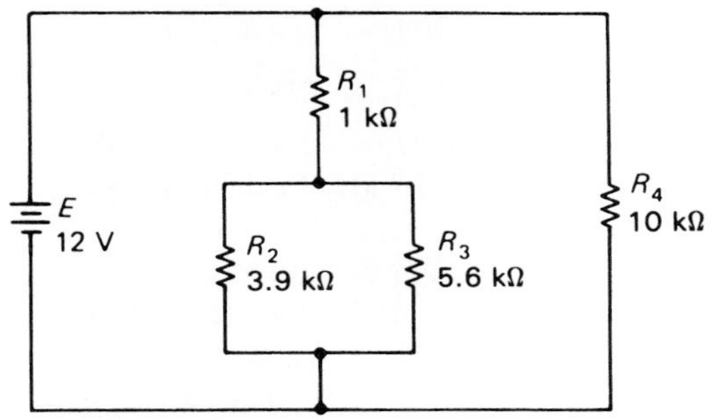

FIGURE 12-12

$$R_B = R_1 + R_A = 1\text{ k}\Omega + 2.3\text{ k}\Omega = 3.3\text{ k}\Omega$$

R_B is in parallel with R_4.

$$R_T = R_B \parallel R_4 = \frac{1}{G_B + G_4} = \frac{1}{403\ \mu\text{S}} = 2.48\text{ k}\Omega$$

Now we can find I_T.

$$I_T = \frac{E}{R_T} = \frac{12\text{ V}}{2.48\text{ k}\Omega} = 4.84\text{ mA}$$

We can see that R_4 is connected across the source. Therefore, $V_4 = E$. Solving for I_4, we get

$$I_4 = \frac{V_4}{R_4} = \frac{12\text{ V}}{10\text{ k}\Omega} = 1.2\text{ mA}$$

Applying Kirchhoff's current law, we get

$$I_T = I_1 + I_4$$

$$I_1 = I_T - I_4 = 4.84\text{ mA} - 1.2\text{ mA} = 3.64\text{ mA}$$

Knowing I_1, we can find I_2 and I_3.

$$I_2 = \frac{I_1 G_2}{G_2 + G_3} = 2.14\text{ mA}$$

$$I_3 = I_T - I_2 = 1.49\text{ mA}$$

We can find the remaining voltage drops now.

$$V_1 = I_1 R_1 = 3.64\text{ mA} \times 1\text{ k}\Omega = 3.64\text{ V}$$

$$V_2 = I_2 R_2 = 2.14\text{ mA} \times 3.9\text{ k}\Omega = 8.36\text{ V}$$

$$V_3 = V_2 = 8.36\text{ V}$$

□

PRACTICE PROBLEMS 12-5

1. Refer to Figure 12-10. Let $E = 25$ V, $R_1 = 2.7$ kΩ, $R_2 = 6.8$ kΩ, and $R_3 = 4.7$ kΩ. Find V_1, V_2, V_3, I_1, I_2, I_3, I_T, and R_T.
2. Refer to Figure 12-11. $E = 40$ V, $R_1 = 680$ Ω, $R_2 = 1.2$ kΩ, and $R_3 = 2.7$ kΩ. Find V_1, V_2, V_3, I_1, I_2, I_3, I_T, and R_T.
3. Refer to Figure 12-10. Let $E = 20$ V, $I_T = 5$ mA, $R_1 = 1.5$ kΩ, and $R_2 = 6.8$ kΩ. Find R_3, R_T, V_1, V_2, V_3, I_2, and I_3.
4. Refer to Figure 12-13. Let $E = 10$ V, $I_T = 4$ mA, $V_2 = 3$ V, $R_3 = 1.5$ kΩ, and $R_4 = 6$ kΩ. Find I_2, I_3, I_4, R_1, R_2, and R_T.
5. Refer to Figure 12-12. Let $E = 9$ V, $R_1 = 9.1$ kΩ, $R_2 = 27$ kΩ, $R_3 = 56$ kΩ, and $R_4 = 120$ kΩ. Find V_1, V_2, V_3, V_4, I_T, I_1, I_2, I_3, I_4, and R_T.
6. Refer to Figure 12-14. Let $E = 9$ V, $R_1 = 330$ Ω, $R_2 = 810$ Ω, $R_3 = 1$ kΩ, and $R_4 = 560$ Ω. Find V_1, V_2, V_3, V_4, I_1, I_2, I_3, I_4, I_T, and R_T.

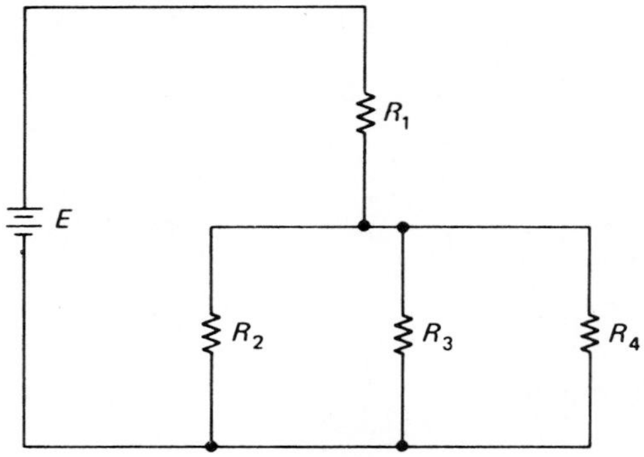

FIGURE 12-13

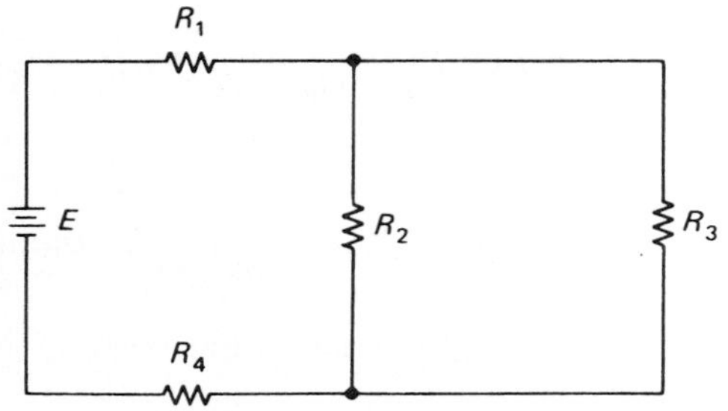

FIGURE 12-14

Solutions

1. $R_T = 5.48$ kΩ, $I_T = 4.56$ mA, $I_1 = 4.56$ mA, $I_2 = 1.86$ mA, $I_3 = 2.7$ mA, $V_1 = 12.3$ V, $V_2 = 12.7$ V, $V_3 = 12.7$ V
2. $R_T = 1.11$ kΩ, $I_T = 36.1$ mA, $I_1 = 21.3$ mA, $I_2 = 21.3$ mA, $I_3 = 14.8$ mA, $V_1 = 14.5$ V, $V_2 = 25.5$ V, $V_3 = 40$ V
3. $R_T = 4$ kΩ, $V_1 = 7.5$ V, $V_2 = 12.5$ V, $V_3 = 12.5$ V, $I_2 = 1.84$ mA, $I_3 = 3.16$ mA, $R_3 = 3.96$ kΩ
4. $R_T = 2.5$ kΩ, $R_1 = 1.75$ kΩ, $R_2 = 2$ kΩ, $I_2 = 1.5$ mA, $I_3 = 2$ mA, $I_4 = 500$ μA
5. $R_T = 22.3$ kΩ, $I_T = 404$ μA, $I_1 = 329$ μA, $I_2 = 223$ μA, $I_3 = 107$ μA, $I_4 = 75$ μA, $V_1 = 2.99$ V, $V_2 = 6.01$ V, $V_3 = 6.01$ V, $V_4 = 9$ V.
6. $R_T = 1.34$ kΩ, $I_T = I_1 = I_4 = 6.73$ mA, $I_2 = 3.72$ mA, $I_3 = 3.01$ mA, $V_1 = 2.22$ V, $V_2 = V_3 = 3.01$ V, $V_4 = 3.77$ V

SELF-TEST 12-3

1. Refer to Figure 12-10. Let $E = 3$ V, $I_T = 70$ μA, $R_2 = 33$ kΩ, and $R_3 = 47$ kΩ. Find V_1, V_2, V_3, I_2, I_3, R_T, and R_1.
2. Refer to Figure 12-11. Let $E = 20$ V, $R_1 = 20$ kΩ, $R_3 = 10$ kΩ, $V_2 = 6.57$ V. Find I_T, I_1, I_2, I_3, V_1, V_3, R_T, and R_2.

3. Refer to Figure 12-13. Let $E = 15$ V, $R_1 = 12$ kΩ, $R_2 = 100$ kΩ, $R_3 = 22$ kΩ, and $R_4 = 68$ kΩ. Find I_T, I_1, I_2, I_3, I_4, V_1, V_2, V_3, V_4, and R_T.

4. Refer to Figure 12-14. Let $E = 25$ V, $R_1 = 2.7$ kΩ, $R_2 = 3.9$ kΩ, $R_3 = 6.8$ kΩ, and $R_4 = 2.2$ kΩ. Find V_1, V_2, V_3, V_4, I_T, I_1, I_2, I_3, I_4, and R_T.

Answers to Self-test 12-3 are at the end of the chapter.

CHAPTER 12 AT A GLANCE

Page	Key Point	Example
301	*Key Point:* In a series circuit the total resistance equals the sum of the individual resistances.	$R_T = R_1 + R_2 + \cdots + R_N$
303	*Key Point:* In a parallel circuit the total conductance equals the sum of the individual branch conductances.	$G_T = G_1 + G_2 + \cdots + G_N$
308	*Key Point:* In a series circuit the greater voltage drops across the greater resistance.	If $R_1 > R_2$, then $V_1 > V_2$.
312	*Key Point:* In a parallel circuit the greater current flows through the greater conductance.	If $G_1 > G_2$, then $I_1 > I_2$.

End of Chapter Problems 12-1

Refer to Figure 12-15(a).

1. Find R_T when (a) $R_1 = 30$ kΩ, $R_2 = 15$ kΩ, (b) $R_1 = 750$ Ω, $R_2 = 1.2$ kΩ, (c) $R_1 = 6.8$ kΩ, $R_2 = 10$ kΩ, and (d) $R_1 = 120$ kΩ, $R_2 = 180$ kΩ.

2. Find R_T when (a) $R_1 = 27$ kΩ, $R_2 = 39$ kΩ, (b) $R_1 = 100$ Ω, $R_2 = 270$ Ω, (c) $R_1 = 680$ kΩ, $R_2 = 470$ kΩ, and (d) $R_1 = 75$ kΩ, $R_2 = 100$ kΩ.

3. Find R_2 when (a) $R_1 = 1.2$ kΩ, $R_T = 3$ kΩ, and (b) $R_1 = 27$ kΩ, $R_T = 66$ kΩ.

4. Find R_2 when (a) $R_1 = 680$ Ω, $R_T = 1.24$ kΩ, and (b) $R_1 = 220$ kΩ, $R_T = 400$ kΩ.

Refer to Figure 12-15(b).

5. Find R_T when (a) $R_1 = 470$ Ω, $R_2 = 820$ Ω, $R_3 = 1$ kΩ, (b) $R_1 = 15$ kΩ, $R_2 = 47$ kΩ, $R_3 = 68$ kΩ, (c) $R_1 = 47$ Ω, $R_2 = 100$ Ω, $R_3 = 150$ Ω, and (d) $R_1 = 330$ kΩ, $R_2 = 470$ kΩ, $R_3 = 560$ kΩ.

6. Find R_T when (a) $R_1 = 1.2$ kΩ, $R_2 = 6.8$ kΩ, $R_3 = 2.2$ kΩ, (b) $R_1 = 82$ kΩ, $R_2 = 68$ kΩ, $R_3 = 100$ kΩ, (c) $R_1 = 47$ Ω, $R_2 = 33$ Ω, $R_3 = 120$ Ω, and (d) $R_1 = 330$ Ω, $R_2 = 200$ Ω, $R_3 = 510$ Ω.

7. Find R_3 when (a) $R_1 = 3.3$ kΩ, $R_2 = 4.7$ kΩ, $R_T = 13.6$ kΩ, and (b) $R_1 = 68$ kΩ, $R_2 = 18$ kΩ, $R_T = 168$ kΩ.

8. Find R_3 when (a) $R_1 = 750$ Ω, $R_2 = 910$ Ω, $R_T = 2.66$ kΩ, and (b) $R_1 = 1.5$ kΩ, $R_2 = 2.7$ kΩ, $R_T = 4.88$ kΩ.

Refer to Figure 12-15(d).

9. Find G_T and R_T when (a) $R_1 = 10$ kΩ, $R_2 = 20$ kΩ, (b) $R_1 = 1.2$ kΩ, $R_2 = 2.7$ kΩ, (c) $R_1 = 47$ kΩ, $R_2 = 18$ kΩ, and (d) $R_1 = 75$ kΩ, $R_2 = 150$ kΩ.

10. Find G_T and R_T when (a) $R_1 = 68$ Ω, $R_2 = 120$ Ω, (b) $R_1 = 560$ Ω, $R_2 = 1.8$ kΩ, (c) $R_1 = 150$ kΩ, $R_2 = 200$ kΩ, and (d) $R_1 = 82$ kΩ, $R_2 = 120$ kΩ.

Refer to Figure 12-15(e).

11. Find G_T and R_T when (a) $R_1 = 2.2$ kΩ, $R_2 = 6.8$ kΩ, $R_3 = 12$ kΩ, (b) $R_1 = 68$ kΩ, $R_2 = 200$ kΩ, $R_3 = 82$ kΩ, (c) $R_1 = 910$ Ω, $R_2 = 2.2$ kΩ, $R_3 = 1.2$ kΩ, and (d) $R_1 = 120$ kΩ, $R_2 = 68$ kΩ, $R_3 = 270$ kΩ.

12. Find G_T and R_T when (a) $R_1 = 18$ kΩ, $R_2 = 18$ kΩ, $R_3 = 18$ kΩ, (b) $R_1 = 330$ Ω, $R_2 = 820$ Ω, $R_3 = 910$ Ω, (c) $R_1 = 47$ kΩ, $R_2 = 22$ kΩ, $R_3 = 8.2$ kΩ, and (d) $R_1 = 220$ kΩ, $R_2 = 1.2$ MΩ, $R_3 = 56$ kΩ.

13. Find R_2 when (a) $R_1 = 56$ kΩ, $R_3 = 33$ kΩ, $R_T = 3.83$ kΩ, and (b) $R_1 = 100$ Ω, $R_3 = 300$ Ω, $R_T = 54.5$ Ω.

14. Find R_3 when (a) $R_1 = 120$ kΩ, $R_2 = 470$ kΩ, $R_T = 78.2$ kΩ, and (b) $R_1 = 27$ kΩ, $R_2 = 39$ kΩ, $R_T = 12.9$ kΩ.

End of Chapter Problems 12-2

1. Refer to Figure 12-16. Find R_T when (a) $R_1 = 560$ Ω, $R_2 = 1$ kΩ, $R_3 = 820$ Ω, (b) $R_1 = 7.5$ kΩ, $R_2 = 18$ kΩ, $R_3 = 27$ kΩ, (c) $R_1 = 27$ kΩ, $R_2 = 12$ kΩ, $R_3 = 47$ kΩ, and (d) $R_1 = 330$ kΩ, $R_2 = 680$ kΩ, $R_3 = 1$ MΩ.

2. Refer to Figure 12-16. Find R_T when (a) $R_1 = 150$ kΩ, $R_2 = 560$ kΩ, $R_3 = 1$ MΩ, (b) $R_1 = 120$ Ω, $R_2 = 470$ Ω, $R_3 = 270$ Ω, (c) $R_1 = 6.8$ kΩ, $R_2 = 12$ kΩ, $R_3 = 1.2$ kΩ, and (d) $R_1 = 68$ kΩ, $R_2 = 120$ kΩ, $R_3 = 56$ kΩ.

3. Refer to Figure 12-17. Find R_T when (a) $R_1 = 4.7$ kΩ, $R_2 = 6.8$ kΩ, $R_3 = 9.1$ kΩ, (b) $R_1 = 1.8$ kΩ, $R_2 = 1.5$ kΩ, $R_3 = 820$ Ω, (c) $R_1 = 27$ kΩ, $R_2 = 18$ kΩ, $R_3 = 56$ kΩ, and (d) $R_1 = 680$ Ω, $R_2 = 470$ Ω, $R_3 = 2.2$ kΩ.

4. Refer to Figure 12-17. Find R_T when (a) $R_1 = 680$ kΩ, $R_2 = 470$ kΩ, $R_3 = 330$ kΩ, (b) $R_1 = 2.2$ kΩ, $R_2 = 8.2$ kΩ, $R_3 = 15$ kΩ, (c) $R_1 = 18$ kΩ, $R_2 = 2$ kΩ, $R_3 = 20$ kΩ, and (d) $R_1 = 27$ Ω, $R_2 = 39$ Ω, $R_3 = 120$ Ω.

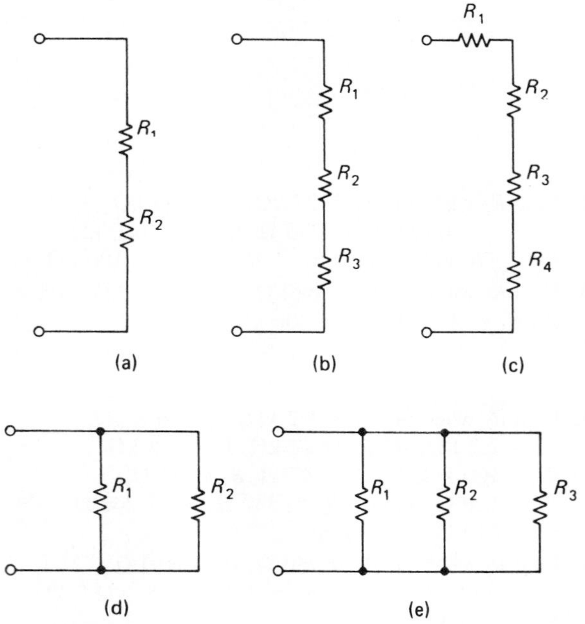

FIGURE 12-15

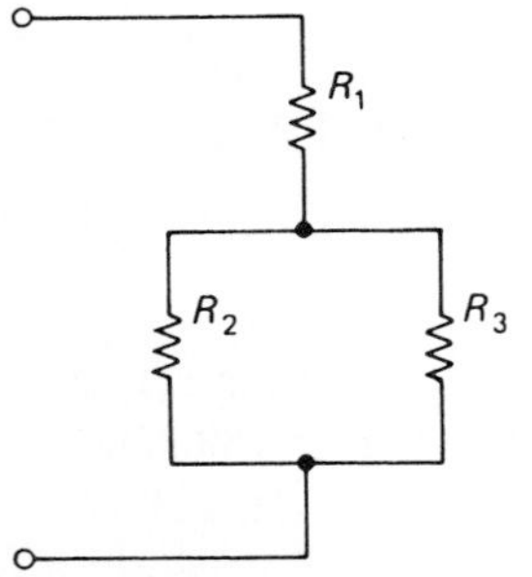

FIGURE 12-16

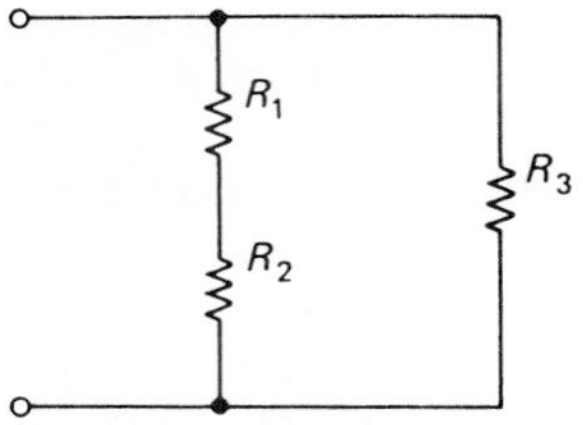

FIGURE 12-17

5. Refer to Figure 12-18. Find R_T when (a) R_1 = 12 kΩ, R_2 = 47 kΩ, R_3 = 33 kΩ, R_4 = 27 kΩ, (b) R_1 = 120 Ω, R_2 = 470 Ω, R_3 = 1 kΩ, R_4 = 560 Ω, (c) R_1 = 390 kΩ, R_2 = 1.2 MΩ, R_3 = 820 kΩ, R_4 = 750 kΩ, and (d) R_1 = 5.1 kΩ, R_2 = 6.8 kΩ, R_3 = 10 kΩ, R_4 = 10 kΩ.

6. Refer to Figure 12-18. Find R_T when (a) R_1 = 910 Ω, R_2 = 2.7 kΩ, R_3 = 10 kΩ, R_4 = 5.6 kΩ, (b) R_1 = 100 kΩ, R_2 = 330 kΩ, R_3 = 680 kΩ, R_4 = 180 kΩ, (c) R_1 = 47 Ω, R_2 = 120 Ω, R_3 = 820 Ω, R_4 = 220 Ω, and (d) R_1 = 100 kΩ, R_2 = 33 kΩ, R_3 = 33 kΩ, R_4 = 39 kΩ.

7. Refer to Figure 12-19. Find R_T when R_1 = 1.2 kΩ, R_2 = 10 kΩ, R_3 = 15 kΩ, R_4 = 6.8 kΩ, R_5 = 4.7 kΩ, and R_6 = 5.6 kΩ.

8. Find R_T when R_1 = 680 Ω, R_2 = 2.2 kΩ, R_3 = 7.5 kΩ, R_4 = 330 Ω, R_5 = 820 Ω, R_6 = 390 Ω.

9. Refer to Figure 12-20. Find R_T when R_1 = 120 kΩ, R_2 = 270 kΩ, R_3 = 91 kΩ, R_4 = 330 kΩ, R_5 = 470 kΩ, R_6 = 560 kΩ, R_7 = 1 MΩ, R_8 = 680 kΩ, and R_9 = 820 kΩ.

10. Find R_T when R_1 = 47 kΩ, R_2 = 100 kΩ, R_3 = 39 kΩ, R_4 = 120 kΩ, R_5 = 220 kΩ, R_6 = 180 kΩ, R_7 = 250 kΩ, R_8 = 82 kΩ, and R_9 = 75 kΩ.

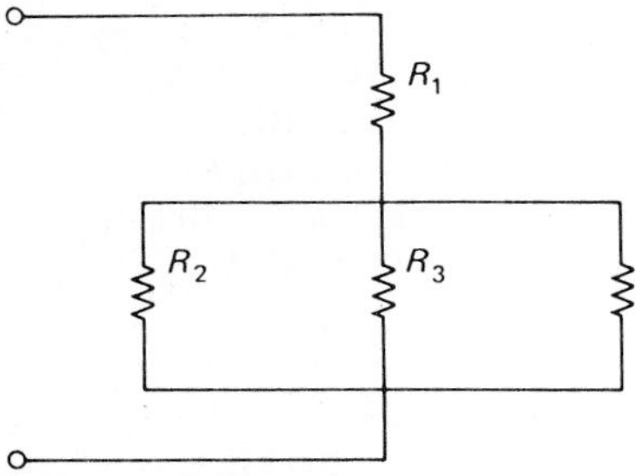

FIGURE 12-18

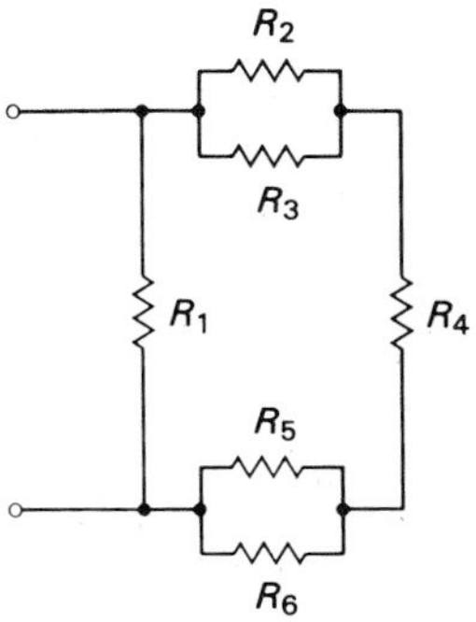

FIGURE 12-19

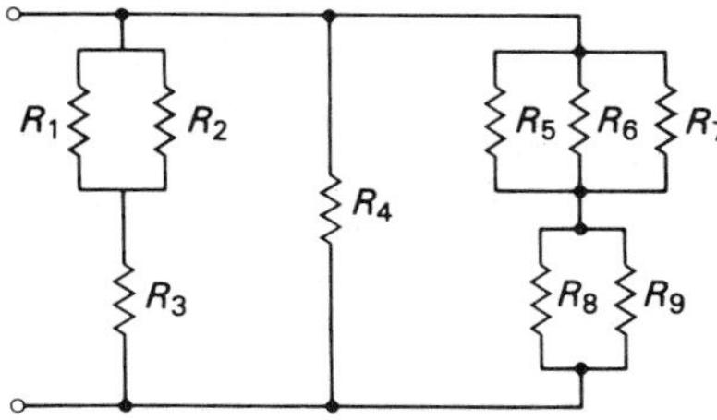

FIGURE 12-20

End of Chapter Problems 12-3

Refer to Figure 12-21.

1. Find I, R_T, V_1, and V_2 when (a) E = 9 V, R_1 = 270 Ω, R_2 = 680 Ω, (b) E = 12 V, R_1 = 1.2 kΩ, R_2 = 3.3 kΩ, (c) E = 40 V, R_1 = 120 kΩ, R_2 = 220 kΩ, and (d) E = 10 V, R_1 = 10 kΩ, R_2 = 3.3 kΩ.

2. Find I, R_T, V_1, and V_2 when (a) E = 25 V, R_1 = 4.7 kΩ, R_2 = 22 kΩ, (b) E = 5 V, R_1 = 68 kΩ, R_2 = 33 kΩ, (c) E = 18 V, R_1 = 180 Ω, R_2 = 1.8 kΩ, and (d) E = 50 V, R_1 = 750 Ω, R_2 = 680 Ω.

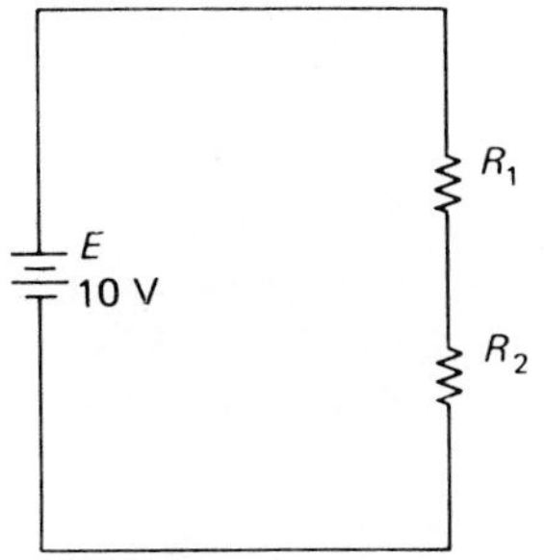

FIGURE 12-21

3. Find I, R_T, V_1, and R_2 when (a) $E = 10$ V, $R_1 = 1$ kΩ, $V_2 = 3$ V, (b) $E = 40$ V, $R_1 = 6.8$ kΩ, $V_2 = 23.5$ V, (c) $E = 15$ V, $R_1 = 8.2$ kΩ, $V_2 = 12$ V, and (d) $E = 100$ V, $R_1 = 220$ kΩ, $V_2 = 29.3$ V.

4. Find I, R_T, V_1, and R_2 when (a) $E = 9$ V, $R_1 = 470$ Ω, $V_2 = 3.71$ V, (b) $E = 35$ V, $R_1 = 68$ kΩ, $V_2 = 22.3$ V, (c) $E = 50$ V, $R_1 = 2.7$ kΩ, $V_2 = 20$ V, and (d) $E = 5$ V, $R_1 = 120$ Ω, $V_2 = 3.13$ V.

5. Find R_1, R_T, V_1, and V_2 when (a) $E = 20$ V, $I = 213$ μA, $R_2 = 33$ kΩ, (b) $E = 25$ V, $I = 1$ mA, $R_2 = 10$ kΩ, (c) $E = 12$ V, $I = 7.69$ mA, $R_2 = 1$ kΩ, and (d) $E = 9$ V, $I = 20$ μA, $R_2 = 180$ kΩ.

6. Find R_1, R_T, V_1, and V_2 when (a) $E = 100$ V, $I = 2.50$ mA, $R_2 = 30$ kΩ, (b) $E = 60$ V, $I = 4.48$ mA, $R_2 = 4.3$ kΩ, (c) $E = 5$ V, $I = 14.7$ mA, $R_2 = 120$ Ω, and (d) $E = 20$ V, $I = 329$ μA, $R_2 = 4.7$ kΩ.

7. Find E, R_1, R_T, and V_2 when (a) $I = 75$ μA, $V_1 = 6.2$ mV, $R_2 = 3$ kΩ, (b) $I = 2$ mA, $V_1 = 14.3$ V, $R_2 = 12$ kΩ, (c) $I = 167$ μA, $V_1 = 4.50$ V, $R_2 = 33$ kΩ, and (d) $I = 14.9$ mA, $V_1 = 1.78$ V, $R_2 = 82$ Ω.

8. Find E, R_1, R_T, and V_2 when (a) $I = 192$ μA, $V_1 = 19.2$ V, $R_2 = 56$ kΩ, (b) $I = 1.32$ mA, $V_1 = 9.87$ V, $R_2 = 3.9$ kΩ, (c) $I = 16.1$ mA, $V_1 = 10.8$ V, $R_2 = 1.5$ kΩ, and (d) $I = 14.5$ mA, $V_1 = 479$ mV, $R_2 = 91$ Ω.

Refer to Figure 12-22.

9. Find I, R_T, V_1, V_2, and V_3 when (a) $E = 9$ V, $R_1 = 33$ kΩ, $R_2 = 10$ kΩ, $R_3 = 18$ kΩ, (b) $E = 40$ V, $R_1 = 220$ kΩ, $R_2 = 100$ kΩ, $R_3 = 68$ kΩ, (c) $E = 25$ V, $R_1 = 820$ Ω, $R_2 = 1.2$ kΩ, $R_3 = 560$ Ω, and (d) $E = 15$ V, $R_1 = 1.8$ kΩ, $R_2 = 4.3$ kΩ, $R_3 = 2.5$ kΩ.

10. Find I, R_T, V_1, V_2, and V_3 when (a) $E = 100$ V, $R_1 = 100$ kΩ, $R_2 = 20$ kΩ, $R_3 = 30$ kΩ, (b) $E = 75$ V, $R_1 = 22$ kΩ, $R_2 = 10$ kΩ, $R_3 = 5.1$ kΩ, (c) $E = 12$ V, $R_1 = 2$ kΩ, $R_2 = 10$ kΩ, $R_3 = 15$ kΩ, and (d)$E = 1$ V, $R_1 = 300$ Ω, $R_2 = 100$ Ω, $R_3 = 200$ Ω.

11. Find V_2, V_3, R_1, R_3, and R_T when (a) $E = 25$ V, $V_1 = 5$ V, $I = 1$ mA, $R_2 = 10$ kΩ, (b) $E = 20$ V, $V_1 = 6.7$ V, $I = 150$ μA, $R_2 = 47$ kΩ, (c) $E = 80$ V, $V_1 = 29.9$ V, $I = 1.66$ mA, $R_2 = 8.2$ kΩ, and (d) $E = 3$ V, $V_1 = 433$ mV, $I = 15.5$ mA, $R_2 = 91$ Ω.

12. Find V_2, V_3, R_1, R_3, and R_T when (a) $E = 9$ V, $V_1 = 2.30$ V, $I = 19.2$ μA, $R_2 = 200$ kΩ, (b) $E = 12$ V, $V_1 = 3.04$ V, $I = 77.9$ μA, $R_2 = 47$ kΩ, (c) $E = 60$ V, $V_1 = 17.6$ V, $I = 58.8$ mA, $R_2 = 330$ Ω, and (d) $E = 10$ V, $V_1 = 1.25$ V, $I = 500$ μA, $R_2 = 7.5$ kΩ.

13. Find E, R_2, R_3, V_1, and V_3 when (a) $I = 3$ mA, $R_1 = 4.7$ kΩ, $V_2 = 7.3$ V, $R_T = 12$ kΩ, (b) $I = 270$ μA, $R_1 = 15$ kΩ, $V_2 = 3.7$ V, $R_T = 43.2$ kΩ, (c) $I = 66.7$ μA, $R_1 = 200$ kΩ, $V_2 = 6.67$ V, $R_T = 450$ kΩ, and (d) $I = 2.30$ mA, $R_1 = 470$ Ω, $V_2 = 690$ mV, $R_T = 870$ Ω.

14. Find E, R_2, R_3, V_1, and V_3 when (a) $I = 220$ μA, $R_1 = 20$ kΩ, $V_2 = 4.40$ V, $R_T = 91$ kΩ, (b) $I = 1.58$ mA, $R_1 = 2.2$ kΩ, $V_2 = 15.8$ V, $R_T = 15.8$ kΩ, (c) $I = 61.7$ mA, $R_1 = 12$ Ω, $V_2 = 1.36$ V, $R_T = 81$ Ω, and (d) $I = 30.6$ μA, $R_1 = 1$ MΩ, $V_2 = 14.4$ V, $R_T = 2.29$ MΩ.

End of Chapter Problems 12-4

Refer to Figure 12-23.

1. Find G_T, R_T, V, I_1, and I_2 when (a) $I_T = 3$ mA, $R_1 = 270$ Ω, $R_2 = 560$ Ω, (b) $I_T = 700$ μA, $R_1 = 68$ kΩ, $R_2 = 39$ kΩ, (c) $I_T = 37.6$ mA, $R_1 = 220$ Ω, $R_2 = 270$ Ω, and (d) $I_T = 100$ μA, $R_1 = 56$ kΩ, $R_2 = 33$ kΩ.

2. Find G_T, R_T, V, I_1, and I_2 when (a) $I_T = 3.55$ mA, $R_1 = 68$ kΩ, $R_2 = 39$ kΩ, (b) $I_T = 3$ mA, $R_1 = 47$ kΩ, $R_2 = 33$ kΩ, (c) $I_T = 22$ mA, $R_1 = 1.8$ kΩ, $R_2 = 3.9$ kΩ, and (d) $I_T = 460$ μA, $R_1 = 2.7$ kΩ, $R_2 = 1.5$ kΩ.

3. Find V, R_2, I_1, G_T, and R_T when (a) I_T = 25 mA, R_1 = 680 Ω, I_2 = 15 mA, (b) I_T = 5 mA, R_1 = 1.8 kΩ, I_2 = 1 mA, (c) I_T = 2.3 mA, R_1 = 4.7 kΩ, I_2 = 1.05 mA, and (d) I_T = 30 mA, R_1 = 4.7 kΩ, I_2 = 12.3 mA.

4. Find V, R_2, I_1, G_T and R_T when (a) I_T = 400 μA, R_1 = 330 Ω, I_2 = 174 μA, (b) I_T = 1 mA, R_1 = 10 kΩ, I_2 = 333 μA, (c) I_T = 10 mA, R_1 = 1.8 kΩ, I_2 = 4 mA, and (d) I_T = 200 μA, R_1 = 22 kΩ, I_2 = 80 μA.

5. Find I_1, I_2, R_2, G_T, and R_T when (a) I_T = 10 mA, V = 20 V, R_1 = 4.7 kΩ, (b) I_T = 350 μA, V = 25 V, R_1 = 180 kΩ, (c) I_T = 20 mA, R_1 = 820 Ω, V = 9.74 V and (d) I_T = 2 mA, R_1 = 43 kΩ, V = 48.6 V.

6. Find I_1, I_2, R_2, G_T, and R_T when (a) I_T = 5 mA, R_1 = 27 kΩ, V = 85.7 V, (b) I_T = 100 μA, R_1 = 220 kΩ, V = 6.88 V, (c) I_T = 200 μA, R_1 = 680 kΩ, V = 81 V, and (d) I_T = 35 mA, R_1 = 100 Ω, V = 2.33 V.

7. Find I_2, I_T, G_T, R_T, and R_1 when (a) V = 12 V, I_1 = 530 μA, R_2 = 20 kΩ, (b) V = 5 V, I_1 = 1 mA, R_2 = 15 kΩ, (c) V = 30.9 V, I_1 = 1.71 mA, R_2 = 10 kΩ, and (d) V = 2.16 V, I_1 = 263 μA, R_2 = 91 kΩ.

8. Find I_2, I, G_T, R_T, and R_1 when (a) V = 99.5 V, I_1 = 23.1 mA, R_2 = 2.7 kΩ, (b) V = 39.5 V, I_1 = 91.9 mA, R_2 = 680 Ω, (c) V = 245 V, I_1 = 164 mA, R_2 = 1.8 kΩ, and (d) V = 11.1 V, I_1 = 503 μA, R_2 = 56 kΩ.

Refer to Figure 12-24.

9. Find G_T, R_T, V, I_1, I_2, and I_3 when (a) I_T = 1.73 mA, R_1 = 2.7 kΩ, R_2 = 1.2 kΩ, R_3 = 1 kΩ, (b) I_T = 500 μA, R_1 = 8.1 kΩ, R_2 = 7.5 kΩ, R_3 = 4.7 kΩ, (c) I_T = 500 μA, R_1 = 22 kΩ, R_2 = 39 kΩ, R_3 = 56 kΩ, and (d) I_T = 2 mA, R_1 = 2.7 kΩ, R_2 = 3.9 kΩ, R_3 = 1.8 kΩ.

10. Find G_T, R_T, V, I_1, I_2, and I_3 when (a) I_T = 1 mA, R_1 = 18 kΩ, R_2 = 8.2 kΩ, R_3 = 22 kΩ, (b) I_T = 100 μA, R_1 = 9.1 kΩ, R_2 = 4.7 kΩ, R_3 = 5.6 kΩ, (c) I_T = 5 mA, R_1 = 910 Ω, R_2 = 2 kΩ, R_3 = 1.2 kΩ, and (d) I_T = 10 mA, R_1 = 820 Ω, R_2 = 680 Ω, R_3 = 430 Ω.

11. Find R_1, R_3, R_T, G_T, I_2, and I_3 when (a) I_T = 4 mA, I_1 = 700 μA, V = 12.6 V, R_2 = 12 kΩ, (b) I_T = 300 μA, I_1 = 75 μA, V = 20 V, R_2 = 270 kΩ, (c) I_T = 300 μA, I_1 = 34.5 μA, V = 23.4 V, R_2 = 750 kΩ, and (d) I_T = 60 mA, I_1 = 15.1 mA, V = 13.8 V, R_2 = 680 Ω.

12. Find R_1, R_3, R_T, G_T, I_2, and I_3 when (a) I_T = 200 μA, I_1 = 36.4 μA, V = 36.4 V, R_2 = 680 kΩ, (b) I_T = 80 mA, I_1 = 19 mA, V = 88.9 V, R_2 = 2.2 kΩ, (c) I_T = 10 mA, I_1 = 4.16 mA, V = 162 V, R_2 = 47 kΩ, and (d) I_T = 2.5 mA, I_1 = 1.39 mA, V = 3.48 V, R_2 = 8.2 kΩ.

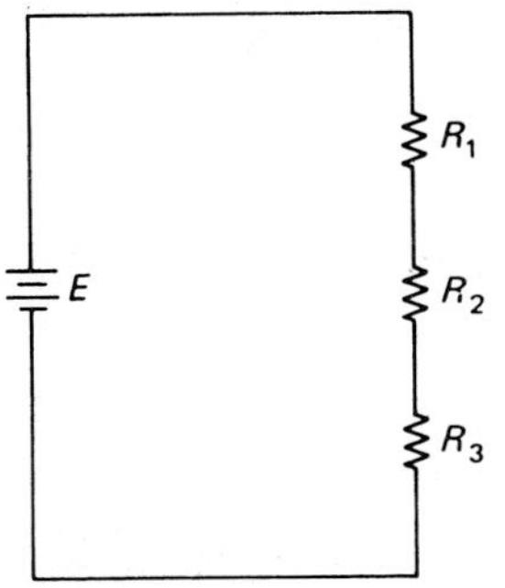

FIGURE 12-22

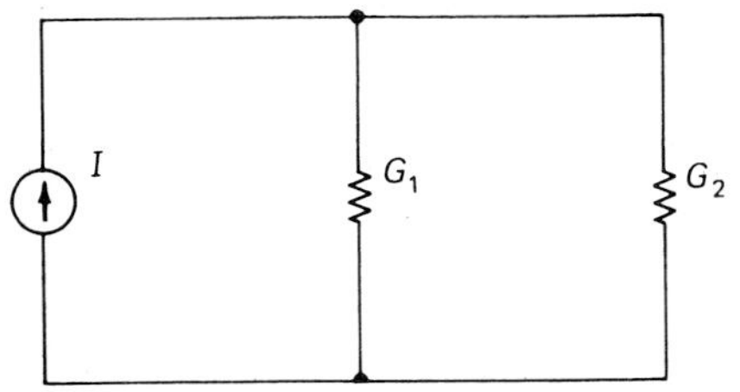

FIGURE 12-23

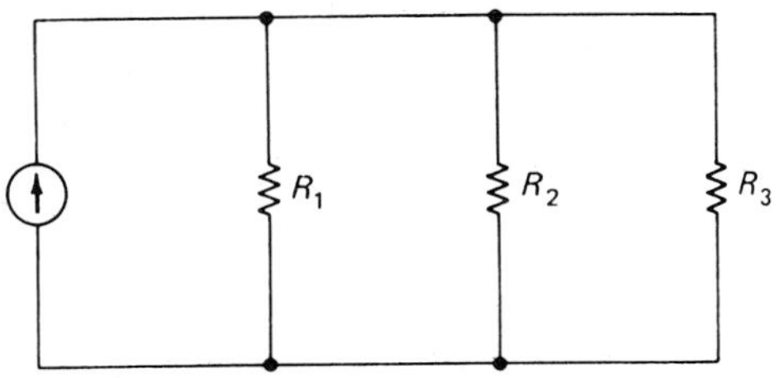

FIGURE 12-24

13. Find I_1, I_3, I_T, R_1, R_2, R_3, and R_T when (a) $V = 15$ V, $G_1 = 213\ \mu$S, $I_2 = 3$ mA, $G_T = 1$ mS, (b) $V = 30$ V, $G_1 = 2$ mS, $I_2 = 60$ mA, $G_T = 8$ mS, (c) $V = 6.09$ V, $I_2 = 10.9$ mA, $G_1 = 1$ mS, $G_T = 3.29$ mS, and (d) $V = 5.29$ V, $G_1 = 13.3\ \mu$S, $I_2 = 53\ \mu$A, $G_T = 28.3\ \mu$S.

14. Find I_1, I_3, I_T, R_1, R_2, R_3, and R_T when (a) $V = 17.6$ V, $G_1 = 4.55\ \mu$S, $I_2 = 31.4\ \mu$A, $G_T = 34.1\ \mu$S, (b) $V = 100$ V, $G_1 = 21.3\ \mu$S, $I_2 = 1.34$ mA, $G_T = 64.9\ \mu$S, (c) $V = 6.87$ V, $I_2 = 123\ \mu$A, $G_1 = 100\ \mu$S, $G_T = 146\ \mu$S, and (d) $V = 35$ V, $G_1 = 455\ \mu$S, $I_2 = 5.14$ mA, $G_T = 858\ \mu$S.

End of Chapter Problems 12-5

Refer to Figure 12-25.

1. Find V_1, V_2, V_3, I_1, I_2, I_3, I_T, and R_T when (a) $E = 15$ V, $R_1 = 10$ kΩ, $R_2 = 33$ kΩ, $R_3 = 47$ kΩ, and (b) $E = 9$ V, $R_1 = 180$ Ω, $R_2 = 560$ Ω, $R_3 = 680$ Ω.

2. Find V_1, V_2, V_3, I_1, I_2, I_3, I_T, and R_T when (a) $E = 50$ V, $R_1 = 8.2$ kΩ, $R_2 = 2.7$ kΩ, $R_3 = 4.7$ kΩ, and (b) $E = 5$ V, $R_1 = 1.2$ kΩ, $R_2 = 30$ kΩ, and $R_3 = 2.2$ kΩ.

3. (a) Let $E = 10$ V, $I_T = 500\ \mu$A, $R_1 = 12$ kΩ, and $R_2 = 47$ kΩ. Find R_3, R_T, V_1, V_2, V_3, I_2, and I_3. (b) Let $E = 12$ V, $I_T = 600\ \mu$A, $R_2 = 39$ kΩ, and $R_3 = 27$ kΩ. Find R_1, R_T, V_1, V_2, V_3, I_2, and I_3.

4. (a) Let $E = 25$ V, $I_T = 1.34$ mA, $R_1 = 12$ kΩ, and $R_2 = 10$ kΩ. Find R_3, R_T, V_1, V_2, V_3, I_2, and I_3. (b) Let $E = 20$ V, $I_T = 48.3\ \mu$A, $R_2 = 470$ kΩ, and $R_3 = 330$ kΩ. Find R_1, R_T, V_1, V_2, V_3, I_2, and I_3.

5. Refer to Figure 12-26. Find V_1, V_2, V_3, I_1, I_2, I_3, I_T, and R_T when (a) $E = 60$ V, $R_1 = 12$ kΩ, $R_2 = 22$ kΩ, $R_3 = 18$ kΩ, and (b) $E = 3$ V, $R_1 = 7.5$ kΩ, $R_2 = 27$ kΩ, $R_3 = 47$ kΩ.

6. Refer to Figure 12-26. Find V_1, V_2, V_3, I_1, I_2, I_3, I_T, and R_T when (a) $E = 25$ V, $R_1 = 12$ kΩ, $R_2 = 10$ kΩ, $R_3 = 20$ kΩ, and (b) $E = 100$ V, $R_1 = 2.5$ kΩ, $R_2 = 4.3$ kΩ, $R_3 = 6.8$ kΩ.

7. Refer to Figure 12-27. (a) Let $E = 40$ V, $R_1 = 680$ Ω, $R_2 = 1.5$ kΩ, $R_3 = 2.7$ kΩ, and $R_4 = 4.7$ kΩ. Find V_1, V_2, V_3, V_4, I_T, I_1, I_2, I_3, I_4, and R_T. (b) Let $E = 30$ V, $R_1 = 150$ kΩ, $R_2 = 470$ kΩ, $R_3 = 680$ kΩ, and $R_4 = 680$ kΩ. Find V_1, V_2, V_3, V_4, I_T, I_1, I_2, I_3, I_4, and R_T.

8. Refer to Figure 12-27. (a) Let $E = 70$ V, $R_1 = 1$ MΩ, $R_2 = 470$ kΩ, $R_3 = 820$ kΩ, and $R_4 = 680$ kΩ. Find V_1, V_2, V_3, V_4, I_T, I_1, I_2, I_3, I_4, and R_T. (b) Let $E = 45$ V, $R_1 = 910$ Ω, $R_2 = 1.5$ kΩ, $R_3 = 470$ Ω, and $R_4 = 1.8$ kΩ. Find V_1, V_2, V_3, V_4, I_T, I_1, I_2, I_3, I_4, and R_T.

9. Refer to Figure 12-28. (a) Let $E = 25$ V, $R_1 = 3$ kΩ, $R_2 = 15$ kΩ, $R_3 = 22$ kΩ, and $R_4 = 18$ kΩ. Find I_T, I_1, I_2, I_3, I_4, R_T, V_1, V_2, V_3, and V_4. (b) Let $E = 25$ V, $I_T = 250\ \mu$A, $V_2 = 10$ V, $R_3 = 150$ kΩ, and $R_4 = 200$ kΩ. Find I_1, I_2, I_3, I_4, R_1, R_2, and R_T.

10. Refer to Figure 12-28. (a) Let $E = 20$ V, $R_1 = 3.3$ kΩ, $R_2 = 10$ kΩ, $R_3 = 15$ kΩ, and $R_4 = 8.2$ kΩ. Find I_T, I_1, I_2, I_3, I_4, R_T, V_1, V_2, V_3, and V_4. (b) Let $E = 9$ V, $I_T = 156\ \mu$A, $V_2 = 4.79$ V, $R_3 = 75$ kΩ, and $R_4 = 68$ kΩ. Find I_1, I_2, I_3, I_4, R_1, R_2, and R_T.

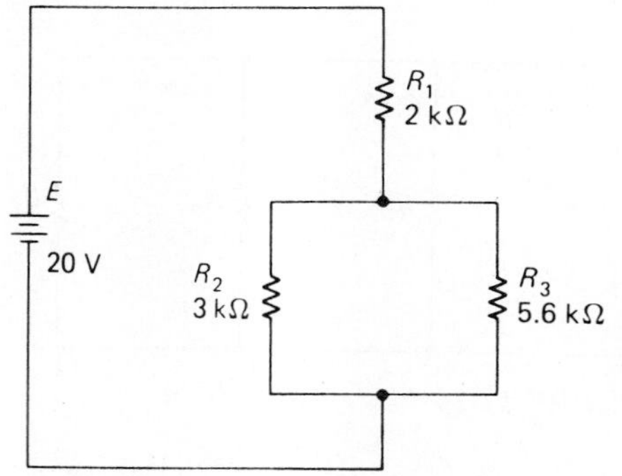

FIGURE 12-25

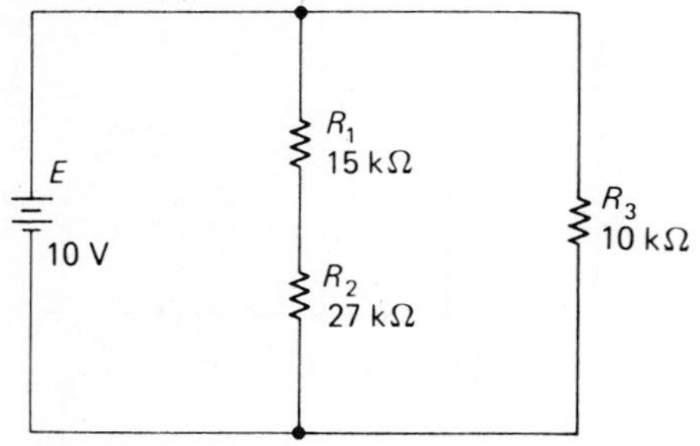

FIGURE 12-26

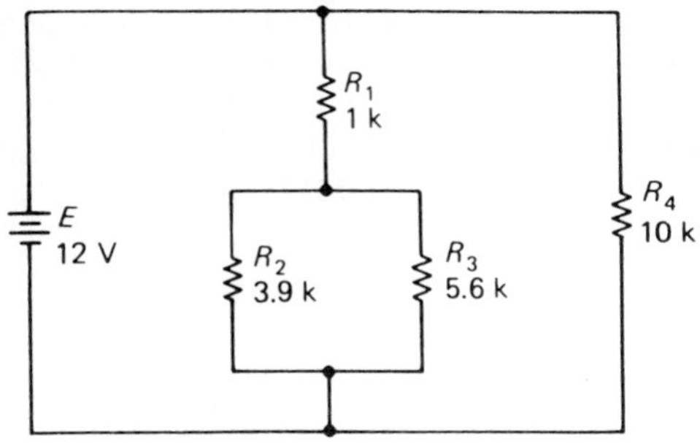

FIGURE 12-27

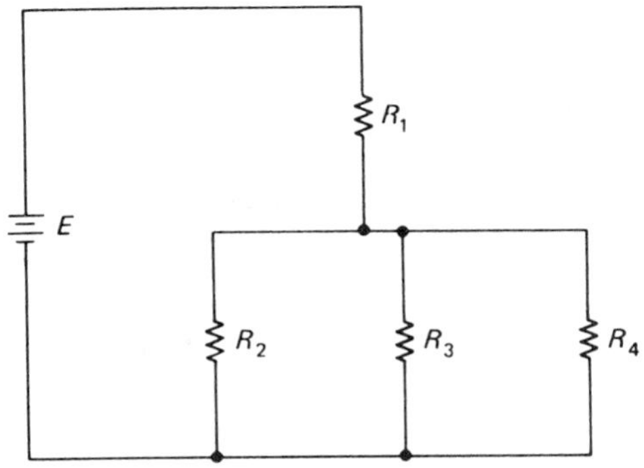

FIGURE 12-28

11. Refer to Figure 12-29. (a) Let $E = 40$ V, $R_1 = 10$ kΩ, $R_2 = 27$ kΩ, $R_3 = 56$ kΩ, and $R_4 = 33$ kΩ. Find V_1, V_2, V_3, V_4, I_1, I_2, I_3, I_4, I_T, and R_T. (b) Let $E = 15$ V, $R_1 = 150$ kΩ, $R_2 = 470$ kΩ, $R_3 = 680$ kΩ, and $R_4 = 150$ kΩ. Find V_1, V_2, V_3, V_4, I_1, I_2, I_3, I_4, I_T, and R_T.

12. Refer to Figure 12-29. (a) Let $E = 15$ V, $R_1 = 1$ kΩ, $R_2 = 12$ kΩ, $R_3 = 22$ kΩ, and $R_4 = 3$ kΩ. Find V_1, V_2, V_3, V_4, I_1, I_2, I_3, I_4, I_T, and R_T. (b) Let $E = 50$ V, $R_1 = 56$ kΩ, $R_2 = 150$ kΩ, $R_3 = 1$ MΩ, and $R_4 = 270$ kΩ. Find V_1, V_2, V_3, V_4, I_1, I_2, I_3, I_4, I_T, and R_T.

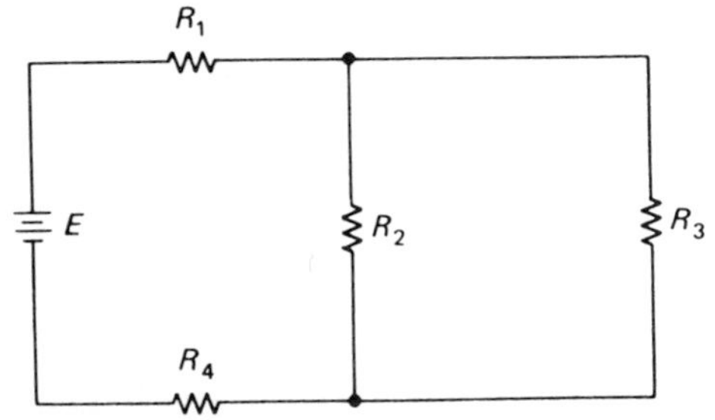

FIGURE 12-29

Answers to Self-tests

SELF-TEST 12-1

1. 122 kΩ

2. 487 Ω

3. 19.0 kΩ

4. 29.1 kΩ

5. 571 Ω

6. 503 kΩ

SELF-TEST 12-2

1. $R_T = 24.3$ kΩ, $I = 1.03$ mA, $R_2 = 6.32$ kΩ, $V_1 = 18.5$ V

2. $E = 160$ V, $V_1 = 54.4$ V, $V_3 = 55.6$ V, $R_3 = 69.5$ kΩ, $R_2 = 62.5$ kΩ

3. $R_T = 33.3$ Ω, $G_T = 30$ mS, $I_1 = 10$ mA, $I_2 = 20$ mA, $R_2 = 50$ Ω

4. $G_T = 100$ μS, $R_T = 10$ kΩ, $I_2 = 370$ μA, $I_3 = 430$ μA, $R_1 = 50$ kΩ, $R_3 = 23.3$ kΩ

SELF-TEST 12-3

1. $R_T = 42.9\ \text{k}\Omega$, $V_1 = 1.64\ \text{V}$, $V_2 = V_3 = 1.36\ \text{V}$, $I_2 = 41.2\ \mu\text{A}$, $I_3 = 28.9\ \mu\text{A}$, $R_1 = 23.4\ \text{k}\Omega$
2. $V_1 = 13.4\ \text{V}$, $V_3 = 20\ \text{V}$, $I_1 = I_2 = 672\ \mu\text{A}$, $I_3 = 2\ \text{mA}$, $I_T = 2.67\ \text{mA}$, $R_T = 7.49\ \text{k}\Omega$, $R_2 = 9.78\ \text{k}\Omega$
3. $I_T = I_1 = 571\ \mu\text{A}$, $I_2 = 81.4\ \mu\text{A}$, $I_3 = 370\ \mu\text{A}$, $I_4 = 120\ \mu\text{A}$, $V_1 = 6.86\ \text{V}$, $V_2 = V_3 = V_4 = 8.14\ \text{V}$, $R_T = 26.3\ \text{k}\Omega$
4. $V_1 = 9.15\ \text{V}$, $V_2 = V_3 = 8.40\ \text{V}$, $V_4 = 7.45\ \text{V}$, $I_T = 3.39\ \text{mA}$, $I_1 = I_4 = 3.39\ \text{mA}$, $I_2 = 2.15\ \text{mA}$, $I_3 = 1.24\ \text{mA}$, $R_T = 7.38\ \text{k}\Omega$

DC CIRCUIT ANALYSIS: CIRCUIT THEOREMS

CHAPTER 13

INTRODUCTION

It is often necessary to determine the internal resistance (or conductance) of an electrical energy source, amplifier, or system. We are also frequently asked to determine current and voltage distribution in electrical circuits for which Ohm's law and Kirchhoff's laws are not adequate. A variety of theorems has been developed over the years to assist the technician and engineer in solving these complex circuit problems. Thévenin's theorem, Norton's theorem, and the superposition theorem are some of the more widely used theorems.

CHAPTER OBJECTIVES

In this chapter you will learn how to:

1. Use the superposition theorem to solve complex circuit problems.
2. Use Thévenin's theorem to solve complex circuit problems.
3. Use Norton's theorem to solve complex circuit problems.

13-1 SUPERPOSITION THEOREM

The superposition theorem is one of the theorems we can use to simplify circuits that contain more than one source. Let's assume that R_2 is the load in Figure 13-1 and find the load current using the superposition theorem. The steps are (1) *short-circuit one supply and calculate the current through the load as a result of the remaining source;* (2) *place the short across the other source and again calculate the current through the load.* The algebraic sum of the two currents is the load current. The polarity of the drop across the load is determined by the direction of the greater current, if they are opposing. (3) *The load current found in step 2 is the load current in the original circuit.* Using this known current, find the other currents and voltage drops using Ohm's law and Kirchhoff's laws.

Using this method, we will find the current through and the voltage drop across R_2 in Figure 13-1. Then we will find the other currents and voltage drops. First, replace one source with a short circuit and determine the magnitude and direction of I_2. If we replace E_2 with a short circuit, the circuit shown in Figure 13-2 results.

$$I_T = \frac{E_1}{R_1 + R_X}$$

where $R_X = R_2 \parallel R_3$.

$$I_T = \frac{20 \text{ V}}{3 \text{ k}\Omega + 667 \ \Omega} = 5.45 \text{ mA}$$

$$I_2 = \frac{I_T R_X}{R_2} = \frac{5.45 \text{ mA} \times 667 \ \Omega}{2 \text{ k}\Omega} = 1.82 \text{ mA}$$

The current is 1.82 mA and has the direction shown. Now let's replace E_1 with a short circuit and compute the current due to E_2. This is done in Figure 13-3.

$$I_T = \frac{E_2}{R_3 + R_X}$$

where $R_X = R_1 \parallel R_2$.

$$I_T = \frac{40 \text{ V}}{1 \text{ k}\Omega + 1.2 \text{ k}\Omega} = 18.2 \text{ mA}$$

$$I_2 = \frac{I_T R_X}{R_2} = \frac{18.2 \text{ mA} \times 1.2 \text{ k}\Omega}{2 \text{ k}\Omega} = 10.9 \text{ mA}$$

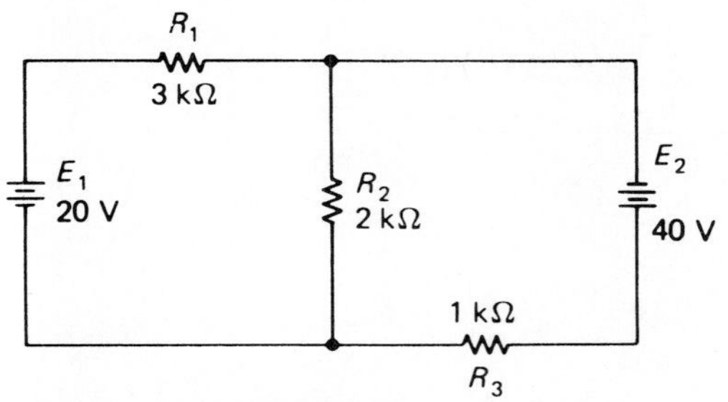

FIGURE 13-1

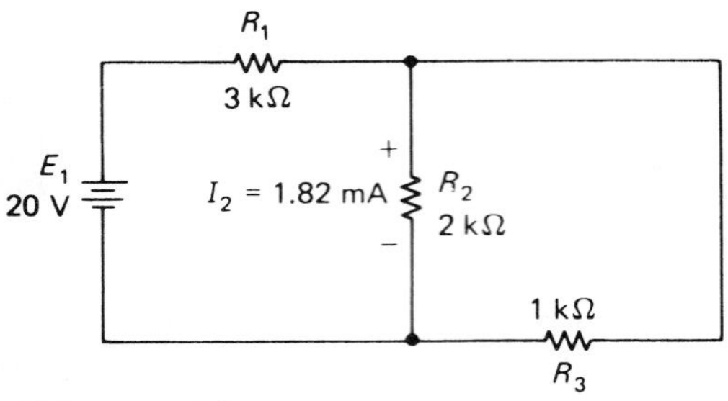

FIGURE 13-2

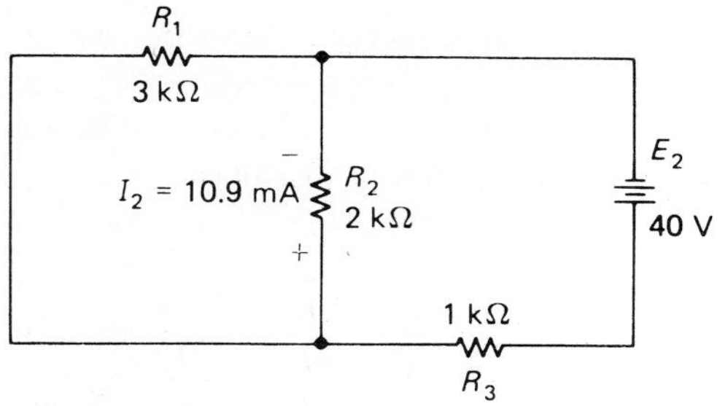

FIGURE 13-3

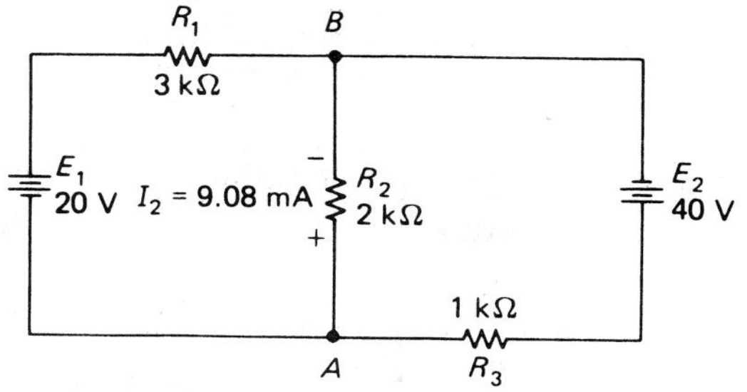

FIGURE 13-4

The current due to E_2 alone is 10.9 mA, and the polarity is as shown in Figure 13-3. Now, if we go back to the original circuit and superimpose the currents through R_2 as in Figure 13-4, we get one current whose value is 1.82 mA, causing a polarity at point A which is negative with respect to point B, and we get a second current of 10.9 mA, causing a polarity at point A which is positive with respect to point B. The currents are opposing. If we add the currents algebraically, we get a resulting current that is the difference between the two: 10.9 mA − 1.82 mA = 9.08 mA. That is, the actual current through R_2 is 9.08 mA, and point A is positive with respect to point B. The voltage drop across R_2 is 9.08 mA × 2 kΩ, which equals 18.2 V.

Knowing V_2, we can determine V_1 and V_3 by using Kirchhoff's voltage law. Using the loop containing E_1, R_1, and R_2 in Figure 13-5(a), we get the general equation $E_1 + V_1 + V_2 = 0$. Starting at point A and moving clockwise, we get

$$-E_1 + V_1 - V_2 = 0$$

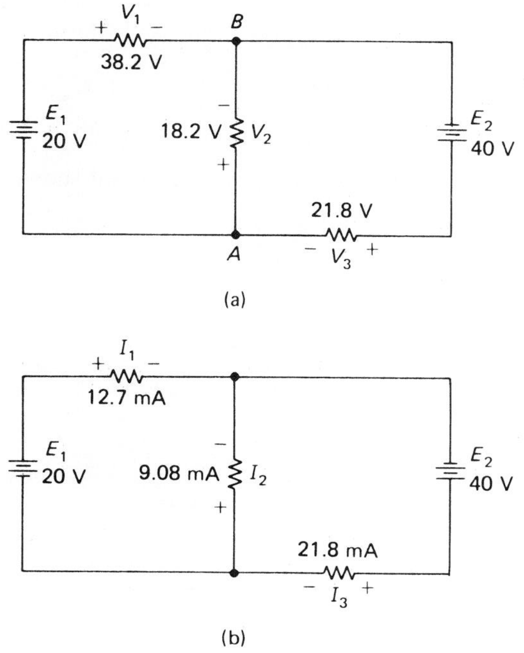

FIGURE 13-5

At this point we don't know the polarity of the drop across V_1; therefore, we assigned positive on the left side. If our assumption is wrong, when we solve for V_1, its magnitude will be correct but its value will be negative. If that turns out to be the case, we simply reverse the assigned polarity. Plugging in known values,

$$-20\text{ V} + V_1 - 18.2\text{ V} = 0$$
$$V_1 - 38.2\text{ V} = 0$$
$$V_1 = 38.2\text{ V}$$

Because the computed value for V_1 is positive, we have assigned the correct polarity and no change is necessary. Then

$$I_1 = \frac{V_1}{R_1} = \frac{38.2\text{ V}}{3\text{ k}\Omega} = 12.7\text{ mA}$$

Now let's look at the loop containing E_2, R_2, and R_3. Referring again to Figure 13-5(a), the general equation is $E_2 + V_2 + V_3 = 0$. Starting at point B and moving counterclockwise, we get $-V_2 + V_3 + E_2 = 0$. Again, the polarity of the drop across V_3 is not known, and we have assumed it to be positive. Plugging in known values and solving for V_3, we get

$$-18.2\text{ V} + V_3 + 40\text{ V} = 0$$
$$V_3 + 21.8\text{ V} = 0$$
$$V_3 = -21.8\text{ V}$$

V_3 equals 21.8 V, but the polarity assigned must be reversed since the computed value was negative. Solving for I_3, we get

$$I_3 = \frac{V_3}{R_3} = \frac{21.8\text{ V}}{1\text{ k}\Omega} = 21.8\text{ mA}$$

The complete circuit showing currents is in Figure 13-5(b).

Let's try another one. Find the currents and voltage drops in the circuit shown in Figure 13-6.

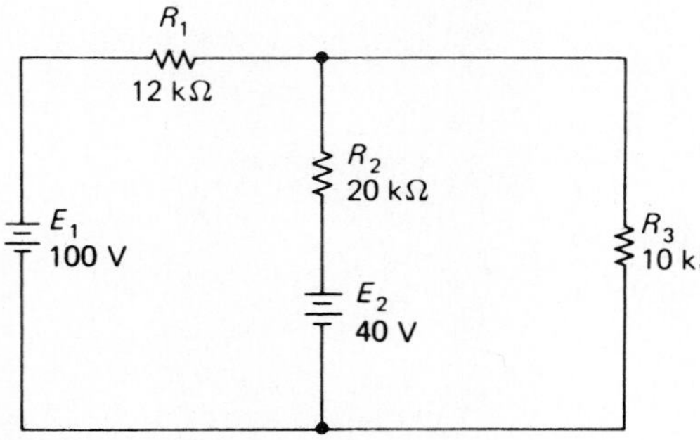

FIGURE 13-6

In Figure 13-7(a), the circuit is redrawn with E_2 replaced by a short circuit. The current through R_3 is calculated:

$$I_T = \frac{E_1}{R_1 + R_X}$$

where $R_X = R_2 \parallel R_3$.

$$I_T = \frac{100 \text{ V}}{12 \text{ k}\Omega + 6.67 \text{ k}\Omega} = 5.36 \text{ mA}$$

$$I_3 = \frac{I_T R_X}{R_3} = \frac{5.36 \text{ mA} \times 6.67 \text{ k}\Omega}{10 \text{ k}\Omega} = 3.57 \text{ mA}$$

In Figure 13-7(b) the circuit is again redrawn, but this time E_1 is replaced with a short circuit. I_3 in this circuit is calculated:

$$I_T = \frac{E_2}{R_2 + R_X}$$

where $R_X = R_1 \parallel R_3$.

$$I_T = \frac{40 \text{ V}}{20 \text{ k}\Omega + 5.45 \text{ k}\Omega} = 1.57 \text{ mA}$$

$$I_3 = \frac{I_T R_X}{R_3} = \frac{1.57 \text{ mA} \times 5.45 \text{ k}\Omega}{10 \text{ k}\Omega} = 856 \ \mu\text{A}$$

The current through R_3 in the original circuit (Figure 13-6) is the algebraic sum of these two currents. Since the two currents are in the same direction, the resulting current is the sum of the two individual currents: $I_3 = 3.57 \text{ mA} + 856 \ \mu\text{A} = 4.43 \text{ mA}$. The voltage drop across R_3 can now be determined:

$$V_3 = I_3 R_3 = 4.43 \text{ mA} \times 10 \text{ k}\Omega = 44.3 \text{ V}$$

In Figure 13-8, polarities have been assigned. The polarity of the voltage drop across R_3 had to be as shown. This was dictated by the resulting direction of I_3. The

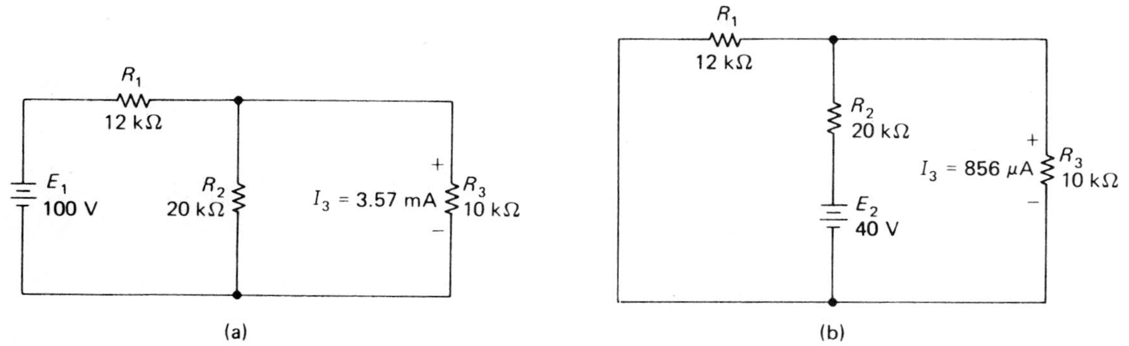

FIGURE 13-7

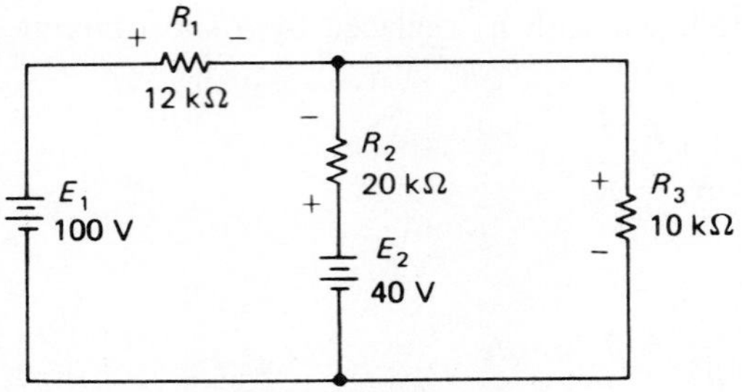

FIGURE 13-8

drops across R_1 and R_2 were chosen arbitrarily. Remember, if we assign the wrong polarity to V_1 and V_2, the calculated values will be correct but they will be negative. Using Kirchhoff's voltage law, we get

$$E_1 + V_1 + V_3 = 0$$
$$100\text{ V} - V_1 - 44.3\text{ V} = 0$$
$$-V_1 + 55.7\text{ V} = 0$$
$$V_1 = 55.7\text{ V}$$
$$E_2 - V_3 - V_2 = 0$$
$$40\text{ V} - 44.3\text{ V} - V_2 = 0$$
$$V_2 = -4.3\text{ V}$$

V_2 is negative, which tells us that the assigned polarity is wrong. The polarity is corrected in Figure 13-9. Now we can find I_1 and I_2 by using Ohm's law:

$$I_1 = \frac{V_1}{R_1} = \frac{55.7\text{ V}}{12\text{ k}\Omega} = 4.63\text{ mA}$$

$$I_2 = \frac{V_2}{R_2} = \frac{4.3\text{ V}}{20\text{ k}\Omega} = 215\ \mu\text{A}$$

Kirchhoff's laws are now used to check our work. I_1 is total current. Therefore,

$$I_1 = I_2 + I_3$$
$$4.64\text{ mA} = 222\ \mu\text{A} + 4.43\text{ mA}$$

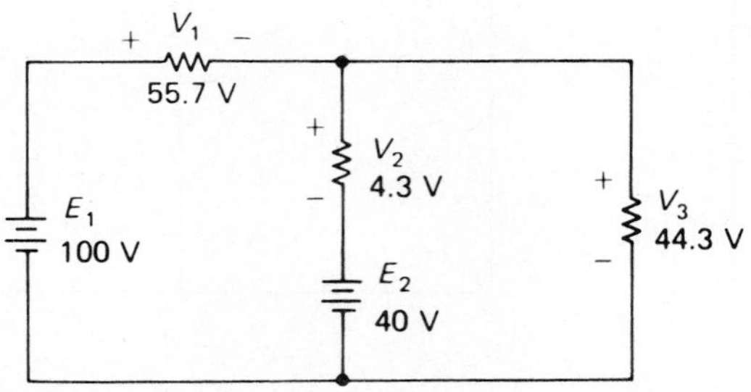

FIGURE 13-9

Also,

$$E_1 - V_3 - V_1 = 0$$

$$100 \text{ V} - 44.3 \text{ V} - 55.7 \text{ V} = 0$$

And

$$E_1 - E_2 - V_2 - V_1 = 0$$

$$100 \text{ V} - 40 \text{ V} - 4.3 \text{ V} - 55.7 \text{ V} = 0$$

PRACTICE PROBLEMS 13-1

Find the various voltage drops and currents in the circuits of Figure 13-10.

Solutions

(a) In Figure 13-10(a) the current through R_2 is 45.3 mA. Replacing E_2 with a short circuit, we get

$$R_T = R_1 + R_2 \parallel R_3 = 100\ \Omega + 132\ \Omega = 232\ \Omega$$

$$I_T = \frac{E_1}{R_T} = \frac{40 \text{ V}}{232\ \Omega} = 172 \text{ mA}$$

$$I_2 = \frac{I_T R_X}{R_2} = \frac{172 \text{ mA} \times 132\ \Omega}{330\ \Omega} = 68.8 \text{ mA}$$

I_2 due to E_1 is 68.8 mA.

Replacing E_1 with a short circuit, we get

$$R_T = R_3 + R_1 \parallel R_2 = 220\ \Omega + 76.7\ \Omega$$
$$= 297\ \Omega$$

$$I_T = \frac{E_2}{R_T} = \frac{30 \text{ V}}{297\ \Omega} = 101 \text{ mA}$$

$$I_2 = \frac{I_T R_X}{R_2} = \frac{101 \text{ mA} \times 76.7\ \Omega}{330\ \Omega} = 23.5 \text{ mA}$$

I_2 due to E_2 is 23.5 mA.

The currents are opposing. Therefore, the resulting current is the difference between the two currents, or 45.3 mA. The polarities are shown in Figure 13-11. Using Ohm's law and Kirchhoff's laws, we can find the rest of the currents and voltage drops.

$$V_2 = I_2 R_2 = 45.3 \text{ mA} \times 330\ \Omega = 14.9 \text{ V}$$

$$V_1 = E_1 - V_2 = 40 \text{ V} - 14.9 \text{ V} = 25.1 \text{ V}$$

$$V_3 = E_2 + V_2 = 30 \text{ V} + 14.9 \text{ V} = 44.9 \text{ V}$$

$$I_1 = \frac{V_1}{R_1} = \frac{25.1 \text{ V}}{100\ \Omega} = 251 \text{ mA}$$

$$I_3 = \frac{V_3}{R_3} = \frac{44.9 \text{ V}}{220\ \Omega} = 204 \text{ mA}$$

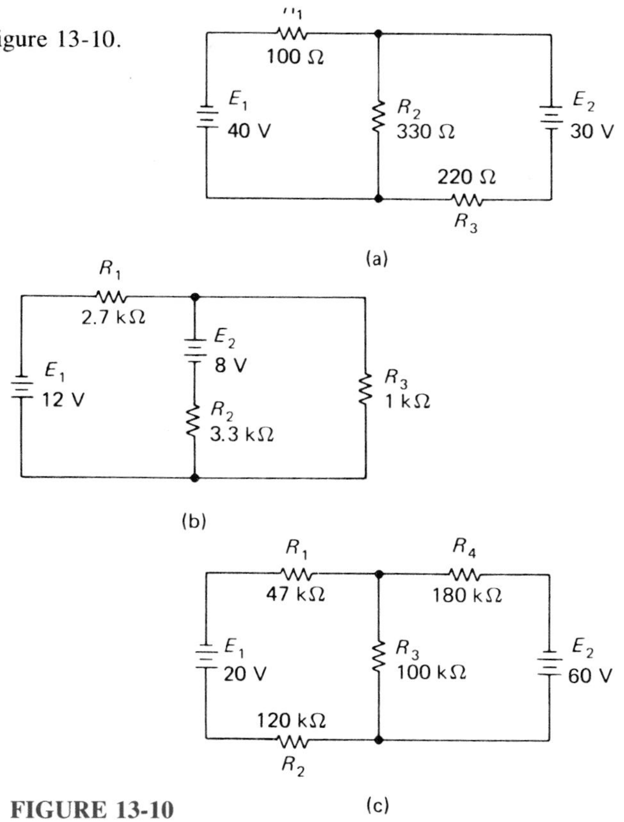

FIGURE 13-10

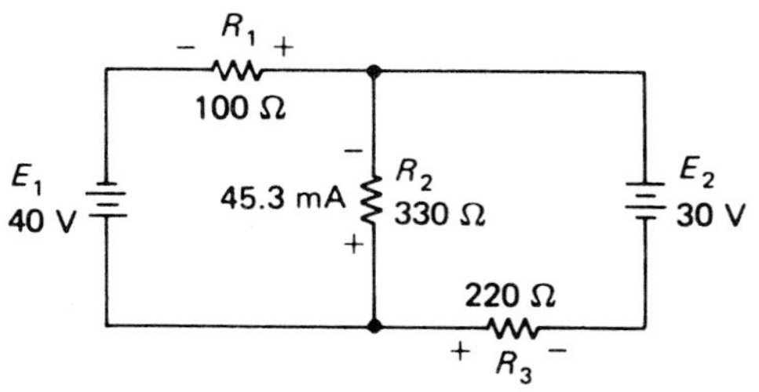

FIGURE 13-11

(b) In Figure 13-10(b) the current through R_3 is 1.2 mA. Replacing E_1 with a short circuit, we get

$$R_T = R_2 + R_1 \parallel R_3 = 3.3\ \text{k}\Omega + 730\ \Omega$$
$$= 4.03\ \text{k}\Omega$$

$$I_T = \frac{E_2}{R_T} + \frac{8\ \text{V}}{4.03\ \text{k}\Omega} = 1.99\ \text{mA}$$

$$I_3 = \frac{I_T R_X}{R_3} = \frac{1.99\ \text{mA} \times 730\ \Omega}{1\ \text{k}\Omega} = 1.45\ \text{mA}$$

I_3 due to E_2 is 1.45 mA.

Replacing E_2 with a short circuit, we get

$$R_T = R_1 + R_2 \parallel R_3 = 2.7\ \text{k}\Omega + 767\ \Omega$$
$$= 3.47\ \text{k}\Omega$$

$$I_T = \frac{E_1}{R_T} = \frac{12\ \text{V}}{3.47\ \text{k}\Omega} = 3.46\ \text{mA}$$

$$I_3 = \frac{I_T R_X}{R_3} = \frac{3.46\ \text{mA} \times 767\ \Omega}{1\ \text{k}\Omega} = 2.65\ \text{mA}$$

I_3 due to E_1 is 2.65 mA.

The two currents are opposing and the resulting current through R_3 is 1.2 mA. The polarities are shown in Figure 13-12. Using Ohm's law and Kirchhoff's laws, we can find the rest of the currents and voltage drops.

$$V_3 = I_3 R_3 = 1.2\ \text{mA} \times 1\ \text{k}\Omega = 1.2\ \text{V}$$
$$V_2 = E_2 + V_3 = 8\ \text{V} + 1.2\ \text{V} = 9.2\ \text{V}$$
$$V_1 = E_1 - V_3 = 12\ \text{V} - 1.2\ \text{V} = 10.8\ \text{V}$$

$$I_2 = \frac{V_2}{R_2} = \frac{9.2\ \text{V}}{3.3\ \text{k}\Omega} = 2.79\ \text{mA}$$

$$I_1 = \frac{V_1}{R_1} = \frac{10.8\ \text{V}}{2.7\ \text{k}\Omega} = 4\ \text{mA}$$

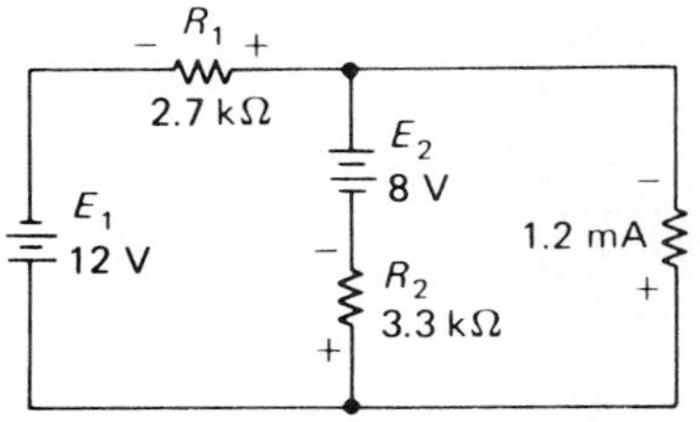

FIGURE 13-12

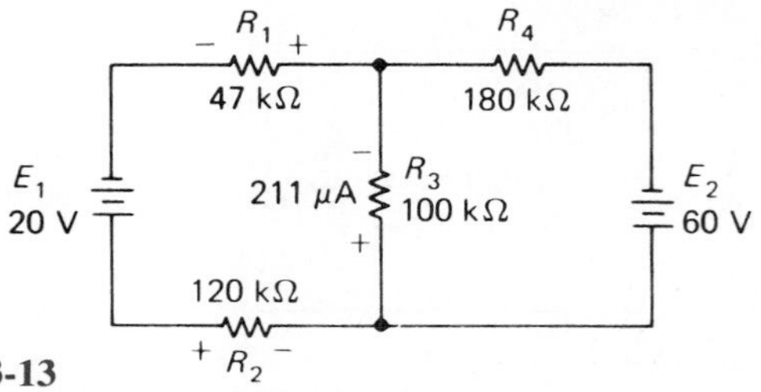

FIGURE 13-13

Additional practice problems are at the end of the chapter.

(c) In Figure 13-10(c) the current through R_3 is 211 μA. Replacing E_1 with a short circuit, we get

$$R_T = R_4 + R_3 \parallel (R_1 + R_2)$$
$$= 180\ \text{k}\Omega + 62.5\ \text{k}\Omega = 243\ \text{k}\Omega$$

$$I_T = \frac{E_2}{R_T} = \frac{60\ \text{V}}{243\ \text{k}\Omega} = 247\ \mu\text{A}$$

$$I_3 = \frac{I_T R_X}{R_3} = \frac{247\ \mu\text{A} \times 62.5\ \text{k}\Omega}{100\ \text{k}\Omega} = 155\ \mu\text{A}$$

I_3 due to E_2 is 155 μA.

Replacing E_2 with a short circuit,

$$R_T = R_1 + R_2 + R_3 \parallel R_4$$
$$= 47\ \text{k}\Omega + 120\ \text{k}\Omega + 64.3\ \text{k}\Omega = 231\ \text{k}\Omega$$

$$I_T = \frac{E_1}{R_T} = \frac{20\ \text{V}}{231\ \text{k}\Omega} = 86.6\ \mu\text{A}$$

$$I_3 = \frac{I_T R_X}{R_3} = \frac{86.6\ \mu\text{A} \times 64.3\ \text{k}\Omega}{100\ \text{k}\Omega} = 55.7\ \mu\text{A}$$

I_3 due to E_1 is 55.7 μA.

The two currents are aiding in making the resulting current through R_3 the sum of the two currents, or 211 μA. The polarities are shown in Figure 13-13. Using Ohm's law and Kirchhoff's laws, the rest of the currents and voltage drops may be found.

$$V_3 = I_3 R_3 = 211\ \mu\text{A} \times 100\ \text{k}\Omega$$
$$= 21.1\ \text{V}$$

$$V_4 = E_2 - V_3 = 60\ \text{V} - 21.1\ \text{V}$$
$$= 38.9\ \text{V}$$

$$I_4 = \frac{V_4}{R_4} = \frac{38.9\ \text{V}}{180\ \text{k}\Omega} = 216\ \mu\text{A}$$

$$V_1 + V_2 = E_2 - V_4 - E_1$$
$$= 60\ \text{V} - 38.9\ \text{V} - 20\ \text{V}$$
$$= 1.1\ \text{V}$$

$$I_1 = I_2 = \frac{V_1 + V_2}{R_1 + R_2} = \frac{1.1\ \text{V}}{167\ \text{k}\Omega}$$
$$= 6.59\ \mu\text{A}$$

$$V_1 = I_1 R_1 = 6.59\ \mu\text{A} \times 47\ \text{k}\Omega$$
$$= 0.31\ \text{V}$$

$$V_2 = I_2 R_2 = 6.59\ \mu\text{A} \times 120\ \text{k}\Omega$$
$$= 0.791\ \text{V}$$

SELF-TEST 13-1

Use the superposition theorem to find the various currents and voltage drops in the circuits in Figure 13-14.
Answers to Self-test 13-1 are at the end of the chapter.

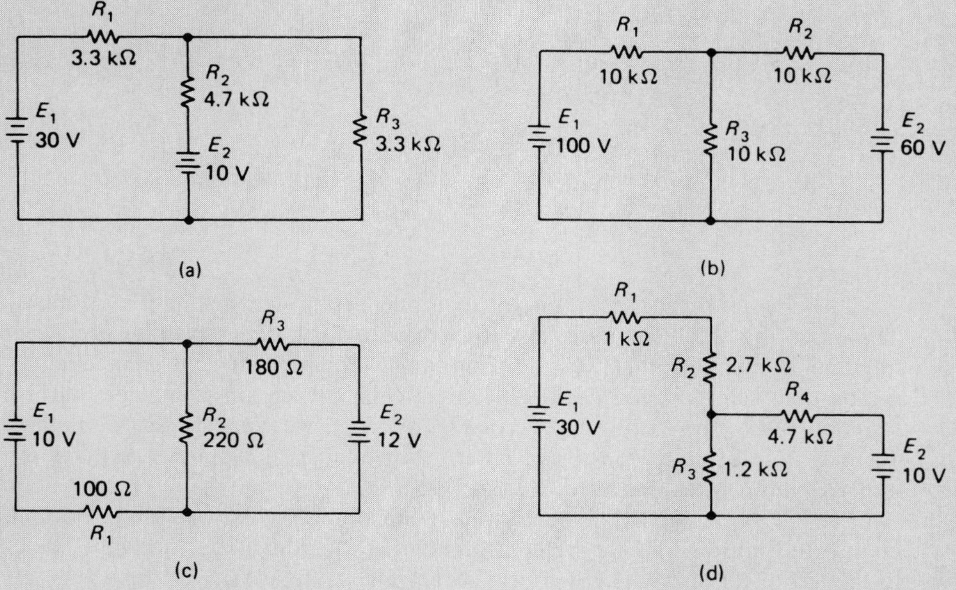

FIGURE 13-14

13-2 THEVENIN'S THEOREM

Thévenin's theorem states that any network, no matter how complex, can be reduced to an equivalent voltage source and series resistance. The voltage source is labeled V_{OC} and the series resistance is R_{TH}, as illustrated in Figure 13-15. Consider the circuit in Figure 13-16. Because this is a simple series–

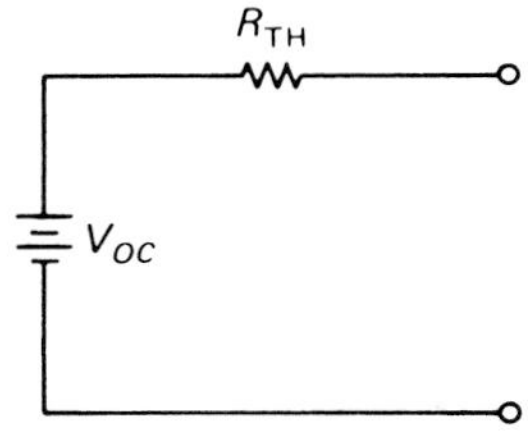

FIGURE 13-15

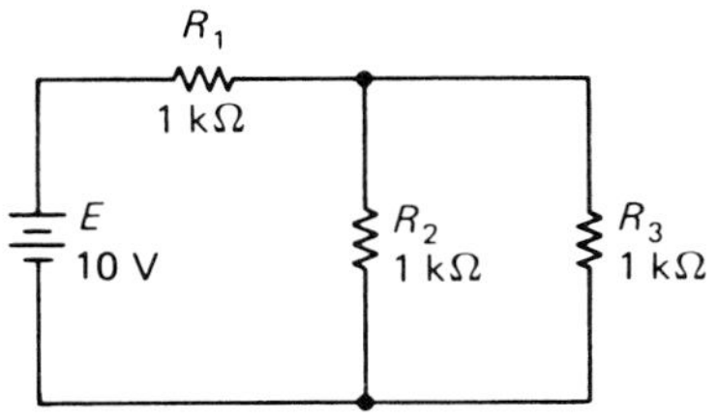

FIGURE 13-16

parallel circuit, the various currents and voltage drops can be found easily by using Ohm's law and Kirchhoff's laws.

$$R_T = R_1 + R_2 \parallel R_3 = 1\ \text{k}\Omega + 500\ \Omega = 1.5\ \text{k}\Omega$$

$$I_T = \frac{E}{R_T} = \frac{10\ \text{V}}{1.5\ \text{k}\Omega} = 6.67\ \text{mA}$$

$$I_1 = I_T = 6.67\ \text{mA}$$

$$V_1 = I_1 R_1 = 6.67\ \text{mA} \times 1\ \text{k}\Omega = 6.67\ \text{V}$$

$$V_2 = V_3 = E - V_1 = 10\ \text{V} - 6.67\ \text{V} = 3.33\ \text{V}$$

$$I_2 = \frac{V_2}{R_2} = \frac{3.33\ \text{V}}{1\ \text{k}\Omega} = 3.33\ \text{mA}$$

$$I_3 = \frac{V_3}{R_3} = \frac{3.33\ \text{V}}{1\ \text{k}\Omega} = 3.33\ \text{mA}$$

In circuits such as this, circuit theorems are not needed. Let's go ahead and Thévenize this circuit anyway just to develop a basic understanding of Thévenin's theorem. Remember that we said Thévenin's circuit is an *equivalent* circuit. This means that the Thévenin equivalent circuit and the circuit it replaces furnish the same energy to the load. Suppose in Figure 13-16 we consider that R_3 is the load. We may consider that E, R_1, and R_2 are the circuit that supplies energy to R_3, just like V_{OC} and R_{TH}, as indicated in Figure 13-17.

In Thévenizing, we first determine from which points we wish to examine the circuit. In Figure 13-17 we are considering that R_3 is the load. Therefore, the circuit to the left of terminals x–y will be Thévenized.

To find Thévenin's equivalent circuit, we use the following procedure:

1. *Remove the load and calculate the difference in potential between the open-circuit terminals (x–y). This is the voltage source V_{OC} in our equivalent circuit.* In Figure 13-18 the load (R_3) has been removed. Looking back from the open-circuit terminals, we see that the circuit has been reduced to a simple series circuit and $V_{OC} = V_2$. Applying Ohm's law, we get $V_{OC} = 5$ V.

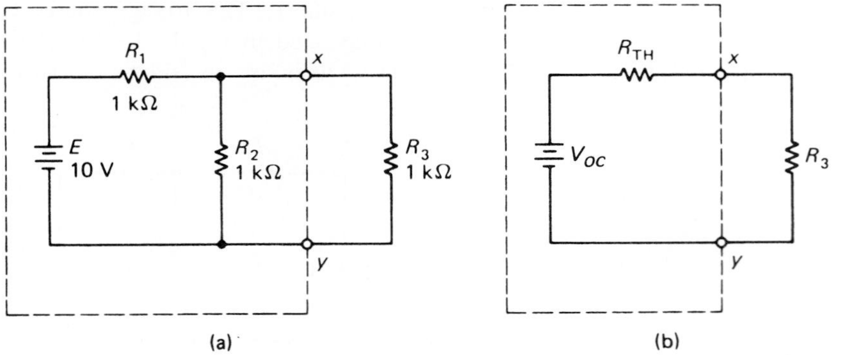

FIGURE 13-17

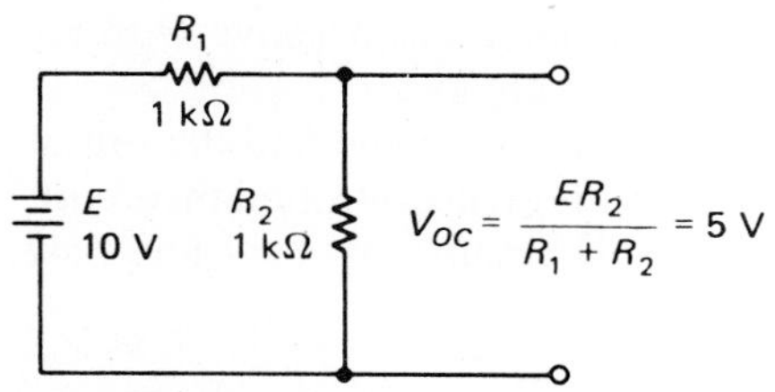

FIGURE 13-18

2. *With the load removed as in step 1, replace the source with a short circuit and determine the resistance at the open-circuit terminals.*[1] This resistance is R_{TH} in our equivalent circuit. Replacing the source with a short circuit as in Figure 13-19 results in a circuit with R_1 and R_2 in parallel. This parallel circuit is seen looking back into the circuit from the open-circuit terminals. The equivalent resistance (R_{TH}) is 500 Ω. V_{OC} and R_{TH} are connected in series to form Thévenin's equivalent circuit in Figure 13-20.
3. *Connect the load across the output terminals of the Thévenin equivalent circuit. Calculate V_L and I_L, the voltage drop across and the current through the load.* Using Ohm's law in Figure 13-21, we get

$$I_L = \frac{V_{OC}}{R_{TH} + R_L} \tag{13-1}$$

$$I_L = \frac{5\text{ V}}{1.5\text{ k}\Omega} = 3.33\text{ mA}$$

$$V_L = \frac{V_{OC} R_L}{R_{TH} + R_L} \tag{13-2}$$

$$V_L = \frac{5\text{ V} \times 1\text{ k}\Omega}{1.5\text{ k}\Omega} = 3.33\text{ V}$$

Thévenin's equivalent circuit causes the same current through R_3 and the same voltage drop across it as does the original circuit. Furthermore, the polarity associated with V_{OC} in Figure 13-18 is the polarity of the voltage drop across R_3 in the

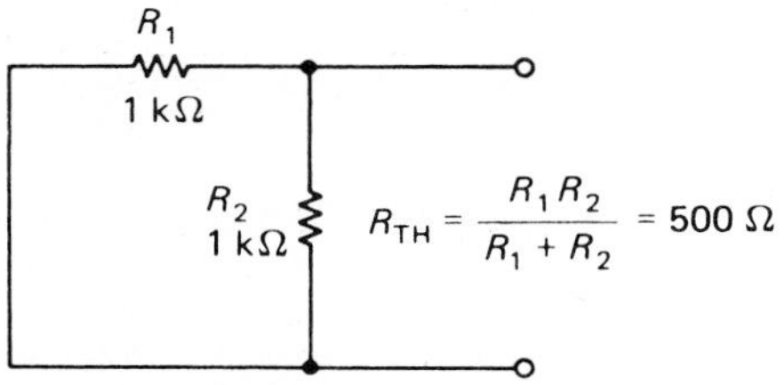

FIGURE 13-19

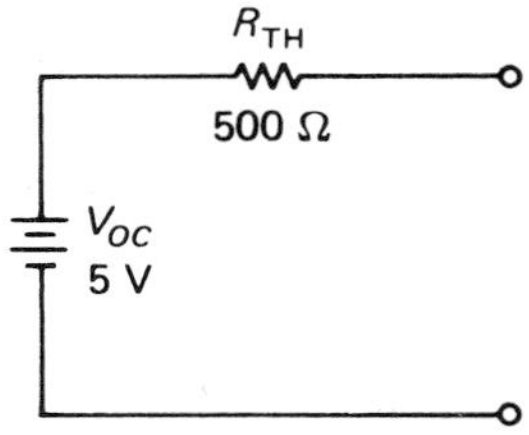

FIGURE 13-20

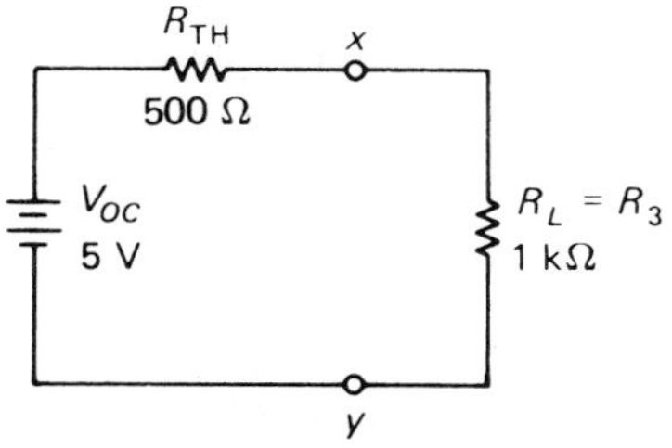

FIGURE 13-21

[1] We have assumed that the source is an ideal voltage source and $R_{int} = 0\ \Omega$. If the source resistance were some finite value, then the source would be replaced with that resistance instead of a short circuit.

original circuit. (In Figure 13-16 it was obvious that point y was negative with respect to point x, but in some complex circuits it is not so obvious.)

Knowing V_3 and I_3 (remember we called R_3 the load), we could find the rest of the voltage drops and currents just as we did in Section 13-1 (superposition theorem).

Let's Thévenize some rather simple series–parallel circuits. We will find only V_L and I_L.

PRACTICE PROBLEMS 13-2

For the problems in Figure 13-22:

1. Develop Thévenin's equivalent circuit.
2. Find V_L and I_L.

Solutions

(a) In Figure 13-22(a), if we open circuit the load, $V_{OC} = V_2$.

$$V_{OC} = V_2 = \frac{ER_2}{R_1 + R_2} = \frac{20\text{ V} \times 20\text{ k}\Omega}{32\text{ k}\Omega}$$

$$= 12.5\text{ V}$$

$$R_{TH} = R_1 \parallel R_2 = 7.5\text{ k}\Omega$$

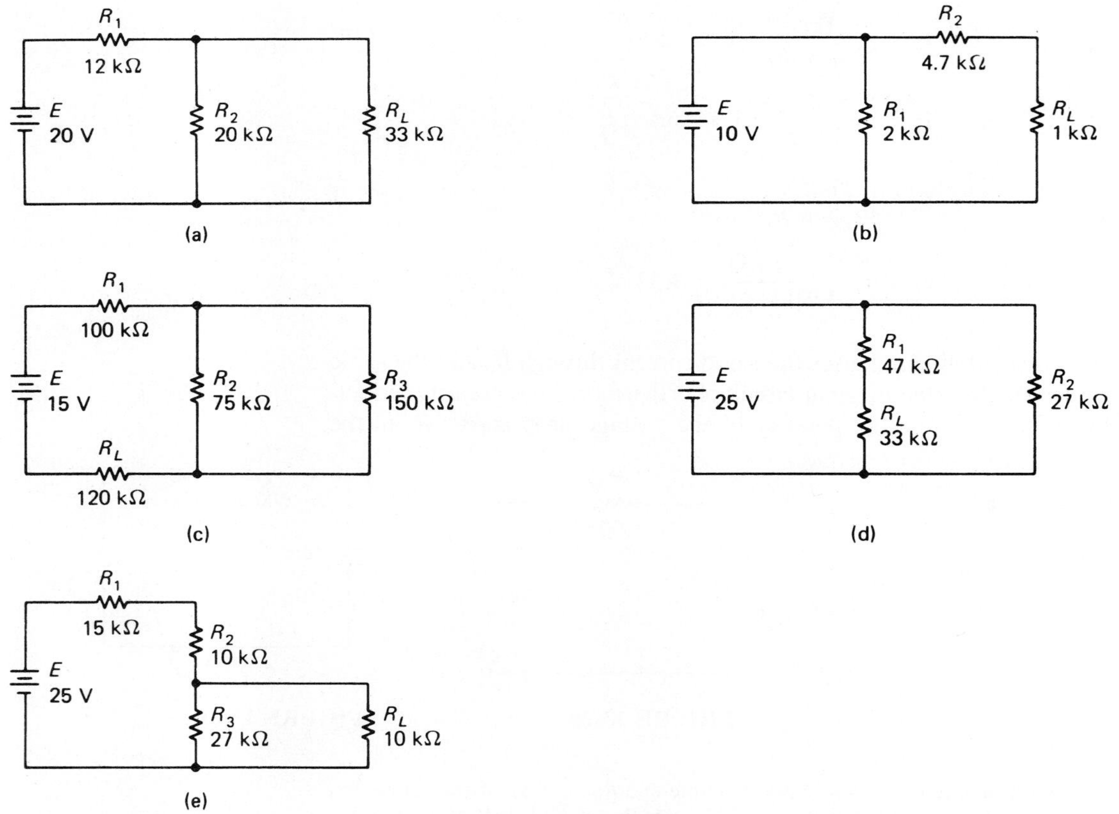

FIGURE 13-22

Thévenin's equivalent circuit with the load connected is shown in Figure 13-23. Applying Ohm's law, we get

$$I_L = \frac{V_{OC}}{R_{TH} + R_L} = \frac{12.5\text{ V}}{40.5\text{ k}\Omega} = 309\ \mu\text{A}$$

$$V_L = \frac{V_{OC}R_L}{R_{TH} + R_L} = \frac{12.5\text{ V} \times 33\text{ k}\Omega}{40.5\text{ k}\Omega} = 10.2\text{ V}$$

or

$$V_L = I_L R_L = 309\ \mu\text{A} \times 33\text{ k}\Omega = 10.2\text{ V}$$

(b) In Figure 13-22(b), if we open circuit the load, $V_{OC} = V_1 = E$. Shorting the source places a short circuit across R_1 so that $R_{TH} = R_2 = 4.7\text{ k}\Omega$. Thévenin's equivalent circuit with the load connected is shown in Figure 13-24. Applying Ohm's law, we get

$$I_L = \frac{V_{OC}}{R_{TH} + R_L} = \frac{10\text{ V}}{5.7\text{ k}\Omega} = 1.75\text{ mA}$$

$$V_L = I_L R_L = 1.75\text{ mA} \times 1\text{ k}\Omega = 1.75\text{ V}$$

(c) In Figure 13-22(c), if we open circuit the load, there is no circuit current. Therefore, $V_{OC} = E = 15$ V.

$$R_{TH} = R_1 + R_2 \parallel R_3 = 100\text{ k}\Omega + 50\text{ k}\Omega = 150\text{ k}\Omega$$

Thévenin's equivalent circuit with the load connected is shown in Figure 13-25.

$$I_L = \frac{V_{OC}}{R_{TH} + R_L} = \frac{15\text{ V}}{270\text{ k}\Omega} = 55.6\ \mu\text{A}$$

$$V_L = I_L R_L = 55.6\ \mu\text{A} \times 120\text{ k}\Omega = 6.67\text{ V}$$

(d) In Figure 13-22(d), if we open circuit the load, V_{OC} again equals E since V_1 would equal 0 V. Shorting the source to find R_{TH} causes a short circuit across R_2. Therefore, $R_{TH} = R_1 = 47\text{ k}\Omega$. Thévenin's equivalent circuit with the load connected is shown in Figure 13-26.

$$I_L = \frac{V_{OC}}{R_{TH} + R_L} = \frac{25\text{ V}}{80\text{ k}\Omega} = 313\ \mu\text{A}$$

$$V_L = I_L R_L = 313\ \mu\text{A} \times 33\text{ k}\Omega = 10.3\text{ V}$$

(e) In Figure 13-22(e), if we open circuit the load, $V_{OC} = V_3$.

$$V_{OC} = V_3 = \frac{ER_3}{R_1 + R_2 + R_3} = \frac{25\text{ V} \times 27\text{ k}\Omega}{52\text{ k}\Omega} = 13\text{ V}$$

$$R_{TH} = R_3 \parallel (R_1 + R_2) = 13\text{ k}\Omega$$

Additional practice problems are at the end of the chapter.

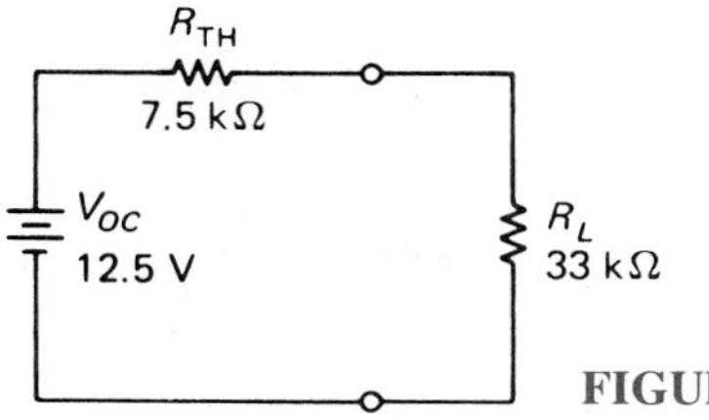

FIGURE 13-23

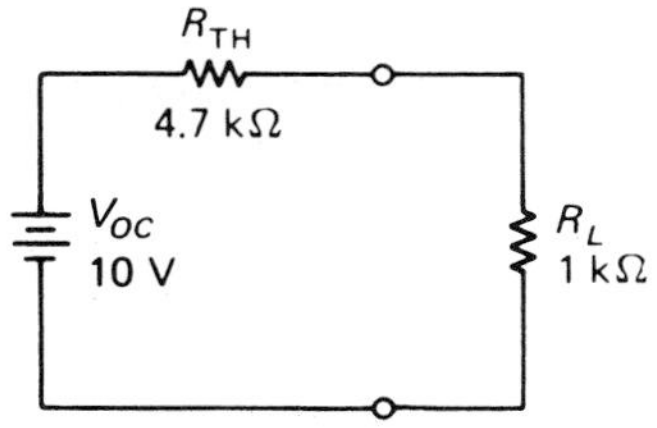

FIGURE 13-24

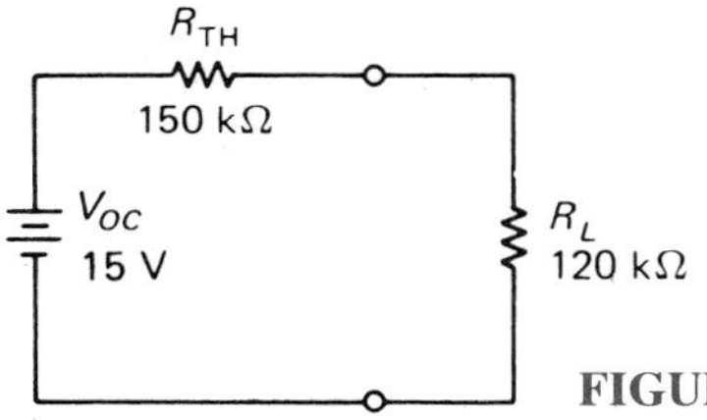

FIGURE 13-25

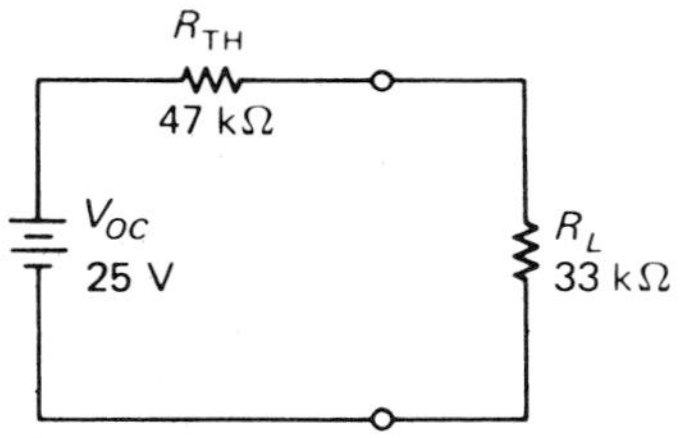

FIGURE 13-26

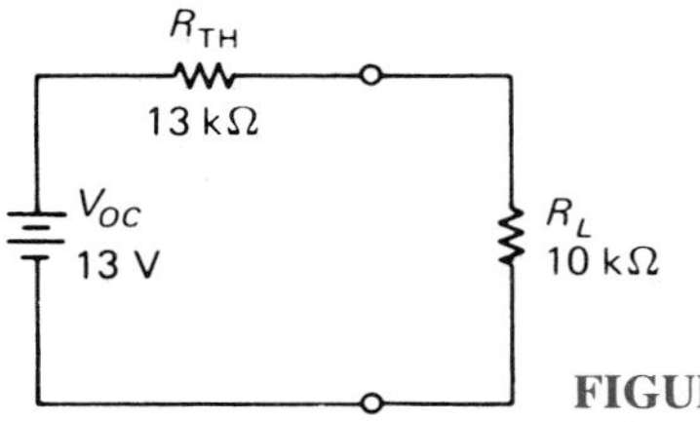

FIGURE 13-27

Thévenin's equivalent circuit with the load connected is shown in Figure 13-27.

$$I_L = \frac{V_{OC}}{R_{TH} + R_L} = \frac{13\text{ V}}{23\text{ k}\Omega} = 565\ \mu\text{A}$$

$$V_L = I_L R_L = 565\ \mu\text{A} \times 10\text{ k}\Omega = 5.65\text{ V}$$

Now let's Thévenize some circuits that are more complex. Consider the circuit in Figure 13-28. We have already found the various currents and voltage drops by using the superposition theorem. We will consider that R_2 is the load and find I_2 and V_2 by using Thévenin's theorem.

First, let's open circuit the load, as in Figure 13-29(a), and calculate V_{OC}. Because E_1 and E_2 are connected series aiding, $E_T = E_1 + E_2 = 60$ V. Therefore, we may consider that there is an emf of 60 V in series with R_1 and R_3.

$$V_1 = \frac{E_T R_1}{R_1 + R_3} = \frac{60 \text{ V} \times 3 \text{ k}\Omega}{4 \text{ k}\Omega} = 45 \text{ V}$$

$$V_3 = E_T - V_1 = 60 \text{ V} - 45 \text{ V} = 15 \text{ V}$$

Using Kirchhoff's voltage law, we get $V_{OC} = E_2 - V_3 = 25$ V (or $V_{OC} = V_1 - E_1 = 25$ V) and the polarity is as shown.

Next, we replace E_1 and E_2 with short circuits as in Figure 13-29(b) and calculate R_{TH}. In Figure 13-29(c) the circuit has been redrawn so that we can see that R_1 and R_3 are in parallel.

$$R_{TH} = R_1 \parallel R_3 = 750 \ \Omega$$

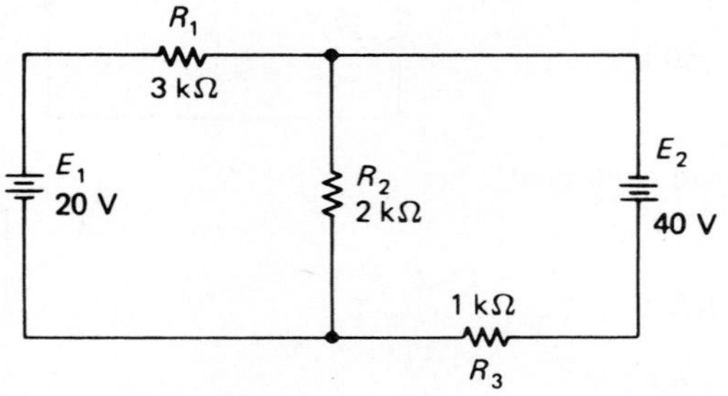

FIGURE 13-28

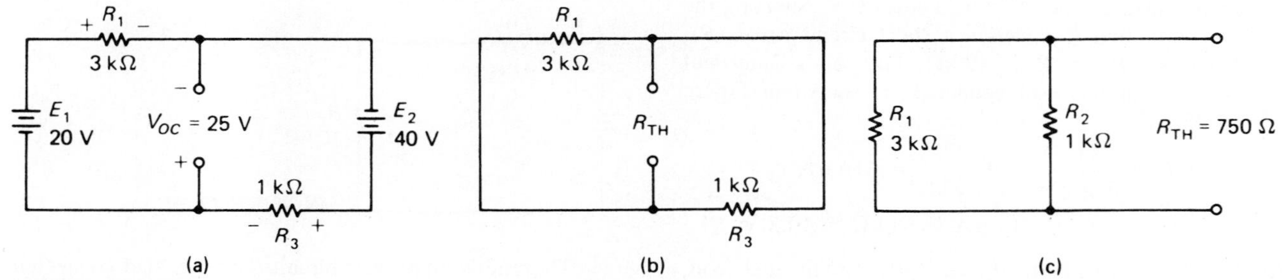

FIGURE 13-29

Now let's draw Thévenin's equivalent circuit and connect the load across it as in Figure 13-30.

$$I_L = \frac{V_{OC}}{R_{TH} + R_L} = \frac{25 \text{ V}}{2.75 \text{ k}\Omega} = 9.09 \text{ mA}$$

$$V_L = I_L R_L = 9.09 \text{ mA} \times 2 \text{ k}\Omega = 18.2 \text{ V}$$

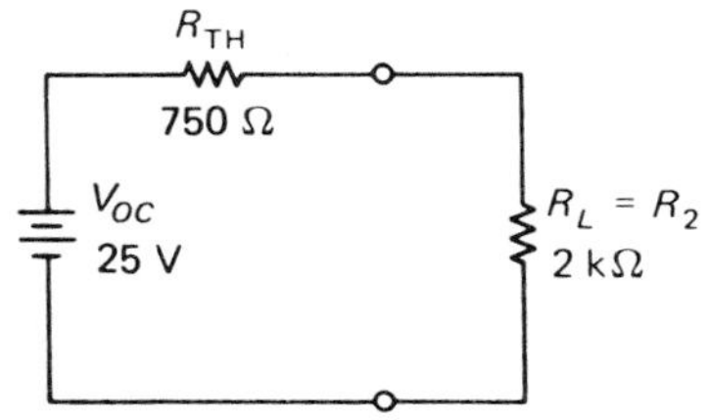

FIGURE 13-30

Knowing V_L and I_L (V_2 and I_2), we could find the rest of the voltage drops and currents as before. Notice that when we found V_{OC} we also found the polarity of the voltage drop across the load in the original circuit.

Let's find I_3 and V_3 in Figure 13-31 by using Thévenin's theorem. In Figure 13-32 the load has been removed, resulting in a series circuit. The sources are connected series opposing, so $E_T = E_1 - E_2 = 60$ V.

$$V_2 = \frac{E_T R_2}{R_1 + R_2} = \frac{60 \text{ V} \times 20 \text{ k}\Omega}{32 \text{ k}\Omega} = 37.5 \text{ V}$$

Using Kirchhoff's voltage law, we get $V_{OC} = E_2 + V_2 = 77.5$ V.

In Figure 13-32(b), E_1 and E_2 are replaced with short circuits and R_{TH} is calculated.

$$R_{TH} = R_1 \parallel R_2 = 7.5 \text{ k}\Omega$$

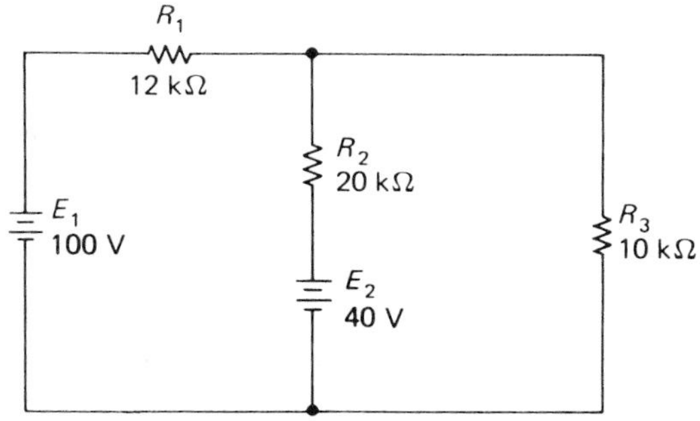

FIGURE 13-31

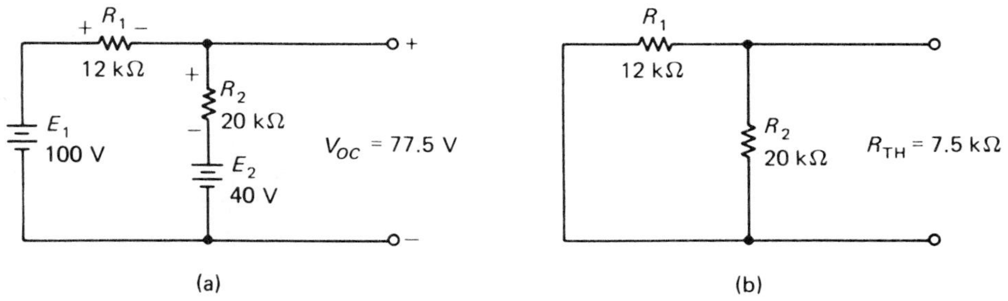

FIGURE 13-32

R_L is connected across Thévenin's equivalent circuit in Figure 13-33.

$$I_L = \frac{V_{OC}}{R_{TH} + R_L} = \frac{77.5 \text{ V}}{17.5 \text{ k}\Omega} = 4.43 \text{ mA}$$

$$V_L = I_L R_L = 4.43 \text{ mA} \times 10 \text{ k}\Omega = 44.3 \text{ V}$$

Consider the bridge circuit in Figure 13-34. Let's remove the load and redraw it slightly to help us see the circuit. With the load removed (Figure 13-35), we can treat the circuit as a series–parallel circuit, where $E = V_1 + V_2 = V_3 + V_4$.

$$V_1 = \frac{ER_1}{R_1 + R_2} = 4.5 \text{ V}$$

$$V_2 = E - V_1 = 7.5 \text{ V}$$

$$V_3 = \frac{ER_3}{R_3 + R_4} = 7.62 \text{ V}$$

$$V_4 = E - V_3 = 4.38 \text{ V}$$

Applying Kirchhoff's voltage law starting at point x and moving clockwise, we get

$$V_{OC} + V_4 - V_2 = 0$$

$$V_{OC} + 4.38 \text{ V} - 7.5 \text{ V} = 0$$

$$V_{OC} = 3.12 \text{ V}$$

We assumed that point x was positive with respect to point y. The fact that V_{OC} is positive in our calculation proves that we were right. If V_{OC} had been -3.12 V in our

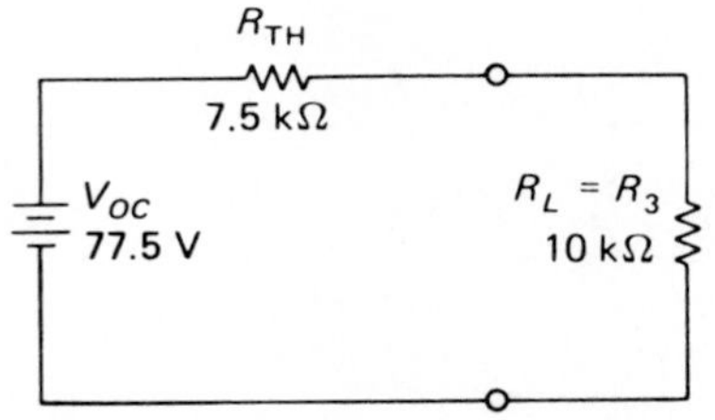

FIGURE 13-33

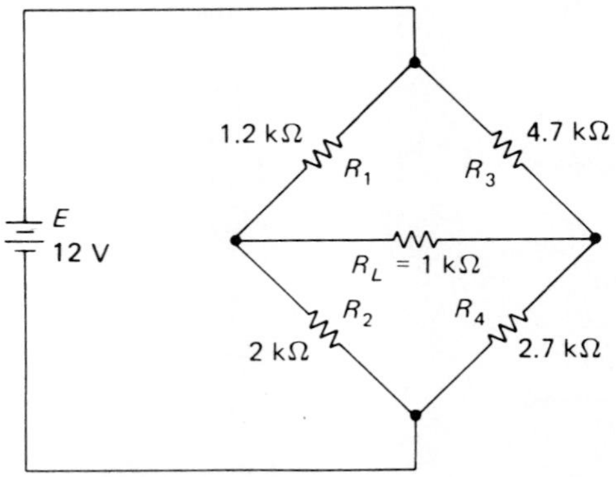

FIGURE 13-34

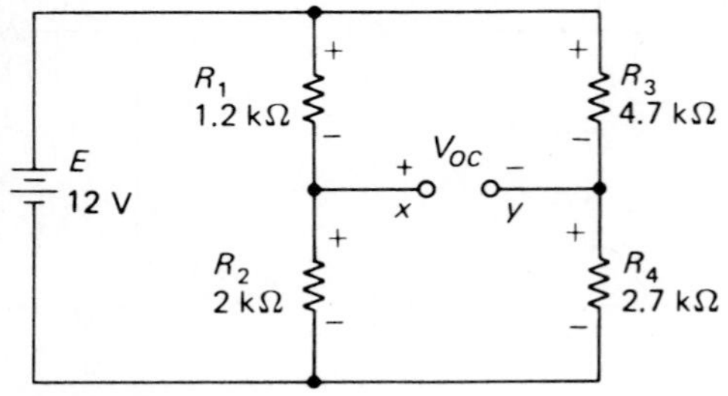

FIGURE 13-35

equation, it would simply have meant that point x was *negative* with respect to point y.

Now let's short E and calculate R_{TH}. Figure 13-36(a) shows the circuit with E shorted. Since it may be difficult to see the series–parallel circuit, let's redraw the circuit as in Figure 13-36(b).

$$R_{TH} = R_1 \parallel R_2 + R_3 \parallel R_4 = 750\ \Omega + 1.71\ \text{k}\Omega = 2.46\ \text{k}\Omega$$

Thévenin's equivalent circuit with the load connected is shown in Figure 13-37.

$$I_L = \frac{V_{OC}}{R_{TH} + R_L} = \frac{3.12\ \text{V}}{3.46\ \text{k}\Omega} = 0.902\ \text{mA}$$

$$V_L = I_L R_L = 0.902\ \text{mA} \times 1\ \text{k}\Omega = 0.902\ \text{V}$$

Let's look at a typical bias arrangement for a bipolar transistor. Such a circuit is drawn in Figure 13-38. Let's find V_C and I_C. If we assume that $I_B = 0\ \mu\text{A}$, then $I_C = I_E$.

$$I_E = \frac{\dfrac{V_{CC}R_2}{R_1 + R_2} - V_{BE}}{R_4} = \frac{\dfrac{25\ \text{V} \times 3.3\ \text{k}\Omega}{47\ \text{k}\Omega + 3.3\ \text{k}\Omega} - 0.6\ \text{V}}{1\ \text{k}\Omega} = 1.04\ \text{mA}$$

$$I_C = I_E = 1.04\ \text{mA}$$

$$V_C = V_{CC} - I_C R_3 = 25\ \text{V} - 7.8\ \text{V} = 17.2\ \text{V}$$

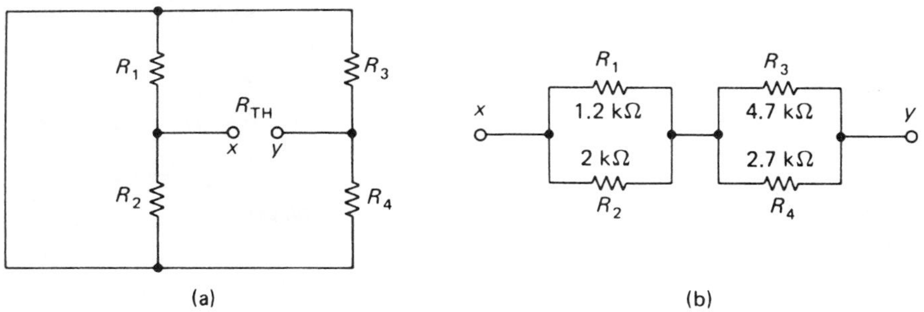

FIGURE 13-36

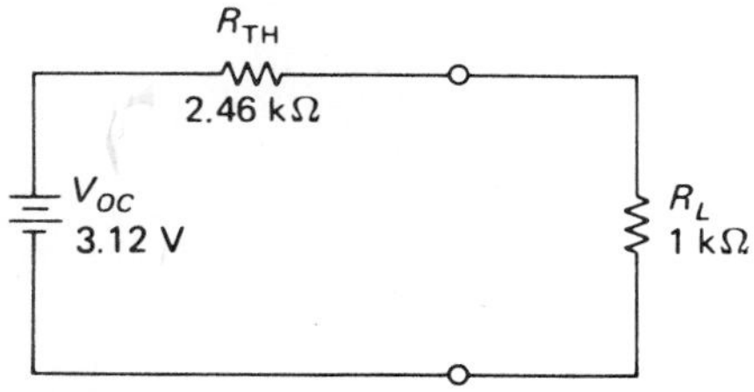

FIGURE 13-37

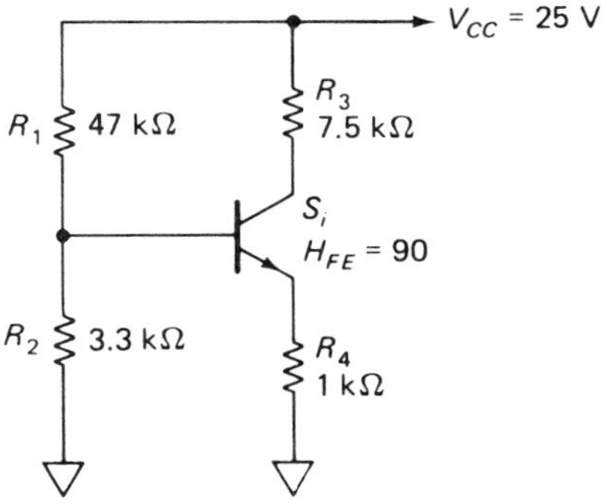

FIGURE 13-38

However, $I_B = 0\ \mu A$ only if H_{FE} is infinitely large. Since H_{FE} is 90, I_B is some finite value and $I_C \neq I_E$. Using Thévenin's theorem, we can find the exact value of I_C for this transistor. Figure 13-39 shows the circuit with the base lead disconnected. That is, the circuit looking into the base is the load.

$$V_{OC} = \frac{V_{CC} R_2}{R_1 + R_2} = \frac{25\text{ V} \times 3.3\text{ k}\Omega}{47\text{ k}\Omega + 3.3\text{ k}\Omega} = 1.64\text{ V}$$

In Figure 13-39(b) the source is shorted and

$$R_{\text{TH}} = R_1 \parallel R_2 = 3.08\text{ k}\Omega$$

Next we will connect the load to the equivalent circuit as in Figure 13-40. We needn't consider the collector since the collector–base junction is reverse biased. Therefore, the load consists of the emitter–base junction in series with some resistance. This resistance, labeled R, is the resistance we see looking across the junction from base to emitter. The base current is the current in this circuit. Therefore,

$$I_B = \frac{V_{OC} - V_{BE}}{R_{\text{TH}} + R_L} = \frac{1.64\text{ V} - 0.6\text{ V}}{3.08\text{ k}\Omega + 91\text{ k}\Omega} = 11.1\ \mu\text{A}$$

$$I_C = H_{FE} I_B = 90 \times 11.1\ \mu\text{A} = 995\ \mu\text{A}$$

$$V_C = V_{CC} - I_C R_3 = 25\text{ V} - 7.46\text{ V} = 17.5\text{ V}$$

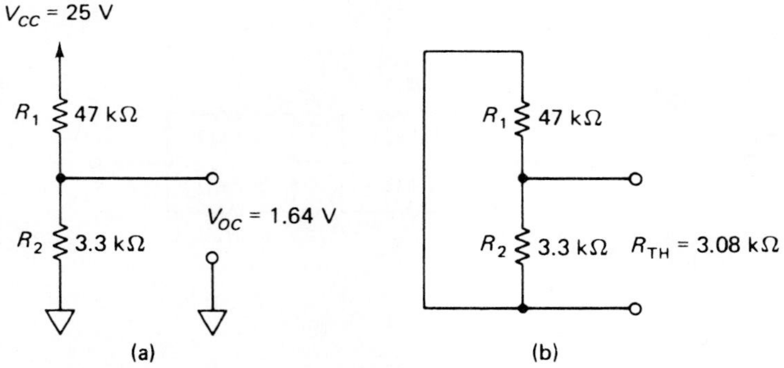

FIGURE 13-39

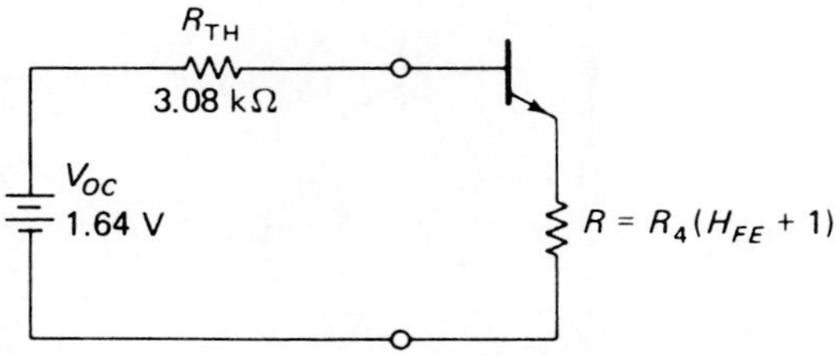

FIGURE 13-40

PRACTICE PROBLEMS 13-3

Develop Thévenin's equivalent circuit for the following problems and find V_L and I_L:

1. Figure 13-41(a). Assume that $R_2 = R_L$.
2. Figure 13-41(b). Assume that $R_3 = R_L$.
3. Figure 13-41(c). Assume that $R_3 = R_L$.
4. Figure 13-42.
5. Figure 13-43. Find I_B, I_C, and V_C.

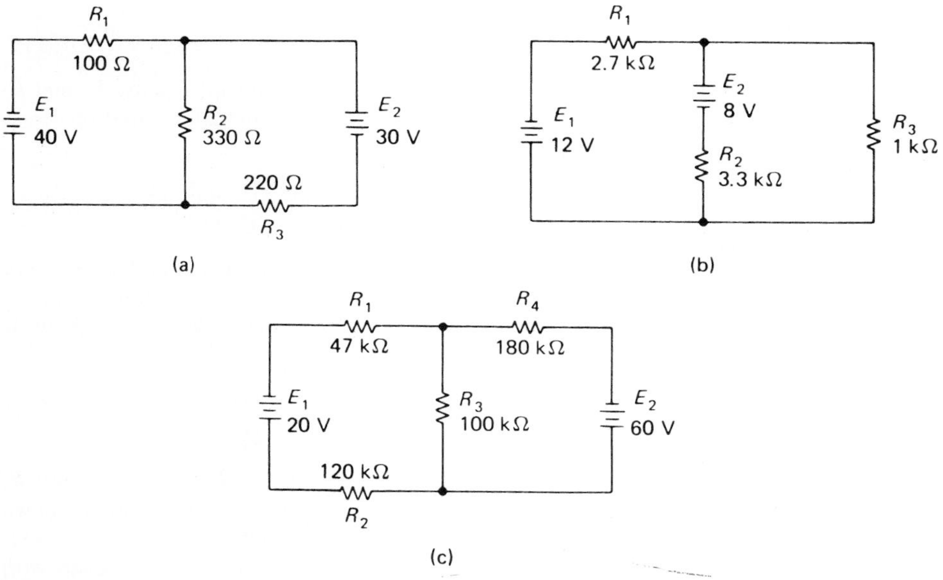

FIGURE 13-41

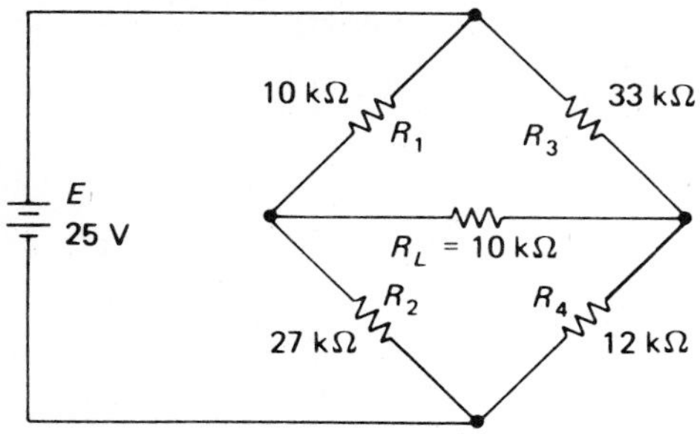

FIGURE 13-42

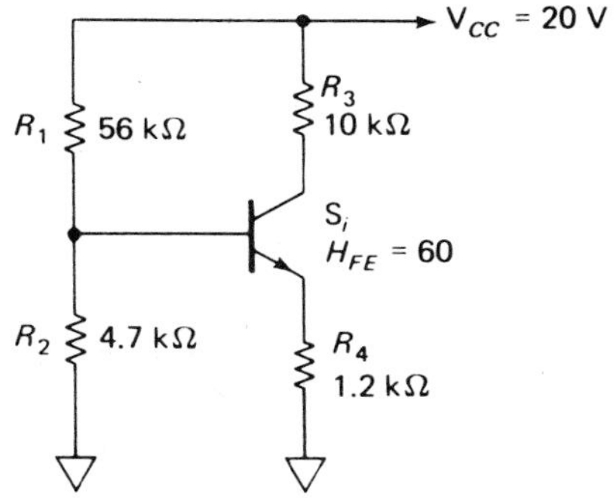

FIGURE 13-43

Solutions

1. With R_2 removed, as in Figure 13-44(a), a series circuit results. E_1 and E_2 are connected series aiding. Therefore, 70 V will drop across R_1 and R_3.

$$V_1 = \frac{70\text{ V} \times R_1}{R_1 + R_3} = \frac{70\text{ V} \times 100\ \Omega}{320\ \Omega} = 21.9\text{ V}$$

Applying Kirchhoff's voltage law to find V_{OC}, we see

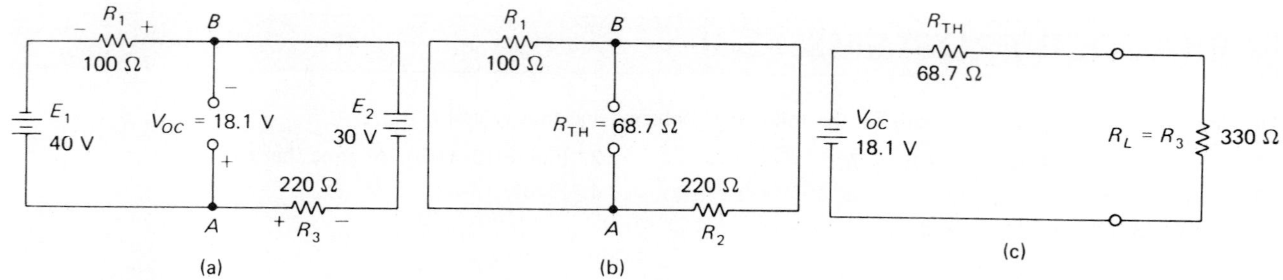

FIGURE 13-44

that the loop equation is $V_{OC} + V_1 + E = 0$. Moving counterclockwise from point A, we get

$$V_{OC} + 21.9\text{ V} - 40\text{ V} = 0$$

$$V_{OC} = 18.1\text{ V}$$

We assumed that the potential at point A was positive with respect to point B. The fact that V_{OC} is positive indicates that we chose the correct polarity. In Figure 13-44(b) the sources have been shorted. Looking into the circuit from the open-circuit terminals, we see that R_1 and R_3 are in parallel and the equivalent resistance (R_{TH}) is 68.7 Ω. The Thévenin's equivalent circuit is shown in series with the load in Figure 13-44(c).

$$I_L = \frac{V_{OC}}{R_{TH} + R_L} = \frac{18.1\text{ V}}{399\ \Omega} = 45.4\text{ mA}$$

$$V_L = I_L R_L = 45.4\text{ mA} \times 330\ \Omega = 15\text{ V}$$

2. With R_3 removed, a series circuit results. E_1 and E_2 are connected series aiding, and 20 V will drop across R_1 and R_2.

$$V_2 = \frac{E_T R_2}{R_1 + R_2} = \frac{20\text{ V} \times 3.3\text{ k}\Omega}{6\text{ k}\Omega} = 11\text{ V}$$

Applying Kirchhoff's voltage law to find V_{OC} in Figure 13-45(a), we see that the loop equation is $V_{OC} + E_2 + V_2 = 0$. Moving counterclockwise from point A, we get:

$$8\text{ V} - 11\text{ V} + V_{OC} = 0$$

$$V_{OC} = 3\text{ V}$$

Shorting E_1 and E_2 results in R_1 in parallel with R_2 looking back from the open-circuit terminals as shown in Figure 13-45(b). This results in an R_{TH} of 1.49 kΩ. Thévenin's equivalent circuit is shown in series with the load in Figure 13-45(c).

$$I_L = \frac{V_{OC}}{R_{TH} + R_L} = \frac{3\text{ V}}{2.49\text{ k}\Omega} = 1.2\text{ mA}$$

$$V_L = I_L R_L = 1.2\text{ mA} \times 1\text{ k}\Omega = 1.2\text{ V}$$

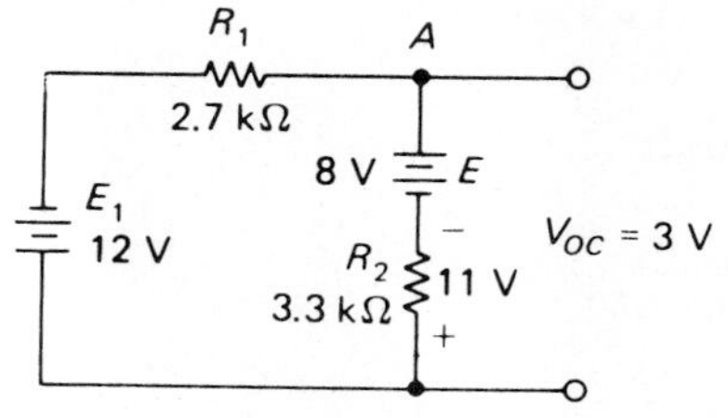

(a)

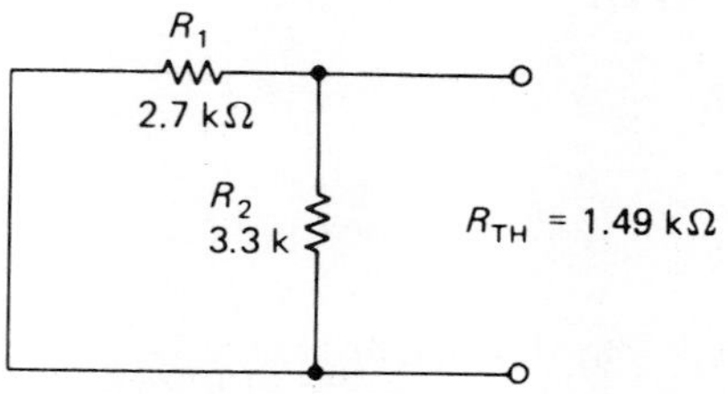

(b)

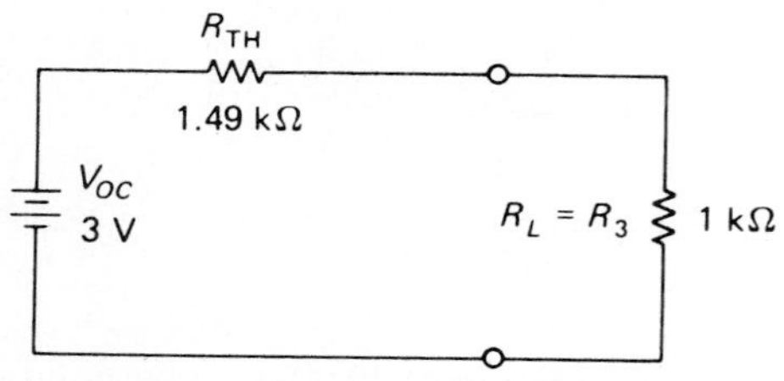

(c)

FIGURE 13-45

3. E_1 and E_2 are connected series opposing in Figure 13-41(c). Therefore, with the load (R_3) removed as in Figure 13-46(a), 40 V will drop across R_1, R_2, and R_4 in series. Let's work with the loop containing V_{OC}, R_4, and E_2.

$$V_4 = \frac{40\text{ V} \times R_4}{R_T} = \frac{40\text{ V} \times 180\text{ k}\Omega}{347\text{ k}\Omega} = 20.7\text{ V}$$

$$V_{OC} = E_2 - V_4 = 60\text{ V} - 20.7\text{ V} = 39.3\text{ V}$$

The source is replaced with a short circuit, and the resulting circuit is shown in Figure 13.46(b).

$$R_{TH} = R_4 \parallel (R_1 + R_2) = 86.6\text{ k}\Omega$$

Connecting the load resistor to Thévenin's equivalent circuit in Figure 13-46(c), we get:

$$I_L = \frac{V_{OC}}{R_{TH} + R_L} = \frac{39.3\text{ V}}{187\text{ k}\Omega} = .210\ \mu\text{A}$$

$$V_L = I_L R_L = 210\ \mu\text{A} \times 100\text{ k}\Omega = 21\text{ V}$$

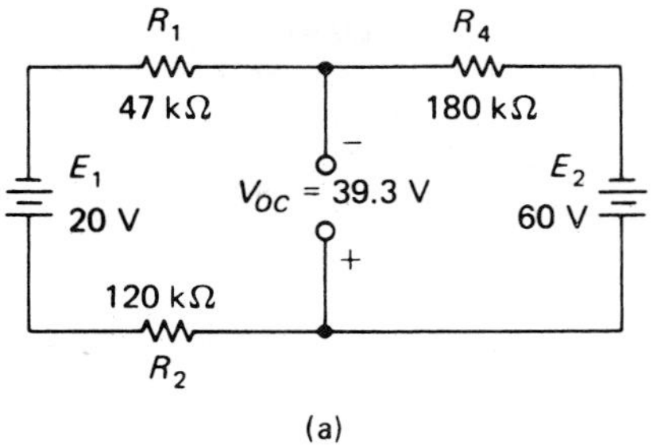

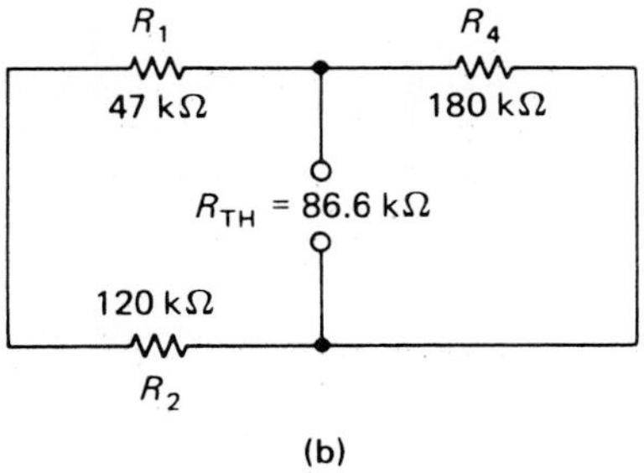

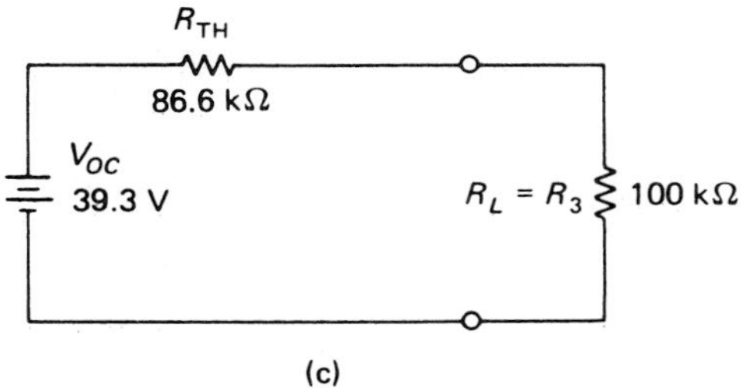

FIGURE 13-46

4. In Figure 13-47(a) the load has been removed and V_2 and V_4 are calculated.

$$V_2 = \frac{ER_2}{R_1 + R_2} = \frac{25\text{ V} \times 27\text{ k}\Omega}{37\text{ k}\Omega} = 18.2\text{ V}$$

$$V_4 = \frac{ER_4}{R_3 + R_4} = \frac{25\text{ V} \times 12\text{ k}\Omega}{45\text{ k}\Omega} = 6.67\text{ V}$$

Using the circuit consisting of V_2, V_4, and V_{OC} in Figure 13-47(b), we see that $V_{OC} = 11.5$ V and the polarity is as shown. The circuit, with the source shorted, is redrawn in Figure 13-47(c) and R_{TH} is calculated.

$$R_{TH} = R_1 \parallel R_2 + R_3 \parallel R_4$$
$$= 7.3\text{ k}\Omega + 8.8\text{ k}\Omega = 16.1\text{ k}\Omega$$

Thévenin's equivalent circuit with the load connected is shown in Figure 13-47(d).

$$I_L = \frac{V_{OC}}{R_{TH} + R_L} = \frac{11.5\text{ V}}{26.1\text{ k}\Omega} = 441\ \mu\text{A}$$

$$V_L = I_L R_L = 441\ \mu\text{A} \times 10\text{ k}\Omega = 4.41\text{ V}$$

5. In Figure 13-48(a), with the load removed, $V_{OC} = V_2$.

$$V_2 = \frac{V_{CC} R_2}{R_1 + R_2} = \frac{20\text{ V} \times 4.7\text{ k}\Omega}{56\text{ k}\Omega + 4.7\text{ k}\Omega} = 1.55\text{ V}$$

With the source shorted as in Figure 13-48(b), $R_{TH} = 4.34$ kΩ.

$$R_{TH} = R_1 \parallel R_2 = 4.34\text{ k}\Omega$$

Thévenin's equivalent circuit with the load connected is shown in Figure 13-48(c).

$$I_L = I_B = \frac{V_{OC} - V_{BE}}{R_{TH} + R_L}$$
$$= \frac{1.55\text{ V} - 0.6\text{ V}}{77.5\text{ k}\Omega} = 12.3\ \mu\text{A}$$

$$I_C = H_{FE} I_B = 60 \times 12.3\ \mu\text{A} = 735\ \mu\text{A}$$

$$V_C = V_{CC} - V_3 = 20\text{ V} - (735\ \mu\text{A} \times 10\text{ k}\Omega)$$
$$= 20\text{ V} - 7.35\text{ V} = 12.6\text{ V}$$

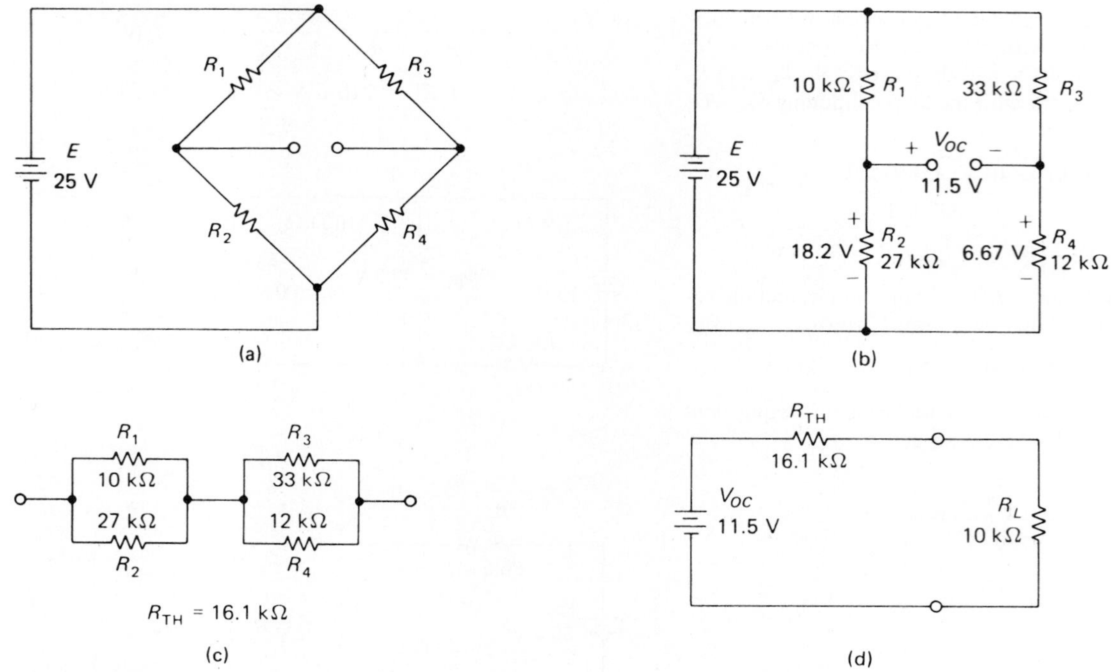

FIGURE 13-47

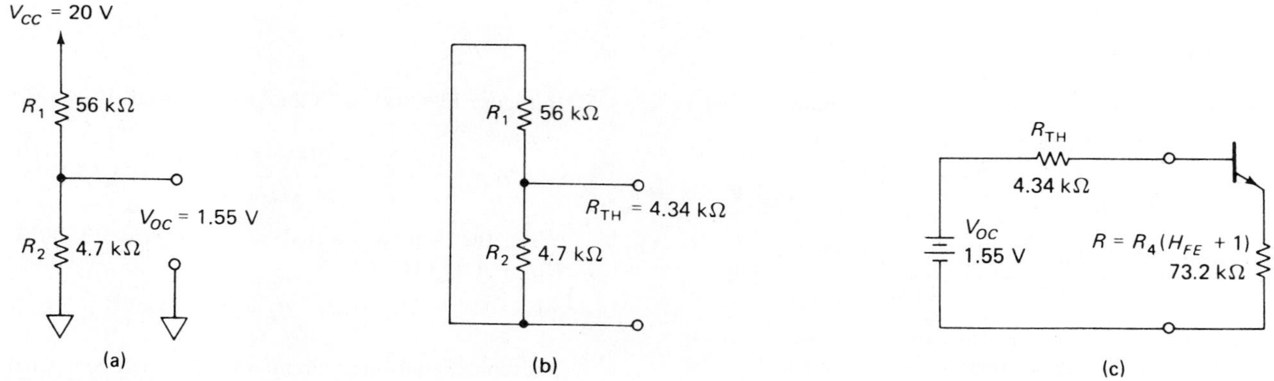

FIGURE 13-48

Additional practice problems are at the end of the chapter.

SELF-TEST 13-2

1. In Figure 13-49, assume that R_4 is the load. Thévenize the circuit and find V_L and I_L.
2. In Figure 13-50, assume that R_3 is the load. Thévenize the circuit and find V_L and I_L.

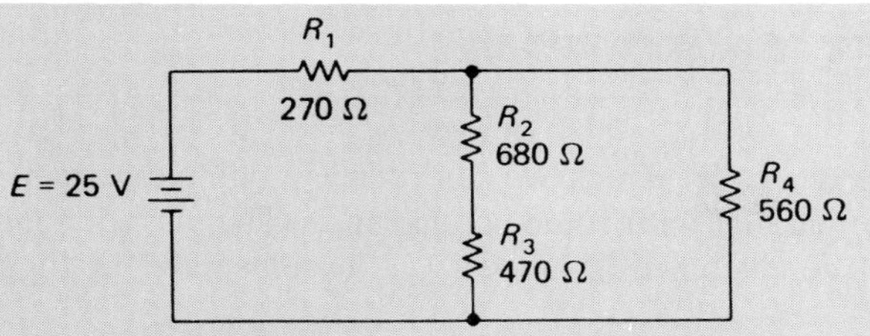

FIGURE 13-49

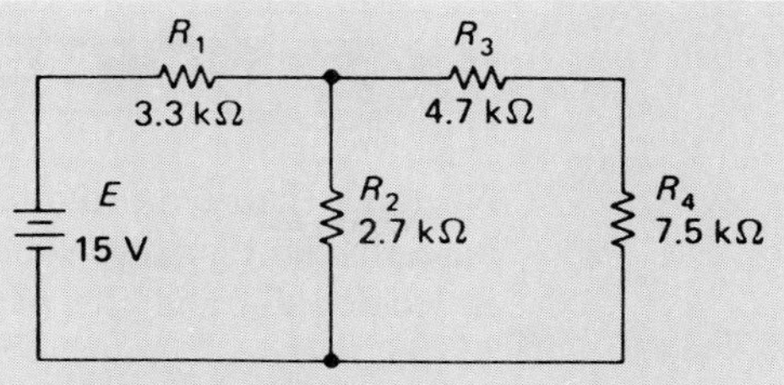

FIGURE 13-50

3. In Figure 13-51, assume that R_3 is the load. Thévenize the circuit and find V_L and I_L.

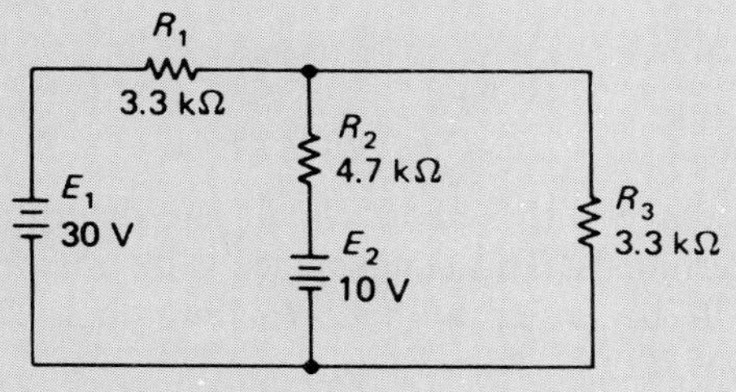

FIGURE 13-51

4. In Figure 13-52, assume that R_3 is the load Thévenize the circuit and find V_L and I_L.

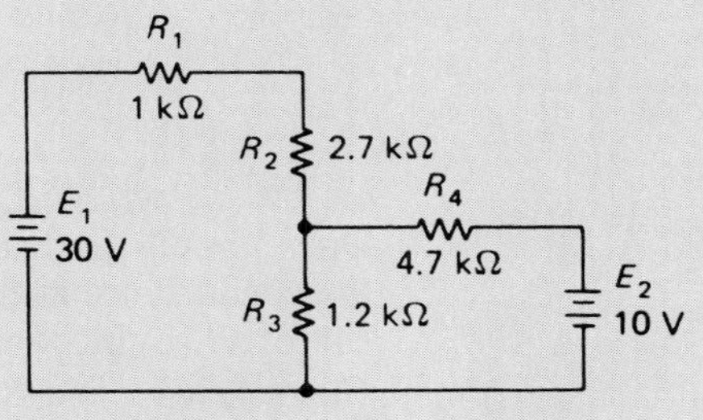

FIGURE 13-52

5. In Figure 13-53, Thévenize the circuit and find V_L and I_L.

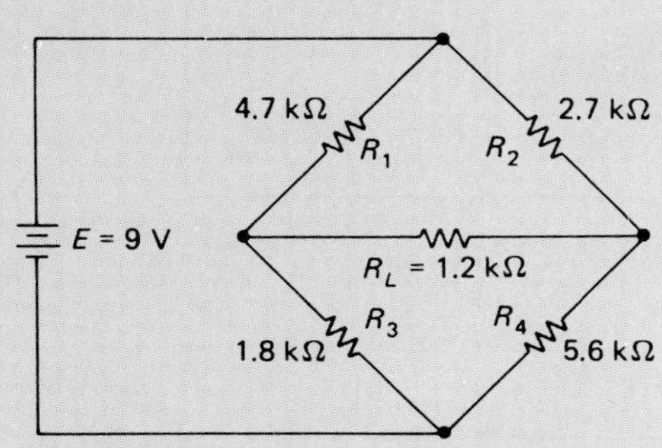

FIGURE 13-53

6. In Figure 13-54, Thévenize the circuit at the base and find I_B, I_C, and V_C.

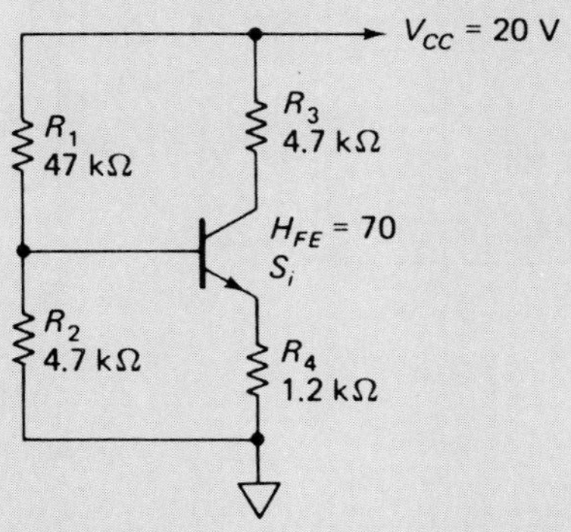

FIGURE 13-54

Answers to Self-test 13-2 are at the end of the chapter.

13-3 NORTON'S THEOREM

Norton's theorem provides the means by which we can reduce a complex circuit to an equivalent current source in parallel with some conductance. The current source is labeled I_{SC} and the parallel conductance is labeled G_N, as illustrated in Figure 13-55.

Consider the circuit in Figure 13-56. This circuit was shown in Figure 13-16 and was Thévenized in Figure 13-17. Assuming that R_3 is the load, Norton's equivalent circuit is found by using the following procedure:

1. Replace the load with a short circuit and calculate the current through the short circuit. This is the current source in our equivalent circuit. In Figure 13-57 the load (R_3) has been replaced with a short circuit. This reduces the circuit resistance to R_1 and the current is 10 mA.

$$I = \frac{V_1}{R_1} = \frac{10\text{ V}}{1\text{ k}\Omega} = 10\text{ mA}$$

2. Open circuit the load. Replace the source with a short circuit and determine the conductance at the open-circuit terminals. This conductance is the reciprocal of R_{TH}. In some circuits it will be easier to find R_{TH} and then G_N. $G_N = 1/R_{TH}$. The conductance, G_N, is 2 mS.

3. Connect the load across the output terminals of the Norton's equivalent circuit. Calculate V_L and I_L. Norton's equivalent circuit of Figure 13-56 with the load connected is shown in Figure 13-58.

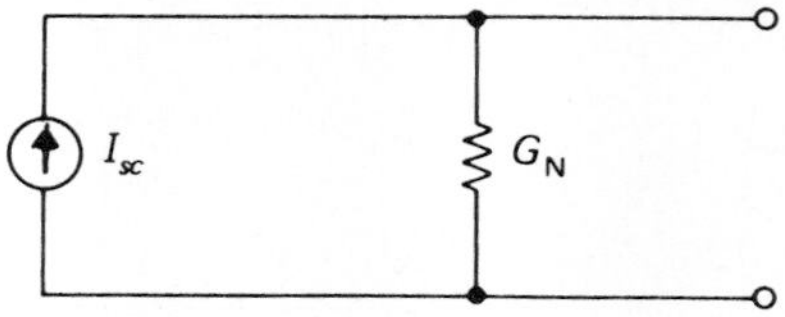

FIGURE 13-55

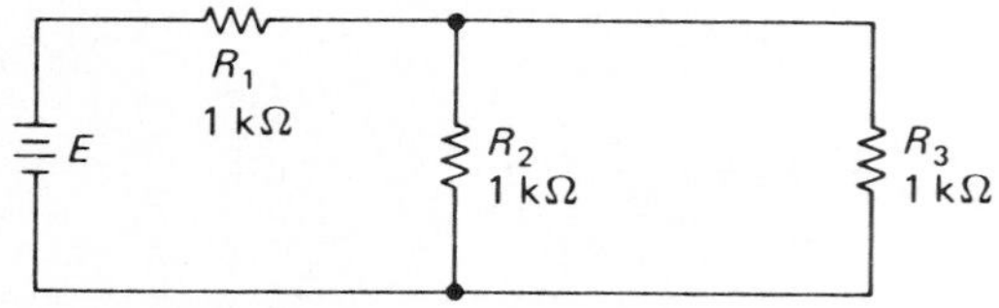

FIGURE 13-56

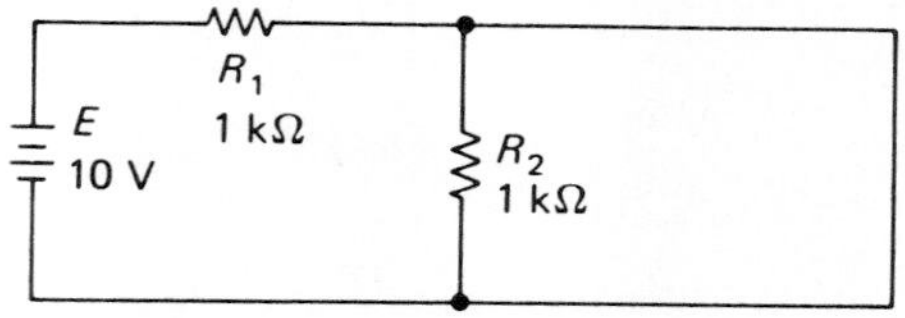

FIGURE 13-57

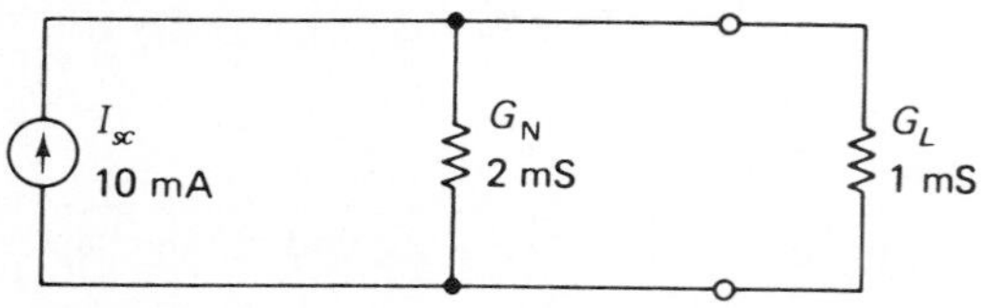

FIGURE 13-58

Using Ohm's law, we get

$$V_L = \frac{I_{SC}}{G_N + G_L}$$

$$= \frac{10 \text{ mA}}{3 \text{ mS}} = 3.33 \text{ V}$$

$$I_L = \frac{I_{SC} G_L}{G_N + G_L}$$

$$= \frac{10 \text{ mA} \times 1 \text{ mS}}{3 \text{ mS}} = 3.33 \text{ mA}$$

or

$$= V_L G_L = 3.33 \text{ V} \times 1 \text{ mS} = 3.33 \text{ mA}$$

Norton's equivalent circuit causes the same current through R_3 and the same voltage drop across it as does the original circuit. Furthermore, the direction of current through the short circuit in Figure 13-57 is the same as in the original circuit.

Knowing V_3 and I_3 (remember we called R_3 the load), we can find the rest of the voltage drops and currents by using Ohm's law and Kirchhoff's laws.

Let's Nortonize some rather simple series–parallel circuits. We will find only V_L and I_L.

PRACTICE PROBLEMS 13-4

For the problems in Figure 13-59:

1. Develop Norton's equivalent circuit.
2. Find V_L and I_L.

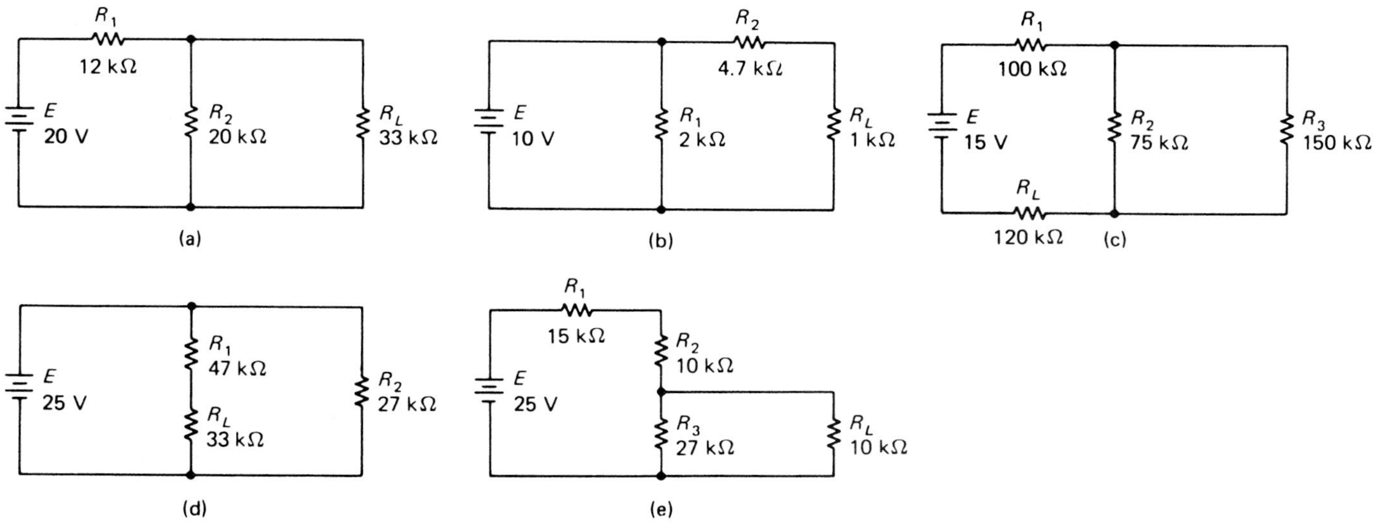

FIGURE 13-59

Solutions

(a) In Figure 13-59(a), if we short circuit the load, we put a short circuit across R_2 and reduce the circuit resistance to R_1. All the circuit current flows through the short circuit.

$$I_{SC} = \frac{V_1}{R_1} = \frac{20\text{ V}}{12\text{ k}\Omega} = 1.67\text{ mA}$$

$$G_N = G_1 + G_2 = 83.3\ \mu\text{S} + 50\ \mu\text{S} = 133\ \mu\text{S}$$

Norton's equivalent circuit with the load connected is shown in Figure 13-60. Applying Ohm's law, we get

$$V_L = \frac{I_{SC}}{G_N + G_L} = \frac{1.67\text{ mA}}{163\ \mu\text{S}} = 10.2\text{ V}$$

$$I_L = V_L G_L = 10.2\text{ V} \times 30.3\ \mu\text{A} = 309\ \mu\text{A}$$

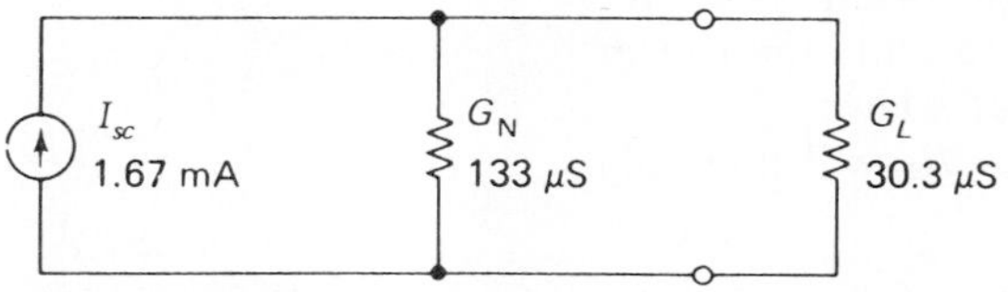

FIGURE 13-60

(b) In Figure 13-59(b), if we short circuit the load, the short circuit current is the current through R_2.

$$I_{SC} = I_2 = \frac{V_2}{R_2} = \frac{10\text{ V}}{4.7\text{ k}\Omega} = 2.13\text{ mA}$$

If we open circuit the load and short circuit the source,

$$R_{TH} = R_2 = 4.7\text{ k}\Omega$$

$$G_N = \frac{1}{R_{TH}} = 213\ \mu\text{S}$$

Norton's equivalent circuit with the load connected is shown in Figure 13-61. Applying Ohm's law, we get

$$V_L = \frac{I_{SC}}{G_N + G_L} = \frac{2.13\text{ mA}}{1.21\text{ mS}} = 1.76\text{ V}$$

$$I_L = V_L G_L = 1.76\text{ mA}$$

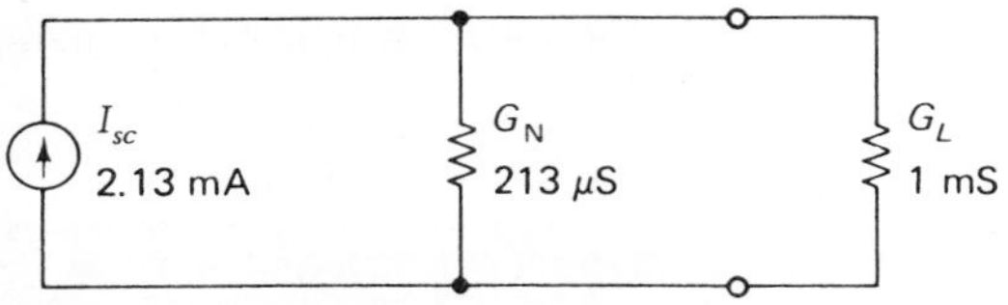

FIGURE 13-61

(c) In Figure 13.59(c), if we short circuit the load, $R_T = R_1 + R_2 \| R_3$ or

$$R_T = 100\text{ k}\Omega + 50\text{ k}\Omega = 150\text{ k}\Omega$$

$$I_{SC} = I_T = \frac{E}{R_T} = \frac{15\text{ V}}{150\text{ k}\Omega} = 100\ \mu\text{A}$$

$$R_{TH} = R_1 + R_2 \| R_3 = 150\text{ k}\Omega$$

$$G_N = \frac{1}{R_{TH}} = \frac{1}{150\text{ k}\Omega} = 6.67\ \mu\text{S}$$

Norton's equivalent circuit with the load connected is shown in Figure 13-62.

$$V_L = \frac{I_{SC}}{G_N + G_L} = \frac{100\ \mu\text{A}}{15\ \mu\text{S}} = 6.67\text{ V}$$

$$I_L = V_L G_L = 6.67\text{ V} \times 8.33\ \mu\text{S} = 55.5\ \mu\text{A}$$

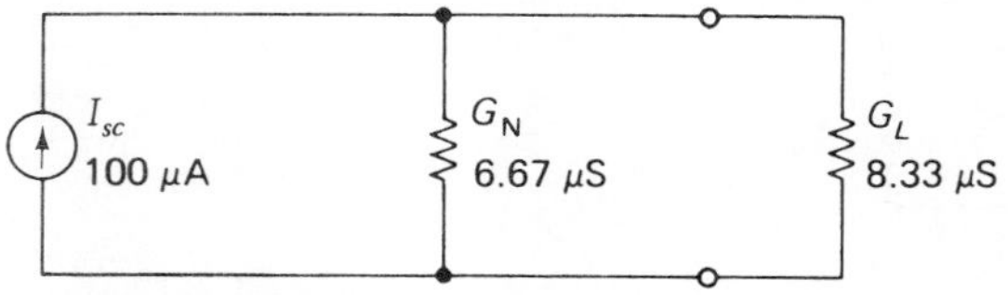

FIGURE 13-62

(d) If we short circuit the load in Figure 13-59(d), the current through R_1 is the short-circuit current.

$$I_{SC} = I_1 = \frac{V_1}{R_1} = \frac{25\text{ V}}{47\text{ k}\Omega} = 532\ \mu\text{A}$$

$$G_N = \frac{1}{R_1} = 21.3\ \mu\text{S}$$

Norton's equivalent circuit with the load connected is shown in Figure 13-63.

$$V_L = \frac{I_{SC}}{G_N + G_L} = \frac{532\ \mu\text{A}}{51.6\ \mu\text{S}} = 10.3\text{ V}$$

$$I_L = V_L G_L = 10.3\text{ V} \times 30.3\ \mu\text{S} = 312\ \mu\text{A}$$

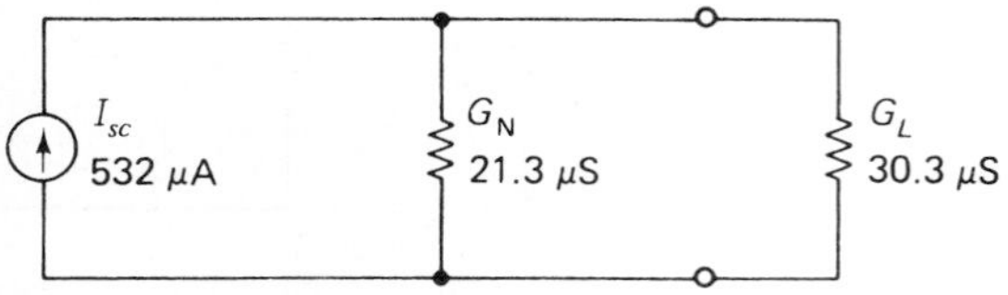

FIGURE 13-63

(e) If we short circuit the load in Figure 13-59(e), R_3 is also shorted. Then $R_T = R_1 + R_2$. I_T in the resulting circuit is I_{SC}.

$$I_{SC} = \frac{E}{R_T} = \frac{25\ \text{V}}{25\ \text{k}\Omega} = 1\ \text{mA}$$

$$G_N = \frac{1}{R_3} + \frac{1}{R_1 + R_2} = 37\ \mu\text{S} + 40\ \mu\text{S}$$

$$= 77\ \mu\text{S}$$

Norton's equivalent circuit with the load connected is shown in Figure 13-64.

$$V_L = \frac{I_{SC}}{G_N + G_L} = \frac{1\ \text{mA}}{177\ \mu\text{S}} = 5.65\ \text{V}$$

$$I_L = V_L G_L = 5.56\ \text{V} \times 100\ \mu\text{S} = 565\ \mu\text{A}$$

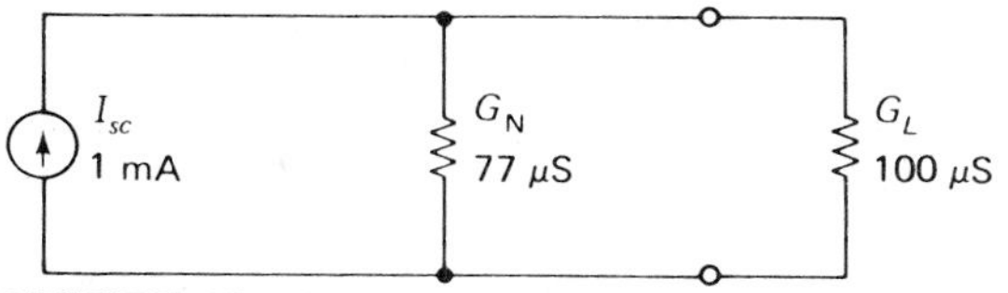

FIGURE 13-64

Additional problems are at the end of the chapter.

SELF-TEST 13-3

1. Refer to Figure 13-59(a). Let $E = 15$, $R_1 = 68\ \text{k}\Omega$, $R_2 = 100\ \text{k}\Omega$, and $R_L = 75\ \text{k}\Omega$. Nortonize the circuit and find V_L and I_L.

2. Refer to Figure 13-59(d). Let $E = 20\ \text{V}$, $R_1 = 680\ \Omega$, $R_2 = 1.2\ \text{k}\Omega$, and $R_L = 560\ \Omega$. Nortonize the circuit and find V_L and I_L.

Answers to Self-test 13-3 are at the end of the chapter.

CHAPTER 13 AT A GLANCE

<table>
<tr><th>Page</th><th>Theorem</th><th>Example</th></tr>
<tr><td>330</td><td>Superposition Theorem: To find currents and voltage drops using the superposition theorem: (1) short circuit one supply and calculate the current through the load as a result of the remaining source; (2) place the short across the other source and again calculate the current through the load; (3) the algebraic sum of load currents in (2) is the load current in the original circuit. Using this current, calculate other currents and voltage drops.</td><td>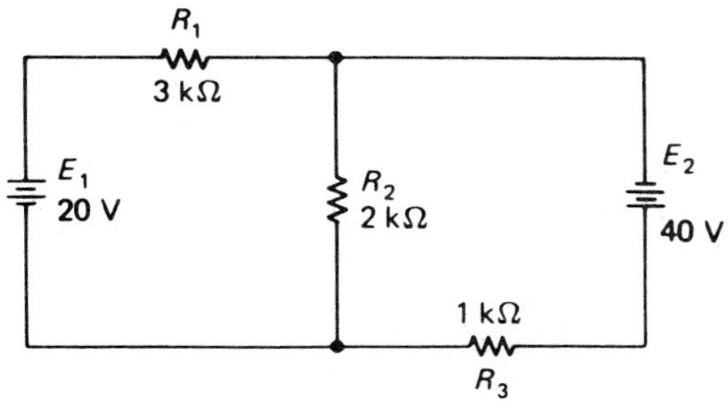

FIGURE 13-65</td></tr>
</table>

Page	Theorem	Example

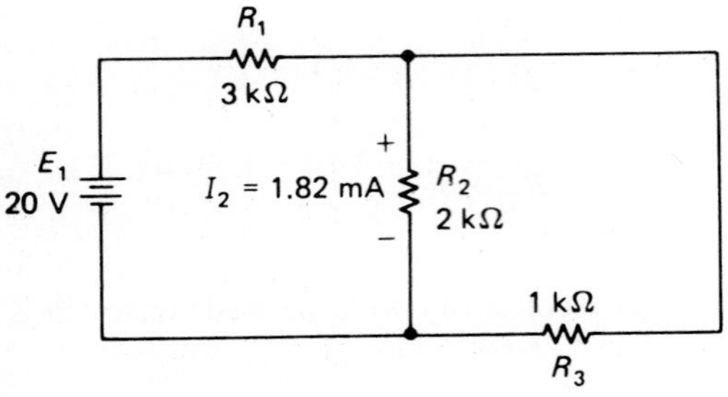

FIGURE 13-66

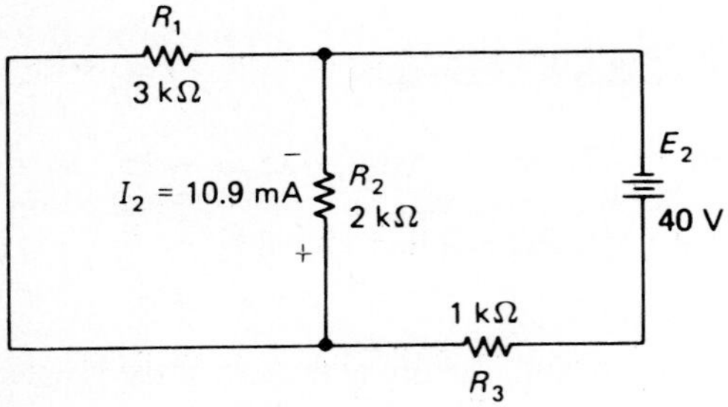

FIGURE 13-67

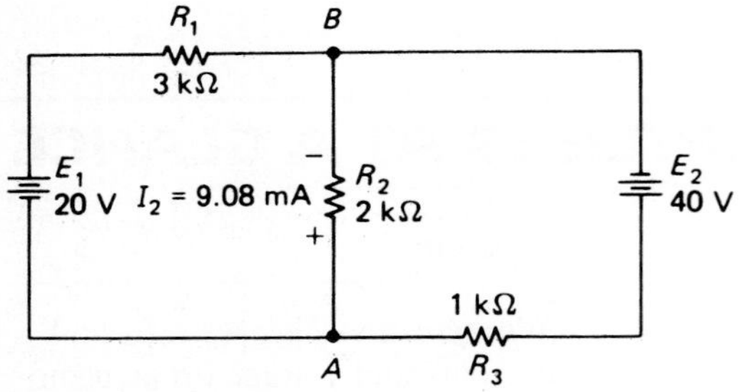

FIGURE 13-68

337 *Thévenin's Theorem:* To find currents and voltage drops using Thévenin's theorem: (1) remove the load and calculate the difference in potential between the open-circuit terminals; (2) with the load removed as in step 1, replace the source with a short cir-

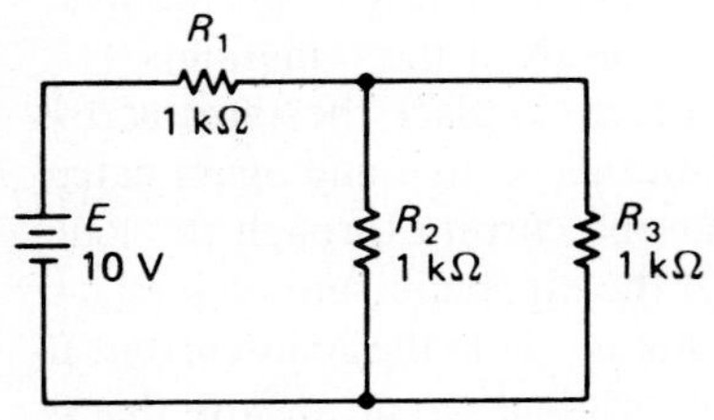

FIGURE 13-69

Page	Theorem	Example
	cuit and determine the resistance at the open-circuit terminals; (3) connect the load across the output terminals of the Thévenin's equivalent circuit. Calculate V_L and I_L	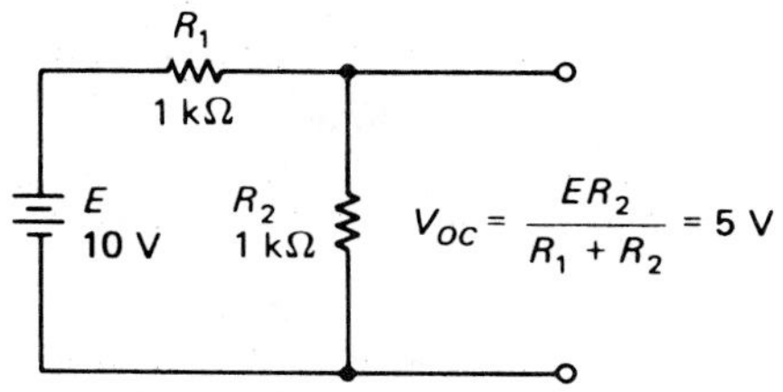 **FIGURE 13-70** 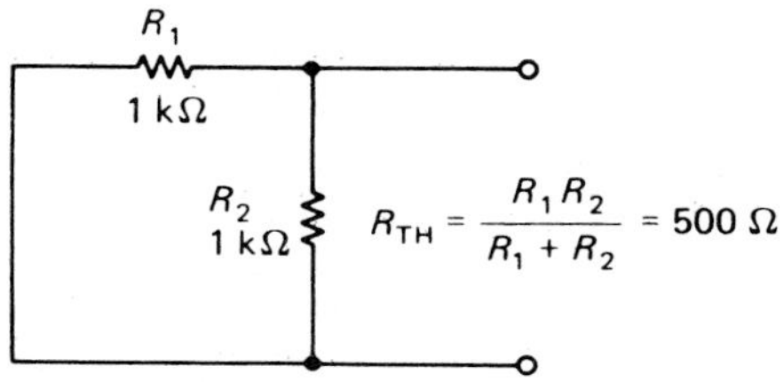 **FIGURE 13-71** 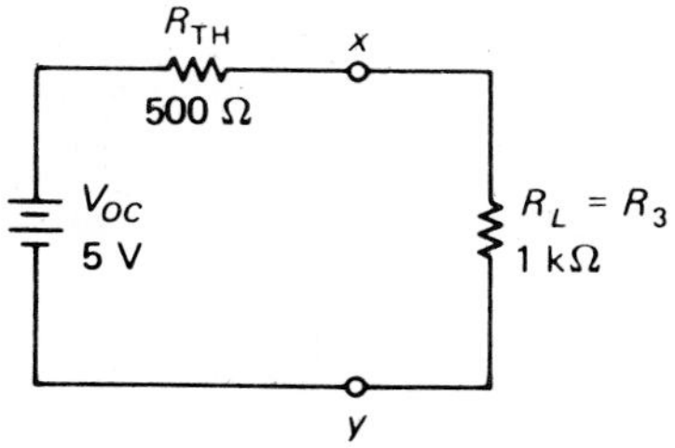 **FIGURE 13-72**
352	*Norton's Theorem:* To find currents and voltage drops using Norton's theorem: (1) replace the load with a short circuit and calculate the current through the short circuit; (2) open circuit the load and replace the source with a short circuit and determine the conductance at the open-circuit terminals; (3) connect the load across the output terminals of the equivalent circuit and calculate V_L and I_L.	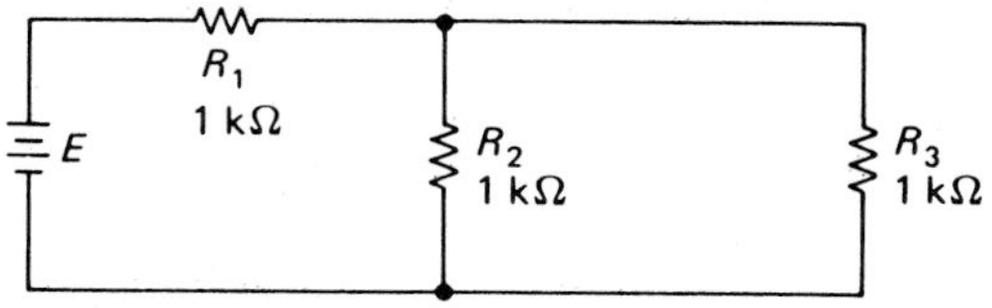 **FIGURE 13-73** 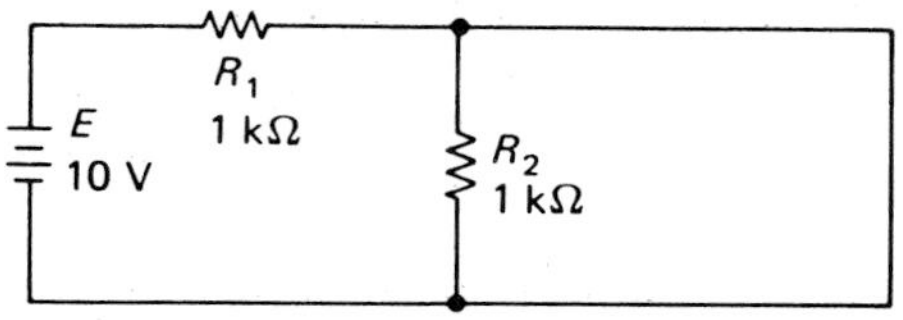 **FIGURE 13-74**

Page	Theorem	Example

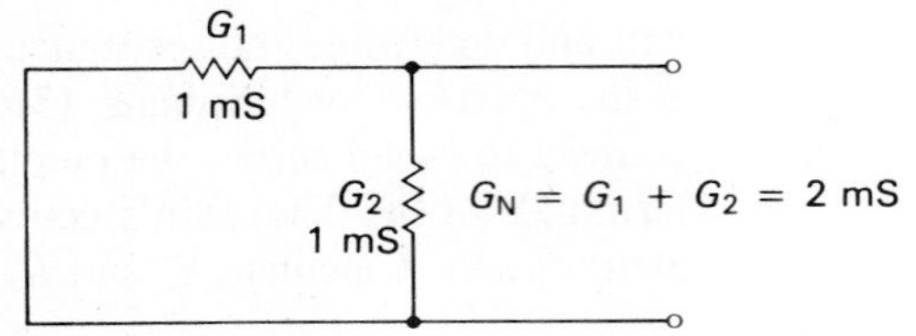

FIGURE 13-75

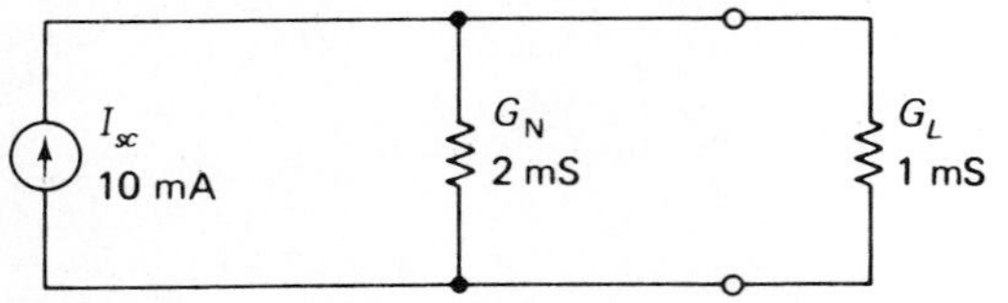

FIGURE 13-76

End of Chapter Problems 13-1

Use the superposition theorem to find the various currents and voltage drops in the following problems:

1. Refer to Figure 13-77(a). Let $E_1 = 25$ V, $E_2 = 50$ V, $R_1 = 2.2$ kΩ, $R_2 = 1.8$ kΩ, and $R_3 = 1.2$ kΩ.
2. Refer to Figure 13-77(b). Let $E_1 = 10$ V, $E_2 = 25$ V, $R_1 = 5.6$ kΩ, $R_2 = 1$ kΩ, and $R_3 = 4.7$ kΩ.
3. Refer to Figure 13-77(c). Let $E_1 = 100$ V, $E_2 = 40$ V, $R_1 = 2.7$ kΩ, $R_2 = 3.3$ kΩ, $R_3 = 1$ kΩ, and $R_4 = 4.7$ kΩ.
4. Refer to Figure 13-77(b). Let $E_1 = 30$ V, $E_2 = 15$ V, $R_1 = 15$ kΩ, $R_2 = 12$ kΩ, and $R_3 = 10$ kΩ.
5. Refer to Figure 13-78(a). Let $E_1 = 9$ V, $E_2 = 18$ V, $R_1 = 3.3$ kΩ, $R_2 = 4.7$ kΩ, and $R_3 = 3.3$ kΩ.
6. Refer to Figure 13-78(b). Let $E_1 = 40$ V, $E_2 = 20$ V, $R_1 = 4.7$ kΩ, $R_2 = 7.5$ kΩ, and $R_3 = 10$ kΩ.
7. Refer to Figure 13-78(c). Let $E_1 = 25$ V, $E_2 = 10$ V, $R_1 = 680$ Ω, $R_2 = 470$ Ω, and $R_3 = 1$ kΩ.
8. Refer to Figure 13-78(c). Let $E_1 = 12$ V, $E_2 = 30$ V, $R_1 = 18$ kΩ, $R_2 = 27$ kΩ, and $R_3 = 47$ kΩ.

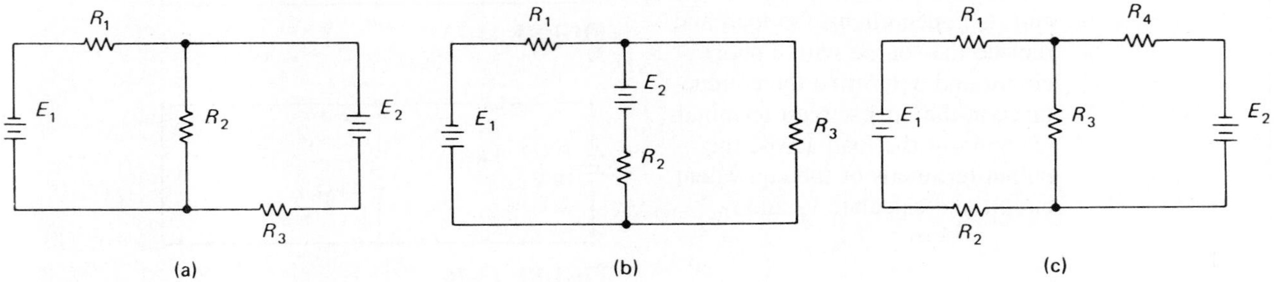

FIGURE 13-77

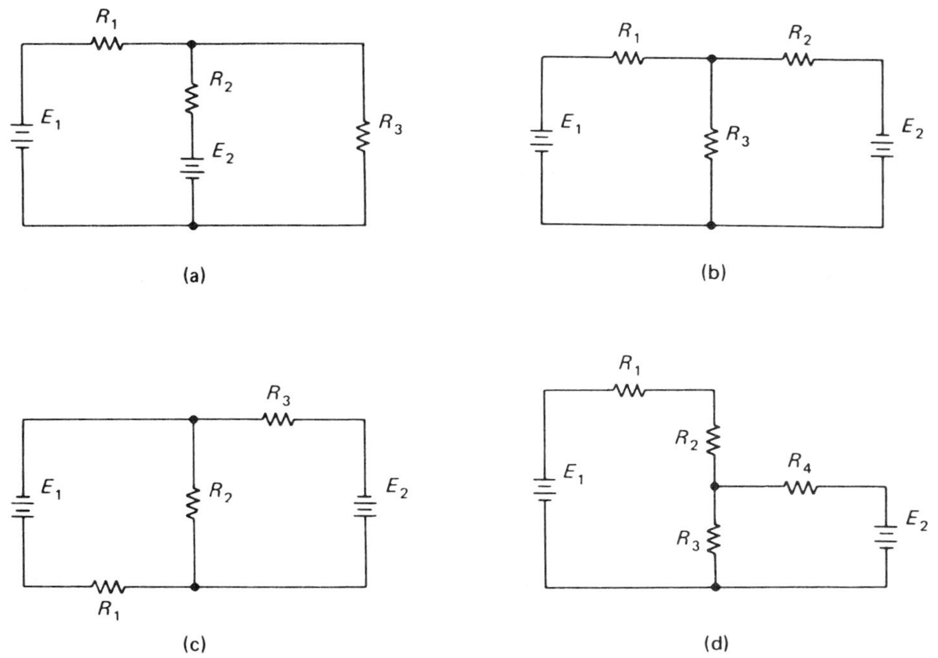

FIGURE 13-78

9. Refer to Figure 13-78(d). Let E_1 = 40 V, E_2 = 15 V, R_1 = 2.2 kΩ, R_2 = 1.2 kΩ, R_3 = 3.3 kΩ, and R_4 = 3.3 kΩ.

10. Refer to Figure 13-78(d). Let E_1 = 50 V, E_2 = 20 V, R_1 = 10 kΩ, R_2 = 20 kΩ, R_3 = 30 kΩ, and R_4 = 30 kΩ.

End of Chapter Problems 13-2

In problems 1 through 10, Thévenize each circuit and find V_L and I_L.

1. Refer to Figure 13-59(a), where E = 30 V, R_1 = 22 kΩ, R_2 = 47 kΩ, and R_L = 10 kΩ.

2. Refer to Figure 13-59(b), where E = 20 V, R_1 = 20 kΩ, R_2 = 56 kΩ, and R_L = 33 kΩ.

3. Refer to Figure 13-59(c), where E = 12 V, R_1 = 680 Ω, R_2 = 470 Ω, R_3 = 1 kΩ, and R_L = 750 Ω.

4. Refer to Figure 13-59(d), where E = 40 V, R_1 = 5.6 kΩ, R_2 = 3.9 kΩ, and R_L = 2.7 kΩ.

5. Refer to Figure 13-59(e), where E = 50 V, R_1 = 22 kΩ, R_2 = 10 kΩ, R_3 = 33 kΩ, and R_L = 15 kΩ.

6. Refer to Figure 13-77(a), where E_1 = 25 V, E_2 = 50 V, R_1 = 2.2 kΩ, R_2 = 1.8 kΩ, and R_3 = 1.2 kΩ. Assume that R_2 is the load.

7. Refer to Figure 13-77(b), where E_1 = 10 V, E_2 = 25 V, R_1 = 5.6 kΩ, R_2 = 1 kΩ, and R_3 = 4.7 kΩ. Assume that R_2 is the load.

8. Refer to Figure 13-77(c), where E_1 = 100 V, E_2 = 40 V, R_1 = 2.7 kΩ, R_2 = 3.3 kΩ, R_3 = 1 kΩ, and R_4 = 4.7 kΩ. Assume that R_3 is the load.

9. Refer to Figure 13-79, where E = 20 V, R_1 = 4.7 kΩ, R_2 = 6.8 kΩ, R_3 = 1.5 kΩ, R_4 = 3 kΩ, and R_L = 1 kΩ.

10. Refer to Figure 13-79, where E = 25 V, R_1 = 680 Ω, R_2 = 1 kΩ, R_3 = 2 kΩ, R_4 = 1.2 kΩ, and R_L = 1.2 kΩ.

11. Refer to Figure 13-80, where V_{CC} = 20 V, R_1 = 47 kΩ, R_2 = 6.8 kΩ, R_3 = 2 kΩ, R_4 = 470 Ω, and H_{FE} = 80. Thévenize and find I_B, I_C, and V_C.

12. Refer to Figure 13-80, where V_{CC} = 25 V, R_1 = 56 kΩ, R_2 = 10 kΩ, R_3 = 8.2 kΩ, R_4 = 2 kΩ, and H_{FE} = 60. Thévenize and find I_B, I_C, and V_C.

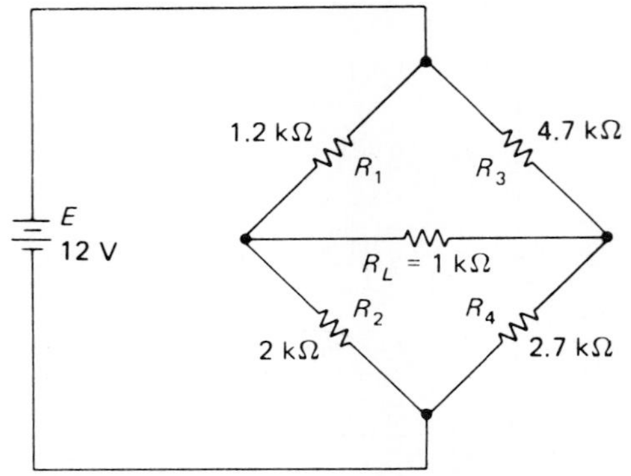

FIGURE 13-79

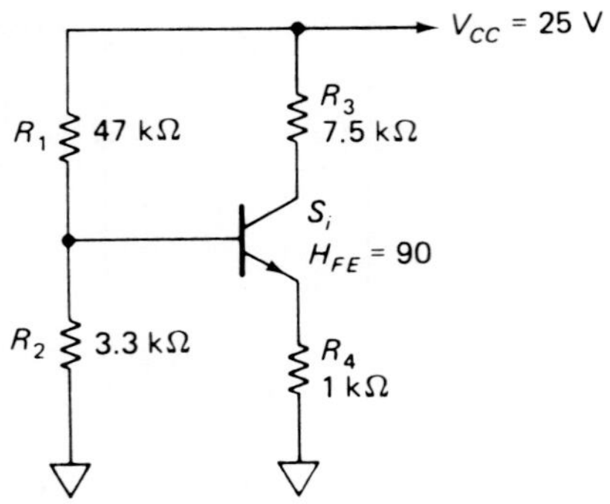

FIGURE 13-80

End of Chapter Problems 13-3

1. Refer to Figure 13-59(a). Nortonize the circuit and find V_L and I_L where (a) $E = 30$ V, $R_1 = 22$ kΩ, $R_2 = 47$ kΩ, and $R_L = 10$ kΩ, and (b) $E = 50$ V, $R_1 = 18$ kΩ, $R_2 = 33$ kΩ, and $R_L = 39$ kΩ.
2. Refer to Figure 13-59(b). Nortonize the circuit and find V_L and I_L where (a) $E = 20$ V, $R_1 = 20$ kΩ, $R_2 = 56$ kΩ, and $R_L = 33$ kΩ, and (b) $E = 25$ V, $R_1 = 330$ Ω, $R_2 = 270$ Ω, and $R_L = 560$ Ω.
3. Refer to Figure 13-59(c). Nortonize the circuit and find V_L and I_L, where (a) $E = 12$ V, $R_1 = 680$ Ω, $R_2 = 470$ Ω, $R_3 = 1$ kΩ, and $R_L = 750$ Ω, and (b) $E = 10$ V, $R_1 = 12$ kΩ, $R_2 = 15$ kΩ, $R_3 = 18$ kΩ, and $R_L = 10$ kΩ.
4. Refer to Figure 13-59(d). Nortonize the circuit and find V_L and I_L where (a) $E = 40$ V, $R_1 = 5.6$ kΩ, $R_2 = 3.9$ kΩ, and $R_L = 2.7$ kΩ, and (b) $E = 9$ V, $R_1 = 330$ Ω, $R_2 = 680$ Ω, and $R_L = 390$ Ω.
5. Refer to Figure 13-59(e). Nortonize the circuit and find V_L and I_L where (a) $E = 50$ V, $R_1 = 22$ kΩ, $R_2 = 10$ kΩ, $R_3 = 33$ kΩ, $R_L = 15$ kΩ, and (b) $E = 10$ V, $R_1 = 3.3$ kΩ, $R_2 = 3.3$ kΩ, $R_3 = 6.8$ kΩ, and $R_L = 2$ kΩ.

Answers to Self-tests

SELF-TEST 13-1

(a) Short E_1:

$$R_T = 4.7\text{ k}\Omega + 1.65\text{ k}\Omega = 6.35\text{ k}\Omega$$

$$I_T = \frac{10\text{ V}}{6.35\text{ k}\Omega} = 1.57\text{ mA}$$

$$I_3 = \frac{1.57\text{ mA} \times 1.65\text{ k}\Omega}{3.3\text{ k}\Omega} = 787\ \mu\text{A}$$

Short E_2:

$$R_T = 3.3\text{ k}\Omega + 1.94\text{ k}\Omega = 5.24\text{ k}\Omega$$

$$I_T = \frac{30\text{ V}}{5.24\text{ k}\Omega} = 5.73\text{ mA}$$

$$I_3 = \frac{5.73\text{ mA} \times 1.94\text{ k}\Omega}{3.3\text{ k}\Omega} = 3.37\text{ mA}$$

In the original circuit:

$$I_3 = 787\ \mu A + 3.37\ mA = 4.15\ mA$$

$$V_3 = 13.7\ V$$

$$V_2 = 3.7\ V$$

$$I_2 = 787\ \mu A$$

$$V_1 = 16.3\ V$$

$$I_1 = 4.94\ mA$$

(b) Short E_1:

$$R_T = 10\ k\Omega + 5\ k\Omega = 15\ k\Omega$$

$$I_T = \frac{60\ V}{15\ k\Omega} = 4\ mA$$

$$I_3 = 2\ mA$$

Short E_2:

$$R_T = 15\ k\Omega$$

$$I_T = \frac{100\ V}{15\ k\Omega} = 6.67\ mA$$

$$I_3 = 3.33\ mA$$

In the original circuit:

$$I_3 = 3.33\ mA - 2\ mA = 1.33\ mA$$

$$V_3 = 13.3\ V$$

$$V_2 = 73.3\ V$$

$$I_2 = 7.33\ mA$$

$$V_1 = 86.7\ V$$

$$I_1 = 8.67\ mA$$

(c) Short E_1:

$$R_T = 180\ \Omega + 68.7\ \Omega = 249\ \Omega$$

$$I_T = \frac{12\ V}{249\ \Omega} = 48.2\ mA$$

$$I_2 = \frac{48.2\ mA \times 68.7\ \Omega}{220\ \Omega} = 15.1\ mA$$

Short E_2:

$$R_T = 100\ \Omega + 99\ \Omega = 199\ \Omega$$

$$I_T = \frac{10\ V}{199\ \Omega} = 50.3\ mA$$

$$I_2 = \frac{50.3\ mA \times 99\ \Omega}{220\ \Omega} = 22.6\ mA$$

In the original circuit:

$$I_2 = 15.1\ mA + 22.6\ mA = 37.7\ mA$$

$$V_2 = 8.29\ V$$

$$V_3 = 3.71\ V$$

$$I_3 = 20.6\ mA$$

$$V_1 = 1.71\ V$$

$$I_1 = 17.1\ mA$$

(d) Short E_2:

$$R_T = R_1 + R_2 + R_3 \parallel R_4$$

$$= 1\ k\Omega + 2.7\ k\Omega + 956\ \Omega = 4.66\ k\Omega$$

$$I_T = \frac{E_1}{R_T} = \frac{30\ V}{4.66\ k\Omega} = 6.44\ mA$$

$$I_3 = \frac{I_T R_X}{R_3} = \frac{6.44\ mA \times 956\ \Omega}{1.2\ k\Omega} = 5.13\ mA$$

Short E_1:

$$R_T = R_4 + R_3 \parallel (R_1 + R_2)$$

$$= 4.7\ k\Omega + 906\ \Omega = 5.61\ k\Omega$$

$$I_T = \frac{E_2}{R_T} = \frac{10\ V}{5.61\ k\Omega} = 1.78\ mA$$

$$I_3 = \frac{I_T R_X}{R_3} = \frac{1.78\ mA \times 906\ \Omega}{1.2\ k\Omega} = 1.35\ mA$$

In the original circuit:

$$I_3 = 1.35\ mA + 5.13\ mA = 6.48\ mA$$

$$V_3 = 6.48\ mA \times 1.2\ k\Omega = 7.77\ V$$

$$V_4 = E_2 - V_3 = 10\ V - 7.77\ V = 2.23\ V$$

$$I_4 = \frac{2.23\ V}{4.7\ k\Omega} = 474\ \mu A$$

$$I_1 = I_2 = I_3 - I_4 = 6.01\ mA$$

$$V_1 = 6.01\ mA \times 1\ k\Omega = 6.01\ V$$

$$V_2 = 6.01\ mA \times 2.7\ k\Omega = 16.2\ V$$

SELF-TEST 13-2

1. The Thévenized circuit with the load connected is shown in Figure 13-81.

$$V_{OC} = V_2 + V_3 = \frac{E(R_2 + R_3)}{R_1 + R_2 + R_3} = 20.2 \text{ V}$$

$$R_{TH} = R_1 \parallel (R_2 + R_3) = 219\ \Omega$$

$$V_L = \frac{V_{OC} R_L}{R_{TH} + R_L} = \frac{20.2 \text{ V} \times 560\ \Omega}{779\ \Omega} = 14.5 \text{ V}$$

$$I_L = \frac{V_{OC}}{R_{TH} + R_L} = \frac{20.2 \text{ V}}{779\ \Omega} = 25.9 \text{ mA}$$

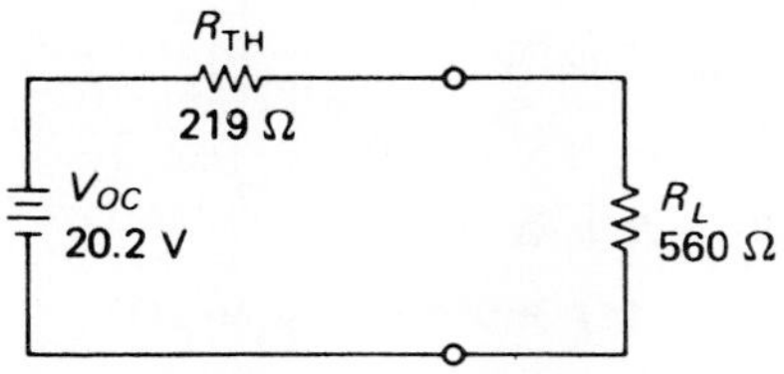

FIGURE 13-81

2. The Thévenized circuit with the load connected is shown in Figure 13-82.

$$V_{OC} = V_2 = \frac{ER_2}{R_1 + R_2} = \frac{15 \text{ V} \times 2.7 \text{ k}\Omega}{6 \text{ k}\Omega}$$

$$= 6.75 \text{ V}$$

$$R_{TH} = R_4 + R_1 \parallel R_2 = 7.5 \text{ k}\Omega + 1.49 \text{ k}\Omega$$

$$= 8.99 \text{ k}\Omega$$

$$V_L = \frac{V_{OC} R_L}{R_{TH} + R_L} = \frac{6.75 \text{ V} \times 4.7 \text{ k}\Omega}{13.7 \text{ k}\Omega}$$

$$= 2.32 \text{ V}$$

$$I_L = \frac{V_{OC}}{R_{TH} + R_L} = \frac{6.75 \text{ V}}{13.7 \text{ k}\Omega} = 493\ \mu\text{A}$$

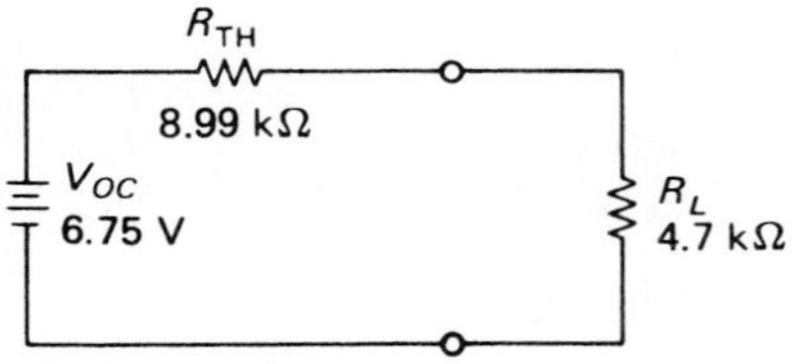

FIGURE 13-82

3. The Thévenized circuit with the load connected is shown in Figure 13-83(b). The circuit used to find V_{OC} is shown in Figure 13-83(a).

$$E_T = E_1 - E_2 = 20 \text{ V}$$

$$V_2 = \frac{20 \text{ V} \times 4.7 \text{ k}\Omega}{8 \text{ k}\Omega} = 11.8 \text{ V}$$

$$V_{OC} = E_2 + V_2 = 21.8 \text{ V}$$

$$R_{TH} = R_1 \parallel R_2 = 1.94 \text{ k}\Omega$$

$$V_L = \frac{V_{OC} R_L}{R_{TH} + R_L} = \frac{21.8 \text{ V} \times 3.3 \text{ k}\Omega}{5.24 \text{ k}\Omega}$$

$$= 13.7 \text{ V}$$

$$I_L = \frac{V_{OC}}{R_{TH} + R_L} = \frac{21.8 \text{ V}}{5.24 \text{ k}\Omega} = 4.15 \text{ mA}$$

4. The Thévenized circuit with the load connected is shown in Figure 13-84(b). The circuit used to find V_{OC} is shown in Figure 13-84(a).

$$E_T = E_1 - E_2 = 20 \text{ V}$$

$$V_4 = \frac{E_T R_4}{R_1 + R_2 + R_4} = \frac{20 \text{ V} \times 4.7 \text{ k}\Omega}{8.4 \text{ k}\Omega}$$

$$= 11.2 \text{ V}$$

$$V_{OC} = V_4 + E_2 = 21.2 \text{ V}$$

$$R_{TH} = R_4 \parallel (R_1 + R_2) = 4.7 \text{ k}\Omega \parallel 3.7 \text{ k}\Omega$$

$$= 2.07 \text{ k}\Omega$$

$$V_L = \frac{V_{OC} R_L}{R_{TH} + R_L} = \frac{21.2 \text{ V} \times 1.2 \text{ k}\Omega}{3.27 \text{ k}\Omega}$$

$$= 7.78 \text{ V}$$

$$I_L = \frac{V_{OC}}{R_{TH} + R_L} = \frac{21.2 \text{ V}}{3.27 \text{ k}\Omega} = 6.48 \text{ mA}$$

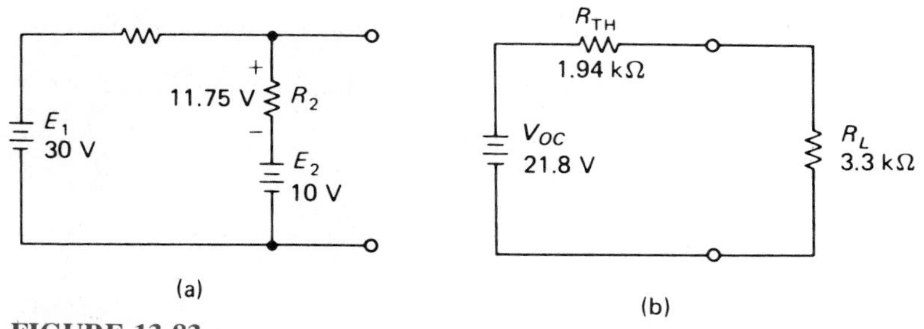

FIGURE 13-83

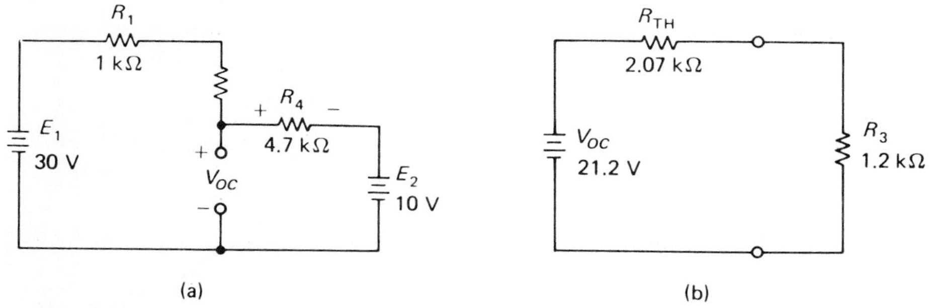

FIGURE 13-84

5. The Thévenized circuit with the load connected is shown in Figure 13-85. Various solutions are possible. We will use a circuit that includes E, R_3, R_L, and R_2 with the load removed:

$$V_3 = \frac{ER_3}{R_1 + R_3} = \frac{9\text{ V} \times 1.8\text{ k}\Omega}{5.5\text{ k}\Omega} = 2.49\text{ V}$$

$$V_2 = \frac{ER_2}{R_2 + R_4} = \frac{9\text{ V} \times 2.7\text{ k}\Omega}{8.3\text{ k}\Omega} = 2.93\text{ V}$$

$$V_{OC} = E - V_2 - V_3 = 9\text{ V} - 5.42 = 3.58\text{ V}$$

$$R_{TH} = R_1 \parallel R_3 + R_2 \parallel R_4$$

$$= 1.3\text{ k}\Omega + 1.82\text{ k}\Omega = 3.12\text{ k}\Omega$$

$$V_L = \frac{V_{OC}R_L}{R_{TH} + R_L} = \frac{3.58\text{ V} \times 1.2\text{ k}\Omega}{4.32\text{ k}\Omega}$$

$$= 0.994\text{ V}$$

$$I_L = \frac{V_{OC}}{R_{TH} + R_L} = \frac{3.58\text{ V}}{4.32\text{ k}\Omega} = 0.829\text{ mA}$$

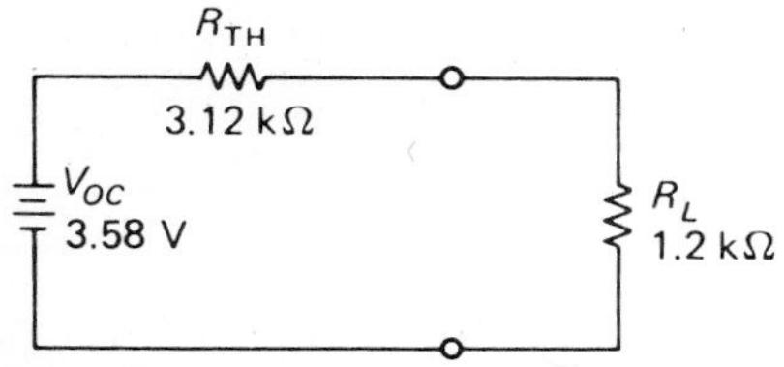

FIGURE 13-85

6. The Thévenized circuit with the load connected is shown in Figure 13-86.

$$V_{OC} = \frac{V_{CC}R_2}{R_1 + R_2} = \frac{20\text{ V} \times 4.7\text{ k}\Omega}{47\text{ k}\Omega + 4.7\text{ k}\Omega} = 1.82\text{ V}$$

$$R_{TH} = R_1 \parallel R_2 = 4.27\text{ k}\Omega$$

$$R = R_4(H_{FE} + 1) = 85.2\text{ k}\Omega$$

$$I_B = \frac{V_{OC} - V_{BE}}{R_{TH} + R} = \frac{1.22\text{ V}}{89.5\text{ k}\Omega} = 13.6\ \mu\text{A}$$

$$I_C = H_{FE}I_B = 955\ \mu\text{A}$$

$$V_C = V_{CC} - V_3 = 20\text{ V} - 4.49\text{ V} = 15.5\text{ V}$$

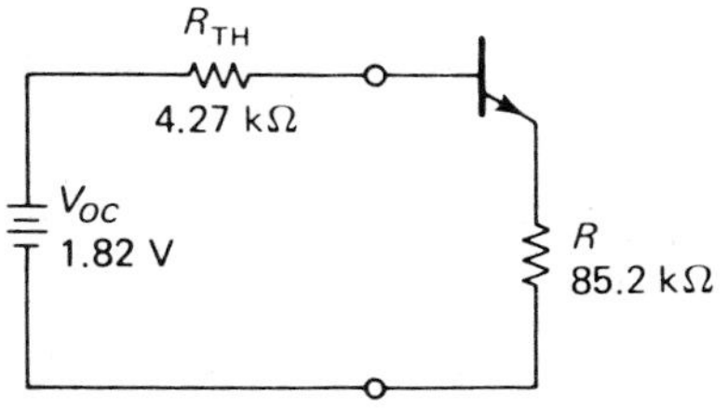

FIGURE 13-86

SELF-TEST 13-3

1. $I_{SC} = 221\ \mu A$
   $G_N = 24.7\ \mu S$
   $V_L = 5.82\ V$
   $I_L = 77.5\ \mu A$

2. $I_{SC} = 29.4\ mA$
   $G_N = 1.47\ mS$
   $V_L = 9.03\ V$
   $I_L = 16.1\ mA$

CHAPTER

14 GRAPHING

INTRODUCTION

Graphs are the result of plotting the relationship of one variable to another. For example, suppose we wanted to plot how temperature changes over a 24-hour period. First, we would collect the data. That is, we would measure the temperature at selected time intervals. Those intervals could be 1 hour, 2 hours, 3 hours, or whatever interval seems appropriate. For our graph, let's select 2-hour time intervals starting at noon. The data collected might look like the data shown in Figure 14-1.

Next we would lay out the graph. Graphs have a vertical part and a horizontal part. It is standard practice to plot the independent variable along the horizontal axis and the dependent variable along the vertical axis. In any graph where time is one of the variables, it will be the independent variable because time is not dependent on temperature or any other variation. Time, then, will be plotted along the horizontal

Time	Temperature (in °F)
12 Noon	88
2 pm	91
4 pm	85
6 pm	81
8 pm	78
10 pm	72
12 am	68
2 am	64
4 am	62
6 am	68
8 am	76
10 am	82

FIGURE 14-1

axis, and temperature will be plotted along the vertical axis. Our graph will look like the graph in Figure 14-2. We have selected graph paper that has five divisions per inch both horizontally and vertically. Next, we need to select scales in both directions. We have selected 4°/division vertically, starting at 60°, and 2 hours/division horizontally, starting at 12 noon. We selected a temperature slightly lower than our lowest temperature and selected temperature variations (4°/division) that would result in a graph that is within the confines of our paper, which has ten divisions in the vertical direction. We selected 2 hours per division horizontally because we chose to measure temperature every 2 hours.

Finally, we plot on the graph paper the data we collected and recorded in Figure 14-1. The resulting graph is displayed in Figure 14-3.

This graph, as is true of most graphs, is not *exactly* correct. For example, we had to interpret where on the graph to plot the points where the temperature did not fall on a division (85°, 91°, and so on). Furthermore, we cannot tell *exactly* when the temperature is at its maximum or minimum value because we only measured the temperature every 2 hours. For *exact* values we would have to provide many more points that we provided here. What the graph *did* do was provide us with a clear picture of how the temperature varied, *generally*, during that 24-hour period.

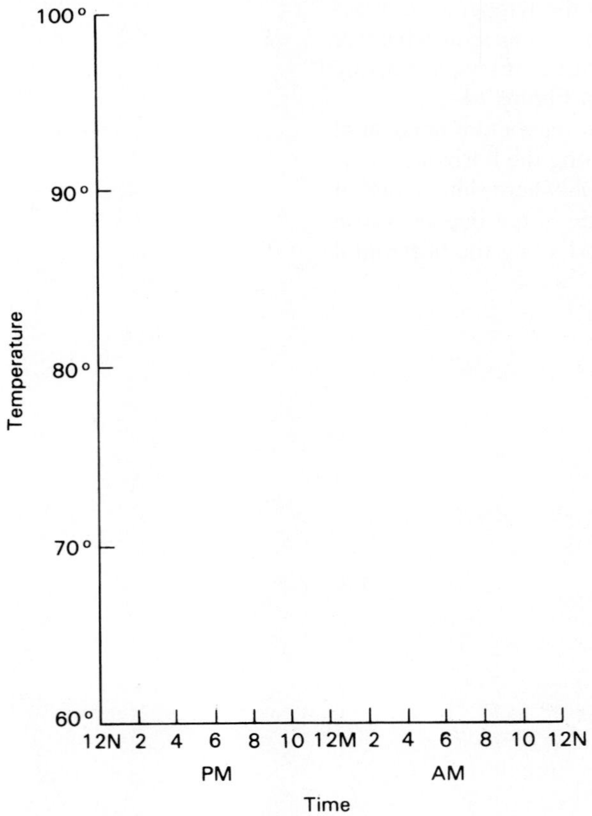

FIGURE 14-2

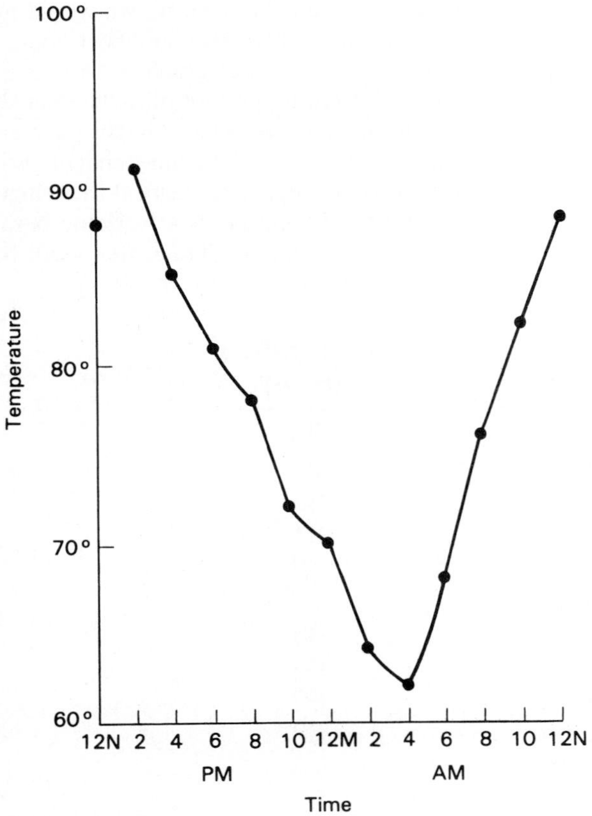

FIGURE 14-3

In electronics, we are constantly noting how one quantity will change as a direct result of changing some other quantity. Any time we collect data about the relationship between variables, we have the potential for creating a graph. The ability to construct and/or read a graph, then, is an important part of our study of electronics.

In this chapter you will learn how to create a graph from collected data. You will also learn how to graph an equation. Finally, you will learn how to interpret a graph that has already been created.

CHAPTER OBJECTIVES

In this chapter you will learn how to:

1. Plot points on a system of rectangular coordinates.
2. Graph linear equations.
3. Put equations in the slope–intercept form.
4. Interpret graphs.
5. Plot curves from given data.

14-1 GRAPHING LINEAR EQUATIONS

Recall from previous chapters that linear equations are equations in which a term contains only *one* variable. Furthermore, the variable exponent is 1. For example, $2x + 1 = 9$, $y - 6 = 0$, and $5 + 6a = 23$ are all linear equations. $6x + 7y = 14$ is also a linear equation. Even though the variables $x + y$ are different, they appear in different terms. $x + 6 + 7y + 14z = 10$ would be a linear equation because no term contains more than one variable. $6xy + 7 = 0$ and $4x^2 + 2y = 14$ are *not* linear equations. In the first equation one term contains two variables (x and y). In the second equation one of the variables is squared ($4x^2$).

KEY POINT *The graph of any linear equation is a straight line.*

Let's see why this is true. Consider the graph in Figure 14-4. This is called a system of *rectangular coordinates*. This kind of graph will be used many times throughout the text. The point where the x-axis (the horizontal axis) and the y-axis (the vertical axis) intersect (cross) is called the *origin*. Notice that the origin is 0 for both vertical and horizontal axes. Along the horizontal or x-axis, positive values are plotted to the right of the origin and negative values are plotted to the left of the origin. Along the vertical, or y-axis, positive values are plotted from the origin upward and negative values are plotted from the origin downward. The graph is divided into equal parts or divisions. The value of each division depends on the requirements of the problem. In Figure 14-4(a), the value of each division both horizontally and vertically is 1. In Figure 14-4(b) the value of each division is 5.

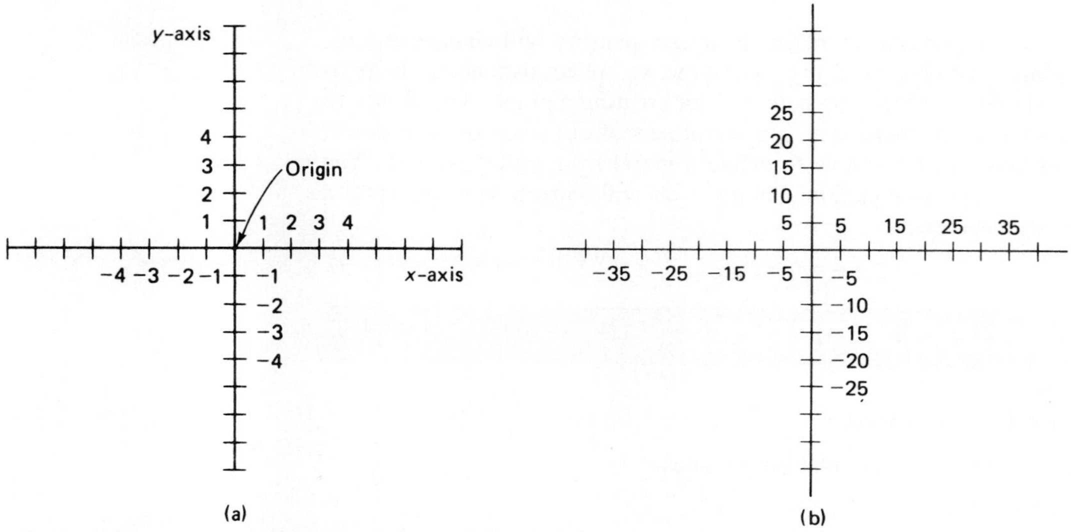

FIGURE 14-4 System of rectangular coordinates.

Let's identify some points using the graph in Figure 14-5. We identify points by giving the x and y values or coordinates. Suppose we were to identify a point whose coordinates are $(5, -2)$. The first number identifies the horizontal or x value and the second number identifies the vertical or y value. The point is located by moving 5 units to the right along the x-axis because the number is positive. We move 2 units down along the y-axis because the number is negative. The point is shown as point A. Let's find the point whose coordinates are $(-4, 6)$. This point is identified as point B in Figure 14-5. The point was found by moving 4 units *left* from the origin along the x-axis. We moved left because $x = -4$. Then we moved upward 6 units. We moved upward because $y = 6$.

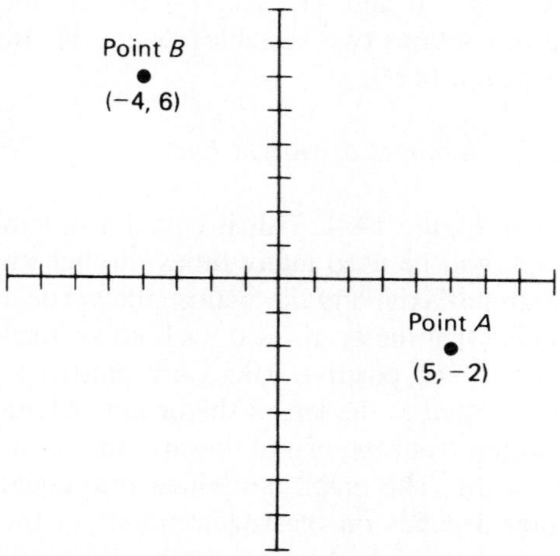

FIGURE 14-5

PRACTICE PROBLEMS 14-1

Identify the following points on the graph of Figure 14-4(a).

1. (2, 6)
2. (−1, 7)
3. (0, −8)
4. (−6, 0)
5. (−1, 2)
6. (6, −3)
7. (−3, −4)

Identify the following points on the graph of Figure 14-4(b).

8. (−15, 0)
9. (5, −10)
10. (25, 25)
11. (−15, −20)
12. (−10, 20)

Solutions

See Figure 14-6(a) for problems 1 through 7; see Figure 14-6(b) for problems 8 through 12.

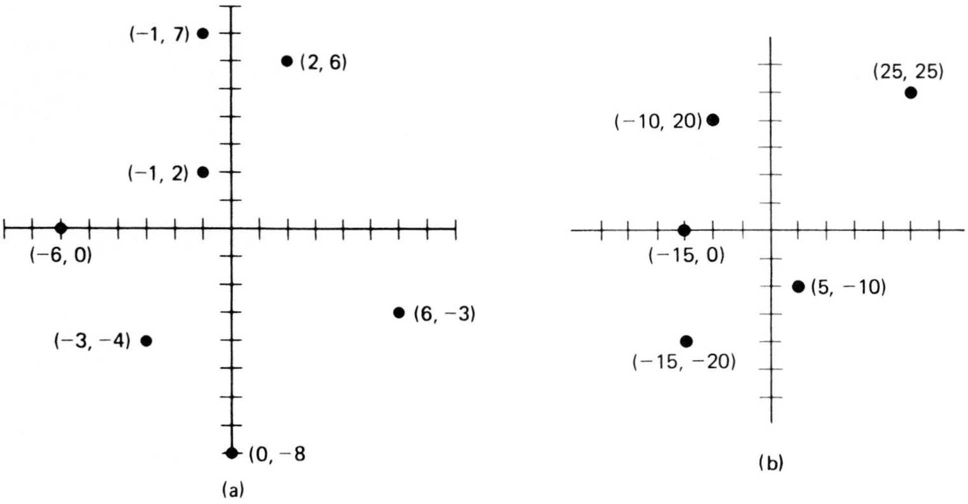

FIGURE 14-6

Additional practice problems are at the end of the chapter.

Consider the equation $2x - y = 6$. Usually, the graph of a linear equation intersects (crosses) both the horizontal and vertical axes. Therefore, we may assume that one point on the graph will be where $y = 0$ and x equals a value to be determined by solving the resulting equation. That is,

$$2x - y = 6$$
$$\text{If } y = 0$$
$$\text{then } 2x = 6$$
$$\text{and } x = 3$$

One point on the graph then is where $y = 0$ and $x = 3$. On the graph the point would be identified as (3, 0). (The x value is always given first.) Another point would be where $x = 0$.

$$2x - y = 6$$
$$\text{If } x = 0$$
$$\text{then } -y = 6$$
$$y = -6$$

A second point on the graph is where $x = 0$ and $y = -6$. On the graph the point would be identified as (0, −6). The graph is shown in Figure 14-7. Point A is called the *x-intercept* because that is where the line crosses the x-axis. Its coordinates are (3, 0). The x-coordinate is called the *abscissa*. The y-coordinate is called the *ordinate*.

Point B is called the *y-intercept*. Its coordinates are (0, −6). Many points along the graph may be plotted or identified. These points may be found by letting x equal some value and then solving for y. Or we could let y equal some value and solve for x. For example, using the same equation, we get

$$2x - y = 6$$
$$\text{If } x = 2$$
$$\text{then } 2(2) - y = 6$$
$$4 - y = 6$$
$$-y = 2$$
$$y = -2$$

The coordinates are (2, −2).

$$\text{If } y = 4$$
$$\text{then } 2x - (4) = 6$$
$$2x = 10$$
$$x = 5$$

The coordinates are (5, 4). These two points are also plotted in Figure 14-7.

A table showing values of x and corresponding values of y is also included in Figure 14-7. Such a table should be drawn whenever points are to be plotted because it helps identify coordinates.

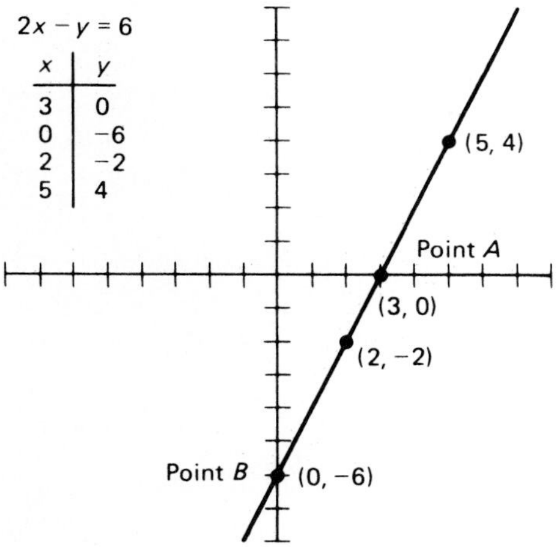

FIGURE 14-7

□ EXAMPLE 14-1

Given the equation $y = 2x + 2$. Draw the graph. Find the x- and y-intercepts. Identify four points on the graph.

Solution First, let's rearrange the terms so that both variables are in the left-hand member. This will make it easier for us to solve for unknowns as we go along.

$$y = 2x + 2$$
$$y - 2x = 2$$

Next, let's find the x-intercept. This is the point where the line passes through the x-axis.

$$y - 2x = 2$$
$$\text{Let } y = 0$$
$$\text{then } -2x = 2$$
$$x = -1$$

The x-intercept is at coordinates $(-1, 0)$. Now find the y-intercept.

$$y - 2x = 2$$
$$\text{Let } x = 0$$
$$\text{then } y = 2$$

The y-intercept is at coordinates $(0, 2)$. Use a straightedge and draw a line extending through the two points as shown in Figure 14-8. A third point is found by letting $x =$ some number. Upon examination of the straight line, we see that the point $(2, 6)$ should fall on the line. Let's see if it does.

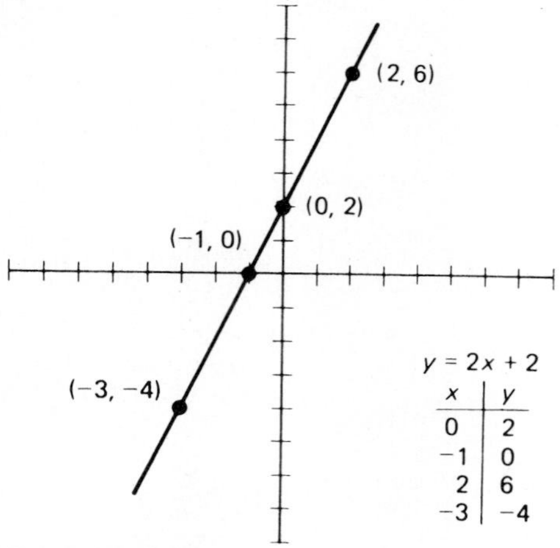

$y = 2x + 2$

x	y
0	2
−1	0
2	6
−3	−4

FIGURE 14-8

$$y - 2x = 2$$

Let $x = 2$. Then

$$y - 2(2) = 2$$
$$y - 4 = 2$$
$$y = 6$$

The fact that the point (2, 6) falls on the line tells us that our line is in the correct location. Let $x = -3$. Then

$$y - 2(-3) = 2$$
$$y + 6 = 2$$
$$y = -4$$

When we plot (−3, −4), we see that this point falls on our line. The point (2, 6) or the point (−3, −4) would not be proof enough that our line is in the correct location. *Three* points are needed to verify our line. If the line passes through all three points, we can assume it to be correct.

□

PRACTICE PROBLEMS 14-2

Plot the following equations on linear graph paper. Identify the x- and y-intercepts. Plot at least three points on the graph:

1. $x - 4y = 8$

2. $3x + 6y = 12$

3. $y = 4x + 4$

4. $x = 3y - 6$

Solutions

1. The graph is shown in Figure 14-9(a). The x-intercept = (8, 0). The y-intercept = (0, −2).

2. The graph is shown in Figure 14-9(b). The x-intercept = (4, 0). The y-intercept = (0, 2).

3. The graph is shown in Figure 14-9(c). The x-intercept $= (-1, 0)$. The y-intercept $= (0, 4)$.

4. The graph is shown in Figure 14-9(d). The x-intercept $= (-6, 0)$. The y-intercept $= (0, 2)$.

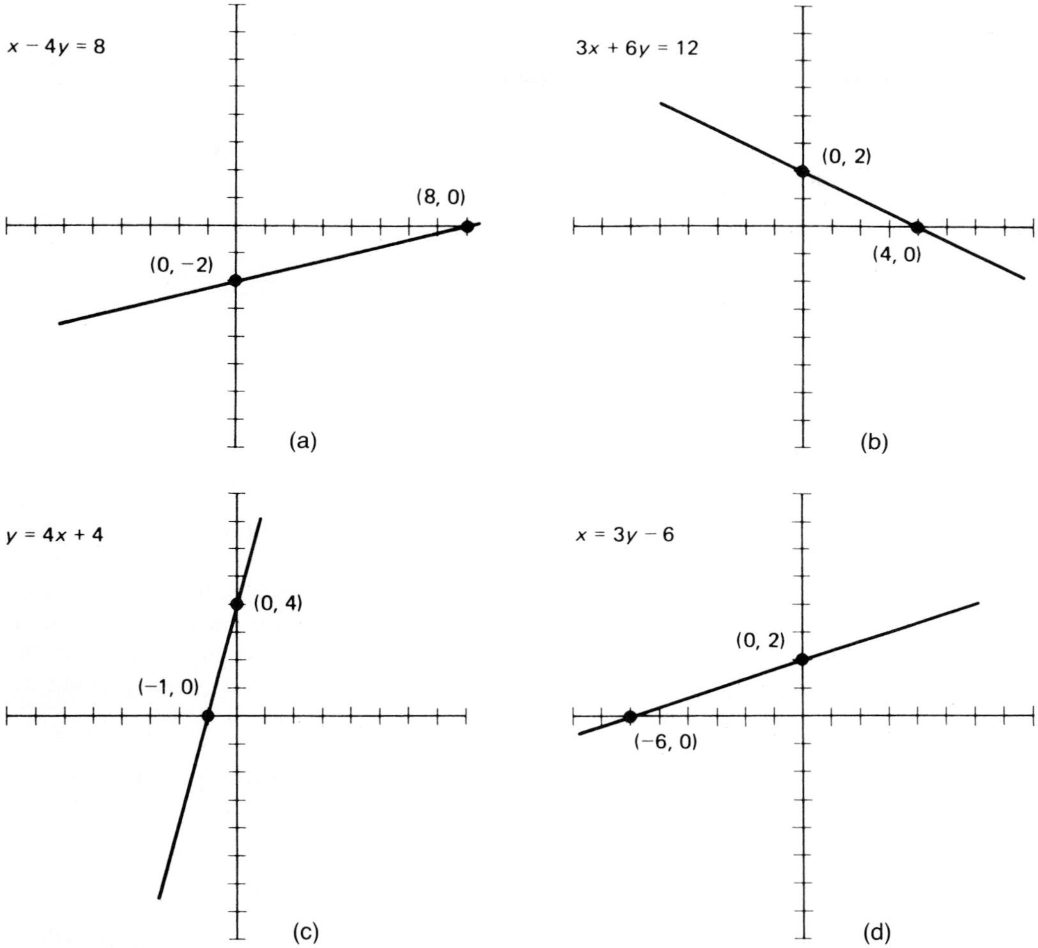

FIGURE 14-9

Additional practice problems are at the end of the chapter.

14-2 SLOPE OF A LINE

The linear equation in Figure 14-7 is redrawn in Figure 14-10. Consider the point (3, 0). If x is changed from 3 to 5, y changes from 0 to 4. The change in x is 2. This would be written $\Delta x = 2$ (*delta* $x = 2$). The change in y is 4. This would be written $\Delta y = 4$ (*delta* $y = 4$). Upon inspection of the curve, we find that for any increase of

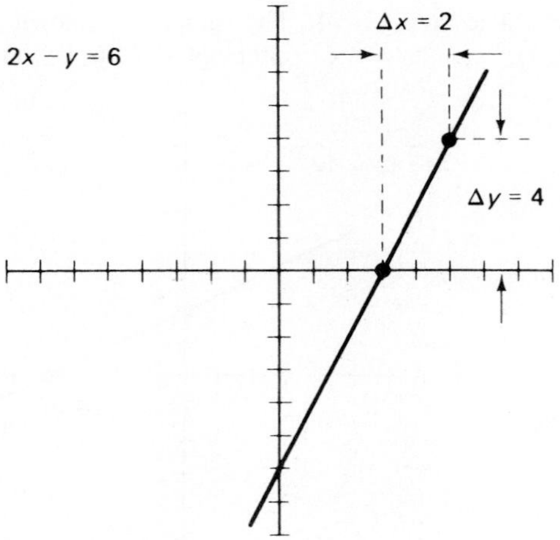

FIGURE 14-10

2 units horizontally ($\Delta x = 2$) y increases 4 units ($\Delta y = 4$). For example, if x increases from -2 units to 0, y increases from -10 units to -6 units. If x increases from 1 unit to 3 units, y increases from -4 units to 0.

Consider the straight line in Figure 14-11, which results from the equation $x + 2y = 4$. If x increases from 0 to 4 ($\Delta x = 4$), y *decreases* from 2 to 0 ($\Delta y = -2$). Any increase in the value of x causes a corresponding decrease in the value of y. Any 4-unit increase in x causes a 2-unit decrease in y. Any 8-unit increase in x causes a 4-unit decrease in y.

KEY POINT The *slope* of a line is defined as the change in y that results from an *increase* in x.

$$\text{slope} = \frac{\Delta y}{\Delta x} = \frac{\text{change in } y}{\text{increase in } x}$$

When we read a graph, we always make Δx positive. The resulting Δy will be either positive or negative depending on whether y increased or decreased as x increased. The slope is positive if Δy is positive and it is negative if Δy is negative. In Figure 14-11 the slope equals

$$\frac{\Delta y}{\Delta x} = \frac{-2}{4} = -\frac{1}{2}$$

When the slope is negative, the line falls from left to right. When the slope is positive, the line rises from left to right.

☐ EXAMPLE 14-2

Find the slopes of the lines in Practice Problems 14-2.

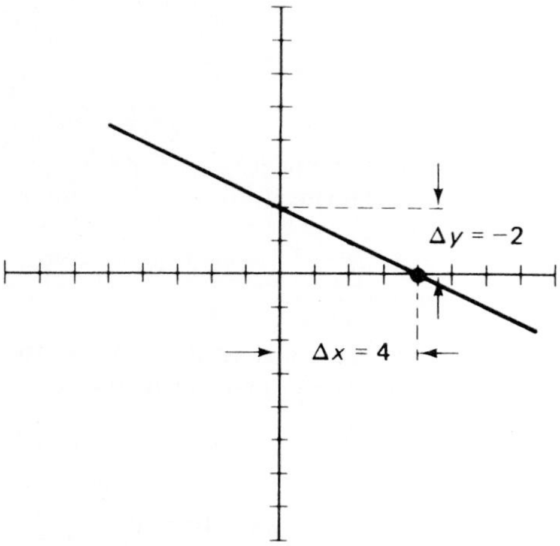

FIGURE 14-11

Solution If we use the x- and y-intercepts to determine the slope, then in problem 1 the slope is

$$\text{slope} = \frac{\Delta y}{\Delta x} = \frac{1}{4}$$

The slope is positive because, when x is increased 4 units, y increases 1 unit. In problem 2

$$\text{slope} = \frac{\Delta y}{\Delta x} = -\frac{1}{2}$$

The slope is negative because, when x is increased 2 units, y decreases 1 unit. In problem 3

$$\text{slope} = \frac{\Delta y}{\Delta x} = \frac{4}{1} = 4$$

The slope is positive because, when x is increased 1 unit, y increases 4 units. In problem 4

$$\text{slope} = \frac{\Delta y}{\Delta x} = \frac{1}{3}$$

The slope is positive because, when x is increased 3 units, y increases 1 unit.

□

PRACTICE PROBLEMS 14-3

Plot the following equations on linear graph paper. Identify the x- and y-intercepts. Determine the slope.

1. $x + y = 4$
2. $x - 3y = -6$
3. $4x + y = -8$
4. $8x - 4y = 16$

Solutions

1. Figure 14-12(a) shows the graph of $x + y = 4$. The x-intercept is at coordinates (4, 0). The y-intercept is at coordinates (0, 4).

$$\text{slope} = \frac{\Delta y}{\Delta x} = \frac{-4}{4} = -1$$

2. Figure 14-12(b) shows the graph of $x - 3y = -6$. The x-intercept is at coordinates $(-6, 0)$. The y-intercept is at coordinate (0, 2).

$$\text{slope} = \frac{\Delta y}{\Delta x} = \frac{2}{6} = \frac{1}{3}$$

3. Figure 14-12(c) shows the graph of $4x + y = -8$. The x-intercept is at coordinates $(-2, 0)$. The y-intercept is $(0, -8)$.

$$\text{slope} = \frac{\Delta y}{\Delta x} = \frac{-8}{2} = -4$$

4. Figure 14-12(d) shows the graph of $8x - 4y = 16$. The x-intercept is (2, 0). The y-intercept is $(0, -4)$.

$$\text{slope} = \frac{\Delta y}{\Delta x} = \frac{4}{2} = 2$$

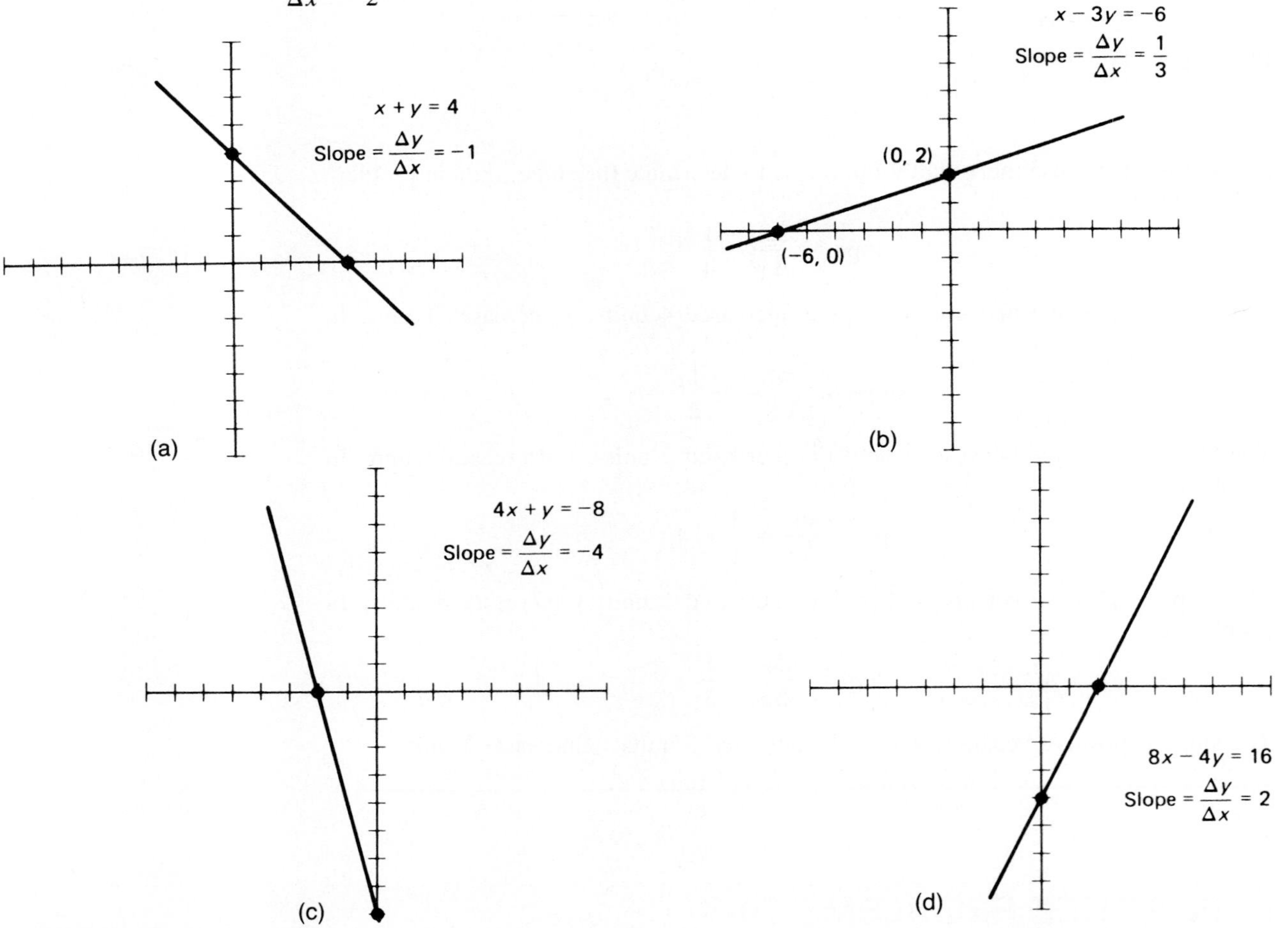

FIGURE 14-12

Additional practice problems are at the end of the chapter.

14-3 SLOPE–INTERCEPT FORM

If a linear equation contains two variables, it can be written like this:

$$y = mx + b$$

y and x are the variables. m and b are constants. When the equation is written in this form, it is said to be written in the *slope–intercept form* because m is the slope and b is the vertical or y-intercept.

■ **KEY POINT** *Any linear equation containing two variables can be written in the slope–intercept form.*

□ EXAMPLE 14-3

Given the equation $3y - 2x = 9$, find the slope and the y-intercept.

Solution Using the rules of algebra, we can change to slope–intercept form:

$$3y - 2x = 9$$

$$3y = 2x + 9$$

$$y = \frac{2}{3}x + 3$$

The slope is $\frac{2}{3}$ and the y-intercept is 3. A slope of $\frac{2}{3}$ means that if x is increased 3 units y increases 2 units. The y-intercept coordinates are (0, 3). With this information we can graph the equation. In Figure 14-13 we first plot the point (0, 3), the y-intercept. Next we find a second point. The slope is $\frac{2}{3}$, which means that, if we change x from 0 to 3, y will increase from 3 to 5. This point is (3, 5) and is plotted. A third point can be plotted by changing x from 3 to 6 and changing y from 5 to 7.

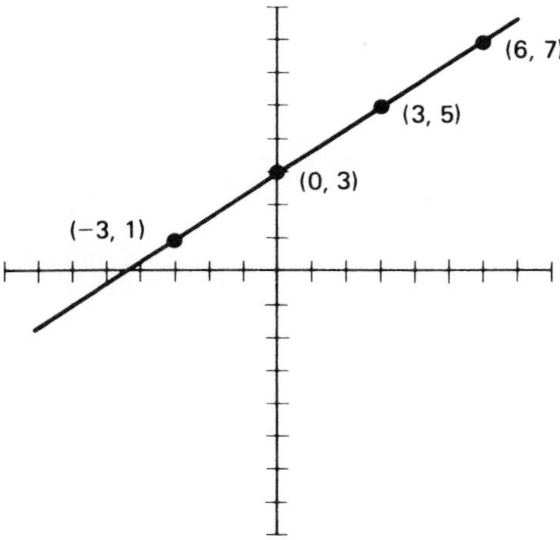

FIGURE 14-13

This point is (6, 7). Starting at the y-intercept, we could have plotted the point $(-3, 1)$. This point results from changing x from 0 to -3, which causes a change in y from 3 to 1.

□

PRACTICE PROBLEMS 14-4

Change the following equations into the slope–intercept form. Determine the slope and y-intercept. Graph the equation.

1. $4x - 3y = -15$
2. $3y - 4x = 18$
3. $3x + 3y = 5$
4. $6y - 5x = 24$
5. $3y - 7x = -12$
6. $4x - y = 5$

Solutions

1. $y = \frac{4}{3}x + 5$. Slope $= \frac{4}{3}$, y-intercept $= (0, 5)$. The graph of the equation is shown in Figure 14-14(a).
2. $y = \frac{4}{3}x + 6$. Slope $= \frac{4}{3}$, y-intercept $= (0, 6)$. The graph of the equation is shown in Figure 14-14(b).
3. $y = -x + \frac{5}{3}$. Slope $= -1$, y-intercept $= (0, \frac{5}{3})$. The graph of the equation is shown in Figure 14-14(c).
4. $y = \frac{5}{6}x + 4$. Slope $= \frac{5}{6}$, y-intercept $= (0, 4)$. The graph of the equation is shown in Figure 14-14(d).
5. $y = \frac{7}{3}x - 4$. Slope $= \frac{7}{3}$, y-intercept $= (0, -4)$. The graph of the equation is shown in Figure 14-14(e).
6. $y = 4x - 5$. Slope $= 4$, y-intercept $= (0, -5)$. The graph of the equation is shown in Figure 14-14(f).

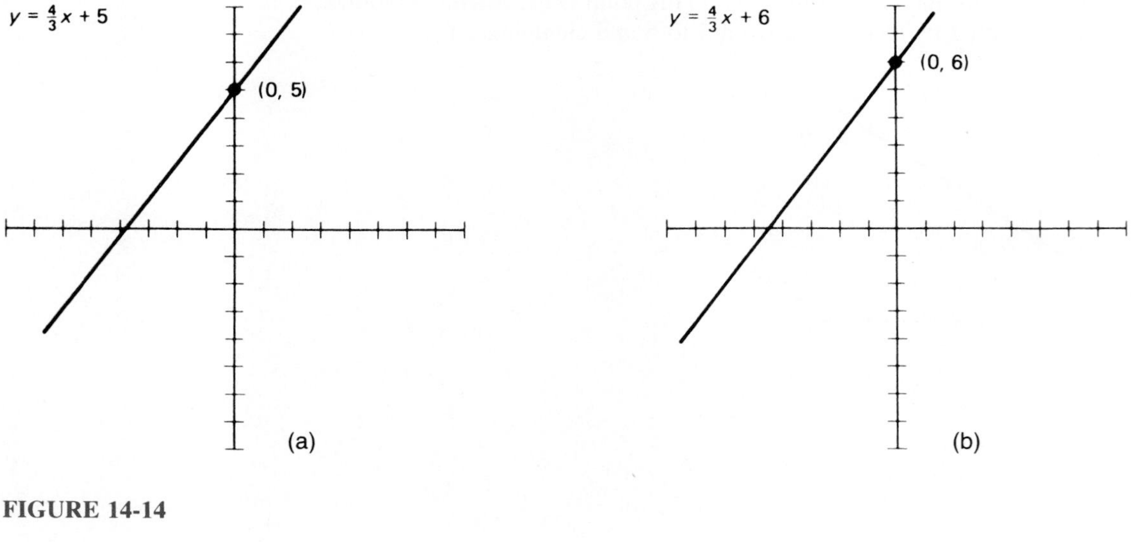

FIGURE 14-14

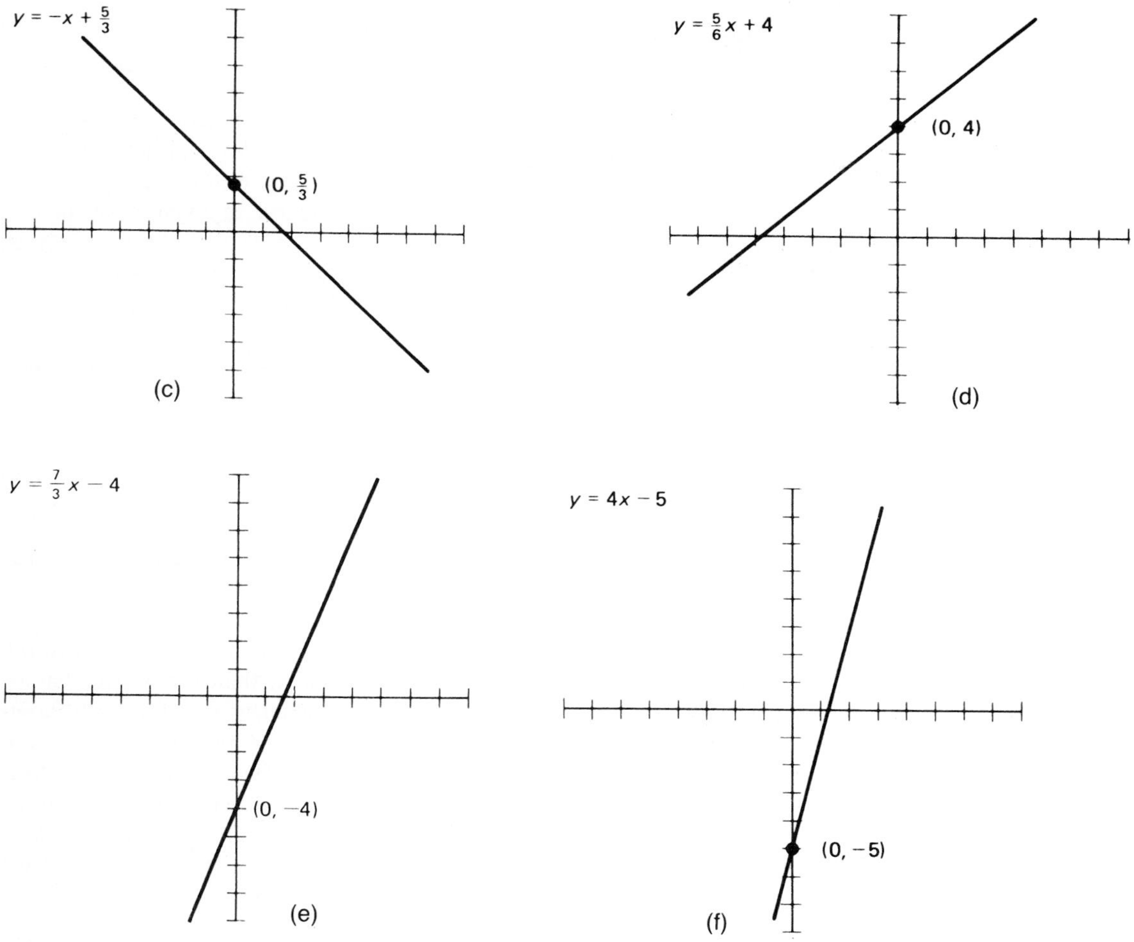

FIGURE 14-14 *(cont.)*

Additional practice problems are at the end of the chapter.

SELF-TEST 14-1

1. Plot the equation $5x + 3y = 15$ on linear graph paper. Identify the x- and y-intercepts.
2. Plot the equation $3x - y = 6$ on linear graph paper. Identify the x- and y-intercepts. Determine the slope.
3. Change the equation $2x - 3y = 6$ into slope–intercept form. Determine the slope. Identify the y-intercept. Graph the equation.

Answers to Self-test 14-1 are at the end of the chapter.

14-4 INTERPRETING GRAPHS

In Figure 14-15 we have plotted current versus voltage for a resistor with values of voltage from 0 to 20 V. The resulting curve is a straight line. We can determine the resistance by using the techniques learned previously. Notice that I is plotted along the y-axis and V is plotted along the x-axis. Relating this curve to a system of rectangular coordinates, we get

$$\text{slope} = \frac{\Delta y}{\Delta x} = \frac{\Delta I}{\Delta V} = \frac{4 \text{ mA}}{4 \text{ V}} = 1 \text{ mS}$$

The slope is equal to the conductance. That is, the slope defines the conductance characteristic of the curve. To find R,

$$R = \frac{1}{G} = \frac{1}{1 \text{ mS}} = 1 \text{ k}\Omega$$

The curve, then, is a curve of current versus voltage for a 1 kΩ resistor. With such a curve we can easily find the resulting current for any value of voltage plotted. Because the curve is linear, the slope is the same anywhere along the curve. Therefore, any convenient place along the curve can be chosen to determine the slope.

We see from the curve in Figure 14-15 that resistors are linear devices. Some devices such as diodes and transistors are nonlinear. That is, their resistance characteristics are not straight line functions. Figure 14-16 is the curve of a forward-biased silicon diode. The resulting curve of current versus voltage is nonlinear. When the curve is nonlinear, the slope of the line depends on where we choose to measure. The most accurate method of determining the slope of such a curve is to first select a point on the curve and then construct a tangent to that point. From geometry a *tangent* is defined as a line that touches a curve at only one point. In Figure 14-17 we

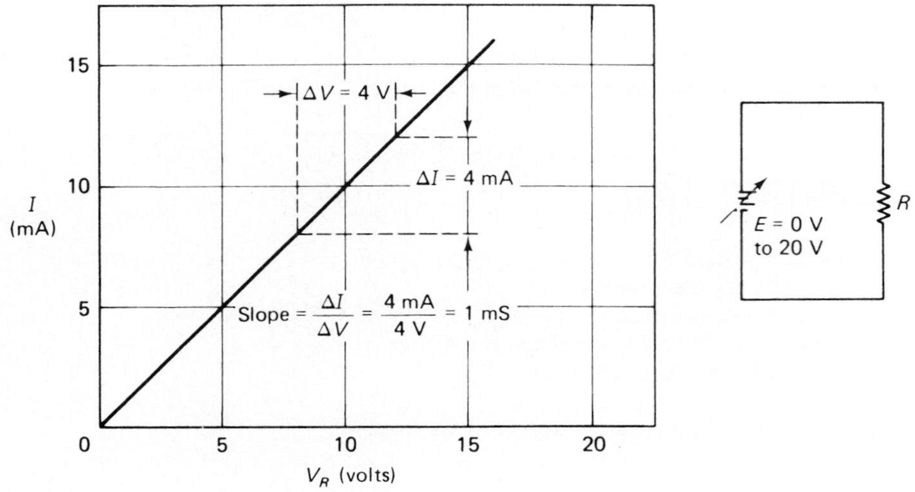

FIGURE 14-15

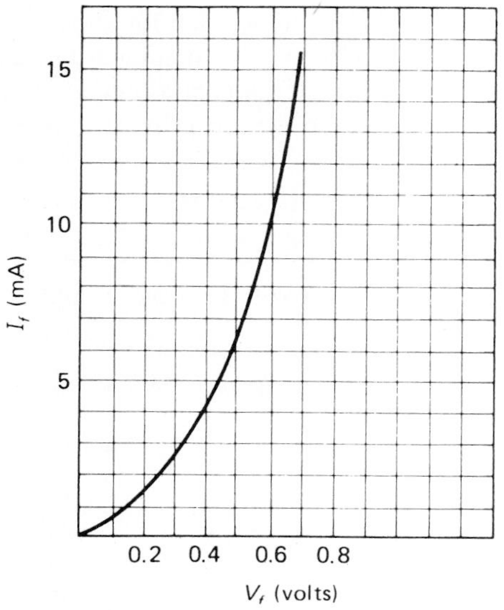

FIGURE 14-16

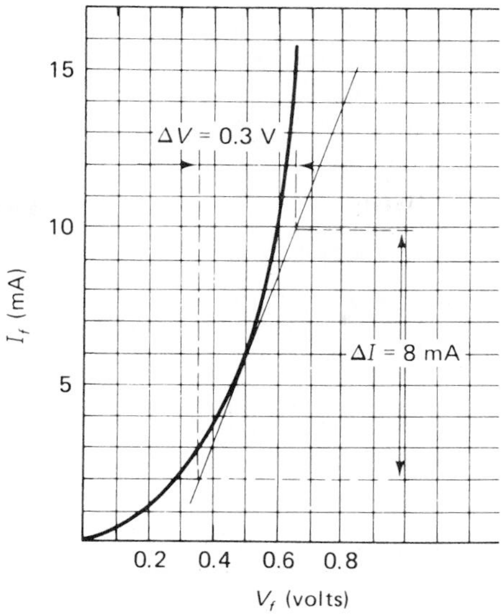

FIGURE 14-17

have constructed a tangent to the curve where $V_f = 0.5$ V. The slope of this line is the same as the slope of the curve at that point.

$$\text{slope} = \frac{\Delta I}{\Delta V} = \frac{8 \text{ mA}}{0.3 \text{ V}} = 26.7 \text{ mS}$$

The resistance determined from the slope is symbolized r_d (lowercase r signifies a dynamic or ac resistance).

$$r_d = \frac{1}{\text{slope}} = \frac{1}{26.7 \text{ mS}} = 37.5\ \Omega$$

In addition to the method discussed previously, r_d can be found by the $\Delta I/\Delta V$ method if we consider that small increments of change along the curve will be nearly a straight line. Notice that r_d changes drastically as V_{dc} changes. This tells us that, when we are required to find r_d, we must specify where on the curve the measurement is to be made.

At the point in question, there are really two resistances. If we simply plot the point, we have a voltage and some resulting current. Since we are examining one point, we consider this to be a dc resistance. This dc resistance is symbolized R_D. The general equation is

$$R_D = \frac{V_{dc}}{I_{dc}}$$

Solving our problem for R_D, we get

$$R_D = \frac{V_{dc}}{I_{dc}} = \frac{0.5 \text{ V}}{6 \text{ mA}} = 83.3\ \Omega$$

Figure 14-18 is a universal time constant (TC) curve. Notice that for about the first 20% of the curve the curve is relatively linear. After that the nonlinearity becomes obvious. This curve is called an *exponential curve*. Exponential curves will be analyzed at length in Chapters 24 and 26. In this curve we are primarily interested in some point on the curve.

☐ EXAMPLE 14-4

In a series *RC* circuit connected to a dc source, if the applied voltage is 50 V, find V_C and V_R after 1.5 TC.

Solution From the curve we see that the capacitor has changed to 78% of maximum in 1.5 TC. V_R is 22% of maximum. Then

$$V_C = 78\% \times 50\text{ V} = 39\text{ V}$$

$$V_R = 22\% \times 50\text{ V} = 11\text{ V}$$

☐ EXAMPLE 14-5

In Figure 14-18, **(a)** how many time constants are required for the capacitor to charge to 60% of maximum, and **(b)** if the maximum voltage is 60 V, how many time constants are required to charge the capacitor to 48 V?

Solution **(a)** From the curve we see that it takes approximately 0.9 TC for the ca-

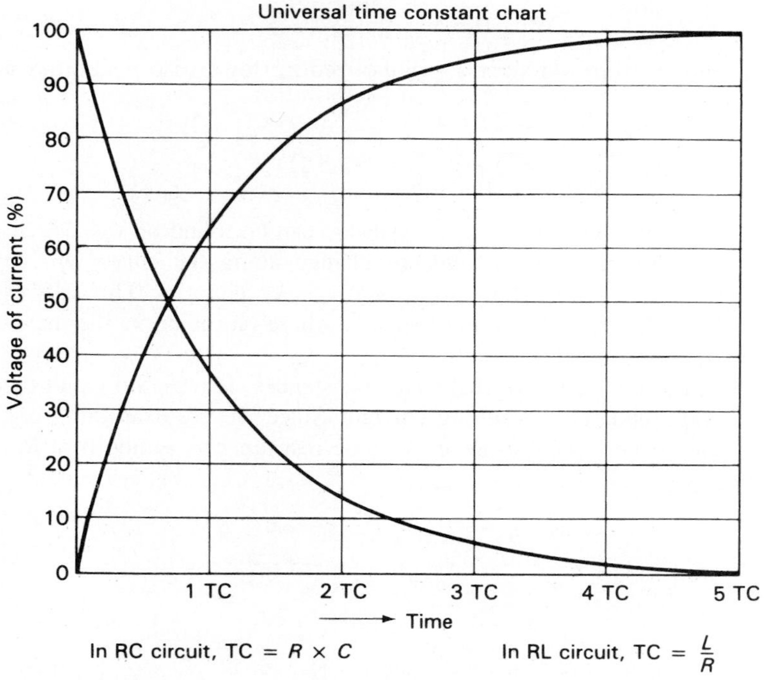

FIGURE 14-18

pacitor to charge to 60% of maximum. **(b)** We must first find what percent 48 V is of 60 V.

$$\frac{48\text{ V}}{60\text{ V}} \times 100 = 80\%$$

Approximately 1.6 TC are required to charge the capacitor to 48 V (80% of maximum).

□

PRACTICE PROBLEMS 14-5

1. Refer to Figure 14-19. Find the slope. Find R.

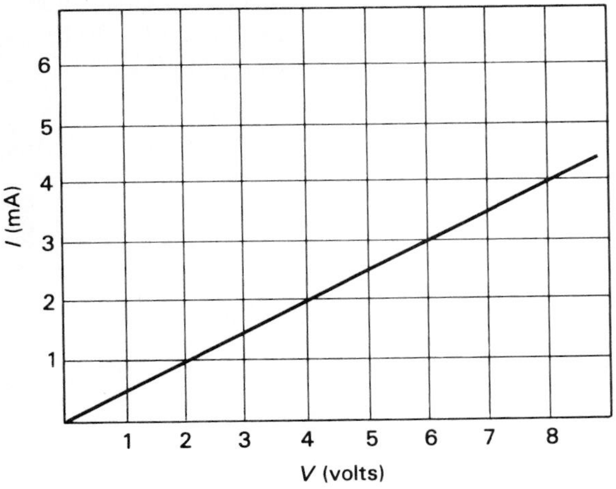

FIGURE 14-19

2. Refer to Figure 14-16. Find R_D and r_d if $V_f = 0.4$ V.
3. Refer to Figure 14-18. Let $E = 10$ V. Find V_C and V_R at 0.7 TC and at 2 TC.

Solutions

1. Slope = 500 μS, $R = 2$ kΩ
2. $R_D \simeq 90\ \Omega$, $r_d \simeq 55\ \Omega$
3. At 0.7 TC, $V_C = 5$ V and $V_R = 5$ V; at 2 TC, $V_C = 8.6$ V and $V_R = 1.4$ V

Additional practice problems are at the end of the chapter.

14-5 PLOTTING CURVES

Refer again to Figure 14-15. A voltage source was connected in series with the resistor. As we vary the applied voltage in the circuit, the current changes. In this circuit the variable V is called the *independent* variable. I is the *dependent* variable. Its value depends on the value of V selected. In plotting curves it is standard practice to plot the independent variable along the x-axis. This leaves the dependent variable to be plotted along the y-axis. Whenever we are required to plot a curve, we must determine the position of the variables in this manner.

□ EXAMPLE 14-6

In a resistive circuit, plot a curve of V versus I where $R = 10\ \text{k}\Omega$. Vary V from 0 to 10 V.

Solution Select an appropriate scale. It would be convenient to use a scale of 1 V/div for voltage and 100 μA/div for current. This is just one of many scales that could be selected. Since this is a resistive circuit, the curve will be linear. We could draw the curve by using two points, but we will use three. The third point is used to check for errors. Using Ohm's law, we can construct a data table of variable values. Figure 14-20(a) is such a table. The resulting curve is shown in Figure 14-20(b).

Let's check our work. The reciprocal of the slope should equal 10 kΩ.

$$\text{slope} = \frac{\Delta I}{\Delta V}$$

$$\text{Let } \Delta V = 2\ \text{V}$$

$$\text{Then } \Delta I = 200\ \mu\text{A}$$

$$\frac{\Delta I}{\Delta V} = \frac{200\ \mu\text{A}}{2\ \text{V}} = 100\ \mu\text{S}$$

$$R = \frac{1}{G} = \frac{1}{100\ \mu\text{S}} = 10\ \text{k}\Omega$$

V (volts)	I (μA)
2	200
5	500
8	800

(a)

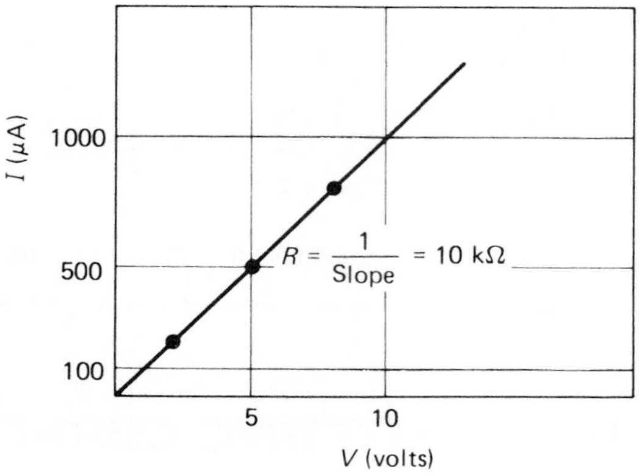

(b)

FIGURE 14-20

☐ EXAMPLE 14-7

Plot a curve of current versus voltage from the given data in Figure 14-21(a). V is the independent variable.

V (volts)	I (mA)
0	0
8	10
20	20
39	30
60	40
85	50

(a)

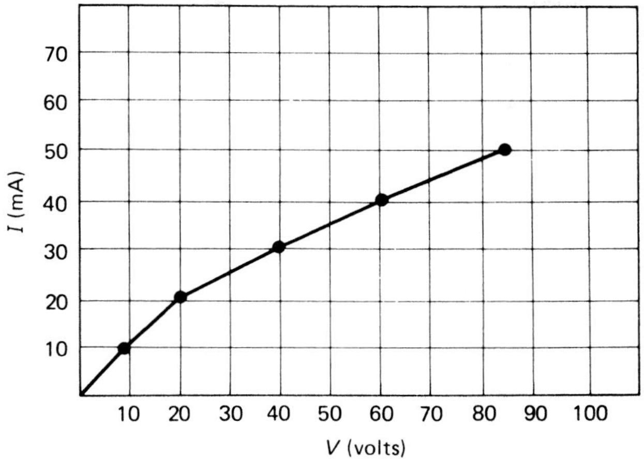

(b)

FIGURE 14-21

Solution Select an appropriate scale. 10 V/div for voltage and 10 mA/div for current were chosen for our solution. The resulting curve is shown in Figure 14-21(b). ☐

PRACTICE PROBLEMS 14-6

1. Plot a curve of current versus voltage for $R = 20\ \text{k}\Omega$ for values of voltage from 0 to 50 V.

2. Plot a curve of current versus voltage from the data in Figure 14-22.

V (volts)	I (mA)
0	0
0.2	0.2
0.3	0.4
0.4	0.65
0.5	1.0
0.6	1.5
0.7	2.4

FIGURE 14-22

Solutions

1. See Figure 14-23.

2. See Figure 14-24.

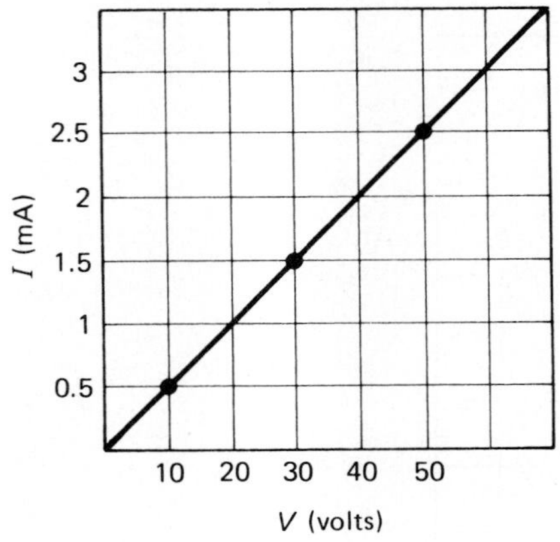

V (volts)	I (mA)
0	0
10	0.5
30	1.5
50	2.5

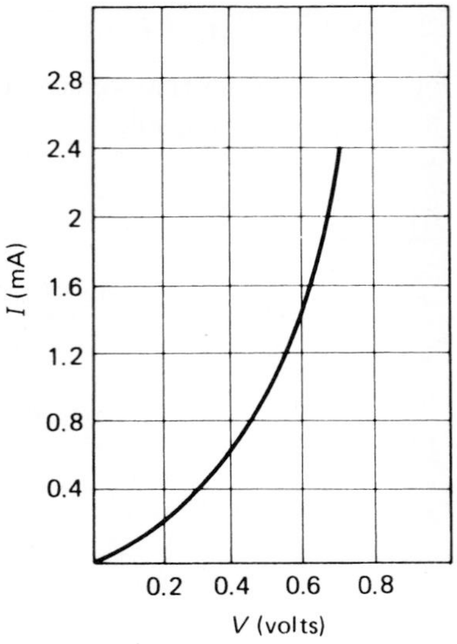

FIGURE 14-23

FIGURE 14-24

Additional practice problems are at the end of the chapter.

SELF-TEST 14-2

1. Refer to Figure 14-25. Find R.
2. Refer to Figure 14-16. Find R_D and r_d if V_f = 0.4 V.
3. Refer to Figure 14-18. Let E = 25 V. Find V_C and V_R 2.3 TC after the capacitor starts to charge.
4. Plot a curve of current versus voltage for R = 100 kΩ. Plot values of voltage from 0 to 10 V.

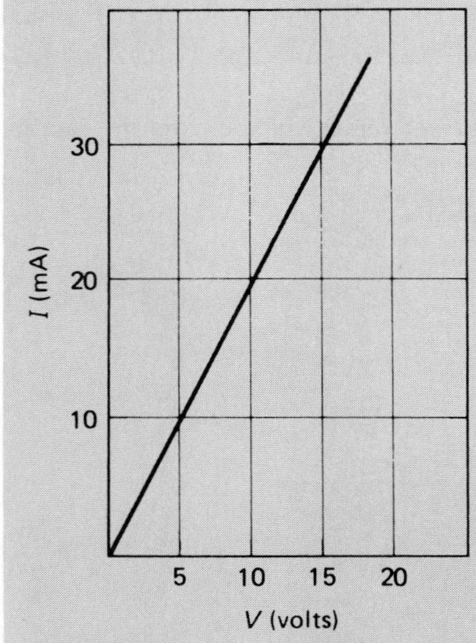

FIGURE 14-25

5. Plot a curve of current versus voltage from the data in Figure 14-26.

V (volts)	I (mA)
0	0
1	1.8
2	3
3	3.8
4	4.3
6	5.3
8	6.2
10	7

FIGURE 14-26

Answers to Self-test 14-2 are at the end of the chapter.

CHAPTER 14 AT A GLANCE

Page	Key Points and definitions	Example
367	*Key point:* The graph of any linear equation is a straight line.	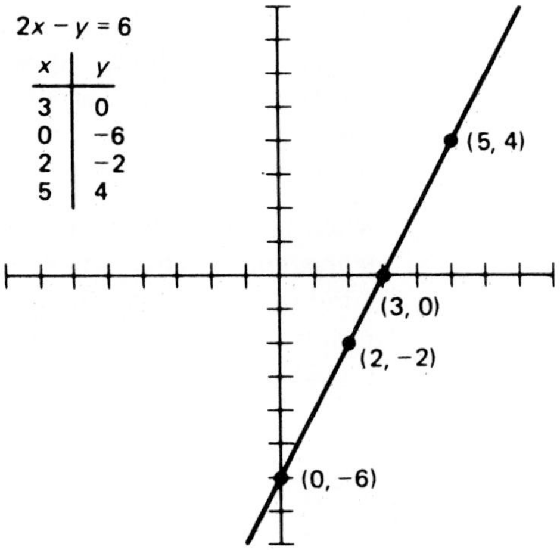 FIGURE 14-27
370	The *x-intercept* is the point where the curve crosses the *x*-axis. The *y-intercept* is the point where the curve crosses the *y*-axis.	

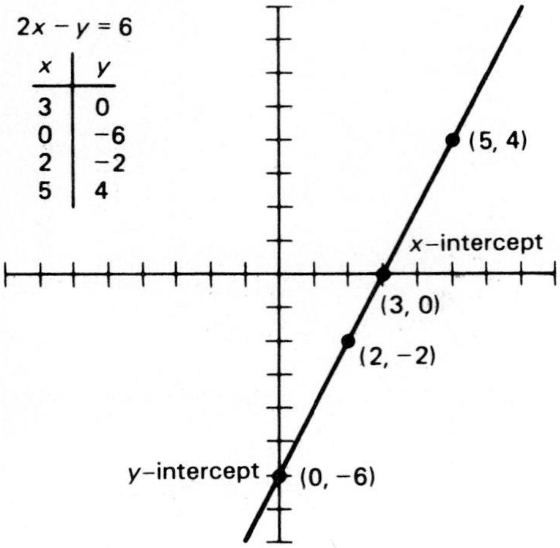

FIGURE 14-28

Page	Key Points and definitions	Example
374	*Key point:* The *slope* of a line is defined as the change in y that results from an *increase* in x.	slope $= \Delta y/\Delta x$
377	*Key point:* Any linear equation containing two variables can be written in the slope–intercept form.	$y = mx + b$. m is the slope. b is the y-intercept.
384	When plotting curves, the *independent* variable is plotted along the x-axis and the *dependent* variable is plotted along the y-axis.	

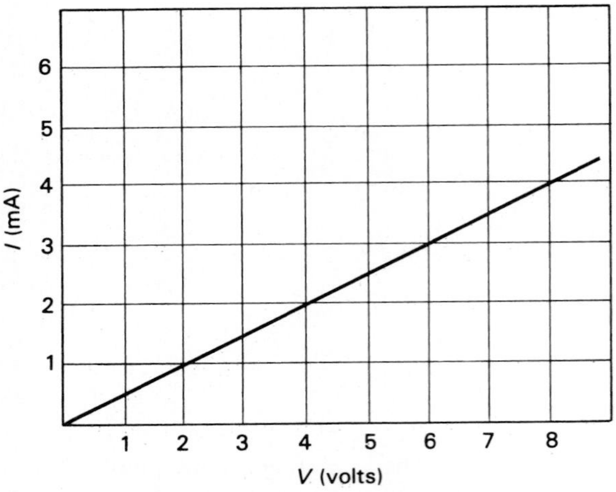

FIGURE 14-29

End of Chapter Problems 14-1

1. Identify the following points on the graph of Figure 14-30:
(a) (0, 6) (b) (−5, 0) (c) (3, −5)
(d) (4, 4) (e) (−2, −5) (f) (−4, 3.5)

2. Identify the following points on the graph of Figure 14-30:
(a) (−5, −5) (b) (−4, 4) (c) (6, 3)
(d) (0, −4) (e) (4, 0) (f) (2, −4.5)

3. Identity the following points on the graph of Figure 14-31:
(a) (30, −20) (b) (−40, 25) (c) (30, 30)
(d) (−10, −30) (e) (0, 45) (f) (−30, 0)
(g) (7, −18) (h) (−23, 47)

4. Identify the following points on the graph of Figure 14-31:
(a) (20, −30) (b) (10, 10) (c) (−40, 15)
(d) (−20, −20) (e) (0, −25) (f) (40, 0)
(g) (−13, 22) (h) (23, −8)

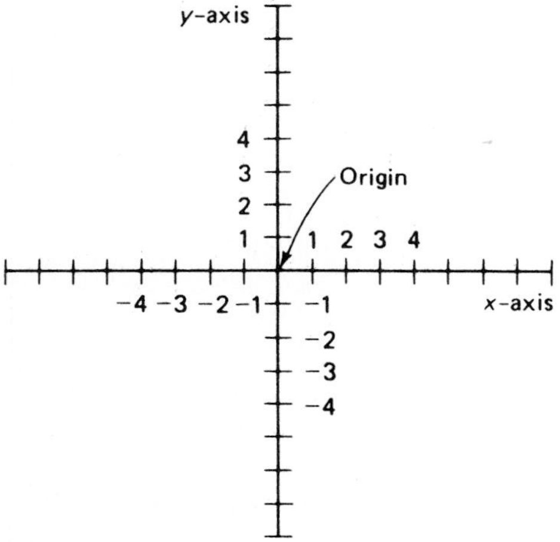

FIGURE 14-30

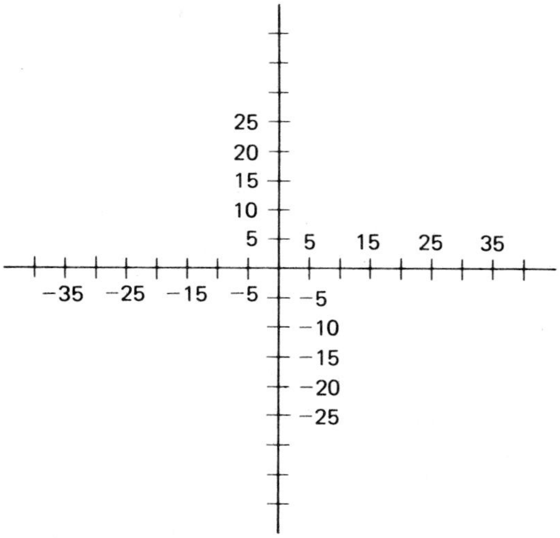

FIGURE 14-31

End of Chapter Problems 14-2

On a system of rectangular coordinates, plot the following equations. Identify the x and y intercepts. Plot at least three points.

1. (a) $x + y = 4$, (b) $2x - y = 6$, and (c) $x + 3y = 3$
2. (a) $2y = 4x + 8$, (b) $x - 2y = -4$, and (c) $x + 2y = -6$
3. $2x + 3y = 8$
4. $x - 4y = 4$
5. $x = -4y + 6$
6. $3x = 2y - 5$
7. $-4x - 3y = 24$
8. $3x - 5y = -20$
9. $-x + 2y = -7$
10. $4x - 5y = -8$

End of Chapter Problems 14-3

Plot the following equations on linear graph paper. Identify the x- and y-intercepts. Determine the slope.

1. $2x + y = 6$
2. $3x - 6y = 12$
3. $2y = 5x + 5$
4. $2x = 2y + 7$
5. $3x + 5y = 15$
6. $-4x - 6y = 24$
7. $2x - 7y = -14$
8. $6x + 3y = -18$
9. $-3x + 4y = -24$
10. $3y = 7x - 21$

End of Chapter Problems 14-4

Change the following equations into the slope–intercept form. Determine the slope and the y-intercept. Graph the equation.

1. $x + 3y = 3$
2. $x - 5y = 7$
3. $3x + 4y = -5$
4. $y - x = -4$
5. $x - 4y = 12$
6. $x + 4y = -5$
7. $5y - 2x = 10$
8. $x - y = 6$
9. $3x - 7y = 21$
10. $8x - 6y = 24$
11. $x + 3y = 6$
12. $2x + y = 8$
13. $2x - 3y = 8$
14. $x - 4y = 9$
15. $2y = x - 6$
16. $3y = x - 9$
17. $3x - 8 = 2y$
18. $4x - 16 = 5y$
19. $3x + y = -9$
20. $4x + y = -12$

End of Chapter Problems 14-5

1. Refer to Figure 14-32. Find R_{dc} and r_{ac} where $V =$ 3 V.
2. Refer to Figure 14-32. Find R_{dc} and r_{ac} where $V =$ 6.5 V.
3. Refer to Figure 14-33. If ΔV_f is measured from 0.4 to 0.6 V, what is ΔI_f? What is the slope? What is the value of r_d?
4. Refer to Figure 14-33. If ΔV_f is measured from 0.2 to 0.6 V, what is ΔI_f? What is the slope? What is the value of r_d?

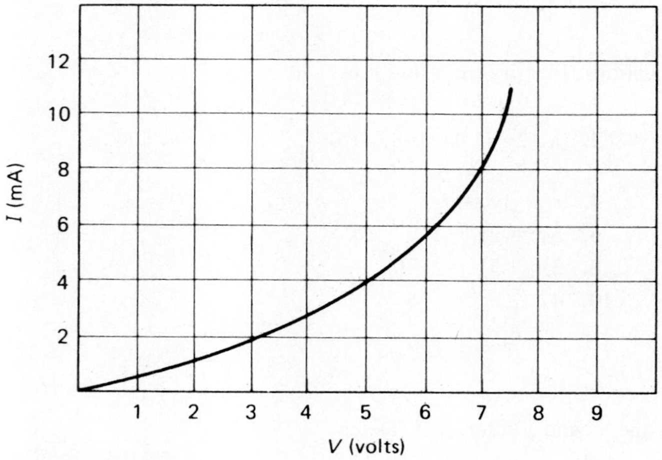

FIGURE 14-32

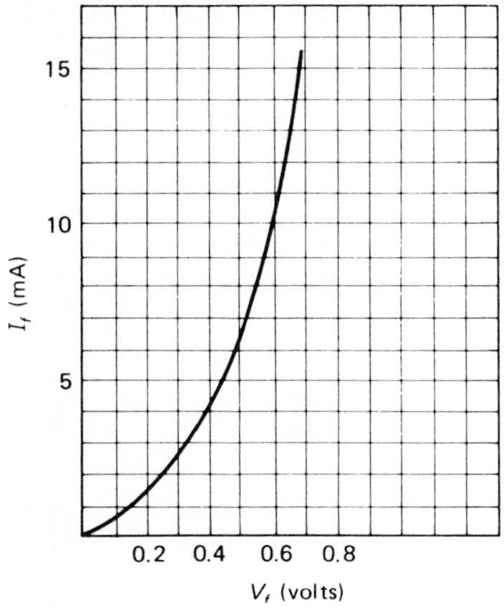

FIGURE 14-33

Refer to Figure 14-34.

5. Let $E = 100$ V. Find V_C and V_R at (a) 0.5 TC and (b) 3 TC.

6. Let $E = 60$ V. Find V_C and V_R at (a) 1 TC and (b) 3.5 TC.

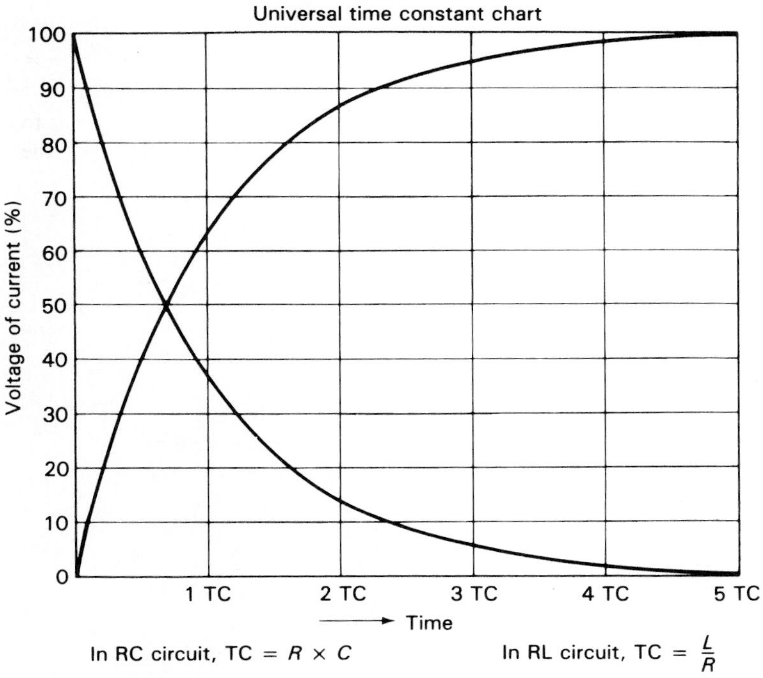

FIGURE 14-34

7. Let $E = 20$ V. Find V_C and V_R at (a) 1.5 TC and (b) 4 TC.

8. Let $E = 50$ V. Find V_C and V_R at (a) 2 TC and (b) 3 TC.

End of Chapter Problems 14-6

Plot a curve of I versus V_R for the following:

1. $R = 100\ \Omega$ for values of V_R from 0 to 50 V.
2. $R = 2\ \text{k}\Omega$ for values of V_R from 0 to 20 V.
3. $R = 1\ \text{k}\Omega$ for values of V_R from 0 to 100 V.
4. $R = 5\ \text{k}\Omega$ for values of V_R from 0 to 80 V.

Plot a curve of I versus V:

5. From the data in Figure 14-35.
6. From the data in Figure 14-36.
7. From the data in Figure 14-37.
8. From the data in Figure 14-38.

V (volts)	I (mA)
0	0
1	15
2	23
3	30
4	34
5	38
6	42

FIGURE 14-35

V (mV)	I (mA)
0	0
40	1
80	2
120	3.2
200	6
240	8
300	11

FIGURE 14-36

V (volts)	I (mA)
0	0
10	20
20	30
30	38
40	40
50	42
70	42

FIGURE 14-37

V (volts)	I (mA)
1	1
2	2
3	4
4	6
5	7

FIGURE 14-38

Answers to Self-tests

SELF-TEST 14-1

1. See Figure 14-39.
2. See Figure 14-40.
3. See Figure 14-41.

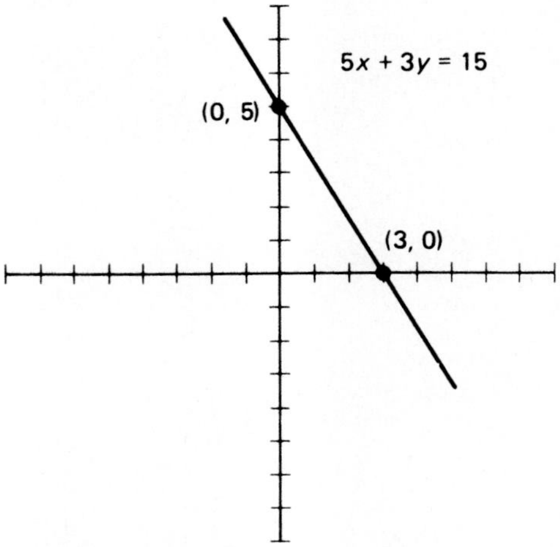

FIGURE 14-39

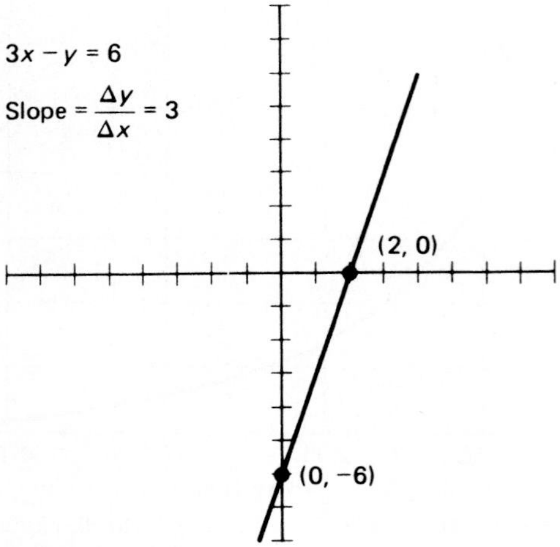

FIGURE 14-40

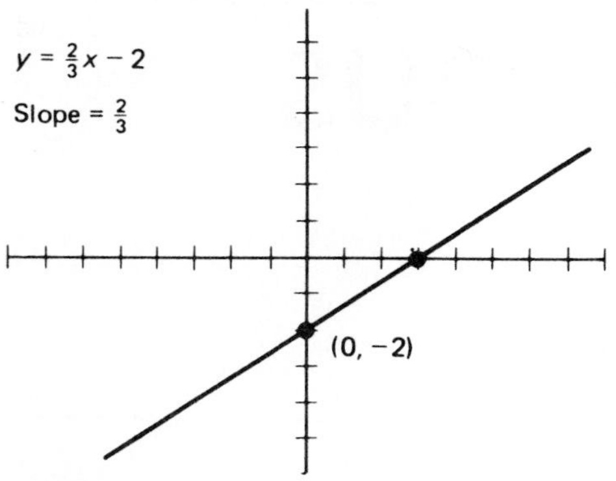

FIGURE 14-41

SELF-TEST 14-2

1. $R = 500\ \Omega$
2. $R_D \simeq 95\ \Omega$, $r_d \simeq 54\ \Omega$
3. $V_C \simeq 22.5$ V, $V_R \simeq 2.5$ V
4. See Figure 14-42.
5. See Figure 14-43.

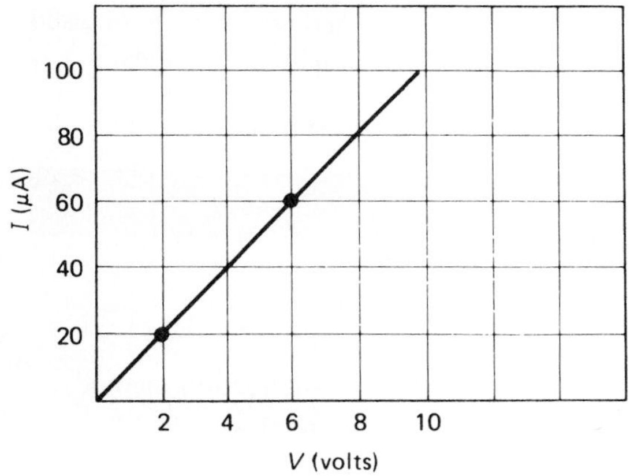

FIGURE 14-42

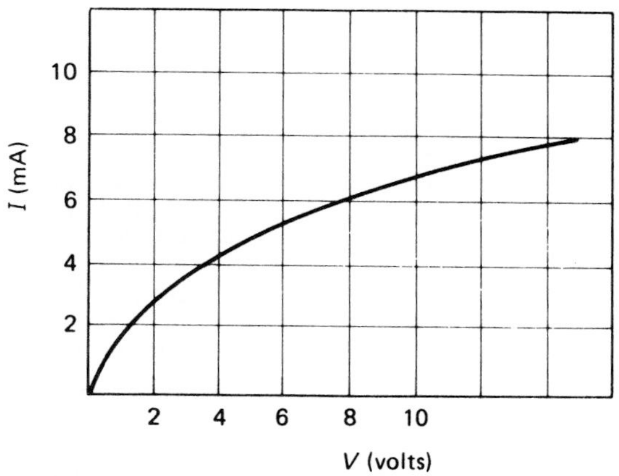

FIGURE 14-43

CHAPTER 15 SIMULTANEOUS LINEAR EQUATIONS

INTRODUCTION

In the chapters on circuit analysis we used various laws and theorems in solving circuit problems. In this chapter we introduce another method of solving complex circuit problems. When there is more than one path for current, we can write an equation in terms of the voltages, currents and resistances for each path. Because parts of each path will be common, we can solve for the unknown values using *simultaneous linear equations*. In this chapter we will show the various methods that can be used to solve sets of linear equations simultaneously, including graphical solutions.

CHAPTER OBJECTIVES

In this chapter you will learn how to:

1. Solve simultaneous linear equations graphically.
2. Solve simultaneous linear equations using the addition or subtraction method.
3. Solve simuitaneous linear equations using the substitution method.
4. Solve simultaneous linear equations using determinants.
5. Use simultaneous linear equations to solve complex circuit problems.

In the chapters on circuit analysis we used various laws and theorems in solving circuit problems. In this chapter, other methods are introduced.

15-1 LINEAR EQUATIONS IN TWO UNKNOWNS: GRAPHICAL SOLUTION

As we saw in Chapter 14, every linear equation in two unknowns has an unlimited number of solutions. If we have two equations, each equation still has an unlimited number of solutions. However, if the two equations are plotted on the same system of rectangular coordinates, they will intersect at some common point provided they (1) are not parallel lines and (2) are not the same line. The coordinates of this common point will satisfy both equations. When such equations have this common point of intersection, they are said to be *simultaneous*.

□ EXAMPLE 15-1

Plot the equations $x - 2y = 4$ and $3x + 2y = 4$ on the same system of rectangular coordinates. Find the point of intersection.

Solution The plot of the equations is shown in Figure 15-1. The point of intersection is at $(2, -1)$. This point is common to both equations. Because the lines intersect, there is a common solution, which in this case is $(2, -1)$. Furthermore, this is the only solution. $x = 2$ and $y = -1$ are the only values common to both equations.

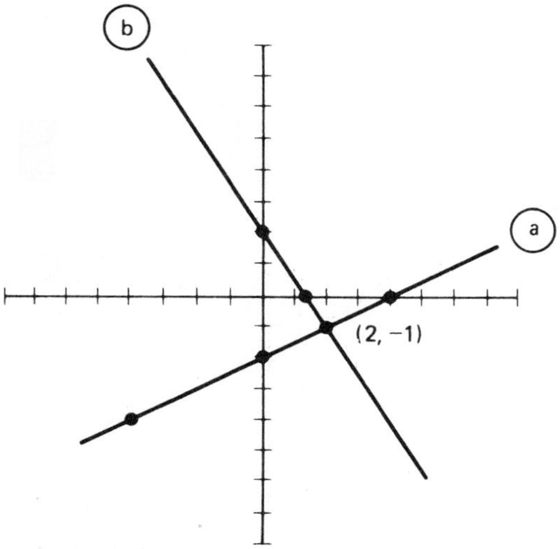

(a) $x - 2y = 4$

x	y
4	0
0	−2
−4	−4

(b) $3x + 2y = 4$

x	y
3	0
0	2
4	−4

FIGURE 15-1

□

PRACTICE PROBLEMS 15-1

Plot the following equations on the same system of rectangular coordinates. Find the coordinates of the point of intersection.

1. $x + y = 5$
$x - y = 3$

2. $x + y = 3$
$2x + y = 7$

3. $x + 2y = 7$
$2x + 2y = 10$

4. $\frac{x}{4} + \frac{y}{3} = \frac{7}{12}$
$\frac{x}{2} - \frac{y}{4} = \frac{1}{4}$

5. $x - y = 1.5$
$0.9375x - y = 3.75$

Solutions

1. (4, 1). The graph is shown in Figure 15-2.

2. (4, −1). The graph is shown in Figure 15-3.

3. (3, 2). The graph is shown in Figure 15-4.

4. (1, 1). The graph is shown in Figure 15-5.

5. (2.6, 1.1). The graph is shown in Figure 15-6.

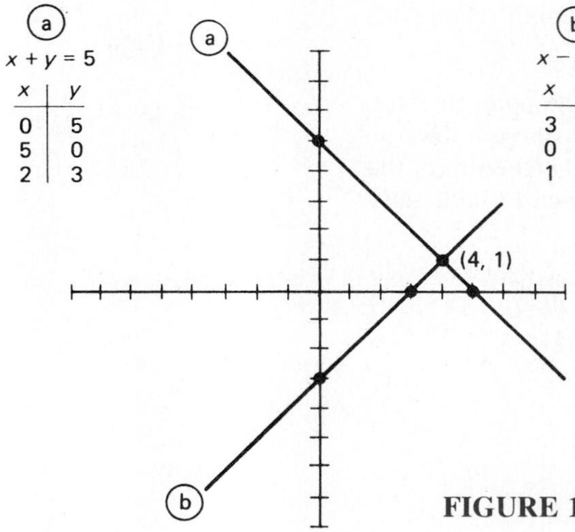

FIGURE 15-2

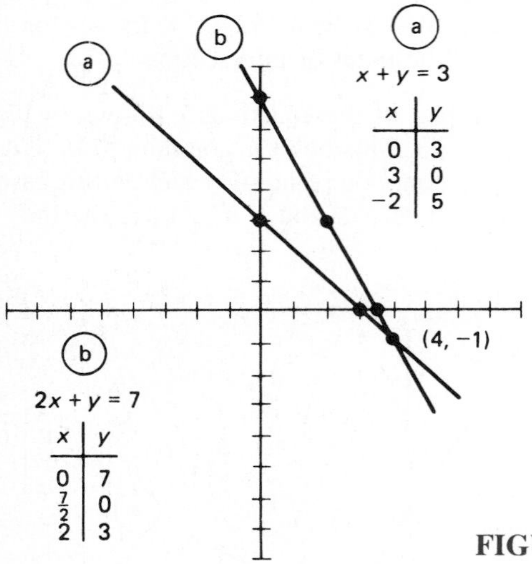

FIGURE 15-3

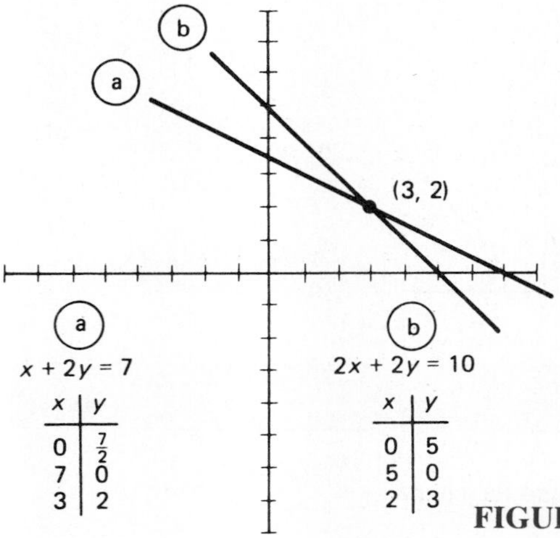

FIGURE 15-4

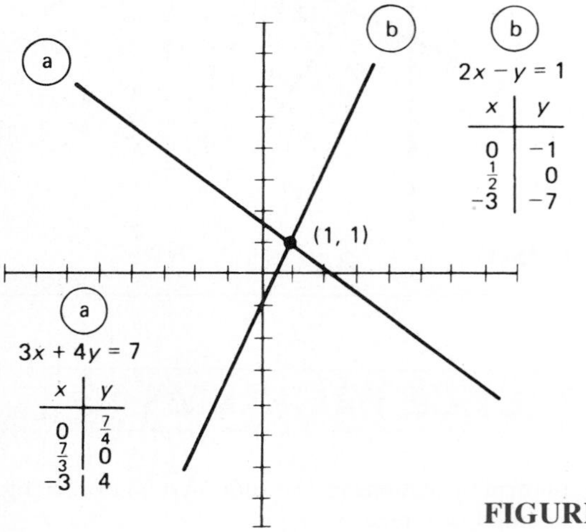

FIGURE 15-5

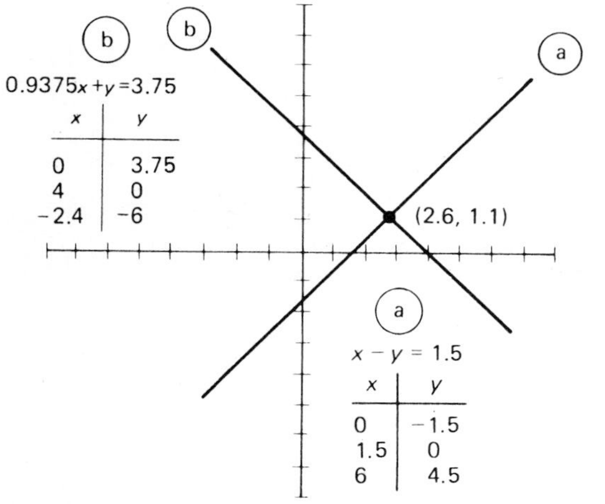

FIGURE 15-6

In graphical analysis all answers are approximations because we cannot read exact values from graphs. We were able to determine accurately the answers to the first four problems because the point of intersection was a whole number and the graphs are scaled in whole numbers. In problem 5 we had to approximate the answer.

Additional practice problems are at the end of the chapter.

15-2 LINEAR EQUATIONS IN TWO UNKNOWNS: ALGEBRAIC SOLUTION

15-2-1 Solution by Addition or Subtraction

Solving the two equations in Example 15-1 graphically resulted in $x = 2$ and $y = -1$. Let's solve the same problem by using the addition or subtraction method. This method uses rules of algebra we have already developed. We will change one or the other of the equations so that both x values, or both y values, have the same coefficient. If the signs of the coefficients are the same, we will subtract one equation from the other. If the signs are different, we will add the equations. The subtraction (or addition) results in a new equation that has only one unknown. We can now find the value of this unknown. This value is common to both original equations. The value of this unknown is then substituted back into one of the original equations and the other unknown is then found. Let's see how it works.

☐ EXAMPLE 15-2

Solve the following set of equations for x and y by the addition or subtraction method:

$$x - 2y = 4 \qquad (15\text{-}1)$$
$$3x + 2y = 4 \qquad (15\text{-}2)$$

Solution We can eliminate the y term by addition.

$$x - 2y = 4 \qquad (15\text{-}1)$$
$$\underline{3x + 2y = 4} \qquad (15\text{-}2)$$
$$\text{(add)}\quad 4x \qquad = 8 \qquad (15\text{-}3)$$

Solve Equation 15-3 for x.

$$4x = 8$$
$$x = 2$$

Substitute $x = 2$ in Equation 15-1 and solve for y.

$$x - 2y = 4 \qquad (15\text{-}1)$$
$$2 - 2y = 4$$
$$-2y = 2$$
$$y = -1$$

The common values of x and y are $x = 2$ and $y = -1$. Check Equation 15-1.

$$x - 2y = 4$$
$$2 - 2(-1) = 4$$
$$2 + 2 = 4$$

Check Equation 15-2.

$$3x + 2y = 4$$
$$3(2) + 2(-1) = 4$$
$$6 - 2 = 4$$

The solution is verified.

☐ EXAMPLE 15-3

Solve the following set of equations for x and y by the addition or subtraction method:

$$2x - y = 4 \qquad (15\text{-}4)$$
$$x - 2y = -1 \qquad (15\text{-}5)$$

Solution

Rewrite 15-4.	$2x - y = 4$	
Multiply 15-5 by 2.	$\underline{2x - 4y = -2}$	(15-6)
Subtract 15-5.	$0 + 3y = 6$	(15-7)

Solve Equation 15-7 for y:

$$3y = 6$$

$$y = 2$$

Substitute $y = 2$ in Equation 15-4 and solve for x.

$$2x - y = 4 \qquad (15\text{-}4)$$

$$2x - 2 = 4$$

$$2x = 6$$

$$x = 3$$

The common values of x and y are $x = 3$ and $y = 2$. Check Equation 15-4.

$$2x - y = 4 \qquad (15\text{-}5)$$

$$2(3) - 2 = 4$$

$$6 - 2 = 4$$

Check Equation 15-5.

$$x - 2y = -1 \qquad (15\text{-}5)$$

$$3 - 2(2) = -1$$

$$3 - 4 = -1$$

□

PRACTICE PROBLEMS 15-2

Solve the following sets of equations for x and y by the addition or subtraction method.

1. $x + y = 5$
$x - y = 3$

2. $x + y = 3$
$2x + y = 7$

3. $x + 2y = 7$
$2x + 2y = 10$

4. $\frac{x}{4} + \frac{y}{3} = \frac{7}{12}$
$\frac{x}{2} - \frac{y}{4} = \frac{1}{4}$

5. $x - y = 1.5$
$x + y = 3.75$

Solutions

1. $x = 4, y = 1$

2. $x = 4, y = -1$

3. $x = 3, y = 2$

4. $x = 1, y = 1$

5. $x = 2.625, y = 1.125$

Additional practice problems are at the end of the chapter.

15-2-2 Solution by Substitution

Let's work our example problems again, but this time we will solve the equations by substitution. This involves finding the value of x (or y) in one equation and then substituting that value back into the other equation. This eliminates one of the unknowns, and we can then find the value of the other. Example 15-4 shows how it works.

☐ EXAMPLE 15-4

Solve the following set of equations for x and y by using the substitution method:

$$x - 2y = 4 \tag{15-1}$$

$$3x + 2y = 4 \tag{15-2}$$

Solution Solve Equation 15-1 for x.

$$x - 2y = 4 \tag{15-1}$$

$$x = 4 + 2y \tag{15-8}$$

Substitute the value of x in Equation 15-8 back into Equation 15-2 and then solve the equation for y.

$$3x + 2y = 4 \tag{15-2}$$

$$3(4 + 2y) + 2y = 4$$

$$12 + 6y + 2y = 4$$

$$8y = -8$$

$$y = -1$$

Now substitute -1 for y in either equation and solve for x. (We will choose Equation 15-1.)

$$x - 2y = 4 \tag{15-1}$$

$$x - 2(-1) = 4$$

$$x + 2 = 4$$

$$x = 2$$

The solution is $x = 2$ and $y = -1$. We could check the solutions as we did in the previous section.

☐ EXAMPLE 15-5

Solve the following set of equations for x and y by using the substitution method:

$$2x - y = 4 \tag{15-4}$$

$$x - 2y = -1 \tag{15-5}$$

Solution Solve for x in Equation 15-4.

$$2x - y = 4 \qquad (15\text{-}4)$$

$$2x = y + 4$$

$$x = \frac{y + 4}{2} \qquad (15\text{-}9)$$

Substitute the value of x in Equation 15-9 for the value of x in Equation 15-5.

$$x - 2y = -1 \qquad (15\text{-}5)$$

$$\frac{y + 4}{2} - 2y = -1$$

Multiply both sides by 2.

$$y + 4 - 4y = -2$$

$$-3y = -6$$

$$y = 2$$

Now substitute this value for y in Equation 15-4.

$$2x - y = 4$$

$$2x - 2 = 4$$

$$2x = 6$$

$$x = 3$$

The solution is $x = 3$ and $y = 2$. We could check the solutions as we did in the previous section.

□

PRACTICE PROBLEMS 15-3

Solve the sets of equations in Practice Problems 15-2 by using the substitution method.

Solutions: The solutions are the same as for Practice Problems 15-2.

Additional practice problems are at the end of the chapter.

15-3 DETERMINANTS

Another way to solve simultaneous equations is by using *determinants*. Determinants are arrays of numbers. These numbers correspond to the coefficients and constants of the different equations.

Consider the following general equations with two unknowns:

$$a_1x + b_1y = k_1 \tag{15-10}$$

$$a_2x + b_2y = k_2 \tag{15-11}$$

a_1, b_1, a_2, and b_2 are the coefficients of the unknowns. k_1 and k_2 are the constants. Notice that the equations are set up the same way we set them up when we were using the addition or subtraction method. The determinant for the denominator looks like this:

$$\begin{vmatrix} a_1 & b_1 \\ a_2 & b_2 \end{vmatrix}$$

This determinant has two rows and two columns and is called a *second-order determinant*. The numbers a_1, a_2, b_1, and b_2 are called the *elements* of the determinants. $a_1b_2 - a_2b_1$ is called the *expansion* of the determinants. The numbers a_1 and b_2 are the *principal diagonal*. The numbers a_2 and b_1 are the *secondary diagonal*. The expansion of the determinant results from the product of the elements in the principal diagonal (a_1b_2) minus the product of the elements in the secondary diagonal (a_2b_1). This process is illustrated as follows:

$$\begin{vmatrix} a_1 & b_1 \\ a_2 & b_2 \end{vmatrix} = a_1b_2 - a_2b_1$$

This is called the *determinant of the denominator*. This is the denominator for both unknowns. The determinant for the numerator of the unknown, x, is written

$$\begin{vmatrix} k_1 & b_1 \\ k_2 & b_2 \end{vmatrix} = k_1b_2 - k_2b_1$$

The determinant for the numerator of the unknown, y, is written

$$\begin{vmatrix} a_1 & k_1 \\ a_2 & k_2 \end{vmatrix} = a_1k_2 - a_2k_1$$

When we put it all together, it looks like this:

$$x = \frac{\begin{vmatrix} k_1 & b_1 \\ k_2 & b_2 \end{vmatrix}}{\begin{vmatrix} a_1 & b_1 \\ a_2 & b_2 \end{vmatrix}} = \frac{k_1b_2 - k_2b_1}{a_1b_2 - a_2b_1}$$

$$y = \frac{\begin{vmatrix} a_1 & k_1 \\ a_2 & k_2 \end{vmatrix}}{\begin{vmatrix} a_1 & b_1 \\ a_2 & b_2 \end{vmatrix}} = \frac{a_1k_2 - a_2k_1}{a_1b_2 - a_2b_1}$$

☐ EXAMPLE 15-6

Solve for x and y in the following set of equations by using determinants. (This is the same set of equations used in Example 15-2.)

$$x - 2y = 4 \tag{15-12}$$

$$3x + 2y = 4 \tag{15-13}$$

Solution Determine the numbers to be used in the following arrays:

$$a_1 = 1, \quad b_1 = -2, \quad k_1 = 4$$

$$a_2 = 3, \quad b_2 = 2, \quad k_2 = 4$$

Now substitute these known values in our general equations for x and y.

$$x = \left| \frac{\begin{matrix} 4 & -2 \\ 4 & 2 \end{matrix}}{\begin{matrix} 1 & -2 \\ 3 & 2 \end{matrix}} \right| = \frac{8 - (-8)}{2 - (-6)} = \frac{16}{8} = 2$$

$$y = \left| \frac{\begin{matrix} 1 & 4 \\ 3 & 4 \end{matrix}}{\begin{matrix} 1 & -2 \\ 3 & 2 \end{matrix}} \right| = \frac{4 - 12}{2 - (-6)} = \frac{-8}{8} = -1$$

The solution is $x = 2$, $y = -1$.

Notice that the denominators are identical. This means that we need calculate the denominator only once. This method may be less laborious than either the addition or subtraction method or the substitution method in problem-solving situations.

☐ EXAMPLE 15-7

Solve for x and y in the following set of equations. (This is the same set of equations used in Example 15-5.)

$$2x - y = 4$$

$$x - 2y = -1$$

Solution Determine the numbers to be used in the following arrays:

$$a_1 = 2, \quad b_1 = -1, \quad k_1 = 4$$

$$a_2 = 1, \quad b_2 = -2, \quad k_2 = -1$$

$$x = \left| \frac{\begin{matrix} 4 & -1 \\ -1 & -2 \end{matrix}}{\begin{matrix} 2 & -1 \\ 1 & -2 \end{matrix}} \right| = \frac{-8 - 1}{-4 - (-1)} = \frac{-9}{-3} = 3$$

$$y = \left| \frac{\begin{matrix} 2 & 4 \\ 1 & -1 \end{matrix}}{\begin{matrix} 2 & -1 \\ 1 & -2 \end{matrix}} \right| = \frac{-2 - 4}{-3} = \frac{-6}{-3} = 2$$

The solution is $x = 3$, $y = 2$.

PRACTICE PROBLEMS 15-4

Solve the sets of equations in Practice Problems 15-2 by using determinants.

Solutions: The solutions are the same as for Practice Problems 15-2.

Additional practice problems are at the end of the chapter.

Let's now consider three equations and three unknowns. Our general equations look like this:

$$a_1x + b_1y + c_1z = k_1$$

$$a_2x + b_2y + c_2z = k_2$$

$$a_3x + b_3y + x_3z = k_3$$

The determinants for the three unknowns look like this:

$$x = \frac{\begin{vmatrix} k_1 & b_1 & c_1 \\ k_2 & b_2 & c_2 \\ k_3 & b_3 & c_3 \end{vmatrix}}{\begin{vmatrix} a_1 & b_1 & c_1 \\ a_2 & b_2 & c_2 \\ a_3 & b_3 & c_3 \end{vmatrix}} = \frac{k_1b_2c_3 + b_1c_2k_3 + c_1k_2b_3 - c_1b_2k_3 - k_1c_2b_3 - b_1k_2c_3}{a_1b_2c_3 + b_1c_2a_3 + c_1a_2b_3 - c_1b_2a_3 - a_1c_2b_3 - b_1a_2c_3} \qquad (15\text{-}14)$$

$$y = \frac{\begin{vmatrix} a_1 & k_1 & c_1 \\ a_2 & k_2 & c_2 \\ a_3 & k_3 & c_3 \end{vmatrix}}{\begin{vmatrix} a_1 & b_1 & c_1 \\ a_2 & b_2 & c_2 \\ a_3 & b_3 & c_3 \end{vmatrix}} = \frac{a_1k_2c_3 + k_1c_2a_3 + c_1a_2k_3 - c_1k_2a_3 - a_1c_2k_3 - k_1a_2c_3}{a_1b_2c_3 + b_1c_2a_3 + c_1a_2b_3 - c_1b_2a_3 - a_1c_2b_3 - b_1a_2c_3} \qquad (15\text{-}15)$$

$$z = \frac{\begin{vmatrix} a_1 & b_1 & k_1 \\ a_2 & b_2 & k_2 \\ a_3 & b_3 & k_3 \end{vmatrix}}{\begin{vmatrix} a_1 & b_1 & c_1 \\ a_2 & b_2 & c_2 \\ a_3 & b_3 & c_3 \end{vmatrix}} = \frac{a_1b_2k_3 + b_1k_2a_3 + k_1a_2b_3 - k_1b_2a_3 - a_1k_2b_3 - b_1a_2k_3}{a_1b_2c_3 + b_1c_2a_3 + c_1a_2b_3 - c_1b_2a_3 - a_1c_2b_3 - b_1a_2c_3} \qquad (15\text{-}16)$$

Various methods are used to find values of x, y, and z. Here is one of them: Rewrite columns 1 and 2 to the right of the determinant. For the unknown, x, the numerator would look like this:

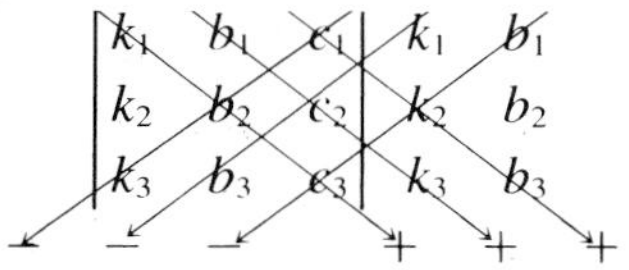

Diagonal lines have been drawn through the elements wherever there are three elements in the diagonal group. Where there are fewer than three elements in the diagonal group, that diagonal group is not used. The result is three diagonal groups running down from left to right and three diagonal groups running down from right to left. The products of each diagonal group running down from left to right are added. The products of the groups that run down from right to left are subtracted. This procedure is used for each numerator and each denominator. Since all the denominators are alike, however, the denominators need to be determined only once.

☐ EXAMPLE 15-8

Solve for the unknowns in the following set of equations by using determinants:

$$x + 2y + 3z = 14$$

$$2x + y + 2z = 10$$

$$3x + 4y - 3z = 2$$

Solution Let's first write down the coefficients:

$$a_1 = 1, \quad b_1 = 2, \quad c_1 = 3, \quad k_1 = 14$$

$$a_2 = 2, \quad b_2 = 1, \quad c_2 = 2, \quad k_2 = 10$$

$$a_3 = 3, \quad b_3 = 4, \quad c_3 = -3, \quad k_3 = 2$$

Next let's find the denominator since it will be common to all solutions. Referring to Equation 15-14 and replacing the literal numbers with real numbers, we get

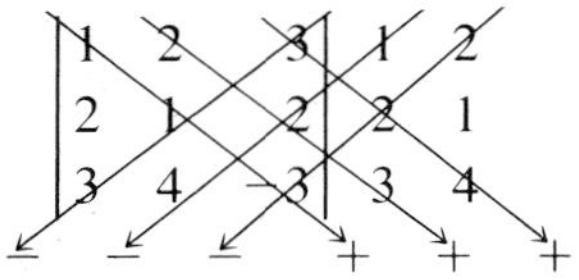

$$= (1)(1)(-3) + (2)(2)(3) + (3)(2)(4) - (3)(1)(3) - (1)(2)(4)$$

$$-(2)(2)(-3)$$

$$= -3 + 12 + 24 - 9 - 8 + 12 = 28$$

Using the same method, let's find x.

$x =$

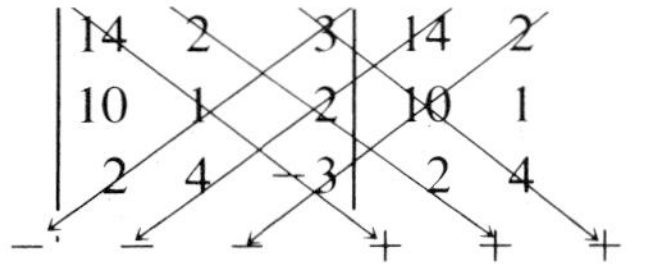

28

$$\frac{(14)(1)(-3) + (2)(2)(2) + (3)(10)(4) - (3)(1)(2) - (14)(2)(4) - (2)(10)(-3)}{28}$$

$$= \frac{-42 + 8 + 120 - 6 - 112 + 60}{28} = \frac{28}{28} = 1$$

Now let's find y the same way.

$$y = \frac{\left|\begin{array}{ccc|cc} 1 & 14 & 3 & 1 & 14 \\ 2 & 10 & 2 & 2 & 10 \\ 3 & 2 & -3 & 3 & 2 \end{array}\right|}{28}$$

$$= \frac{(1)(10)(-3) + (14)(2)(3) + (3)(2)(2) - (3)(10)(3) - (1)(2)(2) - (14)(2)(-3)}{28}$$

$$= \frac{-30 + 84 + 12 - 90 - 4 + 84}{28} = \frac{56}{28} = 2$$

Now we solve for z and are finished.

$$z = \frac{\left|\begin{array}{ccc|cc} 1 & 2 & 14 & 1 & 2 \\ 2 & 1 & 10 & 2 & 1 \\ 3 & 4 & 2 & 3 & 4 \end{array}\right|}{28}$$

$$= \frac{(1)(1)(2) + (2)(10)(3) + (14)(2)(4) - (14)(1)(3) - (1)(10)(4) - (2)(2)(2)}{28}$$

$$= \frac{2 + 60 + 112 - 42 - 40 - 8}{28} = \frac{84}{28} = 3$$

□

PRACTICE PROBLEMS 15-5

Use determinants to solve for the unknowns in the following sets of equations:

1. $x + 2y + z = 6$
$2x - y + 3z = -13$
$3x - 2y + 3z = -16$

2. $3x + 2y = 12$
$4x + 2z = 16$
$4y + 3z = 24$

(Even though only two of the three unknowns appear in each equation, we must use all three in our solution. When the third unknown is missing, we assign a value of zero to that coefficient. For example, $c_1 = 0$ in the first equation of problem 2).

Solutions

1. $x = 1.25$, $y = 4.25$, $z = -3.75$
2. $x = 2$, $y = 3$, $z = 4$

Additional practice problems are at the end of the chapter.

SELF-TEST 15-1

Solve the following sets of equations (a) graphically, (b) by the addition or subtraction method, (c) by the substitution method, and (d) by using determinants.

1. $4x + y = 9$
   $x + y = 6$
2. $2x + 4y = 4$
   $x - y = 5$
3. $3x - 5y = 11$
   $x + 8y = -6$

Answers to Self-test 15-1 are at the end of the chapter.

15-4 APPLICATIONS

In Chapter 11 we learned how to use various circuit theorems to solve complex circuit problems. An alternate method is solution by simultaneous linear equations.

As a first example of solution by simultaneous equations, let's solve for the currents in Figure 15-7. In this example we will solve the problem using all three methods: by addition and subtraction, by substitution, and by determinants.

☐ EXAMPLE 15-9

Find the currents and the voltage drops in the circuit of Figure 15-7.

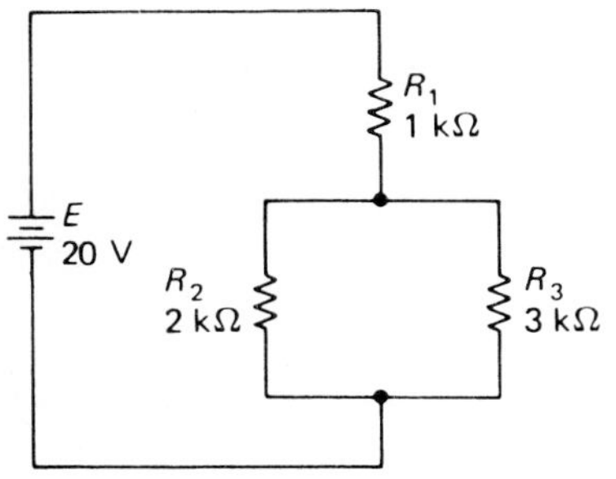

FIGURE 15-7

Solution This circuit can easily be solved by using Ohm's law and Kirchhoff's laws. This solution would yield

$$R_T = R_1 + R_2 \parallel R_3 = 1\text{ k}\Omega + 1.2\text{ k}\Omega = 2.2\text{ k}\Omega$$

$$I_T = \frac{E}{R_T} = \frac{20\text{ V}}{2.2\text{ k}\Omega} = 9.09\text{ mA}$$

$$V_1 = IR_1 = 9.09\text{ mA} \times 1\text{ k}\Omega = 9.09\text{ V}$$

$$V_2 = V_3 = IR_x = 9.09\text{ mA} \times 1.2\text{ k}\Omega = 10.9\text{ V}$$

$$(R_x = R_2 \parallel R_3)$$

$$I_2 = \frac{V_2}{R_2} = \frac{10.9\text{ V}}{2\text{ k}\Omega} = 5.45\text{ mA}$$

$$I_3 = \frac{V_3}{R_3} = \frac{10.9\text{ V}}{3\text{ k}\Omega} = 3.64\text{ mA}$$

$$I_1 = I_T = 9.09\text{ mA}$$

Now let's work the same problem using simultaneous linear equations. Let's first establish the polarities of the voltage drops across each resistor as in Figure 15-8. Using Kirchhoff's voltage law, let's start at point A and move counterclockwise around the loop containing E, R_1, and R_2.

$$E - V_2 - V_1 = 0$$

or

$$E = V_1 + V_2 \tag{15-17}$$

Starting at point A and moving counterclockwise around the loop containing E, R_1, and R_3, we get

$$E - V_3 - V_1 = 0$$

or

$$E = V_1 + V_3 \tag{15-18}$$

If we express these equations in terms of IR drops, then Equation 15-17 would look like this:

$$E = I_1R_1 + I_2R_2$$

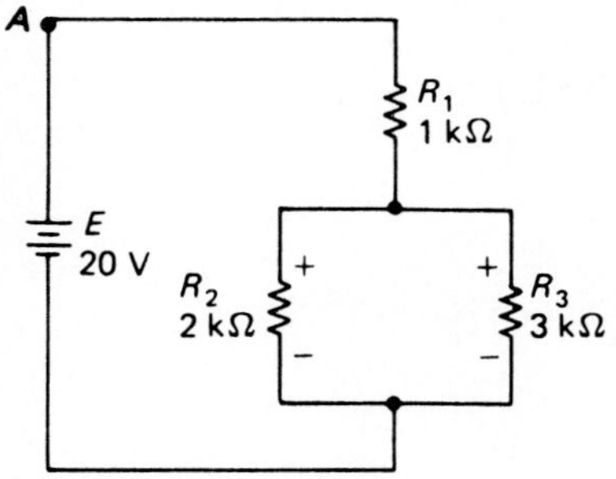

FIGURE 15-8

Substituting known values gives

$$20 = 1kI_1 + 2kI_2 \tag{15-19}$$

Equation 15-18 would look like this:

$$E = I_1R_1 + I_3R_3$$

Substituting known values gives

$$20 = 1kI_1 + 3kI_3 \tag{15-20}$$

Now we have two equations that we can solve simultaneously.

$$20 = 1kI_1 + 2kI_2 \quad (15\text{-}19)$$

$$20 = 1kI_1 + 3kI_3 \quad (15\text{-}20)$$

In Equations 15-19 and 15-20 we have three unknowns. We can simplify the solution by reducing the number of unknowns to two. Since $I_1 = I_2 + I_3$, then $I_3 = I_1 - I_2$. Let's substitute $(I_1 - I_2)$ for I_3 in Equation 15-20.

$$\begin{aligned} 20\text{ V} &= 1kI_1 + 3kI_3 \\ &= 1kI_1 + 3k(I_1 - I_2) \\ &= 1kI_1 + 3kI_1 - 3kI_2 \\ &= 4kI_1 - 3kI_2 \qquad (15\text{-}21) \end{aligned}$$

Now let's look at Equations 15-19 and 15-21. We have two equations and two unknowns and can now solve the problem.

$$20\text{ V} = 1kI_1 + 2kI_2 \quad (15\text{-}19)$$

$$= 4kI_1 - 3kI_2 \quad (15\text{-}21)$$

Solution by Addition and Subtraction: To eliminate one unknown, we will multiply Equation 15-19 by 4.

$$80\text{ V} = 4kI_1 + 8kI_2 \tag{15-22}$$

Subtracting Equation 15-22 from Equation 15-21 gives

$$\begin{aligned} 20\text{ V} &= 4kI_1 - 3kI_2 \quad (15\text{-}21) \\ \underline{80\text{ V}} &\underline{= 4kI_1 + 8kI_2} \quad (15\text{-}22) \\ -60\text{ V} &= 0 \quad - 11kI_2 \end{aligned}$$

Solving for I_2, we get

$$I_2 = \frac{-60\text{ V}}{-11\text{k}} = 5.45\text{ mA}$$

Now let's plug the known value of I_2 back into Equation 15-21 and solve for I_1.

$$20\text{ V} = 4\text{k}I_1 - 3\text{k}I_2$$
$$= 4\text{k}I_1 - 3\text{k}(5.45\text{ mA})$$
$$= 4\text{k}I_1 - 16.4\text{ V}$$
$$36.4\text{ V} = 4\text{k}I_1$$
$$\frac{36.4\text{ V}}{4\text{k}} = 9.09\text{ mA} = I_1$$

Since

$$I_1 = I_2 + I_3$$

then

$$I_3 = I_1 - I_2$$
$$I_3 = 9.09\text{ mA} - 5.45\text{ mA} = 3.64\text{ mA}$$

Knowing I_1, I_2, and I_3, we can solve for V_1, V_2, and V_3:

$$V_1 = I_1R_1 = 9.09\text{ mA} \times 1\text{ k}\Omega = 9.09\text{ V}$$
$$V_2 = I_2R_2 = 5.45\text{ mA} \times 2\text{ k}\Omega = 10.9\text{ V}$$
$$V_3 = I_3R_3 = 3.64\text{ mA} \times 3\text{k}\Omega = 10.9\text{ V}$$

Verifying, we get

$$E = V_1 + V_2 = 9.09\text{ V} + 10.9\text{ V} = 20\text{ V}$$

Solution by Substitution

$$20\text{ V} = 1\text{k}I_1 + 2\text{k}I_2 \qquad (15\text{-}19)$$
$$20\text{ V} = 4\text{k}I_1 - 3\text{k}I_2 \qquad (15\text{-}21)$$

In Equation 15-19, solve for one of the variables. Let's use I_1.

$$1\text{k}I_1 = 20\text{ V} - 2\text{k}I_2$$
$$I_1 = \frac{20\text{ V} - 2\text{k}I_2}{1\text{k}} = 20\text{ mA} - 2I_2$$

Substituting this value of I_1 in Equation 15-21, we get

$$20\text{ V} = 4\text{k}(20\text{ mA} - 2I_2) - 3\text{k}I_2 = 80\text{ V} - 8\text{k}I_2 - 3\text{k}I_2$$
$$20\text{ V} = 80\text{ V} - 11\text{k}I_2$$
$$11\text{k}I_2 = 60\text{ V}$$
$$I_2 = 5.45\text{ mA}$$

Substituting this value for I_2 in Equation 15-19 and solving for I_1,

$$20\text{ V} = 1\text{k}I_1 + 2\text{k}(5.45\text{ mA}) = 1\text{k}I_1 + 10.9\text{ V}$$
$$20\text{ V} - 10.9\text{ V} = 1\text{k}I_1$$
$$I_1 = \frac{9.1}{1\text{k}} = 9.10\text{ mA}$$

I_3 and the voltage drops can be found as before.

Solution by Determinants

$$20\text{ V} = 1\text{k}I_1 + 2\text{k}I_2 \qquad (15\text{-}19)$$

$$20\text{ V} = 4\text{k}I_1 - 3\text{k}I_2 \qquad (15\text{-}21)$$

In the array,

$$a_1 = 1\text{k}, \qquad b_1 = 2\text{k}, \qquad \text{and} \qquad k_1 = 20$$

$$a_2 = 4\text{k}, \qquad b_2 = -3\text{k}, \qquad \text{and} \qquad k_2 = 20$$

$$I_1 = \frac{k_1 b_2 - k_2 b_1}{a_1 b_2 - a_2 b_1} = \frac{-60.0 \times 10^3 - 40.0 \times 10^3}{-11.0 \times 10^6} = 9.09\text{ mA}$$

$$I_2 = \frac{a_1 k_2 - a_2 k_1}{a_1 b_2 - a_2 b_1} = \frac{20.0 \times 10^3 - 80.0 \times 10^3}{-11.0 \times 10^6} = 5.45\text{ mA}$$

I_3 and the voltage drops can be found as before.

Graphical Solution The problem may also be solved graphically. Although this method is seldom used in solving electrical problems such as this, it is useful in some applications and is presented here. Figure 15-9 is the graphical solution.

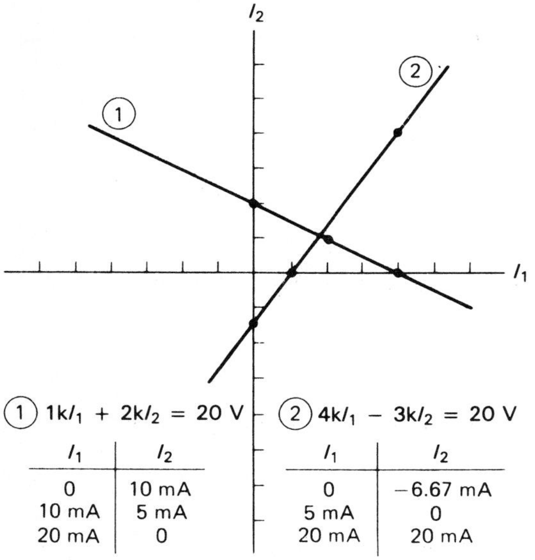

FIGURE 15-9 Graphical solution. Notice that, while other solutions yield exact values for I_1 and I_2, the graphical solution provides only approximate values. That is, the intersect occurs where $I_1 \simeq 5$ mA and $I_2 \simeq 9$ mA.

□

PRACTICE PROBLEMS 15-6

Solve for the currents and voltage drops in Figure 15-10 by using simultaneous linear equations.

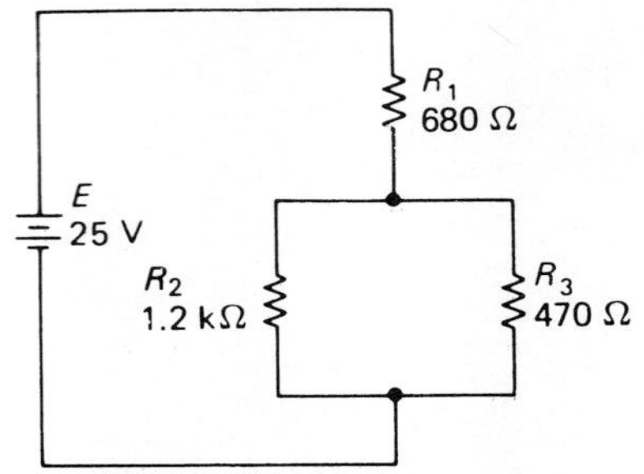

FIGURE 15-10

Solution

The two basic equations should look like this:

$$E = V_1 + V_2 \quad \text{and} \quad E = V_1 + V_3$$

Substituting known values, we get

$$25\text{ V} = 680I_1 + 1.2\text{k}I_2 \qquad (15\text{-}23)$$

and

$$= 680I_1 + 470I_3 \qquad (15\text{-}24)$$

Letting $I_3 = I_1 - I_2$, Equation 15-24 becomes

$$\begin{aligned} 25\text{ V} &= 680I_1 + 470I_3 \\ &= 680I_1 + 470(I_1 - I_2) \\ &= 680I_1 + 470I_1 - 470I_2 \\ &= 1150I_1 - 470I_2 \qquad (15\text{-}25) \end{aligned}$$

Now we need to multiply (or divide) Equations 15-23 and 15-25 by some factor or factors to make I_1 or I_2 equal in both equations. As we know, there are an infinite number of factors we could use. We choose to multiply Equation 15-25 by 2.55 so that the coefficient of I_2 in both equations equals 1.2.

$$\begin{aligned} 25\text{ V} &= 1150I_1 - 470I_2 \qquad (15\text{-}25) \\ 25\text{ V}(2.55) &= 1150(2.55)I_1 - 470(2.55)I_2 \\ 63.8\text{ V} &= 2930I_1 - 1200I_2 \qquad (15\text{-}26) \end{aligned}$$

Putting Equations 15-23 and 15-26 together and adding, we get

$$\begin{aligned} 25\text{ V} &= 680I_1 + 1200I_2 \qquad (15\text{-}23) \\ \underline{63.8\text{ V}} &\underline{= 2930I_1 - 1200I_2} \qquad (15\text{-}26) \\ 88.8\text{ V} &= 3610I_1 \end{aligned}$$

$$\frac{88.8\text{ V}}{3610} = I_1 = 24.6\text{ mA}$$

Substituting 24.6 mA for I_1 in Equation 15-23, we get

$$\begin{aligned} 25\text{ V} &= 680(24.6\text{ mA}) + 1200I_2 \\ 25\text{ V} &= 16.7\text{ V} + 1200I_2 \\ 8.3\text{ V} &= 1200I_2 \end{aligned}$$

$$\frac{8.3\text{ V}}{1200} = I_2 = 6.92\text{ mA}$$

In this problem

$$I_1 = I_2 + I_3$$

$$I_3 = I_1 - I_2$$

Then

$$I_3 = 24.6\text{ mA} - 6.92\text{ mA} = 17.7\text{ mA}$$

Using these currents, we get

$$\begin{aligned} V_1 &= I_1R_1 = 24.6\text{ mA} \times 680\ \Omega = 16.7\text{ V} \\ V_2 &= I_2R_2 = 6.92\text{ mA} \times 1.2\text{ k}\Omega = 8.3\text{ V} \\ V_3 &= I_3R_3 = 17.7\text{ mA} \times 470\ \Omega = 8.32\text{ V} \end{aligned}$$

Let's solve the same problem using determinates:

$$\begin{aligned} 25\text{ V} &= 680I_1 + 1200I_2 \qquad (15\text{-}23) \\ 25\text{ V} &= 1150I_1 - 470I_2 \qquad (15\text{-}25) \end{aligned}$$

In the array

$$a_1 = 680, \qquad b_1 = 1200, \qquad k_1 = 25$$

$$a_2 = 1150, \qquad b_2 = -470, \qquad k_2 = 25$$

$$I_1 = \frac{\begin{vmatrix} 25 & 1200 \\ 25 & -470 \end{vmatrix}}{\begin{vmatrix} 680 & 1200 \\ 1150 & -470 \end{vmatrix}} = \frac{-11.8 \times 10^3 - 30 \times 10^3}{-320 \times 10^3 - 1.38 \times 10^6}$$

$$I_1 = \frac{-41.8 \times 10^3}{-1.70 \times 10^6} = 24.6 \text{ mA}$$

$$I_2 = \frac{\begin{vmatrix} 680 & 25 \\ 1150 & 25 \end{vmatrix}}{\begin{vmatrix} 680 & 1200 \\ 1150 & -470 \end{vmatrix}} = \frac{17.0 \times 10^3 - 28.8 \times 10^3}{-320 \times 10^3 - 1.38 \times 10^6}$$

$$= \frac{-11.8 \times 10^3}{-1.7 \times 10^3} = 6.91 \text{ mA}$$

These values for I_1 and I_2 agree with the values found using the substitution method. The calculations to find I_3 and the voltage drops are the same as when using the substitution method.

Differences between voltage drops or currents using the various methods are due to rounding. You should understand these methods of solution before proceeding to the more complex problems.

Additional practice problems are at the end of the chapter.

In Figure 15-11, other methods than just Ohm's law and Kirchhoff's laws must be used. Let's find the currents and voltage drops by using simultaneous linear equations.

We first have to determine the direction of current in the different branches. If the direction of current is not obvious, we simply assume a direction. If our assumption is wrong, when we solve for current, its magnitude will be correct but it will be negative.

Let's assume that the direction of current results from E_2. The resulting polarities would be as shown in Figure 15-12. There are three loops. One loop includes E_1, R_1, E_2, and R_3. A second loop includes E_1, R_1, and R_2, and a third loop includes E_2, R_2, and R_3. Even though there are three loops, we need use only two. Let's use the loops that contain E_2.

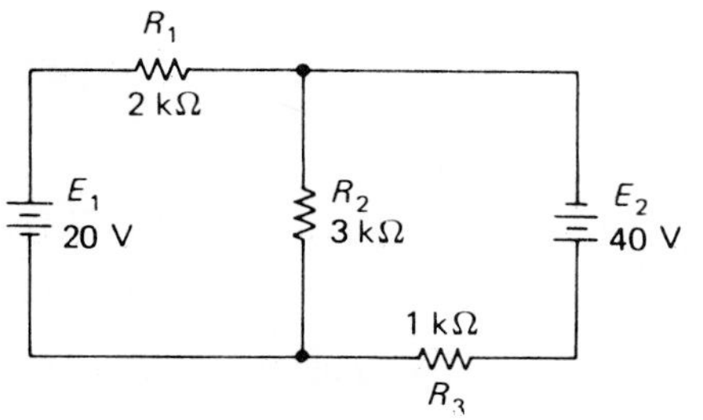

FIGURE 15-11

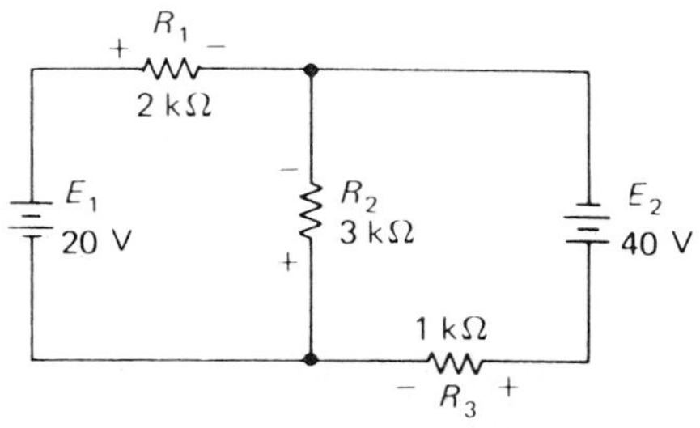

FIGURE 15-12

$$E_2 - V_2 - V_3 = 0$$

$$E_2 = V_2 + V_3 \qquad (15\text{-}27)$$

$$E_2 - V_1 + E_1 - V_3 = 0$$

$$E_2 + E_1 = V_1 + V_3 \qquad (15\text{-}28)$$

E_1 and E_2 are both included in the left member of Equation 15-28 only because they are both known. Substituting known values, we get

$$E_2 = V_2 + V_3 \qquad (15\text{-}27)$$

$$40 \text{ V} = V_2 + V_3 \qquad (15\text{-}29)$$

$$E_1 + E_2 = V_1 + V_3 \qquad (15\text{-}28)$$

$$60 \text{ V} = V_1 + V_3 \qquad (15\text{-}30)$$

Expressing Equations 15-29 and 15-30 in terms of IR drops, we get

$$40 \text{ V} = V_2 + V_3 \qquad (15\text{-}29)$$

$$= I_2R_2 + I_3R_3$$

$$= 3\text{k}I_2 + 1\text{k}I_3 \qquad (15\text{-}31)$$

$$60 \text{ V} = V_1 + V_3 \qquad (15\text{-}30)$$

$$= I_1R_1 + I_3R_3$$

$$= 2\text{k}I_1 + 1\text{k}I_3 \qquad (15\text{-}32)$$

Let's look at Equations 15-31 and 15-32 together.

$$40 \text{ V} = 3000I_2 + 1000I_3 \qquad (15\text{-}31)$$

$$60 \text{ V} = 2000I_1 + 1000I_3 \qquad (15\text{-}32)$$

If we assume that the current through R_3 is the total current, then $I_3 = I_1 + I_2$, which makes $I_1 = I_3 - I_2$. Substituting $I_3 - I_2$ for I_1 in Equation 15-32 to eliminate one unknown yields

$$60 \text{ V} = 2000I_1 + 1000I_3 \qquad (15\text{-}32)$$

$$= 2000(I_3 - I_2) + 1000I_3$$

$$= 2000I_3 - 2000I_2 + 1000I_3$$

$$= -2000I_2 + 3000I_3 \qquad (15\text{-}33)$$

Now let's put Equations 15-31 and 15-33 together and solve.

$$40 \text{ V} = 3000I_2 + 1000I_3 \qquad (15\text{-}31)$$

$$60 \text{ V} = -2000I_2 + 3000I_3 \qquad (15\text{-}33)$$

Solution by addition and subtraction: Multiplying Equation 15-31 by 3 causes I_3 in both equations to be equal.

$$40 \text{ V}(3) = 3000I_2(3) + 1000I_3(3)$$

$$120 \text{ V} = 9000I_2 + 3000I_3 \qquad (15\text{-}34)$$

Now we can put Equations 15-33 and 15-34 together and solve by subtracting.

$$120\text{ V} = 9000I_2 + 3000I_3 \tag{15-34}$$

$$\underline{60\text{ V} = -2000I_2 + 3000I_3} \tag{15-33}$$

$$60\text{ V} = 11{,}000I_2$$

$$\frac{60\text{ V}}{11{,}000\ \Omega} = 5.45\text{ mA} = I_2$$

Substituting 5.45 mA for I_2 in Equation 15-34 yields

$$120\text{ V} = 9000(5.45\text{ mA}) + 3000I_3$$

$$120\text{ V} = 49.1\text{ V} + 3000I_3$$

$$70.9\text{ V} = 3000I_3$$

$$\frac{70.9\text{ V}}{3000\ \Omega} = I_3 = 23.6\text{ mA}$$

If

$$I_3 = 23.6\text{ mA} \quad \text{and} \quad I_2 = 5.45\text{ mA}$$

then

$$I_1 = I_3 - I_2 = 23.6\text{ mA} - 5.45\text{ mA} = 18.2\text{ mA}$$

Referring to Figure 15-11 and solving for the various voltage drops, we get

$$V_1 = I_1R_1 = 18.2\text{ mA} \times 2\text{ k}\Omega = 36.4\text{ V}$$

$$V_2 = I_2R_2 = 5.45\text{ mA} \times 3\text{ k}\Omega = 16.4\text{ V}$$

$$V_3 = I_3R_3 = 23.6\text{ mA} \times 1\text{ k}\Omega = 23.6\text{ V}$$

To verify, we apply Kirchhoff's voltage law to the loops.

Solution by substitution

$$40\text{ V} = 3000I_2 + 1000I_3 \tag{15-31}$$

$$60\text{ V} = -2000I_2 + 3000I_3 \tag{15-33}$$

Solving Equation 15-31 for I_2, we get

$$I_2 = \frac{40\text{ V} - 1000I_3}{3000} = 13.3 \times 10^{-3}\text{ mA} - 3.33 \times 10^{-1}I_3$$

Substituting this value for I_2 in Equation 15-33 yields

$$60\text{ V} = -2000(13.3 \times 10^{-3} - 3.33 \times 10^{-1}I_3) + 3000I_3$$

$$= -26.7\text{ V} + 667I_3 + 3000I_3 = -26.7\text{ V} + 3670I_3$$

$$86.7\text{ V} = 3670I_3$$

$$I_3 = 23.6\text{ mA}$$

Substituting this value back into Equation 15-33 and solving for I_2,

$$60\text{ V} = -2000I_2 + 3000(23.6 \times 10^{-3}) = -2000I_2 + 70.8\text{ V}$$

$$-10.8\text{ V} = -2000I_2$$

$$I_2 = 5.4\text{ mA}$$

I_1 and the voltage drops would be the same as those calculated when the solution was by addition and subtraction (some differences may exist due to rounding).

Solution by determinants

$$40\text{ V} = 3000I_2 + 1000I_3 \qquad (15\text{-}31)$$

$$60\text{ V} = -2000I_2 + 3000I_3 \qquad (15\text{-}33)$$

In the array,

$$a_1 = 3000, \qquad b_1 = 1000, \qquad k_1 = 40$$

$$a_2 = -2000, \qquad b_2 = 3000, \qquad k_2 = 60$$

$$I_2 = \frac{k_1b_2 - k_2b_1}{a_1b_2 - a_2b_1} = \frac{120 \times 10^3 - 60 \times 10^3}{9.00 \times 10^6 - (-2 \times 10^6)} = 5.45\text{ mA}$$

$$I_3 = \frac{a_1k_2 - a_2k_1}{a_1b_2 - a_2b_1} = \frac{180 \times 10^3 - (-80 \times 10^3)}{9.00 \times 10^6 - (-2 \times 10^6)} = 23.6\text{ mA}$$

Values for I_1 and the voltage drops may be calculated as before.

PRACTICE PROBLEMS 15-7

Use simultaneous equations to find the currents and voltage drops of the circuits in Figure 15-13(a), (b), and (c).

Solutions

Remember, your method of solution may be different from the author's because you may not assume the same direction of currents or you may not use the same two loops.

(a) Refer to Figure 15-13(a). This solution uses the loops that contain E_1, R_1, and R_2 and E_1, E_2, R_1, and R_3. Assume that the direction of current results from E_1. Starting at the positive terminal of E_1 and moving in a clockwise direction gives

$$E_1 - V_1 - V_2 = 0$$

$$E_1 = V_1 + V_2$$

$$40\text{ V} = V_1 + V_2$$

$$= 100I_1 + 330I_2 \qquad (15\text{-}35)$$

$$E_1 - V_1 + E_2 - V_3 = 0$$

$$E_1 + E_2 = V_1 + V_3$$

$$40\text{ V} + 30\text{ V} = V_1 + V_3$$

$$70\text{ V} = V_1 + V_3$$

$$= 100I_1 + 220I_3 \qquad (15\text{-}36)$$

Since we assumed that I_1 is the total current, $I_2 = I_1 - I_3$, and we can substitute $I_1 - I_3$ for I_2 in Equation 15-35:

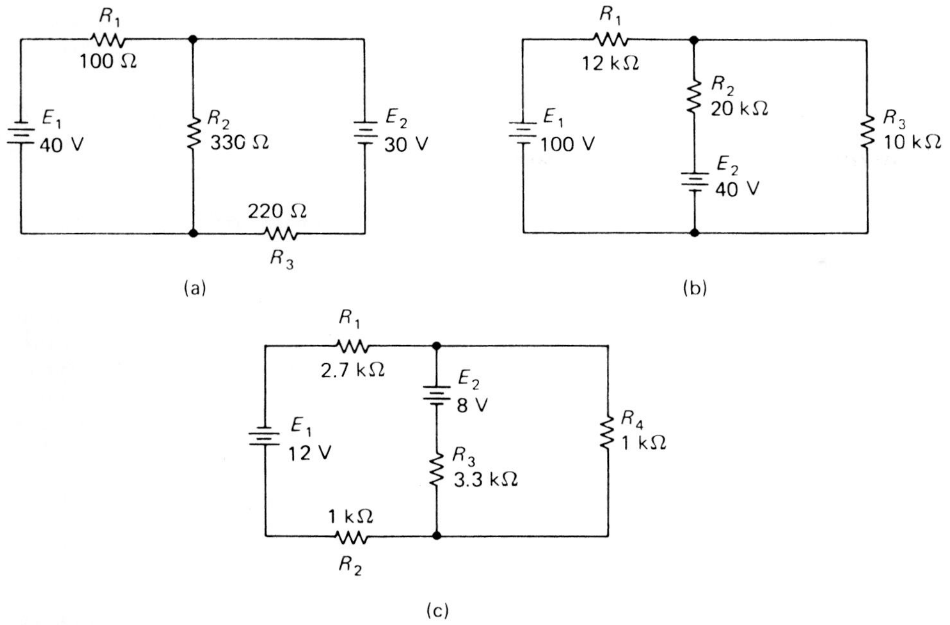

FIGURE 15-13

$$40\text{ V} = 100I_1 + 330I_2 \qquad (15\text{-}35)$$
$$= 100I_1 + 330(I_1 - I_3)$$
$$= 430I_1 - 330I_3 \qquad (15\text{-}37)$$

Multiplying Equation 15-36 by 4.3 yields

$$301\text{ V} = 430I_1 + 946I_3 \qquad (15\text{-}38)$$

Subtracting Equation 15-37 from Equation 15-38 yields

$$40\text{ V} = 430I_1 - 330I_3 \qquad (15\text{-}37)$$
$$\underline{301\text{ V} = 430I_1 + 946I_3} \qquad (15\text{-}38)$$
$$-261\text{ V} = 0 - 1276I_3$$
$$261\text{ V} = 1267I_3$$

Then

$$I_3 = \frac{261\text{ V}}{1276\ \Omega} = 205\text{ mA}$$

Substituting in Equation 15-36 gives

$$70\text{ V} = 100I_1 + 220(205\text{ mA})$$
$$70\text{ V} = 100I_1 + 45.1\text{ V}$$
$$24.9\text{ V} = 100I_1$$
$$I_1 = 249\text{ mA}$$

Substituting in Equation 15-35 gives

$$40\text{ V} = 100I_1 + 330I_2 \qquad (15\text{-}37)$$
$$40\text{ V} = 100(249\text{ mA}) + 330I_2$$
$$40\text{ V} = 24.9\text{ V} + 330I_2$$
$$15.1\text{ V} = 330I_2$$
$$I_2 = 45.8\text{ mA}$$
$$V_1 = I_1R_1 = 249\text{ mA} \times 100\ \Omega = 24.9\text{ V}$$
$$V_2 = I_2R_2 = 45.8\text{ mA} \times 330\ \Omega = 15.1\text{ V}$$
$$V_3 = I_3R_3 = 205\text{ mA} \times 220\ \Omega = 45.1\text{ V}$$

Apply Kirchhoff's laws to verify the solution.

(b) Refer to Figure 15-13(b). This solution uses the loop that includes E_1, E_2, R_1, and R_2 and the loop that includes E_2, R_2, and R_3. Assume that the current in each loop results from E_1. Starting at the negative terminal of E_1 and moving clockwise, for the first loop we get

$$-E_1 + V_1 + V_2 + E_2 = 0$$
$$-100 + V_1 + V_2 + 40\text{ V} = 0$$
$$V_1 + V_2 = 60\text{ V}$$
$$12\text{k}I_1 + 20\text{k}I_2 = 60 \qquad (15\text{-}38)$$

For the second loop, start at the negative terminal of E_2 and move clockwise:

$$-E_2 - V_2 - V_3 = 0$$

$$-40\text{ V} - V_2 + V_3 = 0$$

$$-V_2 + V_3 = 40\text{ V}$$

$$-20\text{k}I_2 + 10\text{k}I_3 = 40\text{ V} \qquad (15\text{-}39)$$

If we assume that I_1 is the total current, then $I_1 = I_2 + I_3$ and we can substitute $I_2 + I_3$ for I_1 in Equation 15-38.

$$12\text{k}(I_2 + I_3) + 20\text{k}I_2 = 60\text{ V}$$

$$12\text{k}I_2 + 12\text{k}I_3 + 20\text{k}I_2 = 60\text{ V}$$

$$32\text{k}I_2 + 12\text{k}I_3 = 60\text{ V} \qquad (15\text{-}40)$$

Using Equations 15-39 and 15-40, we will solve for the currents using the substitution method.

$$-20\text{k}I_2 + 10\text{k}I_3 = 40\text{ V} \qquad (15\text{-}39)$$

$$32\text{k}I_2 + 12\text{k}I_3 = 60\text{ V} \qquad (15\text{-}40)$$

Solving for I_2 in Equation 15-39, we get

$$I_2 = \frac{40\text{ V} - 10\text{k}I_3}{-20\text{k}} = -2\text{ mA} + 0.5I_3$$

Substituting this value into Equation 15-40,

$$32(-2\text{ mA} + 0.5I_3) + 12\text{k}I_3 = 60\text{ V}$$

$$-64\text{ V} + 16\text{k}I_3 + 12\text{k}I_3 = 60\text{ V}$$

$$28\text{k}I_3 = 124\text{ V}$$

$$I_3 = 4.43\text{ mA}$$

Substituting this value of I_3 back into Equation 15-39,

$$-20\text{k}I_2 + 10\text{k}(4.43\text{ mA}) = 40\text{ V}$$

$$-20\text{k}I_2 + 44.3\text{ V} = 40\text{ V}$$

$$-20\text{k}I_2 = -4.3\text{ V}$$

$$I_2 = 215\ \mu\text{A}$$

Knowing I_2 and I_3, we can solve for the rest of the unknowns.

$$I_1 = I_2 + I_3 = 215\ \mu\text{A} + 4.43\text{ mA} = 4.65\text{ mA}$$

$$V_1 = I_1R_1 = 4.65\text{ mA} \times 12\text{ k}\Omega = 55.7\text{ V}$$

$$V_2 = I_2R_2 = 215\ \mu\text{A} \times 20\text{ k}\Omega = 4.3\text{ V}$$

$$V_3 = I_3R_3 = 4.43\text{ mA} \times 10\text{ k}\Omega = 44.3\text{ V}$$

Apply Kirchhoff's laws to verify solutions.

(c) Refer to Figure 15-13(c). This solution uses the loop that includes E_1, E_2, R_1, R_2, and R_3 and the loop that includes E_2, R_3, and R_4. Assume that the current in each loop results from E_2. Starting at the positive terminal of E_2 and moving counterclockwise, for the first loop we get

$$E_2 - I_3R_3 - I_1R_2 + E_1 - I_1R_1 = 0$$

$$8\text{ V} - 3.3\text{k}I_3 - 1\text{k}I_1 + 12\text{ V} - 2.7\text{k}I_1 = 0$$

$$20\text{ V} = 3.7\text{k}I_1 + 3.3\text{k}I_3 \qquad (15\text{-}41)$$

(Because the current is the same through R_1 and R_2, we call that current I_1).

For the second loop, starting again at the positive terminal of E_2 and moving counterclockwise,

$$E_2 - I_3R_3 - I_4R_4 = 0$$

$$8\text{ V} - 3.3\text{k}I_3 - 1\text{k}I_4 = 0$$

$$8\text{ V} = 3.3\text{k}I_3 + 1\text{k}I_4 \qquad (15\text{-}42)$$

If we assume that I_3 is the total current, then $I_3 = I_1 + I_4$ or $I_4 = I_3 - I_1$, and we can substitute $I_3 - I_1$ for I_4 in Equation 15-42.

$$8\text{ V} = 3.3\text{k}I_3 + 1\text{k}(I_3 - I_1)$$

$$= 3.3\text{k}I_3 + 1\text{k}I_3 - 1\text{k}I_1$$

$$8\text{ V} = 4.3\text{k}I_3 - 1\text{k}I_1 \qquad (15\text{-}43)$$

Using Equations 15-41 and 15-43, we will solve for the currents using determinants.

$$20\text{ V} = 3.7\text{k}I_1 = 3.3\text{k}I_3 \qquad (15\text{-}41)$$

$$8\text{ V} = -1\text{k}I_1 + 4.3\text{k}I_3 \qquad (15\text{-}43)$$

In the array

$a_1 = 3.7\text{k}, \quad b_1 = 3.3\text{k}, \quad k_1 = 20 \text{ V}$

$a_2 = -1\text{k}, \quad b_2 = 4.3\text{k}, \quad k_2 = 8 \text{ V}$

$$I_1 = \frac{k_1 b_2 - k_2 b_1}{a_1 b_2 - a_2 b_1}$$

$$= \frac{20 \text{ V} \times 4.3\text{k} - 8 \times 3.3\text{k}}{3.7\text{k} \times 4.3\text{k} - (-1\text{k} \times 3.3\text{k})}$$

$$= \frac{86.0 \times 10^3 - 26.4 \times 10^3}{15.9 \times 10^6 + 3.3 \times 10^3}$$

$$= \frac{59.6 \times 10^3}{19.2 \times 10^6} = 3.1 \text{ mA}$$

$$I_3 = \frac{a_1 k_2 - a_2 k_1}{a_1 b_2 - a_2 b_1}$$

$$= \frac{3.7\text{k} \times 8 - (-1\text{k} \times 20)}{3.7\text{k} \times 4.3\text{k} - (-1\text{k} \times 3.3\text{k})}$$

$$= \frac{29.6 \times 10^3 + 20 \times 10^3}{15.9 \times 10^6 + 3.3 \times 10^6}$$

$$I_3 = \frac{49.6 \times 10^3}{19.2 \times 10^6} = 2.58 \text{ mA}$$

We assumed that I_3 was total current. Then

$$I_4 = I_3 - I_1 = 2.58 \text{ mA} - 3.1 \text{ mA}$$
$$= -520\ \mu\text{A}$$

The negative value for I_4 tells us that we assumed the wrong direction for current through R_4. I_1 is the total current, not I_3. The values calculated are correct. We just have to reverse the direction of current through R_4.

$$V_1 = I_1 R_1 = 3.1 \text{ mA} \times 2.7 \text{ k}\Omega = 8.37 \text{ V}$$

$$V_2 = I_2 R_2 = 3.1 \text{ mA} \times 1 \text{ k}\Omega = 3.1 \text{ V}$$

$$V_3 = I_3 R_3 = 2.58 \text{ mA} \times 3.3 \text{ k}\Omega = 8.51 \text{ V}$$

$$V_4 = I_4 R_4 = 520\ \mu\text{A} \times 1 \text{ k}\Omega = 0.52 \text{ V}$$

Apply Kirchhoff's law to verify the solutions.

Additional practice problems are at the end of the chapter.

SELF-TEST 15-2

1. Use the addition or subtraction method to solve for the currents and voltage drops of the circuit in Figure 15-14.

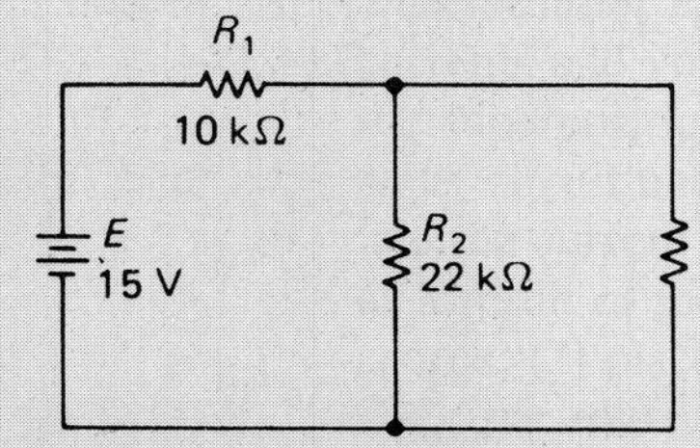

FIGURE 15-14

2. Use the substitution method to solve for the currents and voltage drops of the circuit in Figure 15-15.

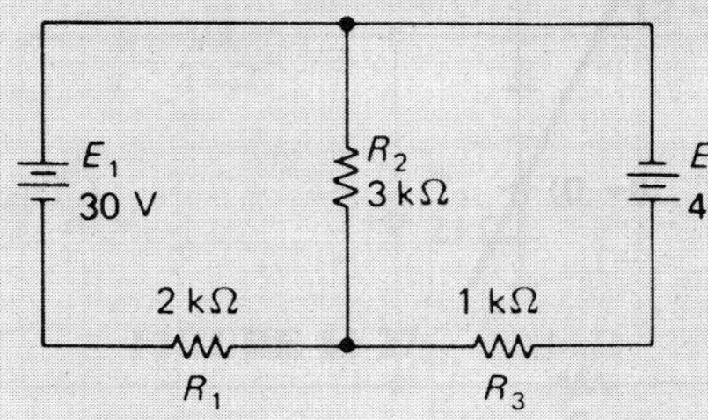

FIGURE 15-15

Answers to Self-test 15-2 are at the end of the chapter.

CHAPTER 15 AT A GLANCE

Page	Method of Solution	Example
397	Sets of equations may be solved by addition or subtraction. In this set, add and solve for x. Knowing x we can then solve for y.	$x - 2y = 4$ (add) $3x + 2y = 4$ $4x = 8$ $x = 2$ Then $y = -1$.
400	Sets of equations may be solved by substitution. Substitute $2y + 4$ for x in (2) and solve for y. Knowing y we can solve for x in either (1) or (2).	$x - 2y = 4$ (1) $3x + 2y = 4$ (2) Solve (1) for x $x = 2y + 4$ $3(2y + 4) + 2y = 4$ $y = -1$ Then $x = 2$.
401	Sets of equations may be solved by using determinants.	See Example 15-6.

End of Chapter Problems 15-1

Solve the following sets of equations graphically:

1. $2x - y = 4$
$2y = x + 1$

2. $3x + 3y = 5$
$x - y = -5$

3. $5x + 2y = 2$
$3x - y = 10$

4. $x + 2y = 6$
$4x + 3y = 4$

5. $8x - 4y = 12$
$x + 2y = 4$

6. $9x - y = 6$
$x + y = 4$

7. $2x - 5y = 1$
$3x - 8y = 2$

8. $3x - 2y = 3$
$2x + y = 2$

9. $2x - 3y = 4$
$x + 2y = 8$

10. $3x + 4y = 12$
$4x - 3y = -12$

11. $\dfrac{x}{3} + \dfrac{y}{2} = \dfrac{5}{12}$
$\dfrac{x}{4} - \dfrac{y}{3} = \dfrac{5}{24}$

12. $\dfrac{x}{2} + \dfrac{y}{5} = \dfrac{7}{20}$
$\dfrac{x}{3} - \dfrac{y}{5} = -\dfrac{4}{15}$

13. $\dfrac{x}{2} + \dfrac{y}{3} = 3$
$\dfrac{x}{2} - \dfrac{y}{3} = 1$

14. $x + 3y = 6$
$2x - 3y = 9$

End of Chapter Problems 15-2

Solve the following equations for x and y by using the addition or subtraction method:

1. $2x - y = 4$
   $2y = x + 1$
2. $3x + 3y = 5$
   $x - y = -5$
3. $5x + 2y = 2$
   $3x - y = 10$
4. $x + 2y = 6$
   $4x + 3y = 4$
5. $8x - 4y = 12$
   $x + 2y = 4$
6. $9x - y = 6$
   $x + y = 4$
7. $2x - 5y = 1$
   $3x - 8y = 2$
8. $3x - 2y = 3$
   $2x + y = 2$
9. $\frac{x}{2} + \frac{y}{3} = 3$
   $\frac{x}{2} - \frac{y}{3} = 1$
10. $x + 3y = 6$
    $2x - 3y = 9$
11. $2x - 3y = 4$
    $x + 2y = 8$
12. $3x + 4y = 12$
    $4x - 3y = -12$
13. $\frac{x}{3} + \frac{y}{2} = \frac{5}{12}$
    $\frac{x}{4} - \frac{y}{3} = \frac{5}{24}$
14. $\frac{x}{2} + \frac{y}{5} = \frac{7}{20}$
    $\frac{x}{3} - \frac{y}{5} = -\frac{4}{15}$

End of Chapter Problems 15-3

Solve the following sets of equations using the substitution method:

1. $x + 2y = 4$
   $2x - y = 2$
2. $3x - y = 6$
   $2x - y = 3$
3. $4x - 2y = 3$
   $3x + 2y = 5$
4. $2x - 5y = 4$
   $3x + 4y = 6$
5. $5x + 3y = 5$
   $3x - 4y = 6$
6. $4x + 5y = 6$
   $5x + 3y = 8$
7. $x + y = 10$
   $4x - 4y = 0$
8. $x - y = 9$
   $3x + 3y = 0$
9. $3x - 3y = -5$
   $-x - y = 4$
10. $5x - y = -4$
    $-x - y = 3$

End of Chapter Problems 15-4

Solve the following sets of equations using determinates:

1. $x + 2y = 4$
   $2x - y = 2$
2. $3x - y = 6$
   $2x - y = 3$
3. $4x - 2y = 3$
   $3x + 2y = 5$
4. $2x - 5y = 4$
   $3x + 4y = 6$

5. $5x + 3y = 5$
$3x - 4y = 6$

6. $4x + 5y = 6$
$5x + 3y = 8$

7. $x + y = 10$
$4x - 4y = 0$

8. $x - y = 9$
$3x + 3y = 0$

9. $3x - 3y = -5$
$-x - y = 4$

10. $5x - y = -4$
$-x - y = 3$

End of Chapter Problems 15-5

Use determinates to solve for the unknowns in the following sets of equations:

1. $x + 2y + 3z = 14$
$2x + y + 2z = 10$
$3x + 4y - 3z = 2$

2. $x - y - z = -22$
$2x - y + 2z = 22$
$3x + 3y - z = 22$

3. $x - 2y + 4z = 4$
$x + 3y - 2z = 6$
$3x + y + 3z = 5$

4. $3x + 2y + 3z = 8$
$4x - 4y - z = 4$
$x + y - 3z = 2$

5. $x + y + z = 12$
$2x - 3y - 4z = 8$
$-3x + 2y + 2z = 5$

6. $10x - 15y + 20z = 12$
$8x + 20y - 30z = 20$
$15x + 20y + 25z = 40$

End of Chapter Problems 15-6

Solve for the currents and voltage drops in the following problems by using simultaneous equations:

1.

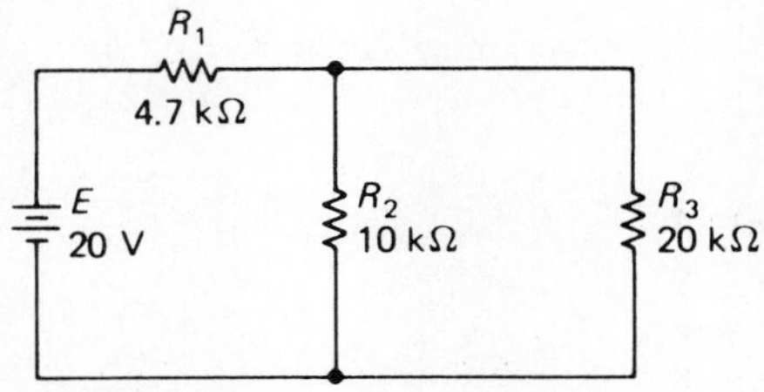

FIGURE 15-16

2.

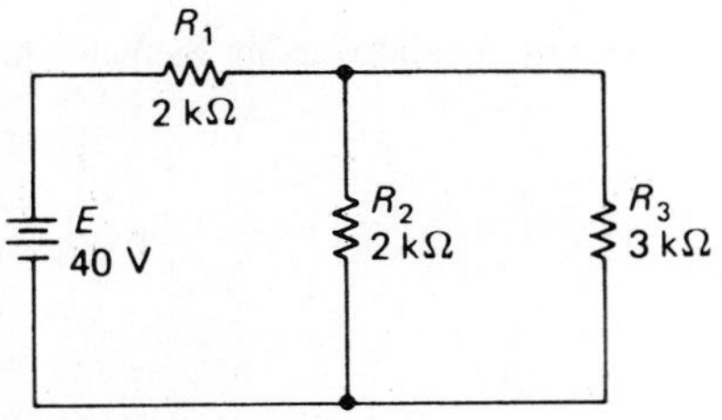

FIGURE 15-17

End of Chapter Problems 15-7

Solve for the currents and voltage drops in the following problems by using simultaneous equations:

1. See Figure 15-18. Let $E_1 = 20$ V, $E_2 = 40$ V, $R_1 = 3$ kΩ, $R_2 = 2$ kΩ, and $R_3 = 1$ kΩ.
2. See Figure 15-18. Let $E_1 = 50$ V, $E_2 = 15$ V, $R_1 = 27$ kΩ, $R_2 = 56$ kΩ, and $R_3 = 47$ kΩ.
3. See Figure 15-19. Let $E_1 = 100$ V, $E_2 = 40$ V, $R_1 = 12$ kΩ, $R_2 = 20$ kΩ, and $R_3 = 10$ kΩ.
4. See Figure 15-19. Let $E_1 = 25$ V, $E_2 = 60$ V, $R_1 = 470$ Ω, $R_2 = 820$ Ω, and $R_3 = 330$ Ω.
5. See Figure 15-20. Let $E_1 = 40$ V, $E_2 = 30$ V, $R_1 = 100$ Ω, $R_2 = 330$ Ω, and $R_3 = 220$ Ω.
6. See Figure 15-20. Let $E_1 = 10$ V, $E_2 = 15$ V, $R_1 = 270$ kΩ, $R_2 = 100$ kΩ, and $R_3 = 150$ kΩ.
7. See Figure 15-21. Let $E_1 = 12$ V, $E_2 = 8$ V, $R_1 = 2.7$ kΩ, $R_2 = 3.3$ kΩ, and $R_3 = 1$ kΩ.
8. See Figure 15-21. Let $E_1 = 80$ V, $E_2 = 50$ V, $R_1 = 68$ kΩ, $R_2 = 22$ kΩ, and $R_3 = 33$ kΩ.

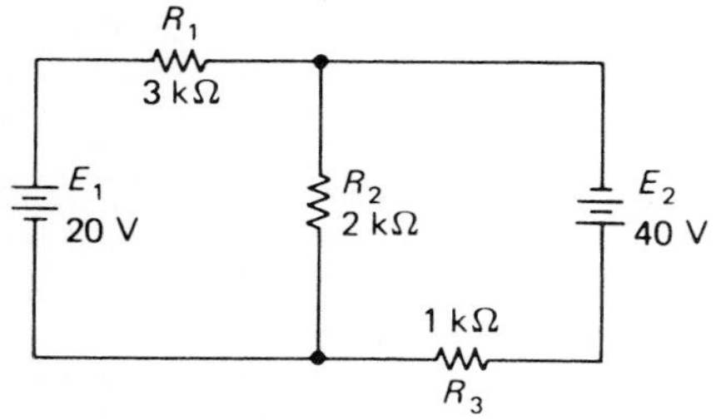

FIGURE 15-18

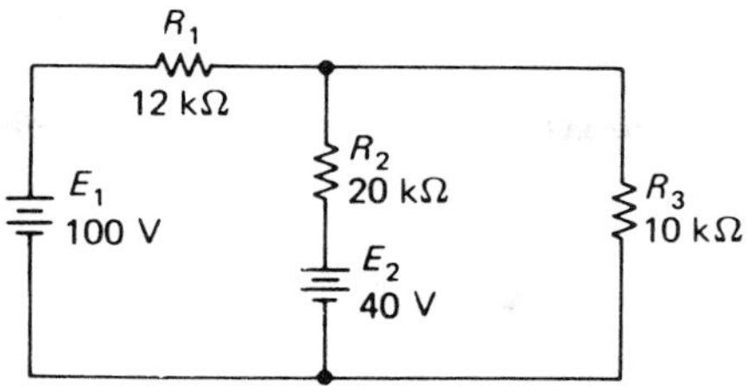

FIGURE 15-19

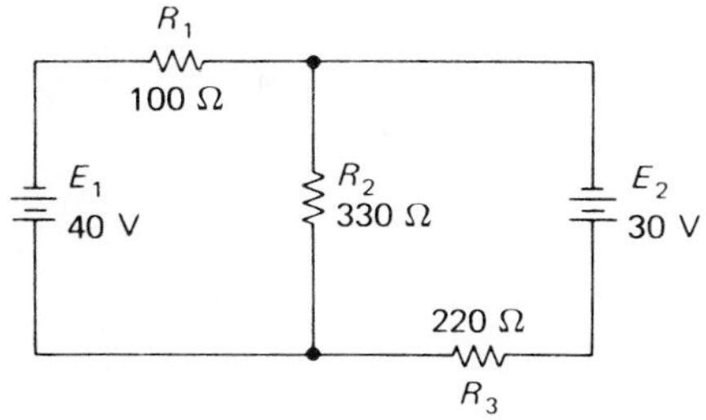

FIGURE 15-20

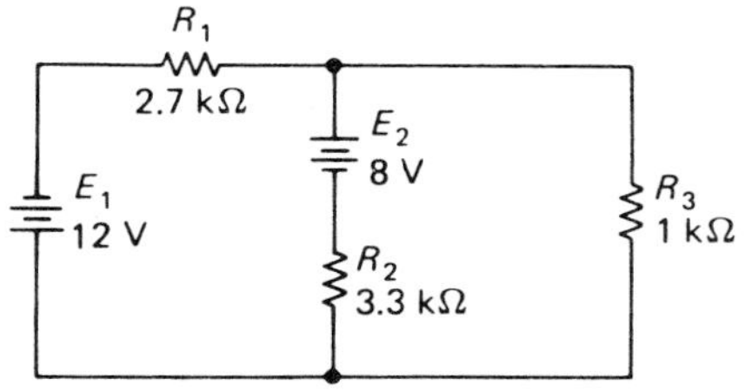

FIGURE 15-21

Answers to Self-tests

SELF-TEST 15-1

1. (a) See Figure 15-22 for graphical solution.

(b) Addition–subtraction method:

$$4x + y = 9 \quad (1)$$
$$x + y = 6 \quad (2)$$

Multiply (2) by 4: $4x + 4y = 24$
Rewrite (1): $4x + y = 9$
Subtract: $3y = 15$
$y = 5$

Substitute in (1).

$$4x + 5 = 9$$
$$4x = 4$$
$$x = 1$$

The solution is $x = 1$, $y = 5$.

(c) Substitution method: Solve for x in (2).

$$x + y = 6$$
$$x = 6 - y$$

Substitute for x in (1).

$$4(6 - y) + y = 9$$
$$24 - 4y + y = 9$$
$$-3y = -15$$
$$y = 5$$

Substitute for y in (1).

$$4x + 5 = 9$$
$$4x = 4$$
$$x = 1$$

The solution is $x = 1$, $y = 5$.

2. (a) See Figure 15-22b for graphical solution.

(b) Addition–subtraction method:

$$2x + 4y = 4 \quad (1)$$
$$x - y = 5 \quad (2)$$

Multiply (2) by 2: $2x - 2y = 10$
Rewrite (1): $2x + 4y = 4$
Subtract: $-6y = 6$
$y = -1$

(c) Substitution method: Solve for x in (2).

$$x - y = 5$$
$$x = y + 5$$

Substitute for x in (1).

$$2(y + 5) + 4y = 4$$
$$2y + 10 + 4y = 4$$
$$6y = -6$$
$$y = -1$$

Substitute for y in (1).

$$2x + 4(-1) = 4$$
$$2x - 4 = 4$$
$$2x = 8$$
$$x = 4$$

The solution is $x = 4$, $y = -1$.

Substitute for y in (2).

$$x - (-1) = 5$$
$$x + 1 = 5$$
$$x = 4$$

The solution is $x = 4$, $y = -1$.

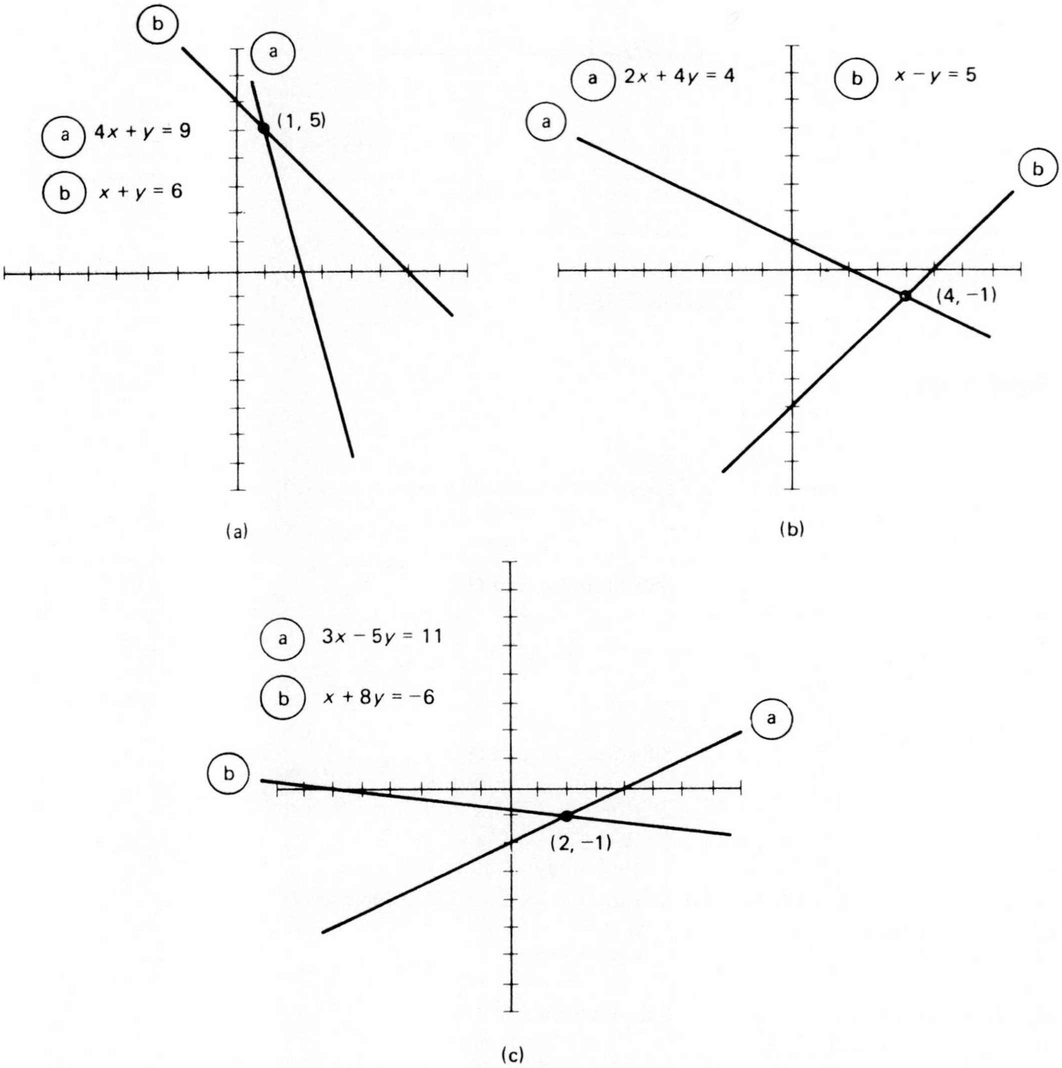

FIGURE 15-22 Answers to Self-Test 15-1: (a) problem 1, (b) problem 2, and (c) problem 3.

3. (a) See Figure 15-22c for graphical solution.

(b) Addition–subtraction method:

$$3x - 5y = 11 \quad (1)$$
$$x + 8y = -6 \quad (2)$$

Multiply (2) by 3: $3x + 24y = -18$

Rewrite (1): $3x - 5y = 11$

Subtract: $29y = -29$

$y = -1$

Substitute for y in (2).

$$x + 8(-1) = -6$$
$$x = 2$$

The solution is $x = 2$, $y = -1$.

(c) Substitution method: Solve for x in (2).

$$x + 8y = -6$$
$$x = -6 - 8y$$

Substitute for x in (1).

$$3(-6 - 8y) - 5y = 11$$
$$-18 - 24y - 5y = 11$$
$$-29y = 29$$
$$y = -1$$

Substitute for y in (2).

$$x + 8(-1) = -6$$
$$x = 2$$

The solution is $x = 2$, $y = -1$.

SELF-TEST 15-2

1. $I_1 = 647\ \mu A$, $I_2 = 388\ \mu A$, $I_3 = 259\ \mu A$, $V_1 = 6.47$ V, $V_2 = V_3 = 8.53$ V

2. $I_1 = 23.2$ mA, $I_2 = 5.47$ mA, $I_3 = 28.6$ mA, $V_1 = 46.4$ V, $V_2 = 16.4$ V, $V_3 = 28.6$ V

CHAPTER 16

COMPLEX NUMBERS

INTRODUCTION

When we have capacitance and inductance, in addition to resistance, in ac circuits, we are dealing with both energy users (resistance) and energy storers (capacitance and inductance). Circuit resistance converts energy to some other form, usually heat or light or both. Capacitance and inductance *store* energy. However, the calculations for voltage drops, current, and power dissipated are the same whether the component has the property of resistance, capacitance, or inductance. Thus, there are two kinds of power dissipation that we calculate: a real part, due to circuit resistance, and an imaginary part, due to circuit capacitance and/or inductance.

When imaginary parts exist, circuit currents and voltages do not lie along the same plane but may be displaced by as little as a few degrees, in circuits that contain mostly real parts, to as much as 180°, in circuits that are made up mostly of imaginary parts. In all such cases we say the variables are *out of phase*. The numbers that represent these variables have both real and imaginary parts and are called *complex* numbers.

This chapter, then, leads us into our study of ac circuits and those circuit variables that are out of phase. We will learn how to add, subtract, multiply, and divide complex numbers and also how to graph them. A thorough understanding of complex numbers is essential as we continue in following chapters to develop the techniques used in solving ac circuit problems.

CHAPTER OBJECTIVES

In this chapter you will learn how to:

1. Solve problems that contain imaginary numbers.
2. Add, subtract, multiply, and divide complex numbers.
3. Graph complex numbers.

16-1 IMAGINARY NUMBERS

What is the result if we find the square root of a negative number? What is $\sqrt{-36}$? The answer cannot be 6 or -6 because the square of either of these numbers results in 36. We cannot take the square root of a negative number and come up with a real-number answer. We can find a solution to the problem in the following manner:

$$\sqrt{-36} = \sqrt{(36)(-1)} = \sqrt{36} \cdot \sqrt{-1} = 6\sqrt{-1}$$

A number in this form is called an *imaginary* number.

KEY POINT Imaginary numbers are numbers that result from taking the square root of negative numbers. In mathematics, the letter i is used in represent $\sqrt{-1}$.

$$\sqrt{-36} = \sqrt{36}\, i = 6i$$

The letter i in electronics denotes current. Therefore, we use the letter j to represent $\sqrt{-1}$. We place the j in front of the number to make it easier to recognize the presence of an imaginary number.

$$\sqrt{-36} = \sqrt{(36)(-1)} = \sqrt{36} \cdot \sqrt{-1} = j6$$

Whenever we see a number preceded by the letter j, we know that the number is multiplied by $\sqrt{-1}$.

The word imaginary is used simply as a means of separating these numbers from real numbers. Imaginary numbers exist, and they are very useful in analyzing ac circuits, as we will see in later chapters.

KEY POINT A *vector* is a line drawn from the origin to some given point. In electronics we call a vector a *phasor*.

The word phasor (*phas*evect*or*) is more descriptive than vector because we are usually describing and/or calculating phase relations between voltages and currents by the use of vectors/phasors.

In Chapter 14 we developed a system of rectangular coordinates. We can develop a similar set of coordinates for plotting imaginary and complex numbers because *the operator j causes a vector to be rotated 90°*. On a system of rectangular coordinates, 0° is plotted along the positive x-axis. Positive rotation is counterclockwise, as shown in Figure 16-1. We start with a point 4 units long, plotted along the positive x-axis. The line drawn from the origin to this point is the phasor.

The operator j rotates the phasor 90°. Thus if we "j" our point, we are at 90° or are at the positive y-axis. Our phasor is now at $j4$. $j4$ tells us that the phasor is 4 units long and falls along the positive y-axis. (The operator j is always written in front of the number to help distinguish j as an operator from j as a literal number.) We can "j" the phasor again, which causes another 90° rotation to 180°. $j \cdot j = j^2$.

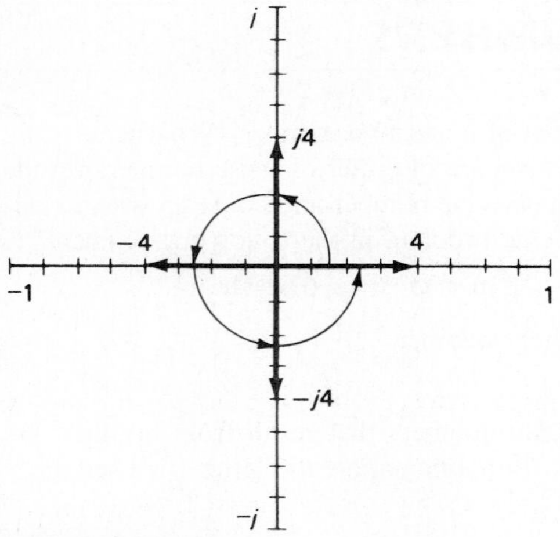

FIGURE 16-1

180° falls along the negative x-axis, so we see that j^2 is also -1. Our phasor is now at $j^2 4$ or -4. Rotating another 90° to 270° yields $j \cdot j \cdot j$ or j^3. This point is also $-j$ since the phasor falls along the negative y-axis. $j \cdot j \cdot j = j^3 = -j$. Our phasor is at $j^3 4$ or $-j4$. (Notice that we could also arrive at $-j$ by rotating the phasor *clockwise* from the x-axis.)

We can complete the circule by j-ing again for another 90° rotation $j \cdot j \cdot j \cdot j = j^4 = 1$. The circle is complete and we are back to the positive x-axis.

Remembering our basic relationship, that $j = \sqrt{-1}$, our graphic illustration helps us see the following relationships:

$$j = \sqrt{-1}$$
$$j^2 = \sqrt{-1} \cdot \sqrt{-1} = -1$$
$$j^3 = \sqrt{-1} \cdot \sqrt{-1} \cdot \sqrt{-1} = -1\sqrt{-1} = -j$$
$$j^4 = \sqrt{-1} \cdot \sqrt{-1} \cdot \sqrt{-1} \cdot \sqrt{-1} = (-1)(-1) = 1$$

Let's use these relationships to work the following problems.

□ EXAMPLE 16-1

Find $\sqrt{-144}$.

Solution

$$\sqrt{-144} = j12$$

□

Imaginary numbers can be added or subtracted together, just as real numbers can be added or subtracted.

☐ EXAMPLE 16-2

Find $\sqrt{-36} + \sqrt{-36}$.

Solution

$$j6 + j6 = j12$$

☐

Multiplication and division of real and imaginary numbers are performed as a simple multiplication of two numbers. The operator j is treated as though it were a literal number.

☐ EXAMPLE 16-3

Find $\sqrt{-36} \cdot \sqrt{36}$.

Solution

$$j6 \cdot 6 = j36$$

☐ EXAMPLE 16-4

Find $\sqrt{-36} \cdot \sqrt{-36}$.

Solution

$$j6 \cdot j6 = j^2 36 = (-1) \cdot 36 = -36$$

☐ EXAMPLE 16-5

Find $j^2 6 \cdot j5$.

Solution

$$j^2 6 = (-1)6 = -6$$

$$j^2 6 \cdot j5 = -6 \cdot j5 = -j30$$

☐ EXAMPLE 16-6

Find $\dfrac{\sqrt{-36}}{3}$.

Solution

$$\frac{\sqrt{-36}}{3} = \frac{j6}{3} = j2$$

☐ EXAMPLE 16-7

Find $\dfrac{j^2 36}{j9}$.

Solution

$$\frac{j^2 36}{j9} = j4$$

□ EXAMPLE 16-8

Find $\dfrac{j24}{j^2 6}$.

Solution

$$\frac{j24}{j^2 6} = -j4$$

When we divide j by j^2, we get $-j$. That is, $\dfrac{1}{j} = j^{-1} = -j$. □

PRACTICE PROBLEMS 16-1

Perform the indicated operations:

1. $\sqrt{-81} + \sqrt{-81}$
2. $\sqrt{-36} + \sqrt{-49}$
3. $\sqrt{-49} - \sqrt{-16}$
4. $j^2 6 \cdot j4$
5. $\dfrac{\sqrt{81}}{\sqrt{-9}}$
6. $\dfrac{j^3 72}{j8}$

Solutions

1. $j9 + j9 = j18$
2. $j6 + j7 = j13$
3. $j7 - j4 = j3$
4. $j^3 24 = -j24$
5. $\dfrac{9}{j3} = -j3$
6. $j^2 9 = -9$

Additional practice problems are at the end of the chapter.

16-2 COMPLEX NUMBERS

■ **KEY POINT** *Complex* numbers are numbers that are part real and part imaginary.

$4 + j3$ is a complex number. 4 is the real-number part and $j3$ is the imaginary-number part. These two parts cannot be added together.

□ EXAMPLE 16-9

Find $\sqrt{36} + \sqrt{-36}$.

Solution

$$\sqrt{36} = 6$$
$$\sqrt{-36} = j6$$
$$\sqrt{36} + \sqrt{-36} = 6 + j6$$

□ EXAMPLE 16-10

Find $j6 + j^2 12 + j^3 2$.

Solution

$$j^2 12 = -12$$
$$j^3 2 = -j2$$

Then

$$j6 + j^2 12 + j^3 2 = j6 - 12 - j2 = -12 + j4$$

□

16-2-1 Graphing Complex Numbers

Consider the complex number $2 + j4$. A look at this number tells us that there is a real part that is 2 units long and an imaginary part that is 4 units long. The point plots as (2,4) on a system of rectangular coordinates. The resulting phasor is shown in Figure 16-2.

In the number $-4 - j6$, the real number part is -4 and the imaginary part is -6. The resulting phasor is also plotted in Figure 16-2.

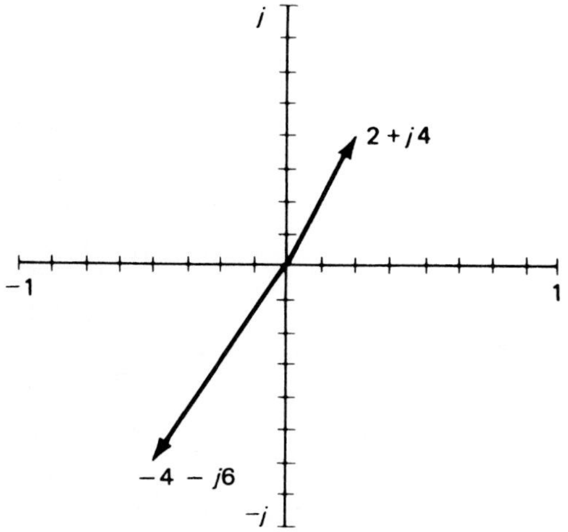

FIGURE 16-2

PRACTICE PROBLEMS 16-2

Perform the indicated operations

1. $j^3 + j^2 3$
2. $j^2 4 + j^3 5 - j^2 6 + j^3$
3. $-j4 + j^3 2 + j^2 4 + 3$
4. $4 + j^2 3 - j^2 4 + j^3 6$

Graph the following complex numbers:

5. $6 + j3$
6. $3 - j6$
7. $-4 + j3$
8. $-3 - j5$
9. $0 - j4$

Solutions

1. $j^3 + j^2 3 = -j + (-1)3 = -3 - j$
2. $j^2 4 + j^3 5 - j^2 6 + j^3 = (-1)4 + (-j)5 - (-1)6 + (-j)$
   $= -4 - j5 + 6 - j = 2 - j6$
3. $-j4 + j^3 2 - j^2 4 + 3 = -j4 + (-j)2 + (-1)4 + 3$
   $= -j4 - j2 - 4 + 3 = -1 - j6$
4. $4 + j^2 3 - j^2 4 + j^3 6 = 4 + (-1)3 - (-1)4 + (-j)6$
   $= 4 - 3 + 4 - j6 = 5 - j6$

See Figure 16-3 for solutions to problems 5 through 9.

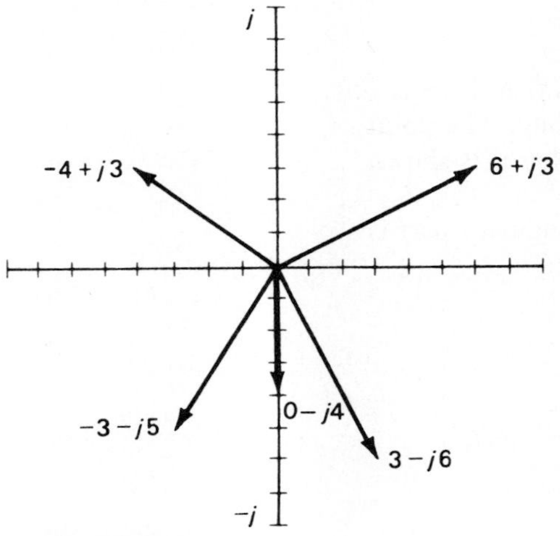

FIGURE 16-3

Additional practice problems are at the end of the chapter.

16-2-2 Addition and Subtraction of Complex Numbers

■ **KEY POINT** Complex numbers can be added or subtracted, but the real and imaginary parts must be added or subtracted separately.

☐ EXAMPLE 16-11

Find the sum of $6 + j3$ and $4 - j7$.

Solution

$$\begin{array}{r} 6 + j3 \\ 4 - j7 \\ \hline 10 - j4 \end{array}$$

$$(6 + j3) + (4 - j7) = 10 - j4$$

☐

Some calculators, will add, subtract, multiply, or divide complex numbers. Consult the users manual for your calculator for this feature.

☐ EXAMPLE 16-12

Find $(25 - j16) - (-15 + j12)$.

Solution

$$\begin{array}{rr} & 25 - j16 \\ (-) & -15 + j12 \\ \hline & 40 - j28 \end{array}$$

$$(25 - j16) - (-15 + j12) = 40 - j28$$

☐

PRACTICE PROBLEMS 16-3

Perform the following additions and subtractions.

1. $(8 + j3) + (7 + j2)$
2. $(10 + j2) + (20 - j14)$
3. $(15 - j4) + (14 + j7)$
4. $(-20 - j6) + (-15 - j10)$
5. $(10 + j10) - (15 + j6)$
6. $(22 + j16) - (17 - j12)$
7. $(6 - j14) - (15 + j8)$
8. $(-40 - j20) - (-25 - j30)$

Solutions

1. $15 + j5$
2. $30 - j12$
3. $29 + j3$
4. $-35 - j16$
5. $-5 + j4$
6. $5 + j28$
7. $-9 - j22$
8. $-15 + j10$

Additional practice problems are at the end of the chapter.

16-2-3 Multiplication and Division of Complex Numbers

KEY POINT We multiply complex numbers in the same way we multiply algebraic terms.

☐ EXAMPLE 16-13

Find the product of $(2 + j3)(2 + j4)$.

Solution

$$(2 + j3)(2 + j4) = 4 + j6 + j8 + j^2 12$$
$$= 4 + j14 + (-1)\,12 = -8 + j14$$

☐

KEY POINT Complex numbers are divided by multiplying the expression by the conjugate of the denominator.

The conjugate of a complex number is the same number, but the sign of the imaginary part is changed. The conjugate of $2 + j3$ is $2 - j3$. The conjugate of $4 - j5$ is $4 + j5$, and so on.

☐ EXAMPLE 16-14

Perform the following division:

$$\frac{2 + j3}{4 - j6}$$

Solution The conjugate of the denominator is $4 + j6$. If we multiply by $\frac{4 + j6}{4 + j6}$, we are really multiplying by 1 since any number divided by itself is 1. We perform this multiplication in order to simplify the denominator.

$$\frac{2 + j3}{4 - j6} \times \frac{4 + j6}{4 + j6} = \frac{(2 + j3)(4 + j6)}{(4 - j6)(4 + j6)}$$
$$= \frac{8 + j12 + j12 + j^2 18}{16 - j24 + j24 - j^2 36} = \frac{-10 + j24}{16 - j^2 36}$$
$$= \frac{-10 + j24}{52} = -\frac{10}{52} + j\frac{24}{52}$$
$$= -0.192 + j0.462$$

☐

PRACTICE PROBLEMS 16-4

Perform the following multiplications. Verify your answer with the calculator.

1. $(3 + j4)(3 + j6)$
2. $(3 + j2)(3 - j3)$
3. $(4 + j6)(4 - j6)$
4. $(-2 + j5)(3 - j6)$

Perform the following divisions. Verify your answers with the calculator.

5. $\dfrac{2 + j3}{3 - j2}$
6. $\dfrac{6 - j2}{2 + j4}$
7. $\dfrac{-6 + j4}{4 + j9}$
8. $\dfrac{7 - j6}{4 - j3}$

Solutions

1. $-15 + j30$
2. $15 - j3$
3. $52 + j0 = 52$
4. $24 + j27$
5. $0 + j1 = j$
6. $0.2 - j1.4$
7. $0.124 + j0.722$
8. $1.84 - j0.120$

Additional practice problems are at the end of the chapter.

SELF-TEST 16-1

Perform the indicated operations:

1. $j^2 + j^2 4$
2. $j^3 4 - j6$
3. $\dfrac{\sqrt{-81}}{\sqrt{9}}$
4. $\dfrac{\sqrt{121}}{\sqrt{-25}}$
5. $\dfrac{j^2 56}{j^3 8}$
6. $(6 - j3)(3 + j7)$
7. $(3 - j6)(4 - j2)$
8. $(4 + j3)(4 - j3)$
9. $\dfrac{3 - j2}{1 + j2}$
10. $\dfrac{2 + j5}{2 - j2}$

Graph the following complex numbers.

11. $6 - j4$
12. $-4 + j6$

Answers to Self-test 16-1 are at the end of the chapter.

CHAPTER 16 AT A GLANCE

Page	Key Points	Example
427	Imaginary numbers are numbers that result from taking the square root of negative numbers.	$\sqrt{-36} = 6\sqrt{-1}$
427	A *vector* is a line drawn form the origin to some given point. In electronics we call a vector a *phasor*. Graphically, the operator j rotates a phasor 90°.	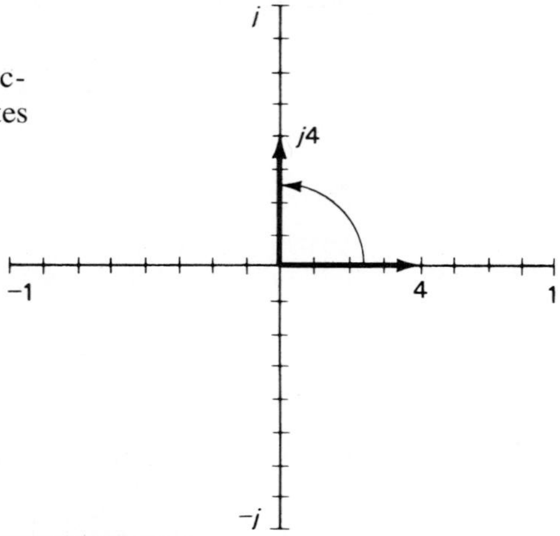 FIGURE 16-4
430	Complex numbers are numbers that are part real and part imaginary.	$4 + j3$
432	Complex numbers can be added or subtracted, but the real and imaginary parts must be added or subtracted separately.	$j6 + j^2 12 + j^3 2$ $= j6 + (-1)12 + (-j)2$ $= -12 + j4$
434	Complex numbers are multiplied the same way we multiply algebraic terms.	$(2 + j3)(2 + j4)$ $= 4 + j6 + j8 + j^2 12$ $= 4 + j14 + (-1)12$ $= -8 + j14$
434	Complex numbers are divided by multiplying the number by the conjugate of the denominator.	$\frac{2 + j3}{4 - j6} \times \frac{4 + j6}{4 + j6}$ $= \frac{(2 + j3)(4 + j6)}{(4 - j6)(4 + j6)}$ $= \frac{8 + j12 + j12 + j^2 18}{16 - j24 + j24 - j^2 36}$ $= \frac{-10 + j24}{16 - j^2 36}$ $= \frac{-10 + j24}{52} = -\frac{10}{52} + j\frac{24}{52}$ $= -0.192 + j0.462$

End of Chapter Problems 16-1

Perform the indicated operations:

1. $\sqrt{-49} + \sqrt{-36}$
2. $\sqrt{-16} + \sqrt{-64}$
3. $\sqrt{-64} - \sqrt{-49}$
4. $\sqrt{-81} - \sqrt{-49}$
5. $\sqrt{-25} \cdot \sqrt{-25}$
6. $\sqrt{-144} \cdot \sqrt{-144}$
7. $\sqrt{-9} \cdot \sqrt{-36}$
8. $\sqrt{-121} \cdot \sqrt{-81}$
9. $\sqrt{25} \cdot \sqrt{-49}$
10. $\sqrt{-100} \cdot \sqrt{4}$
11. $\sqrt{7.02} \cdot \sqrt{-6.2}$
12. $\sqrt{12} \cdot \sqrt{-15}$
13. $j^2 6 \cdot j3$
14. $j^2 4 \cdot j5$
15. $j^3 4 \cdot j3$
16. $j^3 5 \cdot j5$
17. $\dfrac{\sqrt{-144}}{\sqrt{16}}$
18. $\dfrac{\sqrt{-64}}{\sqrt{4}}$
19. $\dfrac{\sqrt{100}}{\sqrt{-25}}$
20. $\dfrac{\sqrt{144}}{\sqrt{-16}}$
21. $\dfrac{j^3 42}{j6}$
22. $\dfrac{j^3 72}{j9}$
23. $\dfrac{j^2 100}{j^3 5}$
24. $\dfrac{j^2 64}{j^3 4}$

End of Chapter Problems 16-2

Perform the indicated operations:

1. $j^3 4 + j^2 4$
2. $j^3 6 + j^2 5$
3. $j^3 8 + j^2 4$
4. $j^3 6 + j^2 5$
5. $j^3 4.5 - j^2 3.1$
6. $j^3 7.3 - j^2 10$
7. $5 + j7 - j^2 6$
8. $4 + j9 - j^2 6$
9. $8.5 + j^3 4.2 + j^2 6.6$
10. $2.75 - j^3 4.55 + j^2 9.2$
11. $-j2 + j^3 5 + j^2 6 + 5$
12. $-j12 + j^3 2 + j^2 3 + 5$
13. $8 + j^2 5 - j^2 7 + j^3 6$
14. $3 + j^2 10 - j^2 2 + j^3 6$
15. $j^2 + j^3 + j + 1$
16. $j^2 2 + j^3 2 + j2 + 2$
17. $-j4 + 2 - j^2 4 - j^3 4$
18. $-j6 + 6 - j^2 3 - j^3 6$
19. $j^3 3 + j^2 4 - j3 + 4$
20. $j^3 2 + j^2 5 - j2 + 5$

Graph the following complex numbers.

21. $0 + j6$
22. $0 + j4$
23. $8 + j0$
24. $4 + j0$
25. $4 + j4$
26. $6 + j6$
27. $3 - j4$
28. $5 - j3$
29. $-4 + j3$
30. $-6 + j$
31. $-20 - j40$
32. $-60 - j80$
33. $-50 + j10$
34. $-100 + j40$

35. $3\text{ k} + j4\text{ k}$

36. $5\text{ k}\Omega + j2\text{ k}\Omega$

37. $-300\Omega + j400\Omega$

38. $-600\ \Omega + j800\ \Omega$

39. $-50\text{ k}\Omega - j70\text{ k}\Omega$

40. $-20\text{ k}\Omega - j40\text{ k}\Omega$

End of Chapter Problems 16-3

Perform the following additions and subtractions. Verify your answers using the calculator.

1. $(10 + j15) + (5 + j12)$
2. $(40 + j16) + (10 + j15)$
3. $(12 + j10) + (13 + j15)$
4. $(17 + j6) + (8 + j6)$
5. $(14 + j6) + (13 - j14)$
6. $(7 + j8) + (15 - j15)$
7. $(25 + j30) + (5 - j20)$
8. $(15 + j10) + (10 - j40)$
9. $(16 - j4) + (4 + j6)$
10. $(7 - j20) + (7 + j14)$
11. $(12 - j16) + (8 + j8)$
12. $(14 - j20) + (6 + j16)$
13. $(4 - j3) + (6 - j3)$
14. $(2 - j5) + (4 - j7)$
15. $(10 - j6) + (7 - j4)$
16. $(7 - j16) + (8 - j14)$
17. $(-4 - j5) + (-6 - j7)$
18. $(-10 - j15) + (-12 - j12)$
19. $(-15 - j10) + (-6 - j10)$
20. $(-12 - j12) + (-9 - j11)$
21. $(10 + j15) - (5 + j12)$
22. $(40 + j16) - (10 + j15)$
23. $(12 + j10) - (13 + j15)$
24. $(17 + j6) - (8 + j6)$
25. $(14 + j6) - (13 - j14)$
26. $(7 + j8) - (15 - j15)$
27. $(25 + j30) - (5 - j20)$
28. $(15 + j10) - (10 - j40)$
29. $(16 - j4) - (4 + j6)$
30. $(7 - j20) - (7 + j14)$
31. $(12 - j16) - (8 + j8)$
32. $(14 - j20) - (6 + j16)$
33. $(4 - j3) - (6 - j3)$
34. $(2 - j5) - (4 - j7)$
35. $(10 - j6) - (7 - j4)$
36. $(7 - j16) - (8 - j14)$
37. $(-4 - j5) - (-6 - j7)$
38. $(-10 - j15) - (-12 - j12)$
39. $(-15 - j10) - (-6 - j10)$
40. $(-12 - j12) - (-9 - j11)$

End of Chapter Problems 16-4

Perform the following multiplications and divisions. Verify your answers using the calculator.

1. $(5 + j2)(3 - j2)$
2. $(6 - j2)(2 - j4)$
3. $(2 + j3)(2 - j3)$
4. $(4 - j3)(4 + j3)$
5. $(4 - j3)(4 - j3)$
6. $(6 + j2)(6 + j3)$
7. $(1 - j5)(3 - j3)$
8. $(2 - j4)(3 - j4)$
9. $(4 + j5)(6 + j2)$
10. $(2 + j2)(5 + j5)$
11. $(5 - j3)(6 + j1)$
12. $(2 - j2)(3 + j3)$
13. $\dfrac{5 + j3}{1 + j4}$
14. $\dfrac{1 + j}{1 - j}$
15. $\dfrac{4 + j4}{3 - j}$
16. $\dfrac{4 - j5}{3 - j5}$

17. $\dfrac{1 + j2}{5 + j2}$

18. $\dfrac{2 - j3}{3 + j2}$

19. $\dfrac{2 - j5}{-3 + j4}$

20. $\dfrac{-8 - j2}{4 - j5}$

21. $\dfrac{3 + j4}{4 + j5}$

22. $\dfrac{1 + j6}{5 + j2}$

Answers to Self-test 16-1

1. -5

2. $-j10$

3. $j3$

4. $-j2.2$

5. $-j7$

6. $39 + j33$

7. $0 - j30$

8. 25

9. $-0.2 - j1.6$

10. $-0.75 + j1.75$

11. See Figure 16-5

12. See Figure 16-5

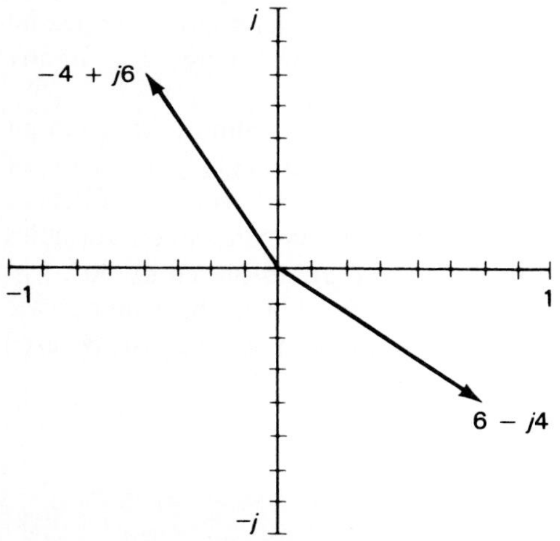

FIGURE 16-5

CHAPTER 17 THE RIGHT TRIANGLE

INTRODUCTION

There are three kinds of triangles: those that contain an obtuse angle and two acute angles, those that contain three acute angles, and those that contain a right angle and two acute angles. Examples of the three triangles are shown in Figure 17-1. In any triangle, the sum of the three angles equals 180°.

We are particularly interested in the right triangle because almost all ac circuit behavior can be evaluated mathematically by use of the right triangle. The effects of inductance and capacitance in ac circuits are at right angles or 90° from the effects of circuit resistance. The result is that we are often asked to compute circuit variables for which voltages or currents are acted on by forces that are displaced by 90°.

In this chapter we will introduce the student to the right triangle and to the trigonometric functions. The skills developed here in problem solving will be used throughout the following five chapters.

CHAPTER OBJECTIVES

In this chapter you will learn how to:

1. Identify the sides and angles of triangles.
2. Recognize the relationship between sides and angles.
3. Use the Pythagorean theorem to find the unknown sides of right triangles.
4. Use trigonometric functions to find the angles and sides of right triangles.
5. Use trigonometric tables.
6. Use the calculator to solve problems using trig functions.

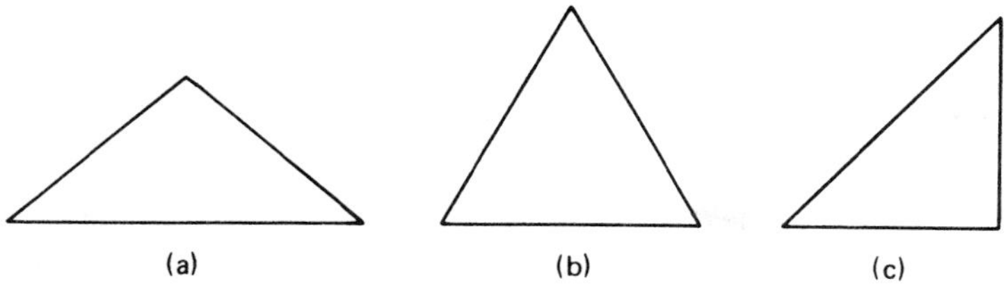

FIGURE 17-1 (a) Triangle with one obtuse and two acute angles; (b) triangle with three acute angles; and (c) triangle with one right angle and two acute angles.

17-1 SIDES AND ANGLES

Figure 17-2 is a right triangle shown in two common positions. For purposes of discussion, the triangles are labeled so that A denotes angle A, B denotes angle B, and C denotes the 90° or right angle. Side a lies opposite angle A, side b lies opposite angle B, and side c lies opposite angle C. In a right triangle the side opposite the right angle is called the *hypotenuse*. In Figure 17-2, side c is the hypotenuse.

Let's develop some basic relationships between the sides and angles in angles of right triangles such as the ones in Figure 17-2.

KEY POINT In any right triangle: (1) the sum of the acute angles equals 90°, (2) the hypotenuse is the side of greatest length, and (3) the greater side lies opposite the greater angle.

When discussing angle A, side b is the *adjacent* side (the side next to angle A) and side a is the opposite side. When discussing angle B, side a is the adjacent side and side b is the opposite side. Side c is always called the hypotenuse.

Let's look at the right triangles in Figure 17-3. In Figure 17-3(a), angle A is 30°; therefore, angle B must equal 60° since the sum of angles A and B must equal 90°. The greater side lies opposite the greater angle. Therefore, since angle B is greater than angle A, side b is greater than side a. Since the hypotenuse is always the side of greatest length, our comparison of length is limited to the relationship between sides a and b. In Figure 17-3(b), angle A is the greater angle; therefore, side a is greater than side b. Finally, in Figure 17-3(c), B is 40°; therefore, angle A must be 50°, making side a the greater side.

Not all triangles are labeled as in Figure 17-3 and not all triangles are shown

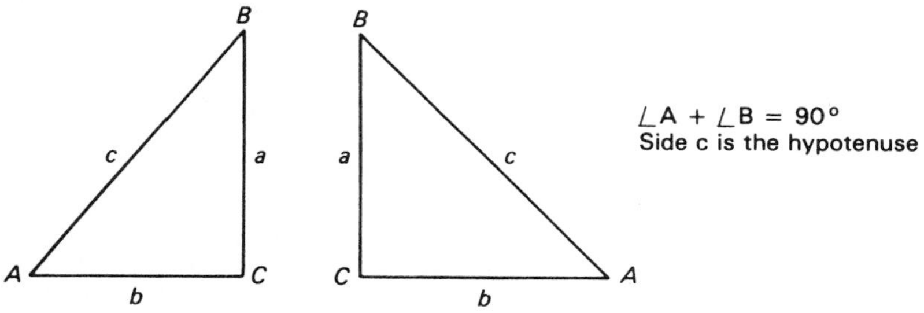

FIGURE 17-2 Right triangles.

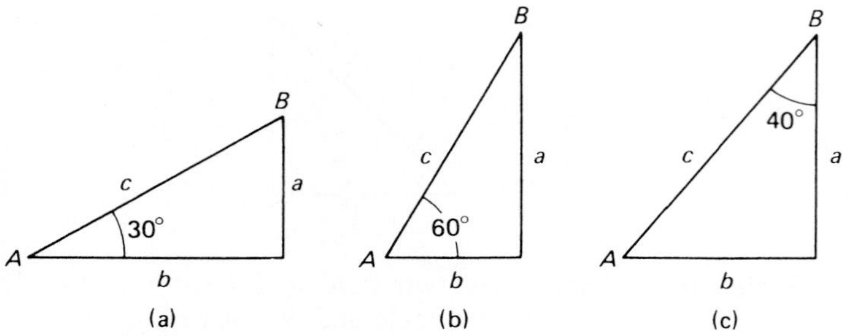

FIGURE 17-3

in this position. Notice in Figure 17-4(a) that the hypotenuse is labeled Z and the angles are θ (the Greek letter *theta*) and ϕ (the Greek letter *phi*). Side R lies opposite angle ϕ and is adjacent to angle θ. Side X lies opposite angle θ and is adjacent to angle ϕ. In Figure 17-4(b), side Y is the hypotenuse. Side G lies opposite angle ϕ and is adjacent to angle θ. Side B lies opposite angle θ and is adjacent to angle ϕ.

In Figure 17-4(c), side T is the hypotenuse. Side W lies opposite angle C and is adjacent to angle D. Side S lies opposite angle D and is adjacent to angle C. You should be able to identify the relationships among sides and angles regardless of the position of the triangle.

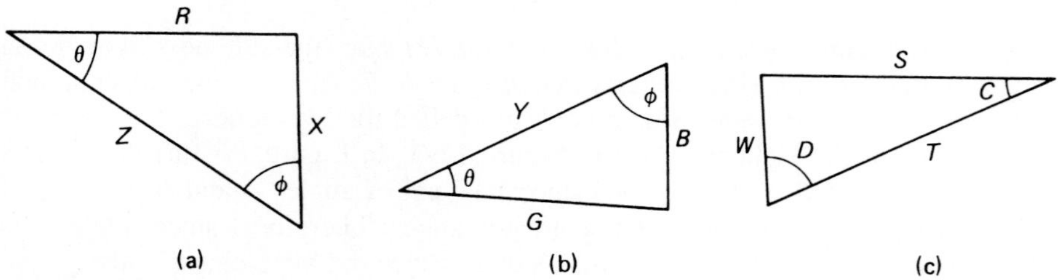

FIGURE 17-4

PRACTICE PROBLEMS 17-1

1. In the triangles in Figure 17-5, identify which of sides a and b is greater. Determine the size of the unknown angles.

2. Refer to Figure 17-6(a).
   (a) Which side is adjacent to angle θ?
   (b) Which side lies opposite angle ϕ?
   (c) Identify the hypotenuse.

3. Refer to Figure 17-6(b).
   (a) Which side lies opposite angle B?
   (b) Side m is adjacent to which angle?
   (c) Identify the hypotenuse.

4. Refer to Figure 17-6(c).
   (a) Side G is opposite which angle?
   (b) Which side is adjacent to angle θ?
   (c) Identify the hypotenuse.

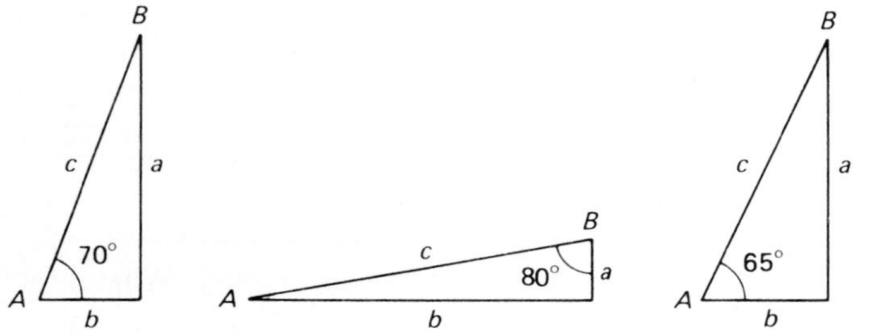

FIGURE 17-5

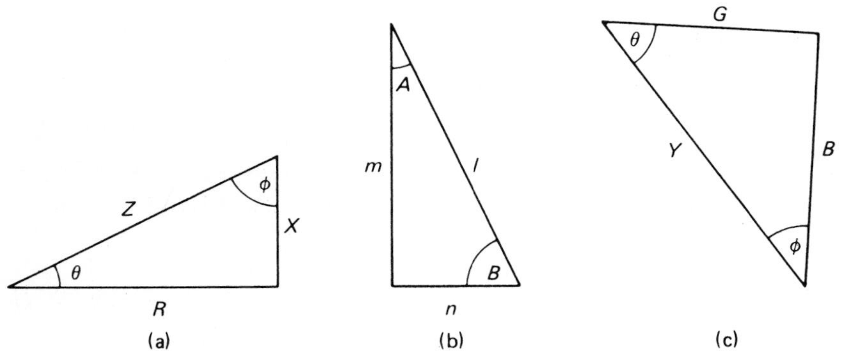

FIGURE 17-6

Solutions

1. In Figure 17-5(a), angle B is 20°. Side a is greater than side b. In (b), angle A is 10°; therefore, side b is greater. In (c), angle B is 25°; therefore, side a is greater.
2. (a) Side R is adjacent to angle θ.
   (b) Side R is opposite angle ϕ.
   (c) Side Z is the hypotenuse.
3. (a) Side m lies opposite angle B.
   (b) Side m is adjacent to angle A.
   (c) Side l is the hypotenuse.
4. (a) Side G is opposite angle ϕ.
   (b) Side G is adjacent to angle θ.
   (c) Side Y is the hypotenuse.

Additional practice problems are at the end of the chapter.

17-2 PYTHAGOREAN THEOREM

If two sides of a right triangle are known, the third side can be found by using the Pythagorean theorem, which is stated as follows:

■ **KEY POINT** *In a right triangle, the square of the hypotenuse is equal to the sum of the squares of the other two sides.*

$$c^2 = a^2 + b^2 \tag{17-1}$$

Taking the square root of both sides, we get

$$c = \sqrt{a^2 + b^2} \tag{17-2}$$

☐ EXAMPLE 17-1

Suppose sides a and b in Figure 17-7 are of length 6 and 8, respectively. What is the length of the hypotenuse?

Solution

$$c = \sqrt{a^2 + b^2} = \sqrt{6^2 + 8^2} = \sqrt{36 + 64} = \sqrt{100} = 10$$

Using the calculator,

6 $\boxed{x^2}$ $\boxed{+}$ 8 $\boxed{x^2}$ $\boxed{=}$ $\boxed{\sqrt{x}}$ *Answer*: 10

☐ EXAMPLE 17-2

If side $a = 5$ and side $c = 8$ in Figure 17-7, what is the value of side b?

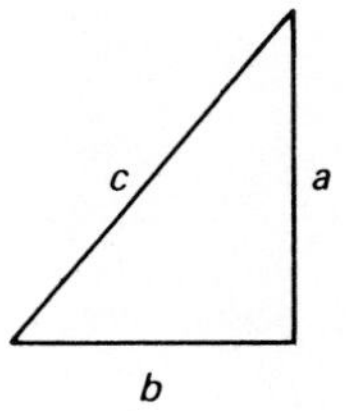

FIGURE 17-7

Solution

$$c^2 = a^2 + b^2 \tag{17-1}$$

$$b^2 = c^2 - a^2$$

$$b = \sqrt{c^2 - a^2} \tag{17-3}$$

$$b = \sqrt{8^2 - 5^2} = \sqrt{64 - 25} = \sqrt{39} = 6.24$$

8 $\boxed{x^2}$ $\boxed{-}$ 5 $\boxed{x^2}$ $\boxed{=}$ $\boxed{\sqrt{x}}$ *Answer*: 6.24

☐ EXAMPLE 17-3

If side $b = 75$ and side $c = 150$ in Figure 17-7, what is the value of side a?

Solution

$$c^2 = a^2 + b^2 \tag{17-1}$$

$$a^2 = c^2 - b^2$$

$$a = \sqrt{c^2 - b^2} \tag{17-4}$$

$$a = \sqrt{150^2 - 75^2} = \sqrt{16{,}875} = 130$$

150 $\boxed{x^2}$ $\boxed{-}$ 75 $\boxed{x^2}$ $\boxed{=}$ $\boxed{\sqrt{x}}$ *Answer*: 130

☐

In an ac circuit the sides of the triangle could represent circuit resistance (R), reactance (X), and impedance (Z), whose unit of measure is the ohm (Ω). Z is the hypotenuse; therefore, the equation is

$$Z^2 = R^2 + X^2 \tag{17-5}$$

☐ EXAMPLE 17-4

Refer to Figure 17-8. Find X if $R = 68\ \text{k}\Omega$ and $Z = 80\ \text{k}\Omega$. Which of angles θ and ϕ is greater?

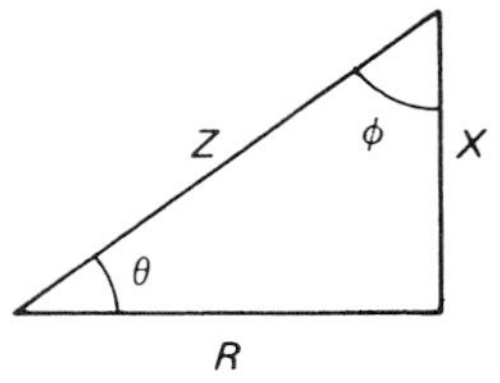

FIGURE 17-8

Solution

$$Z^2 = R^2 + X^2$$

$$X^2 = Z^2 - R^2$$

$$X = \sqrt{Z^2 - R^2} = \sqrt{(80\ \text{k}\Omega)^2 - (68\ \text{k}\Omega)^2} = 42.1\ \text{k}\Omega \qquad \phi > \theta$$

80 [EE] 3 [x^2] [−] 68 [EE] 3 [x^2] [=] [$\sqrt{x}$]   *Answer*: 42.1×10^3

☐

The triangle could also represent circuit conductance (G), susceptance (B), and admittance (Y), whose unit of measure is the siemens (S). Y is the hypotenuse; therefore, the equation is

$$Y^2 = G^2 + B^2 \tag{17-6}$$

☐ EXAMPLE 17-5

Refer to Figure 17-9. Find Y where $G = 400\ \mu\text{S}$ and $B = 600\ \mu\text{S}$.

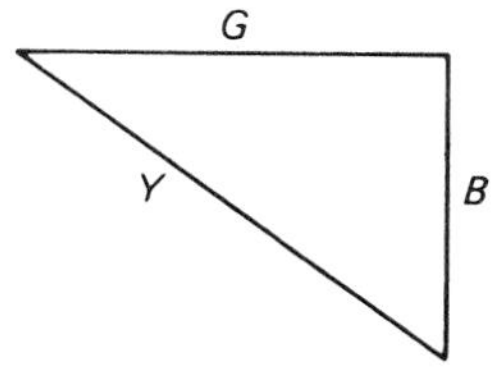

FIGURE 17-9

Solution

$$Y^2 = G^2 + B^2$$

$$Y = \sqrt{G^2 + B^2} = \sqrt{(400\ \mu\text{S})^2 + (600\ \mu\text{S})^2} = 721\ \mu\text{S}$$

400 [EE] [+/−] 6 [x^2] [+] 600 [EE] [+/−] 6 [x^2] [=] [$\sqrt{x}$]   *Answer*:

Answer: 721×10^{-6}

☐

PRACTICE PROBLEMS 17-2

1. Find the length of the unknown side in the triangles in Figure 17-10(a), (b), and (c). Determine whether angle A or angle B is greater.
2. In Figure 17-11, find the unknown side and determine whether θ or ϕ is the greater angle when:
   (a) $Z = 17.3\ \text{k}\Omega$ and $R = 10\ \text{k}\Omega$
   (b) $Z = 800\ \Omega$ and $X = 500\ \Omega$
   (c) $R = 6.8\ \text{k}\Omega$ and $X = 3\ \text{k}\Omega$
3. In Figure 17-12, find the unknown side and determine whether θ or ϕ is the greater angle when:
   (a) $B = 3$ mS and $G = 2$ mS
   (b) $Y = 400\ \mu\text{S}$ and $G = 150\ \mu\text{S}$
   (c) $Y = 1.3$ mS and $B = 600\ \mu\text{S}$

Solutions

1. For triangle (a), use Equation 17-4.

$$a = \sqrt{c^2 - b^2} = \sqrt{40^2 - 10^2} = 38.7$$

Angle A is greater than angle B because side a is greater than side b. Remember, the greater angle lies opposite the greater side. For triangle (b), use Equation 17-2.

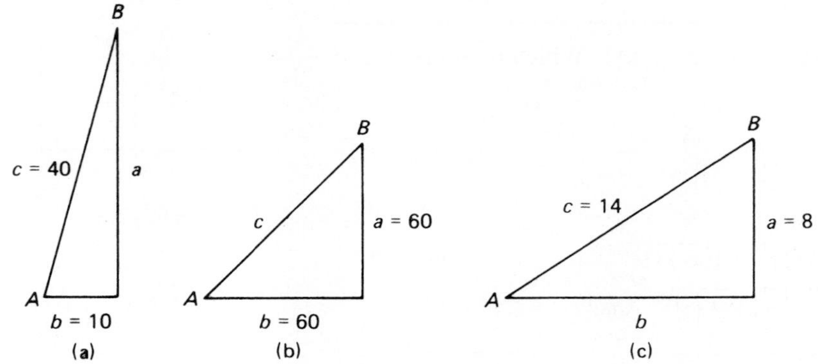

FIGURE 17-10

$$c = \sqrt{a^2 + b^2} = \sqrt{60^2 + 60^2} = 84.9$$

Angles A and B are equal because the sides are equal. For triangle (c), use Equation 17-3.

$$b = \sqrt{c^2 - a^2} = \sqrt{14^2 - 8^2} = 11.5$$

Side b is greater than side a; therefore, angle B is greater than angle A.

3. (a) $Y = 3.61$ mS and θ is the greater angle.
(b) $B = 371\ \mu$S and θ is the greater angle.
(c) $G = 1.15$ mS and ϕ is the greater angle.

Additional practice problems are at the end of the chapter.

2. (a) $X = 14.1$ kΩ and θ is the greater angle.
(b) $R = 624\ \Omega$ and ϕ is the greater angle.
(c) $Z = 7.43$ kΩ and ϕ is the greater angle.

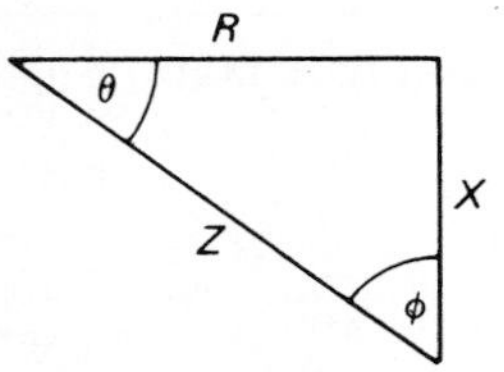

FIGURE 17-11

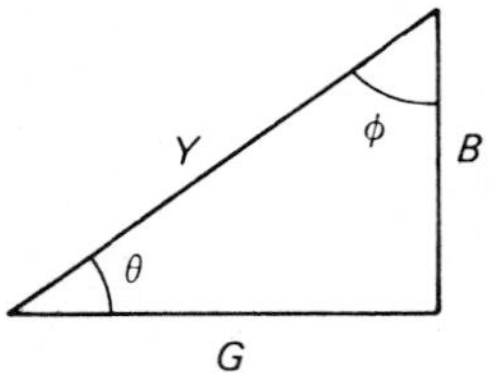

FIGURE 17-12

17-3 TRIGONOMETRIC FUNCTIONS

The Pythagorean theorem has its limitations. We must know two sides in order to find the third side, and we can't determine the actual size of the angles.

We could find the length of the sides *and* the angles by using graph paper, a protractor, and a rule. Let's draw a line 3 units long, as in Figure 17-13. Let this be side a. Another line is drawn at right angles to side a and is 4 units in length (side b). If we complete the triangle and measure the length of side c (the hypotenuse), we will find that side c is 5 units as measured with the rule. $\angle A$ would equal approximately 37° and $\angle B$ would equal approximately 53° as determined by the protractor.

If we doubled sides a and b, as in Figure 17-14, so that side a is 6 units long and side b is 8 units long, how long would the hypotenuse be? What would $\angle A$ and $\angle B$ equal? If the length of a and b is doubled, the length of the hypotenuse would also double. This could be demonstrated either graphically (as in Figure 17-14) or by using the Pythagorean theorem. Again using the protractor, we could measure $\angle A$ and $\angle B$. They would still equal approximately 37° and 53°. They did not change be-

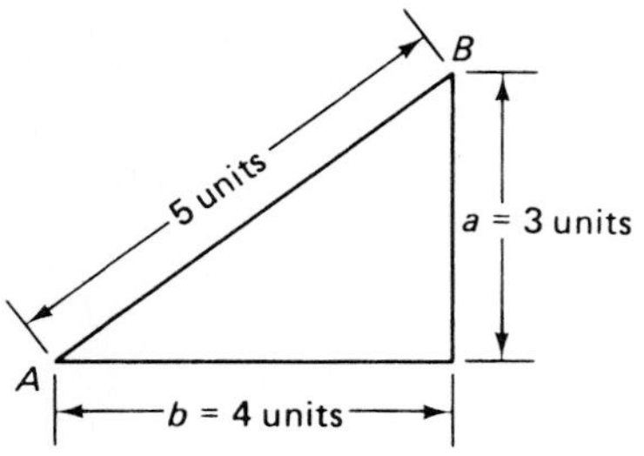

FIGURE 17-13

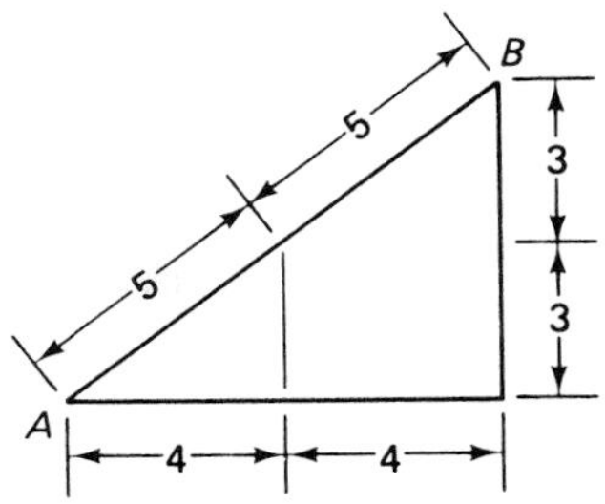

FIGURE 17-14

cause the relative lengths of the sides did not change. In Figure 17-14 the ratio of side a to side b was 3 to 4. When we doubled the length of the sides, the ratio of side a to side b was 6 to 8, which in both cases is 0.75 to 1. 3 to 4 = 0.75 and 6 to 8 = 0.75. As long as the ratio of side a to side b is 0.75 to 1, that is, whenever side a is 0.75 times as long as side b, $\angle A$ is 37°. We could expand on this by selecting other lengths of sides a and b and measuring the angles and the hypotenuse. In this manner, we could determine the approximate value for angles A and B for any ratio of side a to side b. We must consider the angles measured with the protractor as approximate since most protractors are not precision instruments.

It is rarely necessary to solve problems graphically though, because mathematicians have prepared tables of these and other ratios for us to use. These tables are called *trigonometric tables*.

Since a triangle has three sides, there are three possible trigonometric ratios. In Figure 17-15 these ratios would be side a to side b (as previously discussed), side a to side c, and side b to side c. These trigonometric ratios are called *functions* and are identified as *sine, cosine,* and *tangent*. (Actually, there are three additional functions, secant, cosecant, and cotangent, which are reciprocals of the sine, cosine, and tangent functions. But we will confine our discussion to the sine, cosine, and tangent functions since they are the only ones we need to solve for angles and sides of a right triangle.)

$$\text{sine of an angle} = \frac{\text{opposite side}}{\text{hypotenuse}} \tag{17-7}$$

$$\sin\,\theta = \frac{a}{c}$$

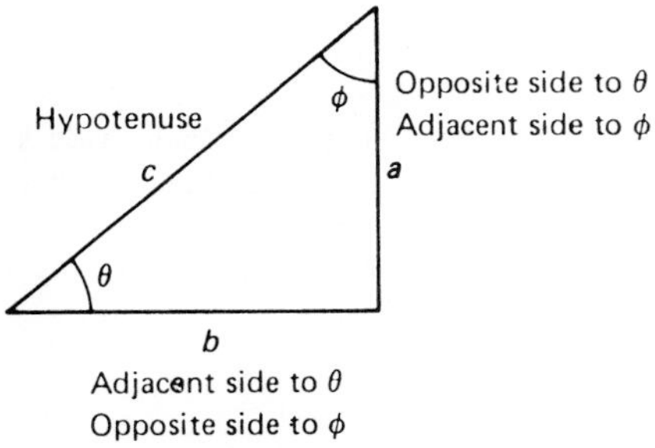

FIGURE 17-15

$$\text{cosine of an angle} = \frac{\text{adjacent side}}{\text{hypotenuse}} \qquad (17\text{-}8)$$

$$\cos \theta = \frac{b}{c}$$

$$\text{tangent of an angle} = \frac{\text{opposite side}}{\text{adjacent side}} \qquad (17\text{-}9)$$

$$\tan \theta = \frac{a}{b}$$

The sine function expresses the ratio of the length of the opposite side to the length of the hypotenuse. The equation would read $\sin \theta = \frac{a}{c}$. The cosine function expresses the ratio of the length of the adjacent side to the length of the hypotenuse. The equation would read $\cos \theta = \frac{b}{c}$. The tangent function expresses the ratio of the length of the opposite side to the length of the adjacent side. The equation would read $\tan \theta = \frac{a}{b}$.

PRACTICE PROBLEMS 17-3

1. Refer to Figure 17-16(a). Write the equation for the three functions with reference to the given angle.
2. Refer to Figure 17-16(b). Write the equation for the three functions with reference to the given angle.
3. Refer to Figure 17-17(a). Write the equations for the three functions with reference to the given angle.
4. Refer to Figure 17-17(b). Write the equations for the three functions with reference to the given angle.

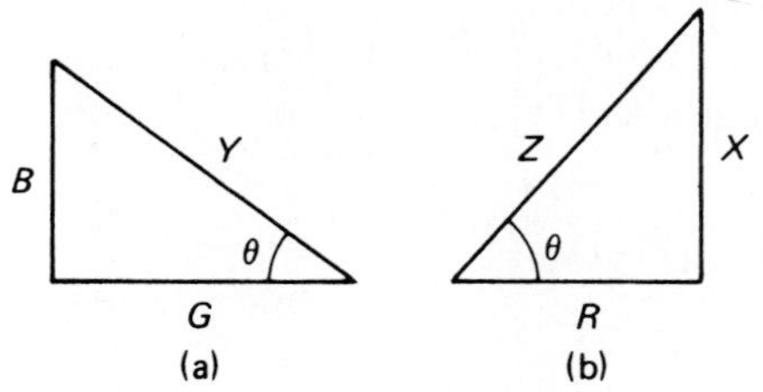

FIGURE 17-16

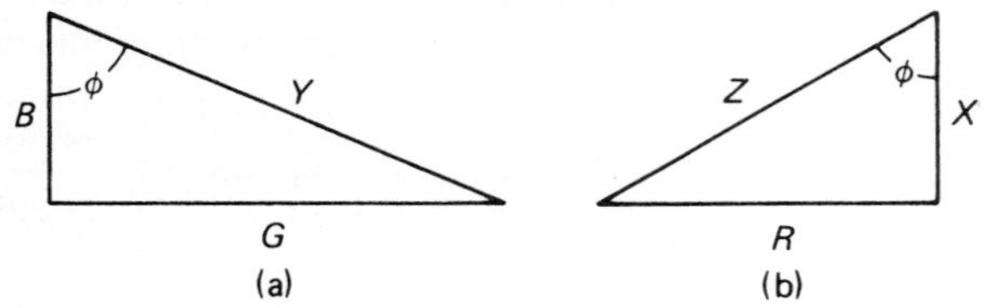

FIGURE 17-17

Solutions

1. In Figure 17-16(a), side G is the adjacent side, side B is the opposite side, and side Y is the hypotenuse when considering the angle θ. The sine function expresses the ratio of the opposite side to the hypotenuse; therefore, $\sin \theta = B/Y$. The cosine function expresses the ratio of the adjacent side to the hypotenuse; therefore, $\cos \theta = G/Y$. Finally, $\tan \theta = B/G$ since the tangent expresses the ratio of the opposite side to the adjacent side.
2. In Figure 17-16(b), side R is the adjacent side, side X is the opposite side, and side Z is the hypotenuse when considering the angle θ. Therefore, $\sin \theta = X/Z$, $\cos \theta = R/Z$, and $\tan \theta = X/R$.

3. If you determined in Figure 17-17(a) that the adjacent side is side B and the opposite side is side G, you are right. Then, $\sin\ \phi = G/Y$, $\cos\ \phi = B/Y$, and $\tan \phi = G/B$.

4. You should have determined in Figure 17-17(b) that the adjacent side is side X and the opposite side is side R. Then, $\sin\ \phi = R/Z$, $\cos\ \phi = X/Z$, and $\tan \phi = R/X$.

Additional practice problems are at the end of the chapter.

SELF-TEST 17-1

Refer to Figure 17-18.

1. $\angle\phi = 47.3°$. Find θ.
2. If $\angle\theta = 37.5°$, which side is longer?
3. If $Z = 73.6\ \text{k}\Omega$ and $R = 47\ \text{k}\Omega$, find X.
4. In problem 3, which angle is greater?
5. Write the equations for the three functions with reference to $\angle\theta$.

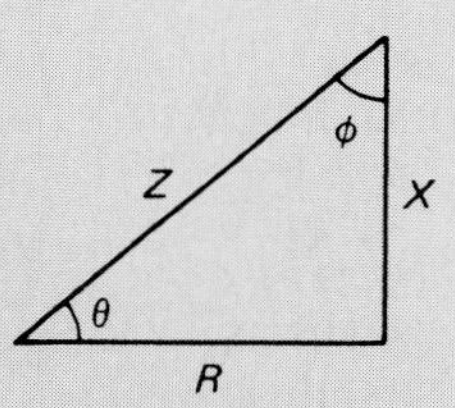

FIGURE 17-18

Answers to Self-test 17-1 are at the end of the chapter.

17-4 TRIGONOMETRIC TABLES

Now that we understand what a trig function is, let's see how we can use trig tables to find unknown angles or ratios.

Table 17-1 is a partial table of the trig functions. Appendix B is a complete table of trig functions.

Let's take a look at Table 17-1. It shows angles from 30° and 60° and the sine, cosine, and tangent functions. What is the tangent of a 40° angle? First, we find 40° and then we find the *tan* column. Tan 40° = 0.8391. What is tan 50°? Tan 50° = 1.1918. What is the sine of 35°? This time we find 35° and locate the sine under the column marked *sin*. Sin 35° = 0.57358. What is the cosine of 54°? The cosine column is marked *cos*. Cos 54° = 0.58779. What is the sine of 33.7°? We note that each degree is divided into tenths of a degree. First we find 33° and then move down to 0.7. This is 7 tenths or 33.7°. Sin 33.7° = 0.55484.

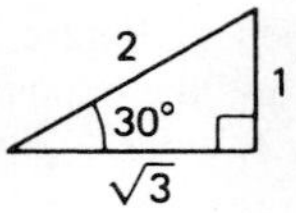

TABLE 17-1 Trigonometric Functions (Partial Table)

Angle°	*sin*	*tan*	*cot*	*cos*	*Angle°*
30.0	.50000	.57735	1.7321	.86603	60.0
.1	.50151	.57968	1.7251	.86515	.9
.2	.50302	.58201	1.7182	.86427	.8
.3	.50453	.58435	1.7113	.86340	.7
.4	.50603	.58670	1.7045	.86251	.6
.5	.50754	.58905	1.6977	.86163	.5
.6	.50904	.59140	1.6909	.86074	.4
.7	.51054	.59376	1.6842	.85985	.3
.8	.51204	.59612	1.6775	.85895	.2
.9	.51354	.59849	1.6709	.85806	.1
31.0	.51504	.60086	1.6643	.85717	59.0
.1	.51653	.60324	1.6577	.85627	.9
.2	.51803	.60562	1.6512	.85536	.8
.3	.51952	.60801	1.6447	.85446	.7
.4	.52101	.61040	1.6383	.85355	.6
.5	.52250	.61280	1.6319	.85264	.5
.6	.52399	.61520	1.6255	.85173	.4
.7	.52547	.61761	1.6191	.85081	.3
.8	.52697	.62003	1.6128	.84989	.2
.9	.52844	.62245	1.6066	.84897	.1
32.0	.52992	.62487	1.6003	.84805	58.0
.1	.53140	.62730	1.5941	.84712	.9
.2	.53288	.62973	1.5880	.84619	.8
.3	.53435	.63217	1.5818	.84526	.7
.4	.53583	.63462	1.5757	.84433	.6
.5	.53730	.63707	1.5697	.84339	.5
.6	.53877	.63953	1.5637	.84245	.4
.7	.54024	.64199	1.5577	.84151	.3
.8	.54171	.64446	1.5517	.84057	.2
.9	.54317	.64693	1.5458	.83962	.1
33.0	.54464	.64941	1.5399	.83867	57.0
.1	.54610	.65189	1.5340	.83772	.9
.2	.54756	.65438	1.5282	.83676	.8
.3	.54902	.65688	1.5224	.83581	.7
.4	.55048	.65938	1.5166	.83485	.6
.5	.55194	.66189	1.5108	.83389	.5
.6	.55339	.66440	1.5051	.83292	.4
.7	.55484	.66692	1.4994	.83195	.3
.8	.55630	.66944	1.4938	.83098	.2
.9	.55775	.67197	1.4882	.83001	.1
34.0	.55919	.67451	1.4826	.82904	56.0
.1	.56064	.67705	1.4770	.82806	.9
.2	.56208	.67960	1.4715	.82708	.8
.3	.56353	.68215	1.4659	.82610	.7
.4	.56497	.68471	1.4605	.82511	.6

TABLE 17-1 (*cont.*)

Angle°	*sin*	*tan*	*cot*	*cos*	*Angle°*
.5	.56641	.68728	1.4550	.82413	.5
.6	.56784	.68985	1.4496	.82314	.4
.7	.56928	.69243	1.4442	.82214	.3
.8	.57071	.69502	1.4388	.82115	.2
.9	.57215	.69761	1.4335	.82015	.1
35.0	.57358	.70021	1.4281	.81915	55.0
.1	.57501	.70281	1.4229	.81815	.9
.2	.57643	.70542	1.4176	.81714	.8
.3	.57786	.70804	1.4124	.81614	.7
.4	.57928	.71066	1.4071	.81513	.6
.5	.58070	.71329	1.4019	.81412	.5
.6	.58212	.71593	1.3968	.81310	.4
.7	.58354	.71857	1.3916	.81208	.3
.8	.58496	.72122	1.3865	.81106	.2
.9	.58637	.72388	1.3814	.81004	.1
36.0	.58779	.72654	1.3764	.80902	54.0
.1	.58920	.72921	1.3713	.80799	.9
.2	.59061	.73189	1.3663	.80696	.8
.3	.59201	.73457	1.3613	.80593	.7
.4	.59342	.73726	1.3564	.80489	.6
.5	.59482	.73996	1.3514	.80386	.5
.6	.59622	.74267	1.3465	.80282	.4
.7	.59763	.74538	1.3416	.80178	.3
.8	.59902	.74810	1.3367	.80073	.2
.9	.60042	.75082	1.3319	.79968	.1
37.0	.60182	.75355	1.3270	.79864	53.0
.1	.60321	.75629	1.3222	.79758	.9
.2	.60460	.75904	1.3175	.79653	.8
.3	.60599	.76180	1.3127	.79547	.7
.4	.60738	.76456	1.3079	.79441	.6
.5	.60876	.76733	1.3032	.79335	.5
.6	.61015	.77010	1.2985	.79229	.4
.7	.61153	.77289	1.2938	.79122	.3
.8	.61291	.77568	1.2892	.79016	.2
.9	.61429	.77848	1.2846	.78908	.1
38.0	.61566	.78129	1.2799	.78801	52.0
.1	.61704	.78410	1.2753	.78694	.9
.2	.61841	.78692	1.2708	.78586	.8
.3	.61978	.78975	1.2662	.78478	.7
.4	.62115	.79259	1.2617	.78369	.6
.5	.62251	.79544	1.2572	.78261	.5
.6	.62388	.79829	1.2527	.78152	.4
.7	.62524	.80115	1.2482	.78043	.3
.8	.62660	.80402	1.2437	.77934	.2
.9	.62796	.80690	1.2393	.77824	.1
	cos	*cot*	*tan*	*sin*	

TABLE 17-1 (*cont.*)

Angle°	*sin*	*tan*	*cot*	*cos*	*Angle°*
39.0	.62932	.80978	1.2349	.77715	51.0
.1	.63068	.81268	1.2305	.77605	.9
.2	.63203	.81558	1.2261	.77494	.8
.3	.63338	.81849	1.2218	.77384	.7
.4	.63473	.82141	1.2174	.77273	.6
.5	.63608	.82434	1.2131	.77162	.5
.6	.63742	.82727	1.2088	.77051	.4
.7	.63877	.83022	1.2045	.76940	.3
.8	.64011	.83317	1.2002	.76828	.2
.9	.64145	.83613	1.1960	.76717	.1
40.0	.64279	.83910	1.1918	.76604	50.0
.1	.64412	.84208	1.1875	.76492	.9
.2	.64546	.84507	1.1833	.76380	.8
.3	.64679	.84806	1.1792	.76267	.7
.4	.64812	.85107	1.1750	.76154	.6
.5	.64945	.85408	1.1708	.76041	.5
.6	.65077	.85710	1.1667	.75927	.4
.7	.65210	.86014	1.1626	.75813	.3
.8	.65342	.86318	1.1585	.75700	.2
.9	.65474	.86623	1.1544	.75585	.1
41.0	.65606	.86929	1.1504	.75471	49.0
.1	.65738	.87236	1.1463	.75356	.9
.2	.65869	.87543	1.1423	.75241	.8
.3	.66000	.87852	1.1383	.75126	.7
.4	.66131	.88162	1.1343	.75011	.6
.5	.66262	.88473	1.1303	.74896	.5
.6	.66393	.88784	1.1263	.74780	.4
.7	.66523	.89097	1.1224	.74664	.3
.8	.66653	.89410	1.1184	.74548	.2
.9	.66783	.89725	1.1145	.74431	.1
42.0	.66913	.90040	1.1106	.74314	48.0
.1	.67043	.90357	1.1067	.74198	.9
.2	.67172	.90674	1.1028	.74080	.8
.3	.67301	.90993	1.0990	.73963	.7
.4	.67430	.91313	1.0951	.73846	.6
.5	.67559	.91633	1.0913	.73728	.5
.6	.67688	.91955	1.0875	.73610	.4
.7	.67816	.92277	1.0837	.73491	.3
.8	.67944	.92601	1.0799	.73373	.2
.9	.68072	.92926	1.0761	.73254	.1
43.0	.68200	.93252	1.0724	.73135	47.0
.1	.68327	.93578	1.0686	.73016	.9
.2	.68455	.93906	1.0649	.72897	.8
.3	.68582	.94235	1.0612	.72777	.7
	cos	*cot*	*tan*	*sin*	

TABLE 17-1 (*cont.*)

Angle°	*sin*	*tan*	*cot*	*cos*	*Angle°*
.4	.68709	.94565	1.0575	.72657	.6
.5	.68835	.94896	1.0538	.72537	.5
.6	.68962	.95229	1.0501	.72417	.4
.7	.69088	.95562	1.0464	.72294	.3
.8	.69214	.95897	1.0428	.72176	.2
.9	.69340	.96232	1.0392	.72055	.1
44.0	.69466	.96569	1.0355	.71934	46.0
.1	.69591	.96907	1.0319	.71813	.9
.2	.69717	.97246	1.0283	.71691	.8
.3	.69842	.97586	1.0247	.71569	.7
.4	.69966	.97927	1.0212	.71447	.6
.5	.70091	.98270	1.0176	.71325	.5
.6	.70215	.98613	1.0141	.71203	.4
.7	.70339	.98958	1.0105	.71080	.3
.8	.70463	.99304	1.0070	.70957	.2
.9	.70587	.99652	1.0035	.70834	.1
45.0	.70711	1.00000	1.0000	.70711	45.0
	cos	*cot*	*tan*	*sin*	

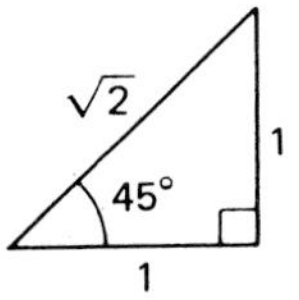

PRACTICE PROBLEMS 17-4

Use the trig table to find the following. Verify your answers by using a calculator.

1. sin 42°
2. cos 58°
3. tan 28°
4. tan 45°
5. sin 45°
6. cos 45°
7. sin 21.6°
8. cos 48.4°
9. tan 50.1°
10. cos 55.6°
11. sin 55.6°
12. tan 31.9°

Solutions

1. 0.66913
2. 0.52992
3. 0.53171
4. 1.0000
5. 0.70711
6. 0.70711
7. 0.36812
8. 0.66393
9. 1.1960
10. 0.56497
11. 0.82511
12. 0.62244

Additional practice problems are at the end of the chapter.

17-4-1 Inverse Trig Functions

Suppose we know the length of the sides and need to find the angles. One of the basic relationships was

$$\sin \theta = \frac{\text{opposite side}}{\text{hypotenuse}}$$

If $\theta = 30°$, then $\sin \theta = \sin 30° = 0.5$. This tells us that if the angle is 30° the ratio of the opposite side to the hypotenuse is 0.5.

Now let's turn it around and say, "Given some ratio of sides, what is the angle?" For example, suppose the ratio of the opposite side to the hypotenuse is 0.74314. What is the angle? The solution requires the use of inverse trig functions. *Inverse trig functions* are written *arc sin, arc cos,* and *arc tan*. Or they are written $\sin^{-1}$, $\cos^{-1}$, and $\tan^{-1}$. Most of the time we will use $\sin^{-1}$, $\cos^{-1}$, and $\tan^{-1}$.

In our example the ratio was opposite side to hypotenuse, which indicates a sine function. Therefore, our inverse trig function is $\sin^{-1}$. $\sin^{-1} 0.74314 = \angle\theta$. The expression $\sin^{-1} 0.74314$ asks the question, "What is the angle whose sin is 0.74314?" To find the answer, we follow along the sin column in Table 17-1 until we find 0.74314, and then we find the angle that corresponds to that value. In this case, the angle is 48°.

$$\sin^{-1} 0.74314 = 48°$$

Using the calculator,

$$.74314 \ \boxed{\sin^{-1}}$$

$$\sin^{-1} 0.74314 = 48.0°$$

On most calculators, $\sin^{-1}$ is the second function of the $\boxed{\sin}$ key.

□ EXAMPLE 17-6

Find the angle whose cosine is 0.90183.

Solution

$$\cos^{-1} 0.90183 = 25.6°$$

$.90183 \ \boxed{\cos^{-1}}$ *Answer*: 25.6°

□ EXAMPLE 17-7

Find the angle whose tangent is 1.4994.

Solution

$$\tan^{-1} 1.4994 = 56.3°$$

$1.4994 \ \boxed{\tan^{-1}}$ *Answer:* 56.3° □

With the calculator,

$$25 \boxed{\times} 40 \boxed{\tan} \boxed{=} \qquad \textit{Answer}: a = 21.0$$

Or

$$\sin \theta = \frac{a}{c} \qquad (17\text{-}7)$$

Solve for a:

$$a = c \sin \theta$$

$$a = 32.6 \sin 40° = 32.6 \times 0.643 = 21.0 \qquad (17\text{-}12)$$

Using the calculator,

$$32.6 \boxed{\times} 40 \boxed{\sin} \boxed{=} \qquad \textit{Answer}: a = 21.0$$

$$\angle\phi = 90° - \angle\theta = 90° - 40° = 50°$$

□ EXAMPLE 17-8

Side $a = 150$ and side $b = 300$ in Figure 17-19. Find the hypotenuse and $\angle\phi$.

Solution Only the tangent function contains one unknown.

$$\tan \phi = \frac{b}{a}$$

$$\tan \phi = \frac{300}{150} = 2$$

Now that we know that $\tan \phi = 2$, our next step is to find the angle whose tangent is 2.

$$\tan^{-1} 2 = \phi = 63.4°$$

We were given sides a and b and we found $\angle\phi$. We can find the hypotenuse by using the Pythagorean theorem:

$$c = \sqrt{a^2 + b^2} = \sqrt{1.125 \times 10^5} = 335$$

Another solution uses the sine function:

$$\sin \phi = \frac{b}{c}$$

$$c = \frac{b}{\sin \phi} = \frac{300}{\sin 63.4°} = 335$$

A third solution uses the cosine function:

$$\cos \phi = \frac{a}{c}$$

$$c = \frac{a}{\cos \phi} = \frac{150}{\cos 63.4°} = 335$$

All solutions yield the same result, and they are all valid.

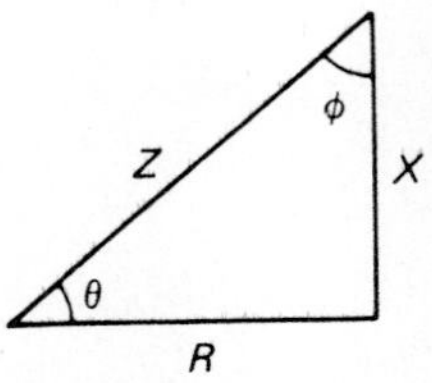

FIGURE 17-20

□ EXAMPLE 17-9

Refer to Figure 17-20. Find R, Z, and $\angle\phi$ if $X = 2.83\text{ k}\Omega$ and $\theta = 33.4°$.

Solution

$$\tan\theta = \frac{X}{R}$$

$$R = \frac{X}{\tan\theta} = \frac{2.83\text{ k}\Omega}{\tan 33.4°} = \frac{2.83\text{ k}\Omega}{0.659} = 4.29\text{ k}\Omega$$

Using the calculator,

2.83 [EE] 3 [÷] 33.4 [tan] [=] *Answer*: $R = 4.29\text{ k}\Omega$

$$\sin\theta = \frac{X}{Z}$$

$$Z = \frac{X}{\sin\theta} = \frac{2.83\text{ k}\Omega}{\sin 33.4°} = 5.14\text{ k}\Omega$$

Using the calculator,

2.83 [EE] 3 [÷] 33.4 [sin] [=] *Answer*: $Z = 5.14\text{ k}\Omega$

$$\phi = 90° - \theta = 90° - 33.4° = 56.6°$$

□ EXAMPLE 17-10

Refer to Figure 17-21. Find B, θ, and ϕ if $Y = 670\ \mu\text{S}$ and $G = 430\ \mu\text{S}$. Find θ.

Solution

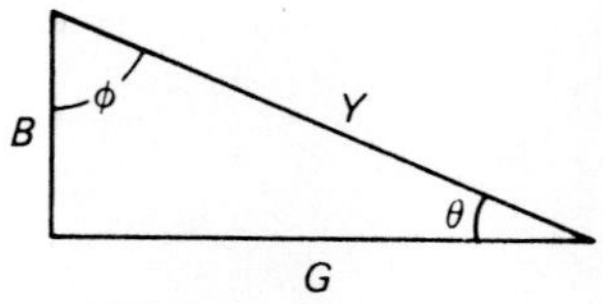

FIGURE 17-21

$$\cos\theta = \frac{G}{Y}$$

$$\cos\theta = \frac{430\ \mu\text{S}}{670\ \mu\text{S}} = 0.642$$

$$\cos^{-1} 0.642 = \theta = 50.1°$$

Using the calculator,

430 [EE] [+/-] 6 [÷] 670 [EE] [+/-] 6 [=] [$\cos^{-1}$] [STO]

Answer: $\theta = 50.1°$ and is stored in memory.

$$\phi = 90° - \theta = 90° - 50.1° = 39.9°$$

Remember, we stored θ, so

90 [−] [RCL] [=] *Answer*: $\phi = 39.9°$

We will use the sine function to find B.

$$\sin\theta = \frac{B}{Y}$$

$$Y\sin\theta = B$$

$$670\ \mu\text{S}\ \sin 50.1° = 670\ \mu\text{S} \times 0.767 = 514\ \mu\text{S}$$

Using the calculator,

670 [EE] [+/-] 6 [×] [RCL] [sin] [=]   *Answer*: $B = 514\ \mu$S

□

PRACTICE PROBLEMS 17-6

1. Refer to Figure 17-19.
   (a) $a = 25.5$, $\theta = 20°$. Find b, c, and ϕ.
   (b) $b = 1400$, $\theta = 65.4°$. Find a, c, and ϕ.
   (c) $a = 350$, $b = 500$. Find θ and c.
   (d) $a = 7.6$, $c = 11.2$. Find θ and b.
   (e) $a = 120$, $\theta = 45°$. Find b and c.
   (f) $\phi = 40°$, $c = 850$. Find a and b.

2. Refer to Figure 17-20.
   (a) $X = 700\ \Omega$, $R = 900\ \Omega$. Find Z and θ.
   (b) $\theta = 68.3°$, $X = 30$ kΩ. Find R and Z.
   (c) $Z = 73.6$ kΩ, $R = 56$ kΩ. Find X and θ.
   (d) $\theta = 27°$, $R = 300\ \Omega$. Find X and Z.
   (e) $Z = 12.3$ kΩ, $X = 8.77$ kΩ. Find R and θ.
   (f) $\theta = 30°$, $Z = 3$ kΩ. Find X and R.

3. Refer to Figure 17-21.
   (a) $G = 2.8$ mS, $B = 3.65$ mS. Find Y and θ.
   (b) $\theta = 30°$, $B = 250\ \mu$S. Find G and Y.
   (c) $G = 10.3$ mS, $Y = 15.8$ mS. Find B and θ.
   (d) $\theta = 52°$, $Y = 650\ \mu$S. Find G and B.
   (e) $B = 140\ \mu$S, $Y = 270\ \mu$S. Find G and θ.
   (f) $\theta = 45°$, $G = 100\ \mu$S. Find B and Y.

Solutions

1. (a) $b = 70$, $c = 74.5$, $\phi = 70°$
   (b) $a = 3060$, $c = 3360$, $\phi = 24.6°$
   (c) $c = 610$, $\theta = 35°$
   (d) $b = 8.23$, $\theta = 42.7°$
   (e) $b = 120$, $c = 170$
   (f) $a = 651$, $b = 546$

2. (a) $Z = 1.14$ kΩ, $\theta = 37.9°$
   (b) $R = 11.9$ kΩ, $Z = 32.3$ kΩ
   (c) $X = 47.8$ kΩ, $\theta = 40.5°$
   (d) $X = 153\ \Omega$, $Z = 337\ \Omega$
   (e) $R = 8.62$ kΩ, $\theta = 45.5°$
   (f) $R = 2.6$ kΩ, $X = 1.5$ kΩ

3. (a) $Y = 4.6$ mS, $\theta = 52.5°$
   (b) $G = 433\ \mu$S, $Y = 500\ \mu$S
   (c) $B = 12$ mS, $\theta = 49.3°$
   (d) $B = 512\ \mu$S, $G = 400\ \mu$S
   (e) $G = 231\ \mu$S, $\theta = 31.2°$
   (f) $B = 100\ \mu$S, $Y = 141\ \mu$S

Additional practice problems are at the end of the chapter.

SELF-TEST 17-2

1. Find the sine, cosine, and tangent of 62.4°.
2. Find $\sin^{-1} 0.74548$.
3. Find $\cos^{-1} 0.75927$.
4. Find $\tan^{-1} 1.42815$.
5. Refer to Figure 17-22. Find b and θ if $a = 27$ and $c = 42.3$.
6. Refer to Figure 17-23. Find Z and θ if $R = 2.7$ kΩ and $X = 3.85$ kΩ.
7. Refer to Figure 17-24. Find B and Y if $G = 220$ μS and $\theta = 38.2°$.

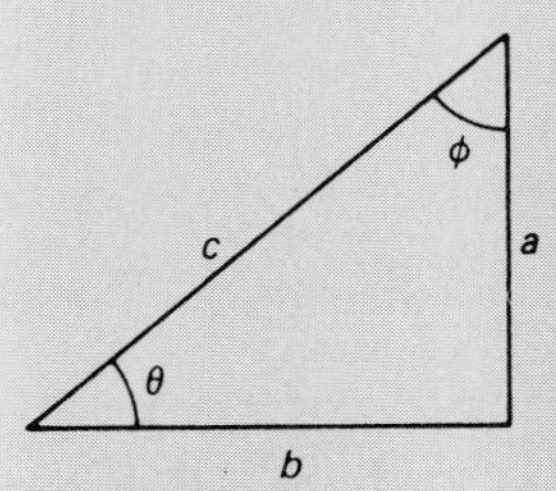

FIGURE 17-22

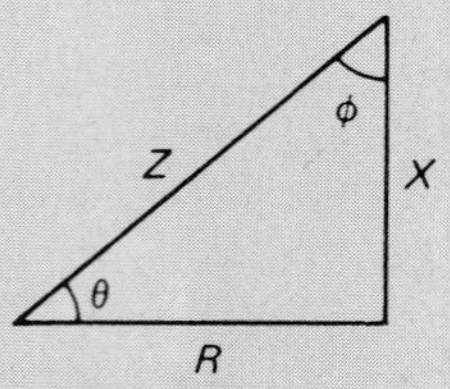

FIGURE 17-23

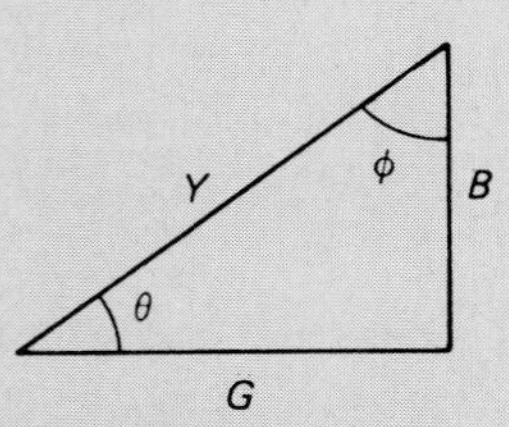

FIGURE 17-24

Answers to Self-test 17-2 are at the end of the chapter.

CHAPTER 17 AT A GLANCE

Page	Key Points, Theorems, and Functions	Example
441	*Key point:* In any right triangle: (1) the sum of the acute angles equals 90°, (2) the hypotenuse is the side of greatest length, and (3) the greater side lies opposite the greater angle.	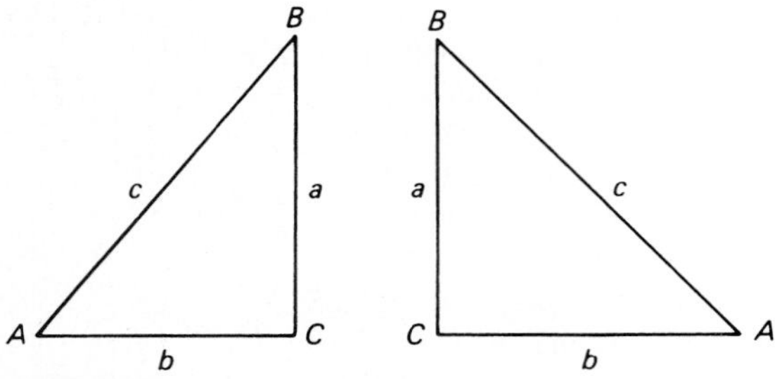 FIGURE 17-25
443	*Key point:* The Pythagorean theorem states that *in a right triangle, the square of the hypotenuse is equal to the sum of the squares of the other two sides.*	$c^2 = a^2 + b^2$

Page	Key Points, Theorems, and Functions	Example
447	*Key point:* Three trigonometric ratios, called *functions,* are the *sine, cosine,* and *tangent* functions. The *sine* function expresses the ratio of the length of the opposite side to the hypotenuse. The *cosine* function expresses the ratio of the length of the adjacent side to the hypotenuse. The *tangent* function expresses the ratio of the length of the opposite side to the adjacent side.	$\sin\theta = \frac{a}{c}$ $\cos\theta = \frac{b}{c}$ $\tan\theta = \frac{a}{b}$ 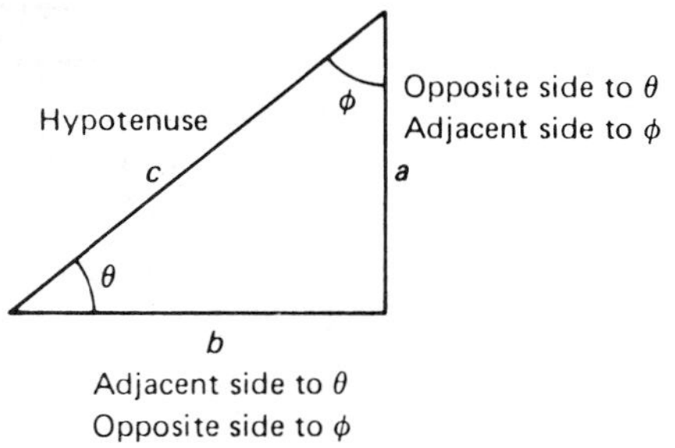 **FIGURE 17-26**
454	*Key point:* Inverse trig functions are used to find the ratio of sides when the angle is known. Three *inverse trig functions* are *arc sin, arc cos,* and *arc tan*. They are usually written sin^{-1}, cos^{-1}, and tan^{-1}.	$\sin^{-1} 0.74314 = 48.0°$ $\cos^{-1} 0.74314 = 42.0°$ $\tan^{-1} 0.74314 = 36.6°$
456	*Key point:* When rearranging the equations of the functions to solve for an unknown side or angle, sin θ (or cos θ or tan θ) is one quantity and cannot be separated.	$\cos\theta = \frac{R}{Z}$ $Z = \frac{R}{\cos\theta}$

End of Chapter Problems 17-1

1. Refer to Figure 17-20:
- (a) If $\angle\theta = 35°$, find $\angle\phi$.
- (b) Which side (R or X) is greater?
- (c) Which side lies opposite $\angle\phi$?
- (d) Which side is adjacent to $\angle\theta$?

2. Refer to Figure 17-21:
- (a) If $\angle\phi = 70°$, find $\angle\theta$.
- (b) Which side (G or B) is greater?
- (c) Which side lies opposite $\angle\theta$? Which side is adjacent to $\angle\phi$?

3. Refer to Figure 17-20:
- (a) Find Z when $R = 2.7\ \text{k}\Omega$ and $X = 4\ \text{k}\Omega$
- (b) Find R when $Z = 600\ \Omega$ and $X = 350\ \Omega$.
- (c) Find X when $Z = 120\ \text{k}\Omega$ and $R = 68\ \text{k}\Omega$.

4. Refer to Figure 17-20:
- (a) Find Z when $R = 20\ \text{k}\Omega$ and $X = 25\ \text{k}\Omega$.
- (b) Find R when $Z = 600\ \text{k}\Omega$ and $X = 280\ \text{k}\Omega$.
- (c) Find X when $Z = 11.7\ \text{k}\Omega$ and $R = 7.8\ \text{k}\Omega$.

5. Refer to Figure 17-21:
- (a) Find Y when $G = 1.83$ mS and B $= 2.37$ mS.
- (b) Find G when $Y = 1$ mS and $B = 600\ \mu$S.
- (c) Find B when $Y = 460\ \mu$S and $G = 175\ \mu$S.

6. Refer to Figure 17-21:
- (a) Find Y when $G = 2.8$ mS and $B = 4.35$ mS.
- (b) Find G when $Y = 170\ \mu$S and $B = 95\ \mu$S.
- (c) Find B when $Y = 10.7$ mS and $G = 6.73$ mS.

End of Chapter Problems 17-2

Refer to Figure 17-26:

1. Let $a = 14$ and $b = 22$. Find c. Which of angles θ and ϕ is greater?
2. Let $a = 40$ and $b = 65$. Find c. Which of angles θ and ϕ is greater?
3. Let $a = 250$ and $c = 350$. Find b.
4. Let $a = 8.93$ and $c = 12.1$. Find b.
5. Let $b = 73.2$ and $c = 100$. Find a.
6. Let $b = 600$ and $c = 935$. Find a.

Refer to Figure 17-27:

7. Let $R = 680\ \Omega$ and $X = 700\ \Omega$. Find Z. Which of angles θ and ϕ is greater?
8. Let $R = 330\ \Omega$ and $X = 200\ \Omega$. Find Z. Which of angles θ and ϕ is greater?
9. Let $R = 470$ kΩ and $X = 560$ kΩ. Find Z. Which of angles θ and ϕ is greater?
10. Let $R = 680$ kΩ and $X = 330$ kΩ. Find Z. Which of angles θ and ϕ is greater?
11. Let $R = 75$ kΩ and $X = 39$ kΩ. Find Z. Which of angles θ and ϕ is greater?
12. Let $R = 12$ kΩ and $X = 18$ kΩ. Find Z. Which of angles θ and ϕ is greater?
13. Let $R = 4.7$ kΩ and $Z = 8.27$ kΩ. Find X. Which of angles θ and ϕ is greater?
14. Let $R = 8.2$ kΩ and $Z = 8.84$ kΩ. Find X. Which of angles θ and ϕ is greater?
15. Let $R = 100$ kΩ and $Z = 180$ kΩ. Find X. Which of angles θ and ϕ is greater?
16. Let $R = 390$ kΩ and $Z = 448$ kΩ. Find X. Which of angles θ and ϕ is greater?
17. Let $X = 43$ kΩ and $Z = 50.8$ kΩ. Find R. Which of angles θ and ϕ is greater?
18. Let $X = 82$ kΩ and $Z = 99.3$ kΩ. Find R. Which of angles θ and ϕ is greater?
19. Let $X = 47\ \Omega$ and $Z = 102\ \Omega$. Find R. Which of angles θ and ϕ is greater?
20. Let $X = 33\ \Omega$ and $Z = 88.4\ \Omega$. Find R. Which of angles θ and ϕ is greater?

Refer to Figure 17-28:

21. Let $G = 450\ \mu$S and $B = 700\ \mu$S. Find Y. Which of angles θ and ϕ is greater?
22. Let $G = 825\ \mu$S and $B = 350\ \mu$S. Find Y. Which of angles θ and ϕ is greater?
23. Let $G = 3.37$ mS and $B = 2.78$ mS. Find Y. Which of angles θ and ϕ is greater?
24. Let $G = 1.08$ mS and $B = 4.33$ mS. Find Y. Which of angles θ and ϕ is greater?
25. Let $G = 37\ \mu$S and $Y = 42.7\ \mu$S. Find B. Which of angles θ and ϕ is greater?
26. Let $G = 30.3\ \mu$S and $Y = 39.7\ \mu$S. Find B. Which of angles θ and ϕ is greater?
27. Let $G = 6.67$ mS and $Y = 8.07$ mS. Find B. Which of angles θ and ϕ is greater?
28. Let $G = 10\ \mu$S and $Y = 12\ \mu$S. Find B. Which of angles θ and ϕ is greater?
29. Let $B = 256\ \mu$S and $Y = 612\ \mu$S. Find G. Which of angles θ and ϕ is greater?
30. Let $B = 23.3\ \mu$S and $Y = 37.5\ \mu$S. Find G. Which of angles θ and ϕ is greater?
31. Let $B = 17.9$ mS and $Y = 22.3$ mS. Find G. Which of angles θ and ϕ is greater?
32. Let $B = 33.3\ \mu$S and $Y = 60.1\ \mu$S. Find G. Which of angles θ and ϕ is greater?

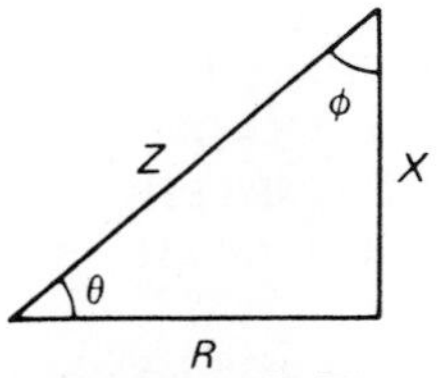

FIGURE 17-27

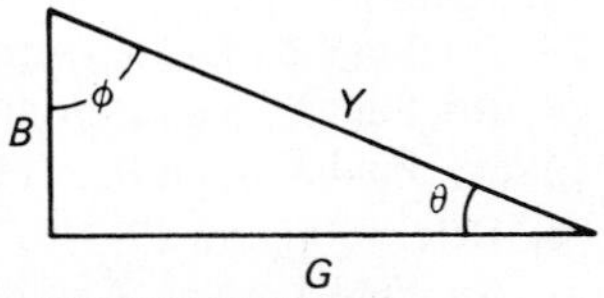

FIGURE 17-28

End of Chapter Problems 17-3

1. Refer to Figure 17-20: Write the equation for the three functions with references to (a) $\angle\theta$ and (b) $\angle\phi$.
2. Refer to Figure 17-21: Write the equation for the three functions with reference to (a) $\angle\theta$ and (b) $\angle\phi$.

End of Chapter Problems 17-4

Use the Trigonometric Functions Table in Appendix B to find the following. Verify your answers using the calculator.

1. sin 27.3°
2. sin 53.2°
3. sin 63.6°
4. sin 37.5°
5. cos 78.3°
6. cos 41.7°
7. cos 23.7°
8. cos 58.9°
9. tan 67.3°
10. tan 12.7°
11. tan 40.3°
12. tan 75.7°

End of Chapter Problems 17-5

Use the Trigonometric Functions Table in Appendix B to find the following. Verify your answers using the calculator.

1. $\sin^{-1} 0.39715$
2. $\sin^{-1} 0.95106$
3. $\sin^{-1} 0.87036$
4. $\sin^{-1} 0.45865$
5. $\cos^{-1} 0.55630$
6. $\cos^{-1} 0.45088$
7. $\cos^{-1} 0.78043$
8. $\cos^{-1} 0.56641$
9. $\tan^{-1} 1.47146$
10. $\tan^{-1} 1.13029$

End of Chapter Problems 17-6

Refer to Figure 17-19:

1. $a = 25$, $b = 40$. Find θ, ϕ, and c.
2. $a = 67.5$, $b = 33$. Find θ, ϕ, and c.
3. $a = 200$, $\phi = 35°$. Find θ, b, and c.
4. $a = 6.35$, $\phi = 61°$. Find θ, b, and c.
5. $c = 17.3$, $\theta = 40°$. Find a and b.
6. $c = 680$, $\theta = 78°$. Find a and b.

Refer to Figure 17-20:

7. $X = 17.6\ \text{k}\Omega$, $\theta = 22.3°$. Find R and Z.
8. $X = 27.8\ \text{k}\Omega$; $\theta = 53°$. Find R and Z.
9. $X = 650\ \Omega$, $\theta = 22.3°$. Find R and Z.
10. $X = 1.87\ \text{k}\Omega$, $\theta = 54.3°$. Find R and Z.
11. $R = 5.6\ \text{k}\Omega$, $X = 10\ \text{k}\Omega$. Find Z and θ.
12. $R = 100\ \text{k}\Omega$, $X = 200\ \text{k}\Omega$. Find Z and θ.
13. $R = 27\ \text{k}\Omega$, $X = 19.7\ \text{k}\Omega$. Find Z and θ.
14. $R = 120\ \Omega$, $X = 150\ \Omega$. Find Z and θ.
15. $R = 20\ \text{k}\Omega$; $\theta = 30°$. Find X and Z.
16. $R = 3.3\ \text{k}\Omega$, $\theta = 60°$. Find X and Z.
17. $R = 270\ \text{k}\Omega$, $\theta = 37.5°$. Find X and Z.
18. $R = 120\ \text{k}\Omega$, $\theta = 64.7°$. Find X and Z.
19. $Z = 40\ \text{k}\Omega$, $\theta = 67.5°$. Find R and X.
20. $Z = 6.2\ \text{k}\Omega$, $\theta = 17.8°$. Find R and X.
21. $Z = 785\ \Omega$, $\theta = 40.3°$. Find R and X.
22. $Z = 310\ \text{k}\Omega$, $\theta = 73.5°$. Find R and X.

23. $R = 12$ kΩ, $Z = 22$ kΩ. Find X and θ.
24. $R = 39$ kΩ, $Z = 65$ kΩ. Find X and θ.
25. $R = 1.8$ kΩ, $Z = 2.55$ kΩ. Find X and θ.
26. $R = 56$ kΩ, $Z = 103$ kΩ. Find X and θ.
27. $X = 17.5$ kΩ, $Z = 38$ kΩ. Find R and θ.
28. $X = 450$ Ω, $Z = 750$ Ω. Find R and θ.
29. $X = 15.3$ kΩ, $Z = 26.8$ kΩ. Find R and θ.
30. $X = 800$ Ω, $Z = 1.28$ kΩ. Find R and θ.

Refer to Figure 17-21:

31. $G = 60$ μS, $\theta = 25°$. Find B and Y.
32. $G = 6.7$ mS, $\theta = 73°$. Find B and Y.
33. $G = 135$ μS, $\theta = 65°$. Find B and Y.
34. $G = 3.45$ mS, $\theta = 32.5°$. Find B and Y.
35. $G = 12.5$ mS, $B = 8.73$ mS. Find Y and θ.
36. $G = 440$ μS, $B = 560$ μS. Find Y and θ.
37. $G = 12.7$ μS, $B = 17.3$ μS. Find Y and θ.
38. $G = 256$ μS, $B = 430$ μS. Find Y and θ.
39. $B = 35$ mS, $\theta = 10.7°$. Find G and Y.
40. $B = 800$ μS, $\theta = 75°$. Find G and Y.
41. $B = 3.32$ mS, $\theta = 27.4°$. Find G and Y.
42. $B = 4.73$ mS, $\theta = 64.5°$. Find G and Y.
43. $G = 6.35$ mS, $Y = 9.35$ mS. Find B and θ.
44. $G = 300$ μS, $Y = 500$ μS. Find B and θ.
45. $G = 2.33$ mS, $Y = 3.30$ mS. Find B and θ.
46. $G = 170$ μS, $Y = 260$ μS. Find B and θ.
47. $Y = 70$ μS, $\theta = 32°$. Find G and B.
48. $Y = 600$ μS, $\theta = 58°$. Find G and B.
49. $Y = 6.57$ mS, $\theta = 55°$. Find G and B.
50. $Y = 200$ μS, $\theta = 34°$. Find G and B.
51. $B = 7.5$ mS, $Y = 11.5$ mS. Find G and θ.
52. $B = 6.35$ mS, $Y = 9.5$ mS. Find G and θ.
53. $B = 170$ μS, $Y = 320$ μS. Find G and θ.
54. $B = 450$ μS, $Y = 670$ μS. Find G and θ.

Answers to Self-tests

SELF-TEST 17-1

1. $\theta = 42.7°$
2. Side R is longer.
3. $X = 56.6$ kΩ
4. $\angle\theta$
5. $\sin\theta = \dfrac{X}{Z}$, $\cos\theta = \dfrac{R}{Z}$, $\tan\theta = \dfrac{X}{R}$

SELF-TEST 17-2

1. $\sin 62.4° = 0.88620$, $\cos 62.4° = 0.46330$, $\tan 62.4° = 1.91282$
2. 48.2°
3. 40.6°
4. 55°
5. $b = 32.6$, $\theta = 39.7°$
6. $Z = 4.7$ kΩ, $\theta = 55°$
7. $B = 173$ μS, $Y = 280$ μS

TRIGONOMETRIC IDENTITIES

CHAPTER 18

INTRODUCTION

We learned in Chapter 17 how to solve triangles when one of the angles is a right angle (90°). In this chapter we will consider triangles other than right triangles.

Let's review a few things we have learned about angles and triangles. Recall that there are three kinds of angles: acute angles (angles that are less than 90°), right angles (90° angles), and obtuse angles (angles that are greater than 90°). In triangles, the sum of the three angles must equal 180°. When we dealt with the right triangle, one angle was 90°. The other two angles, then, had to be acute angles since the sum of the two had to be 90°. Triangles that do not contain a right angle are called *oblique* triangles. An oblique triangle may include an obtuse angle, depending on its shape. In such a case, the other two angles must be acute, since the sum of the three must equal 180°.

In solving right triangles, we used the Pythagorean theorem and/or the trig functions sine, cosine, and tangent. When the triangle is oblique, we must use additional methods of solution. For such triangles, various trigonometric identities have been derived to help in the solutions. In this chapter, we will learn how to use some of these identities in problem solving.

CHAPTER OBJECTIVES

In this chapter you will learn how to:

1. Use the law of sines to find the sides and angles of oblique triangles.
2. Use the law of cosines to find the sides and angles of oblique triangles.

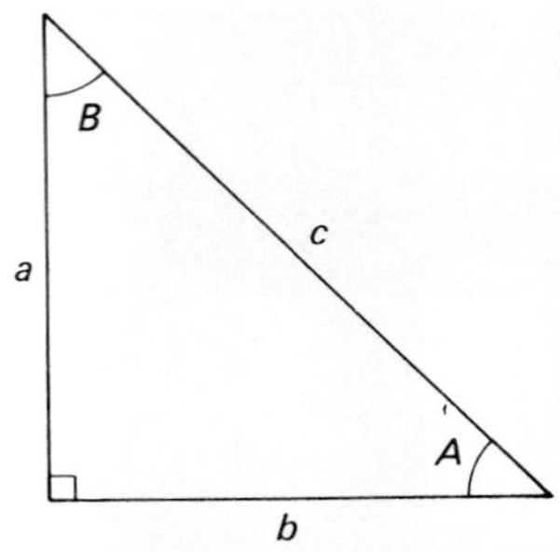

FIGURE 18-1

Several important relationships have been developed between the trigonometric functions and the Pythagorean theorem. These relationships are called trigonometric *identities*. These identities make it possible to simplify complex trigonometric problems so that they can be easily solved. Let's see how we can express the Pythagorean theorem in terms of trig functions: $a^2 + b^2 = c^2$, where a and b are the sides and c is the hypotenuse of the right triangle in Figure 18-1. Dividing both sides by c^2 we get

$$\frac{a^2}{c^2} + \frac{b^2}{c^2} = \frac{c^2}{c^2}, \qquad \frac{c^2}{c^2} = 1$$

$$\frac{a^2}{c^2} + \frac{b^2}{c^2} = 1$$

Using trig functions,

$$\sin A = \frac{a}{c} \quad \text{and} \quad \cos A = \frac{b}{c}$$

Then

$$\frac{b^2}{c^2} = (\cos A)^2$$

$[(\cos A)^2$ is usually written as $\cos^2 A]$ and

$$\frac{a^2}{c^2} = \sin^2 A$$

Therefore,

$$\sin^2 A + \cos^2 A = 1$$

From this identity, other identities have been derived: $\tan^2 A + 1 = \sec^2 A$ and $1 + \cot^2 A = \csc^2 A$. In addition, as stated previously, many other identities have been derived to help in the solution of complex problems. In this chapter we are going to limit ourselves to two of the more useful identities, the law of cosines and the law of sines.

18-1 LAW OF SINES

Consider the oblique triangle in Figure 18-2. We have labeled the sides a, b, and c. These sides lie opposite the angles A, B, and C. We need to develop relationships between the angles and sides so that when we know some of them we can find the others. To do this, let's draw a line from angle B to side b such that we form two right triangles as in Figure 18-3. We have labeled this line h. Now that we have right triangles, we can apply the trig functions that we learned in previous chapters.

$$\sin \theta = \frac{\text{opposite side}}{\text{hypotenuse}} \qquad \text{(general equation)}$$

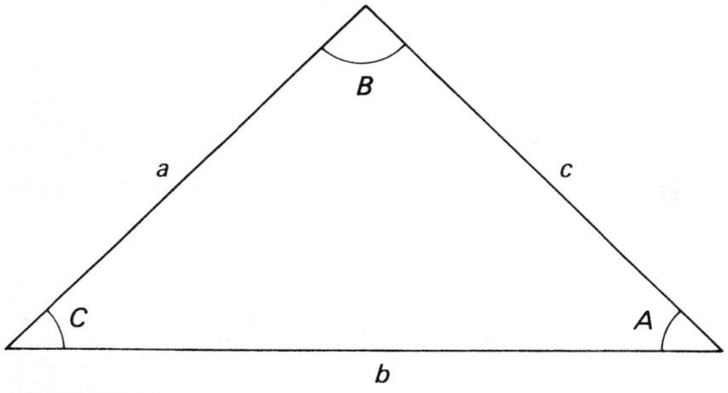

FIGURE 18-2

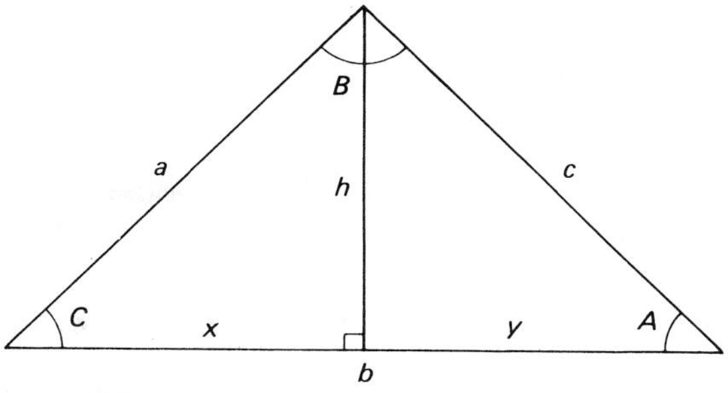

FIGURE 18-3

From the two triangles of Figure 18-3, we can say

$$\sin C = \frac{h}{a} \quad \text{and} \quad \sin A = \frac{h}{c}$$

Moving a and c from the right member to the left member of the equations, we get

$$a \sin C = h \quad \text{and} \quad c \sin A = h$$

If $a \sin C = h$ and $c \sin A = h$, then

$$a \sin C = c \sin A \tag{18-1}$$

Rearranging the equation,

$$\frac{\sin C}{c} = \frac{\sin A}{a} \tag{18-2}$$

Using the same triangles, we could also prove that

$$\frac{\sin C}{c} = \frac{\sin B}{b} \tag{18-3}$$

and we could prove that

$$\frac{\sin A}{a} = \frac{\sin B}{b} \tag{18-4}$$

It follows then that we can combine Equations 18-2, 18-3, and 18-4:

$$\frac{\sin A}{a} = \frac{\sin B}{b} = \frac{\sin C}{c} \tag{18-5}$$

These equations are the trigonometric identities called the *law of sines*. These equations can be used where (1) two angles and a side opposite one of them are given or (2) two sides and an angle opposite one of them are given. In Example 18.1, two angles and a side are given.

□ EXAMPLE 18-1

Refer to Figure 18-2. Using the law of sines, find angle A and sides a and c where angle B is 27°, angle C is 64°, and side b is 700.

Solution Let's first find angle A. $A + B + C = 180°$.

$$A = 180° - B - C = 180° - 27° - 64° = 89°$$

To find side a using the law of sines, we could use Equation 18-4. We couldn't use Equation 18-2 because there are too many unknowns (sides a and c). Let's solve Equation 18-4 for a.

$$a = \frac{b \sin A}{\sin B} = \frac{700 \sin 89°}{\sin 27°} = \frac{700 \times 0.9998}{0.4540} = 1542$$

We will use Equation 18.3 to find side c. Let's rearrange the equation and solve for c.

$$c = \frac{b \sin C}{\sin B} = \frac{700 \times \sin 64°}{\sin 27°} = \frac{700 \times 0.8988}{0.4540} = 1386$$

□

The law of sines can be used to solve problems where two sides and an angle opposite one are given. This is sometimes referred to as the *ambiguous case* because there are three possible results, depending on the information given:

1. There is no solution. This is the case if, in solving the problem, we get values for the sine of the angle that are greater than 1.
2. There is one solution. This is the case if the given angle is greater than 90° (making the other two angles acute angles), or if the side opposite the given angle is larger than the other given side.
3. There are two solutions. This is the case if the given angle is less than 90° and the side opposite the given angle is smaller than the other given side.

In our examples, we will consider only problems with solutions.

□ EXAMPLE 18-2

Refer to Figure 18-2. Using the law of sines, find angles B and C and side c where angle $A = 125°$, side $a = 10$, and side $b = 8$.

Solution Equation 18-4 contains only one unknown (angle A), so let's use that equation and solve for B.

$$\sin B = \frac{b \sin A}{a} = \frac{8 \sin 125°}{10} = \frac{8 \times 0.8192}{10} = 0.6553$$

$$B = \sin^{-1} 0.6553 = 40.94°$$

$$C = 180° - A - B = 180° - 125° - 40.94° = 14.06°$$

Now that we know angle C, we can use Equation 18-2 to find side c (we could also have used Equation 18-3).

$$c = \frac{a \sin C}{\sin A} = \frac{10 \sin 14.06°}{\sin 125°} = \frac{10 \times 0.2429}{0.8192} = 2.966$$

☐ EXAMPLE 18-3

In an oblique triangle, side $a = 70$, side $b = 40$, and angle $A = 65°$. Find the unknown side and angles.

Solution Let's use Equation 18-4 and solve for the unknown, angle B.

$$\sin B = \frac{b \sin A}{a} = \frac{40 \sin 65°}{70} = \frac{40 \times 0.9063}{70} = 0.5179$$

$$\sin^{-1} 0.5179 = B = 31.19°$$

$$C = 180° - A - B = 180° - 65° - 31.19° = 83.81°$$

Now we can solve for side c using either Equation 18-2 or 18-3. We will use 18-2.

$$c = \frac{a \sin C}{\sin A} = \frac{70 \sin 83.81°}{\sin 65°} = \frac{70 \times 0.5179}{0.9063} = 76.79$$

☐ EXAMPLE 18-4

In an oblique triangle, side $b = 10$, side $c = 15$, and angle $B = 30°$. Find the unknown side and angles.

Solution Rearranging Equation 18-3 and solving for angle C, we get

$$\sin C = \frac{c \sin B}{b} = \frac{15 \sin 30°}{10} = \frac{15 \times 0.5}{10} = 0.75$$

$$\sin^{-1} 0.75 = C = 48.59°$$

$$A = 180° - B - C = 180° - 30° - 48.59° = 101.4°$$

But angle C could also be a quadrant II angle since the problem meets the conditions listed above for two solutions. That is, the given angle is an acute angle, and the side opposite the given angle is smaller than the other given side. Then angle C could also be

$$C = 180° - 48.59° = 131.4°$$

which would make angle A:

$$A = 180° - 30° - 131.4° = 18.6°$$

There are two possible answers:

$$C = 48.59° \quad \text{and} \quad A = 101.4°$$

or

$$C = 131.4° \quad \text{and} \quad A = 18.6°$$

Let's find side a for the first case. To avoid compounding possible errors, its better to use equations that contain given values whenever possible, so let's use Equation 18-4.

$$a = \frac{b \sin A}{\sin B} = \frac{10 \sin 101.4°}{\sin 30°} = \frac{10 \times 0.9803}{0.5} = 19.60$$

and for the second case

$$a = \frac{b \sin A}{\sin B} = \frac{10 \sin 18.60°}{\sin 30°} = \frac{10 \times 0.3190}{0.5} = 6.379$$

□

PRACTICE PROBLEMS 18-1

Refer to Figure 18-2. Find the unknown sides and angles in the following problems using the law of sines.

1. $b = 45$, $B = 72°$, and $C = 53°$

2. $b = 110$, $B = 110°$, and $C = 15°$

3. $c = 50$, $A = 35°$, and $C = 70°$

4. $a = 100$, $b = 70$, and $A = 80°$

5. $a = 200$, $b = 150$, and $A = 65°$

6. $a = 200$, $b = 300$, and $A = 20°$

7. $a = 25$, $b = 40$, and $A = 35°$

8. $a = 25$, $b = 15$, and $A = 40°$

9. $b = 150$, $c = 178$, and $C = 80°$

10. $b = 8.5$, $c = 6$, and $C = 32°$

Solutions

1. $a = 38.76$
$c = 37.79$
$A = 55°$

2. $a = 95.89$
$c = 30.30$
$A = 55°$

3. $a = 30.52$
$b = 51.40$
$B = 75°$

4. $c = 84.60$
$B = 43.58°$
$C = 56.42°$

5. $c = 210.1$
$B = 42.82°$
$C = 72.18°$

6. Two solutions:
(1) $c = 453.8$, $B = 30.87°$, $C = 129.1°$
(2) $c = 110.3$, $B = 149.1°$, $C = 10.87°$

7. Two solutions:
(1) $c = 42.70$, $B = 66.60°$, $C = 78.40°$
(2) $c = 22.84$, $B = 113.4°$, $C = 31.60°$

8. $c = 34.56$
$B = 22.69°$
$C = 117.3°$

9. $a = 125.4$
$A = 43.91°$
$B = 56.09°$

10. Two solutions:
(1) $a = 11.17$, $A = 99.35°$, $B = 48.65°$
(2) $a = 3.23$, $A = 16.6°$, $B = 131.4°$

Additional practice problems are at the end of the chapter.

18-2 LAW OF COSINES

The law of sines has its limitations in that we need to know two sides and the included angle, or we need two angles and a side opposite one of them. The law of cosines will provide the means to solve triangles when other information is available.

Let's look again at Figure 18-3. This time let's use the Pythagorean theorem and write equations concerning the right triangle with sides x, h, and a.

$$h^2 = a^2 - x^2$$

Next, write the equation using the right triangle with sides y, h, and c.

$$h^2 = c^2 - y^2$$

If $a^2 - x^2 = h^2$ and $c^2 - y^2 = h^2$, then

$$a^2 - x^2 = c^2 - y^2 \tag{18-6}$$

Now we can manipulate Equation 18-6 so that we can solve for a^2 in terms of b^2 and c^2 in the original triangle. Solving for a^2, we get

$$a^2 = c^2 - y^2 + x^2 \tag{18-7}$$

Since $b = x + y$ in the triangle, then $x = b - y$. Now substitute $b - y$ for x and we get

$$a^2 = c^2 - y^2 + (b - y)^2 \tag{18-8}$$

$$a^2 = c^2 - y^2 + b^2 - 2by + y^2 = c^2 + b^2 - 2by$$

Using trig functions, $y = c \cos A$. Then

$$a^2 = c^2 + b^2 - 2bc \cos A \tag{18-9}$$

In a similar manner, we could prove that

$$b^2 = a^2 + c^2 - 2ac \cos B \tag{18-10}$$

and

$$c^2 = a^2 + b^2 - 2ab \cos C \tag{18-11}$$

Equations 18-9, 18-10, and 18-11 are called the *law of cosines*. These equations and the law of sines will enable us to solve any triangle for which there is a solution.

□ EXAMPLE 18-5

Refer to Figure 18-2. Using the law of cosines, find the unknown angles when $a = 400$, $b = 300$, and $c = 250$.

Solutions Let's rearrange Equation 18-9 and solve for angle A.

$$-2bc \cos A = a^2 - b^2 - c^2$$

$$\cos A = \frac{a^2 - b^2 - c^2}{-2bc} = \frac{400^2 - 300^2 - 250^2}{-2 \times 300 \times 250} = -0.05000$$

$$\cos^{-1} -0.05 = A = 92.87°$$

Cos A was negative because the angle is in the second quadrant (greater than 90° but less than 180°). Using the calculator:

400 [x^2] [−] 300 [x^2] [−] 250 [x^2]
[=] [÷] [(] 2 [+/−] [×] 300 [×] 250 [)] [=] [$\cos^{-1}$] *Answer*: $A = 92.87°$

Now that we know one angle, we can find the other angles by using either the law of sines or the law of cosines. Using the law of sines is less rigorous, so let's rearrange Equation 18-2 and find angle C.

$$\sin C = \frac{c \sin A}{a} = \frac{250 \sin 92.87°}{400} = \frac{250 \times 0.9988}{400} = 0.6242$$

$$\sin^{-1} 0.6242 = C = 38.62°$$

$$\text{angle } B = 180° - A - C = 180° - 38.62° - 92.87° = 48.51°$$

☐ EXAMPLE 18-6

Refer to Figure 18-2. Find the unknown sides and angles when $b = 60$, $c = 80$, and angle $A = 75°$

Solution Solve for side a using Equation 18-9.

$$a = \sqrt{b^2 + c^2 - 2bc \cos A}$$

$$= \sqrt{60^2 + 80^2 - 2 \times 60 \times 80 \times \cos 75°}$$

$$= \sqrt{7515} = 86.69$$

Using the calculator,

60 [x^2] [+] 80 [x^2] [−] [(] 2 [×] 60 [×] 80 [×] 75 [cos] [)] [=] [$\sqrt{x}$]

Answer: $a = 86.69$

Next, find angle B using the law of sines.

$$\sin B = \frac{b \sin A}{a} = \frac{60 \sin 75°}{86.69} = 0.6685$$

$$\sin^{-1} 0.6685 = B = 41.95°$$

$$\text{angle } C = 180° - A - B = 180° - 75° - 41.95° = 63.05°$$

☐

PRACTICE PROBLEMS 18-2

Refer to Figure 18-4. Find the missing sides and angles using the law of cosines and the law of sines where:

1. Side $a = 10$, $b = 15$, and $c = 20$
2. Side $a = 12$, $b = 10$, and $c = 15$
3. Side $b = 55$, $c = 70$, and angle $A = 40°$
4. Side $b = 140$, $c = 100$, and angle $A = 85°$
5. Side $a = 30$, $c = 40$, and angle $B = 75°$
6. Side $a = 85$, $c = 115$, and angle $B = 110°$

Solutions

1. $A = 28.96°$, $B = 46.57°$, $C = 104.5°$
2. $A = 52.89°$, $B = 41.65°$, $C = 85.46°$

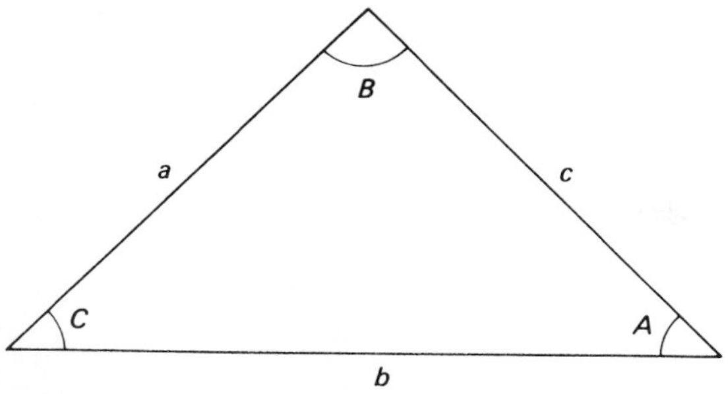

FIGURE 18-4

3. $a = 45.02$, $B = 51.75°$, $C = 88.25°$

4. $a = 164.8$, $B = 57.81°$, $C = 37.19°$

5. $b = 43.35$, $A = 41.95°$, $C = 63.05°$

6. $b = 164.7$, $A = 29.00°$, $C = 41.00°$

Additional practice problems are at the end of the chapter.

SELF-TEST 18-1

Refer to Figure 18-4. Find the unknown sides and angles from the given values, using the law of sines and the law of cosines, where:

1. Side $c = 310$, $A = 55°$, and $C = 80°$

2. Side $a = 65$, $b = 50$, and $A = 100°$

3. Side $a = 50$, $b = 100$, and $c = 60$

4. Side $b = 25$, $c = 15$ and $A = 110°$

Answers to Self-test 18-1 are at the end of the chapter.

CHAPTER 18 AT A GLANCE

Page	Trigonometric Identities	Equations
466	*Law of sines:* these equations can be used when (1) two angles and a side opposite one of them are given or (2) two sides and an angle opposite one of them are given.	$\frac{\sin A}{a} = \frac{\sin B}{b} = \frac{\sin C}{c}$
470	*Law of cosines:* these equations can be used when (1) all three sides are known or (2) when two sides and the angle opposite the unknown side are given.	$a^2 = c^2 + b^2 - 2bc \cos A$ $b^2 = a^2 + c^2 - 2ac \cos B$ $c^2 = a^2 + b^2 - 2ab \cos C$

End of Chapter Problems 18-1

Refer to Figure 18-4. Find the unknown sides and angles using the law of sines when:

1. Side $c = 35$, $A = 50°$, $C = 70°$
2. Side $a = 7.5$, $A = 25°$, $B = 80°$
3. Side $b = 60$, $c = 100$, $C = 35°$
4. Side $b = 45$, $c = 78$, $C = 110°$
5. Side $a = 15$, $b = 22$, $A = 28°$
6. Side $a = 45$, $b = 60$, $A = 35°$
7. Side $a = 10$, $A = 30°$, $B = 70°$
8. Side $a = 300$, $A = 115°$, $B = 25°$
9. Side $b = 160$, $c = 100$, $C = 22°$
10. Side $b = 17$, $c = 12$, $C = 44°$
11. Side $a = 10$, $b = 20$, $A = 25°$
12. Side $a = 40$, $b = 70$, $A = 30°$
13. Side $b = 55$, $B = 50°$, $C = 105°$
14. Side $b = 20$, $B = 60°$, $C = 105°$
15. Side $a = 120$, $b = 65$, $A = 95°$
16. Side $a = 170$, $b = 210$, $A = 40°$
17. Side $b = 6$, $c = 4.7$, $C = 34°$
18. Side $b = 95$, $c = 70$, $C = 15°$
19. Side $c = 14$, $A = 125°$, $C = 45°$
20. Side $c = 65$, $A = 25°$, $C = 40°$
21. Side $c = 350$, $A = 30°$, $C = 110°$
22. Side $c = 38$, $A = 105°$, $C = 32°$
23. Side $b = 40$, $c = 65$, $B = 28°$
24. Side $b = 650$, $c = 900$, $B = 42°$
25. Side $b = 350$, $c = 275$, $B = 100°$
26. Side $b = 55$, $c = 40$, $B = 80°$
27. Side $b = 17$, $B = 95°$, $C = 22°$
28. Side $b = 40$, $B = 60°$, $C = 60°$
29. Side $b = 95$, $B = 20°$, $C = 20°$
30. Side $b = 200$, $B = 80°$, $C = 45°$
31. Side $b = 250$, $c = 200$, $B = 70°$
32. Side $b = 70$, $c = 40$, $B = 105°$
33. Side $b = 150$, $c = 100$, $B = 120°$
34. Side $b = 16$, $c = 12$, $B = 75°$
35. Side $b = 20$, $c = 33$, $B = 33°$
36. Side $b = 70$, $c = 100$, $B = 33°$
37. Side $c = 110$, $A = 47°$, $C = 95°$
38. Side $c = 220$, $A = 25°$, $C = 100°$
39. Side $c = 4.3$, $A = 75°$, $C = 45°$
40. Side $c = 150$, $A = 35°$, $C = 70°$

End of Chapter Problems 18-2

Refer to Figure 18-2. From the given information, find the unknown sides and angles using the law of cosines and the law of sines when:

1. Side $a = 70$, $b = 130$, $c = 90$
2. Side $a = 100$, $b = 200$, $c = 200$
3. Side $a = 250$, $b = 100$, $c = 200$
4. Side $a = 14$, $b = 14$, $c = 20$
5. Side $b = 60$, $c = 80$, $A = 95°$
6. Side $b = 45$, $c = 100$, $A = 80°$
7. Side $b = 12$, $c = 36$, $A = 80°$
8. Side $b = 20$, $c = 70$, $A = 102°$
9. Side $a = 500$, $c = 400$, $B = 100°$
10. Side $a = 130$, $c = 95$, $B = 55°$
11. Side $a = 400$, $b = 300$, $c = 150$
12. Side $a = 500$, $b = 350$, $c = 200$
13. Side $a = 35$, $b = 40$, $c = 25$
14. Side $a = 8$, $b = 9$, $c = 6$
15. Side $a = 8.4$, $c = 10$, $B = 30°$
16. Side $a = 14$, $c = 12$, $B = 40°$
17. Side $a = 40$, $c = 62$, $B = 105°$
18. Side $a = 90$, $c = 110$, $B = 65°$
19. Side $b = 600$, $c = 400$, $A = 95°$
20. Side $b = 120$, $c = 160$, $A = 60°$
21. Side $b = 10$, $c = 15$, $A = 120°$
22. Side $b = 35$, $c = 55$, $A = 63°$
23. Side $a = 45$, $b = 45$, $c = 60$
24. Side $a = 125$, $b = 105$, $c = 140$
25. Side $a = 170$, $b = 140$, $c = 190$
26. Side $a = 50$, $b = 35$, $c = 20$
27. Side $a = 425$, $b = 525$, $c = 300$
28. Side $a = 27$, $b = 40$, $c = 40$
29. Side $a = 5.35$, $b = 8.5$, $c = 8.5$
30. Side $a = 1100$, $b = 950$, $c = 1400$

Answers to Self-test 18-1

1. Side $a = 257.9$, $b = 222.6$, $B = 45°$
2. Side $c = 33.75$, $B = 49.25°$, $C = 30.75°$
3. Angle $A = 22.33°$, $B = 130.5°$, $C = 27.13°$
4. Side $a = 33.26$, $B = 44.93°$, $C = 25.07°$

CHAPTER

19

AC FUNDAMENTALS

INTRODUCTION

An automobile or an airplane traveling in a particular direction has a *linear velocity* measured in miles per hour or kilometers per hour. Other moving objects may have a linear velocity measured in feet per second or centimeters per second. This linear velocity is commonly referred to as the *speed* at which the object is traveling.

Electrical generators have *angular* motion or *angular velocity*. The speed of rotation of the armature of a generator, the angular velocity, is measured in revolutions per minute or, more commonly, in revolutions per second (cycles per second). We can also measure angular velocity in degrees per second or in radians per second. The measurement of angular velocity in degrees or radians per second is necessary in electronics for us to understand the characteristics of alternating voltages and currents.

When alternating values of current and voltage are generated, the value of current or voltage changes sinusoidally. That is, the variation has the characteristics of a sine wave, and we can determine the value of current or voltage at any instant by using the sine function. Since the sine wave repeats itself periodically, we can easily compute the angular velocity.

Because all generated waveforms in electronics are basically sine waves, an understanding of how to compute angular velocity, how to relate time to frequency, and how to distinguish among, rms, peak, and instantaneous values of current and voltage are a necessary part of our studies.

CHAPTER OBJECTIVES

In this chapter you will learn how to:

1. Identify the parts of a system of rectangular coordinates.
2. Generate angles on a system of rectangular coordinates.
3. Convert angles from degrees to radians and from radians to degrees.

4. Calculate angular velocity.
5. Calculate instantaneous values of voltage and current.
6. Convert between effective, peak, and peak-to-peak values of voltage and current.
7. Determine instantaneous values of voltage and current when there is an angle of lead or lag.

19-1 GENERATING ANGLES

To develop an understanding of ac concepts, we need to discuss some basic mathematical and physical ideas. In Figure 19-1, two lines are drawn at right angles to each other. The lines are xx_1 and yy_1. Such a figure has been defined as a system of rectangular coordinates. The origin is the point where the lines cross. A line extending outward from the origin is called a *vector*. Recall from previous chapters that we call a rotating vector a *phasor*. Hereafter, we will use the word *phasor* whenever we are discussing rotating vectors.

Let's consider a line or phasor that extends along the x-axis as in Figure 19-2. If the phasor were rotated either clockwise or counterclockwise until it were again along the x-axis, the phasor would have gone through 360° or a complete circle. It is standard practice to consider the x-axis as 0° of the circle and to rotate the phasor counterclockwise. The y-axis is displaced 90° from the x-axis, the x_1-axis is displaced 180° from the x-axis, and the y_1-axis is displaced 270°. Thus, counterclockwise rotation is considered positive rotation. Clockwise rotation is considered negative rotation. From this we can say that y_1 is displaced 270° or −90° from the x-axis or 0°.

An angle is in its standard position when its initial side is on the positive x-axis and its vertex is at the origin. The x-y coordinates divide the system into fourths or

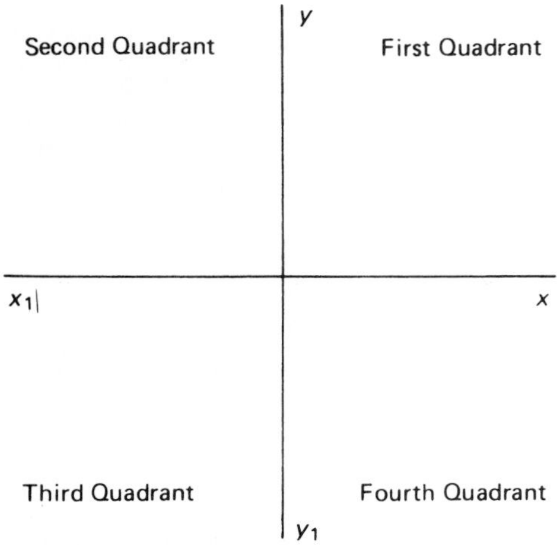

FIGURE 19-1

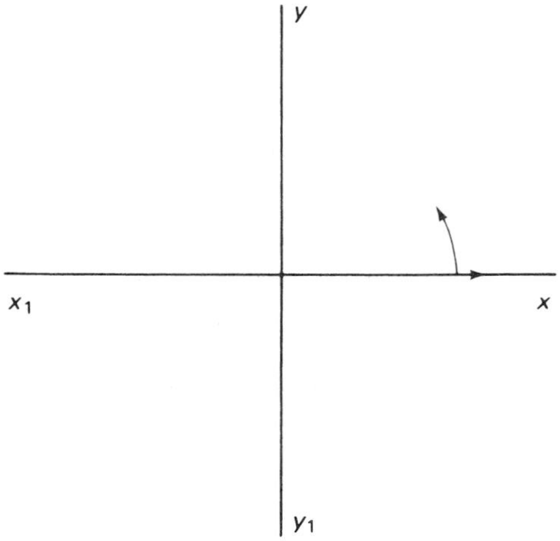

FIGURE 19-2

quadrants, as labeled in Figure 19-1. Note that in the first quadrant both x and y are positive. In the second quadrant, x is negative and y is positive. In the third quadrant, both x and y are negative, and in the fourth quadrant, x is positive and y is negative.

In Figure 19-3 a phasor has been rotated through 30°. The angle is formed by first extending a line from point P, the origin, to point A. In this position the line is in its *initial* position. We then rotate the line through 30° to point B. The line is now in its *terminal* position. Since the resulting angle has its vertex at the origin, the phasor has a length equal to the radius of the circle. Notice that we rotated the line counterclockwise.

Consider Figure 19-4. The line from point P to any point on the circumference is the radius. Let's rotate the line again.

KEY POINT When the length of arc AB equals the length of the radius, the angle is equal to 1 *radian*. In degrees, this angle equals 57.3°. So

$$1 \text{ radian} = 57.3° \tag{19-1}$$

In our studies of ac circuits, we will measure angles in either degrees or radians.[1]

If the line in Figure 19-4 were rotated through 360°, a complete circle would be drawn, as illustrated in Figure 19-5. From such a circle, the relationship between degrees and radians can be determined. In a circle there are 360° or 6.28 radians. It

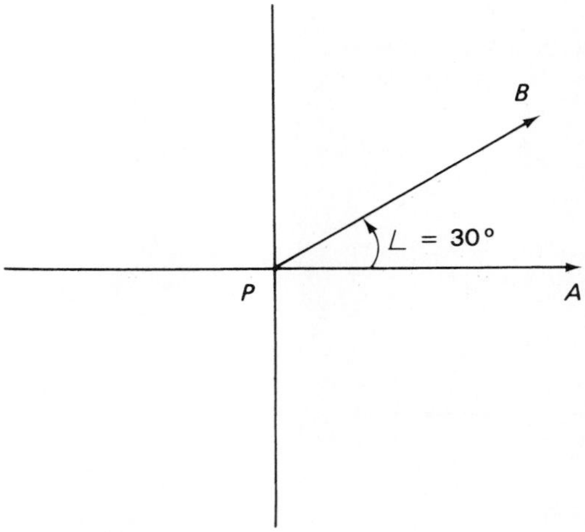

FIGURE 19-3

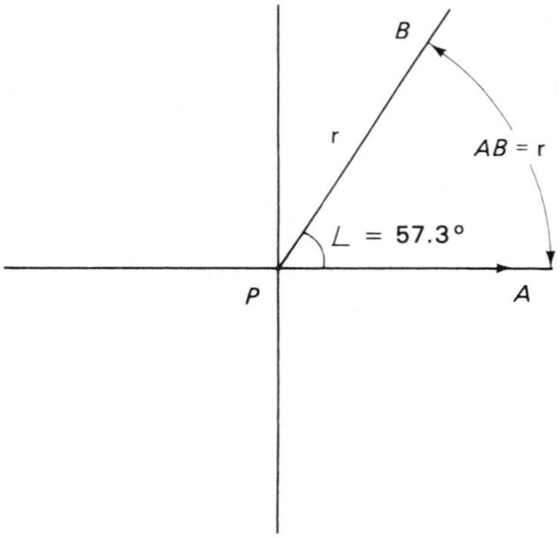

FIGURE 19-4

[1] In some professions the angle is also measured in *grads*. A grad is $\frac{1}{100}$ of a right angle, or, said another way, 90° = 100 grads. We will use only degrees and radians in our discussions of angles.

follows then that there are 3.14 radians in 180°. The greek letter π (*pi*) is used to denote 3.14 rad. π radians = 3.14 radians.

$$180° = \pi \text{ radians} \tag{19-2}$$

We determined previously that there are 57.3° in 1 radian. Rearranging Equation 19-2 is another way of finding the number of degrees in 1 radian.

$$1 \text{ rad} = \frac{180°}{\pi \text{ rad}} = 57.3° \tag{19-3}$$

Since we can use Equation 19-3 to find the number of degrees in 1 radian of rotation, then, to convert from radians to degrees for any angle, we get

$$\text{angle in degrees} = \frac{180°}{\pi \text{ rad}} \times \text{number of radians} \tag{19-4}$$

and to convert from degrees to radians,

$$\text{angle in radians} = \frac{\pi \text{ rad}}{180°} \times \text{number of degrees} \tag{19-5}$$

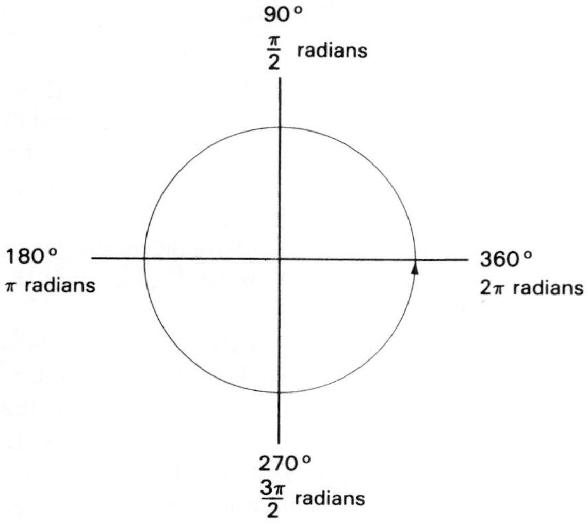

FIGURE 19-5

□ EXAMPLE 19-1

Convert an angle of 85° to radians.

Solution To convert from degrees to radians, we use Equation 19-5.

$$\text{angle in radians} = \frac{\pi}{180°} \times \text{angle in degrees} \qquad (19\text{-}5)$$

$$= \frac{\pi}{180°} \times 85° = 1.48 \text{ rad}$$

Most calculators will convert between degrees, radians, and grads by activating the DRG key. With the calculator in degrees mode and 85 displayed (85°), change to radians mode using the DRG key. 1.48 should be displayed. Consult the users manual for the key combination needed by your calculator.

□ EXAMPLE 19-2

Convert an angle of 3.42 rad to degrees.

Solution To convert from radians to degrees, use Equation 19-4.

$$\text{angle in degrees} = \frac{180°}{\pi} \times \text{angle in radians} \qquad (19\text{-}4)$$

$$= \frac{180°}{\pi} \times 3.42 = 196°$$

□

19-2 GENERATING SINE WAVES

Let's consider the system of rectangular coordinates shown in Figure 19-6(a). If we start at 0° (the positive x-axis) and rotate in a counterclockwise direction, we will have drawn a complete circle. This rotation could represent one complete cycle of a generated sine wave of voltage or current. The radius of the circle could equal the maximum or peak value of the sine wave. Suppose this represents a voltage waveform whose maximum voltage is 100 V. $V = 100\ V_{pk}$. As we rotate our phasor from 0°, the radius, V_{pk}, is the hypotenuse of a right triangle. The vertical component, v, falls along the y-axis and is the instantaneous value of voltage.

The instantaneous voltage is increasing from 0 V at 0° to 100 V at 90°. This is shown graphically in Figure 19-6(b). t_0 on the graph corresponds to 0° on our system of rectangular coordinates. As we increase to 90° ($t = t_1$), we can calculate the voltage generated at any instant by using the equation $v = V_{pk} \sin \angle A$. (Lowercase v and i are used to denote instantaneous values of voltage and current.)

Continuing on from 90° to 180° (t_2), we could calculate instantaneous values and plot our graph using our equation for v ($v = V_{pk} \sin A$). At this point we have completed one half-cycle or one-half revolution and are back to 0 V. As we continue

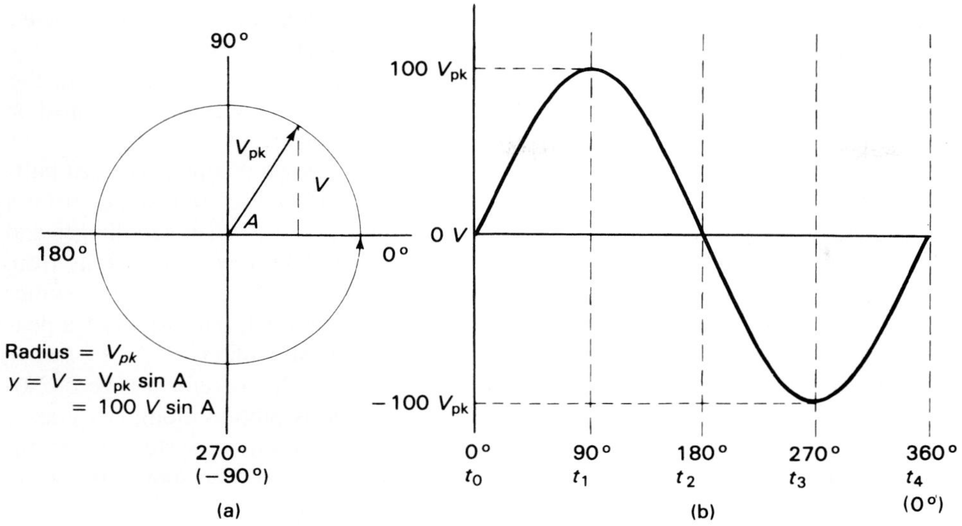

FIGURE 19-6

on past 180°, and using the equation $v = V_{pk} \sin \angle A$, v is negative because $\sin \angle A$ is negative. We can also show graphically why this is so. Look at the angles generated in Figure 19-7. Remember that

$$\sin A = \frac{\text{opposite side}}{\text{hypotenuse}}$$

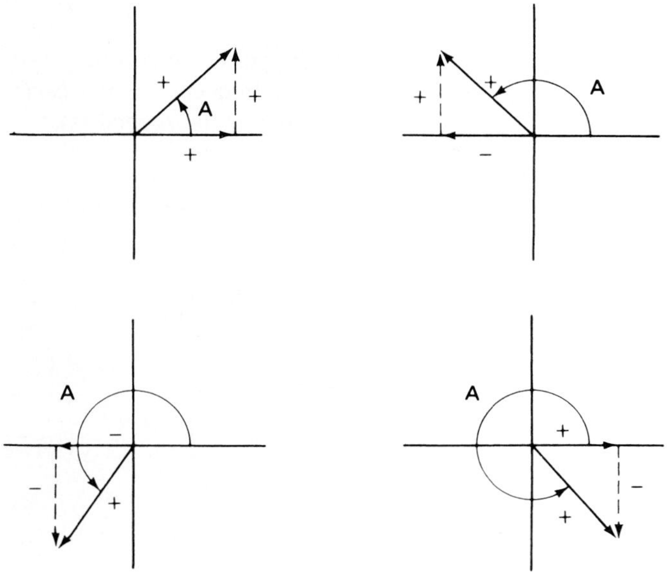

FIGURE 19-7

The side opposite the angle A is the value in the y direction, which we have labeled v. The hypotenuse is the radius vector, which has been labeled V_{pk}. The radius phasor is always considered to be positive. The opposite side, v, is positive in the first and second quadrants because the y-axis is positive. When we are in the third or fourth quadrant, the y-axis is negative; therefore, v is negative.

A look at our graph in Figure 19-6(b) shows that, as we complete the last half-cycle, v goes from 0 to -100 V as we rotate from 180° to 270°. v changes from a value equal to -100 to 0 V as we rotate from 270° back to 0°. When ac voltages and currents are generated, this is the waveform produced. This waveform results from the constant rotation of a generator or from the natural oscillation of an electronics circuit. This sine curve could also represent the back and forth movement of a pendulum or the ripple created when a rock is dropped into still water.

In Figure 19-6(b), at time zero (t_0) the amplitude of the waveform is zero. Notice that time is plotted along the x-axis and amplitude is plotted along the y-axis. This is standard notation; t_0 on our graph corresponds to 0° on our system of rectangular coordinates. The maximum amplitude in the y direction corresponds to the radius of our circle.

19-3 FREQUENCY AND TIME

The number of times per second that we generate the sine wave is called *frequency*. The time it takes to generate one complete sine wave is called the *period*. We can express the relationship between frequency and time as an equation:

$$f = \frac{1}{T} \qquad (19\text{-}6)$$

where f is the frequency in cycles per second and T is the period in seconds. Frequency and time are reciprocals of each other. The unit of frequency is the *hertz*. One hertz (Hz) equals one cycle per second. The unit of time is the second (s).

☐ EXAMPLE 19-3

If the frequency of rotation is 250 Hz, what is the period?

Solution Solve for T:

$$f = \frac{1}{T} \qquad (19\text{-}6)$$

$$T = \frac{1}{f} \qquad (19\text{-}7)$$

$$T = \frac{1}{250\text{ Hz}} = 4\text{ ms}$$

If there are 250 cycles per second (250 Hz), the time for 1 cycle (the period) is 4 ms.

☐ EXAMPLE 19-4

If it takes 100 μs to complete one cycle, what is the frequency?

Solution

$$f = \frac{1}{T}$$

$$f = \frac{1}{100\ \mu\text{s}} = 10\ \text{kHz}$$

If the period is 100 μs, we will generate 10,000 cycles in one second, or 10 kHz.

☐

PRACTICE PROBLEMS 19-1

When using the calculator, conversions between time and frequency are easily solved using the reciprocal key.

1. f = 15 kHz. Find T.
2. f = 500 Hz. Find T.
3. T = 400 μs. Find f.
4. T = 16.7 ms. Find f.
5. Convert 145° to radians.
6. Convert 300° to radians.
7. Convert 2 rad to degrees.
8. Convert 5.73 rad to degrees.

Solutions

1. 66.7 μs
2. 2 ms
3. 2.5 kHz
4. 59.9 Hz
5. 2.53 rad
6. 5.24 rad
7. 115°
8. 328°

Additional practice problems are at the end of the chapter.

19-4 ANGULAR VELOCITY

Let's generate an angle by rotating our radius phasor in Figure 19-3 at a constant speed. In angular motion this speed is called *angular velocity*. Angular velocity is measured in radians per second (rad/s) or in degrees per second. The symbol for angular velocity is ω (the Greek letter *omega*). In radian measure, ω equals 2π times the frequency

$$\boxed{\omega = 2\pi f} \qquad (19\text{-}8)$$

If the angle is measured in degrees, ω equals 360° times the frequency.

$$\boxed{\omega = 360^\circ f} \tag{19-9}$$

anglular velocity is more commonly measured in radians per second.

When the frequency is 1 Hz (one cycle per second), the angular velocity is 2π rad/s or 360°/s. If the frequency is 10 Hz, the angular velocity, in radians or degrees, is

$$\omega = 2\pi f$$

$$\omega = 2\pi \times 10 \text{ Hz} = 20\pi \text{ rad/s}$$

$$= 62.8 \text{ rad/s}$$

or

$$\omega = 360^\circ f = 360^\circ \times 10 \text{ Hz} = 3600^\circ/\text{s}$$

The total angle swept by a phasor in a given time (t) is ωt. In radian measure, the equation would be

$$\boxed{\omega t = 2\pi f t} \tag{19-10}$$

In degrees, the equation would be

$$\boxed{\omega t = 360^\circ f t} \tag{19-11}$$

☐ EXAMPLE 19-5

Find the angle generated after 5 ms when the frequency is 500 Hz. Express your answer in degrees and radians.

Solution

$$\omega t = 2\pi f t \tag{19-10}$$

$$= 2\pi \times 500 \times 5 \text{ ms} = 15.7 \text{ rad}$$

$$= 57.3^\circ \times 15.7 \text{ rad} = 900^\circ$$

☐ EXAMPLE 19-6

Find the frequency if an angle of 2.35 rad is generated in 150 μs.

Solution

$$\omega t = 2\pi f t \tag{19-10}$$

$$f = \frac{\omega t}{2\pi t} = \frac{2.35 \text{ rad}}{2\pi \times 150\ \mu\text{s}} = 2.49 \text{ kHz}$$

☐ EXAMPLE 19-7

How much time is required to generate an angle of 75° when the frequency is 1 kHz?

Solution

$$\omega t = 360°ft \qquad (19\text{-}11)$$

$$t = \frac{\omega t}{360°f} = \frac{75°}{360 \times 1\ \text{kHz}} = 208\ \mu\text{s}$$

☐

PRACTICE PROBLEMS 19-2

1. If the frequency is 2.5 kHz, find the angle in both degrees and radians after (a) 50 μs, (b) 100 μs, (c) 200 μs, and (d) 1 ms.
2. Find the frequency if the angular velocity is (a) 670 rad/s, (b) 2000 rad/s, (c) 7200°/s, and (d) 12,000°/s.
3. What is the frequency if an angle of 0.56 rad is generated in 55 μs?
4. What is the frequency if an angle of 285° is generated in 200 μs?
5. How much time is required to generate an angle of 1.7 rad when the frequency is 25 kHz?
6. How much time is required to generate an angle of 400° when the frequency is 1.5 kHz?

Solutions

1. (a) 0.785 rad or 45° (b) 1.57 rad or 90° (c) 3.14 rad or 180° (d) 15.7 rad or 900°
2. (a) 107 Hz (b) 318 Hz (c) 20 Hz (d) 33.3 Hz
3. 1.62 kHz
4. 3.96 kHz
5. 10.8 μs
6. 741 μs

Additional practice problems are at the end of the chapter.

SELF-TEST 19-1

1. f = 50 kHz. Find T.
2. Convert 160° to radians.
3. Convert 0.628 rad to degrees.
4. If the frequency is 400 Hz, find the angle generated after 750 μs. Express your answer in both degrees and radians.
5. What is the frequency if an angle of 120° is generated in 200 μs?
6. How much time is required to generate an angle of 1.3 rad when the frequency is 5 kHz?

Answers to Self-test 19-1 are at the end of the chapter.

19-5 INSTANTANEOUS VALUES OF VOLTAGE AND CURRENT

In Figure 19-8, at t_0 the voltage is 0 V and at 90° the voltage is at its maximum value, which, in this example, is 100 V. If the frequency is 1 kHz, how long does it take to reach this 100 V? To solve the problem, the first thing we have to do is find the period, the time it takes to complete one cycle:

$$T = \frac{1}{f} \tag{19-7}$$

$$T = \frac{1}{1 \text{ kHz}} = 1 \text{ ms}$$

Since 90° is one-quarter of a complete cycle; then it must take one-quarter the time for a complete cycle to reach 90°:

$$\text{time to complete } 90° = \frac{\text{period}}{4} = \frac{1 \text{ ms}}{4} = 250\ \mu\text{s}$$

During the time from t_0 to $t = 250\ \mu$s, V is increasing from 0 to 100 V. From $t = 250\ \mu$s to $t = 750\ \mu$s, the amplitude is changing from +100 to −100 V. The voltage is 0 V as we pass through 180°, and $t = T/2$ or 500 μs. From $t = 750\ \mu$s, the amplitude is changing from −100 to +100 V. Zero volts is reached when we have completed the cycle, which is 360° or 1 ms. This brings us back to 0°. The cycle then repeats itself 1000 times per second. This change in voltage is not a linear change but a *sinusoidal* change. That is, the amplitude at any instant is a function of the sine of the generated angle. We have already determined that V_{pk} is the hypotenuse of a right triangle, and the opposite side is the magnitude of the voltage at some instant in time. If we display the sine wave on a system of rectangular coordinates, the resulting equation for finding v at any instant is

$$v = V_{pk} \sin \omega t \tag{19-12}$$

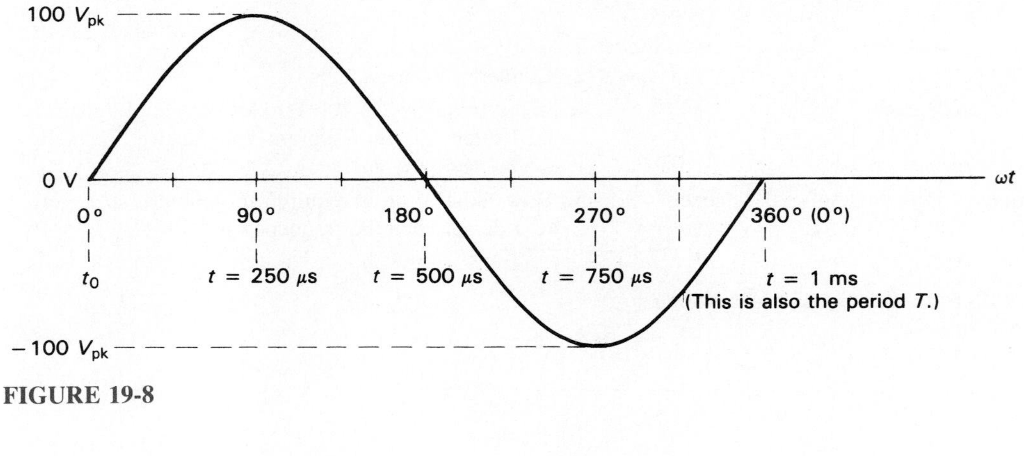

FIGURE 19-8

When working with an emf, our equation would be

$$e = E_{pk} \sin \omega t \tag{19-13}$$

The equation for instantaneous circuit current would be

$$i = I_{pk} \sin \omega t \tag{19-14}$$

□ EXAMPLE 19-8

$e = 100\ V_{pk} \sin \omega t$. If $f = 1$ kHz, find e after **(a)** 50 μs, **(b)** 250 μs, and **(c)** 600 μs.

Solution

$$e = E_{pk} \sin \omega t \tag{19-13}$$

$$= 100 \text{ V} \sin 2\pi f t$$

(a)

$$e = 100 \text{ V}_{pk} \sin 2\pi \times 1 \text{ kHz} \times 50\ \mu\text{s}$$

$$= 100 \text{ V}_{pk} \sin 0.314 \text{ rad}$$

$$= 100 \text{ V}_{pk} \times 0.309 = 30.9 \text{ V}$$

Since the angle ωt is expressed in radians, the calculator must be in the radians mode to find sin ωt; then

100 [×] [(] 2 [×] [π] [×] 1 [EE] 3 [×] 50 [EE] [+/−] 6 [)] [sin] [=]

Answer: $e = 30.9$ V

Notice that in all three problems, (a), (b), and (c), $2\pi f$ is the same. When this happens, we can save keystrokes by storing that product before we start.

2 [×] [π] [×] 1 [EE] 3 [=] [STO]   *Answer*: $2\pi f = 6.28$ kHz

Now we solve the problem:

100 [×] [(] [RCL] [×] 50 [EE] [+/−] 6 [)] [sin] [=]

Answer: $e = 30.9$ V

(b)

$$e = 100 \text{ V}_{pk} \sin (2\pi \times 1 \text{ kHz} \times 250\ \mu\text{s})$$

100 [×] [(] [RCL] [×] 250 [EE] [+/−] 6 [)] [sin] [=]

Answer: $e = 100$ V

We had determined previously that e would equal E_{pk} 250 μs after t_0.

(c)

$$e = 100 \text{ V}_{pk} \sin (2\pi \times 1 \text{ kHz} \times 600\ \mu\text{s})$$

100 [×] [(] [RCL] [×] 600 [EE] [+/−] 6 [)] [sin] [=]

Answer: $e = -58.8$ V

e is negative because 600 μs put us in the third quadrant.

□

PRACTICE PROBLEMS 19-3

1. $V = 60\ V_{pk}$, $f = 500$ Hz. Find v after (a) 100 μs, (b) 250 μs, (c) 500 μs, (d) 800 μs, (e) 1.2 ms, and (f) 1.6 ms.

2. $I = 30\ mA_{pk}$, $f = 250$ Hz. Find i after (a) 400 μs, (b) 1 ms, (c) 1.7 ms, (d) 2.1 ms, (e) 3 ms, and (f) 11.3 ms.

Solutions

1. (a) 18.5 V; (b) 42.4 V; (c) 60 V; (d) 35.3 V; (e) −35.3 V; (f) −57.1 V

2. (a) 17.6 mA; (b) 30 mA; (c) 13.6 mA; (d) −4.69 mA; (e) −30 mA; (f) −26.7 mA

Additional practice problems are at the end of the chapter.

19-6 RMS VALUES OF VOLTAGE AND CURRENT

In Figure 19-8, at t_0 the voltage is 0 V. After some time has lapsed, we reach 90° of rotation and v is at its maximum value. As we continue through the complete cycle, v decreases to 0 at 180°, increases to its maximum negative value at 270°, and decreases to 0 at 360°. If we took an average of all the instantaneous values of voltage for the entire cycle, we would get 0 V. This would be true because one-half the time the voltages are positive and one-half the time they are negative. For every positive instantaneous voltage there is a negative instantaneous voltage. If we consider *half* a cycle, then we could compute an average voltage. This average voltage would be 0.319 of the peak value.

$$V_{AVE} = 0.319\ V_{pk} = \frac{V_{pk}}{\pi} \tag{19-15}$$

If both half-cycles were of the same polarity, then

$$V_{AVE} = 0.637\ V_{pk} = \frac{2\ V_{pk}}{\pi} \tag{19-16}$$

Average values of voltage and current are used primarily in determining component values in rectifier-type power supplies.

In determining currents and voltage drops in ac circuits, *rms* values of current and voltage are used. The rms value of a current or voltage is the value that converts the same energy as does a dc value. For example, in a dc circuit, 100 mA flows through a resistance of 10 Ω. The power dissipated is 100 mW. For an ac current to dissipate 100 mW in that same resistor, the rms current would also have to be 100 mA.

The rms value of an ac voltage or current is found by taking the square root of the average (mean) of the squares of the instantaneous values. Because of the method of calculation, this value is also called the *effective value*. This rms or effective value is the same for all sine waves and equals 0.707 of the maximum or peak value.

$$I_{rms} = I_{eff} = 0.707\ \mathrm{I_{pk}} = \frac{\mathrm{I_{pk}}}{\sqrt{2}} \qquad (19\text{-}17)$$

$$V_{rms} = V_{eff} = 0.707\ \mathrm{V_{pk}} \qquad (19\text{-}18)$$

If we solve for peak values, we get

$$I_{pk} = 1.414\ \mathrm{I_{rms}} = \sqrt{2}\ \mathrm{I_{rms}} \qquad (19\text{-}19)$$

$$V_{pk} = 1.414\ \mathrm{V_{rms}} \qquad (19\text{-}20)$$

Since peak-to-peak values of voltage and current equal twice the peak values ($V_{p\text{-}p} = 2\ V_{pk}$), then

$$V_{p\text{-}p} = 2.828\ \mathrm{V_{rms}} = 2\sqrt{2}\ \mathrm{V_{rms}} \qquad (19\text{-}21)$$

$$I_{p\text{-}p} = 2.828\ \mathrm{I_{rms}} \qquad (19\text{-}22)$$

Solving for effective values, we get

$$V_{eff} = V_{rms} = 0.3535\ \mathrm{V_{p\text{-}p}} = \frac{\mathrm{V_{p\text{-}p}}}{2\sqrt{2}}$$

$$I_{eff} = I_{rms} = 0.3535\ \mathrm{I_{p\text{-}p}}$$

When ac voltages or currents are given, values are always rms values. Thus, if we see an ac voltage written as 120 V, we know that it is 120 $\mathrm{V_{rms}}$. If the voltage or current is either a peak or a peak-to-peak value, the proper subscript must be used. Therefore, ten volts rms is written 10 V. Ten volts peak is written 10 $\mathrm{V_{pk}}$. Ten volts peak-to-peak is written 10 $\mathrm{V_{p\text{-}p}}$.

☐ EXAMPLE 19-9

Convert 12 $\mathrm{V_{pk}}$ to rms and p-p.

Solution

$$V_{rms} = 0.707\ \mathrm{V_{pk}}$$

$$V_{rms} = 0.707 \times 12 = 8.48\ \mathrm{V}$$

$$V_{p\text{-}p} = 2\ \mathrm{V_{pk}}$$

$$V_{p\text{-}p} = 2 \times 12 = 24\ \mathrm{V_{p\text{-}p}}$$

☐ EXAMPLE 19-10

Convert 20 mA to pk and p-p.

Solution

$$I_{pk} = 1.414\ \mathrm{I_{rms}}$$

$$= 1.414 \times 20\ \mathrm{mA} = 28.3\ \mathrm{mA_{pk}}$$

$$I_{p\text{-}p} = 2.828\ \mathrm{I_{rms}}$$

$$I_{p\text{-}p} = 2.828 \times 20\ \mathrm{mA} = 56.6\ \mathrm{mA_{p\text{-}p}}$$

☐ EXAMPLE 19-11

Convert 20 $V_{p\text{-}p}$ to rms and pk.

Solution

$$V_{rms} = 0.3535\ V_{p\text{-}p}$$
$$= 0.3535 \times 20 = 7.07\ V$$
$$V_{pk} = \frac{V_{p\text{-}p}}{2}$$
$$= \frac{20\ V}{2} = 10\ V$$

☐

PRACTICE PROBLEMS 19-4

1. An ac voltage is 30 V. Convert to pk and p-p.
2. An ac current is 150 mA. Convert to pk and p-p.
3. An ac voltage is 50 V_{pk}. Convert to p-p and rms.
4. An ac current is 70 mA_{pk}. Convert to p-p and rms.
5. An ac voltage is 200 $V_{p\text{-}p}$. Convert to pk and rms.
6. An ac current is 1.5 $A_{p\text{-}p}$. Convert to pk and rms.

Solutions

1. 30 V = 42.4 V_{pk} = 84.8 $V_{p\text{-}p}$
2. 150 mA = 212 mA_{pk} = 424 $mA_{p\text{-}p}$
3. 50 V_{pk} = 100 $V_{p\text{-}p}$ = 35.4 V
4. 70 mA_{pk} = 140 $mA_{p\text{-}p}$ = 49.5 mA
5. 200 $V_{p\text{-}p}$ = 100 V_{pk} = 70.7 V
6. 1.5 $A_{p\text{-}p}$ = 750 mA_{pk} = 530 mA

Additional practice problems are at the end of the chapter.

19-7 ANGLE OF LEAD OR LAG

When we have a purely resistive ac circuit, the applied voltage and circuit current equal zero at the same time. They also reach their peak values at the same time. In such a circuit we say that the voltage and current are *in phase*. The voltage drops across the various resistances are also in phase with the current and applied voltage.

■ **KEY POINT** Whenever we have capacitance or inductance in an ac circuit, the current and voltage are *out of phase*. When the phase angle, *theta*(θ), is known, we can calculate the instantaneous values of current and voltage.

The applied voltage either leads or lags the current. If the applied voltage *leads* the current, this means that E reaches its values of 0 V and E_{pk} at an earlier time than does the current. The phase difference can be anywhere from 0° to 90°.

Because the applied voltage is common to each branch, *voltage is the reference in parallel circuits*. In parallel circuits, then, the current either leads or lags the voltage depending on whether the circuit is resistive and capacitive or resistive and inductive. The general equations for instantaneous values of current and voltage in parallel circuits are

$$e = E_{pk} \sin \omega t \tag{19-13}$$

$$i = I_{pk} \sin (\omega t \pm \theta) \tag{19-23}$$

We use the greek letter theta (θ) to donote phase angle.

Because current is the same throughout the circuit, *current is the reference in series circuits*. In series circuits, then, the voltage either leads or lags the current. The general equations for instantaneous values of current and voltage in series circuits are

$$i = I_{pk} \sin \omega t \tag{19-14}$$

$$e = E_{pk} \sin (\omega t \pm \theta) \tag{19-24}$$

Suppose that E leads I by 30° in a series RL circuit as illustrated in Figure 19-9. We could design and construct such a circuit and then look at the current and voltage waveforms on a dual-trace oscilloscope. If we properly calibrate the time base, we can measure the phase angle by noting the time difference between E and I. Because this is a series circuit, current is the reference. The equations for E and I at any instant are

$$i = I_{pk} \sin \omega t \tag{19-14}$$

$$e = E_{pk} \sin (\omega t \pm \theta) \tag{19-24}$$

$$e = E_{pk} \sin (\omega t + 30°)$$

☐ EXAMPLE 19-12

I = 50 mA_{pk}, E = 100 V_{pk}, and the frequency is 2 kHz. Using current as the reference, compute the instantaneous values of i and e 40 μs after t_0 when I leads E by 60°.

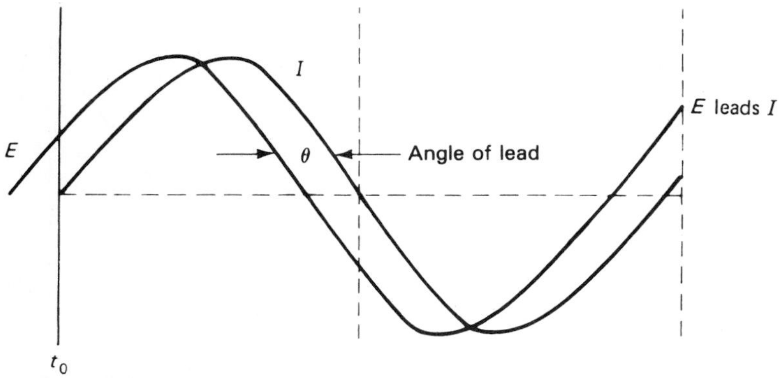

FIGURE 19-9

Solution If we draw the waveforms as in Figure 19-10 with I leading E by 60°, we see that at 40 μs I will be positive and close to one-half its maximum value, or approximately 25 mA. E is negative and moving in a positive direction. Let's estimate its value at about −40 V. (Our freehand graph isn't accurate, but it gets us in the ball park and gives us a visual display of the conditions at a specific time.)

$$\omega t = 360° ft = 360 \times 2 \text{ kHZ} \times 40\ \mu\text{s} = 28.8°$$

360 [×] 2 [EE] 3 [×] 40 [EE] [±] 6 [=] *Answer*: $\omega t = 28.8°$

We found ωt in degrees because the phase angle was given in degrees. Knowing ωt, we can find the instantaneous current using Equation 19-14.

$$i = I_{\text{pk}} \sin \omega t = 50 \text{ mA} \sin 28.8° = 24.1 \text{ mA}$$

50 [EE] [±] 3 [×] 28.8 [sin] [=] *Answer*: $i = 24.1$ mA

E lags I by 60°. Looking at Figure 19-10, we can see that, when the current is at 28.8° (approximately 40 μs from t_0), E is moving from its maximum negative value toward 0 (which would occur 60° after I is zero) and is at an angle of −31.2°. Knowing the phase angle, we can compute e using Equation 19-24.

$$e = E_{\text{pk}} \sin(\omega t \pm \theta) = 100 \text{ V} \sin(28.8° - 60°)$$

$$= 100 \text{ V} \sin -31.2° = -51.8 \text{ V}$$

100 [×] [(] 28.8 [−] 60 [)] [sin] [=] *Answer*: $e = -51.8$ V

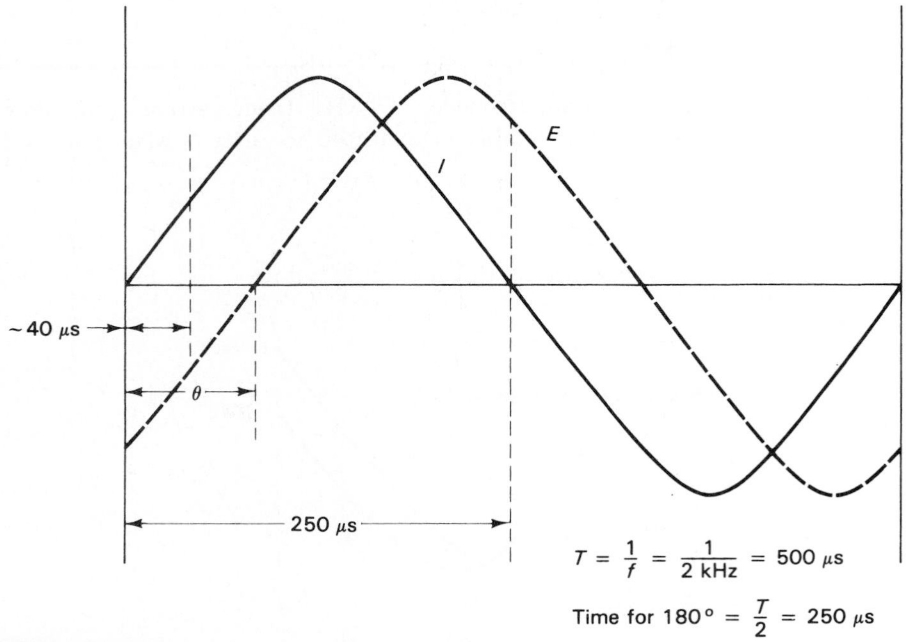

FIGURE 19-10

☐ EXAMPLE 19-13

$I = 20$ mA$_{pk}$, $E = 20$ V$_{pk}$, $f = 12$ kHz, and E leads I by 45°. Compute i and e 15 μs after t_0. Use voltage as the reference.

Solution A graph "ball parking" E and I is drawn in Figure 19-11. The time for 90° is about 20 μs. 15 μs puts us approximately three-quarters of the distance between 0° and 90°, so ωt is about 70°. Its actual value is

$$\omega t = 360° ft = 360° \times 12 \text{ kHz} \times 15\ \mu\text{s} = 64.8°$$

360 [×] 12 [EE] 3 [×] 15 [EE] [±] 6 [=] *Answer*: $\omega t = 64.8°$

E leads I by 45° ($\theta = 45°$). 45° is one-half the distance between 0° and 90°. The graph shows that, at 15 μs after t_0, E is increasing and is near its peak positive value. Using Equation 19-13,

$$e = E_{pk} \sin \omega t = 20 \text{ V} \sin 64.8° = 18.1 \text{ V}$$

20 [×] 64.8 [sin] [=] *Answer*: $e = 18.1$ V

I has just crossed zero and is increasing toward its peak positive value. Using Equation 19-23,

$$i = I_{pk} \sin (\omega t \pm \theta) = 20 \text{ mA} \sin (64.8° - 45°)$$
$$= 20 \text{ mA} \sin 19.8° = 6.77 \text{ mA}$$

20 [EE] [±] 3 [×] [(] 64.8 [−] 45 [)] [sin] [=] *Answer*: $i = 6.77$ mA

☐

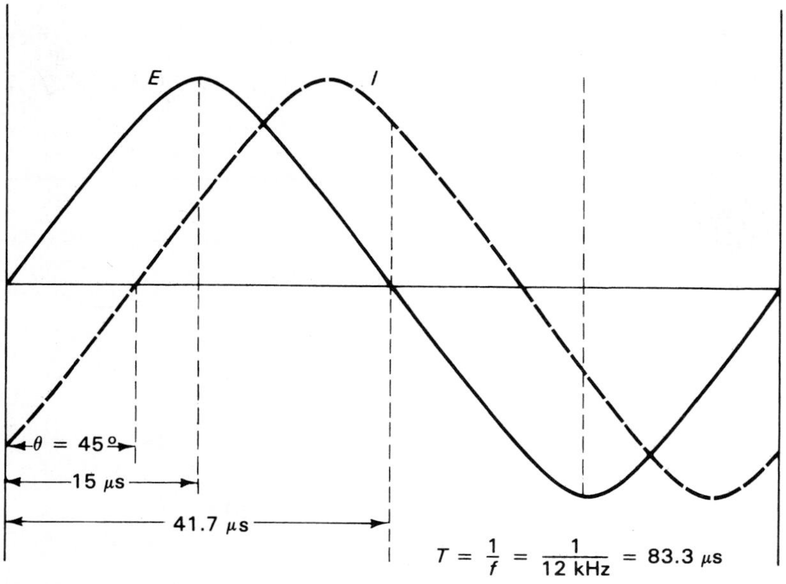

FIGURE 19-11

PRACTICE PROBLEMS 19-5

1. $I = 100\ mA_{pk}$, $E = 30\ V_{pk}$, and $f = 500$ Hz. Use current as the reference. I leads E by 40°. Find i and e after (a) 50 μs, (b) 225 μs, (c) 800 μs, and (d) 1.2 ms.

2. $I = 12\ mA_{pk}$, $E = 15\ V_{pk}$, and $f = 7.5$ kHz. Use voltage as the reference. I leads E by 60°. Find i and e after (a) 10 μs, (b) 40 μs, (c) 100 μs, and (d) 150 μs.

Solutions

1. (a) $i = 15.6$ mA, $e = -15.5$ V
   (b) $i = 64.9$ mA, $e = 0.262$ V
   (c) $i = 58.8$ mA, $e = 29.1$ V
   (d) $i = -58.8$ mA, $e = 2.09$ V

2. (a) $i = 12$ mA, $e = 6.81$ V
   (b) $i = 2.49$ mA, $e = 14.3$ V
   (c) $i = -6$ mA, $e = -15$ V
   (d) $i = 11.6$ mA, $e = 10.6$ V

Additional practice problems are at the end of the chapter.

SELF-TEST 19-2

1. If $V = 25\ V_{pk}$ and $f = 8$ kHz, find v after 25 μs.
2. What is the instantaneous value of current after 2 ms if the frequency is 400 Hz and I_{pk} is 20 mA?
3. Convert 25 V to peak values and peak-to-peak values.
4. Convert 60 $mA_{p\text{-}p}$ to peak values and rms values.
5. I leads E by 25° in an ac circuit. If $E =$ 50 V and the frequency is 25 kHz, find e after 7 μs. Use current as the reference.

Answers to Self-test 19-2 are at the end of the chapter.

CHAPTER 19 AT A GLANCE

Page	Key Points, Definitions, and Equations	Example
478	*Key point:* When the length of an arc equals the length of the radius of a circle, the angle is equal to 1 radian. One radian equals 57.3°.	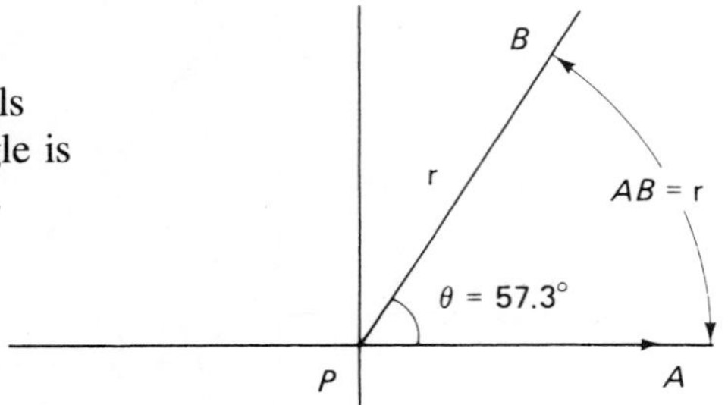 FIGURE 19-12

Page	Key Points, Definitions, and Equations	Example
479	To convert from radians to degrees: $\frac{180°}{\pi \text{ rad}} \times$ number of radians	angle = 1.77 rad $\frac{180°}{\pi} \times 1.77 = 101°$
479	To convert from degrees to radians: $\frac{\pi \text{ rad}}{180°} \times$ number of degrees	angle = 37° $\frac{\pi}{180} \times 37 = 0.646$ rad
482	The *frequency* (f) is the number of times per second that we generate a sine wave. The *period* (T) is the time it takes to generate one complete sine wave.	$f = 1/T$ $T = 1/f$
483	*Angular velocity* (ω) is the angular motion or speed of a rotating vector or phasor. Angular velocity is measured in radians per second or in degrees per second.	$\omega = 2\pi f$ $\omega = 360° f$
484	The generated angle at any instant, t, can be calculated as angle in radians $= 2\pi ft = \omega t$ angle in degrees $= 360° ft = \omega t$	*Given:* $f = 2$ kHz $t = 100\ \mu$s $\omega t = 1.26$ rad $\omega t = 72.0°$
486	In ac circuits, instantaneous values of current and voltage may be found using the following equations: $i = I_{pk} \sin \omega t$ $e = E_{pk} \sin \omega t$	*Let* $I_{pk} = 40$ mA, $E_{pk} = 12$ V, and $f = 5$ kHz Find i and e after 70 μs. $i = 40$ mA $\sin \omega t$ $= 32.4$ mA $e = 12$ V $\sin \omega t$ $= 9.71$ V
489	The *rms* or *effective* value of an ac voltage or current is the value that converts the same energy as does a dc value. The rms value equals 0.707 of the maximum or peak of a sine wave.	$I_{rms} = I_{eff} = 0.707 \quad I_{pk} = \frac{I_{pk}}{\sqrt{2}}$ $V_{rms} = V_{eff} = 0.707\ V_{pk}$ $I_{pk} = 1.414\ I_{rms} = \sqrt{2}\ I_{rms}$ $V_{pk} = 1.414\ V_{rms}$
490	*Key point:* Whenever we have capacitance or inductance in an ac circuit, the current and voltage are out of phase. When the phase angle, *theta* (θ), is known, we can calculate the instantaneous values of current and voltage.	When current is the reference, $i = I_{pk} \sin \omega t$ $e = E_{pk} \sin (\omega t \pm \theta)$ When voltage is the reference, $e = E_{pk} \sin \omega t$ $i = I_{pk} \sin (\omega t \pm \theta)$

End of Chapter Problems 19-1

1. $f = 600$ Hz. Find T.
2. $f = 3.75$ kHz. Find T.
3. $f = 1.2$ kHz. Find T.
4. $f = 150$ Hz. Find T.
5. $f = 1.76$ MHz. Find T.
6. $f = 27.5$ kHz. Find T.
7. $T = 400$ μs. Find f.
8. $T = 74.8$ ms. Find f.
9. $T = 1.33$ ms. Find f.
10. $T = 83.3$ μs. Find f.
11. $T = 25$ μs. Find f.
12. $T = 100$ ms. Find f.
13. Convert 25° to radians.
14. Convert 35° to radians.
15. Convert 90° to radians.
16. Convert 60° to radians.
17. Convert 180° to radians.
18. Convert 120° to radians.
19. Convert 75° to radians.
20. Convert 315° to radians.
21. Convert 600° to radians.
22. Convert 830° to radians.
23. Convert 0.146 rad to degrees.
24. Convert 1.7 rad to degrees.
25. Convert 0.5 rad to degrees.
26. Convert 3.5 rad to degrees.
27. Convert $\pi/2$ rad to degrees.
28. Convert $3\pi/2$ rad to degrees.
29. Convert 1.75 rad to degrees.
30. Convert 3.35 rad to degrees.
31. Convert 10 rad to degrees.
32. Convert 15 rad to degrees.

End of Chapter Problems 19-2

1. Find the angular velocity in both degrees per second and radians per second if the frequency is (a) 60 Hz, (b) 150 Hz, (c) 200 Hz, (d) 300 Hz, (e) 600 Hz, (f) 800 Hz, and (g) 1.25 kHz.
2. Find the angular velocity in both degrees per second and radians per second if the frequency is (a) 30 Hz, (b) 120 Hz, (c) 240 Hz, (d) 1.2 kHz, (e) 3.5 kHz, (f) 10 kHz, (g) 55.7 kHz, and (h) 100 kHz.
3. Find the frequency if the angular velocity is (a) 12 rad/s, (b) 50 rad/s, (c) 400 rad/s, (d) 900 rad/s, (e) 2500 rad/s, (f) 5000 rad/s, and (g) 7000 rad/s.
4. Find the frequency if the angular velocity is (a) 15 rad/s, (b) 47 rad/s, (c) 350 rad/s, (d) 900 rad/s, (e) 2500 rad/s, (f) 4500 rad/s, and (g) 7500 rad/s.
5. Find the frequency if the angular velocity is (a) 10,000°/s, (b) 15,000°/s, (c) 50,000°/s, (d) 120,000°/s, (e) 200,000°/s, (f) 7.35×10^5 degrees/s, and (g) 1.65×10^6 degrees/s.
6. Find the frequency if the angular velocity is (a) 3000°/s, (b) 18,000°/s, (c) 65,000°/s, (d) 2.25×10^4 degrees/s, (e) 5.05×10^5 degrees/s, (f) 8.75×10^5 degrees/s, and (g) 2.75×10^6 degrees/s.
7. If the frequency is 500 Hz, find the angle generated in both degrees and radians after (a) 100 μs, (b) 250 μs, (c) 1 ms, and (d) 2.8 ms.
8. If the frequency is 35 kHz, find the angle generated in both degrees and radians after (a) 5 μs, (b) 12 μs, (c) 40 μs, and (d) 750 μs.
9. If the frequency is 2.55 kHz, find the angle generated in both degrees and radians after (a) 100 μs, (b) 250 μs, (c) 1 ms, and (d) 2.8 ms.
10. If the frequency is 1.75 kHz, find the angle generated in both degrees and radians after (a) 5 μs, (b) 12 μs, (c) 40 μs, and (d) 750 μs.
11. What is the frequency when an angle of (a) 1.8 rad is generated in 20 μs, (b) 3.25 rad is generated in 500 μs, (c) 8.73 rad is generated in 150 μs, and (d) 17.6 rad is generated in 45 μs?
12. What is the frequency when an angle of (a) 0.25 rad is generated in 5 μs, (b) 4.6 rad is generated in 75 μs, (c) 27 rad is generated in 17.5 μs, and (d) 35 rad is generated in 120 ms?
13. What is the frequency when an angle of (a) 145° is generated in 2.3 ms, (b) 310° is generated in 500 μs, (c) 1100° is generated in 50 μs, and (d) 15,000° is generated in 25 μs?
14. What is the frequency when an angle of (a) 75° is generated in 20 μs, (b) 500° is generated in 150 μs, (c) 3000° is generated in 70 μs, and (d) 14,000° is generated in 8 ms?

15. If the frequency is 5 kHz, how much time is required to generate an angle of (a) 1 rad, (b) 0.125 rad, (c) 4.73 rad, (d) 20°, (e) 90°, and (f) 300°?

16. If the frequency is 125 Hz, how much time is required to generate an angle of (a) 0.25 rad, (b) 2.35 rad, (c) 5.6 rad, (d) 300°, (e) 135°, and (f) 270°?

17. If the frequency is 15 kHz, how much time is required to generate an angle of (a) 1 rad, (b) 0.125 rad, (c) 4.73 rad, (d) 20°, (e) 90°, and (f) 300°?

18. If the frequency is 7.5 kHz, how much time is required to generate an angle of (a) 0.25 rad, (b) 2.35 rad, (c) 5.6 rad, (d) 45°, (e) 135°, and (f) 270°?

End of Chapter Problems 19-3

1. $E = 25\ V_{pk}, f = 2.5$ kHz. Find e after (a) 10 μs, (b) 25 μs, (c) 150 μs, (d) 225 μs, (e) 300 μs, and (f) 375 μs.

2. $E = 15\ V_{pk}, f = 18$ kHz. Find e after (a) 2 μs, (b) 10 μs, (c) 20 μs, (d) 45 μs, (e) 70 μs, and (f) 100 μs.

3. $E = 12\ V_{pk}, f = 12.7$ kHz. Find e after (a) 10 μs, (b) 25 μs, (c) 150 μs, (d) 225 μs, (e) 300 μs, and (f) 375 μs.

4. $E = 40\ V_{pk}, f = 5.75$ kHz. Find e after (a) 2 μs, (b) 10 μs, (c) 20 μs, (d) 45 μs, (e) 70 μs, and (f) 100 μs.

5. $I = 700\ \mu A_{pk}, f = 50$ kHz. Find i after (a) 5 μs, (b) 10 μs, (c) 15 μs, (d) 20 μs, (e) 25 μs, and (f) 30 μs.

6. $I = 50\ mA_{pk}, f = 2$ kHz. Find i after (a) 50 μs, (b) 120 μs, (c) 200 μs, (d) 300 μs, (e) 400 μs, and (f) 510 μs.

7. $I = 3.45\ mA_{pk}, f = 800$ Hz. Find i after (a) 5 μs, (b) 10 μs, (c) 15 μs, (d) 20 μs, (e) 25 μs, and (f) 30 μs.

8. $I = 450\ \mu A_{pk}, f = 8.55$ kHz. Find i after (a) 50 μs, (b) 120 μs, (c) 200 μs, (d) 300 μs, (e) 400 μs, and (f) 510 μs.

End of Chapter Problems 19-4

Convert the following voltages and currents to pk and p-p values:

1. (a) 65 V, (b) 110 V

2. (a) 250 mV, (b) 27.5 V

3. (a) 220 V, (b) 440 V

4. (a) 300 V, (b) 600 V

5. (a) 600 μA, (b) 1.2 mA

6. (a) 250 μA, (b) 2.75 mA

7. (a) 37.8 mA, (b) 3 A

8. (a) 850 mA, (b) 6.5 A

Convert the following peak values of voltage and current to p-p and rms values:

9. (a) 350 μV_{pk}, (b) 400 mV_{pk}

10. (a) 800 μV_{pk}, (b) 45 mV_{pk}

11. (a) 80 V_{pk}, (b) 200 V_{pk}

12. (a) 820 mV_{pk}, (b) 15.7 V_{pk}

13. (a) 200 μA_{pk}, (b) 9.45 mA_{pk}

14. (a) 500 μA_{pk}, (b) 14.5 mA_{pk}

15. (a) 500 mA_{pk}, (b) 1.75 A_{pk}

16. (a) 350 mA_{pk}, (b) 2.4 A_{pk}

Convert the following peak-to-peak values of voltage and current to pk and rms:

17. (a) 320 $\mu V_{p\text{-}p}$, (b) 70 $mV_{p\text{-}p}$

18. (a) 450 $\mu V_{p\text{-}p}$, (b) 40 $mV_{p\text{-}p}$

19. (a) 17 $V_{p\text{-}p}$, (b) 283 $V_{p\text{-}p}$

20. (a) 27 $V_{p\text{-}p}$, (b) 141 $V_{p\text{-}p}$

21. (a) 50 $\mu A_{p\text{-}p}$, (b) 700 $\mu A_{p\text{-}p}$

22. (a) 45 $\mu A_{p\text{-}p}$, (b) 825 $\mu A_{p\text{-}p}$

23. (a) 9.38 $mA_{p\text{-}p}$, (b) 650 $mA_{p\text{-}p}$

24. (a) 15 $mA_{p\text{-}p}$, (b) 180 $mA_{p\text{-}p}$

End of Chapter Problems 19-5

1. $I = 4\ mA_{pk}$, $E = 10\ V_{pk}$, and $f = 20$ kHz. Use current as the reference. E leads I by 25°. Find i and e after (a) 5 μs, (b) 12 μs, (c) 25 μs, and (d) 40 μs.

2. $I = 150\ mA_{pk}$, $E = 50$ V, and $f = 3$ kHz. Use voltage as the reference. E leads I by 45°. Find i and e after (a) 25 μs, (b) 60 μs, (c) 160 μs, and (d) 300 μs.

3. $I = 25\ mA_{pk}$, $E = 20$ V, and $f = 100$ Hz. Use voltage as the reference. I leads E by 35°. Find i and e after (a) 1 ms, (b) 2.5 ms, (c) 3.2 ms, and (d) 8.3 ms.

4. $I = 600\ \mu A_{pk}$, $E = 9\ V_{pk}$, and $f = 5$ kHz. Use current as the reference. I leads E by 30°. Find i and e after (a) 15 μs, (b) 40 μs, (c) 140 μs, and (d) 250 μs.

5. $I = 600\ \mu A_{pk}$, $E = 80\ V_{pk}$, and $f = 3.75$ kHz. Use current as the reference. I leads E by 55°. Find i and e after (a) 10 μs, (b) 40 μs, (c) 120 μs, and (d) 200 μs.

6. $I = 1.2\ mA_{pk}$, $E = 12\ V_{pk}$, and $f = 18$ kHz. Use current as the reference. I leads E by 38.5°. Find i and e after (a) 15 μs, (b) 50 μs, (c) 80 μs, and (d) 400 μs.

7. $I = 25\ mA_{pk}$, $E = 65\ V_{pk}$, and $f = 2.35$ kHz. Use voltage as the reference. I leads E by 40°. Find i and e after (a) 25 μs, (b) 70 μs, (c) 200 μs, and (d) 600 μs.

8. $I = 800\ \mu A_{pk}$, $E = 15\ V_{pk}$, and $f = 10$ kHz. Use voltage as the reference. I leads E by 50°. Find i and e after (a) 10 μs, (b) 30 μs, (c) 50 μs, and (d) 180 μs.

Answers to Self-tests

SELF-TEST 19-1

1. $T = 20\ \mu$s

2. 2.79

3. 36°

4. 1.88 rad = 108°

5. 1.67 kHz

6. 41.4 μs

SELF-TEST 19-2

1. 23.8 V

2. −19 mA

3. $35.4\ V_{pk} = 70.7\ V_{p\text{-}p}$

4. $30\ mA_{pk} = 21.2$ mA

5. 43.5 V

AC CIRCUIT ANALYSIS: SERIES CIRCUITS

CHAPTER 20

INTRODUCTION

All electrical circuits contain the properties of resistance, capacitance, and inductance. We didn't consider the effects of capacitance and inductance in the dc circuits in Chapters 11, 12, and 13, because in dc circuits, their effects are negligible. In those chapters we used Kirchhoff's laws and Ohm's law to solve series circuit problems where the only measurable circuit property was resistance and the electrical energy source was some dc source.

In reality, most circuits we work with in electronics contain ac sources and measurable values of inductance and capacitance as well as resistance. Furthermore, the effects of inductance and capacitance are displaced from the effects of circuit resistance by 90°. These effects cause the applied voltage and circuit current to be out of phase. We learned how to solve problems when one angle was a right triangle in the chapters on trig. In this chapter we will use trig functions and the Pythagorean theorem to find the phase angle, the various voltage drops, and the current in ac series circuit problems.

When the effects of both inductance and capacitance are measurable in series circuits, series resonance occurs. *Resonance* is that frequency where the effects of capacitance and inductance cancel each other and the phase angle is 0°. The concept of series resonance is discussed, and the computation of series resonance and the conditions that occur at series resonance are also covered in this chapter.

CHAPTER OBJECTIVES

In this chapter you will learn how to:

1. Use trigonometric functions to find phase angles, reactances, and impedances in series *RC* and *RL* circuits.

2. Express phasors in polar and rectangular form.
3. Express answers to series *RC* and *RL* circuit problems in both polar and rectangular forms.
4. Add and subtract impedances in polar and rectangular form.
5. Find the equivalent series circuit of ac circuits containing all three circuit properties: capacitance, inductance, and resistance.
6. Determine the resonant frequency of series *RCL* circuits.

20-1 SERIES *RC* CIRCUITS

Capacitance in ac circuits causes the current to lead the applied voltage. In a purely capacitive circuit ($R = 0\ \Omega$), current leads the voltage by 90°. At the other extreme, if the circuit is purely resistive, current and applied voltage are in phase. Let's see what happens in a circuit that contains both resistance and capacitance. Consider the circuit in Figure 20-1.

KEY POINT In a series circuit, we use I as the reference. In an *RC* circuit, current leads, so V_C and X_C are plotted along the negative y-axis. E (or Z) is the hypotenuse, and θ is the phase angle.

We have drawn a series *RC* circuit in Figure 20-1(a). As shown in Figure 20-1(b), the current leads the voltage drop across the capacitor by 90°. V_R is the side adjacent to the angle θ and V_C is the opposite side.

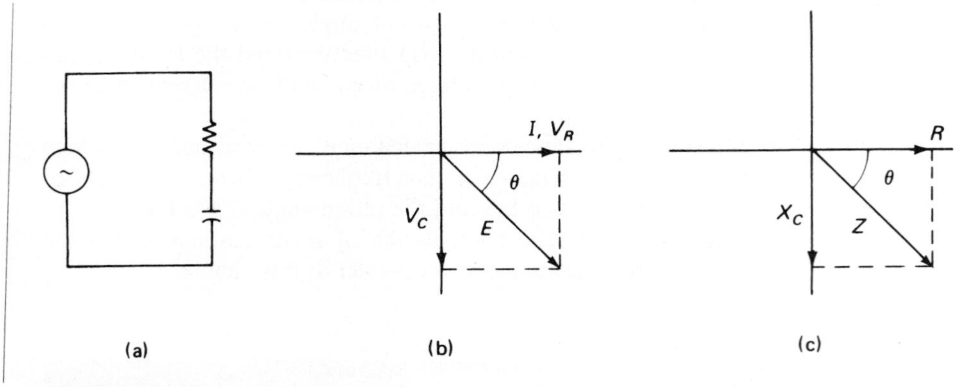

FIGURE 20-1

☐ EXAMPLE 20-1

Refer to Figure 20-1(a). Let $V_R = 10$ V and $V_C = -12$ V. Find **E** and θ.

Solution Refer to Figure 20-2.

$$\tan\theta = \frac{\text{opposite side}}{\text{adjacent side}} = \frac{V_C}{V_R} = \frac{-12\text{ V}}{10\text{ V}} = -1.2$$

$$\tan^{-1} - 1.2 = \theta = -50.2°$$

For the problems in this chapter, let's set up the calculator in engineering mode and display 2 digits to the right of the decimal point.

[ENG] [FIX] 2

Using the calculator,

12 [+/-] [÷] 10 [=] [tan⁻¹] [STO] *Answer:* $\theta = -50.2°$

Store θ because we will use it for the next part of the problem. (Remember, V_C is negative because we plotted V_C along the negative y-axis.) We can now find **E** by using either the sine or cosine function. We will use the sine function.

$$\sin\theta = \frac{\text{opposite side}}{\text{hypotenuse}} = \frac{V_C}{\mathbf{E}}$$

$$\mathbf{E} = \frac{V_C}{\sin\theta} = \frac{-12\text{ V}}{\sin - 50.2°} = \frac{-12\text{ V}}{-0.768} = 15.6\text{ V}$$

12 [+/-] [÷] [RCL] [sin] [=] *Answer*: $\mathbf{E} = 15.6$ V

□ EXAMPLE 20-2

Refer to Figure 20-1(a). Let $\mathbf{E} = 25$ V and $V_R = 18$ V. Find V_C and θ.

Solution

$$\cos\theta = \frac{V_R}{\mathbf{E}} = \frac{18\text{ V}}{25\text{ V}} = 0.72$$

$$\cos^{-1} 0.72 = \theta = -43.9°$$

From the given information we know that θ is negative because we know that we are operating in the fourth quadrant.

$$\sin\theta = \frac{V_C}{\mathbf{E}}$$

$$V_C = \mathbf{E}\sin\theta = 25\text{ V} \sin - 43.9° = -17.3\text{ V}$$

□

PRACTICE PROBLEMS 20-1

Refer to Figure 20-1 for the following problems:

1. $V_R = 6.8$ V, $V_C = -10$ V. Find **E** and θ.
2. $V_R = 30$ V, $\mathbf{E} = 50$ V. Find V_C and θ.
3. $V_R = 10$ V, $\theta = -60°$. Find V_C and **E**.
4. $\mathbf{E} = 20$ V, $\theta = -35°$. Find V_R and V_C.
5. $\mathbf{E} = 50$ V, $V_C = -40$ V. Find V_R and θ.
6. $V_C = -11$ V, $\theta = -65°$. Find **E** and V_R.

Solutions

1. $\mathbf{E} = 12.1$ V; $\theta = -55.8°$
2. $V_C = -40$ V; $\theta = -53.1°$
3. $V_C = -17.3$ V; $\mathbf{E} = 20$ V
4. $V_R = 16.4$ V; $V_C = -11.5$ V
5. $V_R = 30$ V; $\theta = -53.1°$
6. $\mathbf{E} = 12.1$ V; $V_R = 5.11$ V

Additional practice problems are at the end of the chapter.

In Figure 20-1(c), since V_R and I are in phase, R, is plotted along the x-axis. X_C, the capacitive reactance, must be plotted along the negative y-axis to show the 90° phase difference between capacitive values and resistive values. The capacitive reactance is determined by the following equation:

$$X_C = \frac{1}{2\pi f C} \tag{20-1}$$

☐ EXAMPLE 20-3

Refer to Figure 20-1(a). Let $R = 2\ \text{k}\Omega$, $C = 20$ nF, and $f = 3$ kHz. Find X_C, Z, and θ.

Solution

$$X_C = \frac{1}{2\pi f C} = \frac{1}{2\pi \times 3\ \text{k}\Omega \times 20\ \text{nF}} = 2.65\ \text{k}\Omega$$

Using the calculator,

2 [×] [π] [=] [STO] [×] 3 [EE] 3 [×] 20 [EE] [+/-] 9 [=] [1/x]

Answer: $X_C = 2.65\ \text{k}\Omega$

We stored 2π since we will use it often in future problems. X_C will be negative because it is plotted along the negative y-axis (see Figure 20-2).

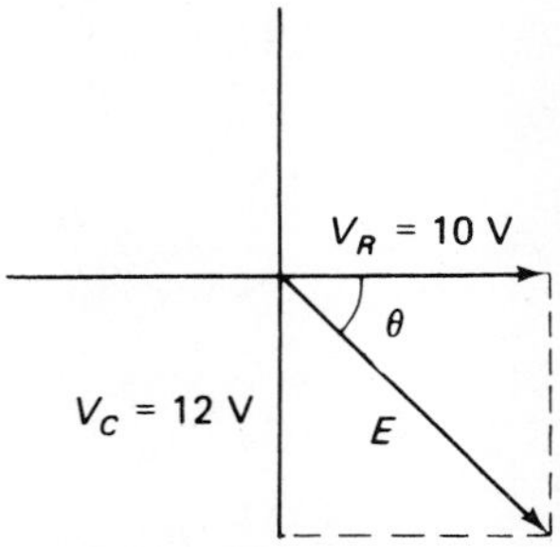

FIGURE 20-2

$$\tan\theta = \frac{\text{opposite side}}{\text{adjacent side}} = \frac{X_C}{R} = \frac{-2.65\ \text{k}\Omega}{2\ \text{k}\Omega} = -1.33$$

$$\tan^{-1} - 1.33 = \theta = -53°$$

The calculator already displays X_C. We need to change its sign and then continue solving the problem:

[+/-] [÷] 2 [EE] 3 [=] [tan⁻¹]   *Answer*: $\theta = -53.0°$

θ is negative because we are in the fourth quadrant.

$$\cos\theta = \frac{R}{Z}$$

$$Z = \frac{R}{\cos\theta} = \frac{2\ \text{k}\Omega}{\cos - 53°} = 3.32\ \text{k}\Omega$$

Now θ is displayed and we need to divide 2 kΩ by the cosine of θ. Dividing by cosine θ is the same as multiplying by the reciprocal of cosine θ.

$$\frac{2\text{ k}\Omega}{\cos\theta} = 2\text{ k}\Omega \times \frac{1}{\cos\theta}$$

So

[cos] [1/x] [×] 2 [EE] 3 [=]   *Answer*: $Z = 3.32$ kΩ

☐ EXAMPLE 20-4

Refer to Figure 20-1(a). Let $R = 10$ kΩ, $C = 50$ nF, $f = 500$ Hz, and $\mathbf{E} = 10$ V. Find X_C, Z, θ, V_R, V_C, and I.

Solution

$$X_C = \frac{1}{2\pi fC} = \frac{1}{2\pi \times 500 \times 50\text{ nF}} = 6.37\text{ k}\Omega$$

$$\tan\theta = \frac{\text{opposite side}}{\text{adjacent side}} = \frac{X_C}{R} = \frac{-6.37\text{ k}\Omega}{10\text{ k}\Omega} = -0.637$$

$$\tan^{-1} - 0.637 = -32.5°$$

$$\sin\theta = \frac{X_C}{Z}$$

$$Z = \frac{X_C}{\sin\theta} = \frac{-6.37\text{ k}\Omega}{\sin -32.5°} = 11.9\text{ k}\Omega$$

With the calculator,

[RCL] [×] 500 [×] 50 [EE] [+/-] 9 [=] [1/x]   *Answer*: $X_C = 6.37$ kΩ

Remember, we had previously stored 2π. Keep X_C displayed but record its value before going to the next step.

[+/-] [÷] 10 [EE] 3 [=] [$\tan^{-1}$]   *Answer*: $\theta = -32.5°$

Change the sign of X_C and then divide by the resistance, 10 kΩ, to get tan θ, which is -0.637. The arctan gives us θ. Keep θ displayed, but record its value before going to the next step.

[sin] [1/x] [×] 6.37 [+/-] [EE] 3 [=]   *Answer*: $Z = 11.9$ kΩ

Find sin θ, take its reciprocal, and multiply by X_C to find the impedance, Z.

Now that we know Z, we can find I.

$$I = \frac{\mathbf{E}}{Z}$$

$$= \frac{10\text{ V}}{11.9\text{ k}\Omega} = 840\ \mu\text{A}$$

To find V_C and V_R, we can use Ohm's law:

$$V_C = IX_C = 840\ \mu\text{A} \times 6.37\ \text{k}\Omega = 5.35\ \text{V}$$

$$V_R = IR = 840\ \mu\text{A} \times 10\ \text{k}\Omega = 8.4\ \text{V}$$

We could also find V_C and V_R by using trig functions since we know **E** and θ.

$$V_R = \mathbf{E} \cos \theta = 10\ \text{V} \cos - 32.5° = 8.43\ \text{V}$$

$$V_C = \mathbf{E} \sin \theta = 10\ \text{V} \sin - 32.5° = -5.37\ \text{V}$$

(Differences in the values of V_R and V_C in the two methods are a result of rounding the value of Z to three places.)

☐ EXAMPLE 20-5

$R = 5\ \text{k}\Omega$, $C = 10\ \text{nF}$, $Z = 8.3\ \text{k}\Omega$, and $\mathbf{E} = 10\ \text{V}$. Find I, θ, f, X_C, V_C, and V_R.

Solution

$$I = \frac{E}{Z} = \frac{10\ \text{V}}{8.3\ \text{k}\Omega} = 1.2\ \text{mA}$$

Since we know R and Z, we can find X_C and θ. We will use the cosine function to find θ and the sine function to find X_C.

$$\cos \theta = \frac{R}{Z} = \frac{5\ \text{k}\Omega}{8.3\ \text{k}\Omega} = 0.602$$

$$\cos^{-1} 0.602 = \theta = -53.1°$$

$$\sin \theta = \frac{X_C}{Z}$$

$$X_C = Z \sin \theta = 8.3\ \text{k}\Omega \times \sin - 53.1° = -6.64\ \text{k}\Omega$$

(Remember, mathematically the value of X_C is negative because we are operating in the fourth quadrant. Electrically, X_C is treated as a positive value.)

Now we can find the frequency because we know C and X_C. If we solve for f in the equation

$$X_C = \frac{1}{2\pi fC}$$

we get

$$f = \frac{1}{2\pi CX_C} = \frac{1}{2\pi \times 10\ \text{nF} \times 6.64\ \text{k}\Omega} = 2.4\ \text{kHz}$$

We can find V_R and V_C by using Ohm's law or by using trig functions. We should solve both ways to check our work. We will use Ohm's law here.

$$V_R = IR = 1.2\ \text{mA} \times 5\ \text{k}\Omega = 6\ \text{V}$$

$$V_C = IX_C = 1.2\ \text{mA} \times 6.64\ \text{k}\Omega = 7.97\ \text{V}$$

☐

PRACTICE PROBLEMS 20-2

Refer to Figure 20-3 for the following problems and find X_C, Z, and θ.

1. $R = 5.6\ \text{k}\Omega$, $C = 50$ nF, and $f = 2.5$ kHz
2. $R = 120\ \Omega$, $C = 0.2\ \mu\text{F}$, and $f = 4.85$ kHz
3. $R = 39\ \text{k}\Omega$, $C = 400$ pF, and $f = 12$ kHz

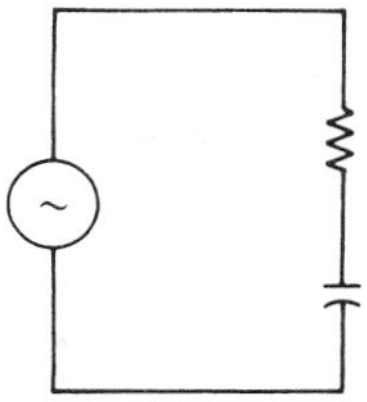

FIGURE 20-3

Refer to Figure 20-3 for the following problems and find X_C, Z, θ, V_R, V_C, and I.

4. $R = 470\ \Omega$, $C = 0.2\ \mu\text{F}$, $f = 1$ kHz, and $\mathbf{E} = 9$ V
5. $R = 2\ \text{k}\Omega$, $C = 30$ nF, $f = 3$ kHz, and $\mathbf{E} = 20$ V
6. $R = 27\ \text{k}\Omega$, $C = 100$ pF, $f = 35$ kHz, and $\mathbf{E} = 15$ V

Refer to Figure 20-3 for the following problems and find I, θ, f, X_C, V_R, and V_C.

7. $R = 12\ \text{k}\Omega$, $C = 200$ pF, $Z = 20\ \text{k}\Omega$, and $\mathbf{E} = 15$ V
8. $R = 680\ \Omega$, $C = 1\ \mu\text{F}$, $Z = 800\ \Omega$, and $\mathbf{E} = 25$ V
9. $R = 3\ \text{k}\Omega$, $C = 50$ nF, $Z = 4.5\ \text{k}\Omega$, and $\mathbf{E} = 30$ V

Solutions

1. $X_C = 1.27\ \text{k}\Omega$, $Z = 5.74\ \text{k}\Omega$, $\theta = -12.8°$
2. $X_C = 164\ \Omega$, $Z = 203\ \Omega$, $\theta = -53.8°$
3. $X_C = 33.2\ \text{k}\Omega$, $Z = 51.2\ \text{k}\Omega$, $\theta = -40.4°$
4. $X_C = 796\ \Omega$, $Z = 924\ \Omega$, $\theta = -59.4°$, $V_R = 4.58$ V, $V_C = 7.75$ V, $I = 9.74$ mA
5. $X_C = 1.77\ \text{k}\Omega$, $Z = 2.67\ \text{k}\Omega$, $\theta = -41.5°$, $V_R = 15$ V, $V_C = 13.2$ V, $I = 7.49$ mA
6. $X_C = 45.5\ \text{k}\Omega$, $Z = 52.9\ \text{k}\Omega$, $\theta = -59.3°$, $V_R = 7.66$ V, $V_C = 12.9$ V, $I = 284\ \mu\text{A}$
7. $I = 750\ \mu\text{A}$, $\theta = -53.1°$, $f = 49.7$ kHz, $X_C = 16\ \text{k}\Omega$, $V_R = 9$ V, $V_C = 12$ V
8. $I = 31.3$ mA, $\theta = -31.8°$, $f = 378$ Hz, $X_C = 421\ \Omega$, $V_R = 21.2$ V, $V_C = 13.2$ V
9. $I = 6.67$ mA, $\theta = -48.2°$, $f = 949$ Hz, $X_C = 3.35\ \text{k}\Omega$, $V_R = 20$ V, $V_C = 22.4$ V

Additional practice problems are at the end of the chapter.

20-2 SERIES *RL* CIRCUITS

KEY POINT In a series circuit, we use I as the reference. In an *RL* circuit, voltage leads, so V_L and X_L are plotted along the positive y-axis. E (or Z) is the hypotenuse, and θ is the phase angle.

We have drawn a series RL circuit in Figure 20-4(a). As shown in Figure 20-4(b), the voltage drop across the inductor leads the current by 90°. V_R is the side adjacent to the angle θ and V_L is the opposite side.

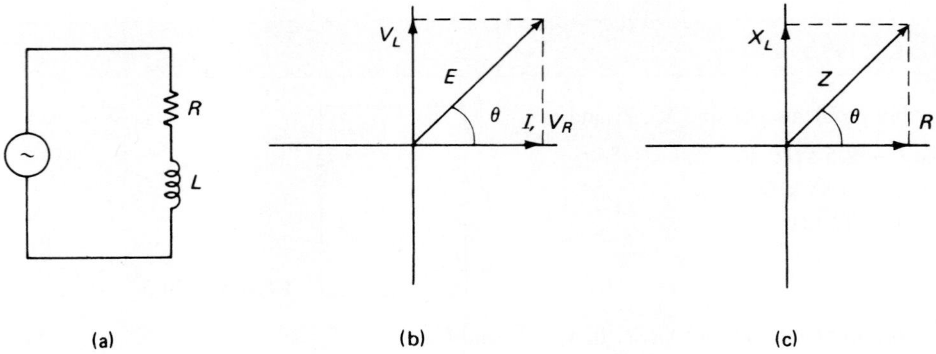

FIGURE 20-4

□ EXAMPLE 20-6

Let $V_R = 15$ V and $V_L = 10$ V in Figure 20-4(a). Find **E** and θ.

Solution Refer to Figure 20-5.

$$\tan\theta = \frac{V_L}{V_R} = \frac{10\text{ V}}{15\text{ V}} = 0.667$$

$$\tan^{-1} 0.667 = \theta = 33.7°$$

E can be found by using either the sine function or the cosine function. Let's use the cosine function.

$$\cos\theta = \frac{V_R}{\mathbf{E}}$$

$$\mathbf{E} = \frac{V_R}{\cos\theta} = \frac{15\text{ V}}{\cos 33.7°} = 18\text{ V}$$

□ EXAMPLE 20-7

Let **E** = 10 V and $V_L = 5$ V in Figure 20-4(a). Find V_R and θ.

Solution

$$\sin\theta = \frac{V_L}{\mathbf{E}} = \frac{5\text{ V}}{10\text{ V}} = 0.5$$

$$\sin^{-1} 0.5 = \theta = 30°$$

Now that we know θ, we can use either the cosine function or the tangent function to find V_R. Let's use the tangent function.

$$\tan\theta = \frac{V_L}{V_R}$$

$$V_R = \frac{V_L}{\tan\theta} = \frac{5\text{ V}}{\tan 30°} = 8.66\text{ V}$$

□

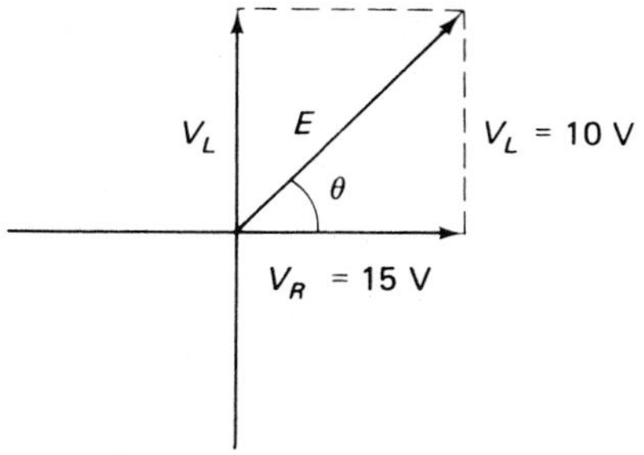

FIGURE 20-5

PRACTICE PROBLEMS 20-3

Refer to Figure 20-4 for the following problems.

1. $V_R = 12$ V, $V_L = 18$ V. Find **E** and θ.
2. $V_R = 27$ V, $\theta = 65°$. Find **E** and V_L.
3. $V_R = 35$ V, **E** $= 50$ V. Find V_L and θ.
4. $V_L = 6$ V, **E** $= 10$ V. Find V_R and θ.
5. $V_L = 15$ V, $\theta = 27°$. Find V_R and **E**.
6. **E** $= 25$ V, $\theta = 33°$. Find V_R and V_L.

Solutions

1. **E** $= 21.6$ V; $\theta = 56.3°$
2. **E** $= 63.9$ V; $V_L = 57.9$ V
3. $V_L = 35.7$ V; $\theta = 45.6°$
4. $V_R = 8$ V; $\theta = 36.9°$
5. $V_R = 29.4$ V; **E** $= 33$ V
6. $V_R = 21$ V; $V_C = 13.6$ V

Additional practice problems are at the end of the chapter.

In Figure 20-4(c), since V_R and I are in phase, R is plotted along the x-axis. X_L, the inductive reactance, must be plotted along the positive y-axis to show the 90° phase difference between inductive and resistive values. The inductive reactance is determined by the equation

$$X_L = 2\pi fL \tag{20-2}$$

☐ EXAMPLE 20-8

Refer to Figure 20-4(a). Let $R = 2.7$ kΩ, $L = 200$ mH, and $f = 3$ kHz. Find X_L, θ, and Z.

Solution

$$X_L = 2\pi fL = 2\pi \times 3\text{ k}\Omega \times 200\text{ mH} = 3.77\text{ k}\Omega$$

Knowing R and X_L, we can use the tangent function to find θ.

$$\tan\theta = \frac{X_L}{R} = \frac{3.77\text{ k}\Omega}{2.7\text{ k}\Omega} = 1.4$$

$$\tan^{-1} 1.4 = \theta = 54.4°$$

We can use either the sine function or the cosine function to find Z. Let's use the sine function.

$$\sin\theta = \frac{X_L}{Z}$$

$$Z = \frac{X_L}{\sin\theta} = \frac{3.77\text{ k}\Omega}{\sin 54.4^\circ} = 4.64\text{ k}\Omega$$

☐ EXAMPLE 20-9

Refer to Figure 20-4. Let $R = 10\text{ k}\Omega$, $L = 100\text{ mH}$, $f = 20\text{ kHz}$, and $\mathbf{E} = 15\text{ V}$. Find X_L, θ, Z, V_R, V_L, and I.

Solution

$$X_L = 2\pi fL = 2\pi \times 20\text{ kHz} \times 100\text{ mH} = 12.6\text{ k}\Omega$$

Knowing R and X_L, we can find Z and θ.

$$\tan\theta = \frac{X_L}{R} = \frac{12.6\text{ k}\Omega}{10\text{ k}\Omega} = 1.26$$

$$\tan^{-1} 1.26 = \theta = 51.5^\circ$$

$$\cos\theta = \frac{R}{Z}$$

$$Z = \frac{R}{\cos\theta} = \frac{10\text{ k}\Omega}{\cos 51.5^\circ} = 16.1\text{ k}\Omega$$

We can find V_R and V_L by using Ohm's law.

$$I = \frac{\mathbf{E}}{Z} = \frac{15\text{ V}}{16.1\text{ k}\Omega} = 932\ \mu\text{A}$$

$$V_R = IR = 932\ \mu\text{A} \times 10\text{ k}\Omega = 9.32\text{ V}$$

$$V_L = IX_L = 932\ \mu\text{A} \times 12.6\text{ k}\Omega = 11.7\text{ V}$$

We can also find V_R and V_L by using trig functions.

$$V_R = \mathbf{E}\cos\theta = 15\text{ V} \times \cos 51.5^\circ = 9.34\text{ V}$$

$$V_L = \mathbf{E}\sin\theta = 15\text{ V} \times \sin 51.5^\circ = 11.7\text{ V}$$

(Differences in the value of V_R are due to rounding.)

☐ EXAMPLE 20-10

Refer to Figure 20-4. Let $L = 30\text{ mH}$, $Z = 12\text{ k}\Omega$, $X_L = 9\text{ k}\Omega$, and $V_R = 3.7\text{ V}$. Find R, θ, $\mathbf{E}$, I, f, and V_L.

Solution Let's find θ since we know Z and X_L.

$$\sin\theta = \frac{X_L}{Z} = \frac{9\text{ k}\Omega}{12\text{ k}\Omega} = 0.75$$

$$\sin^{-1} 0.75 = \theta = 48.6^\circ$$

Having found θ, we can now find R, $\mathbf{E}$, and V_L.

$$\cos\theta = \frac{R}{Z}$$

$$R = Z\cos\theta = 12\text{ k}\Omega\cos 48.6^\circ = 7.94\text{ k}\Omega$$

$$\cos\theta = \frac{V_R}{\mathbf{E}}$$

$$\mathbf{E} = \frac{V_R}{\cos\theta} = \frac{3.7\text{ V}}{\cos 48.6^\circ} = 5.59\text{ V}$$

$$\sin\theta = \frac{V_C}{\mathbf{E}}$$

$$V_C = \mathbf{E}\sin\theta = 5.59\sin 48.6^\circ = 4.19\text{ V}$$

We can use Ohm's law to find I.

$$I = \frac{\mathbf{E}}{Z} = \frac{5.59\text{ V}}{12\text{ k}\Omega} = 466\ \mu\text{A}$$

Solving for f in the equation $X_L = 2\pi fL$, we get

$$f = \frac{X_L}{2\pi L} = \frac{9\text{ k}\Omega}{2\pi\times 30\text{ mH}} = 47.7\text{ kHz}$$

Using the calculator,

9 [EE] 3 [÷] [(] 2 [×] [π] [×] 30 [EE] [+/-] 3 [)] [=]

Answer: $f = 47.7$ kHz

□

PRACTICE PROBLEMS 20-4

Refer to Figure 20-6 for the following problems and find X_L, Z, and θ.

1. $R = 2.7\text{ k}\Omega$, $L = 1$ H, and $f = 600$ Hz
2. $R = 25\text{ k}\Omega$, $L = 200$ mH, and $f = 15$ kHz

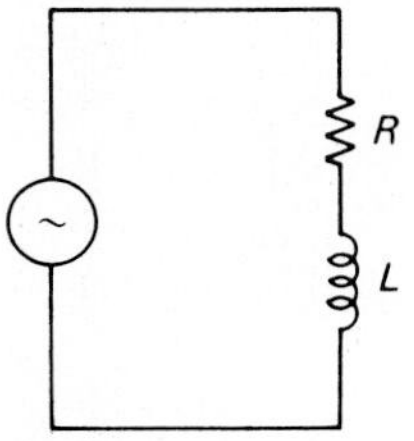

FIGURE 20-6

Refer to Figure 20-6 for the following problems and find X_L, Z, θ, V_R, V_L, and I.

3. $R = 600\ \Omega$, $L = 400$ mH, $f = 200$ Hz, and **E** $= 20$ V
4. $R = 3$ kΩ, $L = 50$ mH, $f = 15$ kHz, and **E** $= 9$ V

Refer to Figure 20-6 for the following problems and find X_L, θ, f, **E**, V_R, and I.

5. $L = 100$ mH, $Z = 7.3$ kΩ, $R = 5$ kΩ, and $V_L = 7.2$ V
6. $L = 20$ mH, $Z = 1$ kΩ, $R = 300\ \Omega$, and $V_L = 3.2$ V

Refer to Figure 20-6 for the following problems and find X_L, L, θ, V_R, **E**, and I.

7. $R = 12$ kΩ, $Z = 22.7$ kΩ, $f = 5.3$ kHz, and $V_L = 14$ V
8. $R = 680\ \Omega$, $Z = 870\ \Omega$, $f = 1.35$ kHz, $V_L = 7.35$ V

Solutions

1. $X_L = 3.77$ kΩ; $Z = 4.64$ kΩ; $\theta = 54.4°$
2. $X_L = 18.8$ kΩ; $Z = 31.3$ kΩ; $\theta = 37°$
3. $X_L = 503\ \Omega$; $\theta = 40°$; $Z = 783\ \Omega$; $V_R = 15.3$ V; $V_L = 12.8$ V; $I = 25.5$ mA
4. $X_L = 4.71$ kΩ; $\theta = 57.5°$; $Z = 5.59$ kΩ; $V_R = 4.83$ V; $V_L = 7.59$ V; $I = 1.61$ mA
5. $X_L = 5.32$ kΩ; $\theta = 46.8°$; $f = 8.47$ kHz; **E** $= 9.88$ V; $V_R = 6.75$ V; $I = 1.35$ mA
6. $X_L = 954\ \Omega$; $\theta = 72.5°$; $f = 7.59$ kHz; **E** $= 3.36$ V; $V_R = 1.01$ V; $I = 3.36$ mA
7. $X_L = 19.3$ kΩ; $L = 579$ mH; $\theta = 58.1°$; $V_R = 8.72$ V; $E = 16.5$ V; $I = 727\ \mu$A
8. $X_L = 543\ \Omega$; $L = 64.0$ mH; $\theta = 38.6°$; $V_R = 9.21$ V; **E** $= 11.8$ V; $I = 13.5$ mA

Additional practice problems are at the end of the chapter.

SELF-TEST 20-1

Refer to Figure 20-7.

1. $V_R = 17$ V, $\theta = -36°$. Find V_C and **E**.
2. $R = 2$ kΩ, $C = 50$ nF, $f = 1$ kHz, and **E** $= 10$ V. Find X_C, Z, θ, V_R, V_C, and I.
3. $R = 330\ \Omega$, $C = 1\ \mu$F, $Z = 600\ \Omega$, and **E** $= 9$ V. Find X_C, f, θ, V_R, V_C, and I.

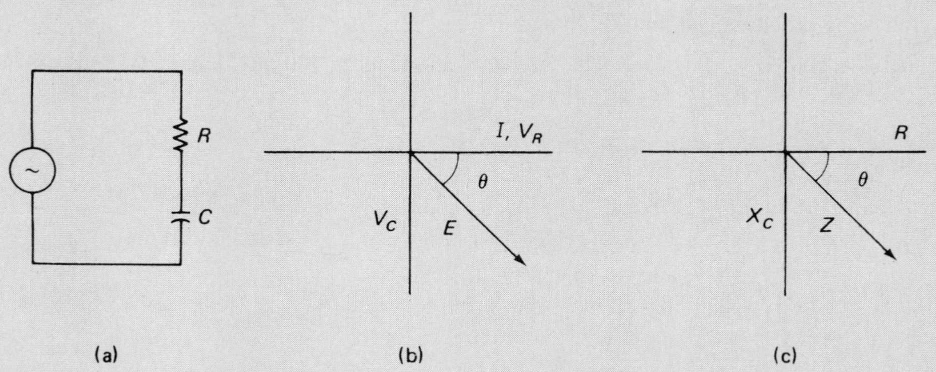

FIGURE 20-7

Refer to Figure 20-6.

4. $V_L = 3.3$ V, $\theta = 65°$. Find V_R and **E**.

5. $R = 4.7$ kΩ, $L = 175$ mH, $f = 5$ kHz, and **E** = 25 V. Find X_L, θ, Z, V_R, V_L, and I.

6. $L = 50$ mH, $Z = 1.2$ kΩ, $R = 750$ Ω, and $V_L = 7.8$ V. Find X_L, θ, f, **E**, V_R, and I.

Answers to Self-test 20-1 are at the end of the chapter.

20-3 POLAR TO RECTANGULAR CONVERSION

Any phasor may be identified by expressing the phasor in terms of its magnitude and direction or in terms of its horizontal and vertical components. For example, if we are given a phasor of magnitude 50 at an angle of −30°, we could represent the phasor graphically as in Figure 20-8. In equation form we could write $r = 50\angle{-30°}$. When the magnitude and direction are given, we say that the quantity is in *polar form*. Polar form notation gives us the magnitude of the hypotenuse of a right triangle and the phase angle. Knowing the hypotenuse and phase angle, we can quickly determine the x and y values necessary to cause such a phasor.

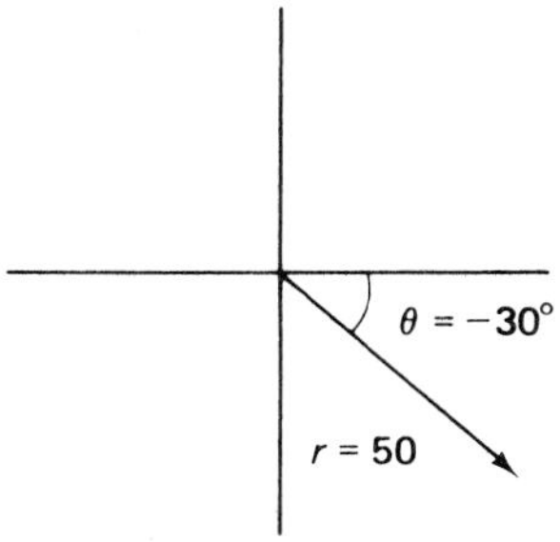

FIGURE 20-8

$$x = r \cos \theta = 50 \cos -30° = 43.3$$

$$y = r \sin \theta = 50 \sin -30° = -25$$

In Chapter 16 we saw how the j operator could be used to indicate 90° rotation from the x-axis. $+j$ indicates 90° rotation in a positive direction and $-j$ indicates 90° rotation in a negative direction. $+j$, then, puts us at 90° and $-j$ puts us at 270° or −90°. We can use the j operator anytime we want to show 90° rotation. This means that we could write the x and y values above in this manner:

$$r = 43.3 - j25$$

We have expressed the magnitude of the hypotenuse in terms of its rectangular coordinates. A phasor expressed in these terms is said to be in *rectangular form*.

$$r_{\text{polar}} = 50\angle{-30°}$$

$$r_{\text{rect}} = 43.3 - j25$$

Either form describes the phasor.

Most scientific or engineering calculators will convert between polar and rectangular forms. In our examples, we will indicate the conversion with the block [R→P] or [P→R] depending on which way we are converting. Consult your users manual for the correct key sequence for your calculator to make the conversion. In all calculators, we must key in the first number, get it into the y register, and then key in the second number. In most calculators, x register and y register changes are done with the [X↔Y] key.

□ EXAMPLE 20-11

Convert $43.3 - j25$ from rectangular to polar form using the calculator.

Solution The calculator sequence is

43.3 [X↔Y] 25 [±] [R→P] *Answer*: 50

[X↔Y] *Answer*: −30

The calculator stores the hypotenuse, 50, in the x register, and the angle, −30°, in the y register. We perform the second x-y interchange so that the angle is displayed.

$$43.3 - j25 = 50\,\underline{/-30^\circ}$$

□ EXAMPLE 20-12

Convert $50\,\underline{/-30^\circ}$ from polar to rectangular form using the calculator.

Solution The calculator sequence is

50 [X↔Y] 30 [±] [P→R] *Answer*: 43.3

The real-number part, which is the side adjacent to the phase angle θ, is displayed. To display the imaginary part, the opposite side, we again use the x-y interchange.

[X↔Y] *Answer*: −25

□

When using the calculator for these conversions, note the following key points:

- **KEY POINT** When converting from rectangular form to polar form, key in the x-axis or real value first. Move that value to the y register using the [X↔Y] key sequence. When finished with the complete calculator sequence, the hypotenuse is displayed. To display the angle, exchange the contents of the registers using the [X↔Y] key sequence.

- **KEY POINT** When converting from polar form to rectangular form, key in the hypotenuse first. Move that value to the y register using the [X↔Y] key sequence. When finished with the complete calculator sequence, the x axis or real value is displayed. To display the y axis or imaginary value, exchange the contents of the registers using the [X↔Y] key sequence.

Polar and rectangular notation is used in ac circuit analysis to define the impedance (Z) and the applied voltage (E) in a series circuit. It is also used to define admittance (Y) and total current (I) in a parallel circuit.

20-3-1 Series *RC* and *RL* Circuits

Consider the circuit in Figure 20-9(a). We could express E in both polar and rectangular forms.

$$E \underline{/\theta} = V_R - jV_C$$

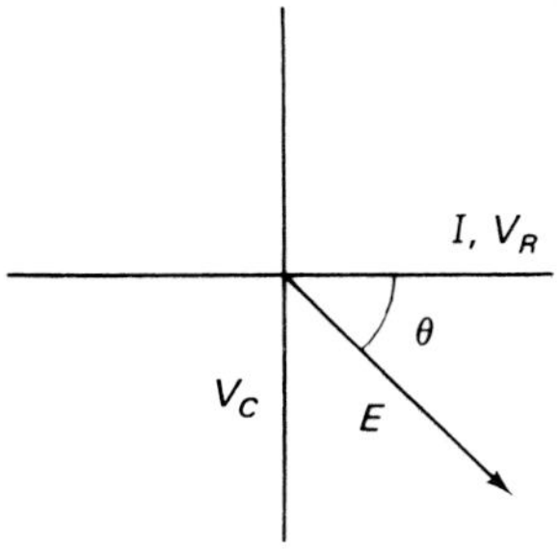

FIGURE 20-9(a)

☐ EXAMPLE 20-13

Let $V_R = 20$ V and $V_C = 30$ V in Figure 20-9(a). Find E and θ. Express E in both polar and rectangular forms.

Solution

$$E_{\text{rect}} = 20\text{ V} - j30\text{ V}$$

$$\tan\theta = \frac{V_C}{V_R} = \frac{-30\text{ V}}{20\text{ V}} = -1.5$$

$$\tan^{-1} - 1.5 = \theta = -56.3^\circ$$

$$\sin\theta = \frac{V_C}{E}$$

$$E = \frac{V_C}{\sin\theta} = \frac{-30\text{ V}}{\sin - 56.3^\circ} = 36.1\text{ V}$$

$$E_{\text{polar}} = 36.1 \underline{/-56.3^\circ}\text{ V}$$

☐

In Figure 20-9(b) we show the relationship between X_C, R, and Z. Z in polar and rectangular forms is

$$Z \underline{/\theta} = R - jX_C$$

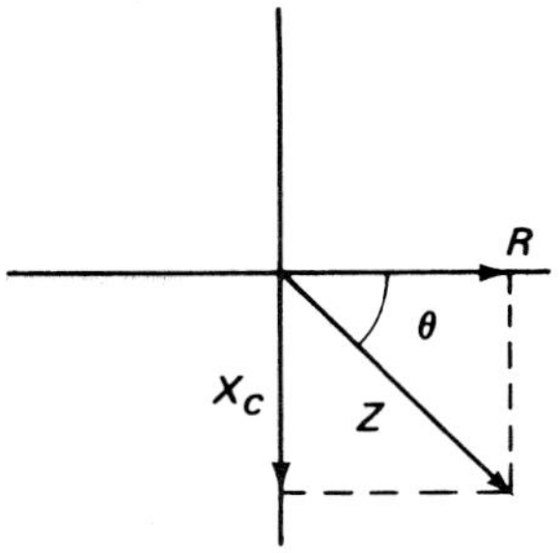

FIGURE 20-9(b)

☐ EXAMPLE 20-14

Let $R = 7.5$ kΩ and $\theta = -27.5^\circ$ in Figure 20-9(b). Express Z in both polar and rectangular forms.

Solution

$$\cos\theta = \frac{R}{Z}$$

$$Z = \frac{R}{\cos\theta} = \frac{7.5\text{ k}\Omega}{\cos -27.5^\circ} = 8.46\text{ k}\Omega$$

$$Z_{\text{polar}} = 8.46 \underline{/-27.5^\circ}\text{ k}\Omega$$

$$X_C = Z\sin\theta = 8.46\text{ k}\Omega \times \sin -27.5^\circ = -3.9\text{ k}\Omega$$

$$Z_{\text{rect}} = 7.5\text{ k}\Omega - j3.9\text{ k}\Omega$$

☐

In the *RL* circuit of Figure 20-10 we could express E in both polar and rectangular forms.

$$E\,\underline{/\theta} = V_R + jV_L$$

□ EXAMPLE 20-15

Let $V_R = 15$ V and $V_L = 10$ V in Figure 20-10(a). Find E and θ. Express E in both polar and rectangular forms.

Solution

$$E_{\text{rect}} = 15\text{ V} + j10\text{ V}$$

$$\tan\theta = \frac{V_L}{V_R} = \frac{10\text{ V}}{15\text{ V}} = 0.667$$

$$\tan^{-1} 0.667 = \theta = 33.7^\circ$$

$$\cos\theta = \frac{V_R}{E}$$

$$E = \frac{V_R}{\cos\theta} = \frac{15\text{ V}}{\cos 33.7^\circ} = 18\text{ V}$$

$$E_{\text{polar}} = 18\,\underline{/33.7^\circ}\text{ V}$$

□

In Figure 20-10(c) we show the relationship between X_L, R, and Z. Z in polar and rectangular forms is

$$Z\,\underline{/\theta} = R + jX_L$$

□ EXAMPLE 20-16

Let $X_L = 27$ kΩ and $\theta = 55^\circ$ in Figure 20-10. Express Z in both polar and rectangular forms.

Solution

$$\sin\theta = \frac{X_L}{Z}$$

$$Z = \frac{X_L}{\sin\theta} = \frac{27\text{ k}\Omega}{\sin 55^\circ} = 33\text{ k}\Omega$$

$$Z_{\text{polar}} = 33\,\underline{/55^\circ}\text{ k}\Omega$$

$$R = Z\cos\theta = 33\text{ k}\Omega\cos 55^\circ = 18.9\text{ k}\Omega$$

$$Z_{\text{rect}} = 18.9\text{ k}\Omega + j27\text{ k}\Omega$$

□

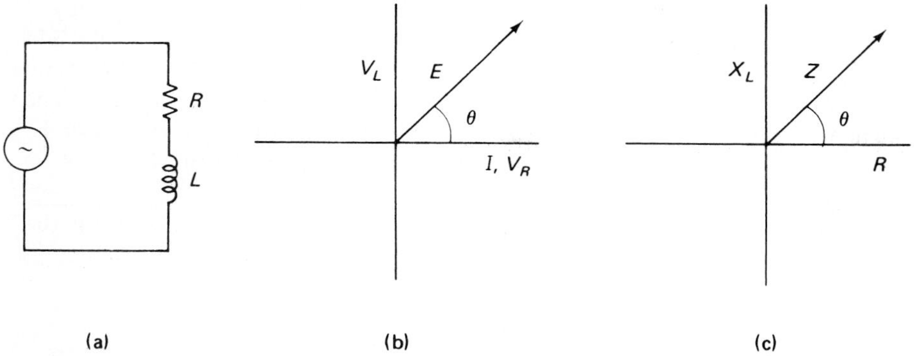

FIGURE 20-10

PRACTICE PROBLEMS 20-5

Express your answers to the following problems in (a) polar form and (b) rectangular form:

1. $R = 3.3\ \text{k}\Omega$
   $\theta = 30°$
2. $X_C = 12.8\ \text{k}\Omega$
   $\theta = -60°$
3. $Z = 17\ \text{k}\Omega$
   $X_L = 10\ \text{k}\Omega$
4. $R = 270\ \Omega$
   $X_L = 500\ \Omega$
5. $V_C = 8.6\ \text{V}$
   $\theta = -35°$
6. $V_R = 7.3\ \text{V}$
   $\theta = 47°$
7. $E = 15\ \text{V}$
   $\theta = -25°$
8. $V_R = 10.7\ \text{V}$
   $V_L = 6.3\ \text{V}$

Solutions

1. $Z_{\text{polar}} = 3.81\ \angle 30°\ \text{k}\Omega$
   $Z_{\text{rect}} = 3.3\ \text{k}\Omega + j1.91\ \text{k}\Omega$
2. $Z_{\text{polar}} = 14.8\ \angle -60°\ \text{k}\Omega$
   $Z_{\text{rect}} = 7.39\ \text{k}\Omega - j12.8\ \text{k}\Omega$
3. $Z_{\text{polar}} = 17\ \angle 36°\ \text{k}\Omega$
   $Z_{\text{rect}} = 13.7\ \text{k}\Omega + j10\ \text{k}\Omega$
4. $Z_{\text{polar}} = 568\ \angle 61.6°\ \Omega$
   $Z_{\text{rect}} = 270\ \Omega + j500\ \Omega$
5. $E_{\text{polar}} = 15\ \angle -35°\ \text{V}$
   $E_{\text{rect}} = 12.3\ \text{V} - j8.6\ \text{V}$
6. $E_{\text{polar}} = 10.7\ \angle 47°\ \text{V}$
   $E_{\text{rect}} = 7.3\ \text{V} + j7.83\ \text{V}$
7. $E_{\text{polar}} = 15\ \angle -25°\ \text{V}$
   $E_{\text{rect}} = 13.6\ \text{V} - j6.34\ \text{V}$
8. $E_{\text{polar}} = 12.4\ \angle 30.5°\ \text{V}$
   $E_{\text{rect}} = 10.7\ \text{V} + j6.3\ \text{V}$

Additional practice problems are at the end of the chapter.

20-3-2 Addition and Subtraction of Phasors

KEY POINT Only phasors that lie along the same line can be added or subtracted directly in polar form. Phasors not on the same line must be added in rectangular form.

For example, in Figure 20-11 we have plotted a number of phasors on a system of rectangular coordinates. These phasors are labeled **A** through **G.** We could add phasors **A** and **B** algebraically. The resultant phasor would be 2 units long and would fall along the positive *x*-axis. Phasors **E** and **G** could be added together. The resultant phasor would be 2 units long and would be in the direction of phasor **E**. Since phasor **F** does not fall on the same plane as any other phasor, it cannot be added to any other phasor.

Consider the phasors labeled Z_1 and Z_2 in Figure 20-12. In polar form the impedances are

$$Z_1 = 1 \angle 30^\circ \text{ k}\Omega$$
$$Z_2 = 1.5 \angle 50^\circ \text{ k}\Omega$$

We cannot find Z_T by adding Z_1 and Z_2 because they don't lie along the same line. However, their rectangular components do fall along the same line. R_1, the side adjacent to θ_1, and R_2, the side adjacent to θ_2, fall on the same line. XL_1, the side opposite θ_1, and XL_2, the side opposite θ_2, fall on the same line. Therefore, if we change Z_1 and Z_2 to rectangular form, we can add the rectangular components. The result of this addition is Z_T in rectangular form. This is easily converted to polar form so that we know the magnitude of Z_T and the phase angle.

$$Z_1 = 1 \angle 30^\circ \text{ k}\Omega = 866\ \Omega + j500\ \Omega$$
$$Z_2 = 1.5 \angle 50^\circ \text{ k}\Omega = 964\ \Omega + j1150\ \Omega$$
$$Z_T = 1830\ \Omega + j1650\ \Omega$$
$$Z_T = 2.46 \angle 42^\circ \text{ k}\Omega$$

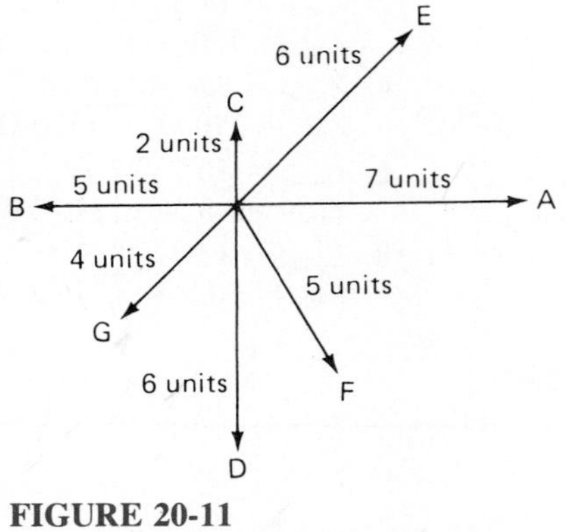

FIGURE 20-11

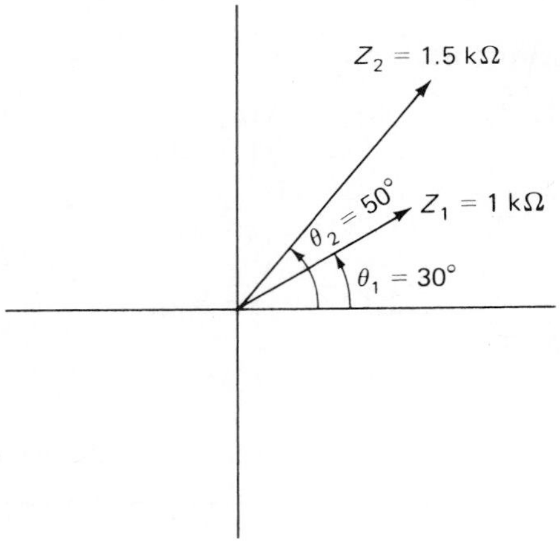

FIGURE 20-12

□ EXAMPLE 20-17

$Z_1 = 3 \angle 30^\circ$ kΩ and $Z_2 = 2 \angle -60^\circ$ kΩ. Find Z_T. Express your answer in polar form.

Solution

$$Z_1 = 2\,\underline{/30^\circ}\text{ k}\Omega \quad = 1.73\text{ k}\Omega + j1.00\text{ k}\Omega$$

$$Z_2 = 2\,\underline{/-60^\circ}\,k\Omega = \underline{1.00\text{ k}\Omega - j1.73\text{ k}\Omega}$$

$$Z_T = 2.73\text{ k}\Omega - j0.73\text{ k}\Omega$$

$$Z_T = 2.83\,\underline{/-15.0^\circ}\text{ k}\Omega$$

□

PRACTICE PROBLEMS 20-6

Find Z_T in the following problems. Express your answers in polar form.

1. $Z_1 = 17\,\underline{/60^\circ}$ kΩ, $Z_2 = 30\,\underline{/20^\circ}$ kΩ
2. $Z_1 = 270\,\underline{/-35^\circ}$ Ω, $Z_2 = 1\,\underline{/65^\circ}$ kΩ
3. $Z_1 = 100\,\underline{/-70^\circ}$ kΩ, $Z_2 = 100\,\underline{/-40^\circ}$ kΩ
4. $Z_1 = 48\,\underline{/20^\circ}$ kΩ, $Z_2 = 70\,\underline{/-65^\circ}$ kΩ
5. $Z_1 = 6.8\,\underline{/70^\circ}$ kΩ, $Z_2 = 3\,\underline{/25^\circ}$ kΩ

Solutions

1. $44.4\,\underline{/34.3^\circ}$ kΩ
2. $990\,\underline{/49.4^\circ}$ Ω
3. $193\,\underline{/-55^\circ}$ kΩ
4. $88.3\,\underline{/-32.2^\circ}$ kΩ
5. $9.17\,\underline{/56.6^\circ}$ kΩ

Additional practice problems are at the end of the chapter.

20-3-3 Equivalent Series Circuit

■ **KEY POINT** The equivalent series circuit (ESC) of Components in Series is the resultant of the vector addition of the circuit *R*, *C*, and *L*.

Any passive ac network may be reduced to a resistance in series with some reactance. The reactance may be either capacitive or inductive depending on the component values. In some circuits the equivalent circuit is purely resistive. Circuits such as these will be discussed later in the chapter.

In a circuit such as the one in Figure 20-13(a), the total resistance is the sum of the individual resistances. $R_T = R_1 + R_2 = 12$ kΩ. Once we have found R_T, we can find Z.

$$Z_{\text{rect}} = 12\text{ k}\Omega + j15\text{ k}\Omega$$

$$Z_{\text{polar}} = 19.2\,\underline{/51.3^\circ}\text{ k}\Omega$$

The equivalent series circuit (ESC) is shown in Figure 20-13(b). In a circuit such as the one in Figure 20-14, we could express Z as the sum of all the resistances and reactances. All the resistances are added together to get R_T. The reactances are added

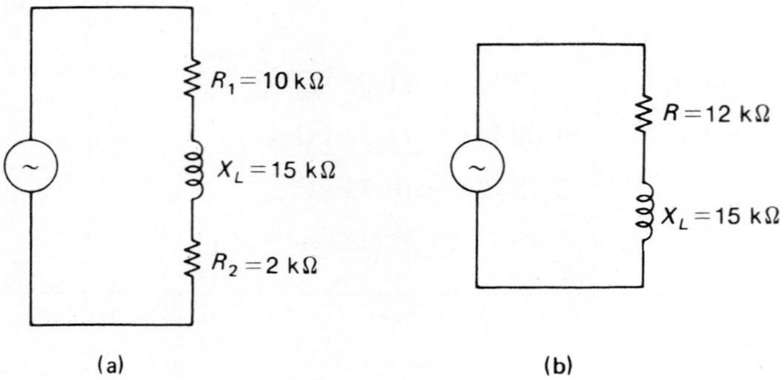

FIGURE 20-13

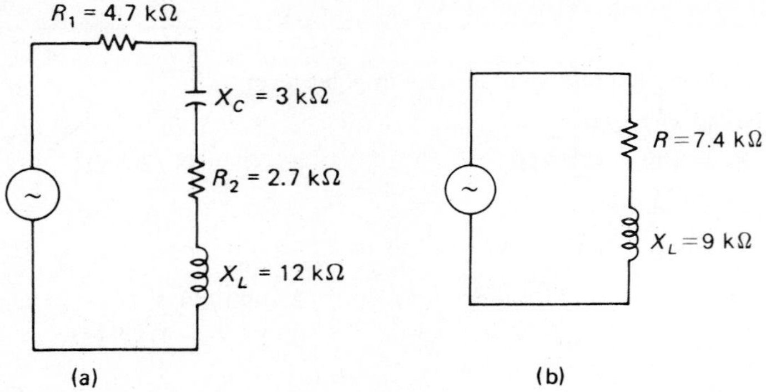

FIGURE 20-14

together to find the resultant reactance. Since X_C and X_L are 180° out of phase, the total reactance will be the difference between X_L and X_C.

$$\begin{aligned} Z_{\text{rect}} &= R_1 + R_2 + jX_L - jX_C \\ &= 7.4\text{ k}\Omega + j12\text{ k}\Omega - j3\text{ k}\Omega \\ &= 7.4\text{ k}\Omega + j9\text{ k}\Omega \\ Z_{\text{polar}} &= 11.7\ \underline{/50.6^\circ}\text{ k}\Omega \end{aligned}$$

The ESC is shown in Figure 20-14(b).

□ EXAMPLE 20-18

In Figure 20-14(a), let $f = 2.5$ kHz, $C = 10$ nF, $L = 600$ mH, $R_1 = 3.3$ kΩ, and $R_2 = 2$ kΩ. Find the ESC. Include component values.

Solution

$$X_L = 2\pi fL = 9.42\ \text{k}\Omega$$

$$X_C = \frac{1}{2\pi fC} = 6.37\ \text{k}\Omega$$

$$Z_{\text{rect}} = 3.3\ \text{k}\Omega + 2\ \text{k}\Omega + j9.42\ \text{k}\Omega - j6.37\ \text{k}\Omega$$

$$= 5.3\ \text{k}\Omega + j3.05\ \text{k}\Omega$$

The ESC includes a resistance of 5.3 kΩ and an inductive reactance of 3.05 kΩ. The equivalent inductance is

$$L = \frac{X_L}{2\pi f} = \frac{3.05\ \text{k}\Omega}{2\pi \times 2.5\ \text{kHz}} = 194\ \text{mH}$$

The ESC is shown in Figure 20-15.

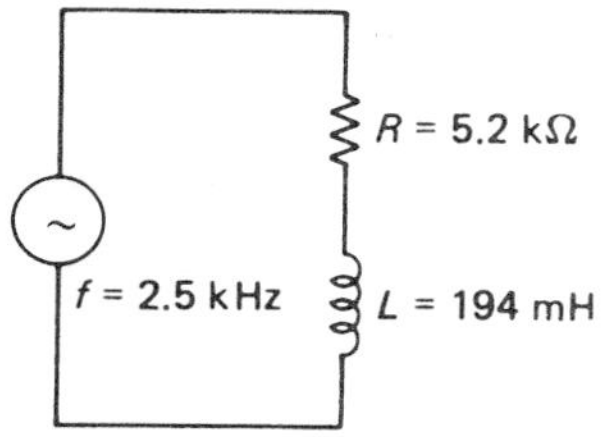

FIGURE 20-15

□

PRACTICE PROBLEMS 20-7

1. Refer to Figure 20-13(a). Find the ESC if R_1 = 27 kΩ, R_2 = 12 kΩ, and X_L = 30 kΩ. Express Z in both polar and rectangular forms.
2. Refer to Figure 20-14(a). Let R_1 = 1.5 kΩ, R_2 = 1.2 kΩ, X_L = 1.5 kΩ, and X_C = 2.7 kΩ. Find the ESC. Express Z in both polar and rectangular forms.
3. Refer to Figure 20-14(a). Let R_1 = 1.5 kΩ, R_2 = 2.7 kΩ, C = 50 nF, L = 800 mH, and f = 400 Hz. Draw the ESC. Include component values.
4. Refer to Figure 20-16. Find Z_{rect}, Z_{polar}, V_R, V_C, V_L, and I. Draw the ESC.

Solutions

1. $Z_{\text{polar}} = 49.2\ \underline{/37.6^\circ}$ kΩ
   Z_{rect} = 39 kΩ + j30 kΩ. See Figure 20-17.
2. Z_{rect} = 2.7 kΩ − j1.2 kΩ
   $Z_{\text{polar}} = 2.95\ \underline{/-24^\circ}$ kΩ. See Figure 20-18.
3. Z_{rect} = 4.2 kΩ − j5.95 kΩ
   $Z_{\text{polar}} = 7.28\ \underline{/-54.8^\circ}$ kΩ. See Figure 20-19.
4. Z_{rect} = 1 kΩ − j2.17 kΩ
   $Z_{\text{polar}} = 2.39\ \underline{/-65.3^\circ}$ kΩ. See Figure 20-20.
   V_R = 4.19 V
   V_C = 9.08 V
   I = 4.18 mA

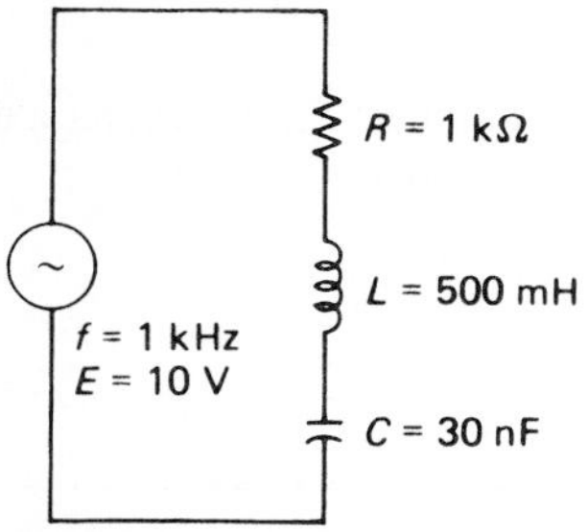

FIGURE 20-16

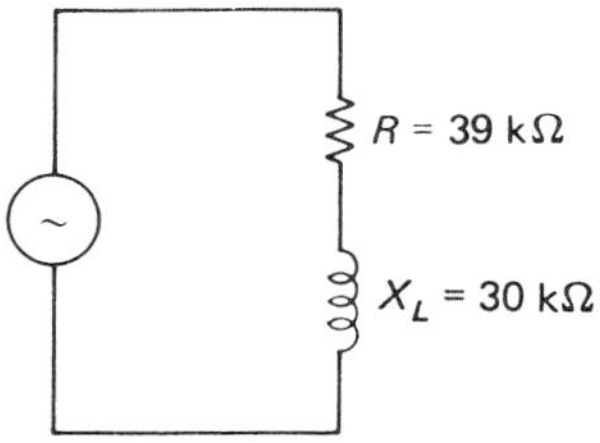

FIGURE 20-17

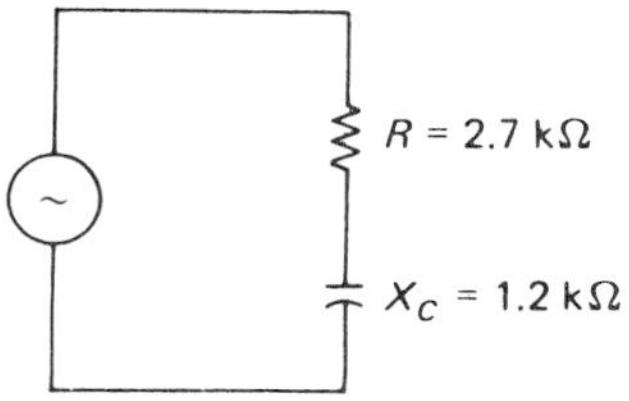

FIGURE 20-18

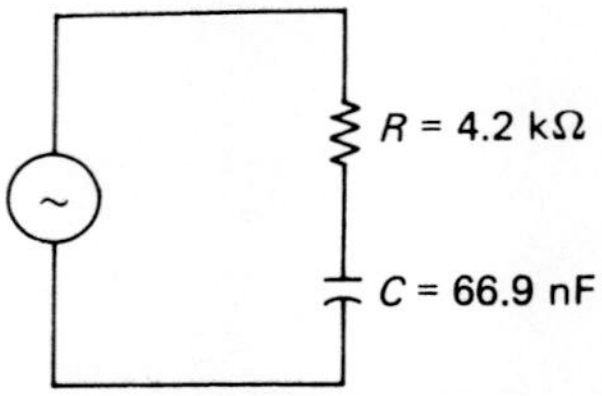

FIGURE 20-19

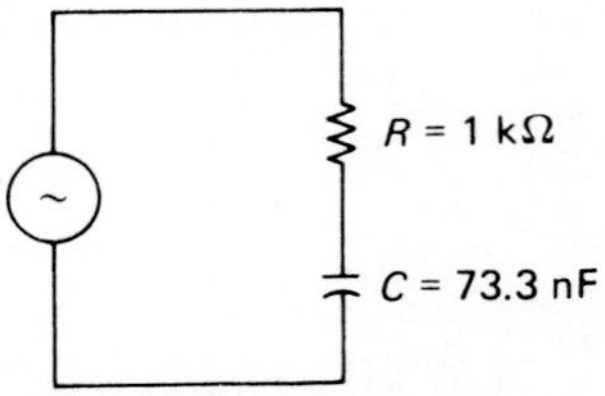

FIGURE 20-20

Additional practice problems are at the end of the chapter.

20-4 SERIES RESONANCE

In circuits that include capacitance and inductance, there is a frequency where $X_L = X_C$. This frequency is called the *resonant frequency*. At this frequency (labeled f_r) the phase angle is 0°. The circuit is purely resistive because the resultant reactance is 0 Ω. The impedance is equal to R. If Z equals R, then Z must be at its minimum value. This makes I equal to its maximum value. So far we have listed the following conditions at resonance:

$$X_C = X_L$$

$$\theta = 0°$$

$$Z = R \text{ and is minimum}$$

$$I \text{ is maximum}$$

The voltage drop across R must equal the applied voltage. The voltage drop across C would equal the voltage drop across L.

$$V_R = E$$

$$V_C = V_L$$

The resonant frequency can be determined in the following manner: Since resonance is defined as that frequency where $X_C = X_L$, then by substitution we can write

$$\frac{1}{2\pi f C} = 2\pi f L$$

Solving for f, we get

$$f = f_r = \frac{1}{2\pi\sqrt{LC}}$$

☐ EXAMPLE 20-19

In Figure 20-21(a) find f_r; at this frequency find V_R, V_L, V_C, X_L, X_C, Z, and I. Draw the ESC.

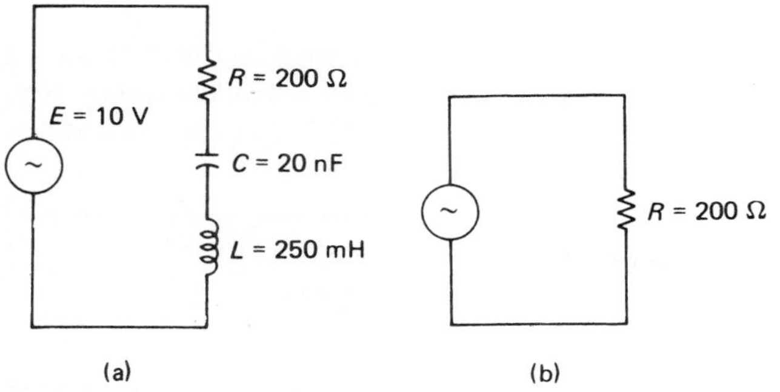

FIGURE 20-21

Solution

$$f_r = \frac{1}{2\pi\sqrt{LC}} = 2.25 \text{ kHz}$$

$$= \frac{1}{2 \times \pi \times \sqrt{250 \times 10^{-3} \times 20 \times 10^{-9}}} = 2.25 \text{ kHz}$$

Using the calculator,

2 [×] [π] [×] [(] 250 [EE] [+/-] 3 [×] 20 [EE] [+/-] 9 [)] [√x] [=] [1/x]

Answer: $f_r = 2.25$ kHz

Knowing f_r, we can solve the rest of the problem by using Ohm's law or trig functions.

$$X_L = 2\pi fL = 3.53 \text{ k}\Omega$$

$$X_C = \frac{1}{2\pi fC} = 3.54 \text{ k}\Omega$$

Our values of X_L and X_C are not exactly equal because we rounded f_r to 2.25 kHz. To five places, $f_r = 2.2508$ kHz, which makes $X_C = X_L = 3.5355$ kΩ. Let's stick to our three-place accuracy and remember that some slight differences may be the result. If

$$X_C = X_L \cong 3.54 \text{ k}\Omega$$

then Z is

$$Z_{\text{rect}} = 200\ \Omega + j3.54 \text{ k}\Omega - j3.54 \text{ k}\Omega$$

$$= 200\ \Omega + j0$$

$$Z_{\text{polar}} = 200 \underline{/0^\circ}\ \Omega$$

$$I = \frac{E}{Z} = \frac{10 \text{ V}}{200\ \Omega} = 50 \text{ mA}$$

$$V_R = IR = 10 \text{ V}$$

$$V_C = IX_C = 177 \text{ V}$$

$$V_L = IX_L = 177 \text{ V}$$

The ESC is shown in Figure 20-21(b).

The fact that V_C and V_L are both greater than E does not mean we got something for nothing, and it does not mean that we have violated Kirchhoff's laws. Remember, V_C and V_L are 180° out of phase. They are not 177 V at the same time. Added vectorially, the resultant voltage drop is 0 V. □

Let's see what conditions exist above and below resonance.

□ EXAMPLE 20-20

Assume that the operating frequency in Figure 20-21 is 2.5 kHz. Find X_L, X_C, V_R, V_L, V_C, Z, θ, and I. Draw the ESC.

Solution

$$X_C = \frac{1}{2\pi fC} = 3.18\ \text{k}\Omega$$

$$X_L = 2\pi fL = 3.93\ \text{k}\Omega$$

$$Z_{\text{rect}} = 200\ \Omega + j3.93\ \text{k}\Omega - j3.18\ \text{k}\Omega$$

$$= 200\ \Omega + j750\ \Omega$$

$$Z_{\text{polar}} = 776\ \underline{/75.1^\circ}\ \Omega$$

$$I = \frac{E}{Z} = \frac{10\ \text{V}}{776\ \Omega} = 12.9\ \text{mA}$$

$$V_L = 50.8\ \text{V}$$

$$V_C = 41.1\ \text{V}$$

$$V_R = 2.59\ \text{V}$$

The ESC is shown in Figure 20-22. The resultant reactance is 750 Ω and is inductive. This inductance is

$$L = \frac{X_L}{2\pi f} = 47.7\ \text{mH}$$

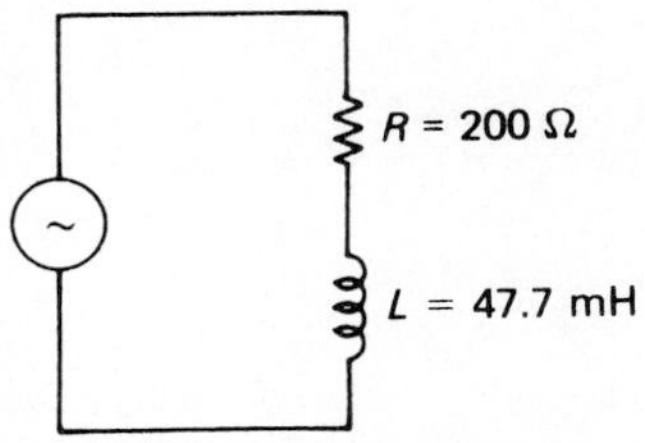

FIGURE 20-22

□ EXAMPLE 20-21

Change the operating frequency to 2 kHz in Figure 20-21. Find X_L, X_C, V_R, V_L, V_C, Z, θ, and I. Draw the ESC.

Solution

$$X_L = 2\pi fL = 3.14\ \text{k}\Omega$$

$$X_C = \frac{1}{2\pi fC} = 3.98\ \text{k}\Omega$$

$$Z_{\text{rect}} = 200\ \Omega + j3.14\ \text{k}\Omega - j3.98\ \text{k}\Omega$$

$$= 200\ \Omega - j840\ \Omega$$

$$Z_{\text{polar}} = 863\ \angle -76.6^\circ\ \Omega$$

$$I = \frac{E}{Z} = \frac{10\ \text{V}}{863\ \Omega} = 11.6\ \text{mA}$$

$$V_R = IR = 2.32\ \text{V}$$

$$V_C = IX_C = 46.1\ \text{V}$$

$$V_L = IX_L = 36.4\ \text{V}$$

The ESC is shown in Figure 20-23. The resultant reactance is 840 Ω and is capacitive. The capacitance is

$$C = \frac{1}{2\pi fX_C} = 94.7\ \text{nF}$$

□

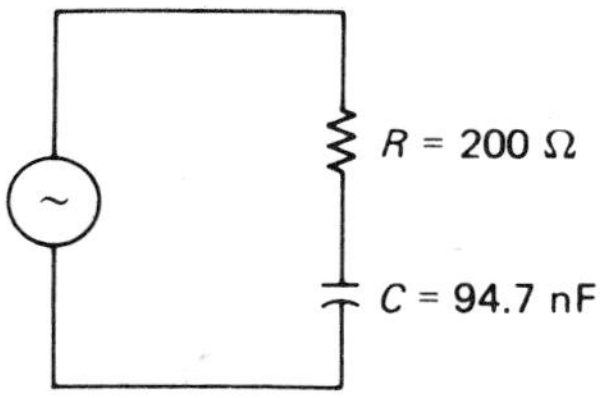

FIGURE 20-23

The examples above show that at frequencies above resonance the equivalent circuit is *RL*. Below resonance, the equivalent circuit is *RC*. The impedance increased as we moved away from resonance, and the current decreased.

PRACTICE PROBLEMS 20-8

Refer to Figure 20-21.

1. $R = 500\ \Omega, C = 30\ \text{nF}, L = 100\ \text{mH}$, and $E = 25$ V. Find f_r. Determine X_L, X_C, V_R, V_C, V_L, Z, θ, I, and the ESC at (a) $f = f_r$, (b) $f = 3.5$ kHz, and (c) $f = 2.7$ kHz.

2. $R = 1\ \text{k}\Omega$, $C = 500\ \text{pF}$, $L = 50\ \text{mH}$, and $E = 15$ V. Find f_r. Determine X_L, X_C, V_R, V_C, V_L, Z, θ, I, and the ESC at (a) $f = f_r$, (b) $f = 33$ kHz, and (c) $f = 29$ kHz.

Solutions

1. $f_r = 2.91$ kHz
(a) $X_L = 1.83\ \text{k}\Omega$, $X_C = 1.83\ \text{k}\Omega$, $Z_{\text{rect}} = 500\ \Omega + j0$, $Z_{\text{polar}} = 500\ \angle 0^\circ\ \Omega$, $I = 50$ mA, $V_R = 25$ V, $V_C = -91.5$ V, and $V_L = 91.5$ V. The ESC consists of 500 Ω of resistance.

2. $f_r = 31.8$ kHz
(a) $X_L = 10\ \text{k}\Omega$, $X_C = 10\ \text{k}\Omega$, $Z_{\text{rect}} = 1\ \text{k}\Omega + j0$, $Z_{\text{polar}} = 1\ \angle 0^\circ\ \text{k}\Omega$, $I = 15$ mA, $V_R = 15$ V, $V_L = 150$ V, and $V_C = -150$ V. The ESC consists of 1 kΩ of resistance.

(b) $X_L = 2.2$ kΩ, $X_C = 1.52$ kΩ, $Z_{rect} = 500\ \Omega + j684\ \Omega$, $Z_{polar} = 847\ \angle 53.8^\circ\ \Omega$, $I = 29.5$ mA, $V_R = 14.8$ V, $V_L = 64.9$ V, and $V_C = -44.9$ V. The ESC consists of 500 Ω of resistance and 31.1 mH of inductance.

(c) $X_L = 1.7$ kΩ, $X_C = 1.96$ kΩ, $Z_{rect} = 500\ \Omega - j265\ \Omega$, $Z_{polar} = 556\ \angle -27.9^\circ\ \Omega$, $I = 44.2$ mA, $V_R = 22.1$ V, $V_L = 75.1$ V, and $V_C = -86.6$ V. The ESC consists of 500 Ω of resistance and 222 nF of capacitance.

(b) $X_L = 10.4$ kΩ, $X_C = 9.65$ kΩ, $Z_{rect} = 1\ \text{k}\Omega + j750\ \Omega$, $Z_{polar} = 1.25\ \angle 36.9^\circ$ kΩ, $I = 12$ mA, $V_R = 12$ V, $V_L = 125$ V, and $V_C = -116$ V. The ESC consists of 1 kΩ of resistance and 3.62 mH of inductance.

(c) $X_L = 9.11$ kΩ, $X_C = 11$ kΩ, $Z_{rect} = 1\ \text{k}\Omega - j1.87$ kΩ, $Z_{polar} = 2.12\ \angle -61.8^\circ$ kΩ, $I = 7.08$ mA, $V_R = 7.08$ V, $V_L = 64.5$ V, and $V_C = -77.8$ V. The ESC consists of 1 kΩ of resistance and 2.93 nF of capacitance.

Additional practice problems are at the end of the chapter.

SELF-TEST 20-2

1. $X_L = 11.3$ kΩ, $\theta = 27.5^\circ$. Find Z in both polar and rectangular forms.
2. $Z_1 = 2.7\ \angle 17.5^\circ$ kΩ, $Z_2 = 4\ \angle -58^\circ$ kΩ. Find Z_T.
3. Refer to Figure 20-24. Let $R_1 = 7.5$ kΩ, $R_2 = 4.7$ kΩ, $X_L = 20$ kΩ, and $X_C = 3.8$ kΩ. Find the ESC. Express Z in both polar and rectangular forms.
4. Refer to Figure 20-25. Let $R = 1.2$ kΩ, $C = 10$ nF, $L = 500$ mH, $E = 20$ V, and $f = 2.1$ kHz. Determine X_L, X_C, V_R, V_L, V_C, Z, θ, I, and the ESC.

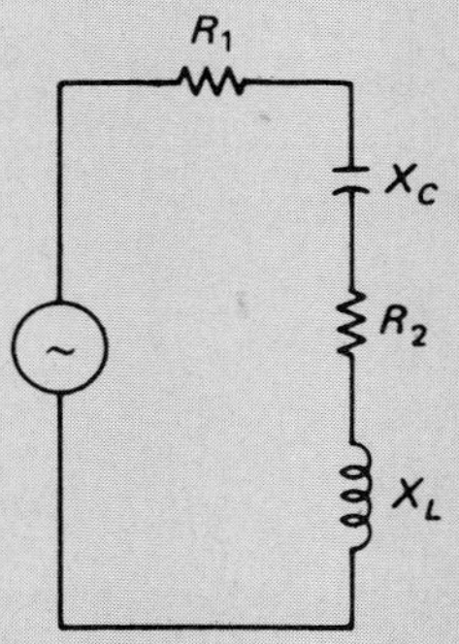

FIGURE 20-24

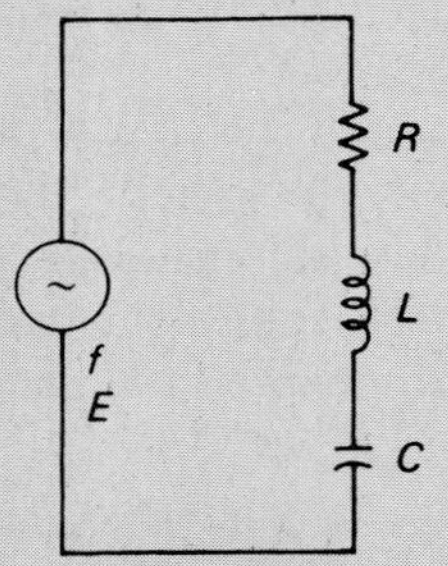

FIGURE 20-25

Answers to Self-test 20-2 are at the end of the chapter.

CHAPTER 20 AT A GLANCE

Page	Key Points and Equations	Example
500	In a series circuit, we use I as the reference. In an RC circuit, current leads, so V_C and	Figure 20-26

Page	Key Points and Equations	Example
	X_C are plotted along the negative y-axis. **E** (or **Z**) is the hypotenuse, and θ is the phase angle.	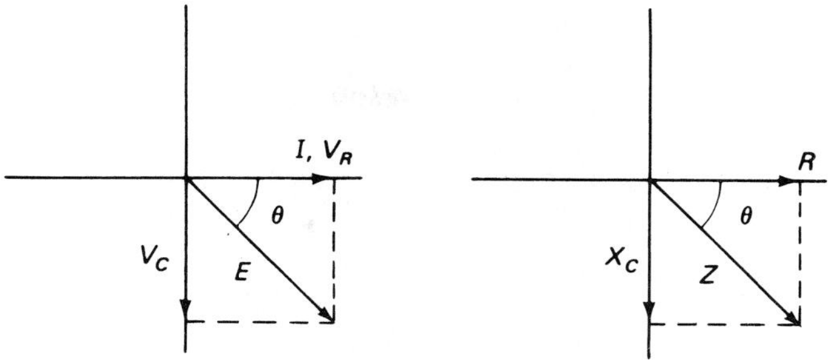 **FIGURE 20-26**
505	*Key point:* In a series circuit, we use I as the reference. In an RL circuit, voltage leads, so V_L and X_L are plotted along the positive y-axis. E (or Z) is the hypotenuse, and θ is the phase angle.	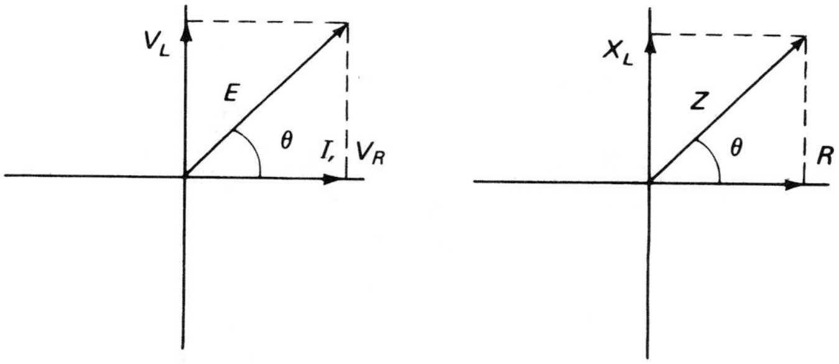 **FIGURE 20-27**
511	Phasors are quantities that have both magnitude and direction. They can be represented in either polar form or rectangular form.	$r_{\text{POLAR}} = 50 \angle -30^\circ$ $r_{\text{RECT}} = 43.3 - j25$
512	To use the calculator in converting between polar and rectangular forms, use the $\boxed{X\leftrightarrow Y}$ and $\boxed{R\rightarrow P}$ keys.	Convert $3 + j4$ to polar form: 3 $\boxed{X\leftrightarrow Y}$ 4 $\boxed{R\rightarrow P}$ *Answer:* 5 $\boxed{X\leftrightarrow Y}$ *Answer:* 53.1 $3 + j4 = 5 \angle 53.1^\circ$
515	Only phasors that lie along the same line can be added or subtracted directly in polar form. Phasors not on the same line must be added in rectangular form.	Add $Z_1 = 1 \angle 30^\circ$ kΩ and $Z_2 = 1.5 \angle 50^\circ$ kΩ. $Z_1 = 1 \angle 30^\circ \text{ k}\Omega = 866\ \Omega + j500\ \Omega$ $Z_2 = 1.5 \angle 50^\circ \text{ k}\Omega = 964\ \Omega + j1150\ \Omega$ $Z_r = 1830\ \Omega + j1650\ \Omega$ $Z_r = 2.46 \angle 42^\circ$ kΩ

Page	Key Points and Equations	Example
517	The equivalent series circuit (ESC) of components in series is the resultant of the vector addition of the circuit R, C, and L.	Find the ESC of the circuit in Figure 20-24. $\text{ESC} = 7.4\ \text{k}\Omega + 12\ \text{k}\Omega - 3\ \text{k}\Omega = 7.4\ \text{k}\Omega + j9\ \text{k}\Omega = 11.7\ \underline{/50.6^\circ}\ \text{k}\Omega$
521	Resonance in a series RCL circuit can be found using the equation $f_r = \dfrac{1}{2\pi\sqrt{LC}}$	Find f_r if $L = 250$ mH and $C = 20$ nF. $f_r = \dfrac{1}{2\pi \times \sqrt{250\ \text{mH} \times 20\ \text{nF}}} = 2.25$ kHz

End of Chapter Problems 20-1

Refer to Figure 20-28 for the following problems:

1. $V_R = 20$ V, $V_C = -20$ V. Find **E** and θ.
2. $V_R = 2.3$ V, $V_C = -1$ V. Find **E** and θ.
3. $V_R = 12$ V, $V_C = -8$ V. Find **E** and θ.
4. $V_R = 36$ V, $V_C = -45$ V. Find **E** and θ.
5. $V_R = 6.35$ V, $E = 10$ V. Find V_C and θ.
6. $V_R = 21.3$ V, **E** $= 30$ V. Find V_C and θ.
7. $V_R = 21$ V, $E = 40$ V. Find V_C and θ.
8. $V_R = 6$ V, **E** $= 10$ V. Find V_C and θ.
9. $V_R = 15$ V, $\theta = -40^\circ$. Find **E** and V_C.
10. $V_R = 28$ V, $\theta = -70^\circ$. Find **E** and V_C.
11. $V_R = 5.5$ V, $\theta = -25^\circ$. Find **E** and V_C.
12. $V_R = 14$ V, $\theta = -60^\circ$. Find **E** and V_C.
13. **E** $= 18$ V, $\theta = -35.6^\circ$. Find V_R and V_C.
14. **E** $= 5$ V, $\theta = -40^\circ$. Find V_R and V_C.
15. **E** $= 70$ V, $\theta = -27^\circ$. Find V_R and V_C.
16. **E** $= 18$ V, $\theta = -50^\circ$. Find V_R and V_C.
17. **E** $= 30$ V, $V_C = -20$ V. Find V_R and θ.
18. **E** $= 9$ V, $V_C = -3$ V. Find V_R and θ.
19. **E** $= 50$ V, $V_C = -40$ V. Find V_R and θ.
20. **E** $= 10$ V, $V_C = -6$ V. Find V_R and θ.
21. $V_C = -5.5$ V, $\theta = -32^\circ$. Find **E** and V_R.
22. $V_C = -14$ V, $\theta = -55^\circ$. Find **E** and V_R.
23. $V_C = -30$ V, $\theta = -70^\circ$. Find **E** and V_R.
24. $V_C = -2.3$ V, $\theta = -20^\circ$. Find **E** and V_R.

End of Chapter Problems 20-2

Refer to Figure 20-28 for the following problems and find X_C, Z, and θ:

1. $R = 10\ \text{k}\Omega$, $C = 50$ nF, $f = 500$ Hz
2. $R = 4.7\ \text{k}\Omega$, $C = 10$ nF, $f = 3.39$ kHz
3. $R = 700\ \Omega$, $C = 1\ \mu$F, $f = 200$ Hz
4. $R = 18\ \text{k}\Omega$, $C = 200$ pF, $f = 50$ kHz
5. $R = 2\ \text{k}\Omega$, $C = 20$ nF, $f = 2.5$ kHz
6. $R = 1\ \text{k}\Omega$, $C = 1$ nF, $f = 200$ kHz
7. $R = 27\ \text{k}\Omega$, $C = 5$ nF, $f = 1.55$ kHz
8. $R = 12\ \text{k}\Omega$, $C = 30$ nF, $f = 720$ Hz
9. $R = 3.3\ \text{k}\Omega$, $C = 0.1\ \mu$F, $f = 2.3$ kHz
10. $R = 680\ \Omega$, $C = 2\ \mu$F, $f = 80$ Hz

Refer to Figure 20-28 for the following problems and find X_C, Z, θ, V_R, V_C, and I:

11. $R = 2\ \text{k}\Omega$, $C = 40$ nF, $f = 3$ kHz, **E** $= 30$ V
12. $R = 25\ \text{k}\Omega$, $C = 470$ pF, $f = 10$ kHz, **E** $= 5$ V
13. $R = 10\ \text{k}\Omega$, $C = 100$ pF, $f = 100$ kHz, **E** $= 10$ V
14. $R = 200\ \Omega$, $C = 2\ \mu$F, $f = 200$ Hz, **E** $= 20$ V

15. $R = 3.3\ k\Omega$, $C = 20\ nF$, $f = 1.5\ kHz$, $\mathbf{E} = 30\ V$
16. $R = 560\ \Omega$, $C = 0.5\ \mu F$, $f = 1\ kHz$, $\mathbf{E} = 50\ V$
17. $R = 10\ k\Omega$, $C = 500\ pF$, $f = 6.5\ kHz$, $\mathbf{E} = 20\ V$
18. $R = 68\ \Omega$, $C = 300\ nF$, $f = 5\ kHz$, $\mathbf{E} = 10\ V$
19. $R = 470\ \Omega$, $C = 200\ nF$, $f = 2\ kHz$, $\mathbf{E} = 100\ V$
20. $R = 2.2\ k\Omega$, $C = 50\ nF$, $f = 2\ kHz$, $\mathbf{E} = 35\ V$

Refer to Figure 20-28 for the following problems and find I, θ, f, X_C, V_R, and V_C:

21. $R = 8.2\ k\Omega$, $C = 200\ pF$, $Z = 12\ k\Omega$, $\mathbf{E} = 40\ V$
22. $R = 1\ k\Omega$, $C = 50\ nF$, $Z = 2\ k\Omega$, $\mathbf{E} = 9\ V$
23. $R = 750\ \Omega$, $C = 10\ nF$, $Z = 1.2\ k\Omega$, $\mathbf{E} = 15\ V$
24. $R = 30\ k\Omega$, $C = 250\ pF$, $Z = 47\ k\Omega$, $\mathbf{E} = 10\ V$
25. $R = 2.7\ k\Omega$, $C = 30\ pF$, $Z = 3.5\ k\Omega$, $\mathbf{E} = 30\ V$
26. $R = 18\ k\Omega$, $C = 1\ nF$, $Z = 25\ k\Omega$, $\mathbf{E} = 5\ V$
27. $R = 56\ k\Omega$, $C = 25\ nF$, $Z = 64.4\ k\Omega$, $\mathbf{E} = 7.5\ V$
28. $R = 2.7\ k\Omega$, $C = 5\ \mu F$, $Z = 2.75\ k\Omega$, $\mathbf{E} = 9\ V$
29. $R = 810\ \Omega$, $C = 1\ \mu F$, $Z = 1.14\ k\Omega$, $\mathbf{E} = 12\ V$
30. $R = 1.8\ k\Omega$, $C = 2\ \mu F$, $Z = 1.97\ k\Omega$, $\mathbf{E} = 15\ V$

End of Chapter Problems 20-3

Refer to Figure 20-29 for the following problems:

1. $V_R = 8\ V$, $V_L = 10\ V$. Find $\mathbf{E}$ and θ.
2. $V_R = 22\ V$, $V_L = 30\ V$. Find $\mathbf{E}$ and θ.
3. $V_R = 14\ V$, $V_L = 20\ V$. Find $\mathbf{E}$ and θ.
4. $V_R = 38.7\ V$, $V_L = 19.3\ V$. Find $\mathbf{E}$ and θ.
5. $V_R = 6.7\ V$, $\theta = 27°$. Find $\mathbf{E}$ and V_L.
6. $V_R = 12.6\ V$, $\theta = 50°$. Find $\mathbf{E}$ and V_L.
7. $V_R = 27.4\ V$, $\theta = 67.4°$. Find $\mathbf{E}$ and V_L.
8. $V_R = 8.35\ V$, $\theta = 34.4°$. Find $\mathbf{E}$ and V_L.
9. $V_R = 8.35\ V$, $\mathbf{E} = 12\ V$. Find V_L and θ.
10. $V_R = 15\ V$, $\mathbf{E} = 30\ V$. Find V_L and θ.
11. $V_R = 43.7\ V$, $\mathbf{E} = 60\ V$. Find V_L and θ.
12. $V_R = 5.36\ V$, $\mathbf{E} = 12\ V$. Find V_L and θ.
13. $V_L = 15\ V$, $\mathbf{E} = 30\ V$. Find V_R and θ.
14. $V_L = 30\ V$, $\mathbf{E} = 35\ V$. Find V_R and θ.
15. $V_L = 10\ V$, $\mathbf{E} = 17\ V$. Find V_R and θ.
16. $V_L = 27.5\ V$, $\mathbf{E} = 38.5\ V$. Find V_R and θ.
17. $V_L = 3\ V$, $\theta = 62°$. Find $\mathbf{E}$ and V_R.
18. $V_L = 10\ V$, $\theta = 33°$. Find $\mathbf{E}$ and V_R.
19. $V_L = 15.5\ V$, $\theta = 30°$. Find $\mathbf{E}$ and V_R.
20. $V_L = 67.4\ V$, $\theta = 55.7°$. Find $\mathbf{E}$ and V_R.
21. $\mathbf{E} = 9\ V$, $\theta = 58°$. Find V_R and V_L.
22. $\mathbf{E} = 22\ V$, $\theta = 38°$. Find V_R and V_L.
23. $\mathbf{E} = 50\ V$, $\theta = 45°$. Find V_R and V_L.
24. $\mathbf{E} = 30\ V$, $\theta = 63°$. Find V_R and V_L.

End of Chapter Problems 20-4

Refer to Figure 20-29 and find X_L, Z, and θ in the following problems:

1. $R = 3.9\ k\Omega$, $L = 600\ mH$, $f = 1\ kHz$
2. $R = 12\ k\Omega$, $L = 150\ mH$, $f = 20\ kHz$
3. $R = 500\ \Omega$, $L = 40\ mH$, $f = 1.5\ kHz$
4. $R = 33\ k\Omega$, $L = 2\ H$, $f = 5\ kHz$

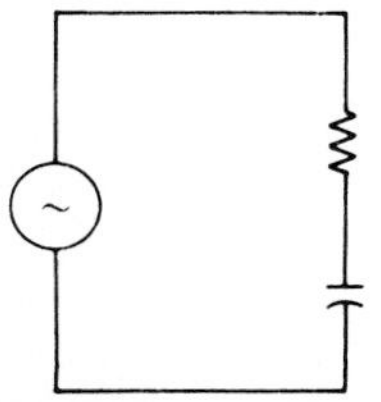

FIGURE 20-28

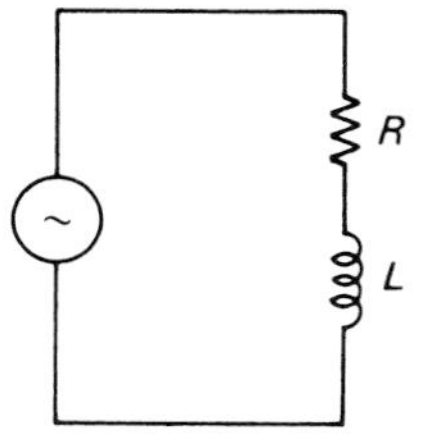

FIGURE 20-29

5. $R = 27\ \text{k}\Omega$, $L = 250\ \text{mH}$, $f = 8.5\ \text{kHz}$
6. $R = 56\ \Omega$, $L = 400\ \mu\text{H}$, $f = 17.5\ \text{kHz}$
7. $R = 1.8\ \text{k}\Omega$, $L = 10\ \text{mH}$, $f = 20\ \text{kHz}$
8. $R = 910\ \Omega$, $L = 650\ \mu\text{H}$, $f = 455\ \text{kHz}$

Refer to Figure 20-29 and find X_L, Z, θ, V_R, V_L, and I in the following problems:

9. $R = 1\ \text{k}\Omega$, $L = 500\ \text{mH}$, $f = 250\ \text{Hz}$, $\mathbf{E} = 15\ \text{V}$
10. $R = 22\ \text{k}\Omega$, $L = 200\ \text{mH}$, $f = 10\ \text{kHz}$, $\mathbf{E} = 5\ \text{V}$
11. $R = 300\ \Omega$, $L = 150\ \text{mH}$, $f = 500\ \text{Hz}$, $\mathbf{E} = 50\ \text{V}$
12. $R = 15\ \text{k}\Omega$, $L = 40\ \text{mH}$, $f = 10\ \text{kHz}$, $\mathbf{E} = 25\ \text{V}$
13. $R = 27\ \text{k}\Omega$, $L = 1.35\ \text{H}$, $f = 2.35\ \text{kHz}$, $\mathbf{E} = 20\ \text{V}$
14. $R = 150\ \text{k}\Omega$, $L = 2.3\ \text{H}$, $f = 12.7\ \text{kHz}$, $\mathbf{E} = 9\ \text{V}$
15. $R = 100\ \Omega$, $L = 500\ \mu\text{H}$, $f = 25\ \text{kHz}$, $\mathbf{E} = 12\ \text{V}$
16. $R = 470\ \Omega$, $L = 65\ \text{mH}$, $f = 1.73\ \text{kHz}$, $\mathbf{E} = 18\ \text{V}$

Refer to Figure 20-29 and find X_L, θ, f, $\mathbf{E}$, V_R, and I in the following problems:

17. $R = 4.7\ \text{k}\Omega$, $Z = 6\ \text{k}\Omega$, $L = 650\ \text{mH}$, $V_L = 7.5\ \text{V}$
18. $R = 56\ \text{k}\Omega$, $Z = 100\ \text{k}\Omega$, $L = 2\ \text{H}$, $V_L = 12\ \text{V}$
19. $R = 18\ \text{k}\Omega$, $Z = 32\ \text{k}\Omega$, $L = 75\ \text{mH}$, $V_L = 3.7\ \text{V}$
20. $R = 750\ \Omega$, $Z = 1.6\ \text{k}\Omega$, $L = 800\ \text{mH}$, $V_L = 8.7\ \text{V}$
21. $R = 22\ \text{k}\Omega$, $Z = 35.7\ \text{k}\Omega$, $L = 800\ \text{mH}$, $V_L = 17\ \text{V}$
22. $R = 33\ \text{k}\Omega$, $Z = 42.7\ \text{k}\Omega$, $L = 760\ \text{mH}$, $V_L = 7.8\ \text{V}$
23. $R = 750\ \Omega$, $Z = 1.77\ \text{k}\Omega$, $L = 100\ \text{mH}$, $V_L = 27\ \text{V}$
24. $R = 560\ \Omega$, $Z = 940\ \Omega$, $L = 250\ \text{mH}$, $V_L = 15.6\ \text{V}$

Refer to Figure 20-29 and find X_L, L, θ, V_R, $\mathbf{E}$, and I in the following problems:

25. $R = 6.8\ \text{k}\Omega$, $Z = 11\ \text{k}\Omega$, $f = 4.3\ \text{kHz}$, $V_L = 6\ \text{V}$
26. $R = 820\ \Omega$, $Z = 1.25\ \text{k}\Omega$, $f = 2\ \text{kHz}$, $V_L = 13.6\ \text{V}$
27. $R = 2.2\ \text{k}\Omega$, $Z = 4.8\ \text{k}\Omega$, $f = 7.5\ \text{kHz}$, $V_L = 4.5\ \text{V}$
28. $R = 560\ \Omega$, $Z = 900\ \Omega$, $f = 25\ \text{kHz}$, $V_L = 14\ \text{V}$
29. $R = 47\ \text{k}\Omega$, $Z = 83\ \text{k}\Omega$, $f = 5.5\ \text{kHz}$, $V_L = 20\ \text{V}$
30. $R = 22\ \text{k}\Omega$, $Z = 34.5\ \text{k}\Omega$, $f = 2.3\ \text{kHz}$, $V_L = 6.7\ \text{V}$
31. $R = 3.9\ \text{k}\Omega$, $Z = 6.75\ \text{k}\Omega$, $f = 12\ \text{kHz}$, $V_L = 8.5\ \text{V}$
32. $R = 5.6\ \text{k}\Omega$, $Z = 7.32\ \text{k}\Omega$, $f = 9.5\ \text{kHz}$, $V_L = 11.3\ \text{V}$

End of Chapter Problems 20-5

In problems 1 through 20, find Z. Express your answers in both *polar* and *rectangular* forms:

1. $X_L = 23\ \text{k}\Omega$, $\theta = 48°$
2. $X_L = 14.5\ \text{k}\Omega$, $\theta = 63.2°$
3. $X_L = 740\ \Omega$, $\theta = -51°$
4. $X_C = 17.5\ \text{k}\Omega$, $\theta = -28°$
5. $R = 820\ \Omega$, $\theta = -17°$
6. $R = 18\ \text{k}\Omega$, $\theta = 72°$
7. $Z = 6.35\ \text{k}\Omega$, $\theta = -45°$
8. $Z = 38.6\ \text{k}\Omega$, $\theta = 24°$
9. $X_C = 30\ \text{k}\Omega$, $R = 40\ \text{k}\Omega$
10. $X_C = 5.85\ \text{k}\Omega$, $R = 6.8\ \text{k}\Omega$
11. $X_L = 120\ \text{k}\Omega$, $R = 180\ \text{k}\Omega$
12. $X_L = 17.5\ \text{k}\Omega$, $R = 11.3\ \text{k}\Omega$
13. $X_C = 3\ \text{k}\Omega$, $Z = 4.6\ \text{k}\Omega$
14. $X_C = 400\ \Omega$, $Z = 560\ \Omega$
15. $X_L = 3.75\ \text{k}\Omega$, $Z = 5.35\ \text{k}\Omega$
16. $X_L = 600\ \Omega$, $Z = 1\ \text{k}\Omega$
17. $X_L = 18\ \text{k}\Omega$, $R = 10\ \text{k}\Omega$
18. $X_L = 46\ \text{k}\Omega$, $R = 68\ \text{k}\Omega$
19. $X_C = 1.45\ \text{k}\Omega$, $R = 1.8\ \text{k}\Omega$
20. $X_C = 18\ \text{k}\Omega$, $R = 33\ \text{k}\Omega$

In problems 21 through 40, find **E**. Express your answers in both *polar* and *rectangular* forms:

21. $V_L = 12.7$ V
$V_R = 10$ V

22. $V_L = 600$ mV
$V_R = 800$ mV

23. $V_C = -6.25$ V
$V_R = 4.32$ V

24. $V_C = -17$ V
$V_R = 25$ V

25. $V_C = -5.7$ V
$E = 12$ V

26. $V_C = -10$ V
$E = 15$ V

27. $V_L = 8.5$ V
$E = 15$ V

28. $V_L = 24$ V
$E = 30$ V

29. $E = 15$ V
$\theta = -40°$

30. $E = 25$ V
$\theta = 72°$

31. $V_L = 7.93$ V
$\theta = 34.6°$

32. $V_L = 14.6$ V
$\theta = 73°$

33. $V_C = -64$ V
$\theta = -28°$

34. $V_C = -28$ V
$\theta = -66°$

35. $V_R = 2.6$ V
$E = 5$ V

36. $V_R = 350$ mV
$E = 1$ V

37. $V_L = 16$ V
$E = 30$ V

38. $V_L = 22$ V
$E = 30$ V

39. $V_C = -6$ V
$E = 10$ V

40. $V_C = -20$ V
$E = 25$ V

End of Chapter Problems 20-6

Find Z_T in the following problems. Express your answers in polar form.

1. $Z_1 = 2\,\underline{/20°}$ kΩ, $Z_2 = 3\,\underline{/65°}$ kΩ

2. $Z_1 = 3\,\underline{/60°}$ kΩ, $Z_2 = 4.5\,\underline{/25°}$ kΩ

3. $Z_1 = 11\,\underline{/-20°}$ kΩ, $Z_2 = 15\,\underline{/-30°}$ kΩ

4. $Z_1 = 600\,\underline{/-35°}$ Ω, $Z_2 = 1\,\underline{/-50°}$ kΩ

5. $Z_1 = 7.3\,\underline{/-25°}$ kΩ, $Z_2 = 3\,\underline{/40°}$ kΩ

6. $Z_1 = 1.73\,\underline{/-60°}$ kΩ, $Z_2 = 3.5\,\underline{/-27°}$ kΩ

7. $Z_1 = 78\,\underline{/38°}$ kΩ, $Z_2 = 50\,\underline{/-65°}$ kΩ

8. $Z_1 = 100\,\underline{/70°}$ Ω, $Z_2 = 400\,\underline{/-15°}$ Ω

9. $Z_1 = 270\,\underline{/30°}$ kΩ, $Z_2 = 150\,\underline{/30°}$ kΩ

10. $Z_1 = 5\,\underline{/90°}$ kΩ, $Z_2 = 3\,\underline{/0°}$ kΩ

11. $Z_1 = 7.5\,\underline{/-90°}$ kΩ, $Z_2 = 13.5\,\underline{/22°}$ kΩ

12. $Z_1 = 10\,\underline{/0°}$ kΩ, $Z_2 = 15\,\underline{/-75°}$ kΩ

End of Chapter Problems 20-7

1. Refer to Figure 20-30. $R_1 = 560$ Ω, $R_2 = 1.2$ kΩ, and $X_L = 3$ kΩ. Draw the ESC. Express Z in both polar and rectangular forms.

2. Refer to Figure 20-30. $R_1 = 1.8$ kΩ, $R_2 = 1.2$ kΩ, and $X_L = 2$ kΩ. Draw the ESC. Express Z in both polar and rectangular forms.

3. Refer to Figure 20-31. $R_1 = 470$ Ω, $R_2 = 680$ Ω, $X_C = 2$ kΩ, and $X_L = 3.3$ kΩ. Draw the ESC. Express Z in both polar and rectangular forms.

4. Refer to Figure 20-31. $R_1 = 2.2$ kΩ, $R_2 = 4.7$ kΩ, $X_C = 6.1$ kΩ, and $X_L = 1.3$ kΩ. Draw the ESC. Express Z in both polar and rectangular forms.

5. Refer to Figure 20-31. $R_1 = 1.2$ kΩ, $R_2 = 2.2$ kΩ, $C = 100$ nF, $L = 800$ mH, and $f = 1.2$ kHz. Draw the ESC. Include component values.

6. Refer to Figure 20-31. $R_1 = 100$ Ω, $R_2 = 330$ Ω, $C = 0.2$ μF, $L = 350$ mH, and $f = 700$ Hz. Draw the ESC. Include component values.

7. Refer to Figure 20-32. $R = 200\ \Omega$, $L = 200$ mH, $C = 350$ nF, $f = 500$ Hz, and $E = 25$ V. Find Z_{rect}, Z_{polar}, V_R, V_C, V_L, and I. Draw the ESC.

8. Refer to Figure 20-32. $R = 500\ \Omega$, $L = 500$ mH, $C = 5$ nF, $f = 3$ kHz, and $E = 20$ V. Find Z_{rect}, Z_{polar}, V_R, V_C, V_L, and I. Draw the ESC.

End of Chapter Problems 20-8

Refer to Figure 20-32.

1. $R = 470\ \Omega$, $C = 5$ nF, $L = 750$ mH, and $E = 10$ V. Find f_r. Determine X_L, X_C, V_R, V_C, V_L, Z, θ, I, and the ESC at (a) $f = f_r$, (b) $f = 2.3$ kHz, and (c) $f = 2.7$ kHz.

2. $R = 1.2$ kΩ, $C = 1$ nF, $L = 50$ mH, and $E = 20$ V. Find f_r. Determine X_L, X_C, V_R, V_C, V_L, Z, θ, I, and the ESC at (a) $f = f_r$, (b) $f = 20$ kHz, and (c) $f = 25$ kHz.

3. $R = 300\ \Omega$, $C = 500$ nF, $L = 1$ H, and $E = 25$ V. Find f_r. Determine X_L, X_C, V_R, V_C, V_L, Z, θ, I, and the ESC at (a) $f = f_r$, (b) $f = 200$ Hz, and (c) $f = 300$ Hz.

4. $R = 680\ \Omega$, $C = 50$ nF, $L = 200$ mH, and $E = 10$ V. Find f_r. Determine X_L, X_C, V_R, V_C, V_L, Z, θ, I, and the ESC at (a) $f = f_r$, (b) $f = 1$ kHz, and (c) $f = 2$ kHz.

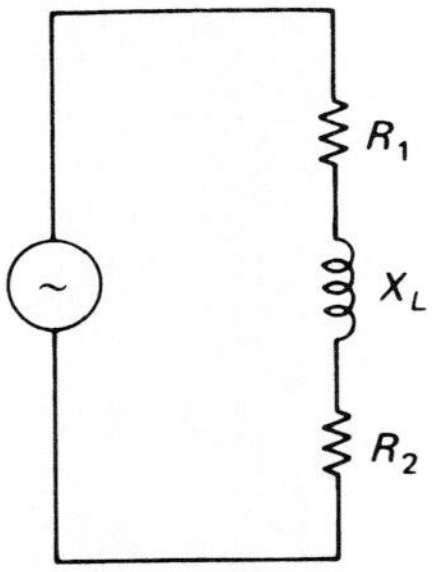

FIGURE 20-30

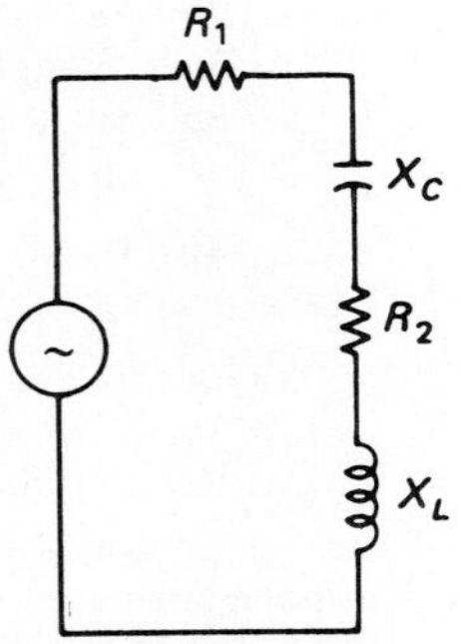

FIGURE 20-31

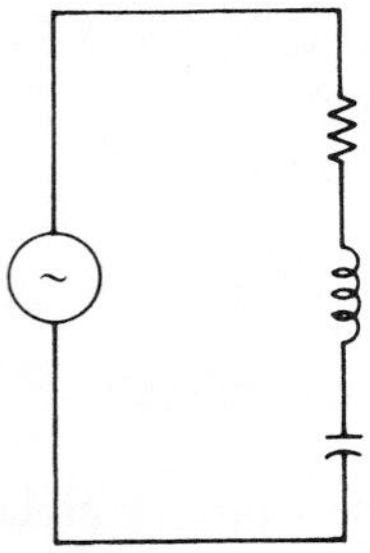

FIGURE 20-32

Answers to Self-tests

SELF-TEST 20-1

1. $V_C = -12.4$ V; $\mathbf{E} = 21$ V

2. $X_C = 3.18$ kΩ; $Z = 3.76$ kΩ; $\theta = 57.9°$; $V_R = 5.32$ V; $V_C = -8.46$ V; $I = 2.66$ mA

3. $X_C = 501\ \Omega$; $f = 318$ Hz; $\theta = -56.6°$; $V_R = 4.95$ V; $V_C = -7.51$ V; $I = 15$ mA

4. $\mathbf{E} = 3.64$ V; $V_R = 1.54$ V

5. $X_L = 5.5$ kΩ: $\theta = 49.5°$; $Z = 7.23$ kΩ; $V_R = 16.2$ V; $V_L = 19$ V; $I = 3.46$ mA

6. $X_L = 937\ \Omega$; $\theta = 51.3°$; $f = 2.98$ kHz; $\mathbf{E} = 9.99$ V; $V_R = 6.69$ V; $I = 8.58$ mA

SELF-TEST 20-2

1. $Z_{rect} = 21.7$ k$\Omega + j11.3$ kΩ
   $Z_{polar} = 24.5\ \underline{/27.5°}$ kΩ

2. $Z_{rect} = 4.7$ k$\Omega - j2.58$ kΩ
   $Z_{polar} = 5.36\ \underline{/-28.8°}$ kΩ

3. $Z_{rect} = 12.2$ k$\Omega + j16.2$ kΩ
   $Z_{polar} = 20.3\ \underline{/53°}$ kΩ
   The ESC consists of 12.2 kΩ of resistance and 16.2 kΩ of inductive reactance.

4. $X_C = 7.58$ kΩ, $X_L = 6.6$ kΩ, $Z_{polar} = 1.55\ \underline{/-39.3°}$ kΩ, $I = 12.9$ mA, $V_R = 15.5$ V, $V_C = -97.7$ V, $V_L = 85.1$ V. The ESC consists of 1.2 kΩ of resistance and 77.1 nF of capacitance.

AC CIRCUIT ANALYSIS: PARALLEL CIRCUITS

CHAPTER 21

INTRODUCTION

Continuing our study of ac circuits, in this chapter we will solve parallel ac circuit problems using trig functions and our calculator. We learned in our study of dc circuits in Chapters 11 and 12 that in series circuits the circuit current is constant and in parallel circuits the applied voltage is constant. These concepts are also true in ac circuits. In ac circuits, resistance, reactance, and impedance are the series circuit parameters. Conductance, susceptance, and admittance are the parallel circuit parameters.

In the chapters on dc circuits, we discussed the duality between current and conductance in parallel circuits and voltage drops and resistance in series circuits. This duality holds true in ac circuits between current and admittance in parallel circuits and voltage and impedance in series circuits. Students should think admittance when dealing with parallel circuits, just as they think impedance when working with series circuits. Familiarity with conductance, susceptance, and admittance is necessary as you further your studies into electronic circuits and systems, especially in the area of microwave circuits and systems.

In this chapter you will also learn how to reduce complex circuits to their equivalent series or parallel circuits in order to simplify their solutions.

CHAPTER OBJECTIVES

In this chapter you will learn how to:

1. Analyze *RC* and *RL* parallel circuits to determine currents, phase angle, and admittance in both polar and rectangular forms.
2. Find the equivalent impedance of impedances in parallel.
3. Convert from parallel to equivalent series circuits and from series to equivalent parallel circuits.
4. Reduce complex circuits to equivalent series circuits or to equivalent parallel circuits.

21-1 *RC* CIRCUIT ANALYSIS

In parallel circuits the voltage drops across the various branches are equal.

■ **KEY POINT** In a parallel circuit, we use E as the reference. In an RC circuit, current leads, so I_C and B_C are plotted along the positive y-axis. I_T (or Y) is the hypotenuse and θ is the phase angle.

As shown in Figure 21-1(b), the current through the capacitor leads the voltage drop across it by 90°. This puts the total current, I_T in the first quadrant. I_R is the side adjacent to the angle θ, and I_C is the opposite side.

□ EXAMPLE 21-1

In the circuit of Figure 21-1(a), let $I_R = 7$ mA and $I_C = 10$ mA. Find I_T and θ. Express I_T in both rectangular and polar forms.

Solution Refer to Figure 21-2.

$$I_{\text{rect}} = 7\text{ mA} + j10\text{ mA}$$

$$\tan\theta = \frac{\text{opposite side}}{\text{adjacent side}}$$

$$\tan\theta = \frac{I_C}{I_R} = \frac{10\text{ mA}}{7\text{ mA}} = 1.43$$

$$\tan^{-1} 1.43 = 55°$$

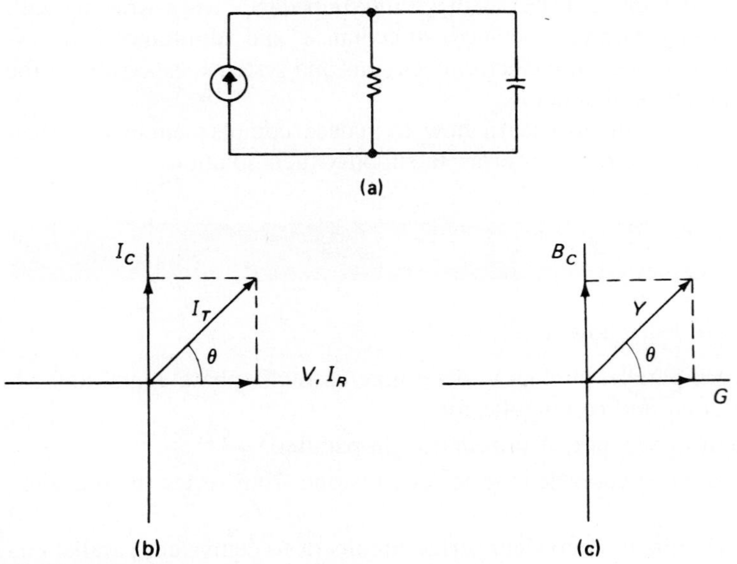

FIGURE 21-1

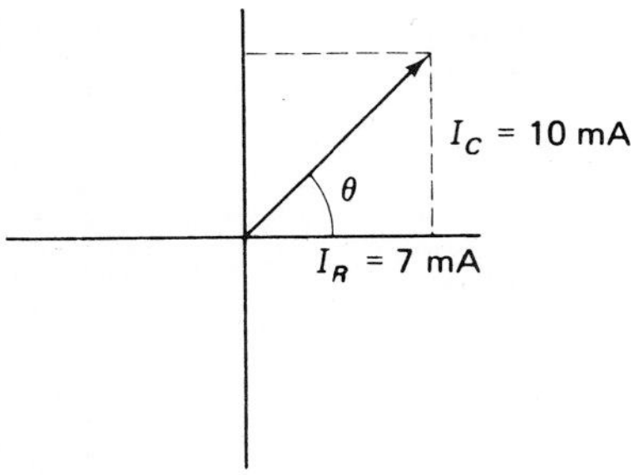

FIGURE 21-2

We can now find I_T by using either the sine or cosine function. We will use the cosine function.

$$\cos\,\theta = \frac{\text{adjacent side}}{\text{hypotenuse}}$$

$$\cos\,\theta = \frac{I_R}{I_T}$$

$$I_T = \frac{I_R}{\cos\,\theta} = \frac{7\text{ mA}}{\cos\,55^\circ} = 12.2\text{ mA}$$

$$I_{\text{polar}} = 12.2\,\underline{/55^\circ}\text{ mA}$$

□ EXAMPLE 21-2

In Figure 21-1(a) let $I_T = 300\ \mu\text{A}$ and $I_C = 100\ \mu\text{A}$. Find I_R and θ. Express I_T in both rectangular and polar forms.

Solution

$$\sin\,\theta = \frac{I_C}{I_T} = \frac{100\ \mu\text{A}}{300\ \mu\text{A}} = 0.333$$

$$\sin^{-1} 0.333 = 19.5^\circ$$

$$\cos\,\theta = \frac{I_R}{I_T}$$

$$I_R = I_T \cos\,\theta = 300\ \mu\text{A}\cos\,19.5^\circ = 283\ \mu\text{A}$$

$$I_{\text{rect}} = 283\ \mu\text{A} + j100\ \mu\text{A}$$

$$I_{\text{polar}} = 300\,\underline{/19.5^\circ}\ \mu\text{A}$$

□

PRACTICE PROBLEMS 21-1

Refer to Figure 21-1(a). From the information given, express I_T in rectangular and polar forms.

1. $I_T = 30$ mA and $\theta = 40°$
2. $I_R = 3.5$ mA and $I_T = 6$ mA
3. $I_C = 40\ \mu$A and $I_T = 80\ \mu$A
4. $I_T = 500$ mA and $\theta = 63.2°$
5. $I_R = 200\ \mu$A and $I_C = 300\ \mu$A

Solutions

1. $I_{\text{rect}} = 23 \text{ mA} + j19.3 \text{ mA}$
   $I_{\text{polar}} = 30 \underline{/40°} \text{ mA}$
2. $I_{\text{rect}} = 3.5 \text{ mA} + j4.87 \text{ mA}$
   $I_{\text{polar}} = 6 \underline{/54.3°} \text{ mA}$
3. $I_{\text{rect}} = 69.3\ \mu\text{A} + j40\ \mu\text{A}$
   $I_{\text{polar}} = 80 \underline{/30°}\ \mu\text{A}$
4. $I_{\text{rect}} = 225 \text{ mA} + j446 \text{ mA}$
   $I_{\text{polar}} = 500 \underline{/63.2°} \text{ mA}$
5. $I_{\text{rect}} = 200\ \mu\text{A} + j300\ \mu\text{A}$
   $I_{\text{polar}} = 361 \underline{/56.3°}\ \mu\text{A}$

Additional practice problems are at the end of the chapter.

We learned in the chapter on Ohm's law that conductance and resistance are reciprocals. Susceptance and reactance are reciprocals, as are admittance and impedance. Conductance, susceptance, and admittance are parallel circuit parameters, just as resistance, reactance, and impedance are series circuit parameters.

The phase relationship between conductance, G, capacitive susceptance, B_C, and admittance, Y, is shown in Figure 21-1(c). The capacitive susceptance is determined by the equation

$$B_C = 2\pi fC \tag{21-1}$$

☐ EXAMPLE 21-3

Refer to Figure 21-1(a). Let $R = 3.3$ kΩ, $C = 75$ nF, and $f = 500$ Hz. Find G, B_C, Y, and θ. Express Y in rectangular and polar forms.

Solution

$$B_C = 2\pi fC = 2\pi \times 500 \text{ Hz} \times 75 \text{ nF} = 236\ \mu\text{S}$$

$$G = \frac{1}{R} = \frac{1}{3.3 \text{ k}\Omega} = 303\ \mu\text{S}$$

$$Y_{\text{rect}} = 303\ \mu\text{S} + j236\ \mu\text{S}$$

$$\tan\theta = \frac{\text{opposite side}}{\text{adjacent side}} = \frac{B_C}{G} = \frac{236\ \mu\text{S}}{303\ \mu\text{S}} = 0.779$$

$$\tan^{-1} 0.779 = 37.9°$$

$$\sin\theta = \frac{B_C}{Y}$$

$$Y = \frac{B_C}{\sin\theta} = \frac{236\ \mu\text{S}}{\sin 37.9°} = 384\ \mu\text{S}$$

$$Y_{\text{polar}} = 384 \underline{/37.9°}\ \mu\text{S}$$

☐ EXAMPLE 21-4

Refer to Figure 21-1(a). Let $R = 680\ \Omega$, $C = 100$ nF, $f = 5.5$ kHz, and $I_T = 25$ mA. Find G, B_C, Y, θ, I_R, I_C, and V. Express Y in rectangular and polar forms.

Solution

$$B_C = 2\pi fC = 2\pi \times 5.5\text{ kHz} \times 100\text{ nF} = 3.46\text{ mS}$$

$$G = \frac{1}{R} = \frac{1}{680\ \Omega} = 1.47\text{ mS}$$

$$Y_{\text{rect}} = 1.47\text{ mS} + j3.46\text{ mS}$$

$$\tan\theta = \frac{B_C}{G} = \frac{3.46\text{ mS}}{1.47\text{ mS}} = 2.35$$

$$\tan^{-1} 2.35 = 67°$$

$$\sin\theta = \frac{B_C}{Y}$$

$$Y = \frac{B_C}{\sin\theta} = \frac{3.46\text{ mS}}{\sin 67°} = 3.76\text{ mS}$$

$$Y_{\text{polar}} = 3.76\ \underline{/67°}\text{ mS}$$

$$I_{\text{polar}} = 25\ \underline{/67°}\text{ mA}$$

$$I_R = I_T\cos\theta = 25\text{ mA}\cos 67° = 9.77\text{ mA}$$

$$I_C = I_T\sin\theta = 25\text{ mA}\sin 67° = 23\text{ mA}$$

$$V = \frac{I_T}{Y} = 6.65\text{ V}$$

☐

PRACTICE PROBLEMS 21-2

Refer to Figure 21-1(a) for problems 1 through 4. Find Y_{rect}, Y_{polar}, I_{rect}, and V.

1. $R = 15\text{ k}\Omega$, $C = 10$ nF, $f = 1.5$ kHz, $I_T = 500\ \mu$A
2. $R = 1.8\text{ k}\Omega$, $C = 200$ nF, $f = 350$ Hz, $I_T = 2$ mA
3. $R = 6.8\text{ k}\Omega$, $C = 50$ nF, $f = 300$ Hz, $I_T = 10$ mA
4. $R = 47\text{ k}\Omega$, $C = 200$ pF, $f = 15$ kHz, $I_T = 50$ mA
5. Refer to Figure 21-1(a). Let $R = 2\text{ k}\Omega$, $f = 2.5$ kHz, $\theta = 35°$, and $I_C = 240\ \mu$A. Find G, B_C, C, I_T, I_R, and V.

Solutions

1. $Y_{rect} = 66.7\ \mu S + j94.2\ \mu S$
   $Y_{polar} = 115 \angle 54.7^\circ\ \mu S$
   $I_{rect} = 289\ \mu A + j408\ \mu A$
   $V = 4.35$ V
2. $Y_{rect} = 556\ \mu S + j440\ \mu S$
   $Y_{polar} = 709 \angle 38.4^\circ\ \mu S$
   $I_{rect} = 1.57$ mA $+ j1.24$ mS
   $V = 2.82$ V
3. $Y_{rect} = 147\ \mu S + j94.2\ \mu S$
   $Y_{polar} = 175 \angle 32.7^\circ\ \mu S$
   $I_{rect} = 8.42$ mA $+ j5.4$ mA
   $V = 57.1$ V
4. $Y_{rect} = 21.3\ \mu S + j18.8\ \mu S$
   $Y_{polar} = 28.4 \angle 41.5^\circ\ \mu S$
   $I_{rect} = 37.4\ \mu A + j33.2\ \mu A$
   $V = 1.76$ V
5. $G = 500\ \mu S$
   $B_C = 350\ \mu S$ $\quad (B_C = G \tan \theta)$
   $C = 22.3$ nF
   $I_T = 418\ \mu A$
   $I_R = 343\ \mu A$
   $V = 686$ mV

Additional practice problems are at the end of the chapter.

21-2 *RL* CIRCUIT ANALYSIS

KEY POINT In a parallel circuit, we use E as the reference. In an *RL* circuit, voltage leads, so I_L and B_L are plotted along the negative y-axis. I_T (or Y) is the hypotenuse and θ is the phase angle.

As shown in Figure 21-3(b), the voltage drop across the inductor leads the current through it by 90°. This puts the total current, I_T in the fourth quadrant. I_R is the side adjacent to the angle θ, and I_L is the opposite side.

☐ EXAMPLE 21-5

In the circuit of Figure 21-3(a), let $I_R = 30$ mA and $I_L = 20$ mA. Find I and θ. Express I_T in rectangular and polar forms.

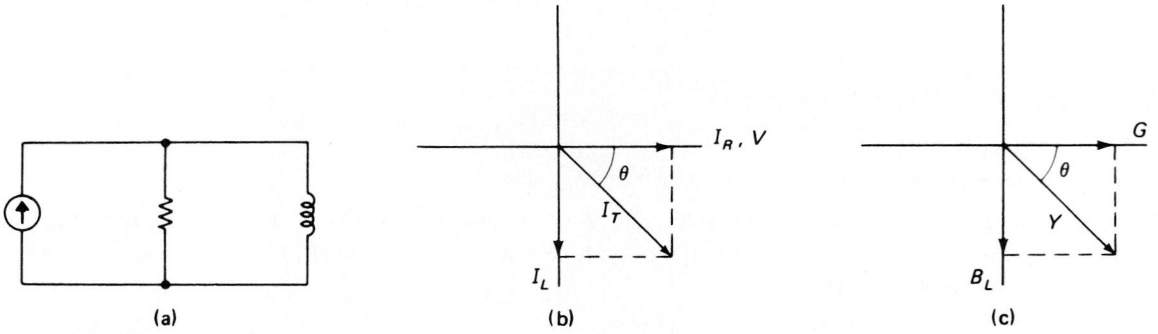

FIGURE 21-3

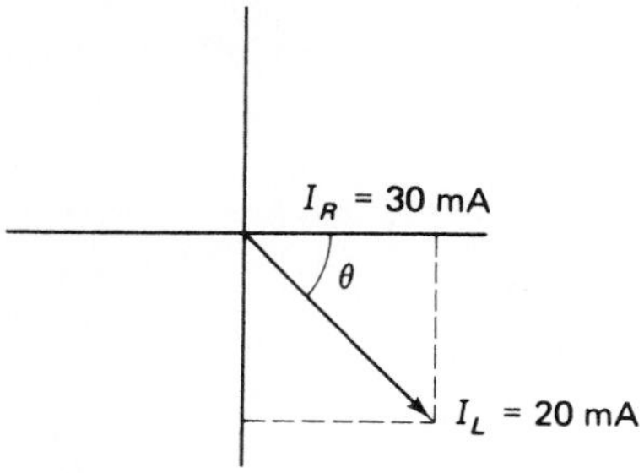

FIGURE 21-4

Solution Refer to Figure 21-4.

$$\tan\theta = \frac{\text{opposite side}}{\text{adjacent side}}$$

$$\tan\theta = \frac{I_L}{I_R} = \frac{-20\text{ mA}}{30\text{ mA}} = -0.667$$

$$\tan^{-1} - 0.667 = -33.7^\circ$$

$$\sin\theta = \frac{\text{opposite side}}{\text{hypotenuse}} = \frac{I_L}{I_T}$$

$$I_T = \frac{I_L}{\sin\theta} = \frac{-20\text{ mA}}{\sin -33.7^\circ} = 36.1\text{ mA}$$

$$I_{\text{rect}} = 30\text{ mA} - j20\text{ mA}$$

$$I_{\text{polar}} = 36.1 \underline{/-33.7^\circ}\text{ mA}$$

Using the calculator to find I_{polar};

30 [EE] [±] 3 [X↔Y] 20 [±] [EE] [±] 3 [R→P] *Answer:* 36.1×10^{-3}

[X↔Y] *Answer*: −33.7, $I = 36.1 \underline{/-33.7^\circ}$ mA

□ EXAMPLE 21-6

In Figure 21-3(a), let $I_T = 7.3$ mA and $I_R = 3.6$ mA. Find I_L and θ. Express I in both rectangular and polar forms.

Solution

$$\cos\theta = \frac{I_R}{I_T} = \frac{3.6\text{ mA}}{7.3\text{ mA}} = 0.493$$

$$\cos^{-1} 0.493 = -60.5^\circ$$

$$I_{\text{polar}} = 7.3 \underline{/-60.5^\circ}\text{ mA}$$

$$I_L = I_T \sin\theta = 7.3\text{ mA} \times \sin -60.5^\circ = -6.35\text{ mA}$$

$$I_{\text{rect}} = 3.6\text{ mA} - j6.35\text{ mA}$$

□

PRACTICE PROBLEMS 21-3

Refer to Figure 21-3(a). From the information given, find I_T in both rectangular and polar forms.

1. $I_R = 400\ \mu A$, $I_L = 300\ \mu A$

2. $I_R = 17$ mA, $I_T = 25$ mA

3. $I_R = 4.5$ mA, $\theta = -65°$

4. $I_T = 50$ mA, $\theta = -36°$

5. $I_L = 50\ \mu A$, $\theta = -17°$

Solutions

1. $I_{rect} = 400\ \mu A - j300\ \mu A$
$I_{polar} = 500\ \underline{/-36.9°}\ \mu A$

2. $I_{rect} = 17\ mA - j18.3\ mA$
$I_{polar} = 25\ \underline{/-47.2°}\ mA$

3. $I_{rect} = 4.5\ mA - j9.65\ mA$
$I_{polar} = 10.6\ \underline{/-65°}\ mA$

4. $I_{rect} = 40.5\ mA - j29.4\ mA$
$I_{polar} = 50\ \underline{/-36°}\ mA$

5. $I_{rect} = 164\ \mu A - j50\ \mu A$
$I_{polar} = 171\ \underline{/-17°}\ \mu A$

Additional practice problems are at the end of the chapter.

The phase relationship between conductance, inductive susceptance, B_L, and admittance is shown in Figure 21-3(c). The inductive susceptance is determined by the equation

$$B_L = \frac{1}{2\pi f L} \tag{21-2}$$

☐ EXAMPLE 21-7

Refer to Figure 21-3(a). Let $R = 6.8\ k\Omega$, $L = 200$ mH, $f = 4$ kHz, and $I_T = 10$ mA. Find Y and I_T in both rectangular and polar forms.

Solution

$$G = \frac{1}{R} = \frac{1}{6.8\ k\Omega} = 147\ \mu S$$

$$B_L = \frac{1}{2\pi f L} = \frac{1}{2\pi \times 4\ kHz \times 200\ mH} = 199\ \mu S$$

$$Y_{rect} = 147\ \mu S - j199\ \mu S$$

$$\tan\theta = \frac{B_L}{G} = \frac{-199\ \mu S}{147\ \mu S} = -1.35$$

$$\tan^{-1} -1.35 = -53.5°$$

$$\sin\theta = \frac{B_L}{Y}$$

$$Y = \frac{B_L}{\sin\theta} = \frac{-199\ \mu S}{\sin -53.5^\circ} = 248\ \mu S$$

$$Y = 248\ \underline{/-53.5^\circ}\ \mu S$$

$$I_{polar} = 10\ \underline{/-53.5^\circ}\ mA$$

$$I_R = I_T \cos\theta = 10\ mA \times \cos -53.5^\circ = 5.95\ mA$$

$$I_L = I_T \sin\theta = 10\ mA \times \sin -53.5^\circ = -8.04\ mA$$

$$I_{rect} = 5.95\ mA - j8.04\ mA$$

We could have found I_R and I_L by using Ohm's law:

$$I_R = \frac{I_T G}{Y} = \frac{10\ mA \times 147\ \mu S}{248\ \mu S} = 5.93\ mA$$

$$I_L = \frac{I_T B_L}{Y} = \frac{10\ mA \times 199\ \mu S}{248\ \mu S} = 8.02\ mA$$

Differences in the answers are due to rounding.

☐ EXAMPLE 21-8

Refer to Figure 21-5. Let $R = 12\ k\Omega$, $L = 100\ mH$, $\theta = -40^\circ$, and $I_R = 50\ \mu A$. Find f. Find Y and I_T in both rectangular and polar forms.

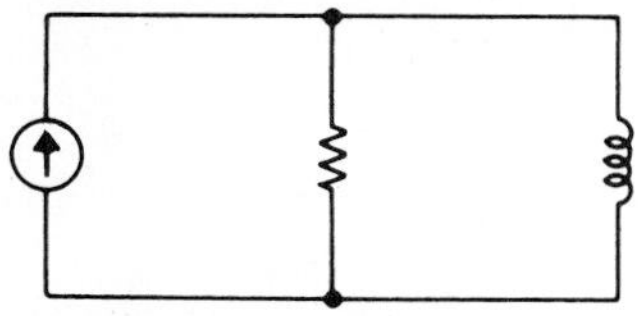

FIGURE 21-5

Solution

$$G = \frac{1}{R} = \frac{1}{12\ k\Omega} = 83.3\ \mu S$$

$$B_L = G \tan\theta = 83.3\ \mu S \times \tan -40^\circ = -69.9\ \mu S$$

$$Y_{rect} = 83.3\ \mu S - j69.9\ \mu S$$

$$\cos\theta = \frac{G}{Y}$$

$$Y = \frac{G}{\cos\theta} = \frac{83.3\ \mu S}{\cos -40^\circ} = 109\ \mu S$$

$$Y_{polar} = 109\ \underline{/-40^\circ}\ \mu S$$

$$f = \frac{1}{2\pi B_L L} = \frac{1}{2\pi \times 69.9\ \mu S \times 100\ mH} = 22.8\ kHz$$

$$I_L = I_R \tan\theta = 50\ \mu A \times \tan -40^\circ = -42\ \mu A$$

$$I_{rect} = 50\ \mu A - j42\ \mu A$$

$$I_T = \frac{I_R}{\cos\theta} = \frac{50\ \mu A}{\cos -40°} = 65.3\ \mu A$$

$$I_{polar} = 65.3\ \angle -40°\ \mu A$$

□

PRACTICE PROBLEMS 21-4

Refer to Figure 21-5. Find Y and I_T in both rectangular and polar forms in problems 1 through 5.

1. $R = 1.2\ k\Omega$, $L = 50\ mH$, $f = 3\ kHz$, $I_T = 400\ \mu A$
2. $R = 15\ k\Omega$, $L = 20\ mH$, $f = 100\ kHz$, $I_T = 1\ mA$
3. $R = 470\ \Omega$, $L = 400\ mH$, $f = 200\ Hz$, $I_T = 20\ mA$
4. $R = 5.6\ k\Omega$, $L = 100\ mH$, $f = 9\ kHz$, $I_T = 50\ mA$
5. $R = 27\ k\Omega$, $L = 100\ mH$, $f = 30\ kHz$, and $I_T = 2.27\ mA$.
6. Refer to Figure 21-5. $L = 10\ mH$, $f = 50\ kHz$, $\theta = -35°$, and $I_L = 60\ \mu A$. Find R. Find Y and I_T in both rectangular and polar forms.

Solutions

1. $Y_{rect} = 833\ \mu S - j1.06\ mS$
   $Y_{polar} = 1.35\ \angle -51.9°\ mS$
   $I_{rect} = 247\ \mu A - j315\ \mu A$
   $I_{polar} = 400\ \angle -51.9°\ \mu A$
2. $Y_{rect} = 66.7\ \mu S - j79.6\ \mu S$
   $Y_{polar} = 104\ \angle -50°\ \mu S$
   $I_{rect} = 642\ \mu A - j766\ \mu A$
   $I_{polar} = 1\ \angle -50°\ mA$
3. $Y_{rect} = 2.13\ mS - j1.99\ mS$
   $Y_{polar} = 2.91\ \angle -43.1°\ mS$
   $I_{rect} = 14.6\ mA - j13.7\ mA$
   $I_{polar} = 20\ \angle -43.1°\ mA$
4. $Y_{rect} = 179\ \mu S - j177\ \mu S$
   $Y_{polar} = 252\ \angle -44.7°\ \mu S$
   $I_{rect} = 35.5\ mA - j35.2\ mA$
   $I_{polar} = 50\ \angle -44.7°\ mA$
5. $Y_{rect} = 37.0\ \mu S - j53.1\ \mu S$
   $Y_{polar} = 64.7\ \angle -55.1°\ \mu S$
   $I_{rect} = 1.3\ mA - j1.86\ mA$
   $I_{polar} = 2.27\ \angle -55.1°\ mA$
6. $R = 22.4\ k\Omega$
   $Y_{rect} = 455\ \mu S - j318\ \mu S$
   $Y_{polar} = 555\ \angle -35°\ \mu S$
   $I_{rect} = 85.7\ \mu A - j60\ \mu A$
   $I_{polar} = 105\ \angle -35°\ \mu A$

Additional practice problems are at the end of the chapter.

SELF-TEST 21-1

1. Refer to Figure 21-6. Let $R = 10\ k\Omega$, $f = 4.3\ kHz$, $C = 2.5\ nF$, and $I_T = 5\ mA$. Find Y_{rect}, Y_{polar}, I_{rect}, I_{polar}, and V.
2. Refer to Figure 21-6. Let $R = 27\ k\Omega$, $f = 800\ Hz$, $C = 10\ nF$, and $I_T = 500\ \mu A$. Find Y_{rect}, Y_{polar}, I_{rect}, I_{polar}, and V.
3. Refer to Figure 21-5. Let $R = 120\ k\Omega$, $f = 50\ kHz$, $L = 0.5\ H$, and $I_T = 70\ \mu A$. Find Y_{rect}, Y_{polar}, I_{rect}, I_{polar}, and V.
4. Refer to Figure 21-5. Let $R = 68\ k\Omega$, $f = 1.5\ kHz$, $L = 5.3\ H$, and $I_T = 2\ mA$. Find Y_{rect}, Y_{polar}, I_{rect}, I_{polar}, and V.

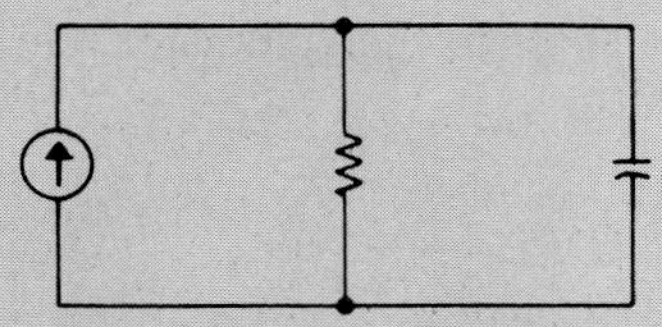

FIGURE 21-6

5. Refer to Figure 21-5. Let $R = 27\ \text{k}\Omega$, $L = 100\ \text{mH}$, $f = 30\ \text{kHz}$, $\theta = 55°$, and $I_T = 2\ \text{mA}$. Find Y and I_T in both rectangular and polar forms.

Answers to Self-test 21-1 are at the end of the chapter.

21-3 EQUIVALENT CIRCUITS

21-3-1 Multiplication and Division of Phasors

We determined in Chapter 16 that addition or subtraction of phasors must be done in rectangular form. When we multiply or divide phasors, the multiplications and divisions can be done in *either* form, but they are simpler to do in polar form. Multiplication of phasors usually occurs when simplifying complex circuits in which impedances are in parallel. Such circuits will be analyzed in Chapter 22. Division of phasors occurs in the simplification of complex circuits and when making impedance–admittance conversions.

When multiplying impedances, we will put the impedances in polar form. The magnitudes are multiplied together. The angles are *added algebraically*. When dividing impedances, the magnitudes are divided. The angles are *subtracted*.

■ **KEY POINT** When multiplying phasors, put in polar form. Multiply the magnitudes and add the angles. When dividing phasors, divide the magnitudes and subtract the angles.

□ EXAMPLE 21-9

Find the circuit impedance when the circuit admittance is $300\ \underline{/-30°}\ \mu\text{S}$.

Solution Impedance and admittance are reciprocals.

$$Z = \frac{1}{Y}$$

$$Z = \frac{1}{300\ \underline{/-30°}\ \mu\text{S}}$$

The rule is that we divide in polar form; therefore, nothing further need be done to solve for Z.

$$Z = 3.33\ \underline{/30°}\ \text{k}\Omega$$

The phase angle is positive because $(-)\,-30° = 30°$. For some students the process of subtraction is simplified by first moving the phase angle to the numerator, changing its sign, and then adding.

$$Z = \frac{1\,\angle 30°}{300\ \mu\text{S}} = 3.33\,\angle 30°\ \text{k}\Omega$$

☐ EXAMPLE 21-10

Two impedances Z_1 and Z_2 are in parallel. $Z_1 = 15\,\angle 60°\ \text{k}\Omega$ and $Z_2 = 10\,\angle -60°\ \text{k}\Omega$. Find Z_T.

Solution

$$Z_T = \frac{Z_1 Z_2}{Z_1 + Z_2}$$

We will simplify the denominator first. Remembering that impedances must be added in rectangular form, our first step is to convert Z_1 and Z_2 to rectangular form.

$$Z_{1(\text{polar})} = 15\,\angle 60°\ \text{k}\Omega$$

$$R_1 = 15\ \text{k}\Omega \cos\theta = 7.5\ \text{k}\Omega$$

$$X_L = 15\ \text{k}\Omega \sin\theta = 13\ \text{k}\Omega$$

$$Z_{2(\text{polar})} = 10\,\angle -60°\ \text{k}\Omega$$

$$R_2 = 10\ \text{k}\Omega \cos -60° = 5\ \text{k}\Omega$$

$$X_C = 10\ \text{k}\Omega \sin -60° = -8.66\ \text{k}\Omega$$

$$Z_{1(\text{rect})} = 7.5\ \text{k}\Omega + j13\ \text{k}\Omega$$

$$Z_{2(\text{rect})} = \underline{5\ \text{k}\Omega - j8.66\ \text{k}\Omega}$$

$$Z_1 + Z_2 = 12.5\ \text{k}\Omega + j4.34\ \text{k}\Omega$$

$$Z_T = \frac{15\,\angle 60°\ \text{k}\Omega \times 10\,\angle -60°\ \text{k}\Omega}{12.5\ \text{k}\Omega + j4.34\ \text{k}\Omega}$$

The division is done in polar form; therefore, we must change the result of $Z_1 + Z_2$ back into polar form.

$$(Z_1 + Z_2)_{\text{rect}} = 12.5\ \text{k}\Omega + j4.34\ \text{k}\Omega$$

$$\tan\theta = \frac{X_L}{R} = \frac{4.34\ \text{k}\Omega}{12.5\ \text{k}\Omega} = 0.347$$

$$\tan^{-1} 0.347 = 19.1°$$

$$Z = \frac{X_L}{\sin\theta} = \frac{4.34\ \text{k}\Omega}{\sin 19.1°} = 13.2\ \text{k}\Omega$$

$$(Z_1 + Z_2)_{\text{polar}} = 13.2\,\angle 19.1°\ \text{k}\Omega$$

Using the calculator,

12.5 [EE] 3 [X↔Y] 4.34 [EE] 3 [R→P]   *Answer*: 13.2×10^3

[X↔Y]   *Answer*: 19.1

$$Z_1 + Z_2 = 13.2 \angle 19.1° \text{ k}\Omega \qquad \text{in polar form}$$

$$Z_T = \frac{15 \angle 60° \text{ k}\Omega \times 10 \angle -60° \text{ k}\Omega}{13.2 \angle 19.1° \text{ k}\Omega}$$

Bringing 19.1° from the denominator results in an angle of −19.1° added to the existing angles.

$$Z_T = \frac{15 \text{ k}\Omega \times 10 \text{ k}\Omega \angle 60° + (-)60° + (-)19.1°}{13.2 \text{ k}\Omega}$$

$$= \frac{15 \text{ k}\Omega \times 10 \text{ k}\Omega \angle -19.1°}{13.2 \text{ k}\Omega} = 11.4 \angle -19.1° \text{ k}\Omega$$

□

PRACTICE PROBLEMS 21-5

1. Find Z_T if $Y_T = 200 \angle 40° \ \mu\text{S}$.
2. Find Y_T if $Z_T = 43.7 \angle -27° \text{ k}\Omega$.
3. Find Z_T if $Z_1 = 2 \angle 20° \text{ k}\Omega$ and $Z_2 = 4 \angle -50° \text{ k}\Omega$.
4. Using the equation in problem 3, find Z_T if $Z_1 = 20 \angle -25° \text{ k}\Omega$ and $Z_2 = 10 \angle -60° \text{ k}\Omega$.

Solutions

1. $Z_T = 5 \angle -40° \text{ k}\Omega$
2. $Y_T = 22.9 \angle 27° \ \mu\text{S}$
3. $Z_T = 1.58 \angle -1.86° \text{ k}\Omega$
4. $Z_T = 6.95 \angle -48.5° \text{ k}\Omega$

Additional practice problems are at the end of the chapter.

21-3-2 Parallel–Series Conversions

In solving complex circuit problems, it is often necessary to reduce circuits to equivalent parallel circuits (EPC) or to equivalent series circuits (ESC). The parallel circuit problems we have been solving can easily be changed to equivalent series circuits. Figure 21-7 shows a parallel circuit with the values of conductance, suscep-

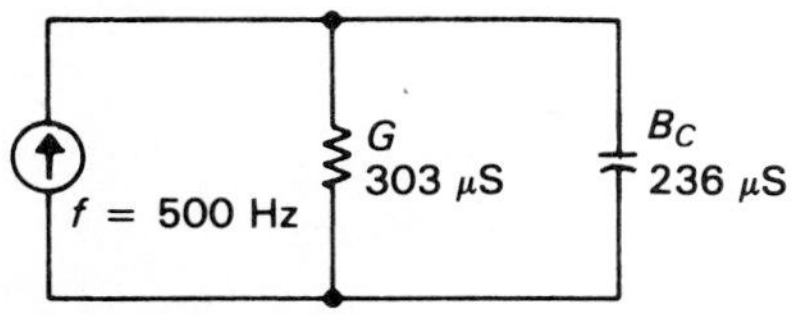

FIGURE 21-7

tance, and frequency given. (See Example 21-3 for the original circuit values.) We determined that $Y_{polar} = 384 \angle 37.9^\circ\ \mu S$. The equivalent series circuit would have an impedance equal to the reciprocal of this admittance.

$$Z_{polar} = \frac{1}{Y_{polar}} = \frac{1}{384 \angle 37.9^\circ\ \mu S} = 2.6 \angle -37.9^\circ\ k\Omega$$

Knowing Z_{polar}, we can determine the values of R and X_C that make up the impedance.

$$R = Z \cos \theta = 2.6\ k\Omega \times \cos -37.9^\circ = 2.05\ k\Omega$$

$$X_C = Z \sin \theta = 2.6\ k\Omega \times \sin -37.9^\circ = -1.6\ k\Omega$$

$$Z_{rect} = 2.05\ k\Omega - j1.6\ k\Omega$$

The frequency in Example 21-3 was 500 Hz. Therefore,

$$C = \frac{1}{2\pi f X_C} = \frac{1}{2\pi \times 500\ Hz \times 1.6\ k\Omega} = 199\ nF$$

The equivalent series circuit consists of a resistance of 2.05 kΩ and a capacitance of 199 nF, as shown in Figure 21-8.

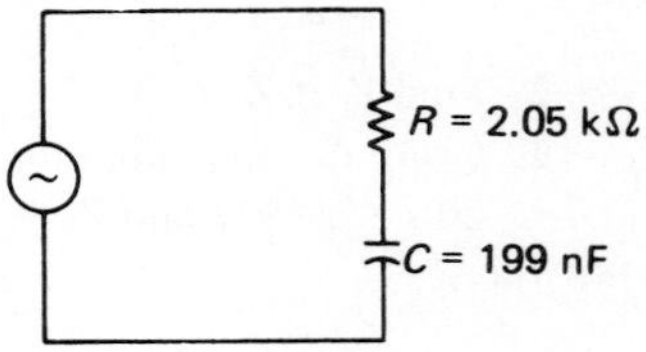

FIGURE 21-8

☐ EXAMPLE 21-11

Find the equivalent series circuit and the branch currents of the parallel circuit in Figure 21-9.

Solution

$$G = \frac{1}{R} = \frac{1}{12\ k\Omega} = 83.3\ \mu S$$

$$B_C = 2\pi f C = 2\pi \times 3.5\ kHz \times 100\ nF = 2.2\ mS$$

$$B_L = \frac{1}{2\pi f L} = \frac{1}{2\pi \times 3.5\ kHz \times 20\ mH} = 2.27\ mS$$

$$Y_{rect} = 83.3\ \mu S + j2.2\ mS - j2.27\ mS = 83.3\ \mu S - j70\ \mu S$$

$$Y_{polar} = 109 \angle -40^\circ\ \mu S$$

For the ESC,

$$Z_{polar} = \frac{1}{Y_{polar}} = \frac{1}{109 \angle -40° \ \mu S} = 9.17 \angle 40° \ k\Omega$$

$$R = Z \cos \theta = 9.17 \ k\Omega \times \cos 40° = 7.01 \ k\Omega$$

$$X_L = Z \sin \theta = 9.17 \ k\Omega \times \sin 40° = 5.9 \ k\Omega$$

$$L = \frac{X_L}{2\pi f} = \frac{5.88 \ k\Omega}{2\pi \times 3.5 \ kHz} = 268 \ mH$$

The ESC is shown in Figure 21-10. In the original circuit,

$$V = \frac{I_T}{Y} = \frac{1 \ mA}{109 \ \mu S} = 9.17 \ V$$

$$I_R = VG = 9.17 \ V \times 83.3 \ \mu S = 764 \ \mu A$$

$$I_C = VB_C = 9.17 \ V \times 2.2 \ mS = 20.2 \ mS$$

$$I_L = VB_L = 9.17 \ V \times 2.27 \ mS = 20.8 \ mS$$

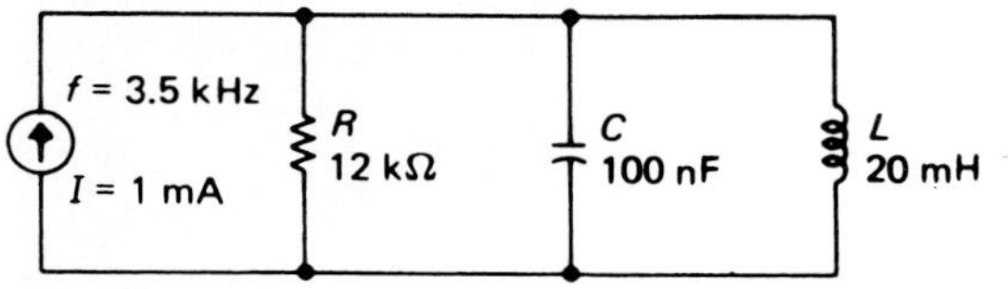

FIGURE 21-9

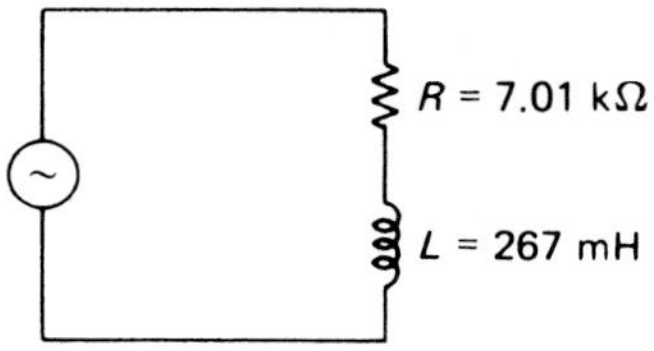

FIGURE 21-10

□

PRACTICE PROBLEMS 21-6

Find the branch currents Y, Z, V, and the ESC in the following problems:

1. Refer to Figure 21-11 f = 1 kHz, C = 20 nF, R = 10 kΩ, and I_T = 10 mA.
2. Refer to Figure 21-11. f = 2.5 kHz, C = 0.5 μF, R = 82 Ω, and I_T = 300 μA.
3. Refer to Figure 21-12. f = 15 kHz, L = 300 mH, R = 22 kΩ, and I_T = 2 mA.
4. Refer to Figure 21-12. f = 2.5 kHz, L = 200 mH, R = 4.7 kΩ, and I_T = 1 mA.
5. Refer to Figure 21-9. f = 3.7 kHz, C = 100 nF, L = 20 mH, R = 12 kΩ, and I_T = 100 μA.
6. Refer to Figure 21-9. f = 7.5 kHz, C = 5 nF, L = 50 mH, R = 20 kΩ, and I_T = 15 mA.

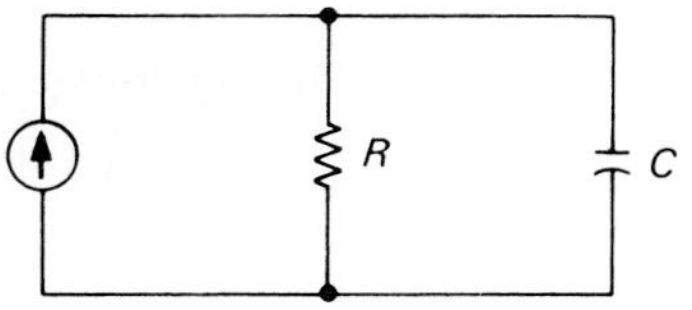

FIGURE 21-11

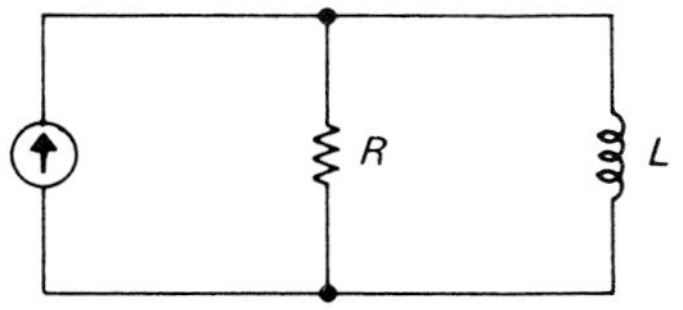

FIGURE 21-12

Solutions

1. $I_R = 6.23$ mA, $I_C = 7.83$ mA, $Y = 161 \angle 51.5° \ \mu$S, $Z = 6.23 \angle -51.5°$ kΩ, and $V = 62.3$ V. In the ESC, $R = 3.88$ kΩ and $C = 32.7$ nF.
2. $I_R = 252 \ \mu$A, $I_C = 163 \ \mu$A, $Y = 14.5 \angle 32.8°$ mS, $Z = 68.9 \angle -32.8° \ \Omega$, and $V = 20.7$ mV. In the ESC, $R = 58 \ \Omega$ and $C = 1.71 \ \mu$F.
3. $I_R = 1.58$ mA, $I_L = 1.23$ mA, $Y = 57.6 \angle -37.9° \ \mu$S, $Z = 17.4 \angle 37.9°$ kΩ, and $V = 34.8$ V. In the ESC, $R = 13.7$ kΩ and $L = 113$ mH.
4. $I_R = 556 \ \mu$A, $I_L = 831 \ \mu$A, $Y = 383 \angle -56.2° \ \mu$S, $Z = 2.61 \angle 56.2°$ kΩ, and $V = 2.61$ V. In the ESC, $R = 1.45$ kΩ and $L = 138$ mH.
5. $I_R = 43.2 \ \mu$A, $I_C = 1.2$ mA, $I_L = 1.11$ mA, $V = 518$ mV, $Y = 193 \angle 64.4° \ \mu$S, and $Z = 5.18 \angle -64.4°$ kΩ. In the ESC, $R = 2.24$ kΩ and $C = 9.21$ nF.
6. $I_R = 3.84$ mA, $I_C = 18.1$ mA, $I_L = 32.6$ mA, $V = 76.8$ V, $Y = 195 \angle -75.2° \ \mu$S, and $Z = 5.12 \angle 75.2°$ kΩ. In the ESC, $R = 1.31$ kΩ and $L = 105$ mH.

SELF-TEST 21-2

1. Find Z_T, where $Z_1 = 7 \angle 25°$ kΩ and $Z_2 = 10 \angle -60°$ kΩ.
2. Refer to Figure 21-11. Find the branch currents, Y, Z, V, and the ESC, where $R = 1.8$ kΩ, $C = 200$ nF, $f = 350$ Hz, and $I_T = 200 \ \mu$A.
3. Refer to Figure 21-12. Find the branch currents, Y, Z, V, and the ESC, where $R = 22$ kΩ, $f = 15$ kHz, $L = 300$ mH, and $I_T = 20$ mA.
4. Refer to Figure 21-9. Find the branch currents, Y, Z, V, and the ESC, where $R = 10$ kΩ, $L = 2$ mH, $C = 500$ pF, $f = 150$ kHz, and $I_T = 500 \ \mu$A.

Answers to Self-test 21-2 are at the end of the chapter.

21-4 AC NETWORKS

As we get into more complex ac circuit problems, many series–parallel conversions become necessary. The methods shown previously may be shortened by the use of equations that we will call *transforms*. These transforms allow us to equate series resistance and reactance to their parallel equivalent conductance and susceptance, or vice versa.

To convert from a series circuit to its equivalent parallel circuit, the equations are

$$G = \frac{R}{R^2 + X^2} \quad \text{and} \quad B = \frac{X}{R^2 + X^2} \tag{21-3}$$

To convert from a parallel circuit to its equivalent series circuit, the equations are

$$R = \frac{G}{G^2 + B^2} \quad \text{and} \quad X = \frac{B}{G^2 + B^2} \tag{21-4}$$

These equations yield absolute values. Signs must be assigned to reactive and susceptive values, depending on circuit components, when problem solving.

Consider the circuit in Figure 21-7 again. In that circuit, the conductance, G, is 303 μS and the susceptance, B, is 236 μS. Let's convert to its equivalent series circuit using Equation 21-4.

$$R = \frac{G}{G^2 + B_C^2} = \frac{303 \times 10^{-6}}{(303 \times 10^{-6})^2 + (236 \times 10^{-6})^2} = 2.05\ \text{k}\Omega$$

303 [EE] [+/-] 6 [÷] [(] 303 [EE] [+/-] 6 [x^2] [+] 236 [EE] [+/-] 6 [x^2] [)] [STO] [=] *Answer*: $R = 2.05\ \text{k}\Omega$

Store the denominator because it is used in the next part.

$$X_C = \frac{B_C}{G^2 + B_C^2} = \frac{236 \times 10^{-6}}{(303 \times 10^{-6})^2 + (236 \times 10^{-6})^2} = 1.6\ \text{k}\Omega$$

236 [EE] [+/-] 6 [÷] [RCL] [=] *Answer*: $X_C = 1.60\ \text{k}\Omega$

In problems of this type, the denominators are the same in the two equations, so store the denominator the first time you compute it to save time in solving the second equation.

The equivalent series circuit in rectangular form for this problem would be

$$Z = R - jX_C = 2.05\ \text{k}\Omega - j1.6\ \text{k}\Omega$$

and the circuit of Figure 21-8 results.

☐ EXAMPLE 21-12

Consider the circuit in Figure 21-13. (It's like the circuit described in Example 20-11.) $X_L = 3.77\ \text{k}\Omega$ and the impedance in rectangular form is

$$Z = 2.7\ \text{k}\Omega + j3.77\ \text{k}\Omega$$

Let's convert to the equivalent parallel circuit using Equation 21-3.

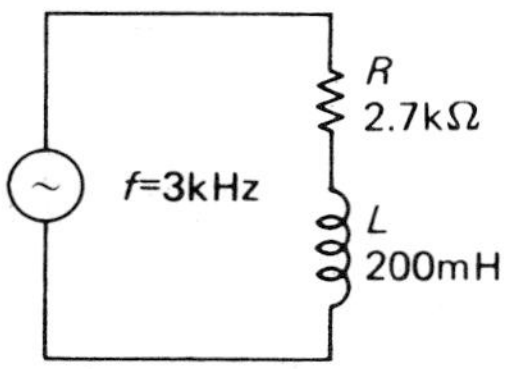

FIGURE 21-13

Solution

$$G = \frac{R}{R^2 + X_L^2} = \frac{2.7 \times 10^3}{(2.7 \times 10^3)^2 + (3.77 \times 10^3)^2} = 126\ \mu\text{S}$$

$$B_L = \frac{X_L}{R^2 + X_L^2} = \frac{3.77 \times 10^3}{(2.7 \times 10^3)^2 + (3.77 \times 10^3)^2} = 175\ \mu\text{S}$$

$$L = \frac{1}{2\pi f B_L} = \frac{1}{6.28 \times 3 \times 10^3 \times 175 \times 10^{-6}} = 303\ \text{mH}$$

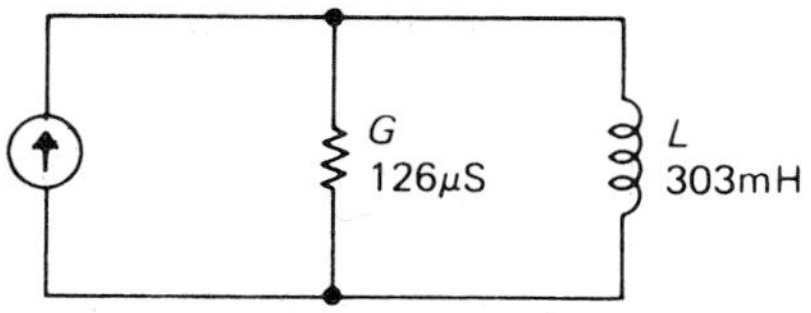

FIGURE 21-14

The equivalent parallel circuit in rectangular form is

$$Y = G - jB_L = 126\ \mu\text{S} - j175\ \mu\text{S}$$

The equivalent parallel circuit is drawn in Figure 21-14. ☐

PRACTICE PROBLEMS 21-7

Find Y_{rect} and C or L in the equivalent parallel circuit for the following problems.

1. In a series RC circuit where $R = 4.7$ kΩ, $C = 0.05$ μF, and $f = 500$ Hz.
2. In a series RL circuit where $R = 27$ kΩ, $L = 200$ mH, and $f = 15$ kHz.

Find Z_{rect} and C or L in the equivalent series circuit for the following problems.

3. In a parallel RC circuit where $G = 66.7$ μS, $C = 500$ pF, and $f = 30$ kHz.
4. In a parallel RL circuit where $G = 500$ μS, $L = 500$ mH, and $f = 500$ Hz.

Solutions

1. $Y = 75.1$ μS $+ j102$ μS, $C = 32.5$ nF
2. $Y = 24.9$ μS $- j17.4$ μS, $L = 610$ mH
3. $Z = 5.00$ kΩ $- j7.07$ kΩ, $C = 750$ pF
4. $Z = 763$ Ω $+ j972$ Ω, $L = 309$ mH

Additional practice problems are at the end of the chapter.

Let's see how we might use these transforms to solve complex circuit problems.

□ EXAMPLE 21-13

Find the equivalent series circuit and the phase angle of the circuit in Figure 21-15.

Solution First, let's convert the series RL circuit to its equivalent parallel circuit.

$$X_L = 2\pi fL = 6.28 \times 10 \text{ kHz} \times 100 \text{ mH} = 6.28 \text{ k}\Omega$$

$$G = \frac{R}{R^2 + X_L^2} = \frac{100}{100^2 + (6.28 \times 10^3)^2} = 2.53\ \mu\text{S}$$

$$B_L = \frac{X_L}{R^2 + X_L^2} = \frac{6.28 \times 10^3}{100^2 + (6.28 \times 10^3)^2} = 159\ \mu\text{S}$$

Next, solve for B_C:

$$B_C = 2\pi fC = 6.28 \times 10 \text{ kHz} \times 3 \text{ nF} = 188\ \mu\text{S}$$

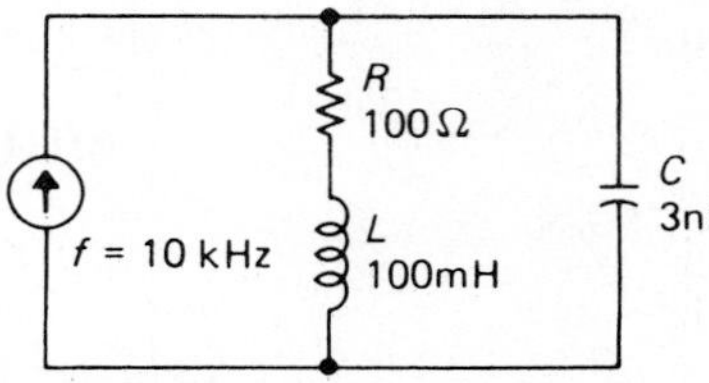

FIGURE 21-15

The parallel circuit in rectangular form is

$$Y = G + jB_C - jB_L = 2.53\ \mu S + j188\ \mu S - j159\ \mu S$$
$$= 2.53\ \mu S + j29.0\ \mu S$$

The equivalent parallel circuit is R and C and is shown in Figure 21-16. Use Equation 21-4 to find the equivalent series circuit.

$$R = \frac{G}{G^2 + B_C^2} = \frac{2.53 \times 10^{-6}}{(2.53 \times 10^{-6})^2 + (29 \times 10^{-6})^2} = 2.99\ k\Omega$$

$$X_C = \frac{B_C}{G^2 + B_C^2} = \frac{29 \times 10^{-6}}{(2.53 \times 10^{-6})^2 + (29 \times 10^{-6})^2} = 34.2\ k\Omega$$

The equivalent series circuit in rectangular form is

$$Z = 2.99\ k\Omega - j34.2\ k\Omega$$

The equivalent series capacitance is

$$C = \frac{1}{2\pi f X_C} = \frac{1}{6.28 \times 10 \times 10^3 \times 34.2 \times 10^3} = 465\ pF$$

The phase angle is

$$\tan\theta = \frac{X_C}{R} = \frac{34.2\ k\Omega}{2.99\ k\Omega} = 11.4$$

$$\tan^{-1} 11.4 = 85°$$

The ESC is shown in Figure 21-17.

This is a circuit encountered whenever L and C are in parallel. The inductance usually contains some measurable series resistance because of the resistance of the wire that makes up the inductor. The result, then, of L and C in parallel is really a circuit such as the one in Figure 21-15, and it is this circuit configuration you must consider when finding the various circuit variables.

☐ EXAMPLE 21-14

Find the ESCs for the circuit in Figure 21-18.

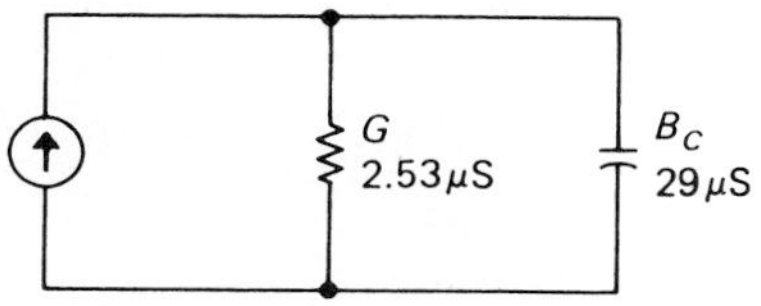

FIGURE 21-16

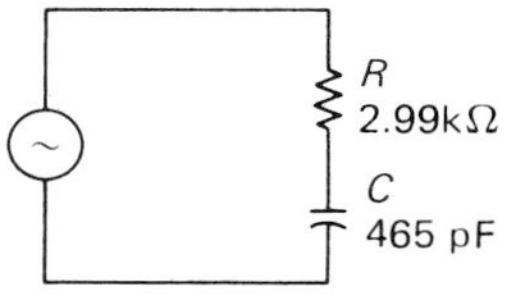

FIGURE 21-17

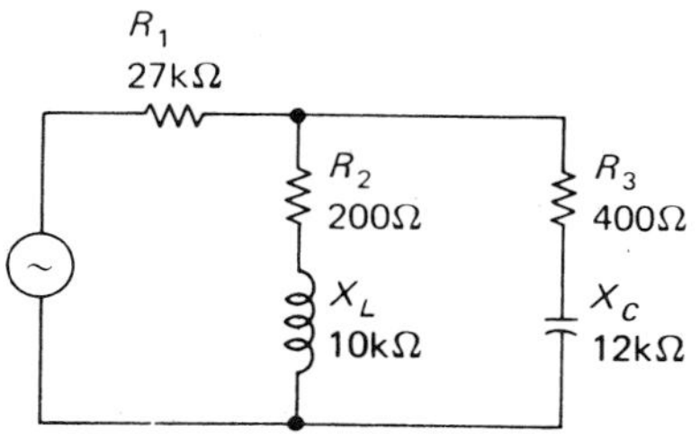

FIGURE 21-18

Solution Convert the *RL* series circuit (call it branch *X*) to its EPC using Equation 21-3.

$$G_X = \frac{R_2}{R_2^2 + X_L^2} = \frac{200}{200^2 + (10 \times 10^3)^2} = 2.00\ \mu\text{S}$$

$$B_L = \frac{X_L}{R_2^2 + X_L^2} = \frac{10 \times 10^3}{200^2 + (10 \times 10^3)^2} = 100\ \mu\text{S}$$

Next, convert the *RC* series circuit (call it branch *Y*) to its EPC using Equation 21-3.

$$G_Y = \frac{R_3}{R_3^2 + X_C^2} = \frac{400}{400^2 + (12 \times 10^3)^2} = 2.77\ \mu\text{S}$$

$$B_C = \frac{X_C}{R_3^2 + X_C^2} = \frac{12 \times 10^3}{400^2 + (12 \times 10^3)^2} = 83.2\ \mu\text{S}$$

Figure 21-19(a) is the circuit at this point. The EPC is

$$\begin{aligned} Y &= G_X + G_Y + jB_C - jB_L \\ &= 2.00\ \mu\text{S} + 2.77\ \mu\text{S} + j83.2\ \mu\text{S} - j100\ \mu\text{S} \\ &= 4.77\ \mu\text{S} - j16.8\ \mu\text{S} \end{aligned}$$

At this point the circuit looks like Figure 21-19(b).

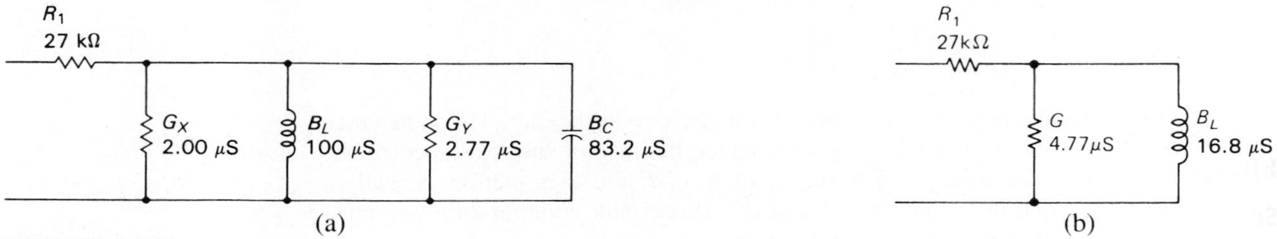

FIGURE 21-19

Next, find the equivalent series circuit of the parallel portion of Figure 21-18.

$$R_{eq} = \frac{G}{G^2 + B_L^2} = \frac{4.77 \times 10^{-6}}{(4.77 \times 10^{-6})^2 + (16.8 \times 10^{-6})^2}$$
$$= 15.6\ \text{k}\Omega$$

$$X_L = \frac{X_L}{G^2 + B_L^2} = \frac{16.8 \times 10^{-6}}{(4.77 \times 10^{-6})^2 + (16.8 \times 10^{-6})^2}$$
$$= 55.1\ \text{k}\Omega$$

The equivalent series circuit in rectangular form is

$$Z = R_1 + R_{eq} + jX_L = 42.6\ \text{k}\Omega + j55.1\ \text{k}\Omega$$

and is drawn in Figure 21-20. □

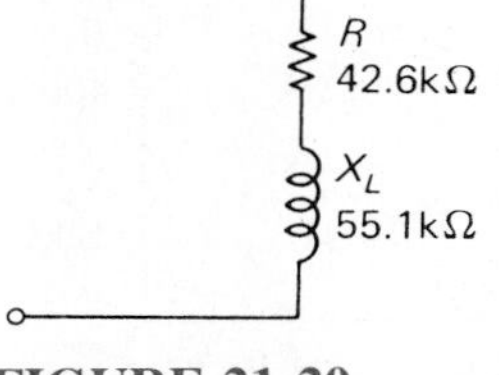

FIGURE 21-20

PRACTICE PROBLEMS 21-8

Find the equivalent series circuit for the following problems. Express answers in rectangular form.

1. Refer to Figure 21-15. Let $R = 2\ \text{k}\Omega$, $X_L = 3\ \text{k}\Omega$, and $X_C = 5\ \text{k}\Omega$.
2. Refer to Figure 21-21.
3. Refer to Figure 21-22.
4. Refer to Figure 21-23. In addition to the ESC, find the component values and the phase angle.

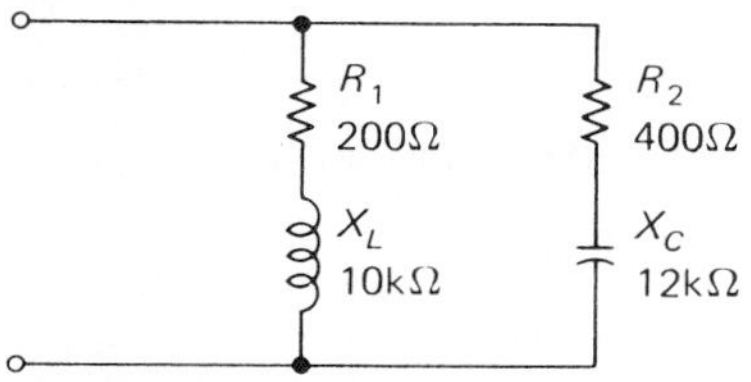

FIGURE 21-21

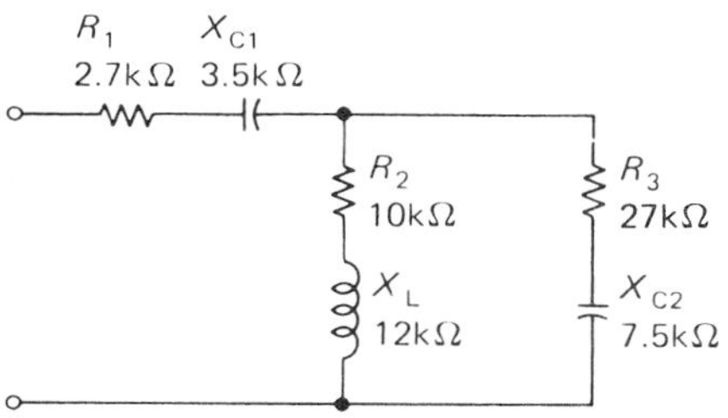

FIGURE 21-22

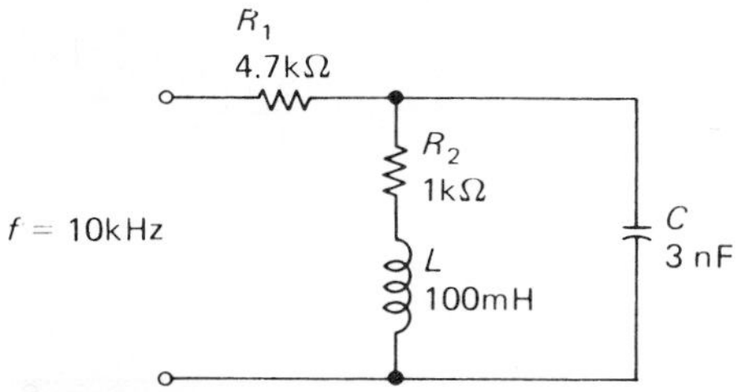

FIGURE 21-23

Solutions

1. The RL branch conversion is

$$Y = 154\ \mu\text{S} - j231\ \mu\text{S}$$

The equivalent parallel circuit in rectangular form is

$$Y = 154\ \mu\text{S} + j200\ \mu\text{S} - j231\ \mu\text{S}$$
$$= 154\ \mu\text{S} - j30.8\ \mu\text{S}$$

Converting to the ESC, we get

$$Z = 6.25\ \text{k}\Omega + j1.25\ \text{k}\Omega$$

The ESC is resistive and inductive.

2. The RL branch conversion is

$$Y = 2.00\ \mu\text{S} - 100\ \mu\text{S}$$

The RC branch conversion is

$$Y = 2.78\ \mu\text{S} + j83.2\ \mu\text{S}$$

The resultant equivalent parallel circuit is

$$Y = G_1 + G_2 + jB_C - jB_L$$
$$= 2.00\ \mu\text{S} + 2.78\ \mu\text{S} + j83.2\ \mu\text{S} - j100\ \mu\text{S}$$
$$= 4.77\ \mu\text{S} - j16.7\ \mu\text{S}$$

Converting to the ESC, we get

$$Z = 15.8\ \text{k}\Omega + j55.3\ \text{k}\Omega$$

The circuit is resistive and inductive.

3. The *RL* branch conversion is

$$Y_L = G_2 - jB_L = 41.0\ \mu S - j49.2\ \mu S$$

The *RC* branch conversion is

$$Y_C = G_3 + jB_C = 34.4\ \mu S + j9.55\ \mu S$$

The EPC of the two branches in rectangular form is

$$\begin{aligned} Y &= G_2 + G_3 + jB_C - jB_L \\ &= 41.0\ \mu S + 34.4\ \mu S + j9.55\ \mu S \\ &\quad - j49.2\ \mu S \\ &= 75.4\ \mu S - j39.6\ \mu S \end{aligned}$$

Equation 21-4 yields an ESC of 10.4 kΩ + j5.47 kΩ. The ESC for the entire circuit is

$$\begin{aligned} Z &= R_1 + R_{eq} + jX_L - jX_C \\ &= 2.7\ k\Omega + 10.4\ k\Omega + j5.47\ k\Omega \\ &\quad - j3.5\ k\Omega \\ &= 13.1\ k\Omega + j1.97\ k\Omega \end{aligned}$$

The equivalent series circuit is resistive and inductive.

4. In the *RL* branch $X_L = 2\pi fL = 6.28$ kΩ and $R =$ 4.7 kΩ. Equation 21-3 yields an EPC of 24.7 μS − j155 μS. In the capacitive branch $B_C = 2\pi fC =$ 189 μS.

$$\begin{aligned} Y_T &= G + jB_C - jB_L \\ &= 24.7\ \mu S - j155\ \mu S + j189\ \mu S \\ &= 24.7\ \mu S + j33.3\ \mu S \end{aligned}$$

(this is the EPC of the two branches)

Equation 21-4 yields an ESC of 14.4 kΩ − j19.4 kΩ.

$$\begin{aligned} Z_{ckt} &= R_1 + R_{eq} - jX_C \\ &= 4.7\ k\Omega + j14.4\ k\Omega - j19.4\ k\Omega \\ &= 19.1\ k\Omega - j19.4\ k\Omega \end{aligned}$$

The circuit is resistive and capacitive. $R =$ 19.1 kΩ.

$$C = \frac{1}{2\pi f X_C} = 820\ pF$$

$$\text{phase angle} = \tan^{-1}\frac{X_C}{R} = \tan^{-1}\frac{19.4\ k\Omega}{19.1\ k\Omega}$$
$$= 45.5°$$

Additional practice problems are at the end of the chapter.

SELF-TEST 21-3

1. Refer to Figure 21-24. Let $f = 7$ kHz, $L =$ 0.3 H, $C = 2$ nF, and $R = 1$ kΩ. Find the ESC, the component values, and the phase angle.

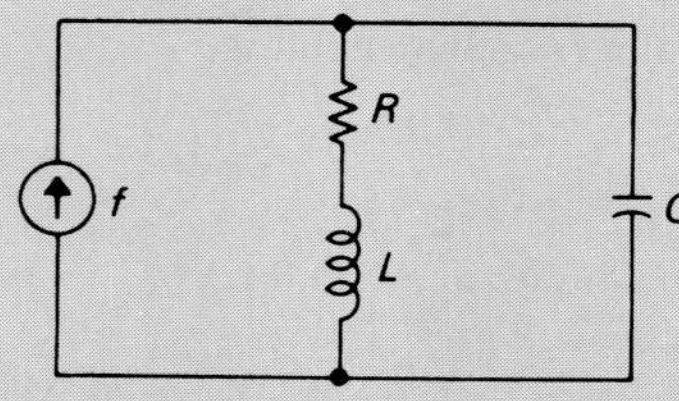

FIGURE 21-24

Answers to Self-test 21-3 are at the end of the chapter.

2. Refer to Figure 21-22. Let $R_1 = 1$ kΩ, $X_{C1} =$ 5 kΩ, $R_2 = 10$ kΩ, $R_3 = 15$ kΩ, $X_L = 2$ kΩ, and $X_{C2} = 7.5$ kΩ. Find the ESC resistance and reactance. Express the answer in rectangular form.

CHAPTER 21 AT A GLANCE

Page	Key Points and Equations	Example
532	*Key point:* In a parallel circuit, we use E as the reference. In an RC circuit, current leads, so I_C and B_C are plotted along the positive y-axis. I_T (or Y) is the hypotenuse and θ is the phase angle.	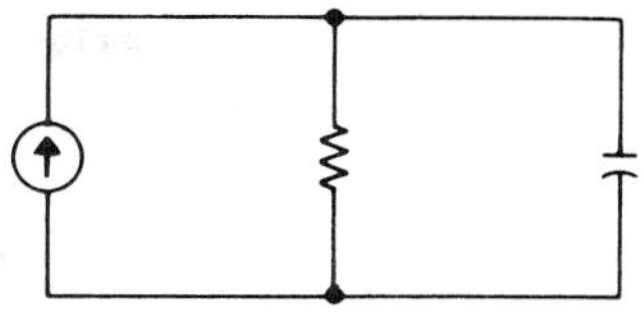 **FIGURE 21-25**
536	*Key point:* In a parallel circuit, we use E as the reference. In an RL circuit, voltage leads, so I_L and B_L are plotted along the negative y-axis. I_T (or Y) is the hypotenuse and θ is the phase angle.	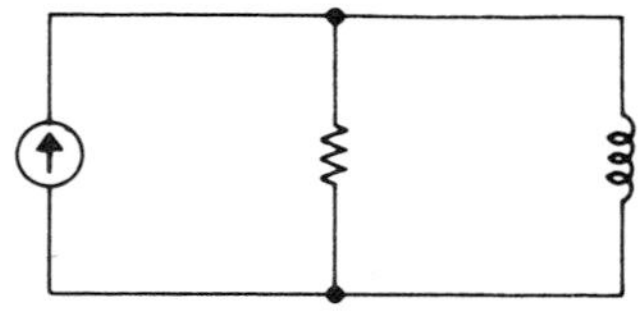 **FIGURE 21-26**
541	*Key point:* When multiplying phasors, put in polar form. Multiply the magnitudes and add the angles.	$3\angle 40^\circ\ \text{k}\Omega \times 4\angle -25.3^\circ\ \text{k}\Omega = 12\angle 14.7^\circ\ \text{k}\Omega$
541	When dividing phasors, divide the magnitudes and subtract the angles.	$3\angle 40^\circ\ \text{k}\Omega \div 4\angle -25.3^\circ\ \text{k}\Omega = 750\angle 65.3^\circ\ \Omega$
546	To convert from a series circuit to its equivalent parallel circuit, the equations are $G = \dfrac{R}{R^2 + X^2}$ and $B = \dfrac{X}{R^2 + X^2}$	Given: $R = 2.7\ \text{k}\Omega$ and $X_L = 3.77\ \text{k}\Omega$. $G = \dfrac{R}{R^2 + X_L^2} = \dfrac{2.7 \times 10^3}{(2.7 \times 10^3)^2 + (3.77 \times 10^3)^2} = 126\ \mu\text{S}$ $B_L = \dfrac{X_L}{R^2 + X_L^2} = \dfrac{3.77 \times 10^3}{(2.7 \times 10^3)^2 + (3.77 \times 10^3)^2} = 175\ \mu\text{S}$
546	To convert from a parallel circuit to its equivalent series circuit, the equations are $R = \dfrac{G}{G^2 + B^2}$ and $X = \dfrac{B}{G^2 + B^2}$	Given: $G = 303\ \mu\text{S}$ and $B_C = 236\ \mu\text{S}$. $X_C = \dfrac{B_C}{G^2 + B_C^2} = \dfrac{236 \times 10^{-6}}{(303 \times 10^{-6})^2 + (236 \times 10^{-6})^2} = 1.60\ \text{k}\Omega$ $R = \dfrac{G}{G^2 + B_C^2} = \dfrac{303 \times 10^{-6}}{(303 \times 10^{-6})^2 + (236 \times 10^{-6})^2} = 2.05\ \text{k}\Omega$

End of Chapter Problems 21-1

Refer to Figure 21-27 and express I_T in *rectangular* and *polar* forms:

1. $I_T = 40$ mA and $\theta = 54°$
2. $I_T = 500\ \mu$A and $\theta = 35°$
3. $I_T = 2.75$ mA and $\theta = 67.4°$
4. $I_T = 7.65$ mA and $\theta = 47.5°$
5. $I_T = 175\ \mu$A and $\theta = 22°$
6. $I_T = 120\ \mu$A and $\theta = 70°$
7. $I_R = 350\ \mu$A and $I_C = 270\ \mu$A
8. $I_R = 63.5$ mA and $I_C = 24$ mA
9. $I_R = 4.77$ mA and $I_C = 7.05$ mA
10. $I_R = 9.5$ mA and $I_C = 9.4$ mA
11. $I_R = 100\ \mu$A and $I_C = 100\ \mu$A
12. $I_R = 1.45$ mA and $I_C = 750\ \mu$A
13. $I_R = 1$ mA and $\theta = 33°$
14. $I_R = 4.5$ mA and $\theta = 57°$
15. $I_R = 640\ \mu$A and $\theta = 20.5°$
16. $I_R = 170\ \mu$A and $\theta = 58.7°$
17. $I_R = 25.3$ mA and $\theta = 62.0°$
18. $I_R = 47.3$ mA and $\theta = 37.5°$
19. $I_C = 45$ mA and $\theta = 71°$
20. $I_C = 33$ mA and $\theta = 18.7°$
21. $I_C = 5.44$ mA and $\theta = 59.5°$
22. $I_C = 9$ mA and $\theta = 45°$
23. $I_C = 275\ \mu$A and $\theta = 55.7°$
24. $I_C = 805\ \mu$A and $\theta = 32.1°$
25. $I_R = 7.5$ mA and $I_T = 12$ mA
26. $I_R = 25$ mA and $I_T = 40$ mA
27. $I_R = 650\ \mu$A and $I_T = 1$ mA
28. $I_R = 275\ \mu$A and $I_T = 400\ \mu$A
29. $I_R = 2.55$ mA and $I_T = 3.15$ mA
30. $I_R = 120\ \mu$A and $I_T = 150\ \mu$A
31. $I_C = 25$ mA and $I_T = 43$ mA
32. $I_C = 13.5$ mA and $I_T = 30$ mA
33. $I_C = 200\ \mu$A and $I_T = 400\ \mu$A
34. $I_C = 900\ \mu$A and $I_T = 2$ mA
35. $I_C = 3.55$ mA and $I_T = 6.08$ mA
36. $I_C = 875\ \mu$A and $I_T = 1.12$ mA

End of Chapter Problems 21-2

Refer to Figure 21-27 for problems 1 through 10. Find Y_{rect}, Y_{polar}, I_{rect}, and V.

1. $R = 3.3$ kΩ, $C = 30$ nF, $f = 2$ kHz, $I_T = 4$ mA
2. $R = 20$ kΩ, $C = 200$ nF, $f = 100$ Hz, $I_T = 500\ \mu$A
3. $R = 750$ Ω, $C = 5$ nF, $f = 60$ kHz, $I_T = 1$ mA
4. $R = 12$ kΩ, $C = 500$ pF, $f = 35$ kHz, $I_T = 25$ mA
5. $R = 6.8$ kΩ, $C = 10$ nF, $f = 1.5$ kHz, $I_T = 20$ mA
6. $R = 47$ kΩ, $C = 0.1\ \mu$F, $f = 50$ Hz, $I_T = 40\ \mu$A
7. $R = 10$ kΩ, $C = 30$ nF, $f = 1$ kHz, $I_T = 10$ mA
8. $R = 82$ Ω, $C = 0.75\ \mu$F, $f = 2.5$ kHz, $I_T = 300\ \mu$A
9. $R = 15$ kΩ, $C = 300$ pF, $f = 35$ kHz, $I_T = 2$ mA
10. $R = 10$ kΩ, $C = 1$ nF, $f = 15$ kHz, $I_T = 7.5$ mA
11. Refer to Figure 21-27. $B_C = 200\ \mu$S, $C = 25$ nF, $\theta = 50°$, and $I_R = 3$ mA. Find R, f, I_T, and V.
12. Refer to Figure 21-27. $B_C = 3.72$ mS, $C = 200$ nF, $\theta = 27°$, and $I_R = 30\ \mu$A. Find R, f, I_T, and V.

End of Chapter Problems 21-3

Refer to Figure 21-28 and express I_T in *rectangular* and *polar* forms.

1. $I_T = 3.8$ mA and $\theta = -30°$
2. $I_T = 7.3$ mA and $\theta = -56°$
3. $I_T = 210\ \mu$A and $\theta = -39°$
4. $I_T = 420\ \mu$A and $\theta = -49.5°$

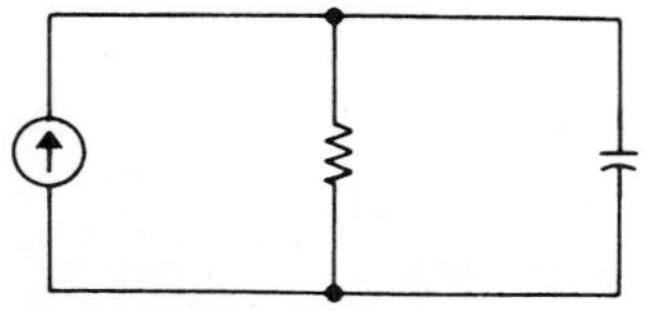
FIGURE 21-27

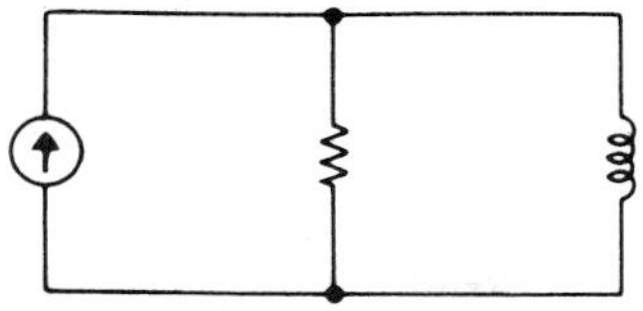
FIGURE 21-28

5. I_T = 1.34 mA and θ = −23.4°
6. I_T = 850 μA and θ = −65.4°
7. I_R = 12 mA and I_L = −8 mA
8. I_R = 18 mA and I_L = −26 mA
9. I_R = 30 mA and I_L = −40 mA
10. I_R = 550 μA and I_L = −200 μA
11. I_R = 150 μA and I_L = −180 μA
12. I_R = 2.35 mA and I_L = −4.73 mA
13. I_R = 2 mA and θ = −28.5°
14. I_R = 3 mA and θ = −75°
15. I_R = 1.3 mA and θ = −73°
16. I_R = 2.4 mA and θ = −15°
17. I_R = 37.6 μA and θ = −27.4°
18. I_R = 350 μA and θ = −62.7°
19. I_L = −730 mA and θ = −45°
20. I_L = −600 mA and θ = −36.7°
21. I_L = −9 mA and θ = −40°
22. I_L = −8 mA and θ = −50°
23. I_L = −475 μA and θ = −43.5°
24. I_L = −370 μA and θ = −55.5°
25. I_L = −20 mA and I_T = 27 mA
26. I_L = −7 mA and I_T = 12 mA
27. I_L = −150 μA and I_T = 300 μA
28. I_L = −225 μA and I_T = 325 μA
29. I_L = −925 μA and I_T = 1.83 mA
30. I_L = −830 μA and I_T = 2.23 mA
31. I_R = 4.7 mA and I_T = 6.7 mA
32. I_R = 3.35 mA and I_T = 7.35 mA
33. I_R = 825 μA and I_T = 1.35 mA
34. I_R = 670 μA and I_T = 1.15 mA
35. I_R = 73.8 mA and I_T = 100 mA
36. I_R = 87.5 mA and I_T = 125 mA

End of Chapter Problems 21-4

Refer to Figure 21-28. Find Y and I_T in both *polar* and *rectangular* forms.

1. R = 20 kΩ, L = 20 mH, f = 200 kHz, I_T = 40 mA
2. R = 1 kΩ, L = 400 mH, f = 600 Hz, I_T = 500 μA
3. R = 10 kΩ, L = 60 mH, f = 20 kHz, I_T = 200 mA
4. R = 3.3 kΩ, L = 5 mH, f = 65 kHz, I_T = 15 mA
5. R = 4.7 kΩ, L = 100 mH, f = 10 kHz, I_T = 100 μA
6. R = 18 kΩ, L = 50 mH, f = 50 kHz, I_T = 1 mA
7. R = 27 kΩ, L = 2.5 H, f = 2 kHz, I_T = 400 μA
8. R = 3.9 kΩ, L = 150 mH, f = 6 kHz, I_T = 2.7 mA
9. R = 750 Ω, L = 600 mH, f = 400 Hz, I_T = 30 mA
10. R = 1 kΩ, L = 400 mH, f = 500 Hz, I_T = 12.5 mA

End of Chapter Problems 21-5

1. Find Z_T if $Y_T = 3.72 \underline{/27^\circ}$ mS.
2. Find Z_T if $Y_T = 37 \underline{/-62^\circ}$ μS.
3. Find Y_T if $Z_T = 6.32 \underline{/-40^\circ}$ kΩ.
4. Find Y_T if $Z_T = 16.7 \underline{/70^\circ}$ kΩ.

Find Z_T in the following problems by using the equation $Z_T = \dfrac{Z_1 Z_2}{Z_1 + Z_2}$.

5. $Z_1 = 700 \angle 70° \ \Omega$, $Z_2 = 1.5 \angle -25° \ \text{k}\Omega$
6. $Z_1 = 12 \angle -65° \ \text{k}\Omega$, $Z_2 = 4 \angle -40° \ \text{k}\Omega$
7. $Z_1 = 35 \angle 20° \ \text{k}\Omega$, $Z_2 = 10 \angle -40° \ \text{k}\Omega$
8. $Z_1 = 5.6 \angle 45° \ \text{k}\Omega$, $Z_2 = 10 \angle -70° \ \text{k}\Omega$
9. $Z_1 = 3.3 \angle 0° \ \text{k}\Omega$, $Z_2 = 4.2 \angle -90° \ \text{k}\Omega$
10. $Z_1 = 12 \angle 0° \ \text{k}\Omega$, $Z_2 = 16 \angle 90° \ \text{k}\Omega$

End of Chapter Problems 21-6

Find the branch currents, Y, Z, V, and the ESC in the following problems:

1. Refer to Figure 21-27. $f = 35$ kHz, $C = 400$ pF, $R = 15$ kΩ, and $I_T = 2$ mA.
2. Refer to Figure 21-27. $f = 15$ kHz, $C = 1$ nF, $R = 7.5$ kΩ, and $I_T = 7.5$ mA.
3. Refer to Figure 21-27. $f = 3$ kHz, $C = 150$ nF, $R = 470$ Ω, and $I_T = 40$ mA.
4. Refer to Figure 21-27. $f = 200$ Hz, $C = 150$ nF, $R = 3.3$ kΩ, and $I_T = 100$ μA.
5. Refer to Figure 21-28. $f = 150$ kHz, $L = 60$ mH, $R = 47$ kΩ, and $I_T = 250$ μA.
6. Refer to Figure 21-28. $f = 500$ Hz, $L = 3$ H, $R = 12$ kΩ, and $I_T = 500$ μA.
7. Refer to Figure 21-28. $f = 5$ kHz, $L = 30$ mH, $R = 680$ Ω, and $I_T = 5$ mA.
8. Refer to Figure 21-28. $f = 10$ kHz, $L = 15$ mH, $R = 2$ kΩ, and $I_T = 100$ μA.
9. Refer to Figure 21-29. $f = 25$ kHz, $L = 30$ mH, $C = 300$ pF, $R = 10$ kΩ, and $I_T = 20$ mA.
10. Refer to Figure 21-29. $f = 6$ kHz, $L = 26.5$ mH, $C = 150$ nF, $R = 300$ Ω, and $I_T = 250$ μA.
11. Refer to Figure 21-29. $f = 15$ kHz, $L = 4$ mH, $C = 30$ nF, $R = 6.8$ kΩ, and $I_T = 400$ μA.
12. Refer to Figure 21-29. $f = 250$ Hz, $L = 70$ mH, $C = 6$ μF, $R = 680$ Ω, and $I_T = 150$ μA.

End of Chapter Problems 21-7

Find Y_{rect} and C or L in the equivalent parallel circuit for the following problems:

1. In a series RC circuit where $R = 680$ Ω, $C = 50$ nF, and $f = 4$ kHz.
2. In a series RC circuit where $R = 1.2$ kΩ, $C = 100$ nF, and $f = 1$ kHz.
3. In a series RC circuit where $R = 15$ kΩ, $C = 100$ pF, and $f = 120$ kHz.
4. In a series RC circuit where $R = 9.1$ kΩ, $C = 1$ nF, and $f = 40$ kHz.
5. In a series RL circuit where $R = 2.2$ kΩ, $L = 100$ mH, and $f = 2.65$ kHz.
6. In a series RL circuit where $R = 5.6$ kΩ, $L = 1.5$ H, and $f = 500$ Hz.
7. In a series RL circuit where $R = 15$ kΩ, $L = 200$ mH, and $f = 10$ kHz.
8. In a series RL circuit where $R = 4.7$ kΩ, $L = 0.5$ H, and $f = 2$ kHz.

Find Z_{rect} and C or L in the equivalent series circuit for the following problems:

9. In a parallel RC circuit where $G = 256$ μS, $C = 2$ nF, and $f = 10$ kHz.
10. In a parallel RC circuit where $G = 556$ μS, $C = 30$ nF, and $f = 1.55$ kHz.

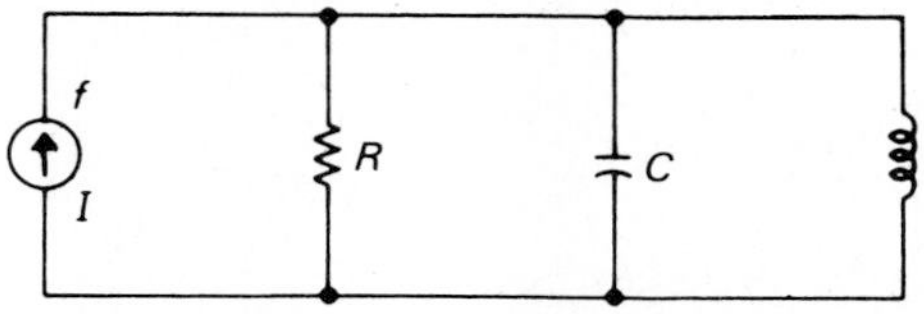

FIGURE 21-29

11. In a parallel RC circuit where $G = 1.22$ mS, $C = 100$ nF, and $f = 3.25$ kHz.

12. In a parallel RC circuit where $G = 833\ \mu$S, $C = 500$ nF, and $f = 800$ Hz.

13. In a parallel RL circuit where $G = 14.7$ mS, $L = 200$ mH, and $f = 100$ Hz.

14. In a parallel RL circuit where $G = 4.55$ mS, $L = 40$ mH, and $f = 3.5$ kHz.

15. In a parallel RL circuit where $G = 110\ \mu$S, $L = 30$ mH, and $f = 8.1$ kHz.

16. In a parallel RL circuit where $G = 40\ \mu$S, $L = 250$ mH, and $f = 7.5$ kHz.

End of Chapter Problems 21-8

For problems 1 to 8, find the ESC. Express answers in rectangular form.

1. Refer to Figure 21-30. Let $R = 500\ \Omega$, $X_L = 10\ k\Omega$, and $X_C = 20\ k\Omega$.

2. Refer to Figure 21-30. Let $R = 100\ \Omega$, $X_L = 250\ \Omega$, and $X_C = 150\ \Omega$.

3. Refer to Figure 21-31. Let $R_1 = 15\ k\Omega$, $R_2 = 10\ k\Omega$, $X_L = 5\ k\Omega$, and $X_C = 5\ k\Omega$.

4. Refer to Figure 21-31. Let $R_1 = 10\ k\Omega$, $R_2 = 4.7\ k\Omega$, $X_L = 2\ k\Omega$, and $X_C = 8\ k\Omega$.

5. Refer to Figure 21-33. Let $R_1 = 5.6\ k\Omega$, $R_2 = 27\ k\Omega$, $X_L = 35\ k\Omega$, and $X_C = 20\ k\Omega$.

6. Refer to Figure 21-33. Let $R_1 = 12\ k\Omega$, $R_2 = 5\ k\Omega$, $X_L = 10\ k\Omega$, and $X_C = 20\ k\Omega$.

7. Refer to Figure 21-32. Let $R_1 = 1.2\ k\Omega$, $R_2 = 12\ k\Omega$, $R_3 = 10\ k\Omega$, $X_L = 12\ k\Omega$, $X_{C1} = 10\ k\Omega$, and $X_{C2} = 3\ k\Omega$.

8. Refer to Figure 21-32. Let $R_1 = 2.7\ k\Omega$, $R_2 = 12\ k\Omega$, $R_3 = 8.2\ k\Omega$, $X_L = 10\ k\Omega$, $X_{C1} = 3.5\ k\Omega$, and $X_{C2} = 7.5\ k\Omega$.

For problems 9 to 16, find the ESC. Include component values and the phase angle.

9. Refer to Figure 21-30. Let $f = 1$ kHz, $R = 1\ k\Omega$, $L = 500$ mH, and $C = 3$ nF.

10. Refer to Figure 21-30. Let $f = 5$ kHz, $R = 1\ k\Omega$, $L = 200$ mH, and $C = 10$ nF.

11. Refer to Figure 21-31. Let $R_1 = 500\ \Omega$, $R_2 = 200\ \Omega$, $L = 250$ mH, $C = 3$ nF, and $f = 5$ kHz.

12. Refer to Figure 21-31. Let $R_1 = 1\ k\Omega$, $R_2 = 470\ \Omega$, $L = 500$ mH, $C = 500$ pF, and $f = 10$ kHz.

13. Refer to Figure 21-32. $R_1 = 3.3\ k\Omega$, $R_2 = 20\ k\Omega$, $R_3 = 6.8\ k\Omega$, $C_1 = 20$ nF, $C_2 = 8$ nF, $L = 3.5$ H, and $f = 1$ kHz.

14. Refer to Figure 21-32. $R_1 = 1\ k\Omega$, $R_2 = 100\ \Omega$, $R_3 = 100\ \Omega$, $C_1 = 1\ \mu$F, $C_2 = 1.2\ \mu$F, $L = 1.6$ H, and $f = 100$ Hz.

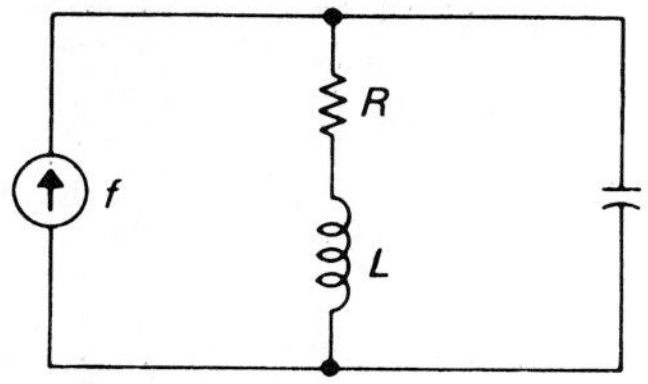

FIGURE 21-30

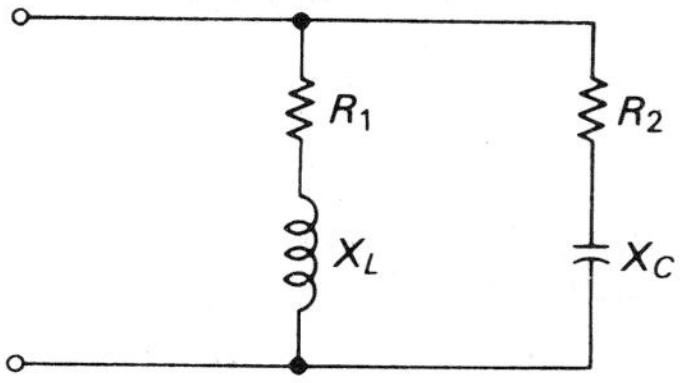

FIGURE 21-31

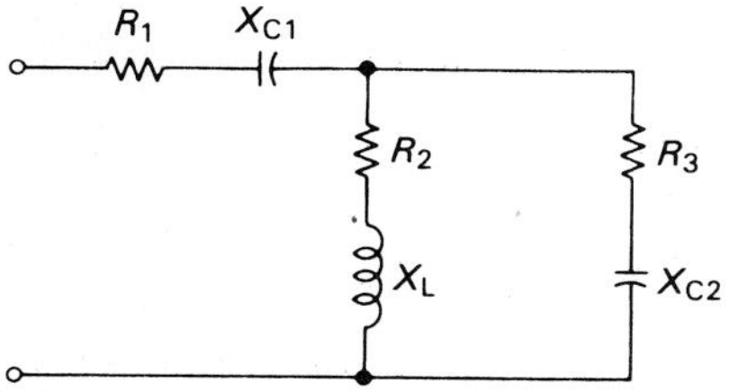

FIGURE 21-32

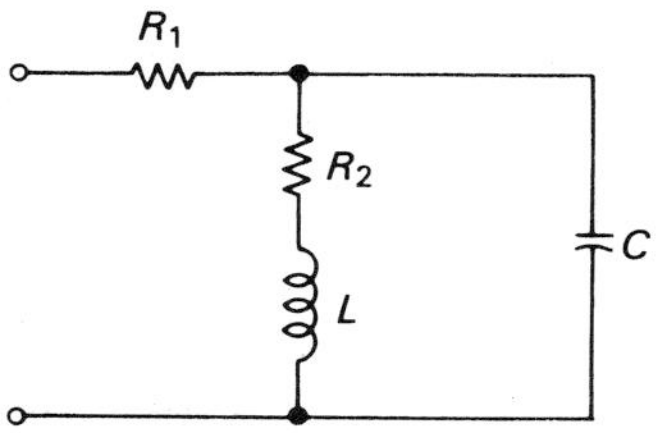

FIGURE 21-33

15. Refer to Figure 21-33. $R_1 = 5.6\ \text{k}\Omega$, $R_2 = 2\ \text{k}\Omega$, $L = 150$ mH, $C = 10$ nF, and $f = 6.7$ kHz.

16. Refer to Figure 21-33. $R_1 = 220\ \Omega$, $R_2 = 470\ \Omega$, $L = 40$ mH, $C = 150$ nF, and $f = 2.7$ kHz.

Answers to Self-tests

SELF-TEST 21-1

1. $Y_{rect} = 100\ \mu S + j67.5\ \mu S$, $Y_{polar} = 121\ \underline{/34°}\ \mu S$, $I_{rect} = 4.14$ mA $+ j2.8$ mA, $I_{polar} = 5\ \underline{/34°}$ mA, $V = 41.3$ V

2. $Y_{rect} = 37\ \mu S + j50.3\ \mu S$, $Y_{polar} = 62.4\ \underline{/53.6°}\ \mu S$, $I_{rect} = 296\ \mu A + j403\ \mu A$, $I_{polar} = 500\ \underline{/53.6°}\ \mu A$, $V = 8.01$ V

3. $Y_{rect} = 8.33\ \mu S - j6.37\ \mu S$, $Y_{polar} = 10.5\ \underline{/-37.4°}\ \mu S$, $I_{rect} = 55.6\ \mu A - j42.5\ \mu A$, $I_{polar} = 70\ \underline{/-37.4°}\ \mu A$, $V = 6.67$ V

4. $Y_{rect} = 14.7\ \mu S - j20\ \mu S$, $Y_{polar} = 24.8\ \underline{/-53.7°}\ \mu S$, $I_{rect} = 1.18$ mA $- j1.61$ mA, $I_{polar} = 2\ \underline{/-53.7°}$ mA, V = 80.6 V

5. $Y_{rect} = 37.1\ \mu S - j53.1\ \mu S$, $Y_{polar} = 64.8\ \underline{/-55°}\ \mu S$, $I_{rect} = 1.15$ mA $- j1.64$ mA, $I_{polar} = 2\ \underline{/-55°}$ mA, $V = 30.9$ V

SELF-TEST 21-2

1. $5.51\ \underline{/-8.3°}\ \text{k}\Omega$

2. $I_R = 157\ \mu A$, $I_C = 124\ \mu A$, $Y_{polar} = 709\ \underline{/38.3°}\ \mu S$, $Z_{polar} = 1.41\ \underline{/-38.3°}\ \text{k}\Omega$, and $V = 282$ mV. In the ESC, $R = 1.11\ \text{k}\Omega$ and $C = 520$ nF.

3. $I_R = 15.8$ mA, $I_L = -12.3$ mA, $Y_{polar} = 57.6\ \underline{/-37.9°}\ \mu S$, $Z_{polar} = 17.4\ \underline{/37.9°}\ \text{k}\Omega$, and $V = 348$ V. In the ESC, $R = 13.7\ \text{k}\Omega$ and $L = 113$ mH.

4. $I_R = 430\ \mu A$, $I_C = 2.03$ mA, $I_L = -2.28$ mA, $Y_{polar} = 116\ \underline{/-30.7°}\ \mu S$, $Z_{polar} = 8.6\ \underline{/30.7°}\ \text{k}\Omega$, and $V = 4.3$ V. In the ESC, $R = 7.4\ \text{k}\Omega$ and $L = 4.65$ mH.

SELF-TEST 21-3

1. The ESC in rectangular form is $Z = 29.8\ \text{k}\Omega - j65.8\ \text{k}\Omega$. $C = 346$ pF and the phase angle is $-65.6°$.

2. The EPC of the RL branch is $Y = 96.15\ \mu S - j19.23\ \mu S$. The EPC of the RC branch is $Y = 53.33\ \mu S + j26.67\ \mu S$. This yields a total admittance of $Y = 149.5\ \mu S + j7.436\ \mu S$. The ESC of the two branches is $Z = 6.673\ \text{k}\Omega - j332$. The circuit ESC in rectangular form is $Z = 6.67\ \text{k}\Omega - j5.33\ \text{k}\Omega$.

FILTERS

CHAPTER 22

INTRODUCTION

In the preceding chapters on ac circuit analysis we calculated voltage drops and currents in many different kinds of circuits. We have seen how the distributions of these voltage drops and currents depend on component values and frequency. If we consider that component values remain fixed in a circuit, then the voltage drops and currents are frequency dependent only.

Typically, in an electronic circuit, or system, the output voltage or current is affected by circuits that are combinations of series and parallel components. Because these circuits are often complex, the analysis is easier if we use circuit theorems and reduce the circuits to their equivalent series or parallel circuits.

In this chapter, then, we will see how frequency affects the output voltage of an *RL*, *RC* or *RCL* circuit. We will also take the output from across different components and note how the voltage changes with frequency.

CHAPTER OBJECTIVES

In this chapter you will learn how to:

1. Identify high-pass and low-pass filters, calculate their cutoff frequencies, and determine the conditions that exist at cutoff.
2. Use circuit theorems to analyze various kinds of band-pass filters.
3. Calculate the various parameters in a series resonant circuit.

22-1 LOW-PASS AND HIGH-PASS FILTERS

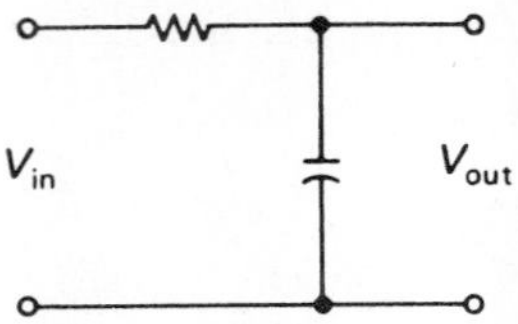

FIGURE 22-1

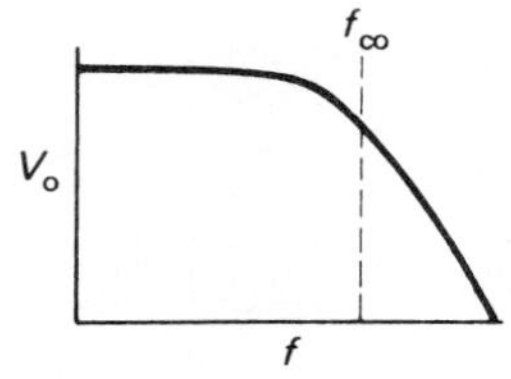

FIGURE 22-2

Consider the circuit in Figure 22-1. The circuit output voltage is developed across the capacitor. Therefore, V_o must vary as V_C varies. Without considering exact values of R and C, let's determine, in general, what happens to V_R and V_C as the frequency changes.

Because X_C is greatest at low frequencies, V_C will be greatest at low frequencies. If we plotted a curve of output voltage versus frequency, we would get a curve like the one in Figure 22-2. The curve shows that at very low frequencies $V_o = V_{in}$. As the frequency increases, X_C decreases and so does V_o.

Again referring to Figure 22-1, as we increase frequency from some low value, we will eventually reach a frequency where $X_C = R$. This frequency is called the *cutoff frequency* (also called the *break frequency* or the *corner frequency*). Because $X_C = R$ at the cutoff frequency (f_{co}), other conditions also exist: $\theta = -45°$, $V_C = V_R = 0.707V_{in}$, $I = 0.707I_{max}$, and $P = 0.5P_{max}$.

□ EXAMPLE 22-1

In Figure 22-1 let $V_{in} = 10$ V, $R = 10$ kΩ, and $X_C \cong 0$ Ω. Find V_o, V_R, I, and P.

Solution

$$V_o = V_C = 0 \text{ V} \qquad (\text{if } X_C = 0\ \Omega,\ V_C = 0 \text{ V})$$

$$V_R = V_{in}$$

$$I = \frac{V_R}{R} = \frac{10 \text{ V}}{10 \text{ k}\Omega} = 1 \text{ mA}$$

$$P = IV_R = 1 \text{ mA} \times 10 \text{ V} = 10 \text{ mW}$$

□

The above example would be the result of operation at some high frequency. I and P are both maximum values here because $Z = R$ and is minimum. At lower frequencies, $Z = R - jX_C$.

□ EXAMPLE 22-2

In Figure 22-1, let $V_{in} = 10$ V, $R = 10$ kΩ, $C = 10$ nF, and $X_C = 10$ kΩ. Find V_o, V_R, I, P, θ, and f_{co}.

Solution

$$Z = R - jX_C = 10 \text{ k}\Omega - j10 \text{ k}\Omega = 14.1 \angle{-45°} \text{ k}\Omega$$

$$I = \frac{V}{Z} = \frac{10 \text{ V}}{14.1 \text{ k}\Omega} = 0.707 \text{ mA}$$

$$V_o = V_C = IX_C = 7.07 \text{ V}$$

or $V_o = V_{in} \sin \theta = 7.07 \text{ V}$

$V_R = IR = 7.07 \text{ V}$

or $V_R = V_{in} \cos \theta = 7.07 \text{ V}$

$P = IV_R = 0.707 \text{ mA} \times 7.07 \text{ V} = 5 \text{ mW}$

$\theta = -45°$

$$f_{co} = \frac{1}{2\pi RC} = 1.59 \text{ kHz}$$

□

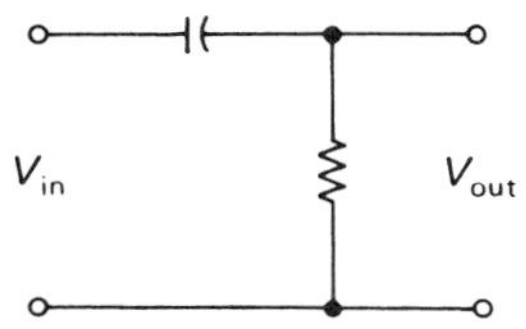

FIGURE 22-3

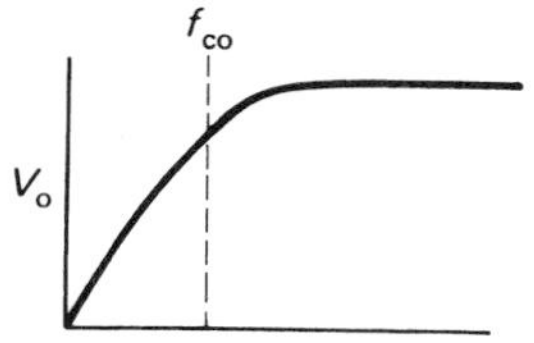

FIGURE 22-4

A circuit such as the one in Figure 22-1 is called a *low-pass filter*. That is, this circuit passes the low frequencies and rejects the high frequencies. Frequencies that result in output voltages that are less than 70.7% of maximum output are considered lost due to their low amplitude.

If we reversed the position of the components as in Figure 22-3, we would have a *high-pass filter*. The curve of output voltage versus frequency for this circuit is shown in Figure 22-4. The relationships discussed in reference to Figure 22-1 are also true here except that the output voltage is taken from across R. This causes V_o to *increase* with frequency. However, the cutoff frequency is the same, and all other conditions at f_{co} are the same.

PRACTICE PROBLEMS 22-1

1. Refer to Figure 22-1. Let $V_{in} = 20$ V and $R = 3$ kΩ. Find I_{max}, P_{max}, and $V_{o(max)}$. Find X_C, I, P, and V_o at f_{co}.
2. Refer to Figure 22-3. Let $V_{in} = 10$ V, $R = 2.2$ kΩ, and $C = 200$ nF. Find f_{co}, I, V_o, and P at f_{co}.

Solutions

1. $I_{max} = 6.67$ mA, $P_{max} = 133$ mW, $V_{o(max)} = 20$ V. At f_{co}, $I = 4.71$ mA, $P = 66.7$ mW, and $V_o = 14.1$ V.
2. $f_{co} = 362$ Hz, $I = 3.21$ mA, $V_o = 7.07$ V, and $P = 22.7$ mW.

Additional practice problems are at the end of the chapter.

22-2 THÉVENIN'S THEOREM

The use of Thévenin's theorem can simplify the solutions of some ac circuits. For example, consider the circuit in Figure 22-5(a). Suppose we wanted to find V_C. If we used trig functions, the problem would be fairly complex. Using Thévenin's theorem reduces the problem to a simple circuit quite easily. If we consider that C is the load, then V_{OC} and R_{TH} can be found. (See Chapter 13 for a review of Thévenin's theorem.) With the load removed as in Figure 22-5(b), $V_{OC} = V_2$ since there is no drop across R_3.

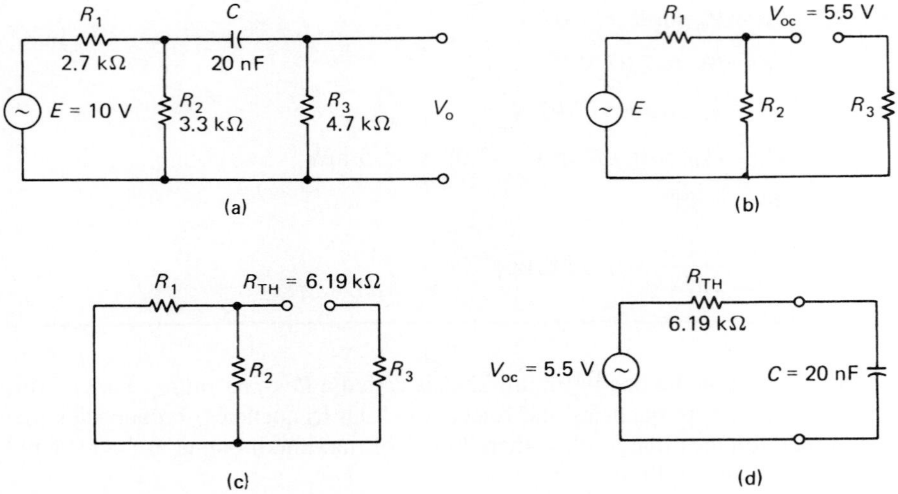

FIGURE 22-5

$$V_{OC} = V_2 = \frac{ER_2}{R_1 + R_2} = \frac{10\text{ V} \times 3.3\text{ k}\Omega}{2.7\text{ k}\Omega + 3.3\text{ k}\Omega} = 5.5\text{ V}$$

Shorting the source results in an R_{TH} of 6.19 kΩ. The circuit is shown in Figure 22-5(c).

$$R_{TH} = R_3 + (R_1 \parallel R_2) = 4.7\text{ k}\Omega + 1.49\text{ k}\Omega = 6.19\text{ k}\Omega$$

Thévenin's equivalent circuit with the load connected is shown in Figure 22-5(d). This equivalent circuit allows us to find the cutoff frequency of the circuit. f_{co} is the frequency where $X_C = R_{TH}$, so

$$f_{co} = \frac{1}{2\pi C R_{TH}} \tag{22-1}$$

$$= \frac{1}{2\pi \times 20\text{nF} \times 6.19\text{ k}\Omega} = 1.29\text{ kHz}$$

The output voltage is taken from across R_3. This results in a *high-pass* filter. The circuit will pass all frequencies above 1.29 kHz.

We can find V_o at f_{co} by using Thévenin's theorem. We know from the equivalent circuit that X_C must equal 6.19 kΩ at f_{co}. Therefore, $V_C = V_{OC} \sin 45° = 3.89$ V. Applying Ohm's law and Kirchhoff's current law, we get

$$I_C = \frac{V_C}{X_C} = \frac{3.89\text{ V}}{6.19\text{ k}\Omega} = 628\ \mu\text{A}$$

$$I_C = I_3 \qquad (C_1 \text{ and } R_3 \text{ are in series})$$

$$V_3 = I_3 R_3 = 628\ \mu\text{A} \times 4.7\text{ k}\Omega = 2.95\text{ V}$$

$$V_o = V_3 = 2.95\text{ V}$$

In a circuit such as the one in Figure 22-6(a), V_o and f_{co} can be found easily by using Norton's theorem. In Figure 22-6(b), the load is replaced with a short circuit. The short circuit current (I_{SC}) would be

$$I_{SC} = \frac{12\ \text{V}}{2.7\ \text{k}\Omega} = 4.44\ \text{mA}$$

(The short circuit reduces the resistance of the parallel circuit to 0 Ω.) G_N is found from the circuit in Figure 22-6(c).

$$G_N = G_1 + G_2 = 370\ \mu\text{S} + 303\ \mu\text{S} = 673\ \mu\text{S}$$

The equivalent circuit with the load connected is shown in Figure 22-6(d).

The circuit is a low-pass filter. C provides a low-impedance path for current at high frequencies. f_{co} may be calculated from the equivalent circuit:

$$f_{co} = \frac{G_N}{2\pi C} \qquad (22\text{-}2)$$

$$f_{co} = \frac{673\ \mu\text{S}}{2\pi \times 20\ \text{nF}} = 5.36\ \text{kHz}$$

The output voltage at f_{co} can be determined by first determining $V_{o\,(max)}$. $V_{o\,(max)}$ would occur where $G_N \gg B_C$.

$$V_{o\,(max)} = \frac{I_{SC}}{G_N} = 6.6\ \text{V}$$

At f_{co}

$$V_o = 0.707\ V_{o\,(max)} = 4.66\ \text{V}$$

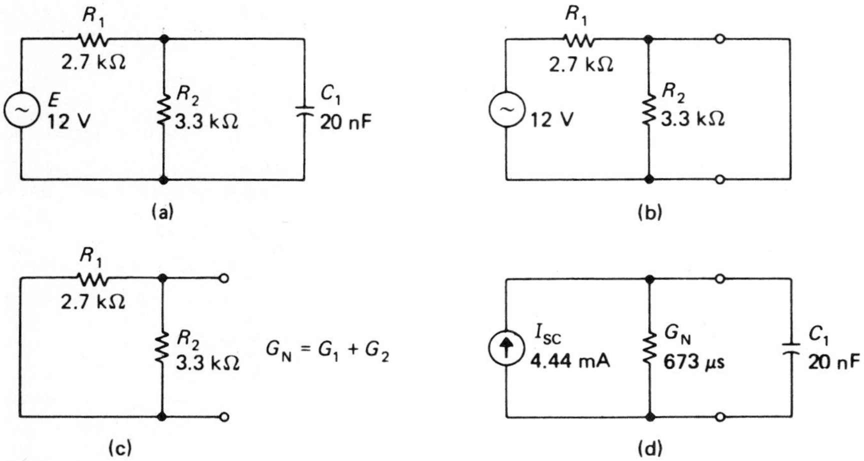

FIGURE 22-6

PRACTICE PROBLEMS 22-2

1. Refer to Figure 22-5. Let $R_1 = 1$ kΩ, $R_2 = 2$ kΩ, $R_3 = 3$ kΩ, $C = 100$ nF, and $E = 10$ V. Thévenize the circuit. Find f_{co} and V_o at f_{co}.
3. Refer to Figure 22-6. Let $R_1 = 2$ kΩ, $R_2 = 2.7$ kΩ, $C = 500$ pF, and $E = 12$ V. Nortonize the circuit. Find f_{co} and V_o at f_{co}.

2. Refer to Figure 22-7. Thévenize the circuit. Find f_{co} and V_o at f_{co}.
4. Refer to Figure 22-8. Nortonize the circuit. Find f_{co} and V_o at f_{co}.

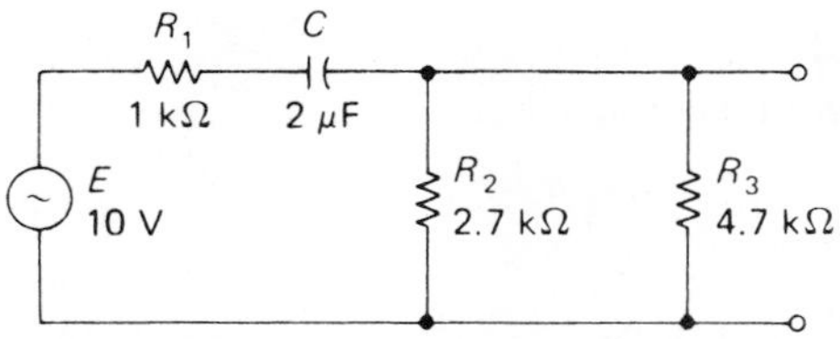

FIGURE 22-7

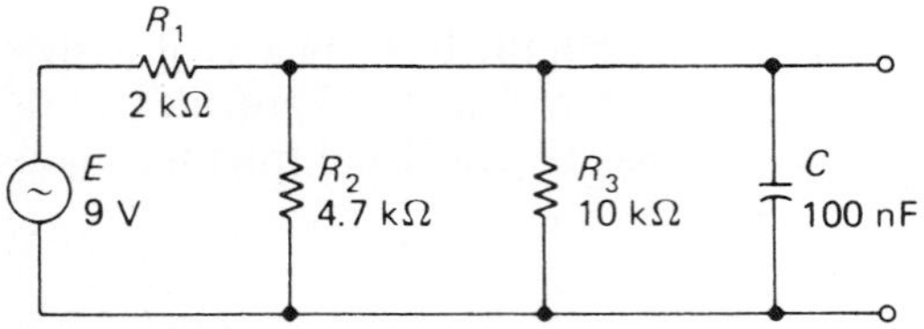

FIGURE 22-8

Solutions

1. $V_{oc} = 6.67$ V, $R_{TH} = 3.67$ kΩ, $f_{co} = 434$ Hz, V_o at $f_{co} = 3.86$ V
3. $I_{SC} = 6$ mA, $G_N = 870$ μS, $f_{co} = 277$ kHz, V_o at $f_{co} = 4.88$ V

2. $V_{OC} = 10$ V, $R_{TH} = 2.71$ kΩ, $f_{co} = 29.4$ Hz, V_o at $f_{co} = 4.46$ V
4. $I_{SC} = 4.5$ mA, $G_N = 813$ μS, $f_{co} = 1.29$ kHz, V_o at $f_{co} = 3.91$ V

Additional practice problems are at the end of the chapter.

SELF-TEST 22-1

1. Refer to Figure 22-9. $V_{in} = 25$ V, $R = 2.2$ kΩ, and $C = 200$ nF. Find f_{co}. Find I, V_o, and P at f_{co}.
3. Refer to Figure 22-7. Let $R_1 = 3.3$ kΩ, $R_2 = 1.8$ kΩ, $R_3 = 4.7$ kΩ, $C = 500$ nF, and $E = 12$ V. Thévenize the circuit. Find f_{co}, and V_o at f_{co}.
5. Refer to Figure 22-8. Let $R_1 = 680$ Ω, $R_2 = 2.7$ kΩ, $R_3 = 6.8$ kΩ, $C = 10$ nF, and $E = 15$ V. Nortonize the circuit. Find f_{co}, and V_o at f_{co}.

2. Refer to Figure 22-5. Let $R_1 = 4.7$ kΩ, $R_2 = 1$ kΩ, $R_3 = 1.8$ kΩ, $C = 250$ nF, and $E = 12$ V. Thévenize the circuit. Find f_{co}.
4. Refer to Figure 22-6. Let $R_1 = 680$ Ω, $R_2 = 2.7$ kΩ, $C = 5$ nF, and $E = 20$ V. Nortonize the circuit. Find f_{co}, and V_o at f_{co}.

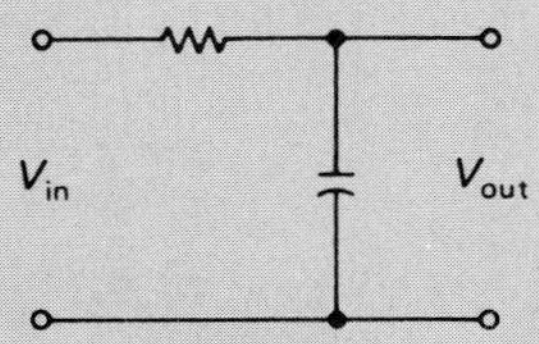

FIGURE 22-9

Answers to Self-test 22-1 are at the end of the chapter.

22-3 BAND-PASS FILTERS

Band-pass filters are filters that pass a band of frequencies. That is, the filter rejects both high frequencies and low frequencies but passes a range of frequencies between. Figure 22-10 shows a curve of such a filter. Various circuit configurations could produce such a curve, but, in general, the curve usually results from both series and parallel capacitances in the circuit. The curve has two cutoff frequencies. f_1 is the lower cutoff frequency and f_2 is the upper cutoff frequency.

Consider the circuit in Figure 22-11. This circuit could be the equivalent circuit of an amplifier or passive network. At low frequencies the capacitive reactances will be quite large compared to the circuit resistances. Because C_2 is connected in parallel, its high reactance (low susceptance) will have negligible effect on the circuit. C_1, though, is connected in series, thus resulting in low output voltages at these low frequencies. We can better see the low-frequency circuit if we Thévenize it. Since C_2 will have little effect on the circuit at low frequencies, we can treat it as an open circuit. With C_2 open-circuited, the new circuit is shown in Figure 22-12.

The Thévenin's equivalent circuit is shown in Figure 22-13. $R_{TH} = R_1 + R_2$. At f_1, $X_C = R_{TH}$.

$$f_1 = \frac{1}{2\pi R_{TH} C_1} \tag{22-3}$$

At high frequencies we can ignore the low reactance of C_1. The high susceptance of C_2 (low reactance) causes a reduction in output voltage at high frequencies. The resulting circuit at high frequencies is shown in Figure 22-14. Norton's equivalent circuit is shown in Figure 22-15. At f_2, $G_N = B_C$.

$$f_2 = \frac{G_N}{2\pi C_2} \tag{22-4}$$

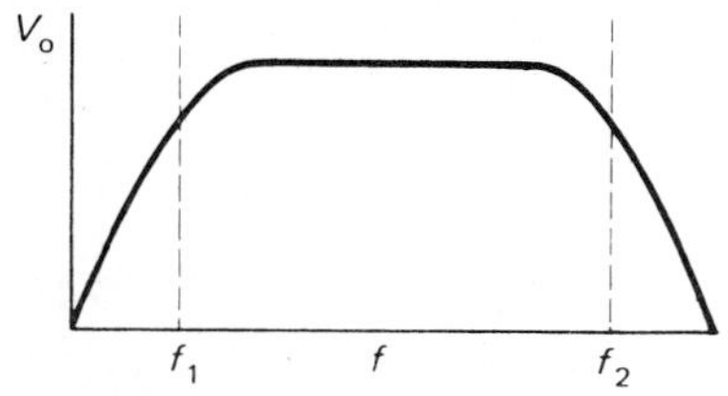

FIGURE 22-10

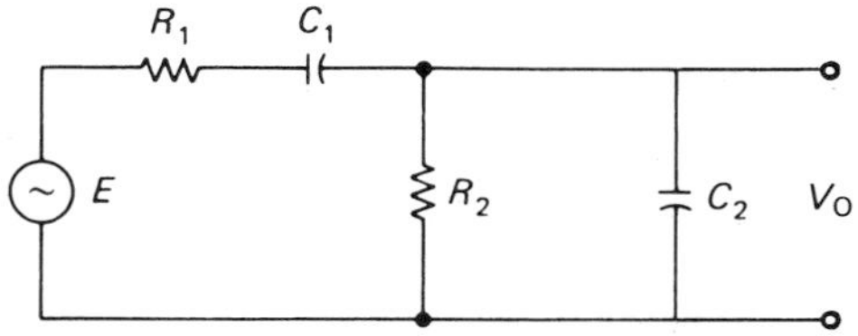

FIGURE 22-11

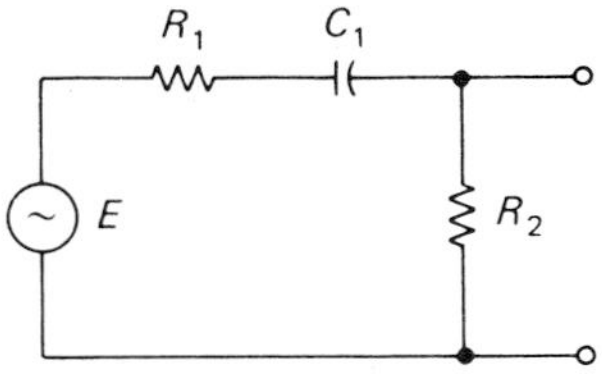

FIGURE 22-12

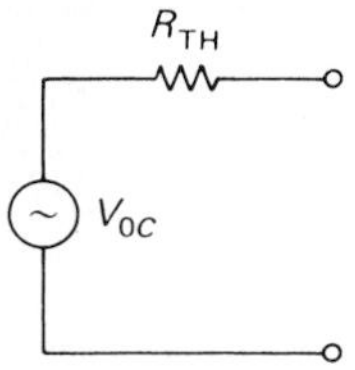

FIGURE 22-13

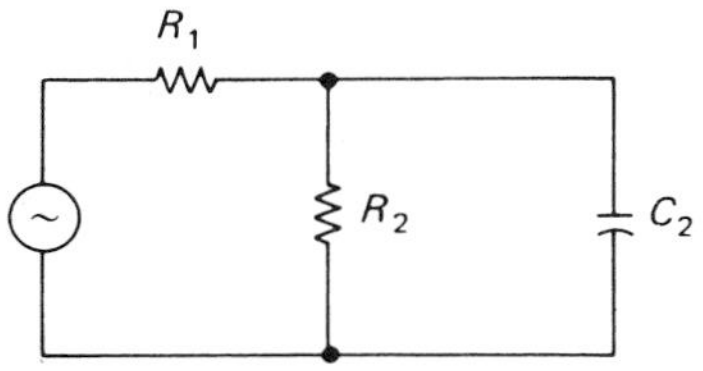

FIGURE 22-14

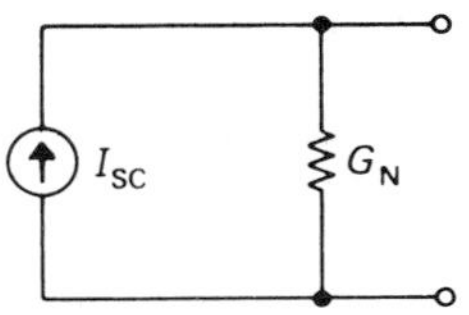

FIGURE 22-15

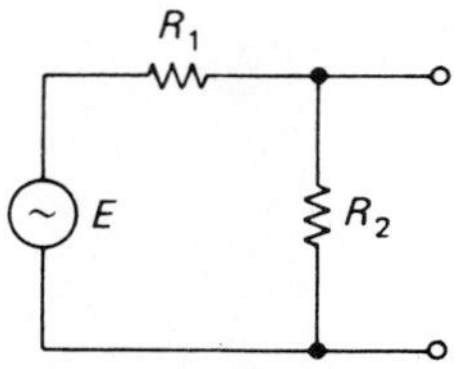

FIGURE 22-16

In the mid-frequency range (the range of frequencies between f_1 and f_2) we ignore the effects of both capacitances. The reactance of C_1 is negligible compared to $R_{TH}(R_1 + R_2)$. The susceptance of C_2 is negligible compared to $G_N(G_1 + G_2)$. The mid-frequency circuit consists of R_1 and R_2, as in Figure 22-16. Even though we consider that the circuit passes a band or range of frequencies, a *middle* frequency (f_m) may be found by using the following equation:

$$f_m = \sqrt{f_1 f_2}$$

Theoretically, this middle frequency is at the mid-point of the mid-frequency range.

☐ EXAMPLE 22-3

Refer to Figure 22-11. Let $R_1 = 2\ k\Omega$, $R_2 = 3\ k\Omega$, $C_1 = 1\ \mu F$, $C_2 = 500\ pF$, and $E = 20\ V$. Find f_1, f_2, V_o at f_m and V_o at f_{co}.

Solution

$$f_1 = \frac{1}{2\pi R_{TH} C_1}$$

$$R_{TH} = R_1 + R_2 = 2\ k\Omega + 3\ k\Omega = 5\ k\Omega$$

$$f_1 = \frac{1}{2\pi \times 5\ k\Omega \times 1\ \mu F} = 31.8\ Hz$$

$$f_2 = \frac{G_N}{2\pi C_2} \tag{22-1}$$

$$G_N = G_1 + G_2 = 500\ \mu S + 333\ \mu S = 833\ \mu S$$

$$f_2 = \frac{833\ \mu S}{2\pi \times 500\ pF} = 265\ kHz$$

At f_m

$$V_o = \frac{ER_2}{R_1 + R_2} = \frac{20\ V \times 3\ k\Omega}{5\ k\Omega} = 12\ V$$

At f_1 and f_2 the phase angle is 45°. At these frequencies, $V_o = 0.707$ times $V_{o\,(max)}$. $V_{o\,(max)}$ is the output voltage in the mid-frequency range.

$$V_o \text{ at } f_1 = V_o \text{ at } f_2 = 12\ V \times 0.707 = 8.48\ V$$

☐ EXAMPLE 22-4

Refer to Figure 22-17. Find f_1 and f_2.

Solution Ignoring C_2 and treating C_1 as the load, we can determine the value

$$R_{TH} = R_3 + R_1 \parallel R_2 = 5.24\ k\Omega$$

$$f_1 = \frac{1}{2\pi C_1 R_{TH}} = 30.4\ Hz$$

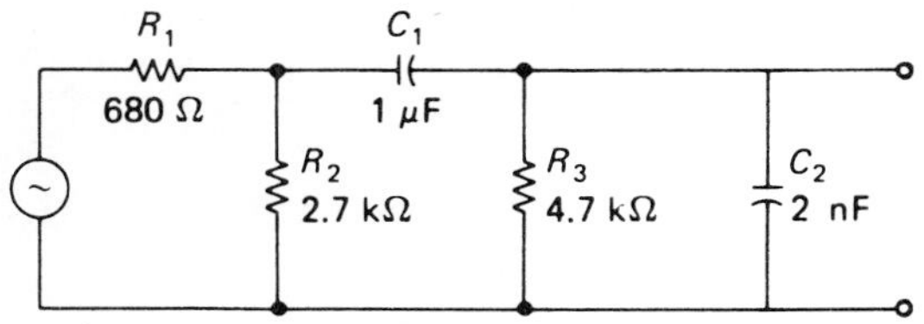

FIGURE 22-17

Treating C_1 as a short circuit and assuming that C_2 is the load, we can determine the value of f_2:

$$G_N = G_1 + G_2 + G_3 = 1.47 \text{ mS} + 370\ \mu\text{S} + 213\ \mu\text{S} = 2.05 \text{ mS}$$

$$f_2 = \frac{G_N}{2\pi C_2} = 163 \text{ kHz}$$

Consider the circuit in Figure 22-18. We recognize a series *LCR* band-pass circuit. We have determined previously that the resonant frequency (f_r) is that frequency where $X_L = X_C$. Other conditions at resonance are $\theta = 0°$, $V_{in} = V_R = V_o$, $V_C = V_L$, $Z = R$, I is max, and P is max. The curve of output voltage versus frequency is shown in Figure 22-19.

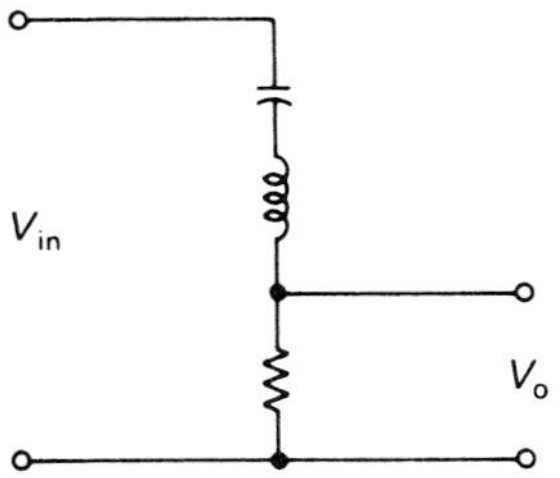

FIGURE 22-18

As we increase the frequency above resonance, X_L increases and X_C decreases. Eventually, we reach a frequency where the difference between X_L and X_C equals R. This is the *upper cutoff frequency*, f_2.

As we decrease the frequency below resonance, X_C increases and X_L decreases. Eventually we reach a frequency where the difference between X_C and X_L equals R. This is the *lower cutoff frequency*, f_1. In *RCL* circuits the range of frequencies between f_1 and f_2 is called the *bandwidth* (BW).

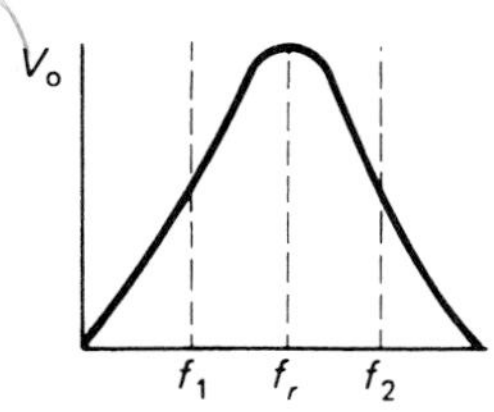

FIGURE 22-19

$$\text{BW} = f_2 - f_1$$

$$f_1 = f_r - \frac{\text{BW}}{2}$$

$$f_2 = f_r + \frac{\text{BW}}{2}$$

There is a direct relationship between f_r, BW, and circuit Q.

$$\text{BW} = \frac{f_r}{Q}$$

□ EXAMPLE 22-5

In Figure 22-18, let $V_{in} = 3$ V, $R = 100\ \Omega$, $f_r = 1$ kHz, and $Q = 20$. Find the bandwidth, f_1, f_2, I_{max}, P_{max}, V_R at resonance, and V_R, I, and P at f_1 and f_2.

Solution

$$\text{BW} = \frac{f_r}{Q} = \frac{1\text{ kHz}}{20} = 50\text{ Hz}$$

$$f_1 = f_r - \frac{\text{BW}}{2} = 1\text{ kHz} - 25\text{ Hz} = 975\text{ Hz}$$

$$f_2 = f_r + \frac{\text{BW}}{2} = 1\text{ kHz} + 25\text{ Hz} = 1025\text{ Hz}$$

$V_R = V_{in} = 3$ V at resonance. At other frequencies, V_R would be less than 3 V. Therefore,

$$I_{max} = \frac{V_R}{R} = \frac{3\text{ V}}{100\ \Omega} = 30\text{ mA}$$

$$P_{max} = V_R I = 3\text{ V} \times 30\text{ mA} = 90\text{ mW}$$

At f_1, $\theta = -45°$. At f_2, $\theta = 45°$. Then, at either f_1 or f_2,

$$V_R = V_{in}\cos 45° = 3\text{ V} \times 0.707 = 2.12\text{ V}$$

$$I = \frac{V_R}{R} = \frac{2.12\text{ V}}{100\ \Omega} = 21.2\text{ mA}$$

$$P = IV_R = 21.2\text{ mA} \times 2.12\text{ V} = 45\text{ mW}$$

This circuit (Figure 22-18) is a *band-pass* filter. The circuit passes the band of frequencies between f_1 and f_2 and rejects all others.

□ EXAMPLE 22-6

In Figure 22-18, let $L = 40$ mH, $C = 200$ nF, and $R = 30\ \Omega$. Find f_r, f_1, f_2, BW, and Q.

Solution

$$f_r = \frac{1}{2\pi\sqrt{LC}} = 1.78\text{ kHz}$$

$$Q = \frac{X_L}{R} = \frac{2\pi f_r L}{R} = \frac{447\ \Omega}{30\ \Omega} = 14.9$$

$$\text{BW} = \frac{f_r}{Q} = \frac{1.78\text{ kHz}}{14.9} = 119$$

$$f_1 = f_r - \frac{\text{BW}}{2} = 1.78\text{ kHz} - 60\text{ Hz} = 1.72\text{ kHz}$$

$$f_2 = f_r + \frac{\text{BW}}{2} = 1.78\text{ kHz} + 60\text{ Hz} = 1.84\text{ kHz}$$

□

The circuit in Figure 22-20 is a *band-reject* filter. The curve of output voltage versus frequency is shown in Figure 22-21. The parallel LC circuit offers a maximum impedance to the circuit at resonance. Therefore, circuit current is minimum and V_o is minimum. The frequencies between f_1 and f_2 are rejected.

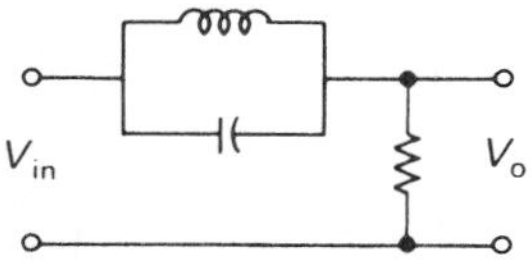

FIGURE 22-20

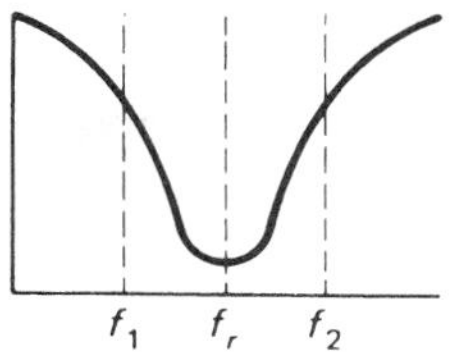

FIGURE 22-21

PRACTICE PROBLEMS 22-3

1. Refer to Figure 22-22. Let R_1 = 4.7 kΩ, R_2 = 1.8 kΩ, C_1 = 100 nF, C_2 = 1 nF, and E = 20 V. Find f_1, f_2, V_o at f_m, and V_o at f_{co}.
3. Refer to Figure 22-18. V_{in} = 9 V, R = 200 Ω, f_r = 800 Hz, and Q = 20. Find BW, f_1, and f_2. Find V_R, P, and I at f_r, f_1, and f_2.
2. Refer to Figure 22-17. Let R_1 = 1.5 kΩ, R_2 = 3.3 kΩ, R_3 = 2.7 kΩ, C_1 = 200 nF, and C_2 = 700 pF. Find f_1 and f_2.
4. Refer to Figure 22-18. L = 200 mH, C = 200 nF, and R = 100 Ω. Find f_r, f_1, f_2, BW, and Q.

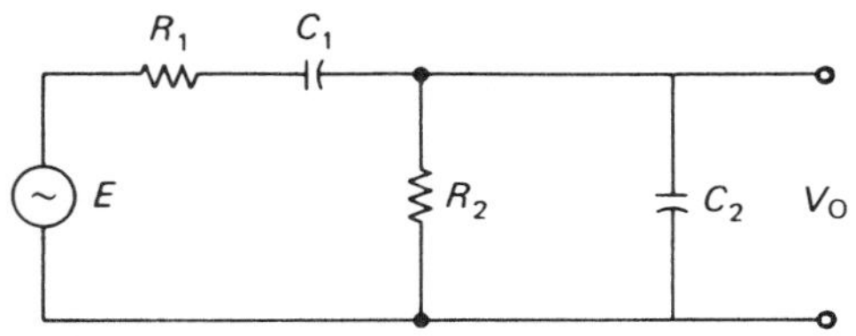

FIGURE 22-22

Additional practice problems are at the end of the chapter.

Solutions

1. f_1 = 245 Hz, f_2 = 122 kHz, V_o at f_m = 5.54 V, V_o at f_{co} = 3.92 V
3. BW = 40 Hz, f_1 = 780 Hz, f_2 = 820 Hz. At f_r, V_R = 9 V, I = 45 mA, and P = 405 mW. At f_{co}, V_R = 6.36 V, I = 31.8 mA, and P = 202 mW.
2. f_1 = 213 Hz, f_2 = 305 kHz
4. f_r = 796 Hz, f_1 = 756 Hz, f_2 = 836 Hz, BW = 79.6 Hz, Q = 10

SELF-TEST 22-2

1. Refer to Figure 22-22. Let R_1 = 2 kΩ, R_2 = 1.2 kΩ, C_1 = 200 nF, C_2 = 2 nF, and E = 9 V. Find f_1, f_2, V_o at f_m, and V_o at f_{co}.
3. Refer to Figure 22-18. L = 200 mH, C = 2 nF, and R = 1 kΩ. Find f_r, f_1, BW, and Q.
2. Refer to Figure 22-17. Let R_1 = 3.9 kΩ, R_2 = 4.7 kΩ, R_3 = 2.2 kΩ, C_1 = 100 nF, and C_2 = 2 nF. Find f_1 and f_2.

Answers to Self-test 22-2 are at the end of the chapter.

CHAPTER 22 AT A GLANCE

Page	Definitions	Example
560	The *cutoff frequency* is that frequency where the reactance equals the resistance.	$R = X$

561 A *low-pass* filter passes low frequencies and rejects those frequencies above the cutoff frequency.

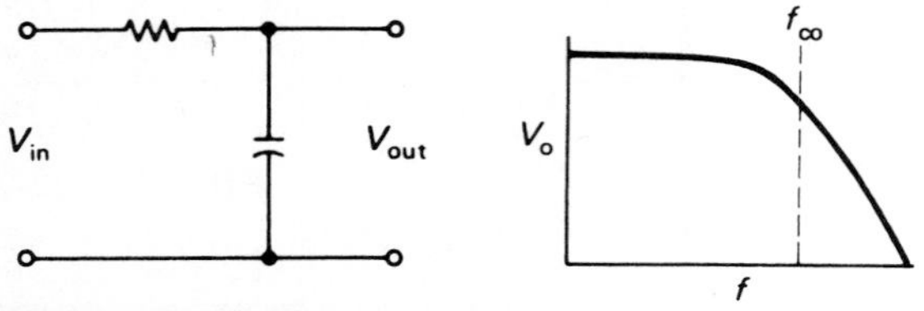

FIGURE 22-23

561 A *high-pass* filter passes high frequencies and rejects those frequencies below the cutoff frequency.

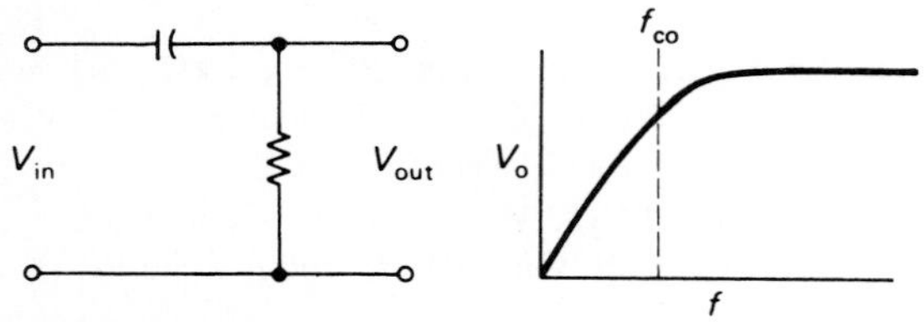

FIGURE 22-24

565 A *band-pass* filter passes that band of frequencies between the lower and upper cutoff frequencies.

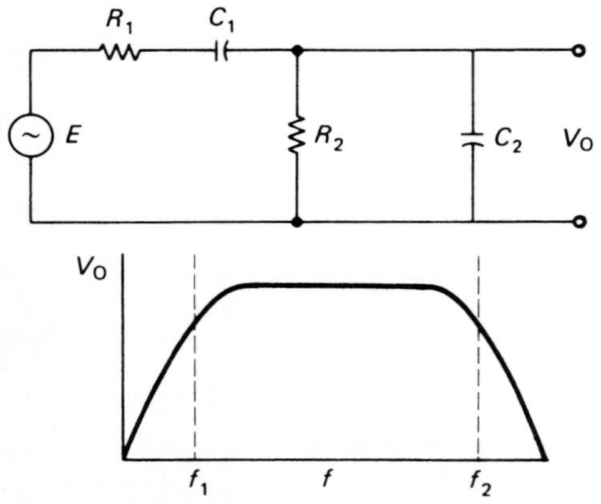

FIGURE 22-25

569 A *band-reject* filter rejects that band of frequencies between the cutoff frequencies and passes all other frequencies

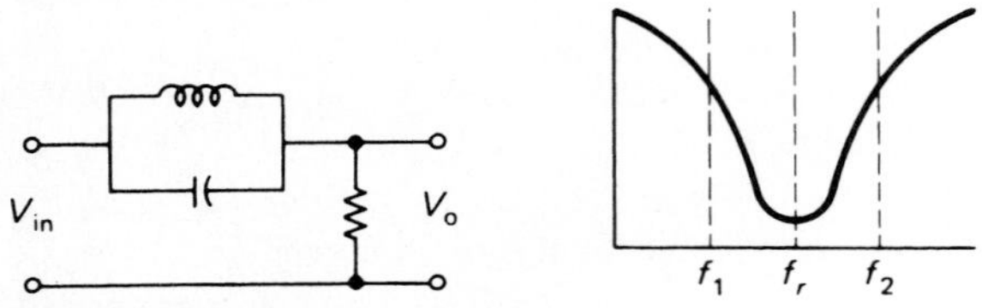

FIGURE 22-26

End of Chapter Problems 22-1

Refer to Figure 22-27.

1. Let V_{in} = 15 V and R = 470 Ω. Find I_{max}, P_{max}, and $V_{o(max)}$. Find X_C, I, P, and V_o at f_{co}.
2. Let V_{in} = 25 V and R = 5.6 kΩ. Find I_{max}, P_{max}, and $V_{o(max)}$. Find X_C, I, P, and V_o at f_{co}.
3. Let V_{in} = 300 mV and R = 5.6 kΩ. Find I_{max}, P_{max}, and $V_{o(max)}$. Find X_C, I, P, and V_o at f_{co}.
4. Let V_{in} = 12 V and R = 1.2 kΩ. Find I_{max}, P_{max}, and $V_{o(max)}$. Find X_C, I, P, and V_o at f_{co}.
5. Let V_{in} = 20 V, R = 1.2 kΩ, and C = 500 nF. Find f_{co}. Find I, V_o, and P at f_{co}.
6. Let V_{in} = 25 V, R = 300 Ω, and C = 1 μF. Find f_{co}. Find I, V_o, and P at f_{co}.
7. Let V_{in} = 40 V, R = 2.2 kΩ, and C = 200 nF. Find f_{co}. Find I, V_o, and P at f_{co}.
8. Let V_{in} = 600 mV, R = 100 Ω, and C = 2 μF. Find f_{co}. Find I, V_o, and P at f_{co}.

End of Chapter Problems 22-2

1. Refer to Figure 22-28. Let R_1 = 1.8 kΩ, R_2 = 3.3 kΩ, R_3 = 1 kΩ, C = 500 nF, and E = 10 V. Thévenize the circuit and find f_{co} and V_o at f_{co}.
2. Refer to Figure 22-28. Let R_1 = 2.7 kΩ, R_2 = 1.2 kΩ, R_3 = 3 kΩ, C = 1 μF, and E = 15 V. Thévenize the circuit and find f_{co} and V_o at f_{co}.
3. Refer to Figure 22-29. Thévenize the circuit and find f_{co} and V_o at f_{co}.
4. Refer to Figure 22-29. Let R_1 = 5.6 kΩ, R_2 = 2.7 kΩ, C = 1 μF, and E = 10 V. Thévenize the circuit and find f_{co} and V_o at f_{co}.
5. Refer to Figure 22-30. Let R_1 = 2.7 kΩ, R_2 = 3.9 kΩ, R_3 = 5.6 kΩ, C = 1 μF, and E = 10 V. Thévenize the circuit. Find f_{co} and V_o at f_{co}.
6. Refer to Figure 22-30. Let R_1 = 750 Ω, R_2 = 1 kΩ, R_3 = 1.5 kΩ, C = 800 nF, and E = 20 V. Thévenize the circuit. Find f_{co} and V_o at f_{co}.
7. Refer to Figure 22-31. Thévenize the circuit and find f_{co}.
8. Refer to Figure 22-31. Let R_1 = 7.5 kΩ, R_2 = 2.7 kΩ, R_3 = 4.7 kΩ, R_4 = 3 kΩ, C = 500 nF, and E = 15 V. Thévenize the circuit and find f_{co}.
9. Refer to Figure 22-32. Let R_1 = 470 Ω, R_2 = 1.2 kΩ, C = 50 nF, and E = 15 V. Nortonize the circuit and find f_{co} and V_o at f_{co}.
10. Refer to Figure 22-32. Let R_1 = 1.5 kΩ, R_2 = 3.3 kΩ, C = 100 nF, and E = 20 V. Nortonize the circuit and find f_{co} and V_o at f_{co}.

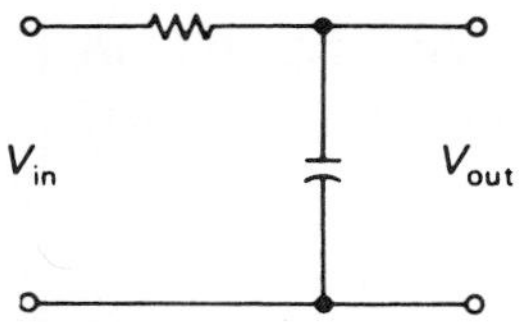

FIGURE 22-27

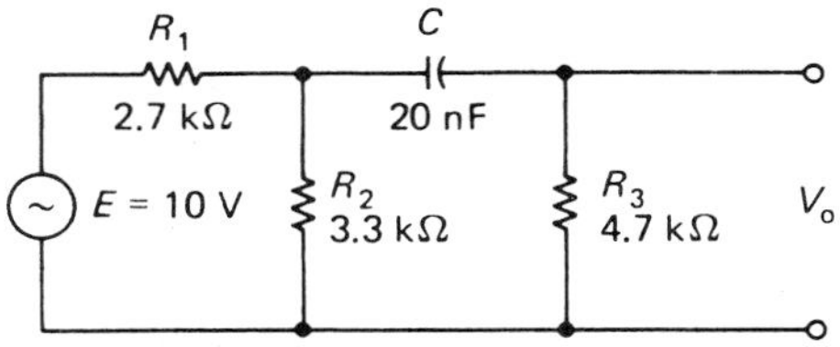

FIGURE 22-28

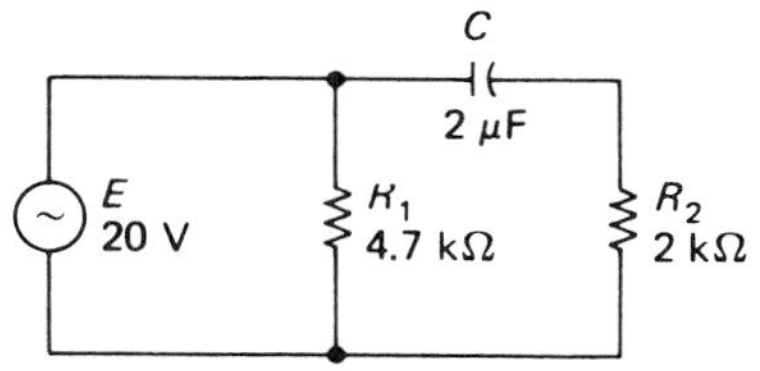

FIGURE 22-29

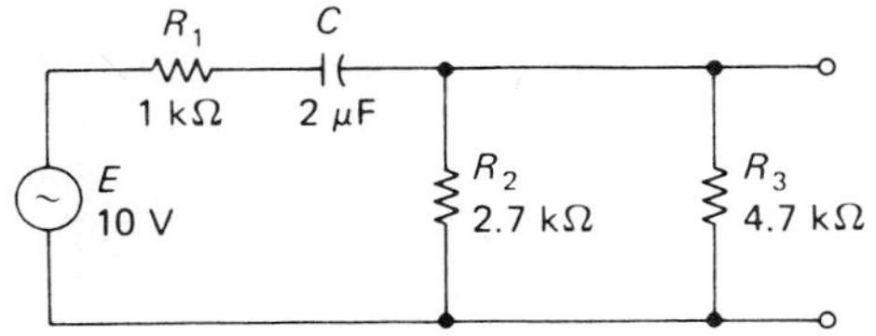

FIGURE 22-30

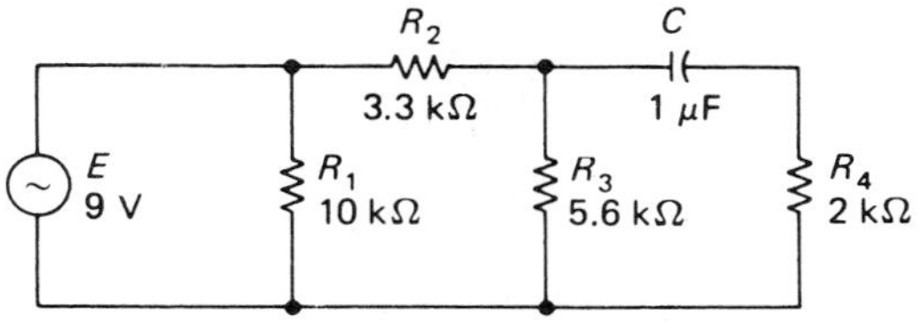

FIGURE 22-31

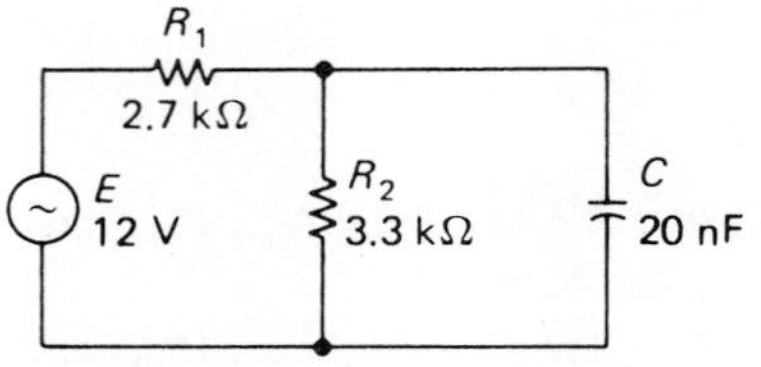

FIGURE 22-32

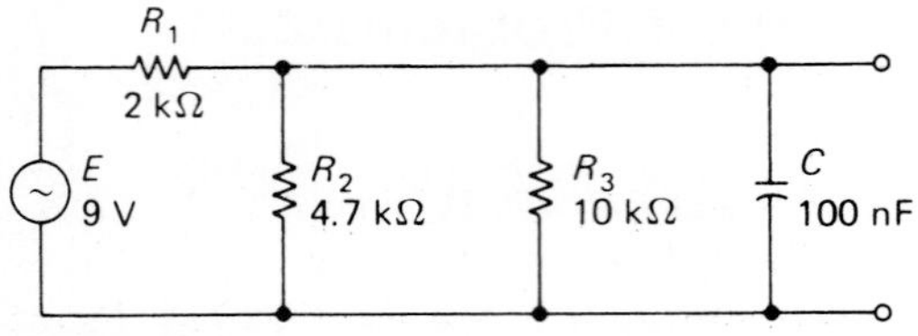

FIGURE 22-33

11. Refer to Figure 22-33. Let $R_1 = 2$ kΩ, $R_2 = 4.7$ kΩ, $R_3 = 5.6$ kΩ, $C = 1$ nF, and $E = 10$ V. Nortonize the circuit and find f_{co} and V_o at f_{co}.

12. Refer to Figure 22-33. Let $R_1 = 3$ kΩ, $R_2 = 2.7$ kΩ, $R_3 = 3.3$ kΩ, $C = 2$ nF, and $E = 12$ V. Nortonize the circuit and find f_{co} and V_o at f_{co}.

End of Chapter Problems 22-3

Refer to Figure 22-22.

1. Let $R_1 = 680$ Ω, $R_2 = 1.2$ kΩ, $C_1 = 500$ nF, $C_2 = 2$ nF, and $E = 9$ V. Find f_1, f_2, V_o at f_m, and V_o at f_{co}.

2. Let $R_1 = 2.7$ kΩ, $R_2 = 1.5$ kΩ, $C_1 = 100$ nF, $C_2 = 100$ pF, and $E = 10$ V. Find f_1, f_2, V_o at f_m, and V_o at f_{co}.

3. Let $R_1 = 3.3$ kΩ, $R_2 = 2$ kΩ, $C_1 = 2$ μF, $C_2 = 1$ nF, and $E = 15$ V. Find f_1, f_2, V_o at f_m, and V_o at f_{co}.

4. Let $R_1 = 1.8$ kΩ, $R_2 = 2.2$ kΩ, $C_1 = 750$ nF, $C_2 = 500$ pF, and $E = 5$ V. Find f_1, f_2, V_o at f_m, and V_o at f_{co}.

5. Let $R_1 = 2.2$ kΩ, $R_2 = 3.9$ kΩ, $C_1 = 1.2$ μF, $C_2 = 20$ nF, and $E = 2$ V. Find f_1, f_2, V_o at f_m, and V_o at f_{co}.

6. Let $R_1 = 5.6$ kΩ, $R_2 = 2.7$ kΩ, $C_1 = 0.5$ μF, $C_2 = 100$ nF, and $E = 600$ mV. Find f_1, f_2, V_o at f_m, and V_o at f_{co}.

7. Let $R_1 = 330$ Ω, $R_2 = 820$ Ω, $C_1 = 200$ nF, $C_2 = 500$ pF, and $E = 3$ V. Find f_1, f_2, V_o at f_m, and V_o at f_{co}.

8. Let $R_1 = 910$ Ω, $R_2 = 2$ kΩ, $C_1 = 500$ nF, $C_2 = 1$ nF, and $E = 7.5$ V. Find f_1, f_2, V_o at f_m, and V_o at f_{co}.

Refer to Figure 22-34.

9. Let $R_1 = 5.6$ kΩ, $R_2 = 2.7$ kΩ, $R_3 = 1.8$ kΩ, $C_1 = 1$ μF, and $C_2 = 2$ nF. Find f_1 and f_2.

10. Let $R_1 = 10$ kΩ, $R_2 = 2$ kΩ, $R_3 = 4.7$ kΩ, $C_1 = 1$ μF, and $C_2 = 2$ nF. Find f_1 and f_2.

11. Let $R_1 = 1.2$ kΩ, $R_2 = 4.3$ kΩ, $R_3 = 6.8$ kΩ, $C_1 = 2$ μF, and $C_2 = 800$ pF. Find f_1 and f_2.

12. Let $R_1 = 220$ Ω, $R_2 = 560$ Ω, $R_3 = 1$ kΩ, $C_1 = 1$ μF, and $C_2 = 1$ nF. Find f_1 and f_2.

Refer to Figure 22-35.

13. $V_{in} = 20$ V, $R = 1$ kΩ, $f_r = 2$ kHz, and $Q = 25$. Find BW, f_1, and f_2. Find V_R, I, and P at f_r, f_1, and f_2.

14. $V_{in} = 10$ V, $R = 810$ Ω, $f_r = 1$ kHz, and $Q = 10$. Find BW, f_1 and f_2. Find V_R, I, and P at F_r, f_1, and f_2.

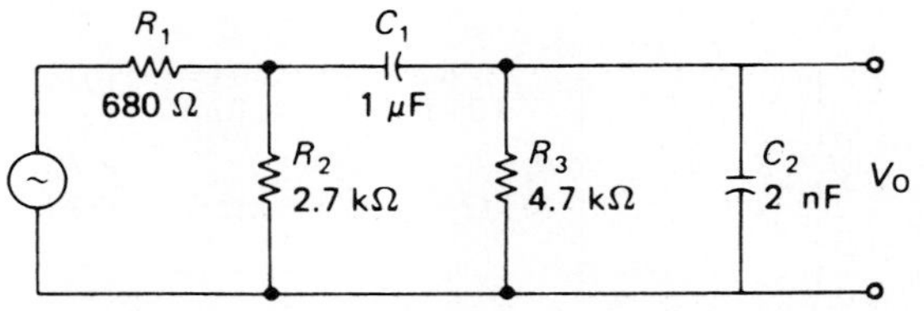

FIGURE 22-34

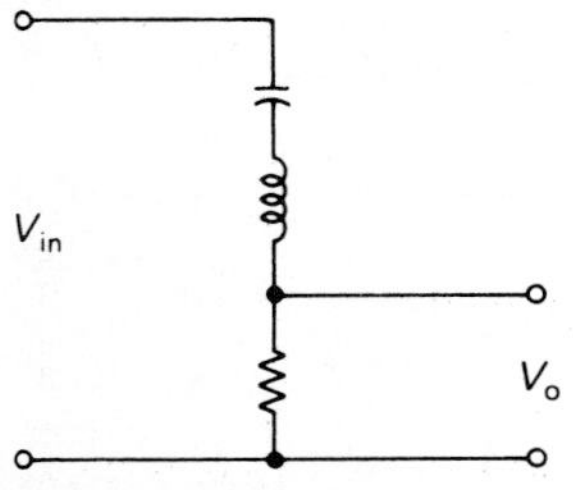

FIGURE 22-35

15. $V_{in} = 5$ V, $R_1 = 120\ \Omega$, $f_r = 5.5$ kHz, and $Q = 14$. Find BW, f_1, and f_2. Find V_R, I, and P at f_r, f_1, and f_2.

16. $V_{in} = 200$ mV, $R_1 = 300\ \Omega$, $f_r = 2.73$ kHz, and $Q = 20$. Find BW, f_1, and f_2. Find V_R, I, and P at f_r, f_1, and f_2.

17. $L = 100$ mH, $C = 250$ nF, and $R = 45\ \Omega$. Find f_r, f_1, f_2, BW, and Q.

18. $L = 500$ mH, $C = 1$ nF, and $R = 470\ \Omega$. Find f_r, f_1, f_2, BW, and Q.

19. $L = 20$ mH, $C = 100$ nF, and $R = 75\ \Omega$. Find f_r, f_1, f_2, BW, and Q.

20. $L = 500$ mH, $C = 2$ nF, and $R = 100\ \Omega$. Find f_r, f_1, f_2, BW, and Q.

Answers to Self-tests

SELF-TEST 22-1

1. $f_{co} = 362$ Hz, $I = 8.03$ mA, $V_o = 17.7$ V, $P = 142$ mW

2. $V_{OC} = 2.11$ V, $R_{TH} = 2.62$ kΩ, $f_{co} = 243$ Hz

3. $V_{OC} = 12$ V, $R_{TH} = 4.6$ kΩ, $f_{co} = 69.2$ Hz, V_o at $f_{co} = 2.4$ V

4. $I_{SC} = 29.4$ mA, $G_N = 1.84$ mS, $f_{co} = 58.6$ kHz, V_o at $f_{co} = 11.3$ V

5. $I_{SC} = 22.1$ mA, $G_N = 1.99$ mS, $f_{co} = 31.7$ kHz, V_o at $f_{co} = 7.85$ V

SELF-TEST 22-2

1. $f_1 = 249$ Hz, $f_2 = 106$ kHz, V_o at $f_m = 3.38$ V, V_o at $f_{co} = 2.39$ V

2. $f_1 = 367$ Hz, $f_2 = 73.5$ kHz

3. $f_r = 7.96$ kHz, $f_1 = 7.56$ kHz, $f_2 = 8.36$ kHz, BW = 796 Hz, $Q = 10$

CHAPTER 23

LOGARITHMS

INTRODUCTION

Logarithms are used in solving many electrical and electronics problems. The use of logarithms simplifies the solution of some complex problems. Their use reduces multiplication to addition, division to subtraction, raising to a power to simple multiplication, and extracting square roots to simple division.

Multiplication, division, or raising a number to some power can be accomplished with a calculator. However, in some problem-solving situations an understanding of the solution to such problems using logarithms is necessary. The following chapters dealing with logarithmic equations and applications of logarithms require this understanding.

CHAPTER OBJECTIVES

In this chapter you will learn how to:

1. Identify logarithmic expressions and exponential expressions.
2. Convert between logarithmic and exponential forms.
3. Find common logarithms of numbers using log tables and then using a calculator.
4. Multiply and divide numbers by using logarithms.
5. Raise a number to some power by using logarithms.
6. Find natural logarithms of numbers using ln tables and then using a calculator.

23-1 LOGARITHMS DEFINED

Consider the expression $10^2 = 100$. Recall from previous chapters that 10 is called the base, 2 is the exponent, and 100 is its numerical value. The *logarithm* of the expression is the exponent 2.

☐ EXAMPLE 23-1

The logarithm is the exponent.

exponent → $10^2 = 100$ ← number; base → 10

logarithm → $\log_{10} 100 = 2$; base → 10; number → 100

We can write the equations generically, where b is the base, N is the number, and x is the exponent.

$\log_b N = x$	is the same as	$b^x = N$

☐

That is, the logarithm is the power to which 10 (the base) must be raised in order for it to equal 100 (the number). Simply stated, *a logarithm is an exponent*.

Ten is the base of this system of logarithms. When 10 is the base, the system is called a system of *common logarithms*. In the system of logarithms, this is one of the two bases used in problem solving. The other base is 2.718. This system is referred to as a system of *natural logarithms*. Natural logarithms will be discussed later in the chapter.

23-2 COMMON LOGARITHMS

When dealing with common logarithms, the logarithmic expression for $10^2 = 100$ would be $\log_{10} 100 = 2$. This is read, "The log, to the base 10, of 100 is 2." When using common logs, it is not necessary to indicate the base because it is assumed to be 10. The expression "$\log 100 = 2$" is more often used. Notice that we have expressed the quantity two different ways. The first expression, $10^2 = 100$, is called an *exponential expression*. We have expressed the quantity in *exponential form*. The second expression, $\log 100 = 2$, is called a *logarithmic expression*. We have written the quantity in *logarithmic form*.

Let's find the log of 1000. First, we express the quantity in exponential form.

To do this, we merely find the power of 10 that equals 1000. In exponential form, $10^3 = 1000$. The logarithm is the exponent to which 10 is raised and equals 3. In logarithmic form, $\log 1000 = 3$.

Find the log of 0.0001. Again, we first express the quantity in exponential form: $10^{-4} = 0.0001$. The logarithm is the exponent and equals -4. In logarithmic form, $\log 0.0001 = -4$.

PRACTICE PROBLEMS 23-1

Find the logarithms of the following numbers. Express your answers first in logarithmic form and then in exponential form.

1. 0.00001
2. 0.01
3. 10,000
4. 1,000,000

Solutions

	Logarithmic form	*Exponential form*
1.	$\log 0.00001 = -5$	$10^{-5} = 0.00001$
2.	$\log 0.01 = -2$	$10^{-2} = 0.01$
3.	$\log 10{,}000 = 4$	$10^4 = 10{,}000$
4.	$\log 1{,}000{,}000 = 6$	$10^6 = 1{,}000{,}000$

Additional practice problems are at the end of the chapter.

23-2-1 Finding the Logarithms of Numbers between 1 and 10

What about the log of numbers that are not multiples of 10? Let's consider numbers between 1 and 10. Since $10^0 = 1$ and $10^1 = 10$, then numbers between 1 and 10 would have logs or exponents between 0 and 1. For example (to four significant figures) $5.32 = 10^{0.7259}$ and $7.63 = 10^{0.8825}$. In logarithmic form, $\log 5.32 = 0.7259$ and $\log 7.63 = 0.8825$.

We can find the logarithms of all numbers between 1 and 10 in a *table of common logarithms*. Your calculator also contains a table of common logarithms. The table in your calculator contains the logarithms of *all* numbers. The calculator has made tables of common logarithms obsolete in almost all problem-solving situations because it is faster and easier to use.

□ EXAMPLE 23-2

Find log 4.23.

Solution

$$\log 4.23 = 0.6263$$

To find logarithms using the calculator, let's first clear and set up to display 4 digits to the right of the decimal point.

[AC] [FIX] 4 4.23 [LOG] *Answer*: log 4.23 = 0.6263

In exponential form,

$$10^{0.6263} = 4.23$$

10^x is usually the second function of the LOG key; so

0.6263 [10^x] *Answer*: $10^{0.6263} = 4.23$

□ EXAMPLE 23-3

Find log 5.03.

Solution

$$\log 5.03 = 0.7016$$

In exponential form,

$$10^{0.7016} = 5.03$$

□

PRACTICE PROBLEMS 23-2

Using the calculator, find the logs of the following numbers to four significant figures. Express your answers in both logarithmic and exponential form.

1. 1.76
2. 2.03
3. 9.83
4. 4.35
5. 3.16

Solutions

Logarithmic form	*Exponential form*
1. log 1.76 = 0.2455	$10^{0.2455} = 1.76$
2. log 2.03 = 0.3075	$10^{0.3075} = 2.03$
3. log 9.83 = 0.9926	$10^{0.9926} = 9.83$
4. log 4.35 = 0.6385	$10^{0.6385} = 4.35$
5. log 3.16 = 0.4997	$10^{0.4997} = 3.16$

Additional practice problems are at the end of the chapter.

23-2-2 Finding the Logarithms of Numbers Greater Than 10

We can see that if we wanted to find the logarithm of 50, for example, the exponent (or log) would be greater than 1 but less than 2. (The number is greater than 10 but less than 100.) The log will consist of a whole number and a decimal fraction.

■ **KEY POINT** *Only numbers that are multiples of 10 will have logarithms that are whole numbers.*

Notice that in Practice Problems 23-2 the logs of the numbers were all decimal fractions. This was because all the numbers were between 1 and 10. Remember, the log of 1 is 0 and the log of 10 is 1. If problems contained numbers between 1 and 10, each logarithm would be a number between 0 and 1.

If problems contained numbers between 10 and 100, the logarithms would be numbers between 1 and 2 because

$$\log 10 = 1 \quad \text{and} \quad \log 100 = 2$$

If the numbers were between 100 and 1000, the logarithms would be numbers between 2 and 3 because

$$\log 100 = 2 \quad \text{and} \quad \log 1000 = 3$$

If the numbers were between 1000 and 10,000, the logarithms would be numbers between 3 and 4, and so on.

What is the log of 50? Log 50 = 1.6990. What is the log of 500? 5000? 50,000? Log 500 = 2.6990; log 5000 = 3.6990; log 50,000 = 4.6990. Notice that in each case the fractional part of the logarithm is 0.6990. The whole-number part depends on the multiples of 10 contained in the number. For example, let's express the number 5000 in scientific notation: $5000 = 5 \times 10^3$. We know from previous discussions that $\log 10^3 = \log 1000 = 3$. Log 5 = 0.6990. If we express $5 \times 10^3 = 5000$ by using the laws of exponents, $10^{0.6990+3} = 10^{3.6990} = 5000$. When finding logarithms of numbers, we are finding a fractional part that results from finding the log of a number between 1 and 10 and a whole-number part that is the power of 10. What is log 276,000? Log 276,000 = 5.4409. When we consider that $276{,}000 = 2.76 \times 10^5$, we really found the log of 2.76 and the log of 10^5. Log 2.76 = 0.4409 and $\log 10^5 = 5$. $10^5 \times 10^{0.4409} = 10^{5.4409}$. The logarithm is the exponent to the base 10. We found the power to which 10 is raised so that its value is 276,000.

□ EXAMPLE 23-4

Find log 37,500.

Solution

$$\log 37{,}500 = 4.574$$

In exponential form,

$$10^{4.574} = 37{,}500$$

□

PRACTICE PROBLEMS 23-3

Find the logarithms of the following numbers. Express your answers in both logarithmic and exponential forms.

1. 67.3
2. 67,300
3. 742
4. 103,000
5. 8730

Solutions

Logarithmic form	*Exponential form*
1. log 67.3 = 1.8280	$10^{1.8280} = 67.3$
2. log 67,300 = 4.8280	$10^{4.8280} = 67{,}300$
3. log 742 = 2.8704	$10^{2.8704} = 742$
4. log 103,000 = 5.0128	$10^{5.0128} = 103{,}000$
5. log 8730 = 3.9410	$10^{3.9410} = 8730$

Additional practice problems are at the end of the chapter.

23-2-3 Logarithms of Numbers Less Than 1

If problems contained numbers between 0 and 0.1, the logarithms would be numbers between 0 and −1. If the numbers were between 0.1 and 0.01, the logarithms would be numbers between −1 and −2, and so on. Therefore, we can make the following statement:

KEY POINT *For any number less than 1, the logarithm (or exponent) is a negative number.*

Find the log of 0.0275. Log 0.0275 = −1.5607. In exponential form, $10^{-1.5607} =$ 0.0275. Using scientific notation gives $0.0275 = 2.75 \times 10^{-2}$. Log 2.75 = 0.4393 and $\log 10^{-2} = -2$. The resulting exponent is $-2 + 0.4393 = -1.5607$.

□ EXAMPLE 23-5

Find log 0.432.

Solution

$$\log 0.432 = -0.3645$$

In exponential form,

$$10^{-0.3645} = 0.432$$

□

PRACTICE PROBLEMS 23-4

Find the logs of the following numbers. Express your answers in both logarithmic and exponential forms.

1. 0.0463
2. 0.000935
3. 0.00205
4. 0.127
5. 0.00000812

Solutions

Logarithmic form	*Exponential form*
1. log 0.0463 = − 1.3344	$10^{-1.3344} = 0.0463$
2. log 0.000935 = −3.0292	$10^{-3.0292} = 0.000935$
3. log 0.00205 = −2.6882	$10^{-2.6882} = 0.00205$
4. log 0.127 = −0.8962	$10^{-0.8962} = 0.127$
5. log 0.00000812 = −5.0904	$10^{-5.0904} = 0.00000812$

Additional practice problems are at the end of the chapter.

SELF-TEST 23-1

Find the logarithms of the following numbers. Express your answers in both logarithmic and exponential form.

1. 7
2. 98.3
3. 0.00420
4. 7,650,000
5. 5762
6. 0.273
7. 0.00017
8. 0.00000983
9. 1786
10. 0.0933

Answers to Self-test 23-1 are at the end of the chapter.

23-3 MULTIPLICATION OF NUMBERS BY USING LOGARITHMS

Numbers can be multiplied by using logarithms. When we covered exponents in Chapter 2, we learned that, when multiplying like bases, the exponents are added. For example,

$$10^2 \times 10^3 = 10^{2+3} = 10^5$$

$$y^3 \times y^4 = y^{3+4} = y^7$$

$$10^{2.34} \times 10^{1.72} = 10^{4.06}$$

Since this is true, when multiplying numbers together we could change the numbers by using logarithms so that the bases are alike. Then we could add the logs (exponents) together. The number in exponential form would be the answer. For instance, suppose we wanted to multiply 37 × 763. We would first change the numbers to base 10. Log 37 = 1.568 and log 763 = 2.883. In exponential form,

$$37 = 10^{1.568} \quad \text{and} \quad 763 = 10^{2.883}$$

Then

$$37 \times 763 = 10^{1.568} \times 10^{2.883} = 10^{4.451}$$

$$10^{4.451} = 2.82 \times 10^4$$

Since logarithms are exponents, we can make the following statement:

KEY POINT *Multiplication of numbers, using logarithms, is reduced to addition.*

To find the product of numbers using logs:

1. Find the log of each number.
2. Change each number to base 10.
3. Add the exponents.

When we found the product of 37 × 763, we could have written the problem in this way:

$$\log (37 \times 763) = \log 37 + \log 763$$

The general equation is

$$\log ab = \log a + \log b$$

Remember, the logs are the exponents to the base 10 and, according to the laws of exponents, are added.

☐ EXAMPLE 23-6

Find the product of 7600 × 230. Round answer to three places.

Solution

$$\log (7600 \times 230) = \log 7600 + \log 230 = 3.881 + 2.362 = 6.243$$

$$7600 \times 230 = 10^{3.881} \times 10^{2.362} = 10^{6.243} = 1.75 \times 10^6$$

☐ EXAMPLE 23-7

Find the product of 1760 × 0.0273. Round answer to three places.

Solution

$$\log(1760 \times 0.0273) = \log 1760 + \log 0.0273$$
$$= 3.246 + (-1.564) = 1.682$$
$$1760 \times 0.0273 = 10^{3.246} \times 10^{-1.564} = 10^{1.682} = 4.8 \times 10$$

□ EXAMPLE 23-8

Find the product of 0.0783 × 0.0000342.

Solution

$$\log(0.0783 \times 0.0000342) = \log 0.0783 + \log 0.0000342$$
$$= -1.106 + (-4.466) = -5.572$$
$$10^{-5.572} = 2.68 \times 10^{-6}$$

□

PRACTICE PROBLEMS 23-5

Find the product of the following numbers by using logs. Round your answers to three digits.

1. 76 × 2730
2. 0.783 × 6500
3. 0.00376 × 0.107
4. 7,600,000 × 9800
5. 10,700 × 0.0083

Solutions

1. $\log(76 \times 2730) = \log 76 + \log 2730$
$= 1.881 + 3.436 = 5.317$
$10^{5.317} = 2.07 \times 10^5$

2. $\log(0.783 \times 6500) = \log 0.783 + \log 6500$
$= -0.106 + 3.813 = 3.707$
$10^{3.707} = 5.09 \times 10^3$

3. $\log(0.00376 \times 0.107) = \log 0.0037 + \log 0.107$
$= -2.425 + (-0.971)$
$= -3.395$
$10^{-3.395} = 4.02 \times 10^{-4}$

4. $\log(7{,}600{,}000 \times 9800) = \log 7{,}600{,}000 + \log 9800$
$= 6.881 + 3.991$
$= 10.872$
$10^{10.872} = 7.45 \times 10^{10}$

5. $\log(10{,}700 \times 0.0083) = \log 10{,}700 + \log 0.0083$
$= 4.029 + (-2.081) = 1.948$
$10^{1.948} = 8.88 \times 10$

Additional practice problems are at the end of the chapter.

23-4 DIVISION OF NUMBERS BY USING LOGARITHMS

Division of numbers by using logarithms is possible by again observing the laws of exponents. When like bases are divided, the exponents are subtracted. For example,

$$\frac{10^7}{10^4} = 10^{7-4} = 10^3$$

$$\frac{8^2}{8^5} = 8^{2-5} = 8^{-3}$$

$$\frac{a^5}{a^{-3}} = a^{5-(-3)} = a^{5+3} = a^8$$

$$\frac{10^{4.384}}{10^{2.732}} = 10^{4.384-2.732} = 10^{1.652}$$

Let's find the quotient of $\frac{6270}{220}$ by using logs. To do this:

1. Find the log of both dividend and divisor.
2. Subtract the log of the divisor from the log of the dividend.
3. Change to exponential form.

We would solve the problem this way:

$$\log\left(\frac{6270}{220}\right) = \log 6270 - \log 220 = 3.797 - 2.342 = 1.455$$

$$10^{1.455} = 2.85 \times 10$$

As with multiplication, finding the logs of the numbers gives us the opportunity to add or subtract exponents of like bases:

$$\frac{6270}{220} = \frac{10^{3.797}}{10^{2.342}} = 10^{3.797-2.342} = 10^{1.455} = 2.85 \times 10$$

☐ EXAMPLE 23-9

Find the quotient of $\frac{1.73}{96{,}700}$.

Solution

$$\log\left(\frac{1.73}{96{,}700}\right) = \log 1.73 - \log 96{,}700 = 0.238 - 4.985$$

$$= -4.747$$

$$10^{4.747} = 1.79 \times 10^{-5}$$

☐ EXAMPLE 23-10

Find the quotient of $\frac{0.0735}{0.00034}$.

Solution

$$\log\left(\frac{0.0735}{0.00034}\right) = \log 0.0735 - \log 0.00034 = -1.134 - (-3.469)$$

$$= 2.335$$

$$10^{2.335} = 2.16 \times 10^2$$

☐

PRACTICE PROBLEMS 23-6

Find the quotients of the following problems by using logs. Round your answers to three places.

1. $\frac{76{,}000}{42}$
2. $\frac{207}{56{,}300}$
3. $\frac{1670}{0.0073}$
4. $\frac{0.000426}{0.862}$
5. $\frac{0.0000942}{6040}$

Solutions

1. $\log\left(\frac{76{,}000}{42}\right) = \log 76{,}000 - \log 42$
   $= 4.881 - 1.623 = 3.258$
   $10^{3.258} = 1.81 \times 10^3$

2. $\log\left(\frac{207}{56{,}300}\right) = \log 207 - \log 56{,}300$
   $= 2.316 - 4.751 = -2.435$
   $10^{-2.435} = 3.68 \times 10^{-3}$

3. $\log\left(\frac{1670}{0.0073}\right) = \log 1670 - \log 0.0073$
   $= 3.223 - (-2.137) = 5.359$
   $10^{5.359} = 2.29 \times 10^5$

4. $\log\left(\frac{0.000426}{0.862}\right) = \log 0.000426 - \log 0.862$
   $= -3.371 - (-0.064)$
   $= -3.306$
   $10^{-3.306} = 4.94 \times 10^{-4}$

5. $\log\left(\frac{0.0000942}{6040}\right) = \log 0.0000942 - \log 6040$
   $= -4.026 - 3.781 = -7.807$
   $10^{-7.807} = 1.56 \times 10^{-8}$

Additional practice problems are at the end of the chapter.

23-5 RAISING A NUMBER TO A POWER BY USING LOGARITHMS

KEY POINT *Raising a number to a power is a process of multiplication.* In general, to find a^x using logarithms,

$$\log a^x = x(\log a)$$

Let's find 4^5 by using logarithms. To solve problems of this type:

1. Find the log of the number.
2. Multiply the log by the exponent.
3. Change to exponential form.

We can express the problem this way: $\log 4^5 = 5 \log 4$. Remember, we are going to multiply the log of the number by the exponent.

$$5 \log 4 = 5 \times 0.602 = 3.01$$

$$10^{3.01} = 1.02 \times 10^3$$

$$4^5 = (10^{0.602})^5 = 10^{(0.602)(5)} = 10^{3.01} = 1.02 \times 10^3$$

□ EXAMPLE 23-11

What does 0.0745^3 equal?

Solution

$$\log 0.0745^3 = 3 \log 0.0745 = (3)(-1.128) = -3.384$$

$$10^{-3.384} = 4.13 \times 10^{-4}$$

$$0.0745^3 = (10^{-1.128})^3 = 10^{-3.384} = 4.13 \times 10^{-4}$$

□ EXAMPLE 23-12

What does $74.6^{-2.5}$ equal?

Solution

$$\log 74.6^{-2.5} = -2.5 \log 74.6 = -2.5 \times 1.873 = -4.682$$

$$10^{-4.682} = 2.08 \times 10^{-5}$$

$$74.6^{-2.5} = (10^{1.873})^{-2.5} = 10^{-4.682} = 2.08 \times 10^{-5}$$

□

PRACTICE PROBLEMS 23-7

Find the answers to the following problems by using logarithms. Round to three places.

1. 6.7^5

2. $2.3^{-3.6}$

3. 0.00403^4

4. $0.6^{4.2}$

5. 50^3

Solutions

1. $\log 6.7^5 = 5 \log 6.7 = 5 \times 0.826 = 4.13$
$10^{4.13} = 1.35 \times 10^4$

2. $\log 2.3^{-3.6} = -3.6 \log 2.3 = -3.6 \times 0.362$
$= -1.302$
$10^{-1.302} = 4.99 \times 10^{-2}$

3. $\log 0.00403^4 = 4 \log 0.00403 = 4 \times (-2.395)$
$= -9.579$
$10^{-9.579} = 2.64 \times 10^{-10}$

4. $\log 0.6^{4.2} = 4.2 \log 0.6 = 4.2 \times (-0.222)$
$= -0.932$
$10^{-0.932} = 0.117 = 1.17 \times 10^{-1}$

5. $\log 50^3 = 3 \log 50 = 3 \times 1.699 = 5.097$
$10^{5.097} = 1.25 \times 10^5$

Additional practice problems are at the end of the chapter.

SELF-TEST 23-2

Find the answers to the following problems by using logarithms. Round to three places.

1. 463×0.027

2. $\dfrac{34.3}{8700}$

3. $7^{4.6}$

4. $\dfrac{0.0063}{2.73}$

5. 0.463^4

6. $97{,}000 \times 170$

7. 0.0128×0.000055

8. $27.3^{2.6}$

9. $\dfrac{43.2}{0.000772}$

10. 0.00193×1930

Answers to Self-test 23-2 are at the end of the chapter.

23-6 NATURAL LOGARITHMS

Common logarithms are exponents where 10 is the base. *Natural logarithms* are exponents where e is the base. The numerical value of e, rounded to three digits, is 2.72. In electronics, the instantaneous charge or discharge of a capacitor or the instantaneous current in a coil can be determined by equations using natural logarithms.

23-6-1 Finding Natural Logarithms of Numbers

Just as we have tables of common logarithms, we have tables of natural logarithms. We used the abbreviation "log" for common logarithms. We use the abbreviation "ln" for natural logarithms.

As with common logarithms, the calculator stores a table of natural logarithms. What is the natural log of 5? Using the calculator,

5 [Ln] *Answer*: $\ln 5 = 1.6094$

In exponential form, $e^{1.6094} = 5$. e^x is usually the second function of the [Ln] key, so

1.6094 [e^x] *Answer*: $e^{1.6094} = 5$

Notice that the change from logarithmic form to exponential form is the same for natural logs as for common logs. The only difference is in the base. Note the following examples:

Logarithmic form	***Exponential form***
$\log 5 = 0.6990$	$10^{0.6990} = 5$
$\ln 5 = 1.6094$	$e^{1.6094} = 5$
$\log 1 = 0$	$10^0 = 1$
$\ln 1 = 0$	$e^0 = 1$
$\log 10 = 1$	$10^1 = 10$
$\ln e = 1$	$e^1 = e$

Find ln 45. $\ln 45 = 3.8067$. In exponential form, $e^{3.8067} = 45$. Find the ln of 0.832. $\ln 0.832 = -0.1839$. In exponential form, $e^{-0.1839} = 0.832$. As with common logarithms, the natural logarithm of numbers greater than 1 are *positive*, and the natural logarithms of numbers less than 1 are *negative*.

PRACTICE PROBLEMS 23-8

Find the natural logarithms of the following numbers by using the calculator.

1. 7.4
2. 300
3. 0.8
4. 3.4
5. 100

Solutions

1. $\ln 7.4 = 2.0015$
2. $\ln 300 = 5.7038$
3. $\ln 0.8 = -0.2231$
4. $\ln 3.4 = 1.2238$
5. $\ln 100 = 4.6052$

Additional practice problems are at the end of the chapter.

If we know the exponent to the base e, we can find the number by using the calculator, just as we found numbers when we were using base 10. What does $e^{3.434}$ equal? $e^{3.434} = 31$. What does e^2 equal? $e^2 = 7.39$.

PRACTICE PROBLEMS 23-9

Find the number whose natural logarithm is:

1. 1.946
2. 3
3. 4.489
4. 0.0953
5. 1.435

Solutions

1. $e^{1.946} = 7.00$
2. $e^3 = 20.1$
3. $e^{4.489} = 89$
4. $e^{0.0953} = 1.10$
5. $e^{1.435} = 4.20$

Additional practice problems are at the end of this chapter.

SELF-TEST 23-3

Find the natural logarithm of the following numbers. Convert your answers to exponential form.

1. 14
2. 0.0073
3. 740
4. 1073
5. 10
6. 0.00001
7. 0.463
8. 0.0673
9. 4.73
10. 73

Answers to Self-test 23-3 are at the end of the chapter.

CHAPTER 23 AT A GLANCE

Page	Key Points and Definitions	Example
575	A logarithm is an exponent.	
578	*Key Point:* Only numbers that are multiples of 10 will have logarithms that are whole numbers.	log 10 = 1 log 100 = 2

Page	Key Points and Definitions	Example
579	*Key Point:* For any number less than 1, the logarithm is a negative number.	$\log 1 = 0$ $\log 0.1 = -1$
581	*Key Point:* Multiplication of numbers, using logarithms, is reduced to addition.	$\log ab = \log a + \log b$
585	*Key Point:* Raising a number to a power is a process of multiplication.	$\log a^x = x(\log a)$
586	*Natural* logarithms are exponents where e is the base. $e = 2.718$.	$\ln 10 = 2.30$

End of Chapter Problems 23-1

Find the logarithms of the following numbers. Express answers in both logarithmic and exponential forms.

1. 1000
2. 100,000
3. 1
4. 10
5. 10,000,000
6. 100
7. 0.1
8. 0.001
9. 0.000001
10. 0.0001

End of Chapter Problems 23-2

Find the logarithms of the following numbers accurate to four significant figures. Express answers in both logarithmic and exponential forms.

1. 3.42
2. 8.45
3. 4.05
4. 6.33
5. 6.44
6. 1.73
7. 9.45
8. 7.85
9. 2.50
10. 5.66

End of Chapter Problems 23-3

Find the logarithms of the following numbers accurate to four significant figures. Express answers in both logarithmic and exponential forms.

1. 14
2. 376
3. 743
4. 5650
5. 1070
6. 7070
7. 43.4
8. 74.7

9. 10,000
10. 400,000
11. 23,400
12. 74,500
13. 47,700
14. 175
15. 14.6
16. 7.84
17. 37,000
18. 301,000
19. 2,000,000
20. 84.7

End of Chapter Problems 23-4

Find the logarithms of the following numbers accurate to four significant figures: Express answers in both logarithmic and exponential forms.

1. 0.00377
2. 0.000248
3. 0.037
4. 0.000637
5. 0.746
6. 0.332
7. 0.407
8. 0.001
9. 0.000614
10. 0.000276
11. 0.49
12. 0.00699
13. 0.00814
14. 0.00419
15. 0.301
16. 0.000301
17. 0.975
18. 0.00075
19. 0.0275
20. 0.00875

End of Chapter Problems 23-5

Find the product of the following numbers by using logarithms:

1. 27×56
2. 83×720
3. $107 \times 243{,}000$
4. 110×3070
5. 43×0.143
6. 65×0.243
7. 0.043×0.0106
8. 0.00274×0.0976
9. $0.0000176 \times 0.000357$
10. 0.000555×0.00477
11. 256×0.00246
12. 670×0.0178
13. 78×0.473
14. 48.7×0.00942
15. 4730×0.000506
16. 2780×0.000225
17. 760×7360
18. 0.146×0.0356
19. 2560×0.00457
20. 1650×0.00735

End of Chapter Problems 23-6

Find the quotients of the following numbers by using logarithms:

1. $\dfrac{1760}{43}$
2. $\dfrac{6530}{3.75}$

3. $\frac{65}{467}$

4. $\frac{27}{730}$

5. $\frac{0.00273}{172}$

6. $\frac{0.000825}{70}$

7. $\frac{4300}{0.15}$

8. $\frac{5430}{0.0335}$

9. $\frac{0.0675}{0.00103}$

10. $\frac{0.000565}{0.0093}$

11. $\frac{0.00467}{73.2}$

12. $\frac{0.00675}{62.3}$

13. $\frac{0.147}{0.0432}$

14. $\frac{0.473}{0.0563}$

15. $\frac{4730}{0.443}$

16. $\frac{7430}{0.257}$

17. $\frac{0.445}{635}$

18. $\frac{0.655}{425}$

19. $\frac{0.00275}{4.35}$

20. $\frac{0.000375}{6.45}$

End of Chapter Problems 23-7

Express the following numbers by using logarithms:

1. 3.4^{3}
2. 8.26^{8}
3. $6^{-1.3}$
4. $14^{-2.43}$
5. $0.43^{2.5}$
6. 0.007^{4}
7. $6^{4.3}$
8. $4^{2.73}$
9. $0.27^{-4.4}$
10. $0.083^{-2.1}$
11. $8.7^{6.3}$
12. $5.67^{5.3}$
13. $3.4^{3.2}$
14. $2.75^{4.3}$
15. $6.35^{-2.3}$
16. $1.73^{-4.3}$
17. $0.045^{3.6}$
18. $0.00563^{4.3}$
19. $0.00735^{-2.5}$
20. $0.00245^{-3.6}$

End of Chapter Problems 23-8

Find the natural logarithms of the following numbers. Express your answers in both logarithmic and exponential forms.

1. 2
2. 6.78
3. 0.932
4. 0.0273
5. 60
6. 23

7. 0.073
8. 0.345
9. 247
10. 783
11. 0.00146
12. 0.00541
13. 1100
14. 1500
15. 0.00027
16. 0.000481
17. 8.43
18. 12.3
19. 0.346
20. 0.000092
21. 3.67
22. 73.8
23. 0.107
24. 0.0081
25. 176
26. 500
27. 0.293
28. 0.372
29. 873
30. 1800

End of Chapter Problems 23-9

Find the number whose natural logarithm is:

1. 2.34
2. 1.07
3. 3.77
4. 4
5. 4.25
6. 2.7
7. 5.3
8. 3.89
9. 1.55
10. 3.9
11. 0.43
12. 0.732
13. −0.0045
14. −0.0078
15. 0.15
16. 0.225
17. −0.315
18. −0.41
19. −0.142
20. −0.914

Answers to Self-tests

SELF-TEST 23-1

	Logarithmic form	*Exponential form*
1.	log 7 = 0.8451	$10^{0.8451} = 7$
2.	log 98.3 = 1.993	$10^{1.993} = 98.3$
3.	log 0.0042 = −2.377	$10^{-2.377} = 0.0042$
4.	log 7,650,000 = 6.883	$10^{6.883} = 7{,}650{,}000$
5.	log 5762 = 3.76	$10^{3.76} = 5762$
6.	log 0.273 = −0.564	$10^{-0.564} = 0.273$
7.	log 0.00017 = −3.77	$10^{-3.77} = 0.00017$
8.	log 0.00000983 = −5.007	$10^{-5.007} = 0.00000983$
9.	log 1786 = 3.253	$10^{3.253} = 1786$
10.	log 0.0933 = −1.03	$10^{-1.03} = 0.0933$

SELF-TEST 23-2

1. $\log(463 \times 0.027) = \log 463 + \log 0.027 = 2.666 + (-1.569) = 1.097$
   $10^{1.097} = 1.25 \times 10$
2. $\log\left(\frac{34.3}{8700}\right) = \log 34.3 - \log 8700 = 1.535 - 3.94 = -2.404$
   $10^{-2.404} = 3.94 \times 10^{-3}$
3. $\log 7^{4.6} = 4.6 \log 7 = 4.6 \times 0.845 = 3.89$
   $10^{3.89} = 7.72 \times 10^3$
4. $\log\left(\frac{0.0063}{2.73}\right) = \log 0.0063 - \log 2.73 = -2.201 - 0.436 = -2.637$
   $10^{-2.637} = 2.31 \times 10^{-3}$
5. $\log 0.463^4 = 4 \log 0.463 = 4 \times (-0.334) = -1.34$
   $10^{-1.34} = 4.6 \times 10^{-2}$
6. $\log(97{,}000 \times 170) = \log 97{,}000 + \log 170 = 4.987 + 2.23 = 7.217$
   $10^{7.217} = 1.65 \times 10^7$
7. $\log(0.0128 \times 0.000055) = \log 0.0128 + \log 0.000055 = -1.893 + (-4.26) = -6.152$
   $10^{-6.152} = 7.04 \times 10^{-7}$
8. $\log 27.3^{2.6} = 2.6 \log 27.3 = 2.6 \times 1.436 = 3.734$
   $10^{3.734} = 5.42 \times 10^3$
9. $\log \frac{43.2}{0.000772} = \log 43.2 - \log 0.000772 = 1.635 - (-3.112) = 4.748$
   $10^{4.748} = 5.60 \times 10^4$
10. $\log(0.00193 \times 1930) = \log 0.00193 + \log 1930 = -2.714 + 3.286 = 0.571$
    $10^{0.571} = 3.72$

SELF-TEST 23-3

	Logarithmic form	*Exponential form*
1.	$\ln 14 = 2.639$	$e^{2.639} = 14$
2.	$\ln 0.0073 = -4.92$	$e^{-4.92} = 0.0073$
3.	$\ln 740 = 6.607$	$e^{6.607} = 740$
4.	$\ln 1073 = 6.978$	$e^{6.978} = 1073$
5.	$\ln 10 = 2.303$	$e^{2.303} = 10$
6.	$\ln 0.00001 = -11.51$	$e^{-11.51} = 0.00001$
7.	$\ln 0.463 = -0.770$	$e^{-0.770} = 0.463$
8.	$\ln 0.0673 = -2.699$	$e^{-2.699} = 0.0673$
9.	$\ln 4.73 = 1.554$	$e^{1.554} = 4.73$
10.	$\ln 73 = 4.29$	$e^{4.29} = 73$

CHAPTER 24

LOGARITHMIC EQUATIONS

INTRODUCTION

This chapter continues our study of logarithms. Because of the logarithmic nature of the way the human ear perceives changes in sound, we must develop circuits to deal with this characteristic. Whenever we analyze circuits, we work with equations. If we are working with amplifiers and developing curves and collecting data that describe the amplifiers' characteristics, we will be working with some sort of logarithmic equation. In this chapter we will work with the kinds of logarithmic equations you are most likely to encounter in your electronics studies.

CHAPTER OBJECTIVES

In this chapter you will learn how to:

1. Write equations in both logarithmic and exponential forms.
2. Solve equations involving both common and natural logarithms.
3. Change equations from logarithmic to exponential form and from exponential to logarithmic form.
4. Use the calculator to help solve logarithmic equations.

24-1 LOGARITHMIC AND EXPONENTIAL FORMS

We have discussed logarithmic and exponential expressions in previous chapters. Now let's write them again in the form of simple equations.

Logarithmic	*Exponential*	
(1) $\log N = x$	$10^x = N$	(24-1)
(2) $\ln a = b$	$e^b = a$	(24-2)

In (24-1), x is the exponent and N is the number. The base is 10. In (24-2), b is the exponent and a is the number. The base is e.

KEY POINT Logarithmic equations can either be solved in logarithmic form or in exponential form.

Consider the following equation:

$$\log 73 = t$$

This equation is in logarithmic form. Seventy-three is the number and t is the exponent. The equation is solved when we find log 73.

KEY POINT Logarithmic equations can be solved directly when the exponent is unknown.

$$\log 73 = t$$
$$t = 1.86$$

In the following logarithmic equation, the number is unknown.

$$\log y = 4.73$$

We can't solve this equation in its present form because we don't know the value of y. However, the equation is easily solved if we change to exponential form.

$$\log y = 4.73 \qquad \text{(logarithmic form)}$$
$$10^{4.73} = y \qquad \text{(exponential form)}$$
$$y = 5.37 \times 10^4$$

KEY POINT Change from logarithmic to exponential form when the number is unknown.

PRACTICE PROBLEMS 24-1

Change the following equations from logarithmic form to exponential form:

1. $\log 7600 = a$

2. $\log x = -3.74$

3. $\log y = 0.742$

4. $\log 0.00432 = x$

Change the following equations from exponential form to logarithmic form:

5. $10^{4.2} = y$

6. $10^{t} = 0.043$

7. $10^{y} = 173$

8. $10^{-2.43} = y$

Solutions

1. $10^{a} = 7600$

2. $10^{-3.74} = x$

3. $10^{0.742} = y$

4. $10^{x} = 0.00432$

5. $\log y = 4.2$

6. $\log 0.043 = t$

7. $\log 173 = y$

8. $\log y = -2.43$

Additional practice problems are at the end of the chapter.

24-2 LOGARITHMIC EQUATIONS: COMMON LOGS

Let's consider some of the equations we might encounter in problem solving that involve the use of common logs.

☐ EXAMPLE 24-1

$x = \log 250$. Solve for x.

$$x = 2.40$$

☐ EXAMPLE 24-2

$\log a = 7.83$. Solve for a.

Solution Change to exponential form:

$$10^{7.83} = a$$

$$a = 6.76 \times 10^{7}$$

7.83 [10^x] *Answer*: 6.76×10^{7}

■ **KEY POINT** *We solve logarithmic equations using rules developed in Chapter 8: We may perform any mathematical operation on one side of an equation provided we perform the same operation on the other side.*

☐ EXAMPLE 24-3

$3^x = 104$. Solve for x.

Solution Take the log of both sides to get rid of the exponent:

$$x \log 3 = \log 104$$

$$x = \frac{\log 104}{\log 3} = \frac{2.02}{0.477}$$

$$= 4.23$$

(We could have found log 3 and log 104 first and then solved for x.)

104 [log] [÷] 3 [log] [=]   *Answer:* 4.23

☐ EXAMPLE 24-4

$x^{3.6} = 85$. Solve for x.

Solution Take the log of both sides to get rid of the exponent:

$$3.6 \log x = \log 85$$

$$\log x = \frac{\log 85}{3.6} = \frac{1.93}{3.6} = 0.536$$

Change to exponential form:

$$10^{0.536} = x$$

$$x = 3.44$$

85 [log] [÷] 3.6 [=] [10^x]   *Answer:* 3.44

☐ EXAMPLE 24-5

$6.3^4 = x$. Solve for x.

Solution Take the log of both sides to get rid of the exponent:

$$4 \log 6.3 = \log x$$

$$4(0.799) = \log x$$

$$3.2 = \log x$$

Change to exponential form:

$$10^{3.2} = x$$

$$x = 1.58 \times 10^3$$

☐

PRACTICE PROBLEMS 24-2

Solve for the unknown in the following problems:

1. $\log 473 = x$
2. $\log 0.0783 = x$
3. $\log t = 2.74$
4. $\log x = -3.74$
5. $\log a = 0.274$
6. $\log 10{,}700 = y$
7. $4^x = 256$
8. $x^{2.7} = 56$
9. $3.6^3 = x$
10. $x^{-1.4} = 0.073$
11. $3.4^{-3} = x$
12. $0.087^x = 4$

Solutions

1. $x = 2.67$
2. $y = -1.11$
3. $t = 5.5 \times 10^2$
4. $x = 1.82 \times 10^{-4}$
6. $y = 4.03$
7. $x = 4$
8. $x = 4.44$
9. $x = 46.7$
10. $x = 6.488$
11. $x = 0.0254$
12. $x = -0.5678$

Additional practice problems are at the end of the chapter.

☐ EXAMPLE 24-6

$3^{1-x} = 12$. Solve for x.

Solution Take the log of both sides to get rid of the exponent:

$$(1 - x) \log 3 = \log 12$$

$$1 - x = \frac{\log 12}{\log 3} = \frac{1.08}{0.477} = 2.26$$

$$1 - 2.26 = x$$

$$x = -1.26$$

☐ EXAMPLE 24-7

$27^{x/3} = 18$. Solve for x.

Solution Take the log of both sides:

$$\frac{x}{3} \log 27 = \log 18$$

$$\frac{x}{3} = \frac{\log 18}{\log 27} = \frac{1.255}{1.431} = 0.877$$

$$x = 2.63$$

☐ EXAMPLE 24-8

$4 = \log \frac{x}{100}$. Solve for x.

Solution Change to exponential form:

$$10^4 = \frac{x}{100}$$

$$x = 100 \times 10^4$$

$$x = 10^6$$

☐ EXAMPLE 24-9

$8 = \log \frac{200}{x}$. Solve for x.

Solution Change to exponential form:

$$10^8 = \frac{200}{x}$$

$$10^8 x = 200$$

$$x = \frac{200}{10^8}$$

$$x = 2 \times 10^{-6}$$

☐ EXAMPLE 24-10

$65 = 10 \log \frac{x}{0.03}$. Solve for x.

Solution

$$\frac{65}{10} = \log \frac{x}{0.03}$$

$$6.5 = \log \frac{x}{0.03}$$

Change to exponential form:

$$10^{6.5} = \frac{x}{0.03}$$

$$0.03 \times 10^{6.5} = x$$

$$x = 9.49 \times 10^4$$

☐

PRACTICE PROBLEMS 24-3

Solve for x in the following problems:

1. $4^{3-x} = 128$
2. $5.4^{x+2} = 200$
3. $53^{x/2} = 35$
4. $3.6 = \log \frac{x}{25}$
5. $7.3 = \log \frac{75}{x}$
6. $x = \log \frac{36}{2}$
7. $43 = 10 \log \frac{x}{12}$
8. $85 = 20 \log \frac{40}{x}$

Solutions

1. $x = -0.5$
2. $x = 1.14$
3. $x = 1.79$
4. $x = 9.95 \times 10^4$
5. $x = 3.76 \times 10^{-6}$
6. $x = 1.26$
7. $x = 2.39 \times 10^5$
8. $x = 2.25 \times 10^{-3}$

Additional practice problems are at the end of the chapter.

SELF-TEST 24-1

Solve for x in the following problems:

1. $x = \log 400$
2. $4^{3.5} = x$
3. $12^{-4} = x$
4. $x^{1.7} = 40$
5. $5^{x+2} = 625$
6. $9^{1-x} = 36$
7. $65^{x/3} = 10$
8. $3^{4/x} = 0.506$
9. $5 = \log \frac{x}{40}$
10. $14 = \log \frac{200}{x}$
11. $\log x^4 = 2$
12. $\log 3^x = 0.93$

Answers to Self-test 24-1 are at the end of the chapter.

24-3 LOGARITHMIC EQUATIONS: NATURAL LOGS

■ **KEY POINT** *Equations involving natural logarithms are solved just like equations involving common logarithms.*

Here are some problems involving the use of natural logs.

□ EXAMPLE 24-11

$e^{1.7} = y$. Solve for y.

Solution $y = 5.47$

□ EXAMPLE 24-12

$e^y = 83$. Solve for y.

Solution Take the ln of both sides:

$$y \ln e = \ln 83$$
$$y(1) = \ln 83 \qquad \text{(remember: } \ln e = 1\text{)}$$
$$y = 4.42$$

□ EXAMPLE 24-13

$89 = e^{-x}$. Solve for x.

Solution

$$\ln 89 = -x \ln e$$
$$4.49 = -x$$
$$x = -4.49$$

□ EXAMPLE 24-14

$\ln x = 1.7$. Solve for x.

Solution Change to exponential form:

$$e^{1.7} = x$$
$$x = 5.47$$

□

PRACTICE PROBLEMS 24-4

Solve for the unknown in the following problems:

1. $e^{3.4} = x$

2. $e^x = 49$

3. $89 = e^t$

4. $y = e^{4.5}$

5. $e^y = 58$

6. $e^x = 150$

7. $\ln x = 3$

8. $\ln x = -2.7$

Solutions

1. $x = 30$

2. $x = 3.89$

3. $t = 4.49$

4. $y = 90$

5. $y = 4.06$

6. $x = 5.01$

7. $x = 20.1$

8. $x = 6.72 \times 10^{-2}$

Additional practice problems are at the end of the chapter.

☐ EXAMPLE 24-15

$x = 20(1 - e^{-0.5})$. Solve for x.

Solution

$$\begin{aligned} x &= 20(1 - 0.606) \\ &= 20(0.393) \\ &= 7.87 \end{aligned}$$

☐ EXAMPLE 24-16

$30 = x(1 - e^{-4})$. Solve for x.

Solution

$$\begin{aligned} 30 &= x(1 - 0.0183) \\ 30 &= x(0.982) \\ x &= 30.6 \end{aligned}$$

☐ EXAMPLE 24-17

$5 = 15(1 - e^{-1/x})$. Solve for x.

Solution

$$\begin{aligned} 0.333 &= 1 - e^{-1/x} \\ 0.333 - 1 &= -e^{-1/x} \\ -0.667 &= -e^{-1/x} \\ 0.667 &= e^{-1/x} \\ \ln 0.667 &= -\frac{1}{x} \ln e \\ -0.405 &= \frac{-1}{x} \\ 0.405x &= 1 \\ x &= 2.47 \end{aligned}$$

☐

PRACTICE PROBLEMS 24-5

Solve for the unknown in the following problems:

1. $x = 25(1 - e^{-1.75})$

2. $40 = 60(1 - e^{-x})$

3. $15 = 45(1 - e^{-2/3x})$

4. $0.45 = e^{-x/3.2}$

5. $0.175 = 1 - e^{-3x/4}$

6. $22 = x(1 - e^{-2.3})$

Solutions

1. $x = 20.7$

2. $x = 1.1$

3. $x = 1.65$

4. $x = 2.56$

5. $x = 0.256$

6. $x = 24.4$

SELF-TEST 24-2

Solve for the unknown in the following problems:

1. $x = \ln 40$

2. $\ln x = 14$

3. $e^x = 20$

4. $x = 1 - e^{-0.6}$

5. $36 = 100e^{-x}$

6. $x = 50(1 - e^{-2.75})$

7. $86 = x(1 - e^{-1.5})$

8. $40 = 80e^{-2/x}$

9. $0.45 = e^{-x/4}$

10. $0.2 = 1 - e^{-2/6x}$

Answers to Self-test 24-2 are at the end of the chapter.

CHAPTER 24 AT A GLANCE

Page	Key Points and Rules	Example
595	*Key Point:* Logarithmic equations can either be solved in logarithmic form or in exponential form.	$\log N = x$ $10^x = N$
595	*Key Point:* Logarithmic equations can be solved directly when the exponent is unknown.	$x = \log 250$ $= 2.40$
595	*Key Point:* Change from logarithmic to exponential form when the number is unknown.	$\log a = 7.83$ $a = 10^{7.83}$ $= 6.76 \times 10^7$
597	We solve logarithmic equations using rules developed in Chapter 8.	$3^x = 104$. Solve for x. Take the log of both sides. $x \log 3 = \log 104$ $x = \dfrac{\log 104}{\log 3} = \dfrac{2.02}{0.477} = 4.23$

Page	Key Points, and Rules	Example
600	*Key Point:* Equations involving natural logarithms are solved just like equations involving common logarithms.	$e^y = 83$. Solve for y. $y \ln e = \ln 83$ $y(1) = \ln 83$ $y = 4.42$

End of Chapter Problems 24-1

Change the following equations from logarithmic form to exponential form:

1. $\log 340 = x$
2. $\log 4760 = x$
3. $\log 14 = x$
4. $\log 25 = x$
5. $\log 7.4 = x$
6. $\log 6.3 = x$
7. $\log 0.014 = y$
8. $\log 0.0073 = y$
9. $\log 0.734 = y$
10. $\log 0.417 = y$
11. $\log -0.00178 = y$
12. $\log 0.00427 = y$
13. $\log x = -2.43$
14. $\log x = -0.44$
15. $\log x = -3.643$
16. $\log x = -2.743$
17. $\log x = -0.143$
18. $\log x = -0.173$
19. $\log y = 1.44$
20. $\log y = 0.042$
21. $\log y = 2.834$
22. $\log y = 1.738$
23. $\log y = 4.146$
24. $\log y = 2.75$

Change the following equations from exponential form to logarithmic form:

25. $10^{3.6} = y$
26. $10^{1.7} = y$
27. $10^{4.73} = y$
28. $10^{2.66} = y$
29. $10^{2.73} = y$
30. $10^{1.732} = y$
31. $10^a = 732$
32. $10^a = 1.73$
33. $10^a = 47.3$
34. $10^a = 176$
35. $10^a = 276$
36. $10^a = 27.4$
37. $10^y = 0.043$
38. $10^y = 0.736$
39. $10^y = 0.1778$
40. $10^y = 0.4663$
41. $10^y = 0.00178$
42. $10^y = 0.00556$
43. $10^{-3.4} = x$
44. $10^{-1.02} = x$
45. $10^{-2.76} = x$
46. $10^{-3.45} = x$
47. $10^{-1.17} = x$
48. $10^{-3.14} = x$

End of Chapter Problems 24-2

Find the unknown in the following problems:

1. $\log y = 7.36$
2. $\log y = 0.0276$
3. $\log x = 0.000736$
4. $\log x = 5.6$

5. $\log a = -4.23 \times 10^{-1}$
6. $\log a = -2.73$
7. $\log x = -3.64 \times 10^{-3}$
8. $\log x = -8.44 \times 10^{-2}$
9. $10^x = 462$
10. $10^x = 2700$
11. $10^x = 0.0173$
12. $10^x = 0.00906$
13. $10^y = 2.5 \times 10^{-3}$
14. $10^y = 2.6 \times 10^{-6}$
15. $10^y = 4.66 \times 10^4$
16. $10^y = 7.4 \times 10^3$
17. $10^{4.16} = x$
18. $10^{7.23} = x$
19. $10^{0.033} = x$
20. $10^{0.0056} = x$
21. $10^{-3.6} = y$
22. $10^{-1.73} = y$
23. $10^{-0.173} = y$
24. $10^{-0.00066} = y$
25. $\log 746 = x$
26. $\log 217 = x$
27. $\log 0.00716 = x$
28. $\log 0.0000106 = x$
29. $7.26^4 = x$
30. $12.4^{3.2} = x$
31. $0.026^x = 7$
32. $0.173^x = 0.037$
33. $x^{4.3} = 7300$
34. $x^{-4.3} = 0.45$
35. $\log y^3 = 9$
36. $\log 5^x = 5$
37. $\log y = 7.36$
38. $\log y = 0.0276$
39. $x^7 = 14$
40. $x^{4.5} = 75$
41. $70^x = 12$
42. $200^x = 6000$

End of Chapter Problems 24-3

Solve for the unknown in the following problems:

1. $6^{2-y} = 68$
2. $8^{1+y} = 50$
3. $75^{y/4} = 3.54$
4. $38^{y/3} = 6.73$
5. $1.7 = \log \dfrac{x}{42}$
6. $1 = \log \dfrac{x}{0.5}$
7. $6 = \log \dfrac{100}{x}$
8. $4 = \log \dfrac{0.43}{x}$
9. $3^{x-2} = 50$
10. $5^{x-1} = 80$
11. $16^{3/x} = 200$
12. $7^{2/x} = 0.763$
13. $6 = \log \dfrac{x}{100}$
14. $2 = \log \dfrac{x}{1000}$
15. $-4 = \log \dfrac{20}{x}$
16. $-3 = \log \dfrac{10}{x}$
17. $14^{x/2} = 43$
18. $8^{x/4} = 27$
19. $35^{3-x} = 470$
20. $15^{4-x} = 127$
21. $3 = \log \dfrac{0.005}{x}$
22. $7 = \log \dfrac{0.043}{x}$
23. $4.3 = \log \dfrac{y}{9}$
24. $2.7 = \log \dfrac{y}{4}$
25. $45^{x-2} = 3.75$
26. $30^{x-4} = 2.36$
27. $8.7^{x+2} = 150$
28. $7.3^{x+3} = 220$
29. $60^{y/3} = 1.78$
30. $100^{y/4} = 2.4$

End of Chapter Problems 24-4

Solve for the unknown in the following problems:

1. $x = \ln 11$
2. $x = \ln 47$
3. $\ln 15 = x$
4. $\ln 0.635 = x$
5. $e^x = 200$
6. $e^{-x} = 0.73$
7. $\ln 0.0004 = x$
8. $\ln 0.146 = x$
9. $e^{-x} = 15$
10. $e^x = 14.7$
11. $e^{2.7} = x$
12. $e^{4.6} = x$
13. $46 = e^y$
14. $70 = e^y$
15. $e^{-4} = y$
16. $e^{-1.6} = y$

End of Chapter Problems 24-5

1. $0.6254 = 1 - e^{-x}$
2. $x = 1 - e^{-0.5}$
3. $50 = 80e^{-x}$
4. $45 = 60e^{-x}$
5. $x = 25(1 - e^{-0.75})$
6. $x = 100(1 - e^{-2})$
7. $30 = x(1 - e^{-3})$
8. $20 = x(1 - e^{-1.5})$
9. $10 = 25e^{-x/3}$
10. $2.7 = 10e^{-x/1.7}$
11. $3 = 9e^{-4/x}$
12. $6 = 12e^{-0.27/x}$
13. $0.176 = e^{-2x/3}$
14. $0.93 = e^{-x/4.7}$
15. $0.5 = e^{-2/x}$
16. $0.37 = e^{-3.2/x}$
17. $x = 1 - e^{-3}$
18. $1 - x = e^{-1.7}$
19. $0.376 = 1 - e^{-x}$
20. $0.673 = e^{-x}$
21. $e^{-x/2} = 0.76$
22. $1 - x = e^{-2}$

Answers to Self-tests

SELF-TEST 24-1

1. $x = 2.6$
2. $x = 128$
3. $x = 4.82 \times 10^{-5}$
4. $x = 8.76$
5. $x = 2$
6. $x = -0.63$
7. $x = 1.66$
8. $x = -6.45$
9. $x = 4 \times 10^6$
10. $x = 2 \times 10^{-12}$
11. $x = 3.16$
12. $x = 1.95$

SELF-TEST 24-2

1. $x = 3.69$
2. $x = 1.2 \times 10^6$
3. $x = 3$
4. $x = 0.451$
5. $x = 1.02$
6. $x = 46.8$
7. $x = 111$
8. $x = 2.89$
9. $x = 3.19$
10. $x = 1.49$

APPLICATIONS OF LOGARITHMS

CHAPTER 25

INTRODUCTION

A parameter of many electronic circuits or systems is *gain*. The symbol for gain is A. Gain is the ratio of an output quantity to an input quantity. We may analyze a device in terms of power gain (A_p), voltage gain (A_v), current gain (A_i), or resistance gain (A_r). In ac circuits, gain will vary as the frequency varies. In this chapter we will calculate the gain of different kinds of circuits, note how the gain will change with frequency, and then plot curves of gain versus frequency so that we may have a visual display of these characteristics.

CHAPTER OBJECTIVES

In this chapter you will learn how to:

1. Compute both power and voltage gains of electrical systems.
2. Determine the frequency response of an electrical network or system.
3. Construct a Bode plot of the frequency response of an electrical network or system.
4. Compute the charge and discharge of a capacitor in a series RC circuit.

25-1 GAIN MEASUREMENTS

It has been demonstrated experimentally that the human ear responds logarithmically to changes in sound levels. It has also been found that it requires about a 1-decibel change in a sound level before we are aware that a change has taken place. Because of this logarithmic characteristic, power gains are usually expressed as the logarithm of the ratio of power levels.

$$A_{p(\text{bels})} = \log \frac{P_{\text{out}}}{P_{\text{in}}}$$

The unit of measure is the *bel* in honor of Alexander Graham Bell.

The bel is seldom used because it is considered too large a unit. Instead, the *decibel* is the unit of measure we normally use. A decibel (dB) equals $\frac{1}{10}$ bel. Therefore,

$$A_{p(\text{dB})} = 10 \log \frac{P_{\text{out}}}{P_{\text{in}}} \tag{25-1}$$

☐ EXAMPLE 25-1

The power out of an amplifier is 30 W. The input power is 100 mW. What is the power gain in decibels?

Solution

$$A_p = 10 \log \frac{P_{\text{out}}}{P_{\text{in}}}$$

$$A_p = 10 \log \frac{30\ \text{W}}{100\ \text{mW}} = 10 \log 300 = 24.8\ \text{dB}$$

☐

☐ EXAMPLE 25-2

The power gain of a power amplifier is 40 dB. If the input power is 50 μW, what is the output power?

Solution

$$A_p = 10 \log \frac{P_{\text{out}}}{P_{\text{in}}}$$

$$40\ \text{dB} = 10 \log \frac{P_{\text{out}}}{50\ \mu\text{W}}$$

Divide both sides by 10:

$$4 = \log \frac{P_{\text{out}}}{50\ \mu\text{W}}$$

Change to exponential form:

$$10^4 = \frac{P_{out}}{50\ \mu W}$$

Multiply both sides by 50 μW:

$$50\ \mu W \times 10^4 = P_{out}$$

$$P_{out} = 0.5\ W$$

If we consider that

$$P_{out} = \frac{V_{out}^2}{R_{out}} \quad \text{and} \quad P_{in} = \frac{V_{in}^2}{R_{in}}$$

then we can write the equation for voltage gain in decibels.

$$A_v = 10 \log \left(\frac{V_{out}^2}{R_{out}} \div \frac{V_{in}^2}{R_{in}}\right)$$

$$= 10 \log \left(\frac{V_{out}^2}{V_{in}^2} \times \frac{R_{in}}{R_{out}}\right)$$

$$= 10 \log \frac{V_{out}^2}{V_{in}^2} + 10 \log \frac{R_{in}}{R_{out}}$$

If $R_{out} = R_{in}$, then

$$A_v = 10 \log \left(\frac{V_{out}}{V_{in}}\right)^2 + 10 \log 1 = 20 \log \frac{V_{out}}{V_{in}} + 0$$

$$A_{v(dB)} = 20 \log \frac{V_{out}}{V_{in}} \tag{25-2}$$

□ EXAMPLE 25-3

The output voltage of an amplifier is 1.73 V. The input voltage is 50 mV. Assuming that $R_{out} = R_{in}$, find the voltage gain in dB.

Solution

$$A_v = 20 \log \frac{V_{out}}{V_{in}} = 20 \log \frac{1.73\ V}{50\ mV} = 20 \log 34.6 = 30.8\ dB$$

1.73 [÷] 50 [EE] [±] 3 [=] [LOG] [×] 20 [=]

Answer: $A_v = 30.8$ dB

□ EXAMPLE 25-4

The voltage gain of a device is −6 dB. If the output voltage is 700 mV, find V_{in}.

Solution

$$A_v = 20 \log \frac{V_{out}}{V_{in}}$$

$$-6 = 20 \log \frac{700 \text{ mV}}{V_{in}}$$

Divide by 20:

$$-0.3 = \log \frac{700 \text{ mV}}{V_{in}}$$

Change to exponential form:

$$10^{-0.3} = \frac{700 \text{ mV}}{V_{in}}$$

$$0.5 = \frac{700 \text{ mV}}{V_{in}}$$

Multiply by V_{in}:

$$0.5\ V_{in} = 700 \text{ mV}$$

Divide by 0.5:

$$V_{in} = \frac{700 \text{ mV}}{0.5}$$

$$= 1.4 \text{ V}$$

□

25-2 REFERENCE LEVELS

In the preceding examples the output powers and output voltages were given with reference to input values. Sometimes other reference levels are used. One common reference level is 6 mW. In other words, 0 dB equals 6 mW. Another reference level is 1 mW. When 1 mW is the reference, the decibel measurement is called a *dBm*. If a power gain is given in dBm, we assume that P_{in} is 1 mW.

□ EXAMPLE 25-5

The power gain of an amplifier is 33 dBm. Determine the output power.

Solution

$$A_p = 10 \log \frac{P_{out}}{P_{in}}$$

$$33 = 10 \log \frac{P_{out}}{1 \text{ mW}}$$

Divide by 10:

$$3.3 = \log \frac{P_{out}}{1 \text{ mW}}$$

Change to exponential form:

$$10^{3.3} = \frac{P_{out}}{1 \text{ mW}}$$

Multiply by 1 mW:

$$1 \text{ mW} \times 10^{3.3} = P_{out}$$

1 [EE] [±] 3 [×] 3.3 [10^x] [=] *Answer*: $P_{out} = 2.00$ W

□

PRACTICE PROBLEMS 25-1

1. $P_{in} = 20$ mW, $P_{out} = 35$ W. Find $A_{p(dB)}$.
2. $A_p = 38$ dB, $P_{in} = 5$ mW. Find P_{out}.
3. $A_p = 25$ dB, $P_{out} = 15$ W. Find P_{in}.
4. $A_p = -3$ dB, $P_{out} = 1$ W. Find P_{in}.

Assume that $R_{out} = R_{in}$ for problems 5 through 7:

5. $V_{in} = 50\ \mu$V, $V_{out} = 1$ V. Find $A_{v(dB)}$.
6. $A_v = 60$ dB, $V_{out} = 3$ V. Find V_{in}.
7. $A_v = 75$ dB, $V_{in} = 2$ mV. Find V_{out}.
8. $A_v = -10$ dB, $V_{in} = 2.7$ V. Find V_{out}.
9. $A_p = 15$ dBm. Find P_{out}.
10. $P_{out} = 10$ W. Find A_p in dBm.

Solutions

1. 32.4 dB
2. 31.5 W
3. 47.4 mW
4. 2 W
5. 86 dB
6. 3 mV
7. 11.2 V
8. 854 mV
9. 31.6 mW
10. 40 dBm

Additional practice problems are at the end of the chapter.

25-3 FREQUENCY RESPONSE

25-3-1 Bode Plot

A Bode plot is a useful tool for approximating the frequency response of a circuit or system. Let's construct a Bode plot of the frequency response characteristics of the *RC* circuit in Figure 25-1. Recall from ac circuit analysis that X_C varies inversely

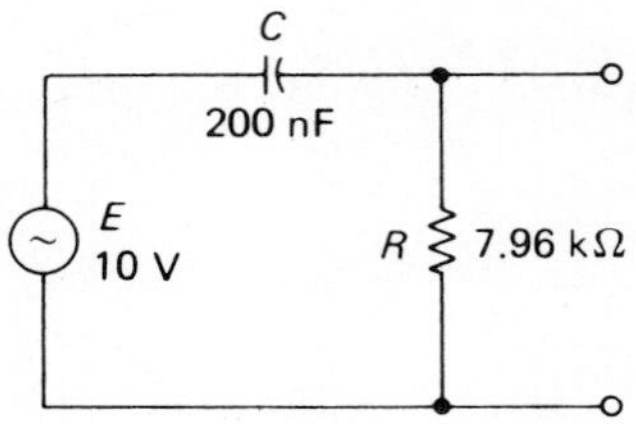

FIGURE 25-1

with frequency. Furthermore, the cutoff frequency f_{co} occurs where $X_C = R$. Because the output is taken from across R, $V_{out} = V_R$. We can compute the voltage gain of the circuit by using the equation learned previously.

$$A_{v(\mathrm{dB})} = 20 \log \frac{V_{out}}{V_{in}}$$

At high frequencies $V_{out} = V_{in}$. The voltage gain is

$$A_{v(\mathrm{dB})} = 20 \log \frac{10\ \mathrm{V}}{10\ \mathrm{V}} = 0\ \mathrm{dB}$$

The maximum possible gain for this circuit is 0 dB.

The Bode plot is a plot of gain versus frequency. Semilog graph paper is used in such a plot. Frequency is the independent variable and is plotted along the logarithmic scale. A scale for A_v is chosen that will allow us to change the gain by at least -20 dB. Once our scale is chosen, a horizontal line is drawn across our graph at 0 dB as shown in Figure 25-2. Four-cycle semilog paper has been chosen with a minimum frequency of 1 Hz. This allows us to examine the gain at frequencies from 1 Hz to 10 kHz. The cutoff frequency is 100 Hz.

$$f_{co} = \frac{1}{2\pi CR} = 100\ \mathrm{Hz}$$

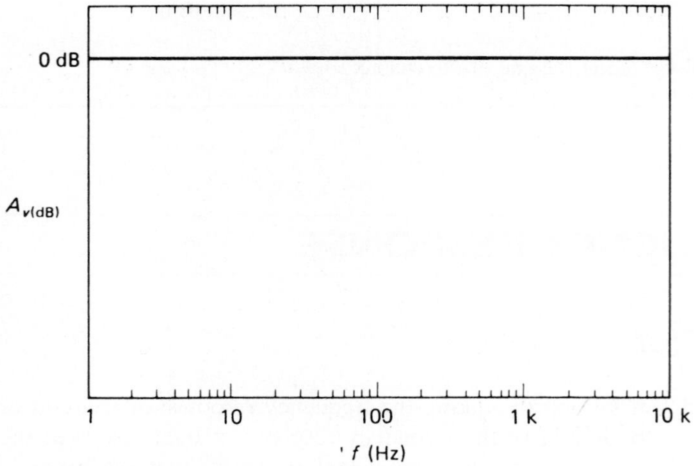

FIGURE 25-2

Figure 25-2 has been expanded in Figure 25-3 and the 100-Hz point has been plotted on our horizontal line. This becomes one point on our Bode plot.

■ **KEY POINT** As a straight-line approximation, the gain will change at the rate of 6 *dB per octave* and 20 *dB per decade*.

An octave change in frequency is a halving or doubling of the frequency. A decade change is a change in frequency of one-tenth or ten times.

Because gain decreases as frequency decreases, a second point is plotted in Figure 25-3 where the gain is down 6 dB and the frequency is 50 Hz, one-half the cutoff frequency. A third point is plotted where the gain is down 20 dB and the frequency is 10 Hz, 0.1 times f_{co}. In Figure 25-4 a straight line is drawn to connect these three points. This is our Bode plot. All Bode plots are plotted in a like manner. When the gain decreases with frequency, the line slopes downward and to the *left* as in Figure 25-4. When the gain decreases as frequency *increases*, the line slopes downward and to the right.

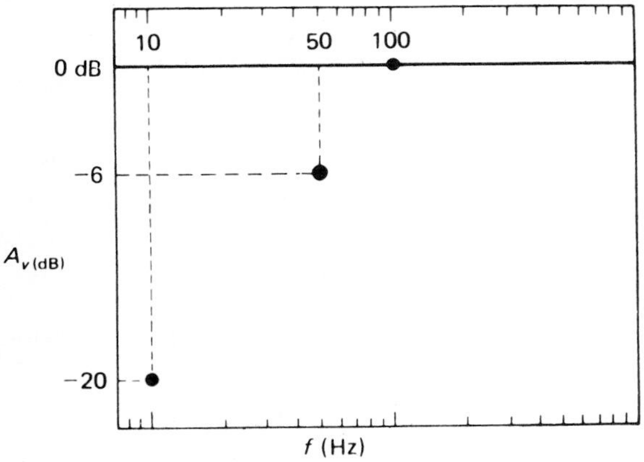

FIGURE 25-3

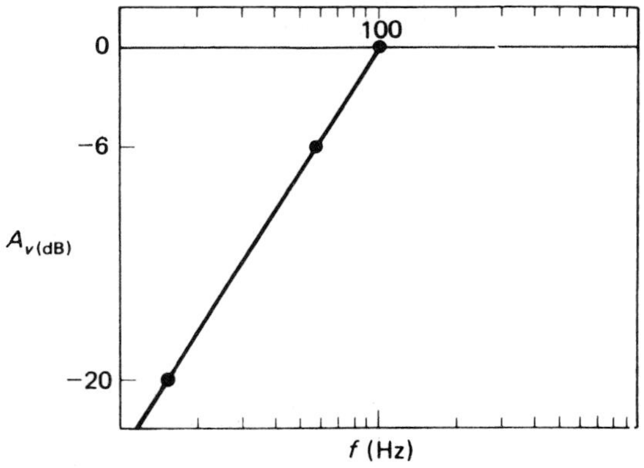

FIGURE 25-4

PRACTICE PROBLEMS 25-2

1. Refer to Figure 25-1. Construct a Bode plot for $C = 100$ nF and $R = 2.7$ kΩ.

2. Refer to Figure 25-5. Construct a Bode plot of the frequency response.

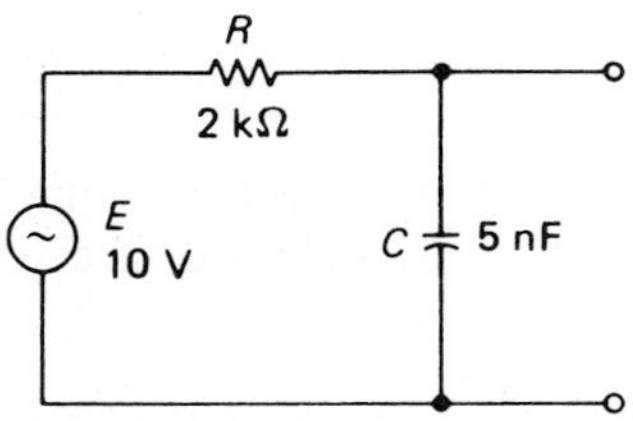

FIGURE 25-5

Solutions

1. f_{co} = 589 Hz. The Bode plot is shown in Figure 25-6.

2. f_{co} = 15.9 kHz. See Figure 25-7 for the Bode plot.

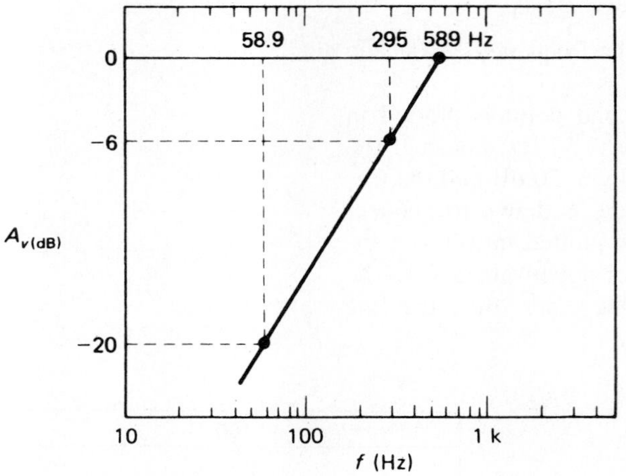

FIGURE 25-6

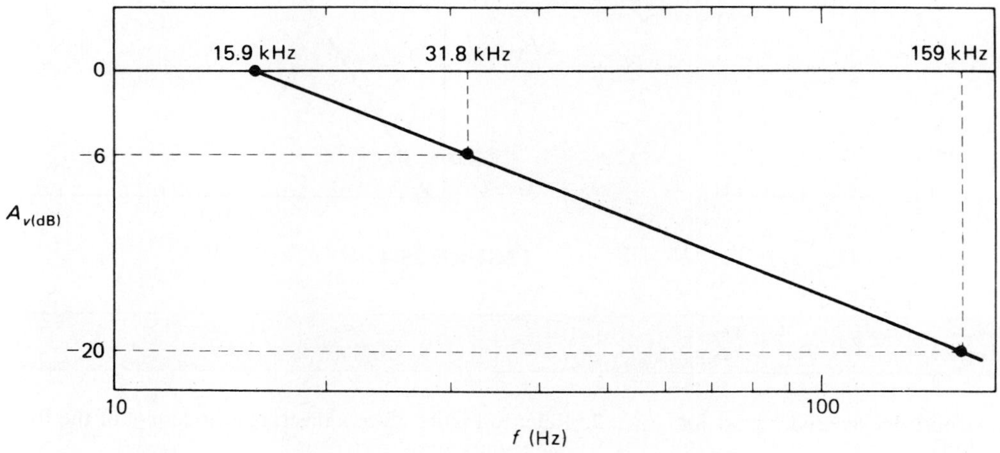

FIGURE 25-7

☐ EXAMPLE 25-6

An amplifier has a mid-frequency gain of 25 dB. The lower cutoff frequency (symbolized f_1) is 90 Hz. The upper cutoff frequency (symbolized f_2) is 12 kHz. Construct a Bode plot of the frequency response.

Solution We will use five-cycle semilog graph paper for this example. Choose a practical scale for A_v and draw a horizontal line at 25 dB. Plot the lower cutoff frequency as before. One point is f_1 and 25 dB. $\frac{f_1}{2}$ and 19 dB is a second point (6 dB down). $\frac{f_1}{10}$ and 5 dB is a third point (20 dB down). The upper cutoff frequency is plotted by using the points f_2 and 25 dB, $2f_2$ and 19 dB, and $10f_2$ and 5 dB. The complete plot is shown in Figure 25-8.

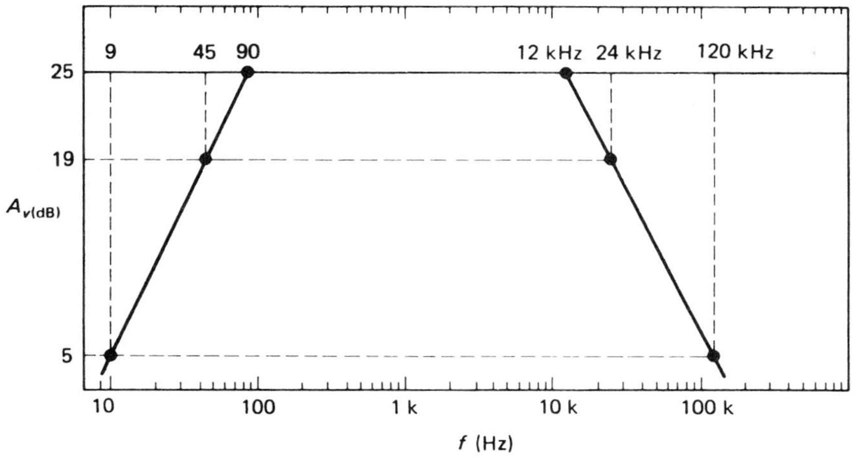

FIGURE 25-8

Consider the circuit in Figure 25-9. In circuits like this, the mid-frequency gain can be determined (if f_1 and f_2 are separated by at least one decade) by ignoring the effects of C_1 and C_2 and using the ratio of R_2 to the total resistance $R_1 + R_2$.

$$\frac{R_2}{R_1 + R_2} = \frac{V_{out}}{V_{in}}$$

The mid-frequency equivalent circuit is shown in Figure 25-10(a). (In practice there may be other factors affecting response, such as source impedance and cable capacitance.)

Mid-frequency gain would be determined by

$$A_v = 20 \log \frac{R_2}{R_1 + R_2} \tag{25-3}$$

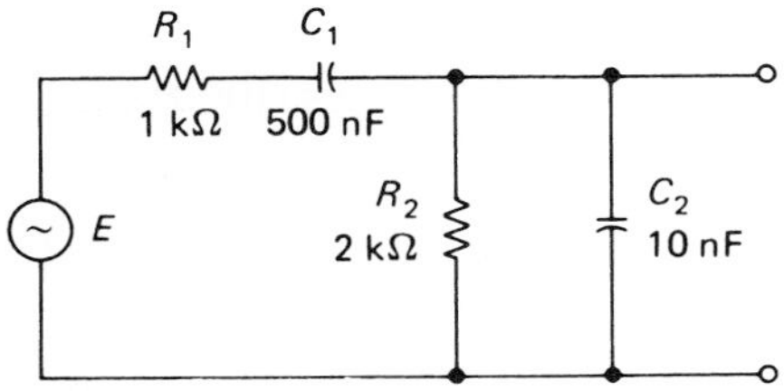

FIGURE 25-9

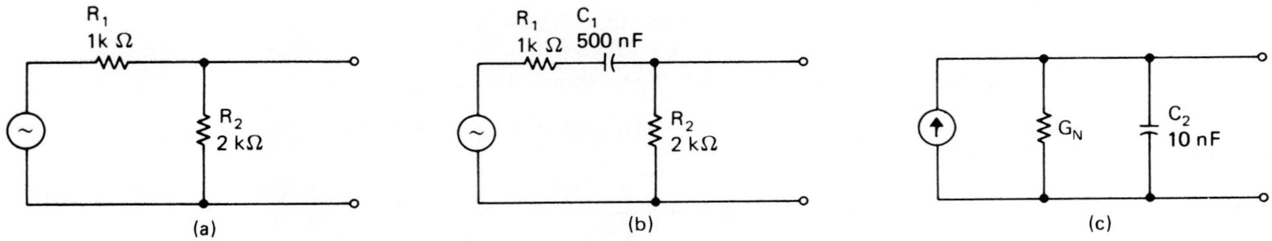

FIGURE 25-10 (a) Mid-frequency equivalent circuit of Figure 25-9; (b) low-frequency equivalent circuit of Figure 25-9; (c) high-frequency equivalent circuit of Figure 25-9.

In this circuit

$$A_v = 20 \log \frac{2\ \text{k}\Omega}{1\ \text{k}\Omega + 2\ \text{k}\Omega} = -3.52\ \text{dB}$$

2 [EE] 3 [÷] [(] 1 [EE] 3 [+] 2 [EE] 3 [)] [=] [LOG] [×] 20 [=]

Answer: $A_v = -3.52$ dB

This is the Bode limit on circuit gain. If the frequency is decreased, the reactance of C_1 becomes larger and some voltage drops across it. This causes a decrease in the output voltage. If the frequency is increased, the susceptance of C_2 increases, which again causes a decrease in output voltage. The series reactive components (C_1 in our circuit) affect the response as we lower frequency. The effect on the circuit due to C_2 is negligible, which is the case if f_1 and f_2 are at least a decade apart. The equivalent low-frequency circuit looks like the circuit in Figure 25-10(b).

At high frequencies, the reactance of C_1 becomes negligible (compared to R_T), but the susceptance of C_2 becomes a factor. Since C_2 is a parallel component, if we Nortonize the circuit at R_2, we get an equivalent circuit consisting of C_2, R_1, and R_2 all in parallel, as shown in Figure 25-10(c).

□ EXAMPLE 25-7

Find f_1 and f_2 in the circuit of Figure 25-9. Construct a Bode plot.

Solution We previously determined that the mid-frequency gain is −3.52 dB. Let's compute f_1 and f_2.

$$f_1 = \frac{1}{2\pi C_1(R_1 + R_2)} = \frac{1}{2 \times 3.14 \times 500 \times 10^{-9} \times 3 \times 10^3} = 106\ \text{Hz}$$

$$f_2 = \frac{G_1 + G_2}{2\pi C_2} = \frac{1.5 \times 10^{-3}}{2 \times 3.14 \times 10 \times 10^{-9}} = 23.9\ \text{kHz}$$

The Bode plot is constructed in Figure 25-11.

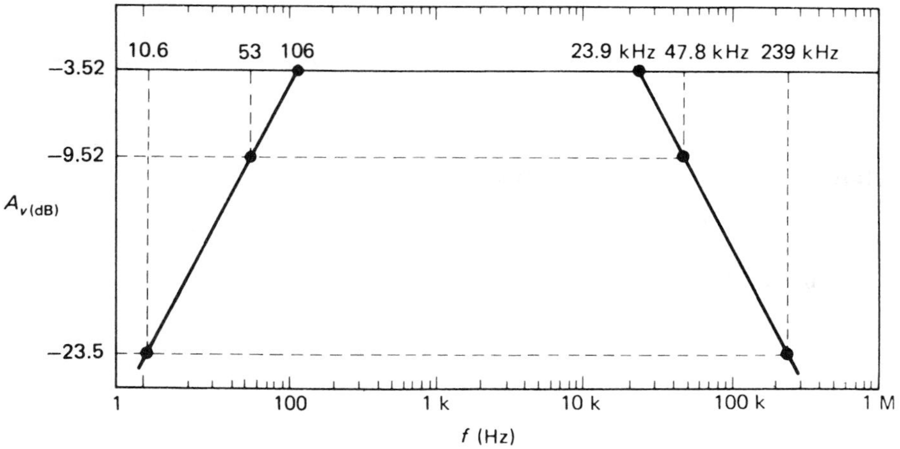

FIGURE 25-11

25-3-2 Frequency Response Curve

The Bode plot results in some error, particularly around the cutoff frequency. As we know, the gain is really down 3 dB at f_{co}. At frequencies other than cutoff, the error is less than 3 dB. The error is practically 0 one decade above and below f_{co}. The reduction in gain (from maximum gain) at any frequency can be found by using the following equations:

For an *RC* high-pass filter:

$$A_v = 20 \log \frac{1}{\sqrt{1 + \left(\frac{f_1}{f}\right)^2}} \tag{25-4}$$

where f_1 is the lower cutoff frequency and f is the frequency in question.

For an *RC* low-pass filter:

$$A_v = 20 \log \frac{1}{\sqrt{1 + \left(\frac{f}{f_2}\right)^2}} \tag{25-5}$$

where f_2 is the upper cutoff frequency and f is the frequency in question. The general shape of the frequency response curve results from plotting gain versus frequency using these equations. The actual curve and the Bode plot are drawn together in Figure 25-12.

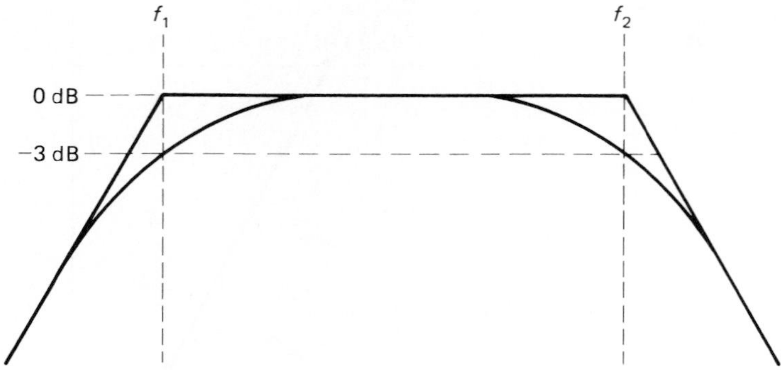

FIGURE 25-12

☐ EXAMPLE 25-8

Refer to Figure 25-9. Find the gain at 50 Hz and at 100 kHz by constructing a Bode plot and finding those frequencies on the curve. Verify your answers by using Equations 25-4 and 25-5.

Solution We previously determined that the mid-frequency gain (the maximum gain) is −3.52 dB. At 50 Hz

$$A_v = 20 \log \frac{1}{\sqrt{1 + \left(\dfrac{f_1}{f}\right)^2}} = 20 \log \frac{1}{\sqrt{1 + \left(\dfrac{106\text{ Hz}}{50\text{ Hz}}\right)^2}} = -7.4\text{ dB}$$

106 [÷] 50 [=] [x^2] [+] 1 [=] [$\sqrt{x}$] [$1/x$] [LOG] [×] 20 [=]

Answer: $A_v = -7.4$ dB

This tells us that we are down 7.4 dB from −3.52 dB. The resultant gain is

$$A_{v(\text{dB})} = -3.52\text{ dB} + (-7.4\text{ dB}) = -10.9\text{ dB}$$

At 100 kHz

$$A_v = 20 \log \frac{1}{\sqrt{1 + \left(\dfrac{f}{f_2}\right)^2}} = 20 \log \frac{1}{\sqrt{1 + \left(\dfrac{100\text{ kHz}}{23.9\text{ kHz}}\right)^2}}$$

$$= -12.7\text{ dB}$$

100 [EE] 3 [÷] 23.9 [EE] 3 [=] [x^2] [+] 1 [=] [$\sqrt{x}$] [$1/x$] [LOG] [×] 20 [=]

Answer: $A_v = -12.7$ dB

The resultant gain is

$$A_{v(\text{dB})} = -3.52\text{ dB} + (-12.7\text{ dB}) = -16.2\text{ dB}$$

The Bode plot is shown in Figure 25-11. An inspection of the Bode plot shows that our gain calculations at 50 Hz and at 100 kHz were correct.

☐

PRACTICE PROBLEMS 25-3

1. An amplifier has a mid-frequency gain of 40 dB. $f_1 = 200$ Hz and $f_2 = 20$ kHz. Construct a Bode plot of the frequency response.

2. Refer to Figure 25-9. Let $R_1 = 330\ \Omega$ and $R_2 = 680\ \Omega$. Compute the gain at 100 Hz and 200 kHz. Construct a Bode plot of the frequency response. Do the computed gains at 100 Hz and 200 kHz fall on the Bode plot? If they do not, either the plot is wrong or your computations are wrong.

3. The mid-frequency gain of an amplifier is 27 dB. $f_1 = 100$ Hz and $f_2 = 15$ kHz. Find the gain at f_1, f_2, 25 Hz, and 30 kHz.

Solutions

1. See Figure 25-13 for the Bode plot.

2. The mid-frequency gain is -3.44 dB. $f_1 = 315$ Hz and $f_2 = 72.3$ kHz. At 100 Hz, $A_v = -13.8$ dB (-3.44 dB + -10.4 dB). At 200 kHz, $A_v = -12.8$ dB (-3.44 dB + -9.37 dB). The Bode plot is constructed in Figure 25-14. An inspection of the plot shows that both 100 Hz and 200 kHz fall on the curve. This verifies the accuracy of our curve and our calculations.

3. At f_1, $A_v = 24$ dB. At f_2, $A_v = 24$ dB. At 25 Hz, $A_v = 14.7$ dB. At 30 kHz, $A_v = 20$ dB.

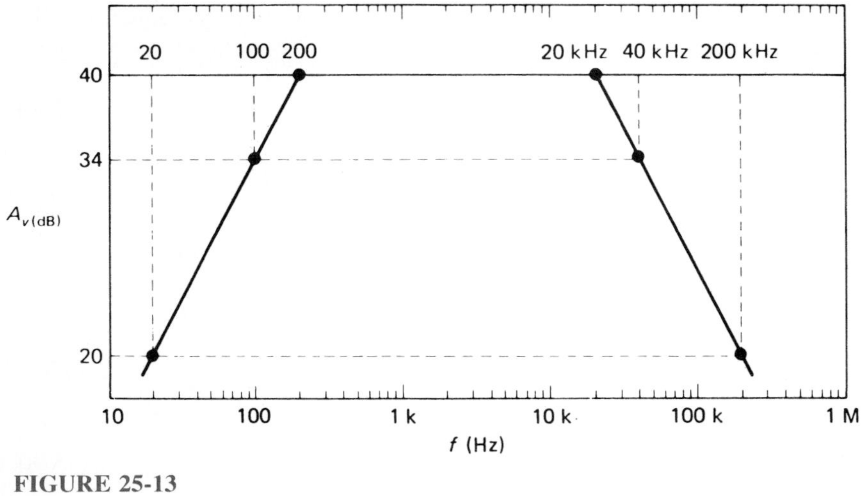

FIGURE 25-13

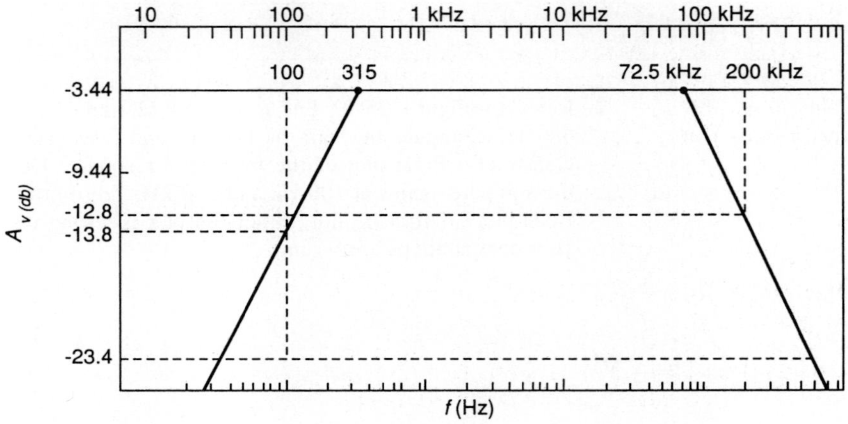

FIGURE 25-14

Additional practice problems are at the end of the chapter.

25-4 *RC* CIRCUITS

In Chapter 14 we used a universal time constant curve to determine the charge on a capacitor at some time after power was applied to the circuit. In this section we will compute the charge or discharge of a capacitor by using logarithmic equations. To determine the charge on a capacitor in a series *RC* circuit, the following equation is used:

$$v_c = E(1 - e^{-t/RC}) \tag{25-6}$$

where v_c = charge on the capacitor
E = applied voltage
t = time in seconds
R = resistance in ohms
C = capacitance in farads

The equation for v_R is

$$v_R = Ee^{-t/RC} \tag{25-7}$$

When the capacitor is discharging, the equation is

$$v_C = Ee^{-t/RC} \tag{25-8}$$

□ EXAMPLE 25-9

In the circuit of Figure 25-15, find v_R and v_C 1 ms after the switch is closed.

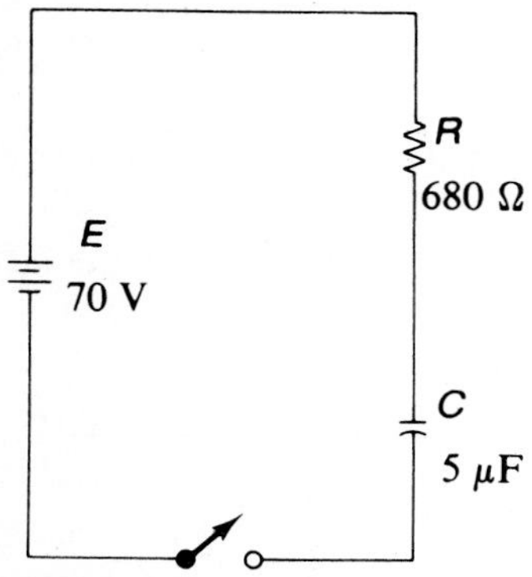

FIGURE 25-15

Solution

$$v_C = E(1 - e^{-t/RC}) = 70(1 - e^{-1\text{ms}/3.4\times 10^{-3}})$$

1 [±] [EE] [±] 3 [÷] [(] 680 [×] 5 [EE] [±] 6 [)]

[=] [e^x] [±] [+] 1 [=] [×] 70 [=]   *Answer:* 17.8 V

An alternate method is

1 [±] [EE] [±] 3 [÷] [(] 680 [×] 5 [EE] [±] 6 [)] [=] [STO]

(store the exponent to which e will be raised)

Then

70 [×] [(] 1 [−] [RCL] [e^x] [)] [=]   *Answer:* v_c = 17.8 V

v_R can be found by applying Kirchhoff's voltage law:

$$v_R = E - v_C = 70\text{ V} - 17.8\text{ V} = 52.2\text{ V}$$

☐ EXAMPLE 25-10

Refer to Figure 25-15. If V_C = 40 V, how much time has lapsed?

Solution

$$v_C = E(1 - e^{-t/RC})$$

$$40 = 70(1 - e^{-t/3.4\times 10^{-3}})$$

Divide by 70:

$$0.571 = 1 - e^{-t/3.4\times 10^{-3}}$$

Subtract 1:

$$-0.429 = -e^{-t/3.4\times 10^{-3}}$$

Change signs:

$$0.429 = e^{-t/3.4\times 10^{-3}}$$

Take ln: $\ln 0.429 = -\frac{t}{3.4 \times 10^{-3}} \ln e$

$$-0.846 = -\frac{t}{3.4 \times 10^{-3}} \qquad (\ln e = 1)$$

Multiply by -3.4×10^{-3}:

$$2.88 \times 10^{-3} = t$$

$$t = 2.88 \text{ ms}$$

□

PRACTICE PROBLEMS 25-4

Refer to Figure 25-15.

1. $t = 2$ ms. Find v_C and v_R.
2. $V_R = 10$ V. Find t.
3. $i = 20$ mA. Find t. (The equation for i is of the same form as Equation 25-7).
4. $E = 50$ V, $R = 4.7$ kΩ, $v_C = 10$ V, and $t = 50$ μs. Find C.
5. In a circuit like Figure 25-15, $E = 50$ V, $C = 0.2$ μF, $v_R = 20$ V, and $t = 500$ ms. Find R.

Solutions

1. $v_C = 31.1$ V, $v_R = 38.9$ V
2. 6.62 ms
3. 5.57 ms
4. 47.7 nF
5. 2.73 MΩ

SELF-TEST 25-1

1. Refer to Figure 25-16. Let $C_1 = 1$ μF, $C_2 = 470$ pF, $R_1 = 4.7$ kΩ, and $R_2 = 2$ kΩ. Determine $A_{v(\text{dB})}$ at 50 Hz and 200 kHz. Construct a Bode plot.
2. $A_p = 25$ dB, $P_{\text{out}} = 3$ W. Find P_{in}.
3. $A_p = 45$ dBm. Find P_{out}.
4. Refer to Figure 25-15. $E = 60$ V, $v_C = 20$ V, $C = 500$ nF, and $t = 3$ ms. Find R.

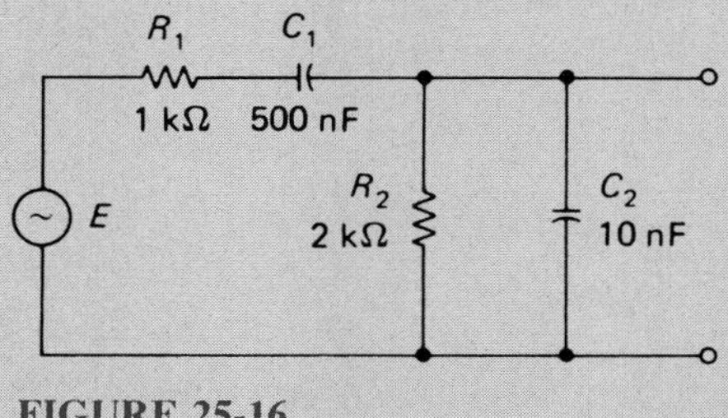

FIGURE 25-16

Answers to Self-test 25-1 are at the end of the chapter.

CHAPTER 25 AT A GLANCE

Page	Key Points, Definitions, and Equations	Example
608	The decibel is the unit of measure we use when computing power gain. $$A_{p(\text{dB})} = 10 \log \frac{P_{\text{out}}}{P_{\text{in}}}$$	$P_{\text{in}} = 100$ mW, $P_o = 30$ W. Find A_p. $A_p = 10 \log \frac{P_{\text{out}}}{P_{\text{in}}}$ $= 10 \log \frac{30 \text{ W}}{100 \text{ mW}}$ $= 10 \log 300 = 24.8$ dB
609	The equation for voltage gain is $$A_{v(\text{dB})} = 20 \log \frac{V_{\text{out}}}{V_{\text{in}}}$$	$V_{\text{in}} = 50$ mV, $V_o = 1.73$ V. Find A_v. $A_v = 20 \log \frac{V_{\text{out}}}{V_{\text{in}}} = 20 \log \frac{1.73 \text{ V}}{50 \text{ mV}}$ $= 20 \log 34.6 = 30.8$ dB
613	*Key Point:* As a straight line approximation, the gain will change at the rate of 6 dB per octave and 20 dB per decade.	
612	A Bode plot is a plot of gain versus frequency. As a straight-line approximation, the gain will change at the rate of 6 dB per octave and 20 dB per decade.	If the mid-frequency gain is 25 dB, f_1 is 90 Hz and f_2 is 12 kHz. Construct a Bode plot.

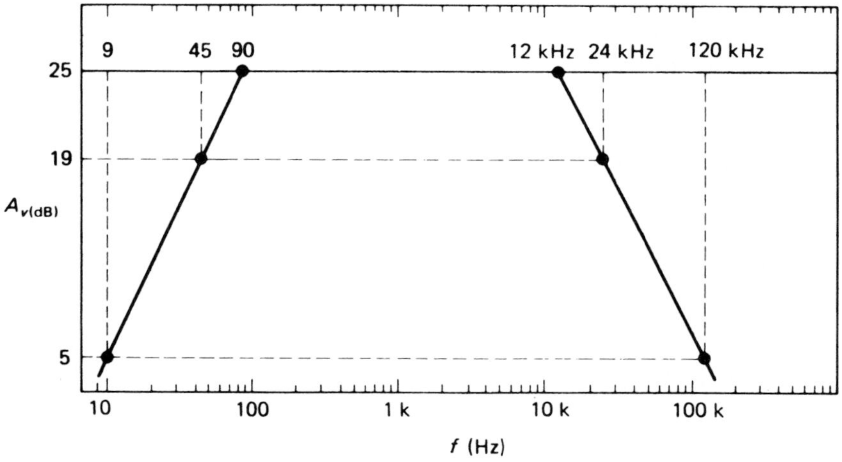

FIGURE 25-17

End of Chapter Problems 25-1

1. $P_{\text{in}} = 30\ \mu$W, $P_{\text{out}} = 30$ W. Find A_p in dB.
2. $P_{\text{in}} = 10$ mW, $P_{\text{out}} = 3.16$ W. Find A_p in dB.
3. $P_{\text{in}} = 1$ mW, $P_{\text{out}} = 20$ W. Find A_p in dB.
4. $P_{\text{in}} = 100\ \mu$W, $P_{\text{out}} = 10$ W. Find A_p in dB.

5. $P_{in} = 2$ mW, $P_{out} = 50$ W. Find A_p in dB.
6. $P_{in} = 5$ mW, $P_{out} = 40$ W. Find A_p in dB.
7. $A_p = 35$ dB, $P_{out} = 12$ W. Find P_{in}.
8. $A_p = 30$ dB, $P_{out} = 1$ W. Find P_{in}.
9. $A_p = 20$ dB, $P_{out} = 100$ W. Find P_{in}.
10. $A_p = 40$ dB, $P_{out} = 2$ W. Find P_{in}.
11. $A_p = 30$ dB, $P_{out} = 5$ W. Find P_{in}.
12. $A_p = 50$ dB, $P_{out} = 60$ W. Find P_{in}.
13. $A_p = 22$ dB, $P_{in} = 10$ mW. Find P_{out}.
14. $A_p = 40$ dB, $P_{in} = 150$ mW. Find P_{out}.
15. $A_p = 38$ dB, $P_{in} = 500$ μW. Find P_{out}.
16. $A_p = 17$ dB, $P_{in} = 100$ mW. Find P_{out}.
17. $A_p = 15$ dB, $P_{in} = 10$ mW. Find P_{out}.
18. $A_p = 45$ dB, $P_{in} = 50$ mW. Find P_{out}.
19. $A_p = 20$ dBm, $P_{in}= 10$ mW. Find P_{out}.
20. $A_p = 43$ dBm. Find P_{out}.
21. $A_p = 32$ dBm. Find P_{out}.
22. $A_p = 7.5$ dBm. Find P_{out}.
23. $A_p= 50$ dBm. Find P_{out}.
24. $A_p = 25$ dBm. Find P_{out}.

Assume $R_{out} = R_{in}$ for the following problems:

25. $V_{in} = 100$ μV, $V_{out} = 2.7$ V. Find A_v in dB.
26. $V_{in} = 5.7$ V, $V_{out} = 3.4$ V. Find A_v in dB.
27. $V_{in} = 10$ mV, $V_{out} = 3$ V. Find A_v in dB.
28. $V_{in} = 600$ μV, $V_{out} = 10$ V. Find A_v in dB.
29. $V_{in} = 50$ μV, $V_{out} = 10$ V. Find A_v in dB.
30. $V_{in} = 4$ mV, $V_{out} = 12$ V. Find A_v in dB.
31. $A_v = 40$ dB, $V_{in} = 5$ mV. Find V_{out}.
32. $A_v = 70$ dB, $V_{in} = 10$ mV. Find V_{out}.
33. $A_v = 22.7$ dB, $V_{in} = 2$ mV. Find V_{out}.
34. $A_v = 37.5$ dB, $V_{in} = 20$ mV. Find V_{out}.
35. $A_v = 73$ dB, $V_{in} = 23$ mV. Find V_{out}.
36. $A_v = 44$ dB, $V_{in} = 4.7$ mV. Find V_{out}.
37. $A_v = 90$ dB, $V_{out} = 4$ V. Find V_{in}.
38. $A_v = 45$ dB, $V_{out} = 1.73$ V. Find V_{in}.
39. $A_v = 75$ dB, $V_{out} = 15$ V. Find V_{in}.
40. $A_v = 63$ dB, $V_{out} = 7.5$ V. Find V_{in}.
41. $A_v = 80$ dB, $V_{out} = 600$ mV. Find V_{in}.
42. $A_v = 32$ dB, $V_{out} = 200$ mV. Find V_{in}.
43. $A_v = -3$ dB, $V_{out} = 500$ mV. Find V_{in}.
44. $A_v = -3$ dB, $V_{out} = 10$ V. Find V_{in}.
45. $A_v = -6$ dB, $V_{out} = 1$ V. Find V_{in}.
46. $A_v = -6$ dB, $V_{out} = 3$ V. Find V_{in}.
47. $A_v = -3$ dB, $V_{out} = 5$ V. Find V_{in}.
48. $A_v = -3$ dB, $V_{out} = 7$ V. Find V_{in}.
49. $A_v = -30$ dB, $V_{out} = 100$ mV. Find V_{in}.
50. $A_v = 20$ dB, $V_{out} = 500$ μV. Find V_{in}.

End of Chapter Problems 25-2

Refer to Figure 25-18 and construct a Bode plot of the frequency response.

1. $C = 0.3$ μF and $R = 1$ kΩ.
2. $C = 0.5$ μF and $R = 470$ Ω.
3. $C = 1$ μF and $R = 470$ Ω.
4. $C = 100$ nF and $R = 12$ kΩ.
5. $C = 300$ nF and $R = 3.9$ kΩ.
6. $C = 50$ nF and $R = 2.5$ kΩ.

Refer to Figure 25-19 and construct a Bode plot of the frequency response.

7. $C = 50$ nF and $R = 2$ kΩ.
8. $C = 20$ nF and $R = 1.5$ kΩ.
9. $C = 200$ nF and $R = 750$ Ω.
10. $C = 400$ nF and $R = 270$ Ω.
11. $C = 40$ nF and $R = 2$ kΩ.
12. $C = 120$ nF and $R = 1.2$ kΩ.

End of Chapter Problems 25-3

Refer to Figure 25-16 and construct a Bode plot of the frequency response.

1. $C_1 = 250$ nF, $C_2 = 40$ nF, $R_1 = 680$ Ω, and $R_2 = 1.2$ kΩ.
2. $C_1 = 400$ nF, $C_2 = 25$ nF, $R_1 = 1$ kΩ, and $R_2 = 1$ kΩ.

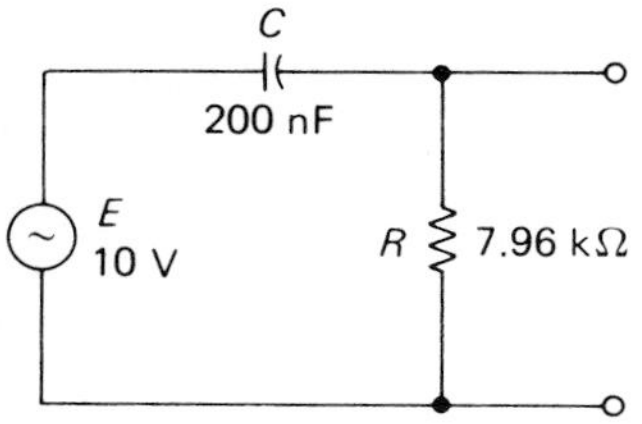

FIGURE 25-18

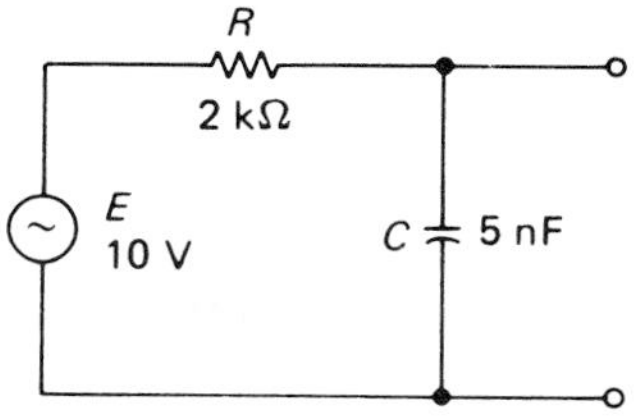

FIGURE 25-19

3. $C_1 = 500$ nF, $C_2 = 50$ nF, $R_1 = 470\ \Omega$, and $R_2 = 1.8$ kΩ.

4. $C_1 = 1\ \mu$F, $C_2 = 400$ pF, $R_1 = 1.2$ kΩ, and $R_2 = 2$ kΩ.

5. $C_1 = 200$ nF, $C_2 = 30$ nF, $R_1 = 3.3$ kΩ, and $R_2 = 2.2$ kΩ.

6. $C_1 = 450$ nF, $C_2 = 2$ nF, $R_1 = 750\ \Omega$, and $R_2 = 2.7$ kΩ.

7. An amplifier has a mid-frequency gain of 20 dB. $f_1 = 50$ Hz and $f_2 = 18$ kHz. Construct a Bode plot of the frequency response. Find the gain at 40 Hz and 25 kHz.

8. An amplifier has a mid-frequency gain of 35 dB. $f_1 = 80$ Hz and $f_2 = 30$ kHz. Construct a Bode plot of the frequency response. Find the gain at 20 Hz and 100 kHz.

9. An amplifier has a mid-frequency gain of 10 dB. $f_1 = 80$ Hz and $f_2 = 4$ kHz. Construct a Bode plot of the frequency response. Find the gain at 40 Hz and 12 kHz.

10. An amplifier has a mid-frequency gain of 40 dB. $f_1 = 20$ Hz and $f_2 = 50$ kHz. Construct a Bode plot of the frequency response. Find the gain at 5 Hz and 100 kHz.

11. Refer to Figure 25-16. If $R_1 = 470\ \Omega$, $R_2 = 680\ \Omega$, $C_1 = 1\ \mu$F, and $C_2 = 20$ nF, find the gain at 30 Hz and at 50 kHz.

12. Refer to problem 11. Find the gain at 200 Hz and at 20 kHz.

End of Chapter Problems 25-4

Refer to Figure 25-15 for the following problems. Let $E = 100$ V, $R = 47$ kΩ, and $C = 50$ nF for problems 1 through 8.

1. $t = 1$ ms. Find v_R and v_C.

2. $t = 2$ ms. Find v_R and v_C.

3. $v_C = 65$ V. Find t.

4. $v_C = 40$ V. Find t.

5. $i = 2$ mA. Find t.

6. $i = 1$ mA. Find t.

7. $v_R = 10$ V. Find t.

8. $v_R = 80$ V. Find t.

9. $E = 25$ V, $C = 10$ nF, $v_R = 15$ V, and $t = 50\ \mu$s. Find R.

10. $E = 40$ V, $C = 200$ nF, $v_R = 20$ V, and $t = 200\ \mu$s. Find R.

11. $E = 100$ V, $R = 47$ kΩ, $v_C = 20$ V, and $t = 100\ \mu$s. Find C.

12. $E = 100$ V, $R = 2.7$ kΩ, $v_C = 50$ V, and $t = 50\ \mu$s. Find C.

Answers to Self-test 25-1

1. At 50 Hz, $A_v = -11.4$ dB. At 200 kHz, $A_v = -12.8$ dB. See Figure 25-20 for the Bode plot.

2. $P_{in} = 9.49$ mW

3. $P_{out} = 31.6$ W

4. $R = 14.8$ kΩ

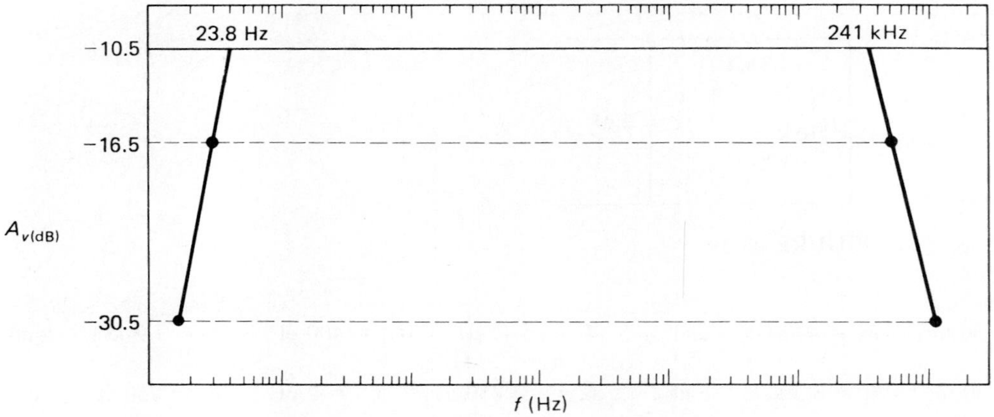

FIGURE 25-20

BOOLEAN ALGEBRA

CHAPTER 26

INTRODUCTION

In today's world, technicians must have some background in digital logic circuits. In Chapter 4 we introduced the various kinds of number systems used in computers. In this chapter we will examine the basic building blocks of any logic system, the AND, OR, and NOT functions. We will learn how to relate between logic diagrams and logic expressions. We will simplify logic circuits and construct truth tables to verify the accuracy of our simplifications.

CHAPTER OBJECTIVES

In this chapter you will learn how to:

1. Identify the three basic logic operations.
2. Draw the logic symbol and construct truth tables for the three basic logic operations.
3. Write the Boolean expression for various logic circuits.
4. Use Boolean postulates and theorems to simplify logic expressions.

26-1 LOGIC VARIABLES

Boolean algebra is used in manipulating logic variables. A variable is either completely true or completely false; partly true or partly false values are not allowed. When a variable is not true, by implication it must be false. Conversely, if the variable is not false, it must be true. Because of this characteristic, Boolean algebra is ideally suited to variables that have two states or values, such as YES and NO, or for a number system that has two values, 1 and 0 (for example, the binary number system). For example, B (a variable) could represent the presence of Bob. B has two values: if Bob is present, B equals "true"; if Bob is absent, B equals "false." Note that Bob is not the variable; B is the variable that represents the presence of Bob.

A switch is ideally suited to represent the value of any two-state variable because it can only be "off" or "on." Consider the SPST switch in Figure 26-1. When the switch is in the closed position, it indicates that Bob is present (B = true). When it is in the open position, it then represents that Bob is absent (B = false).

A closed switch could also represent values such as true, yes, one (1), HIGH (H), go, and so on; an open switch could represent values such as false, no, zero (0), LOW (L), no go, and so on.

26-2 LOGIC OPERATIONS

There are only three basic logic operations:

1. The *conjunction* (logical product), commonly called AND, symbolized by (·).
2. The *disjunction* (logical sum), commonly called OR, symbolized by (+).
3. The *negation,* commonly called NOT, symbolized by ($\overline{\ \ }$).

These operations are performed by logic circuits. All functions within a computer can be performed by combinations of these three basic logic operations.

26-3 THE AND FUNCTION

■ **KEY POINT** *For the AND function, the output is true only when ALL inputs are true. Otherwise, the output is false.*

FIGURE 26-1 SPST switch: (a) closed; (b) open.

Figure 26-2 shows the logic symbol and truth table for a two-input AND gate. A and B are the input variables. The function (A·B) is produced when A is true AND B is true. That is, if both inputs are true, the output is true. If either input, or both, is false, the output is false. To show all the conditions that may exist at the input and output of a logic gate, truth tables are used. Since there are two inputs and each input has two possible states (true and false), the number of possible conditions at the input would be 2 raised to the second power (2^2), or 4. Note in the truth table that condition 4 is the only one in which all the inputs are true so that the AND function is produced and a true output appears. If A is true AND B is true, the output will be true.

AND gates may have two or more inputs. Figure 26-3 shows a three-input AND gate and its truth table. The function (A·B·C) is produced when A is true AND B is true AND C is true. Ones (1) and zeros (0) are used for the values of the variables (true = 1 and false = 0). Because three variables are used, eight possible conditions exist ($2^3 = 8$). Condition 8 is the only one that will produce a true output (1) because all the input variables are true. Remember, in an AND gate, *all* the inputs must be true in order to produce a true output.

The (·) symbol used in the expression A·B·C is the AND operator and indicates logical multiplication. For example, condition 8 in Figure 26-3 can be interpreted as $1 \times 1 \times 1 = 1$, whereas condition 7 can be interpreted as $1 \times 1 \times 0 = 0$.

Using the same approach, we find that all other conditions (1 through 6) also produce a zero. It is important to remember that in the binary system a 2 cannot exist, only 1's and 0's. As with ordinary algebra, the operator symbol (·) can be omitted; thus, A·B·C = ABC and is read "A AND B AND C."

The AND function can be illustrated by the following analogy. The members of group A are Bob, Charley, and Dick. Note that the names in this group are com-

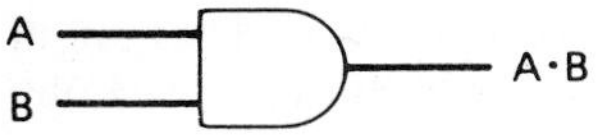

(a)

Condition	A	B	A·B
1	F	F	F
2	F	T	F
3	T	F	F
4	T	T	T

(b)

FIGURE 26-2 Two-input AND gate: (a) logic symbol; (b) truth table.

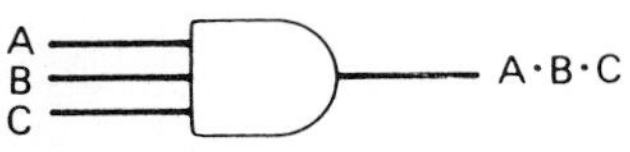

(a)

Condition	A	B	C	A·B·C
1	0	0	0	0
2	0	0	1	0
3	0	1	0	0
4	0	1	1	0
5	1	0	0	0
6	1	0	1	0
7	1	1	0	0
8	1	1	1	1

(b)

FIGURE 26-3 Three-input AND gate: (a) logic symbol; (b) truth table.

bined by the conjunction "and." That means group A equals the presence of Bob AND Charley AND Dick. This may be symbolized as

$$A = BCD$$

A is true (group A is present) when B is true (Bob is present), AND C is true (Charley is present), AND D is true (Dick is present). A is not true if any one or more of the members are absent.

The circuit in Figure 26-4 can be used to produce the AND function of the above example. The light indicates that group A is present only when all members of the group are present. (All switches are closed.) A logic circuit producing this AND function is symbolized in Figure 26-5.

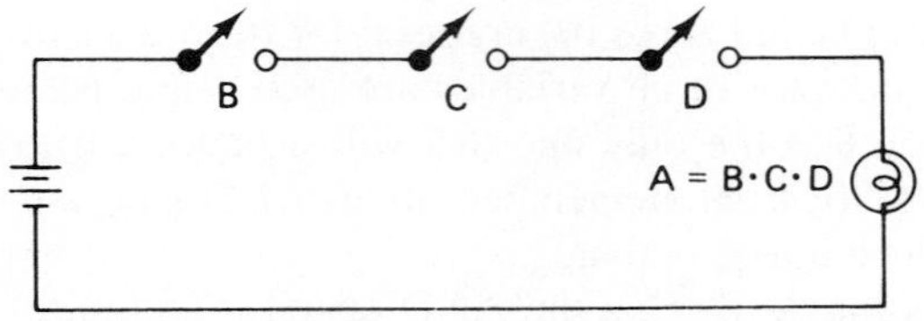

FIGURE 26-4 Switches connected in series to produce the AND function.

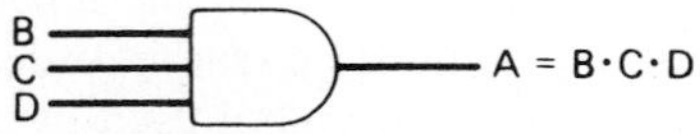

FIGURE 26-5 Logic symbol for the AND function BCD.

26-4 THE OR FUNCTION

■ **KEY POINT** *For the OR function, the output is true when any one of the inputs is true. Otherwise, the output is false.*

Figure 26-6 shows the logic symbol and truth table for a two-input OR gate. The function (A + B) is produced when A is true OR B is true. That is, if either input is true, the output is true. If both inputs are false, the output is false. Because there are two input variables, a total of four conditions is possible ($2^2 = 4$). Note in the truth table that conditions 2, 3, and 4 produce a true output because at least one of the input variables is true. Condition 1 produces a false output because neither of the input variables is true.

OR gates may have two or more inputs. Figure 26-7 shows a three-input OR gate and its truth table. Ones (1) and zeros (0) are used for the values of the variables. Because three variables are used, eight possible conditions occur ($2^3 = 8$). Conditions 2 through 8 produce a true (1) output because at least one of the input

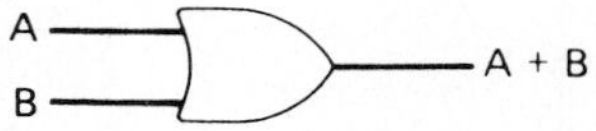

(a)

Condition	A	B	A + B
1	F	F	F
2	F	T	T
3	T	F	T
4	T	T	T

(b)

FIGURE 26-6 Two-input OR gate: (a) logic symbol; (b) truth table.

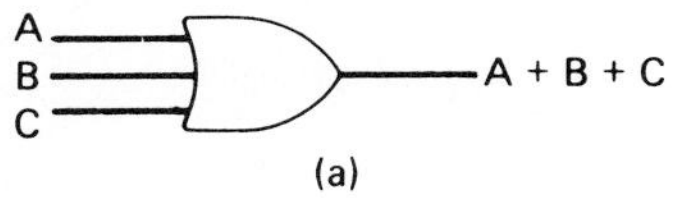

(a)

Condition	A	B	C	A + B + C
1	0	0	0	0
2	0	0	1	1
3	0	1	0	1
4	0	1	1	1
5	1	0	0	1
6	1	0	1	1
7	1	1	0	1
8	1	1	1	1

(b)

FIGURE 26-7 Three-input OR gate: (a) logic symbol; (b) truth table.

variables is true (1). Condition 1 is the only condition that does not produce a true output.

The + symbol used in the expression A + B + C is the OR operator and indicates logical addition. In logical addition $1 + 0 = 1$ and $1 + 1 = 1$. One OR one does not equal two! Condition 8 in Figure 26-7 may be interpreted as $1 + 1 + 1 = 1$, whereas condition 5 can be interpreted as $1 + 0 + 0 = 1$. Condition 1 produces a zero output because $0 + 0 + 0 = 0$.

The OR function is illustrated by the following analogy. The members of group A are Bob, Charley, and Dick. A representative of this group could be Bob OR Charley OR Dick, or any combination of them. This expression may be symbolized as

$$R = B + C + D$$

where R is a representative of group A. R is true (a representative of group A is present) when B is true (Bob is present) OR C is true (Charley is present) OR D is true (Dick is present). Only one of the members needs to be present for group A to be represented. However, group A is also represented when more than one member is present. Group A will not be represented (false) when all members are absent.

The circuit in Figure 26-8 can be used to produce the OR function of the above example. The light (R) indicates that a representative of group A is present when one OR more members of the group are present. (At least one switch is closed.) A logic circuit producing this OR function is symbolized in Figure 26-9.

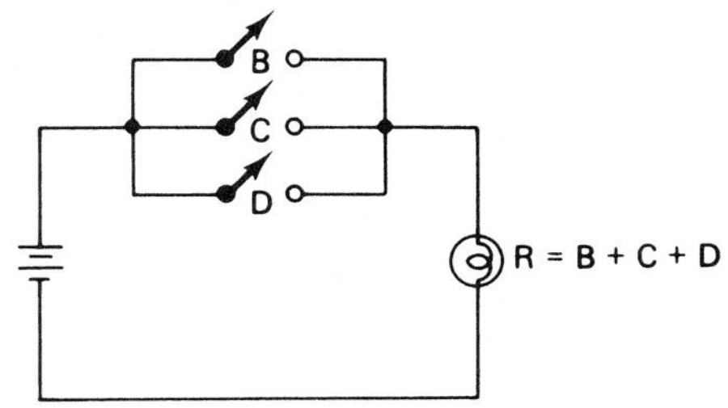

FIGURE 26-8 Switches connected in parallel to produce the OR function.

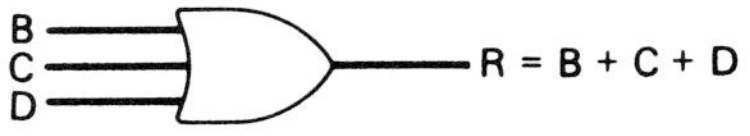

FIGURE 26-9 Logic symbol for the OR function B + C + D.

26-5 THE NOT FUNCTION

■ **KEY POINT** *For the NOT function, the output is the complement of the input.*

The concept of the NOT function can be illustrated by the circuit in Figure 26-10. When the switch is closed, the indicator lamp lights (true). Opening the switch will break the circuit and the lamp will go out (false). Only two conditions may exist. Either the lamp is on (true) or it is off (false). No other condition may exist. The conditions are *complements* of each other.

A logic circuit producing the NOT function is called an *inverter;* its symbol appears in Figure 26-11. An inverter converts the state or value of a variable to its complement. Thus, if variable A appears at the input, $\overline{A}$ (not A) is produced at the output; and, conversely, when $\overline{A}$ appears at the input, A is produced at the output. When performing the NOT function, a 1 will be changed to a 0, and vice versa, as shown in the truth table.

26-6 BOOLEAN EXPRESSIONS

The three basic logic functions discussed, AND, OR, and NOT, either individually or in various combinations, are the basic building blocks for all computer logic circuits. A few of these combinations will be illustrated by our group A analogy consisting of the presence of Bob (B), Charley (C), and Dick (D). Suppose we wish to describe a situation in which the group (A) is represented by the presence of Bob (B) and Charley (C), but not Dick (D). This situation could be expressed as $X = BC\overline{D}$, which states that Bob and Charley are present but Dick is NOT present; it is illustrated in Figure 26-12. Note the use of the inverter.

How would you describe a condition in which the group (A) is represented by at least two members—in other words, a majority? This situation could be expressed in Boolean algebra as $X = BC + BD + CD$ and is symbolized in Figure 26-13. The output is true if B AND C is true (Bob AND Charley are present), OR B AND D is true (Bob AND Dick are present), OR C AND D is true (Charley AND Dick are present).

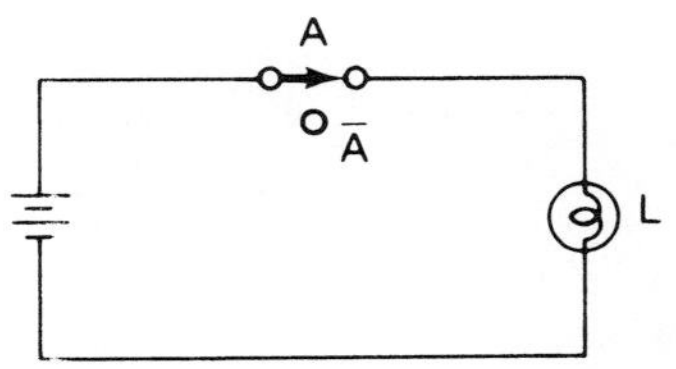

FIGURE 26-10 Representation of the NOT function.

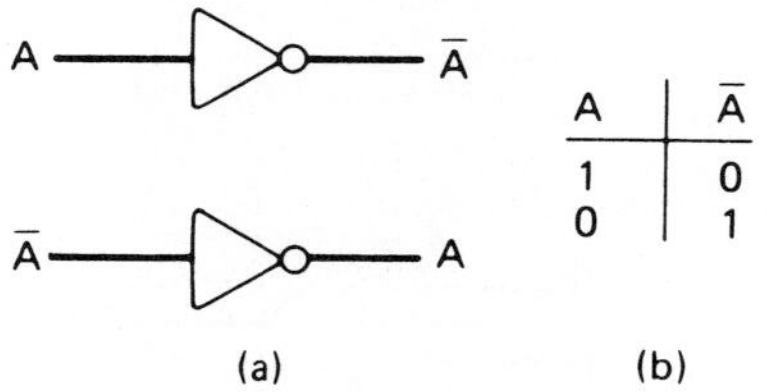

FIGURE 26-11 Inverter: (a) logic symbol; (b) truth table.

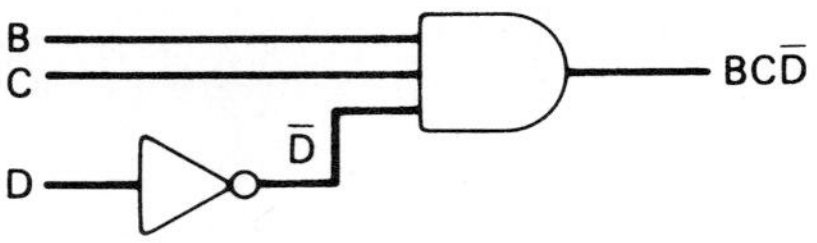

FIGURE 26-12 Logic circuit used to represent the expression $A = BC\overline{D}$.

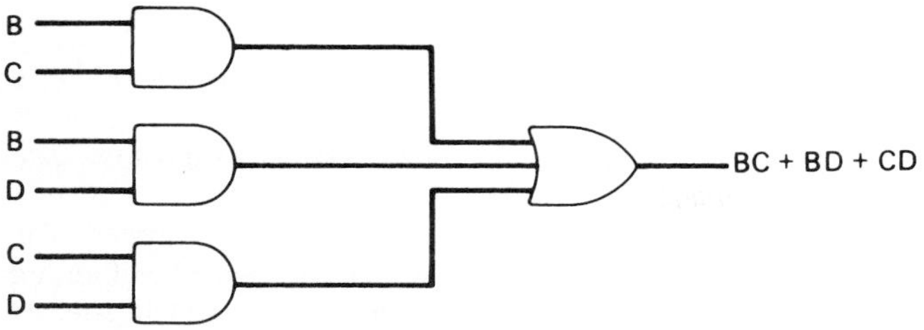

FIGURE 26-13 Logic circuit used to represent the expression BC + BD + CD.

How could the group be represented by the presence of Charley and the absence of both Bob and Dick? One possible method would be $\overline{B}C\overline{D}$, as shown in Figure 26-14.

Suppose we wish to express the situation in which the entire group is NOT present. In other words, one or more members are absent. The Boolean expression would be $\overline{BCD}$ and is illustrated in Figure 26-15. Note that the entire group is affected by the inverter.

Finally, how would we express a situation in which the group is represented by either Bob and Charley or by Dick alone? One possible method is shown in Figure 26-16.

Logic diagrams are drawn to symbolize logic expressions. In the following problems, logic diagrams are to be interpreted or logic diagrams are to be drawn to symbolize logic expressions.

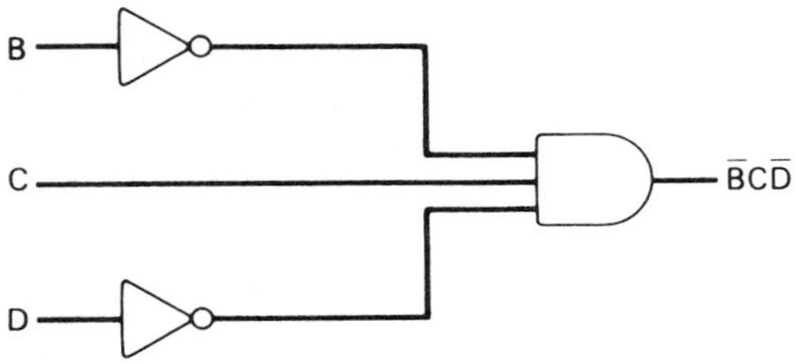

FIGURE 26-14 Logic circuit used to represent the expression $\overline{B}C\overline{D}$.

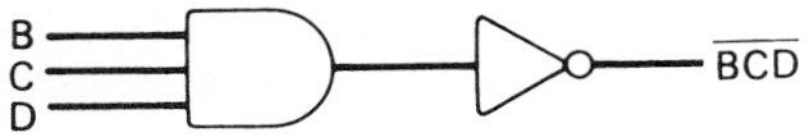

FIGURE 26-15 Logic circuit used to represent the expression $\overline{BCD}$.

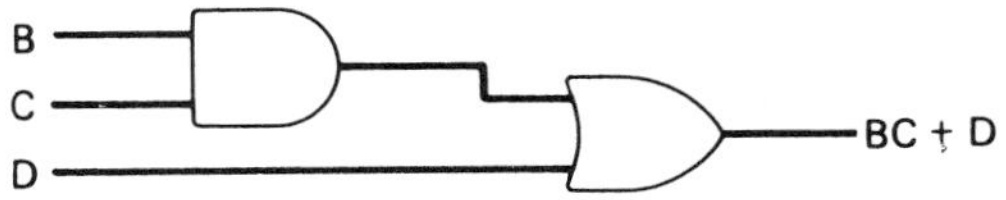

FIGURE 26-16 Logic circuit used to represent the expression BC + D.

PRACTICE PROBLEMS 26-1

Draw the logic diagrams that represent the following Boolean expressions:

1. $B + \overline{C}$

2. $\overline{A}B$

3. $AB + \overline{C}$

4. $A(\overline{B} + C)$

5. $\overline{B + \overline{C} + D}$

Solutions

In problem 1, Figure 26-17, the OR function is indicated; therefore, an OR gate is drawn. To realize the input $\overline{C}$, an inverter is used.

In problem 2 the AND function is indicated; therefore, an AND gate is drawn. To realize the input $\overline{A}$, an inverter is used.

In problem 3 both AND and OR functions are indicated. A and B are the inputs to the AND gate. The output of the AND gate (AB) and $\overline{C}$ are the inputs to the OR gate. An inverter is used to produce the input variable $\overline{C}$.

In problem 4 parentheses are used to indicate logical multiplication as in any algebraic expression. Therefore, a two-input AND gate is indicated. The inputs are A and $\overline{B}$ OR C. The input $\overline{B}$ OR C results from a two-input OR gate whose inputs are $\overline{B}$ and C.

In problem 5 the entire output function is negated. The bar across the entire expression tells us this. Therefore, an inverter must be connected to the output of the OR gate. At the output of the OR gate is the expression $B + \overline{C} + D$. Because there are three variables, the gate is a three-input OR gate.

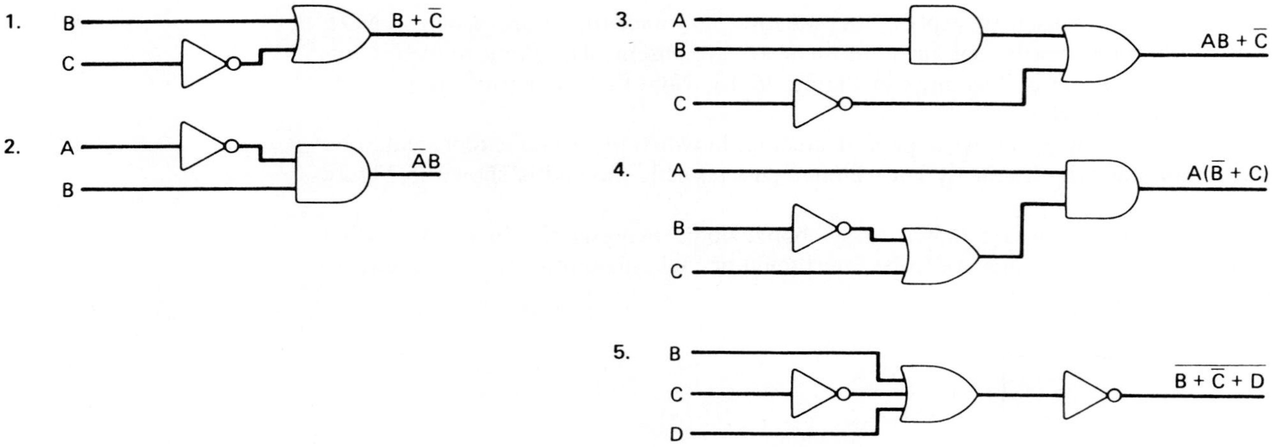

FIGURE 26-17 Answers to Practice Problems 26-1.

Now let's turn it around.

PRACTICE PROBLEMS 26-2

Write the Boolean expression represented by the logic circuits in Figure 26-18.

Solutions

1. $F = \overline{AB}$

2. $F = \overline{A} + B$

3. $A = \overline{XY} + Z$

4. $X = \overline{\overline{A}BC}$

5. $F = \overline{C(\overline{A} + B)}$

In problem 1 the AND function is indicated, followed by an inverter. Because the inverter is on the output side of the gate, the entire expression is complemented.

In problem 2 the inverter complements one input to the OR gate and results in the output $\overline{A} + B$.

The inverter complements the output of the AND gate

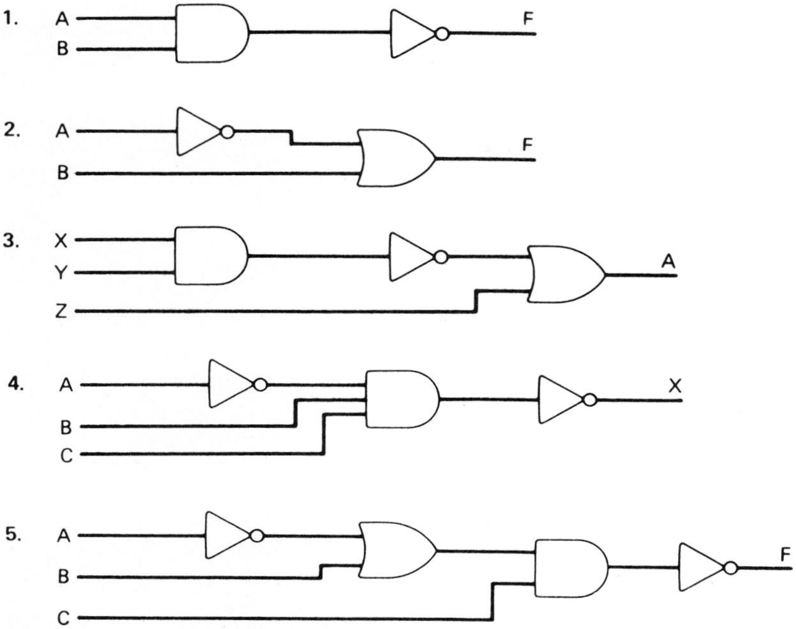

FIGURE 26-18 Logic circuits for Practice Problems 26-2.

in problem 3 to give the expression $\overline{XY}$. This expression is one input to the OR gate. The other input is Z. The output of the OR gate then is $\overline{XY} + Z$.

In problem 4 the three inputs to the AND gate are $\overline{A}$, B, and C. The output is $\overline{A}BC$. This output is complemented by the inverter and results in $\overline{\overline{A}BC}$ at its output.

In problem 5 the output of the OR gate ($\overline{A} + B$) is one input to the AND gate. C is the other input. The output of the AND gate is $C(\overline{A} + B)$. This output is complemented and results in $\overline{C(\overline{A} + B)}$ at the output of the inverter.

SELF-TEST 26-1

Problems **1** through **10**: Write the Boolean expressions for the logic circuits in Figure 26-19. Draw the logic diagrams that represent the following Boolean expressions:

11. $A\overline{B}$
12. $\overline{A} + B$
13. $\overline{F + G}$
14. $\overline{C}(B + \overline{D})$
15. $A\overline{B} + C$
16. AB + AC
17. (X + Z)(X + Y)
18. $A\overline{B} + \overline{CD}$
19. $\overline{A(B + C)}$
20. $\overline{B\overline{C} + D}$

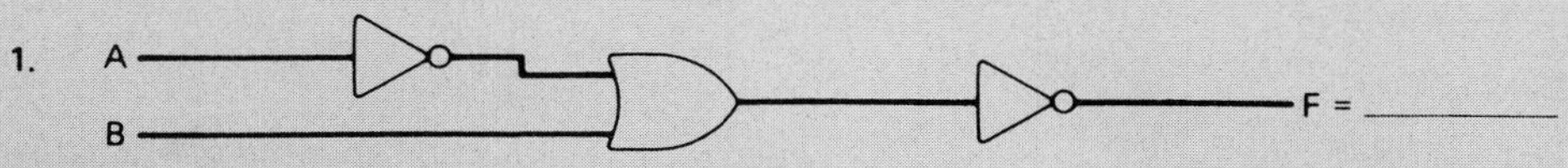

FIGURE 26-19 Logic circuits for problems 1 through 10 for Self-test 26-1.

2.

3.

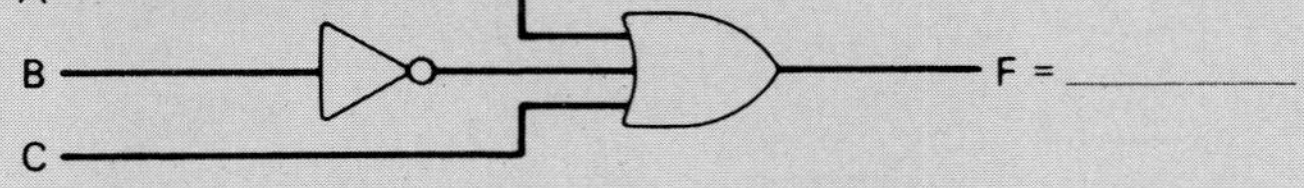

4.

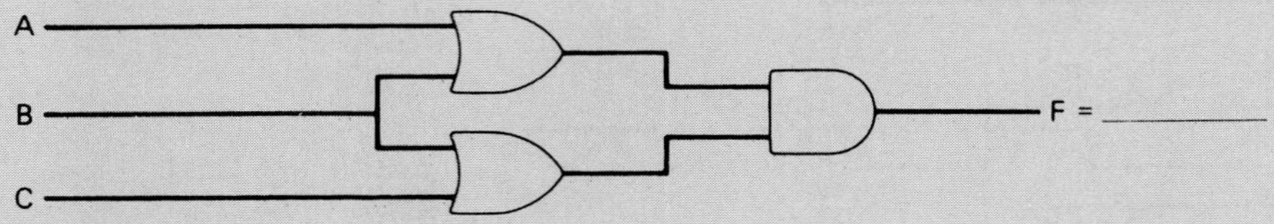

5.

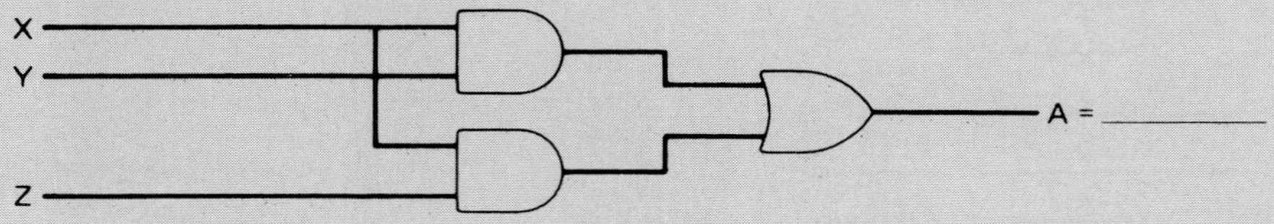

6.

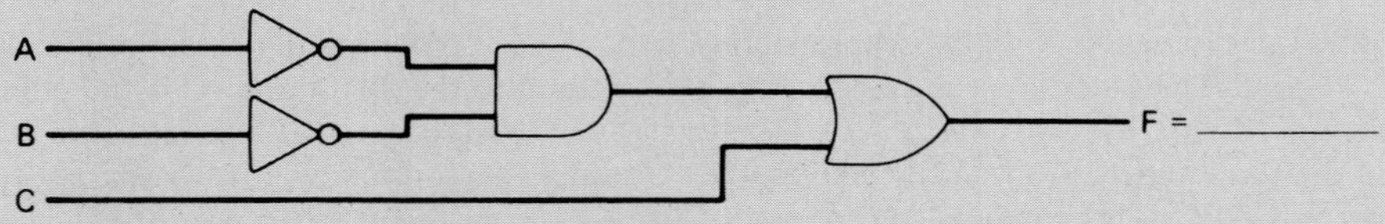

7.

8.

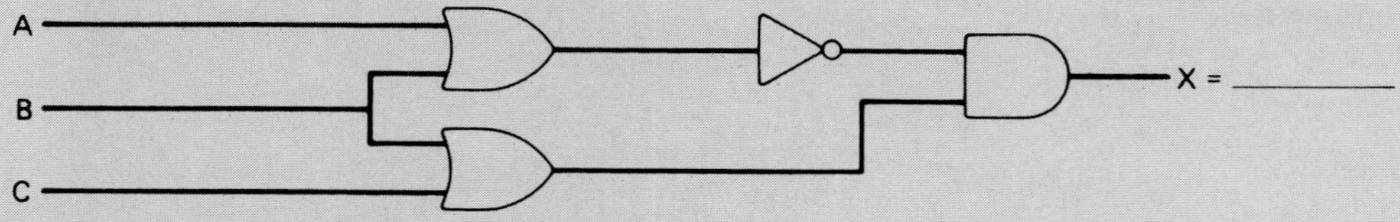

9.

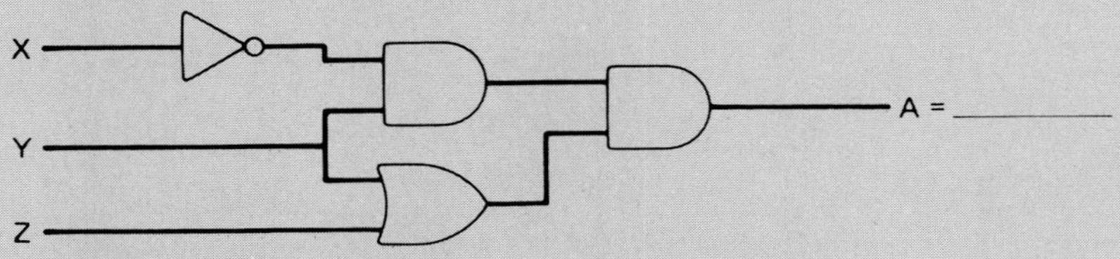

FIGURE 26-19 ***(cont.)***

10. 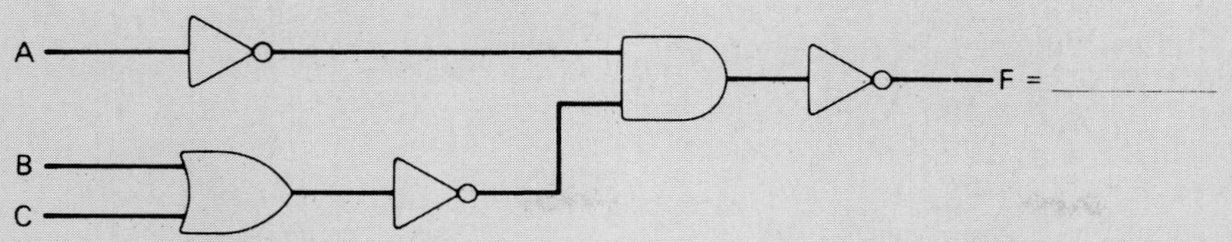

FIGURE 26-19 *(cont.)*

Answers to Self-test 26-1 are at the end of the chapter.

26-7 BOOLEAN POSTULATES AND THEOREMS

The following postulates, laws, and theorems are important in the simplification and manipulation of logic expressions. The student is encouraged to become familiar with these because they will be used throughout this chapter. Note that each postulate, law, or theorem is described in two parts. These are *duals* of each other; that is, the dual of the OR operator is the AND, while the dual of a given variable is its complement.

26-7-1 Postulates

Postulates are self-evident truths. Consider postulates 1a and 1b. A variable is either true (1) or false (0). Postulates 2a, 3a, and 4a represent the conjunctive (AND) form and define the function of the AND operator, as shown in Figure 26-20. Postulates 2b, 3b, and 4b represent the disjunctive (OR) form and define the function of the OR operator, as shown in Figure 26-21. Postulates 5a and 5b define the function of the NOT operator, as illustrated in Figure 26-22.

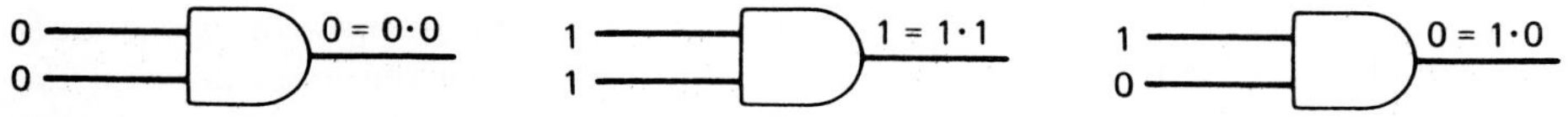

FIGURE 26-20 Logic symbolization of postulates 2a, 3a, and 4a.

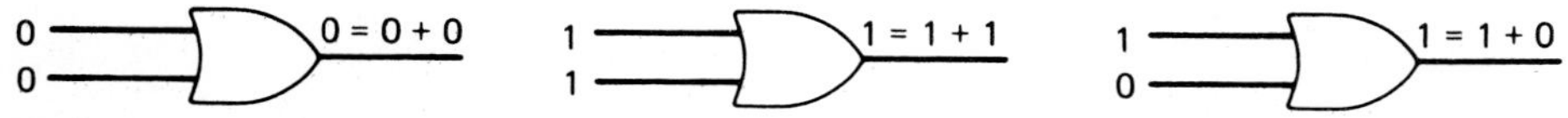

FIGURE 26-21 Logic symbolization of postulates 2b, 3b, and 4b.

FIGURE 26-22 Logic symbolization of postulates 5a and 5b.

Postulates

1a. $A = 1$ (if $A \neq 0$)	1b. $A = 0$ (if $A \neq 1$)
2a. $0 \cdot 0 = 0$	2b. $0 + 0 = 0$
3a. $1 \cdot 1 = 1$	3b. $1 + 1 = 1$
4a. $1 \cdot 0 = 0$	4b. $1 + 0 = 1$
5a. $\overline{1} = 0$	5b. $\overline{0} = 1$

26-7-2 Algebraic Properties

The following are properties of ordinary algebra that also apply to Boolean algebra. Remember, Boolean expressions contain variables having only two possible values.

Algebraic Properties

Commutative

6a. $AB = BA$ — 6b. $A + B = B + A$

Associative

7a. $A(BC) = AB(C)$ — 7b. $A + (B + C) = (A + B) + C$

Distributive

8a. $A(B + C) = AB + AC$ — 8b. $A + BC = (A + B)(A + C)$

The *commutative* property simply means that the circuit is not affected by the order or sequence of the variables. This is illustrated in Figure 26-23.

The *associative* property pertains to the parentheses. It shows that a sequence exclusively of AND functions (7a) or a sequence exclusively of OR functions (7b) is not affected by the placement of the parentheses, as indicated in Figure 26-24.

The *distributive* property for 8a and 8b may be proved by performing the algebraic multiplication or by factoring. Figure 26-25 shows that the application of the distributive property produces two forms of an expression and, thus, two circuits may be realized for each.

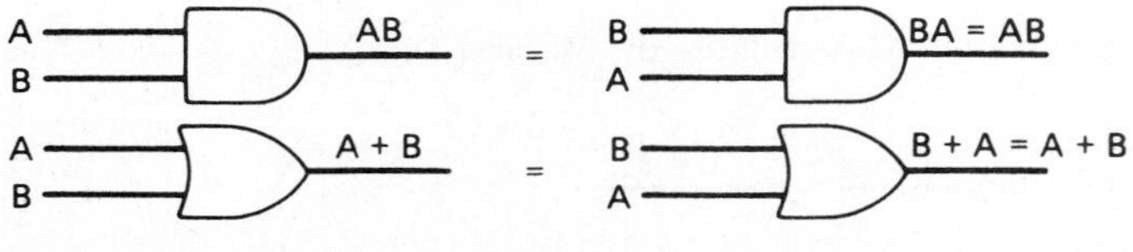

FIGURE 26-23 Logic symbols used to demonstrate commutative property.

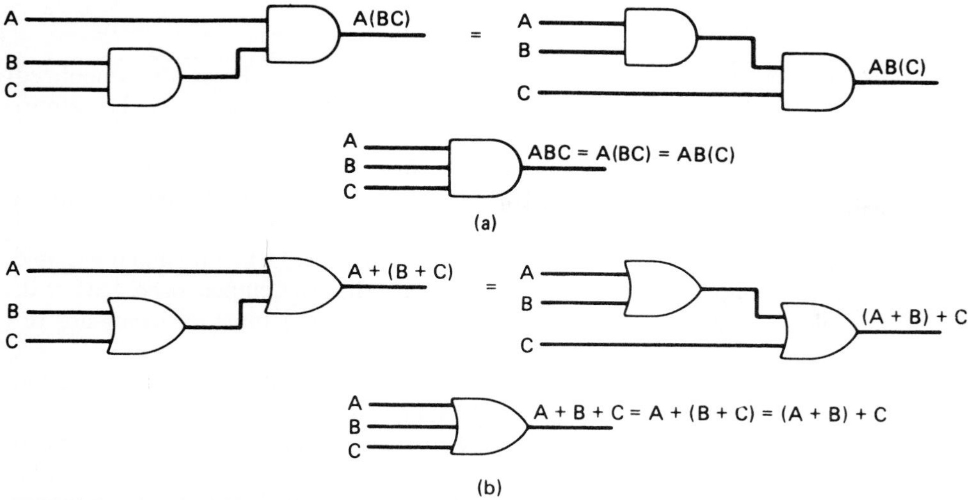

FIGURE 26-24 **Logic symbols and circuits used to demonstrate the associative property: (a) logic circuit to demonstrate AND; (b) logic circuit to demonstrate OR.**

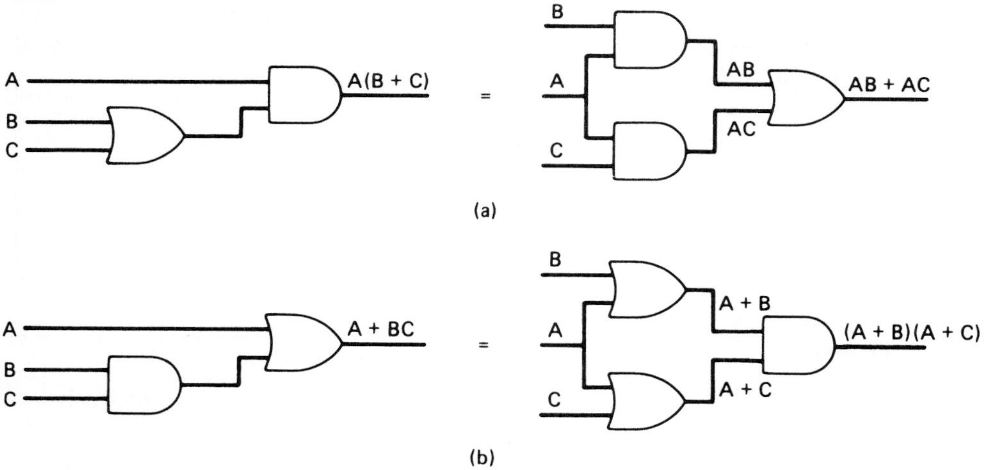

FIGURE 26-25 **Logic symbols and circuits used to demonstrate distributive property: (a) logic circuit to demonstrate A(B + C) = AB + AC; (b) logic circuit to demonstrate A + BC = (A + B)(A + C).**

26-7-3 Theorems

The following theorems define the application of the operators to variables:

Theorems

9a. $A \cdot 0 = 0$	9b. $A + 0 = A$
10a. $A \cdot 1 = A$	10b. $A + 1 = 1$
11a. $A \cdot A = A$	11b. $A + A = A$
12a. $A \cdot \overline{A} = 0$	12b. $A + \overline{A} = 1$
13a. $\overline{\overline{A}} = A$	13b. $A = \overline{\overline{A}}$

Theorems 9a, 10a, 11a, and 12a pertain to the AND function. These are symbolized in Figure 26-26. For theorem 9a, one of the variables is always a zero. Therefore, the output will be a zero regardless of the value of A. Theorem 10a tells us the output will be determined by the input variable A. If A = 1, then $1 \cdot 1 = 1$. If A = 0, then $1 \cdot 0 = 0$. For Theorem 11a, if the input variable A = 1, the output will be $1 \cdot 1 = 1$. If the input variable A = 0, the output will be $0 \cdot 0 = 0$.

For Theorem 12a, the output will always be 0 because, when the input variable A = 1, the other input variable will be 0 ($\bar{A}$), causing the output to be $1 \cdot 0 = 0$. The same output will be produced when the values of the input variables are reversed. Theorem 12a is called a *self-contradiction*.

Theorem 13a expresses double negation. An expression that has been inverted twice equals its original value, as shown in Figure 26-27.

Theorems 9b, 10b, 11b, and 12b pertain to the OR function and are illustrated in Figure 26-28. For the OR function, one or more of the input variables must be true in order that a true output be present. For theorems 9b and 11b, the output is determined by the input variable A. For 10b and 12b, either A or $\bar{A}$ must always equal 1.

Consider the logic diagram in Figure 26-29. The Boolean expression at its output is (AB)(BC). By applying various postulates, laws, and theorems, we can simplify the expression to ABC. Here's how we could do it.

1.	(AB)(BC)	Original expression
2.	ABBC	7a
3.	ABC	11a

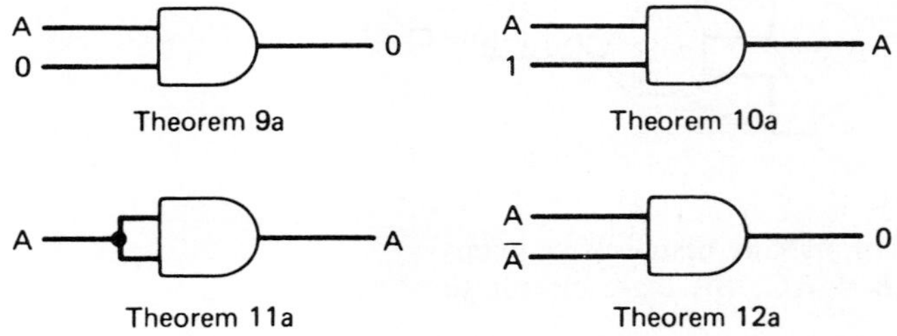

FIGURE 26-26 Logic symbolization of the AND function.

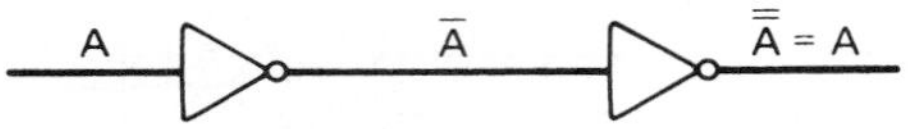

FIGURE 26-27 Logic symbolization of double negation.

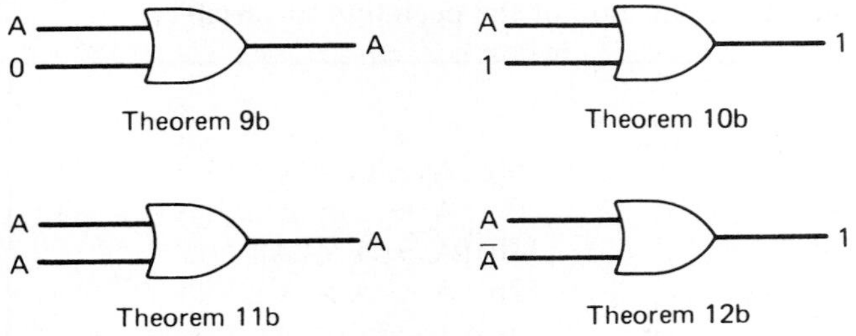

FIGURE 26-28 Logic symbolization of the OR function.

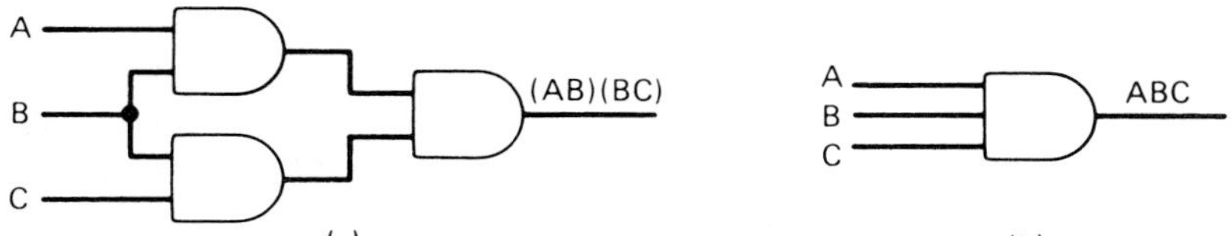

(a) (b)

A	B	C	AB	BC	(AB)(BC)	ABC
0	0	0	0	0	0	0
0	0	1	0	0	0	0
0	1	0	0	0	0	0
0	1	1	0	1	0	0
1	0	0	0	0	0	0
1	0	1	0	0	0	0
1	1	0	1	0	0	0
1	1	1	1	1	1	1

(c)

FIGURE 26-29 (a) Original circuit; (b) simplified circuit; (c) truth table.

Step 1 shows the original expression. In step 2 we remove the parentheses. Our justification for this step is the associative property (7a). The simplified expression in step 3 results from applying theorem 11a. Any variable AND that same variable equals that variable ($B \cdot B = B$). The simplified expression results in 3 two-input AND gates being replaced by 1 three-input AND gate, a substantial savings of both space and money to the designer.

Let's prove that the expressions are equal by using a truth table as in Figure 26-29(c). Notice that there are eight conditions. There are three input variables (A, B, and C) and each variable has two possible states ($2^3 = 8$). When we construct a truth table, it is important that we keep the order of the original statement and not skip any steps along the way. Thus, we list the variables first, then we list the function of the first two AND gates (AB and BC), and then we write the original expression and the simplified expression. The truth table shows that (AB)(BC) = ABC.

The process of using truth tables to prove the equality of two expressions is called *proof by perfect induction*. The student is encouraged to use truth tables to evaluate logic expressions or to prove the results of logic operations. Truth tables can also be used to define the functions of various logic circuits.

The distributive property (8a) states that $A(B + C) = AB + AC$. Let's prove that they are equal by constructing a truth table, as in Figure 26-30. First we list the input variables. Again, because there are three variables, eight conditions are possible. Next we list each intermediate step. $B + C$ is the result of ORing the input variables B and C. The columns AB and AC are the result of ANDing the input variables A and B and the input variables B and C. Finally, there are columns showing the expressions $A(B + C)$ and $AB + AC$.

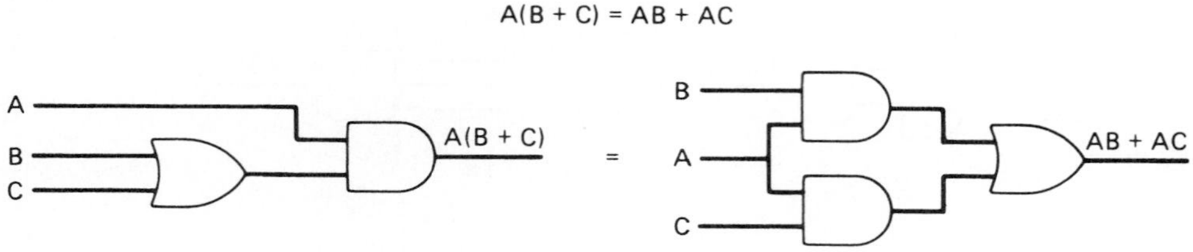

A	B	C	B + C	AB	AC	A(B + C)	AB + AC
0	0	0	0	0	0	0	0
0	0	1	1	0	0	0	0
0	1	0	1	0	0	0	0
0	1	1	1	0	0	0	0
1	0	0	0	0	0	0	0
1	0	1	1	0	1	1	1
1	1	0	1	1	0	1	1
1	1	1	1	1	1	1	1

Equal (A(B + C) and AB + AC columns)

FIGURE 26-30 Logic diagrams and truth tables for A(B + C) = AB + AC.

Now let's do the same thing with the distributive property 8b: (A + B)(A + C). The logic diagrams and truth table are shown in Figure 26-31. Notice again how each part of each expression is listed in the truth table.

Let's see if it is possible to simplify the logic diagram in Figure 26-32. The step-by-step simplification and the justification for each step are shown by the following:

1. A + B + AC + AB	Original expression
2. A + AB + B + AC	6b
3. A(1 + B) + B + AC	8a
4. A(1) + B + AC	10b
5. A + B + AC	10a
6. A + AC + B	6b
7. A(1 + C) + B	8a
8. A(1) + B	10b
9. A + B	10a

When many steps are involved in the simplification, there is usually more than one way to arrive at the final answer. The student may find alternate methods to the above simplification, but in each case the final simplification should be A + B. The original logic diagram, simplified logic diagram, and truth table are found in Figure 26-32.

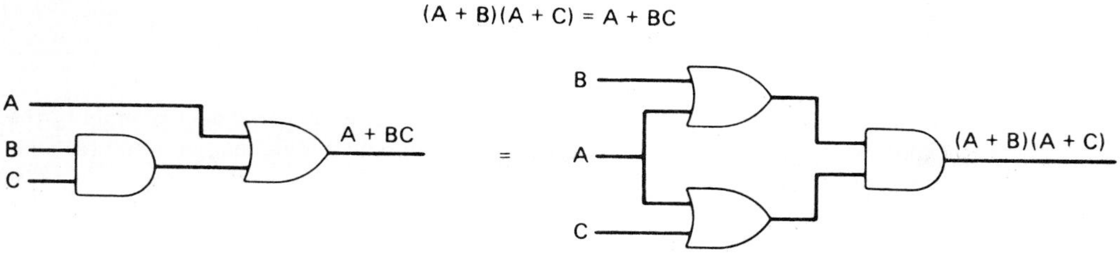

A	B	C	BC	A + B	A + C	A + BC	(A + B)(A + C)
0	0	0	0	0	0	0	0
0	0	1	0	0	1	0	0
0	1	0	0	1	0	0	0
0	1	1	1	1	1	1	1
1	0	0	0	1	1	1	1
1	0	1	0	1	1	1	1
1	1	0	0	1	1	1	1
1	1	1	1	1	1	1	1

Equal (A + BC and (A + B)(A + C))

FIGURE 26-31 **Logic diagrams and truth table for A + BC = (A + B)(A + C).**

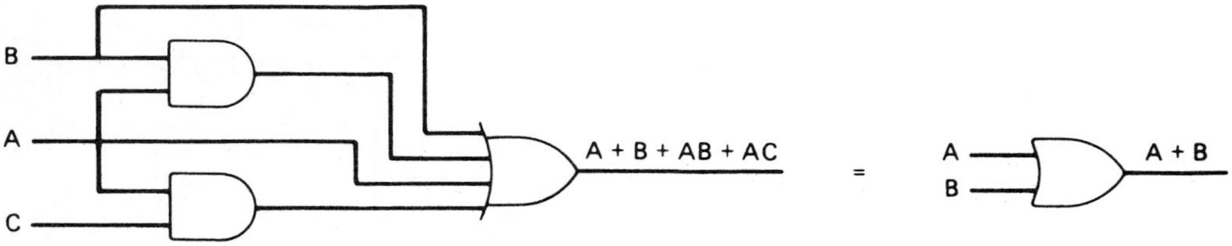

A	B	C	AC	AB	A + B + AC + AB	A + B
0	0	0	0	0	0	0
0	0	1	0	0	0	0
0	1	0	0	0	1	1
0	1	1	0	0	1	1
1	0	0	0	0	1	1
1	0	1	1	0	1	1
1	1	0	0	1	1	1
1	1	1	1	1	1	1

Equal (A + B + AC + AB and A + B)

FIGURE 26-32 **Logic symbols and truth table for A + B + AB + AC = A + B.**

PRACTICE PROBLEMS 26-3

For each logic diagram in Figure 26-33, (a) write the Boolean expression, (b) simplify the Boolean expression and justify each step, (c) draw the simplified logic diagram, and (d) construct a truth table to prove that the two expressions are equal.

1.
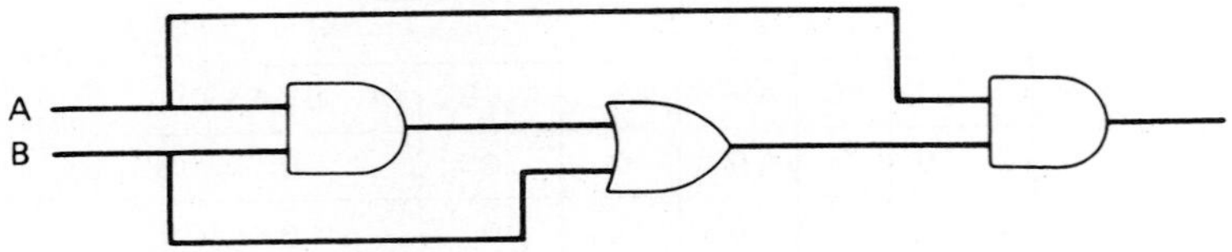

2.
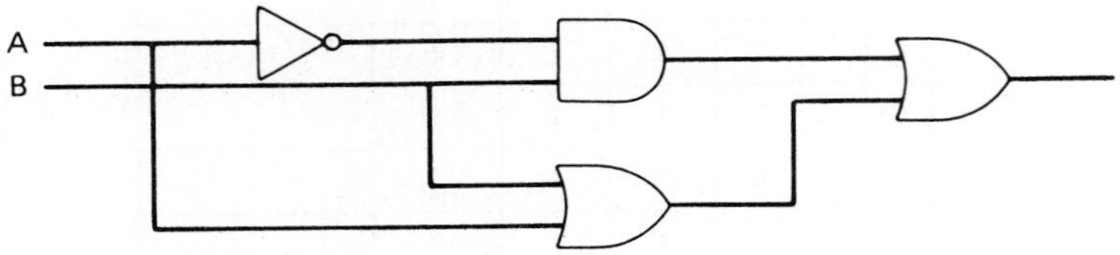

3.
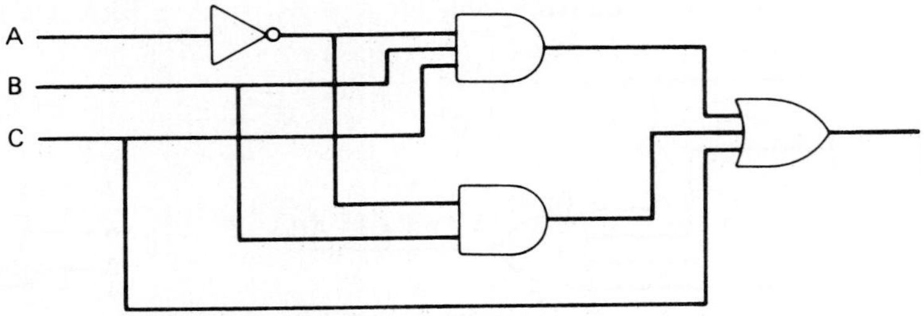

4.
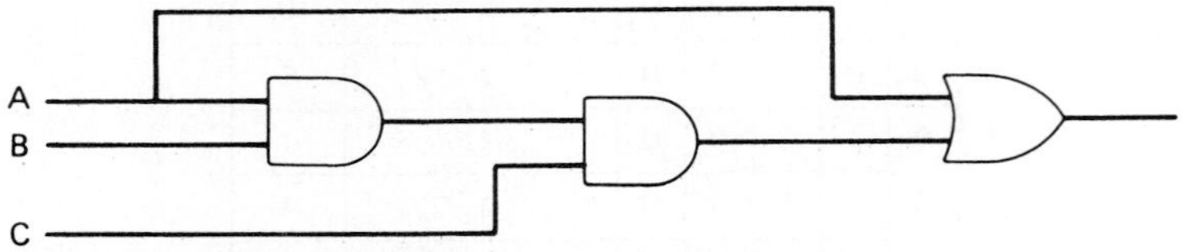

FIGURE 26-33 **Logic circuits for Practice Problems 26-3.**

Solutions

1. See Figure 26-34.
2. See Figure 26-35.
3. See Figure 26-36.
4. See Figure 26-37.

A(AB + B)	Original expression
AAB + AB	8a
AB + AB	11a
AB	11b

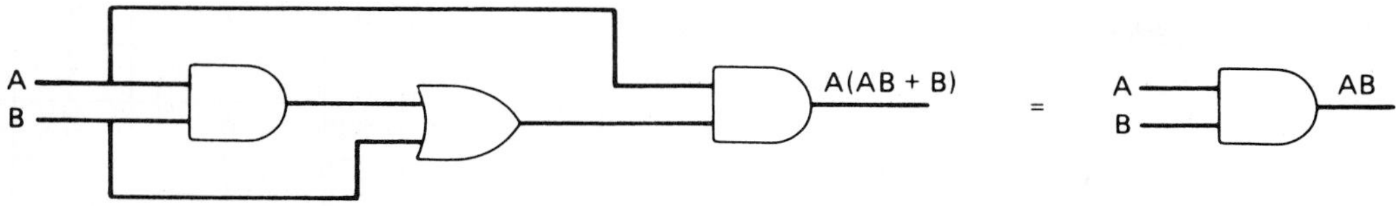

A	B	AB	AB + B	A(AB + B)
0	0	0	0	0
0	1	0	1	0
1	0	0	0	0
1	1	1	1	1

Equal

FIGURE 26-34 **Answer to problem 1, Practice Problems 26-3.**

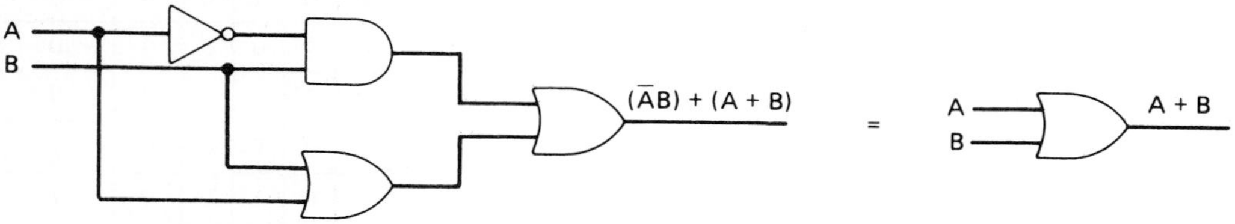

$(\bar{A}B) + (A + B)$	Original expression
$\bar{A}B + A + B$	7b
$\bar{A}B + B + A$	6b
$B(\bar{A} + 1) + A$	8a
B(1) + A	10b
B + A	10a

A	B	$\bar{A}$	$\bar{A}B$	A + B	$\bar{A}B + (A + B)$
0	0	1	0	0	0
0	1	1	1	1	1
1	0	0	0	1	1
1	1	0	0	1	1

FIGURE 26-35 **Answer to problem 2, Practice Problems 26-3.**

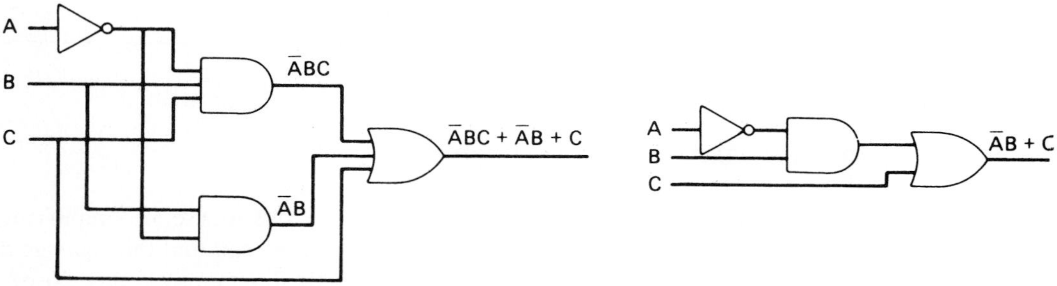

FIGURE 26-36 **Answer to problem 3, Practice Problems 26-3.**

$\bar{A}BC + \bar{A}B + C$	Original expression
$\bar{A}B\,(C + 1) + C$	8a
$\bar{A}B\,(1) + C$	10b
$\bar{A}B + C$	10a

A	B	C	$\bar{A}$	$\bar{A}B$	$\bar{A}BC$	$\bar{A}BC + \bar{A}B + C$	$\bar{A}B + C$
0	0	0	1	0	0	0	0
0	0	1	1	0	0	1	1
0	1	0	1	1	0	1	1
0	1	1	1	1	1	1	1
1	0	0	0	0	0	0	0
1	0	1	0	0	0	1	1
1	1	0	0	0	0	0	0
1	1	1	0	0	0	1	1

Equal

FIGURE 26-36 *(cont.)*

A + ABC	Original expression
A(1 + BC)	8a
A(1)	10b
A	10a

A	B	C	AB	ABC	A + ABC
0	0	0	0	0	0
0	0	1	0	0	0
0	1	0	0	0	0
0	1	1	0	0	0
1	0	0	0	0	1
1	0	1	0	0	1
1	1	0	1	0	1
1	1	1	1	1	1

Equal

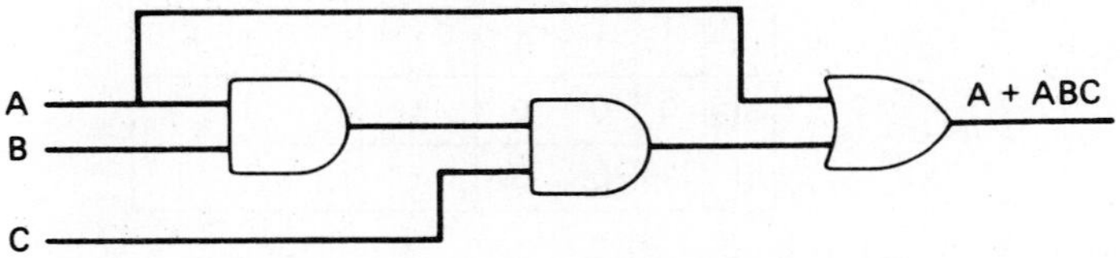

FIGURE 26-37 **Answer to problem 4, Practice Problems 26-3.**

26-7-4 De Morgan's Theorem

De Morgan's theorem provides us with an invaluable tool to use in simplifying Boolean expressions. By applying De Morgan's theorem, we can find the equivalent of a Boolean expression. For example, consider the logic diagram in Figure 26-38. The Boolean expression at the output is $\overline{\bar{A}\,\bar{B}}$. We would not consider $\overline{\bar{A}\,\bar{B}}$ to be in its simplest form because of the bar extending over the entire expression. By applying De Morgan's theorem, the equivalent Boolean expression A + B is realized. This expression is much easier to understand and much more simple to construct.

Three steps are required in De Morganizing a Boolean expression: First, we replace all the OR operator symbols (+) with AND operator symbols ($\cdot$) and all the AND operator symbols with OR operator symbols; next, we replace all variables with their complements; finally, we complement the entire expression. To De Morganize the expression $\overline{\overline{A}\,\overline{B}}$ above, we use the following steps:

- □ Step 1. Replace the AND operator with the OR operator: $\overline{\overline{A} + \overline{B}}$.
- □ Step 2. Complement each variable: $\overline{\overline{\overline{A}} + \overline{\overline{B}}} = \overline{A + B}$.
- □ Step 3. Complement the entire expression: $\overline{\overline{A + B}} = A + B$.

The logic diagrams and truth table are found in Figure 26-39.

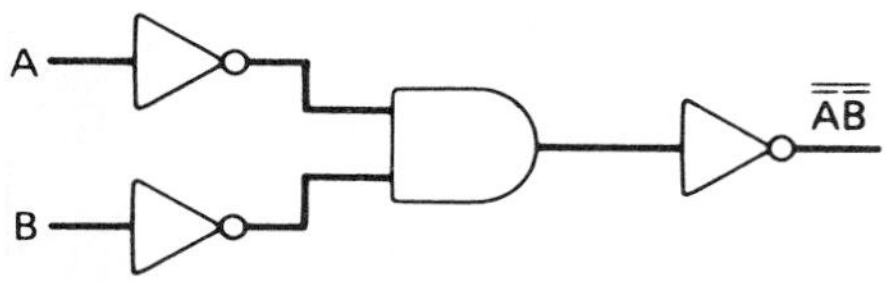

FIGURE 26-38 Logic diagram for the expression $\overline{\overline{A}\,\overline{B}}$.

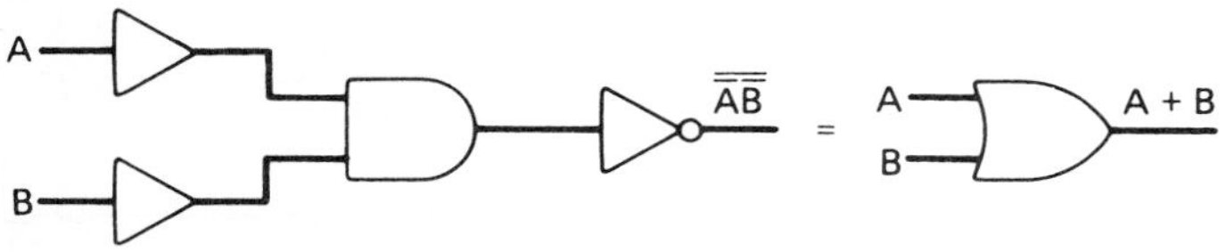

A	B	$\overline{A}$	$\overline{B}$	$\overline{A}\,\overline{B}$	$\overline{\overline{A}\,\overline{B}}$	A + B
0	0	1	1	1	0	0
0	1	1	0	0	1	1
1	0	0	1	0	1	1
1	1	0	0	0	1	1

Equal

FIGURE 26-39 Logic diagrams and truth table for $\overline{\overline{A}\,\overline{B}}$ = A + B.

Let's determine De Morgan's equivalent of the expression $\overline{\overline{A} + \overline{B}}$.

- □ Step 1. Replace the OR operator with the AND operator: $\overline{\overline{A} \cdot \overline{B}}$.
- □ Step 2. Complement each variable: $\overline{\overline{\overline{A}} \cdot \overline{\overline{B}}} = \overline{AB}$.
- □ Step 3. Complement the entire expression: $\overline{\overline{AB}} = AB$.

Therefore,

$$\overline{\overline{A} + \overline{B}} = AB$$

and is symbolized in Figure 26-40.

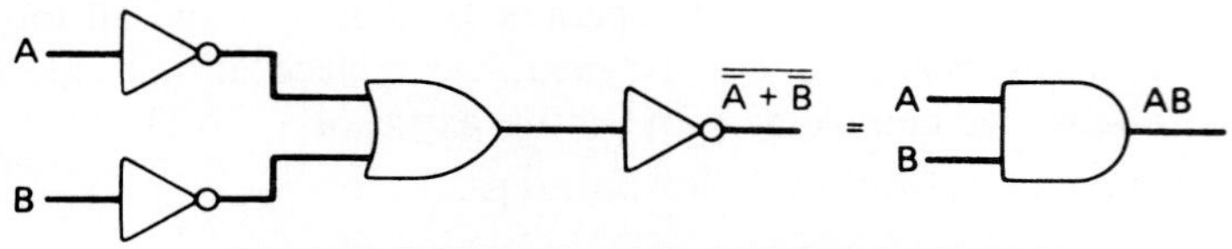

A	B	$\overline{A}$	$\overline{B}$	$\overline{A}+\overline{B}$	$\overline{\overline{A}+\overline{B}}$	AB
0	0	1	1	1	0	0
0	1	1	0	1	0	0
0	0	0	1	1	0	0
1	1	0	0	0	1	1

FIGURE 26-40 Logic diagrams and truth table for $\overline{\overline{A} + \overline{B}} = AB$.

Let's find the De Morgan's equivalent of the expression in theorem 14a, $\overline{A}\overline{B}\overline{C}$.

- □ Step 1. $\overline{A} + \overline{B} + \overline{C}$
- □ Step 2. $\overline{\overline{A}} + \overline{\overline{B}} + \overline{\overline{C}} = A + B + C$
- □ Step 3. $\overline{A + B + C}$

Therefore,

$$\overline{A}\overline{B}\overline{C} = \overline{A + B + C}$$

De Morgan's Theorem

14a. $\overline{A}\overline{B}\overline{C} = \overline{A + B + C}$ 14b. $\overline{A} + \overline{B} + \overline{C} = \overline{ABC}$

The symbols for the equivalent expressions above are given in Figure 26-41.

Now let's try finding De Morgan's equivalent of the expression in theorem 14b, $\overline{A} + \overline{B} + \overline{C}$.

- □ Step 1. $\overline{A} \cdot \overline{B} \cdot \overline{C}$
- □ Step 2. $\overline{\overline{A}} \cdot \overline{\overline{B}} \cdot \overline{\overline{C}} = A \cdot B \cdot C$
- □ Step 3. $\overline{A \cdot B \cdot C}$

Therefore,

$$\overline{A} + \overline{B} + \overline{C} = \overline{A \cdot B \cdot C}$$

The logic symbols for these expressions are given in Figure 26-42.

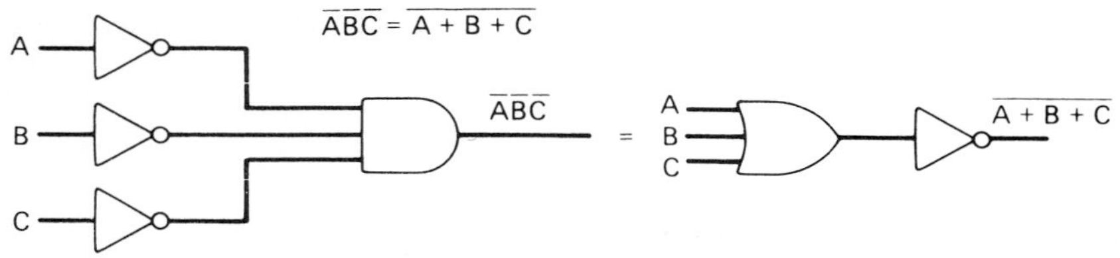

A	B	C	$\overline{A}$	$\overline{B}$	$\overline{C}$	$\overline{A}\,\overline{B}\,\overline{C}$	A + B + C	$\overline{A + B + C}$
0	0	0	1	1	1	1	0	1
0	0	1	1	1	0	0	1	0
0	1	0	1	0	1	0	1	0
0	1	1	1	0	0	0	1	0
1	0	0	0	1	1	0	1	0
1	0	1	0	1	0	0	1	0
1	1	0	0	0	1	0	1	0
1	1	1	0	0	0	0	1	0

Equal

FIGURE 26-41 **Logic diagrams and truth table for $\overline{A}\,\overline{B}\,\overline{C} = \overline{A + B + C}$.**

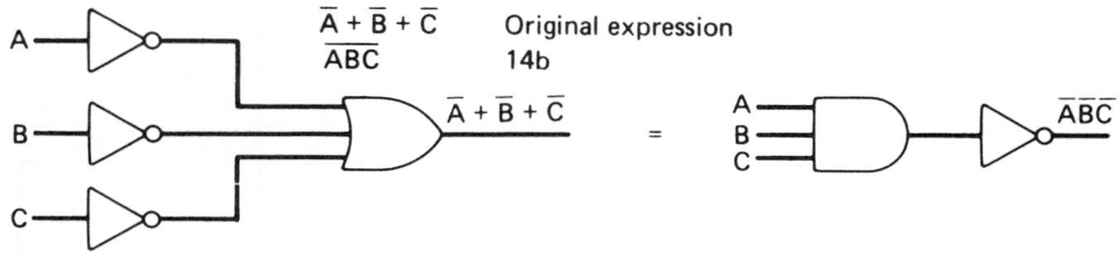

A	B	C	$\overline{A}$	$\overline{B}$	$\overline{C}$	$\overline{A} + \overline{B} + \overline{C}$	ABC	$\overline{ABC}$
0	0	0	1	1	1	1	0	1
0	0	1	1	1	0	1	0	1
0	1	0	1	0	1	1	0	1
0	1	1	1	0	0	1	0	1
1	0	0	0	1	1	1	0	1
1	0	1	0	1	0	1	0	1
1	1	0	0	0	1	1	0	1
1	1	1	0	0	0	0	1	0

Equal

FIGURE 26-42 **Logic diagrams and truth table for $\overline{A} + \overline{B} + \overline{C} = \overline{ABC}$.**

The three-step solution process can be simplified. Consider the expression $\overline{\overline{A} + \overline{B}}$ that we simplified earlier. To simplify, we just break the bar and change the operator.

$$\overline{\overline{A} + \overline{B}} = \overline{\overline{A}} \cdot \overline{\overline{B}} = AB$$

This is an example of theorem 14a. When breaking the bar results in a double negation, we apply theorem 13a ($\overline{\overline{A}} = A$).

Let's simplify the Boolean expression $\overline{\overline{A} + B + \overline{C}}$.

$\overline{\overline{A} + B + \overline{C}}$	Original expression
$\overline{\overline{A}} \cdot \overline{B} \cdot \overline{\overline{C}}$	14b
$A\overline{B}C$	13a

Notice we broke the bar between each variable and changed each OR operator to AND. The logic diagrams and truth table are shown in Figure 26-43.

Let's simplify the expression $\overline{\overline{A} + \overline{B + C}}$. Again we use De Morgan's theorem since simplification requires removing the signs of negation (the bars). Let's start by breaking the bar between $\overline{A}$ and $\overline{B + C}$ and changing the operator.

$\overline{\overline{A} + \overline{B + C}}$	Original expression
$\overline{\overline{A}} \cdot \overline{\overline{B + C}}$	14a
$A \cdot (B + C)$	13a

Or

$$A(B + C)$$

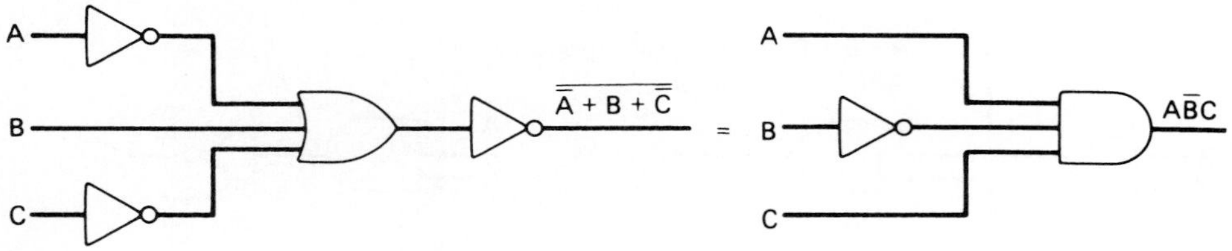

A	B	C	$\overline{A}$	$\overline{B}$	$\overline{C}$	$\overline{A} + B + \overline{C}$	$\overline{\overline{A} + B + \overline{C}}$	$A\overline{B}C$
0	0	0	1	1	1	1	0	0
0	0	1	1	1	0	1	0	0
0	1	0	1	0	1	1	0	0
0	1	1	1	0	0	1	0	0
1	0	0	0	1	1	1	0	0
1	0	1	0	1	0	0	1	1
1	1	0	0	0	1	1	0	0
1	1	1	0	0	1	1	0	0

Equal

FIGURE 26-43 Logic diagrams and truth table for $\overline{\overline{A} + B + \overline{C}} = A\overline{B}C$.

Notice the parentheses (sign of grouping) around B + C. We must treat $\overline{B + C}$ in the original expression as a single term (in algebra the bar is a sign of grouping). Therefore, when we De Morganize, we must include the sign of grouping around B + C to avoid an invalid expression. Without the sign of grouping, we get $A \cdot B + C$ or AB + C, which is not the same as A(B + C). We could have included the sign of grouping in the original expression:

$$\overline{\overline{A} + \overline{B + C}} = \overline{\overline{A} + (\overline{B + C})}$$

It isn't necessary though because the bar ($\overline{B + C}$) tells us to treat B + C as a single term or variable.

We could have De Morganized $\overline{B + C}$ first:

$\overline{\overline{A} + \overline{B + C}}$	Original expression	$\overline{\overline{A} + \overline{B}\,\overline{C}}$	14a

Notice this did not affect the bar that extends over the entire expression since we were De Morganizing just $\overline{B + C}$. Next

$\overline{\overline{A}} \cdot \overline{\overline{B}\,\overline{C}}$	14a (again)	$A \cdot \overline{\overline{B}\,\overline{C}}$	13a
$A \cdot (\overline{\overline{B}} + \overline{\overline{C}})$	14b	A(B + C)	13a

We arrive at the same answer by a much more difficult route. Notice the parentheses when theorem 14b was used. Only the expression BC was De Morganized. Therefore, we include the sign of grouping to indicate that B + C is a single variable (in this case one input to an AND gate). Figure 26-44 includes the logic diagrams and truth table.

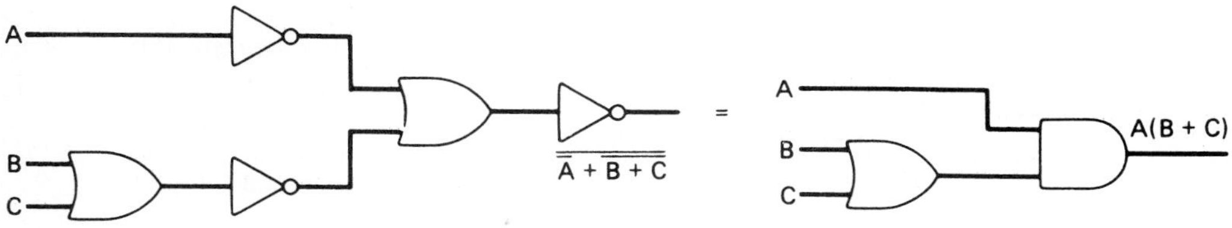

A	B	C	$\overline{A}$	B + C	$\overline{B + C}$	$\overline{A} + \overline{B + C}$	$\overline{\overline{A} + \overline{B + C}}$	A(B + C)
0	0	0	1	0	1	1	0	0
0	0	1	1	1	0	1	0	0
0	1	0	1	1	0	1	0	0
0	1	1	1	1	0	1	0	0
1	0	0	0	0	1	1	0	0
1	0	1	0	1	0	0	1	1
1	1	0	0	1	0	0	1	1
1	1	1	0	1	0	0	1	1

Equal

FIGURE 26-44 Logic diagrams and truth table for $\overline{\overline{A} + \overline{B + C}}$ = A(B + C).

26-7-5 Absorption Theorem

Sometimes variables appear in a certain pattern that can obviously be simplified. The *absorption* theorem shows how substitution may be used on some commonly derived patterns.

Absorption Theorem

15a. $A(A + B) = A$	15b. $A + AB = A$
16a. $A(\overline{A} + B) = AB$	16b. $A + \overline{A}B = A + B$

Theorems 15a and 15b show that the value of B is redundant. A true output occurs only when A is true, as shown in Figure 26-45.

Theorem 16a states that only when A AND B are true is the statement true, as shown in Figure 26-46. Recall that A and $\overline{A}$ cannot be true at the same time.

Theorem 16b is the dual of theorem 16a and is shown in Figure 26-47. For the expression to be true, either A must be true or B must be true. $\overline{A}$ is redundant.

FIGURE 26-45 Logic symbolization of (a) theorem 15a and (b) theorem 15b.

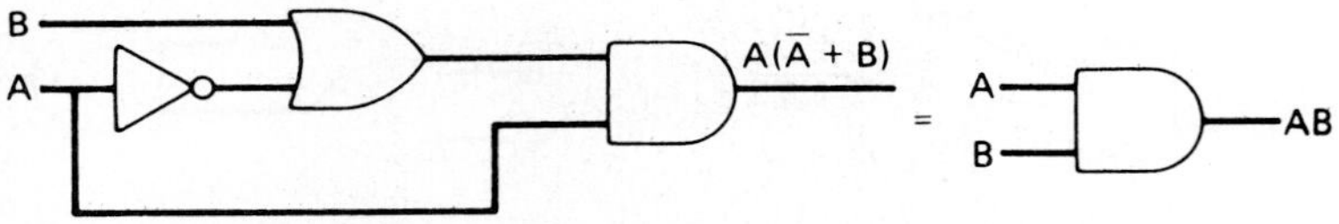

FIGURE 26-46 Logic symbolization of theorem 16a.

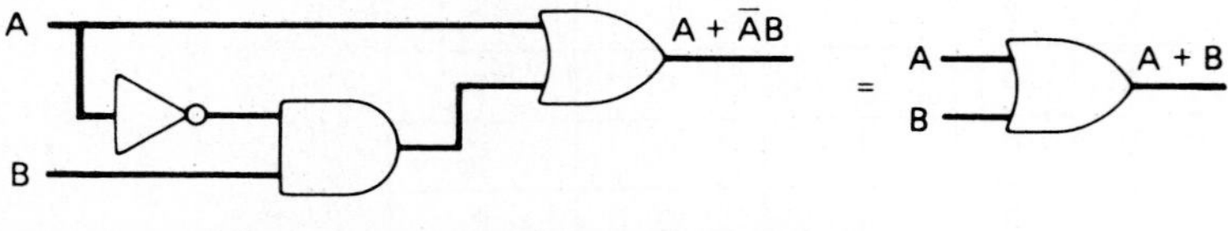

FIGURE 26-47 Logic symbolization of theorem 16b.

PRACTICE PROBLEMS 26-4

For each logic diagram in Figure 26-48, (a) write the Boolean expression, (b) simplify the Boolean expression and justify each step, (c) draw the simplified logic diagram, and (d) construct a truth table to prove that the two expressions are equal.

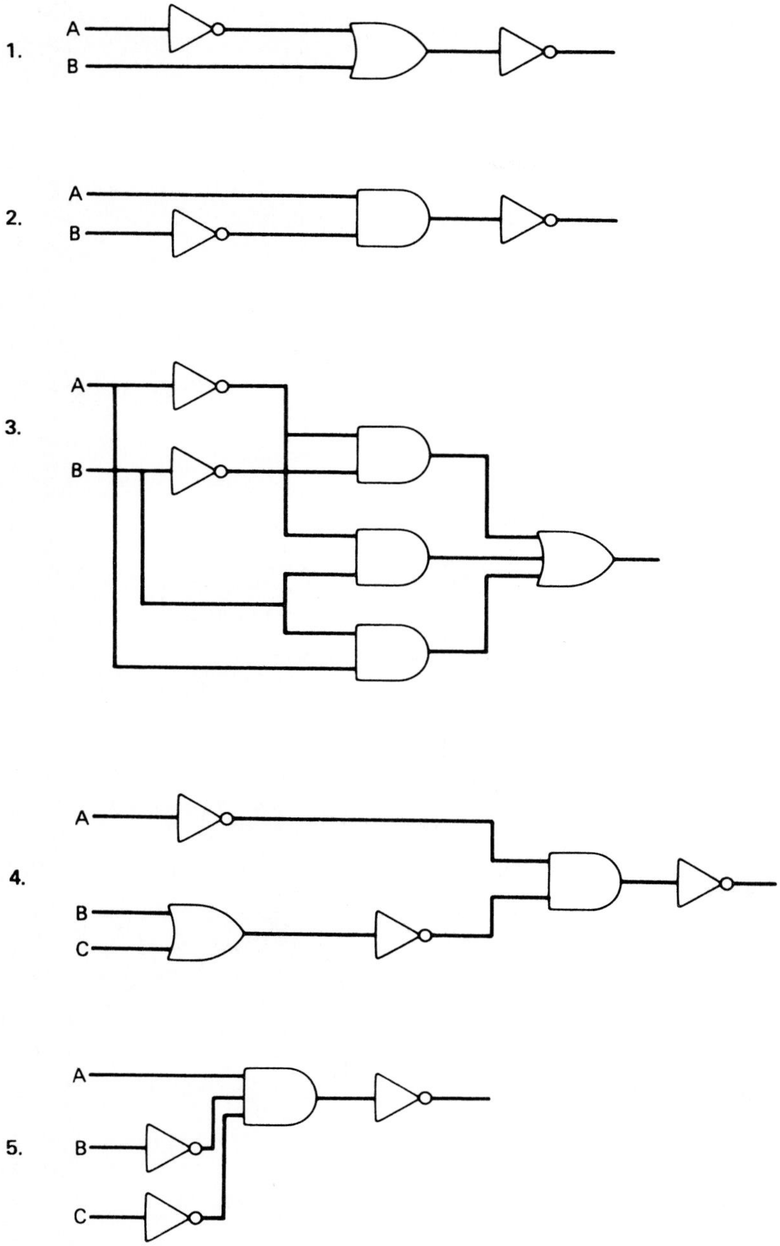

FIGURE 26-48 **Practice Problems 26-4.**

Solutions

1. See Figure 26-49.
2. See Figure 26-50.
3. See Figure 26-51.
4. See Figure 26-52.
5. See Figure 26-53.

$\overline{\overline{A}+B}$	Original expression
$\overline{\overline{A}}\cdot\overline{B}$	14a
$A\overline{B}$	13a

A	B	$\overline{A}$	$\overline{B}$	$\overline{A}+B$	$\overline{\overline{A}+B}$	$A\overline{B}$
0	0	1	1	1	0	0
0	1	1	0	1	0	0
1	0	0	1	0	1	1
1	1	0	0	1	0	0

└Equal┘ (last two columns)

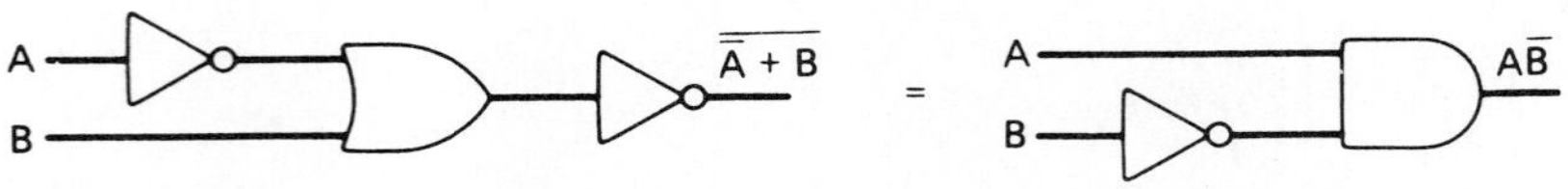

FIGURE 26-49 **Answer to problem 1, Practice Problems 26-4.**

$\overline{A\overline{B}}$	Original expression
$\overline{A}+\overline{\overline{B}}$	14b
$\overline{A}+B$	13a

A	B	$\overline{A}$	$\overline{B}$	$A\overline{B}$	$\overline{A\overline{B}}$	$\overline{A}+B$
0	0	1	1	0	1	1
0	1	1	0	0	1	1
1	0	0	1	1	0	0
1	1	0	0	0	1	1

└Equal┘ (last two columns)

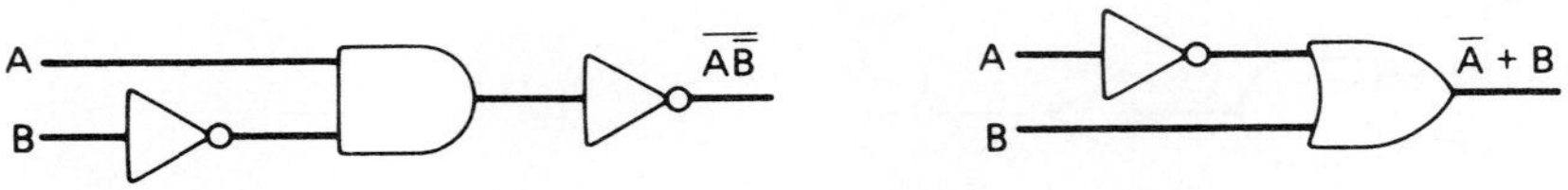

FIGURE 26-50 **Answer to problem 2, Practice Problems 26-4.**

$\overline{A}\,\overline{B}+\overline{A}B+AB$	Original expression
$\overline{A}(\overline{B}+B)+AB$	8a
$\overline{A}(1)+AB$	12b
$\overline{A}+AB$	10a
$\overline{A}+B$	16b

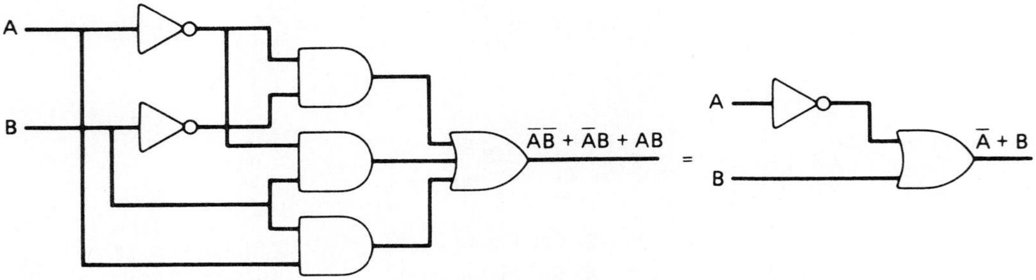

FIGURE 26-51 **Answer to problem 3, Practice Problems 26-4.**

A	B	$\overline{A}$	$\overline{B}$	$\overline{A}\,\overline{B}$	$\overline{A}B$	AB	$\overline{A}\,\overline{B} + \overline{A}B + AB$	$\overline{A} + B$
0	0	1	1	1	0	0	1	1
0	1	1	0	0	1	0	1	1
1	0	0	1	0	0	0	0	0
1	1	0	0	0	0	1	1	1

Equal

FIGURE 26-51 (*cont.*)

$\overline{\overline{A}\,(\overline{B+C})}$	Original expression
$\overline{\overline{A}} + (\overline{\overline{B+C}})$	14b
$A + (B + C)$	13a
$A + B + C$	7b

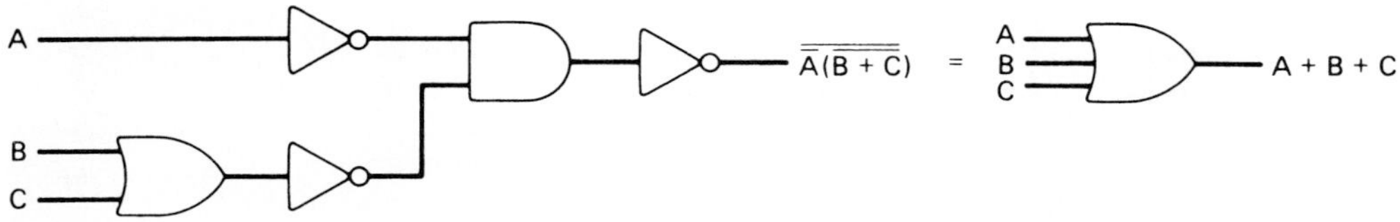

A	B	C	$\overline{A}$	B + C	$\overline{B+C}$	$\overline{A}(\overline{B+C})$	$\overline{\overline{A}\,(\overline{B+C})}$	A + B + C
0	0	0	1	0	1	1	0	0
0	0	1	1	1	0	0	1	1
0	1	0	1	1	0	0	1	1
0	1	1	1	1	0	0	1	1
1	0	0	0	0	1	0	1	1
1	0	1	0	1	0	0	1	1
1	1	0	0	1	0	0	1	1
1	1	1	0	1	0	0	1	1

Equal

FIGURE 26-52 Answer to problem 4, Practice Problems 26-4.

$\overline{A\overline{B}\,\overline{C}}$	Original expression
$\overline{A} + B + C$	14b

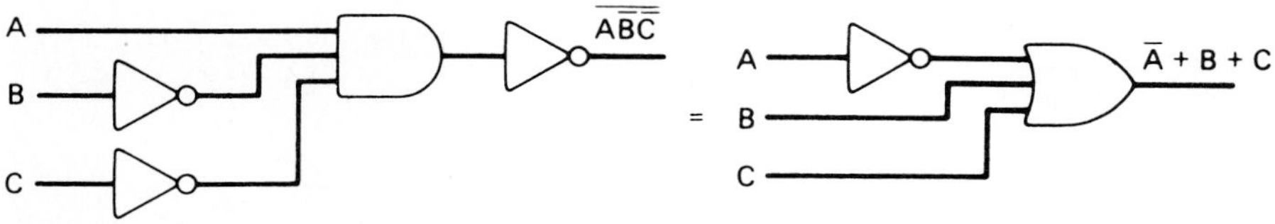

FIGURE 26-53 Answer to problem 5, Practice Problems 26-4.

A	B	C	$\bar{A}$	$\bar{B}$	$\bar{C}$	$A\bar{B}\bar{C}$	$\overline{A\bar{B}\bar{C}}$	$\bar{A}+B+C$
0	0	0	1	1	1	0	1	1
0	0	1	1	1	0	0	1	1
0	1	0	1	0	1	0	1	1
0	1	1	1	0	0	0	1	1
1	0	0	0	1	1	1	0	0
1	0	1	0	1	0	0	1	1
1	1	0	0	0	1	0	1	1
1	1	1	0	0	0	0	1	1

FIGURE 26-53 *(cont.)*

SELF-TEST 26-2

Problems **1** through **5**: For the logic diagrams in Figure 26-54, (a) write the Boolean expression at the output, (b) simplify and justify each step, (c) draw the logic diagram of the simplified expression, and (d) construct a truth table to prove that the expressions are equal.

For the following Boolean expressions, (a) draw the logic diagram, (b) simplify and justify each step, (c) draw the logic diagram of the simplified expression, and (d) construct a truth table to prove that the expressions are equal.

6. $C(\overline{B + D})$

7. $\overline{A\bar{B} + \bar{C}}$

8. $\overline{\bar{A} + \bar{\bar{B}}} + C$

9. $(AB + BC)(AC + BC)$

10. $\overline{\overline{\bar{A}B + A\bar{C}} + \overline{BC}}$

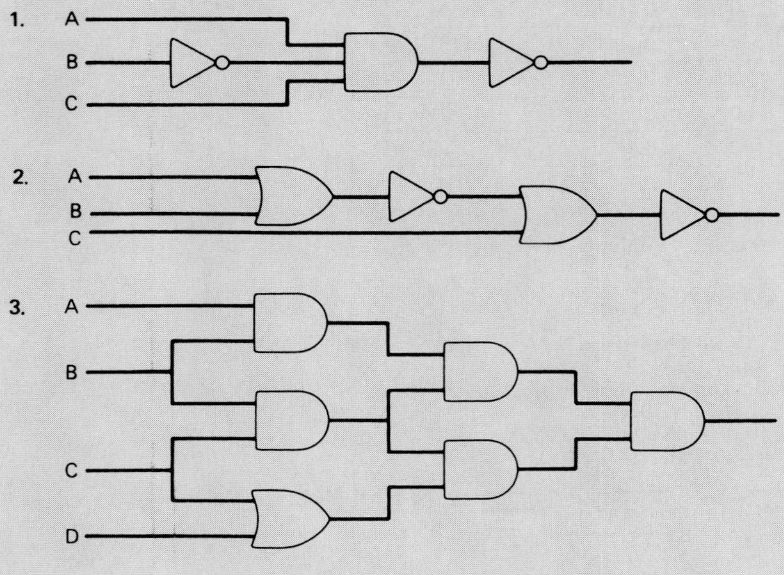

FIGURE 26-54 Problems 1 through 5, Self-test 26-2.

4.

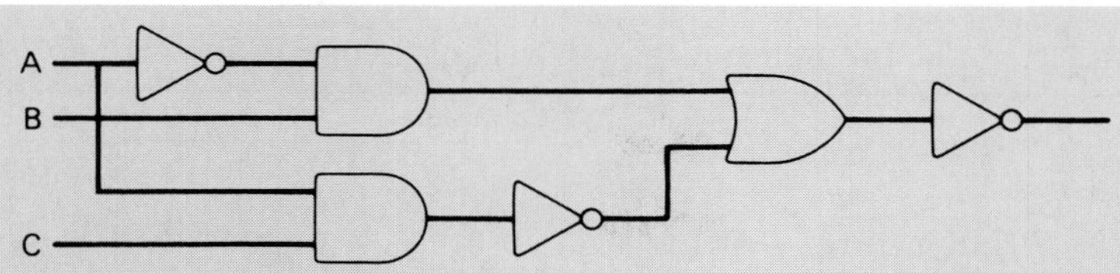

5.

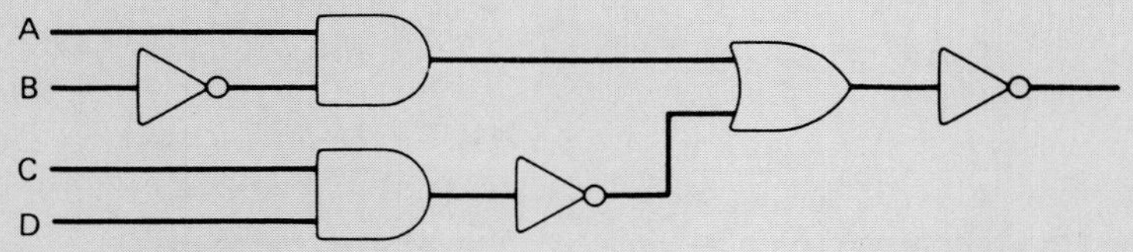

FIGURE 26-54 (*cont.*)

Answers to Self-test 26-2 are at the end of the chapter.

CHAPTER 26 AT A GLANCE

Page	Logic Functions, Postulates, and Theorems	Logic Symbols and Circuits
628	For the AND function, the output is true only when ALL inputs are true. Otherwise, the output is false.	See Figure 26-55 below
630	For the OR function, the output is true when any one of the inputs is true. Otherwise, the output is false.	See Figure 26-56 below

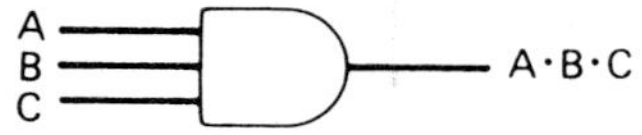

(a)

Condition	A	B	C	A•B•C
1	0	0	0	0
2	0	0	1	0
3	0	1	0	0
4	0	1	1	0
5	1	0	0	0
6	1	0	1	0
7	1	1	0	0
8	1	1	1	1

(b)

FIGURE 26-55 Three-input AND gate: (a) logic symbol; (b) truth table.

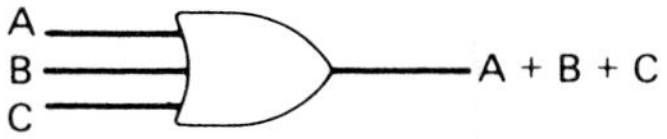

(a)

FIGURE 26-56 Three-input OR gate: (a) logic symbol; (b) truth table.

Page	Logic Functions, Postulates, and Theorems	Logic Symbols and Circuits

Condition	A	B	C	A + B + C
1	0	0	0	0
2	0	0	1	1
3	0	1	0	1
4	0	1	1	1
5	1	0	0	1
6	1	0	1	1
7	1	1	0	1
8	1	1	1	1

(b)

FIGURE 26-56 ***(cont.)***

632 For the NOT function, the output is the complement of the input.

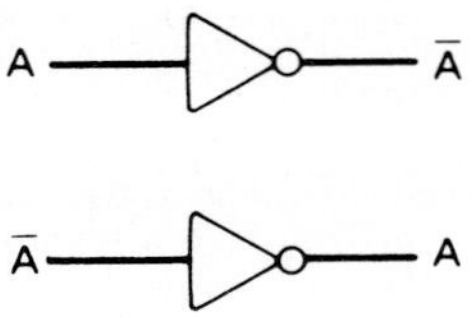

(a)

A	$\overline{A}$
1	0
0	1

(b)

FIGURE 26-57 Inverter: (a) logic symbol; (b) truth table.

637 Postulates, laws, and theorems are used in the simplification and manipulation of logic expressions.

End of Chapter Problems 26-1

Draw the logic diagrams that represent the following Boolean expressions:

1. $\overline{A + \overline{B}}$
2. $A\overline{B} + \overline{AC}$
3. $ABC + D$
4. $X(\overline{Y + Z})$
5. $\overline{A + B\overline{C}}$
6. $\overline{AB} + \overline{A}C$
7. $A\overline{B} + \overline{A}B$
8. $\overline{B}(\overline{C + D})$
9. $\overline{AB} + C + D$
10. $\overline{BC\overline{D} + E}$
11. $AB + A\overline{C} + B\overline{C}$
12. $(\overline{B + C})(C + D)$
13. $\overline{A\overline{B} + BC}$
14. $\overline{\overline{B}(\overline{A} + \overline{C})}$
15. $\overline{A\overline{B}C} + D$
16. $\overline{\overline{A} + B + C}$
17. $(\overline{X + Y})(\overline{Y + Z})$
18. $\overline{\overline{AC} + D}$
19. $\overline{\overline{AB} + C\overline{D}}$
20. $\overline{A\overline{\overline{B}} + C}$

End of Chapter Problems 26-2

Write the Boolean expression represented by the logic diagrams in Figures 26-58 through 26-60.

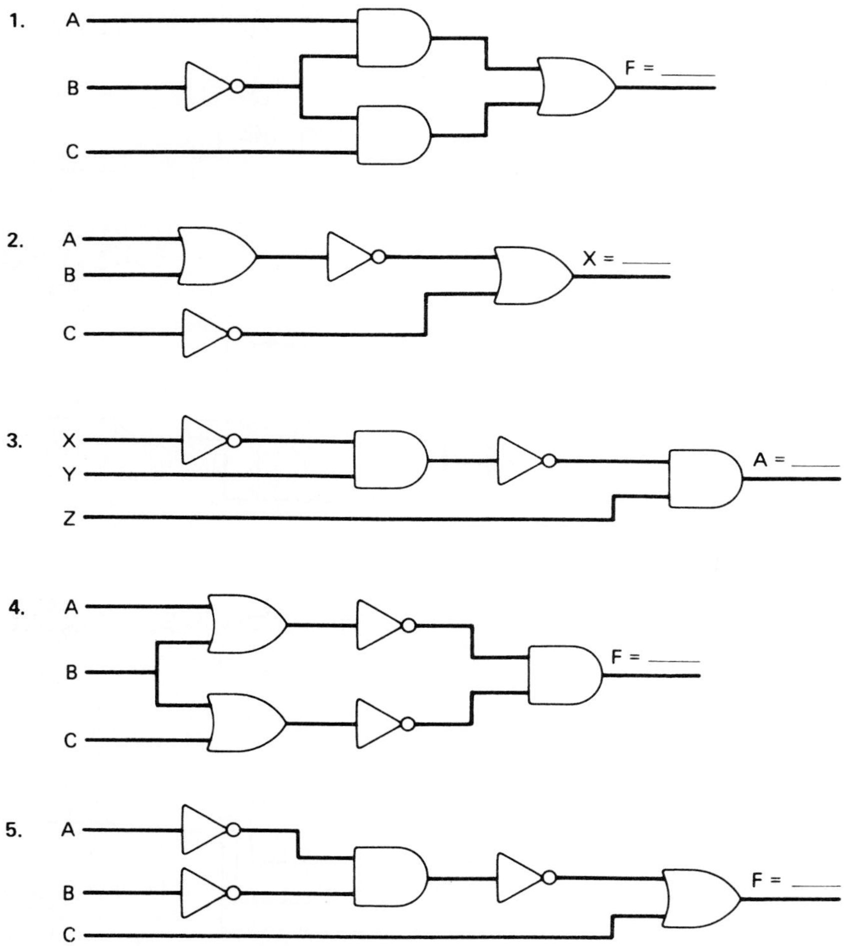

FIGURE 26-58 Logic diagrams for End of Chapter Problems 26-2, problems 1 through 5.

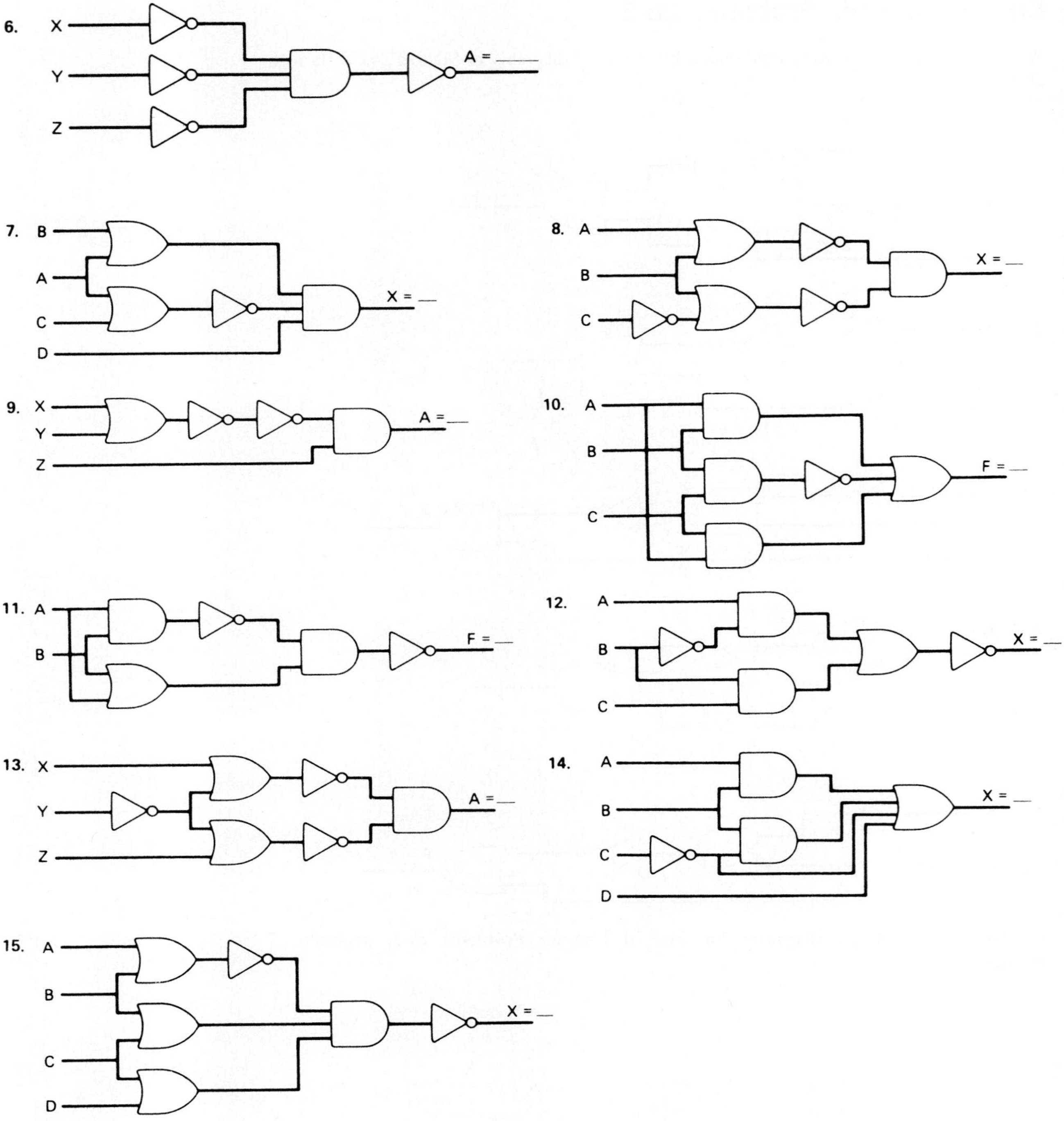

FIGURE 26-59 Logic diagrams for End of Chapter Problems 26-2, problems 6 through 15.

16.

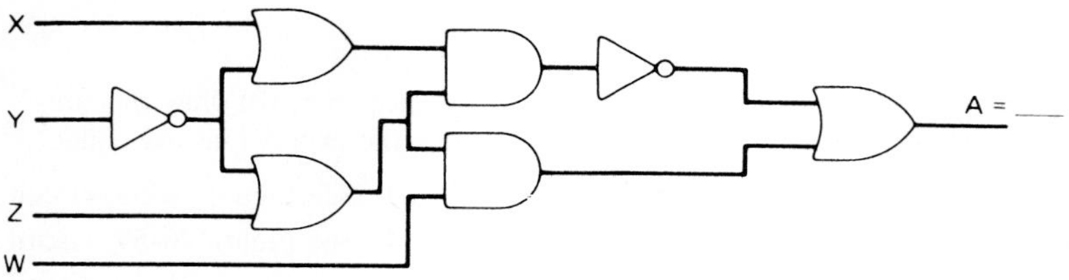

17.

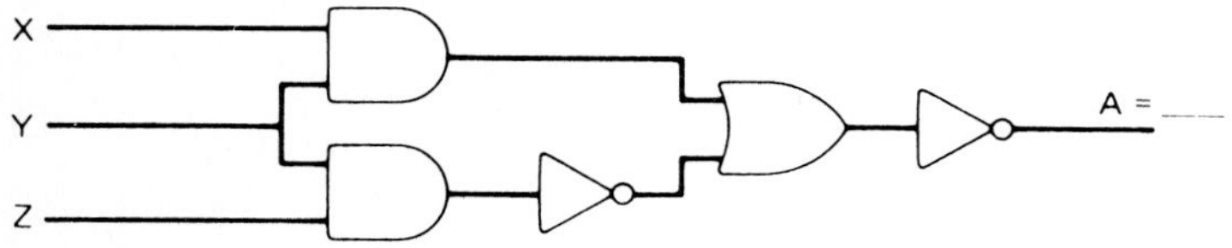

18.

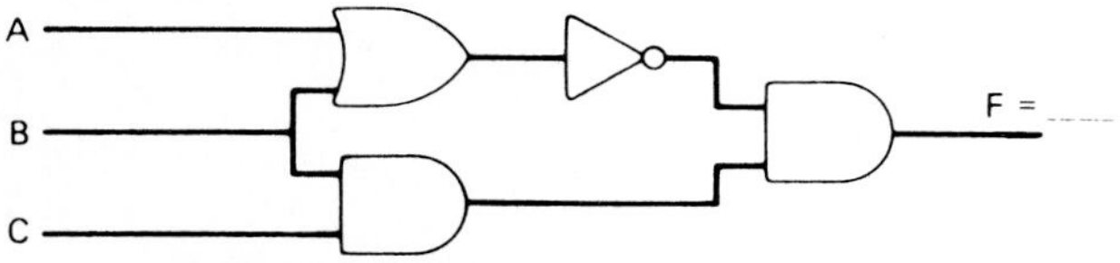

19.

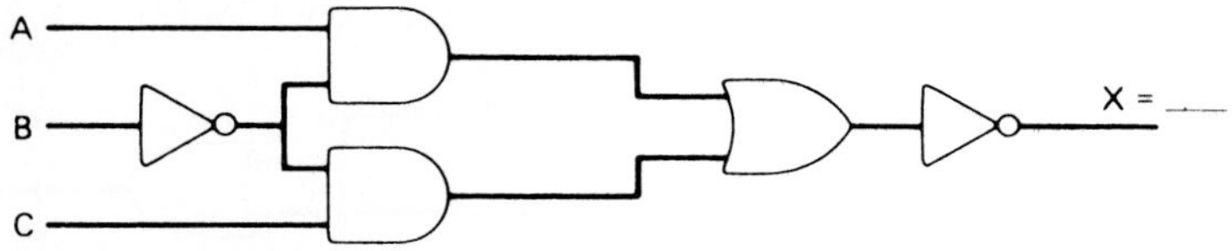

20.

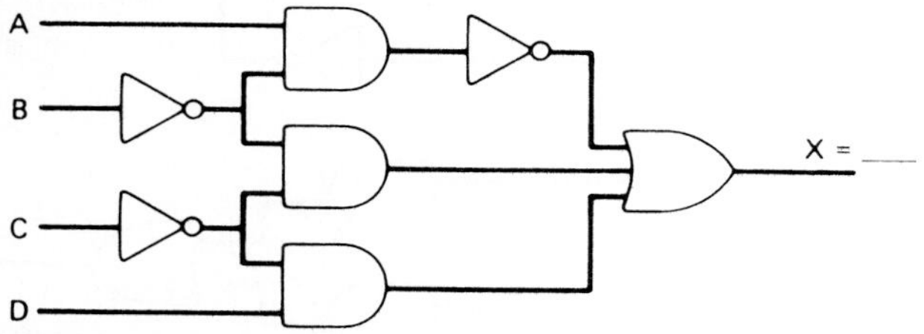

FIGURE 26-60 Logic diagrams for End of Chapter Problems 26-2, problems 16 through 20.

End of Chapter Problems 26-3

In each problem, (a) simplify the output expression and justify each step, (b) draw the simplified logic diagram, and (c) construct a truth table to prove that the expressions are equal.

1. See Figure 26-58, circuit 2.
2. See Figure 26-58, circuit 5.
3. See Figure 26-59, circuit 7.
4. See Figure 26-59, circuit 10.
5. See Figure 26-60, circuit 19.
6. See Figure 26-60, circuit 20.
7. $\overline{A + B\overline{C}}$
8. $\overline{\overline{AB} + \overline{A}C}$
9. $(\overline{B + C})(C + D)$
10. $\overline{\overline{B}(\overline{A} + \overline{C})}$
11. $\overline{\overline{AB} + C\overline{D}}$
12. $\overline{A\overline{\overline{B}} + C}$

Answers to Self-tests

SELF-TEST 26-1

1. $F = \overline{\overline{A} + B}$
2. $F = \overline{A\overline{B}}$
3. $X = A + \overline{B} + C$
4. $X = (A + B)(B + C)$
5. $A = XY + XZ$
6. $\overline{A}\,\overline{B} + C$
7. $F = C(\overline{A + \overline{B}})$
8. $X = (\overline{A + B})(B + C)$
9. $(\overline{X}Y)(Y + Z)$
10. $F = \overline{\overline{A}(\overline{B + C})}$

11 through 20. See Figure 26-61.

SELF-TEST 26-2

1. See Figure 26-62.
2. See Figure 26-63.
3. See Figure 26-64.
4. See Figure 26-65.
5. See Figure 26-66.
6. See Figure 26-67.
7. See Figure 26-68.
8. See Figure 26-69.
9. See Figure 26-70.
10. See Figure 26-71.

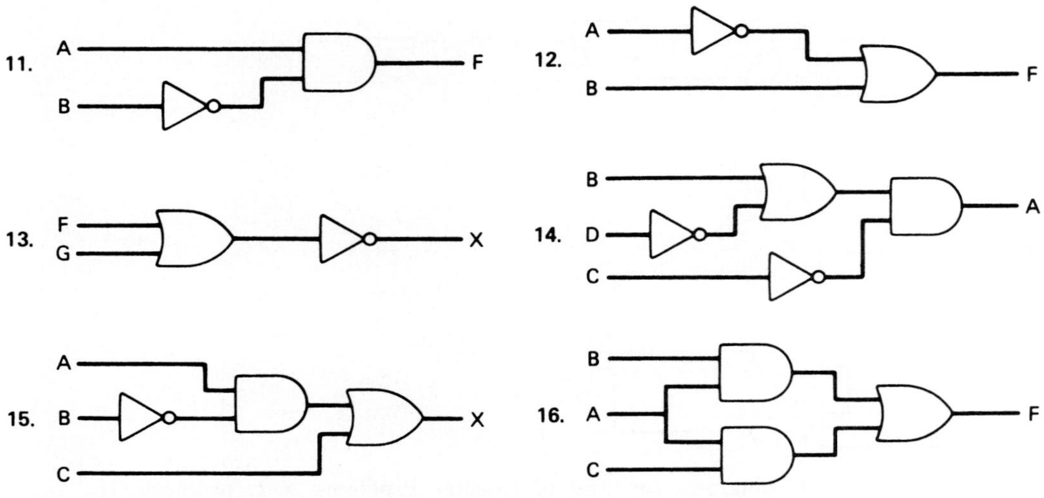

FIGURE 26-61 **Answers to questions 11 through 20 for Self-test 26-1.**

17. 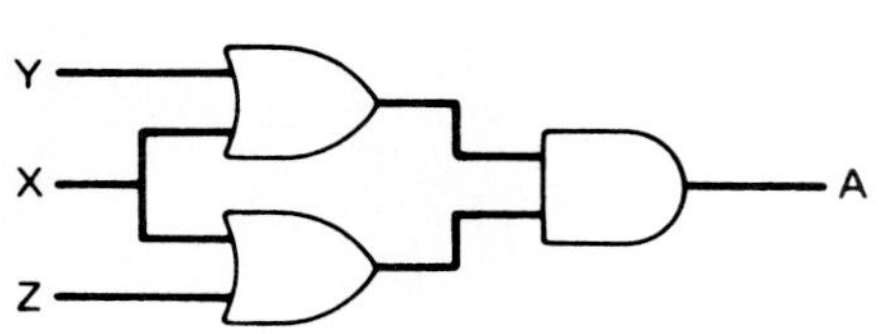

18. 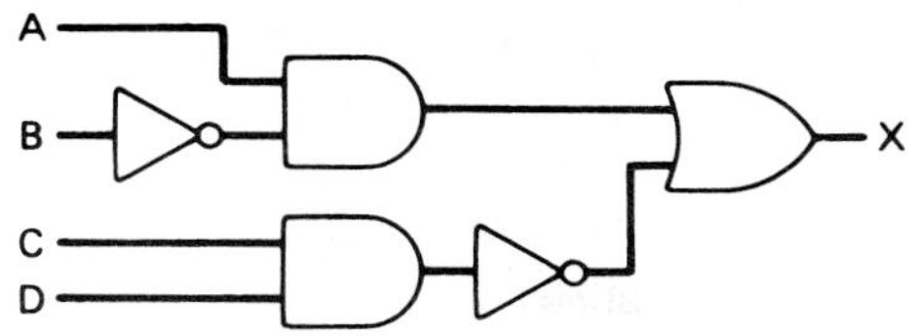

19. 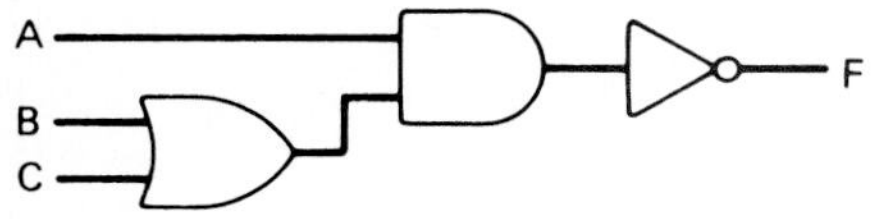

20. 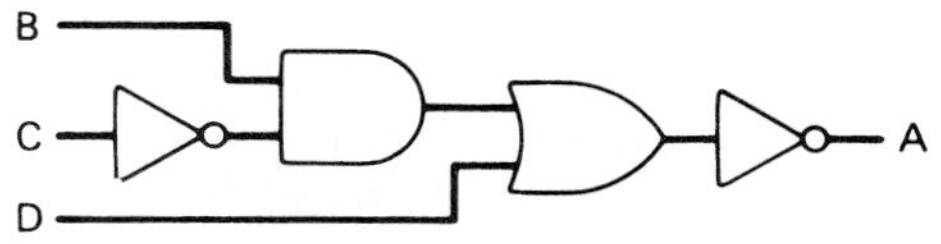

FIGURE 26-61 *(cont.)*

$\overline{A\overline{B}C}$	Original expression
$\overline{A} + B + \overline{C}$	14b

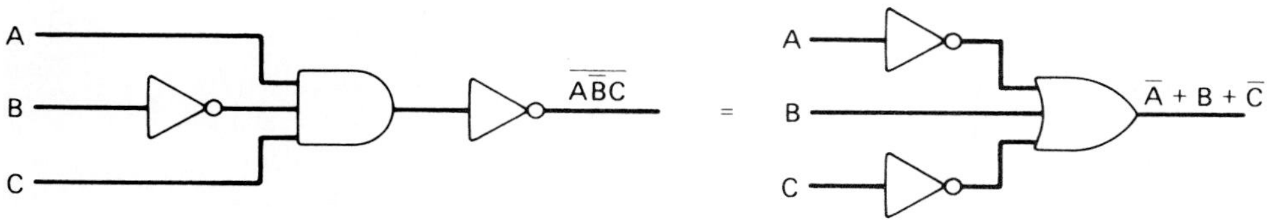

A	B	C	$\overline{A}$	$\overline{B}$	$\overline{C}$	$A\overline{B}C$	$\overline{A\overline{B}C}$	$\overline{A} + B + \overline{C}$
0	0	0	1	1	1	0	1	1
0	0	1	1	1	0	0	1	1
0	1	0	1	0	1	0	1	1
0	1	1	1	0	0	0	1	1
1	0	0	0	1	1	0	1	1
1	0	1	0	1	0	1	0	0
1	1	0	0	0	1	0	1	1
1	1	1	0	0	0	0	1	1

Equal

FIGURE 26-62 Logic diagrams and truth table for $A\overline{B}C = \overline{A} + B + \overline{C}$. Answer to problem 1, Self-test 26-2.

$\overline{\overline{A+B}+C}$	Original expression
$\overline{\overline{\overline{A+B}}}\cdot\overline{C}$	14a
$(A+B)\overline{C}$	13a
$\overline{C}(A+B)$	6a

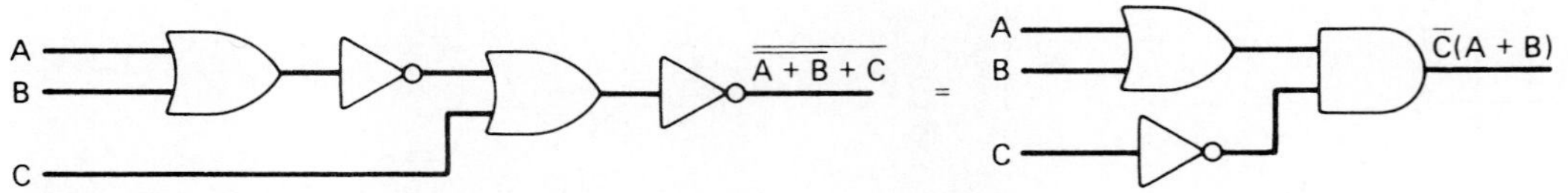

A	B	C	$\overline{C}$	A + B	$\overline{A+B}$	$\overline{A+B}+C$	$\overline{\overline{A+B}+C}$	$\overline{C}(A+B)$
0	0	0	1	0	1	1	0	0
0	0	1	0	0	1	1	0	0
0	1	0	1	1	0	0	1	1
0	1	1	0	1	0	1	0	0
1	0	0	1	1	0	0	1	1
1	0	1	0	1	0	1	0	0
1	1	0	1	1	0	0	1	1
1	1	1	0	1	0	1	0	0

Equal

FIGURE 26-63 Logic diagram and truth table for $\overline{\overline{A+B}+C} = \overline{C}(A+B)$. Answer to problem 2, Self-test 26-2.

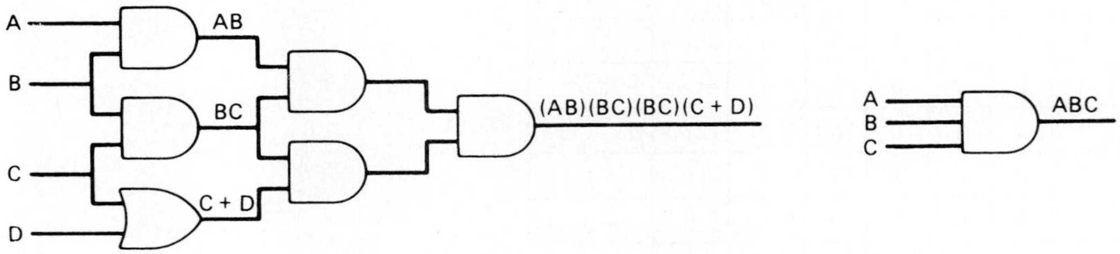

(AB)(BC)(BC)(C + D)	Original expression
(AB)(BC)(C + D)	11b
ABC (C + D)	11a
ABCC + ABCD	8a
ABC + ABCD	11a
ABC(1 + D)	8a
ABC	10b

FIGURE 26-64 Answer to problem 3, Self-test 26-2.

A	B	C	D	AB	BC	C + D	(AB)(BC)	(BC)(C + D)	(AB)(BC)(BC)(C + D)	ABC
0	0	0	0	0	0	0	0	0	0	0
0	0	0	1	0	0	1	0	0	0	0
0	0	1	0	0	0	1	0	0	0	0
0	0	1	1	0	0	1	0	0	0	0
0	1	0	0	0	0	0	0	0	0	0
0	1	0	1	0	0	1	0	0	0	0
0	1	1	0	0	1	1	0	1	0	0
0	1	1	1	0	1	1	0	1	0	0
1	0	0	0	0	0	0	0	0	0	0
1	0	0	1	0	0	1	0	0	0	0
1	0	1	0	0	0	1	0	0	0	0
1	0	1	1	0	0	1	0	0	0	0
1	1	0	0	1	0	0	0	0	0	0
1	1	0	1	1	0	1	0	0	0	0
1	1	1	0	1	1	1	1	1	1	1
1	1	1	1	1	1	1	1	1	1	1

Equal

FIGURE 26-64 *(cont.)*

$\overline{\overline{A}B + \overline{AC}}$	Original expression
$\overline{\overline{A}B} \cdot AC$	14a
$(A + \overline{B})AC$	14b
$AAC + A\overline{B}C$	8a
$AC + A\overline{B}C$	11a
$AC(1 + \overline{B})$	8a
$AC(1)$	10b
AC	10a

A	B	C	$\overline{A}$	$\overline{A}B$	AC	$\overline{AC}$	$\overline{A}B + \overline{AC}$	$\overline{\overline{A}B + \overline{AC}}$
0	0	0	1	0	0	1	1	0
0	0	1	1	0	0	1	1	0
0	1	0	1	1	0	1	1	0
0	1	1	1	1	0	1	1	0
1	0	0	0	0	0	1	1	0
1	0	1	0	0	1	0	0	1
1	1	0	0	0	0	1	1	0
1	1	1	0	0	1	0	0	1

Equal

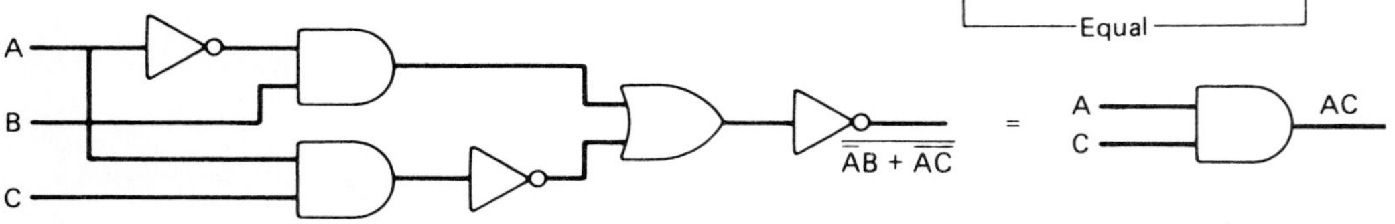

FIGURE 26-65 Answer to problem 4, Self-test 26-2.

$\overline{A\overline{B} + \overline{CD}}$	Original expression
$\overline{A\overline{B}} \cdot \overline{\overline{CD}}$	14a
$\overline{A\overline{B}} \cdot CD$	13a
$(\overline{A} + B)\ CD$	14b
$CD\ (\overline{A} + B)$	7a

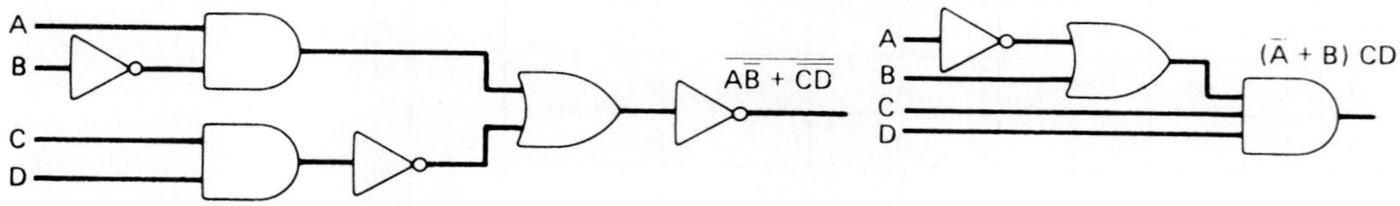

A	B	C	D	$\overline{A}$	$\overline{B}$	CD	$\overline{CD}$	$A\overline{B}$	$A\overline{B} + \overline{CD}$	$\overline{A\overline{B} + \overline{CD}}$	$\overline{A} + B$	$CD(\overline{A} + B)$
0	0	0	0	1	1	0	1	0	1	0	1	0
0	0	0	1	1	1	0	1	0	1	0	1	0
0	0	1	0	1	1	0	1	0	1	0	1	0
0	0	1	1	1	1	1	0	0	0	1	1	1
0	1	0	0	1	0	0	1	0	1	0	1	0
0	1	0	1	1	0	0	1	0	1	0	1	0
0	1	1	0	1	0	0	1	0	1	0	1	0
0	1	1	1	1	0	1	0	0	0	1	1	1
1	0	0	0	0	1	0	1	1	1	0	0	0
1	0	0	1	0	1	0	1	1	1	0	0	0
1	0	1	0	0	1	0	1	1	1	0	0	0
1	0	1	1	0	1	1	0	1	1	0	0	0
1	1	0	0	0	0	0	1	0	1	0	1	0
1	1	0	1	0	0	0	1	0	1	0	1	0
1	1	1	0	0	0	0	1	0	1	0	1	0
1	1	1	1	0	0	1	0	0	0	1	1	1

Equal

FIGURE 26-66 Answer to problem 5, Self-test 26-2.

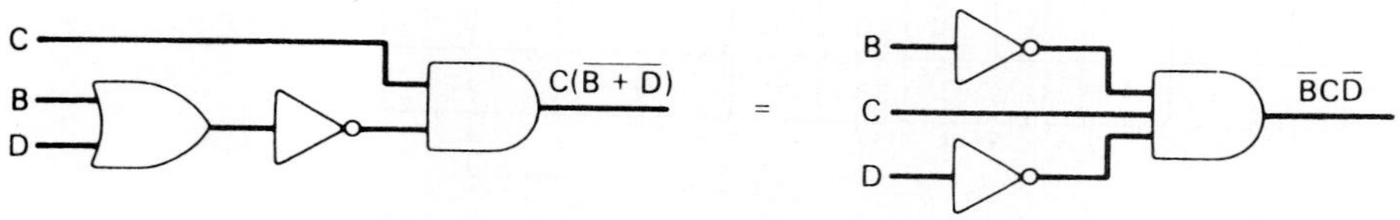

$C(\overline{B + D})$	Original expression
$C(\overline{B}\,\overline{D})$	14a
$C\overline{B}\,\overline{D}$	7a
$\overline{B}C\overline{D}$	6a

FIGURE 26-67 Answer to problem 6, Self-test 26-2.

B	C	D	$\overline{B}$	$\overline{D}$	B + D	$\overline{B + D}$	$C(\overline{B + D})$	$\overline{B}C\overline{D}$
0	0	0	1	1	0	1	0	0
0	0	1	1	0	1	0	0	0
0	1	0	1	1	0	1	1	1
0	1	1	1	0	1	0	0	0
1	0	0	0	1	1	0	0	0
1	0	1	0	0	1	0	0	0
1	1	0	0	1	1	0	0	0
1	1	1	0	0	1	0	0	0

Equal

FIGURE 26-67 *(cont.)*

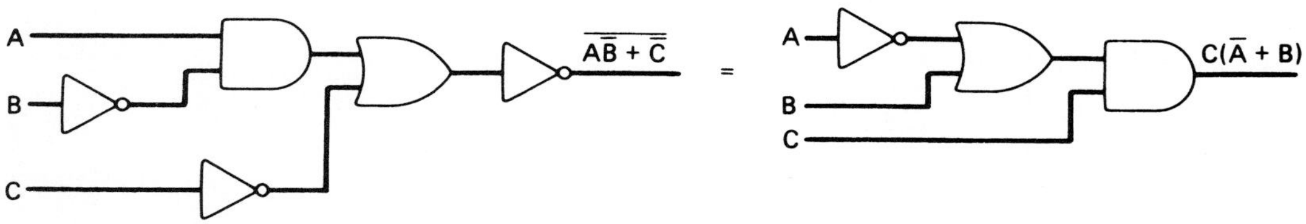

$\overline{A\overline{B} + \overline{C}}$	Original expression
$\overline{A\overline{B}} \cdot \overline{\overline{C}}$	14a
$\overline{A\overline{B}} \cdot C$	13a
$(\overline{A} + \overline{\overline{B}})C$	14b
$(\overline{A} + B)C$	13a
$C(\overline{A} + B)$	7a

A	B	C	$\overline{A}$	$\overline{B}$	$\overline{C}$	$A\overline{B}$	$A\overline{B} + \overline{C}$	$\overline{A\overline{B} + \overline{C}}$	$\overline{A} + B$	$C(\overline{A} + B)$
0	0	0	1	1	1	0	1	0	1	0
0	0	1	1	1	0	0	0	1	1	1
0	1	0	1	0	1	0	1	0	1	0
0	1	1	1	0	0	0	0	1	1	1
1	0	0	0	1	1	1	1	0	0	0
1	0	1	0	1	0	1	1	0	0	0
1	1	0	0	0	1	0	1	0	1	0
1	1	1	0	0	0	0	0	1	1	1

Equal

FIGURE 26-68 **Answer to problem 7, Self-test 26-2.**

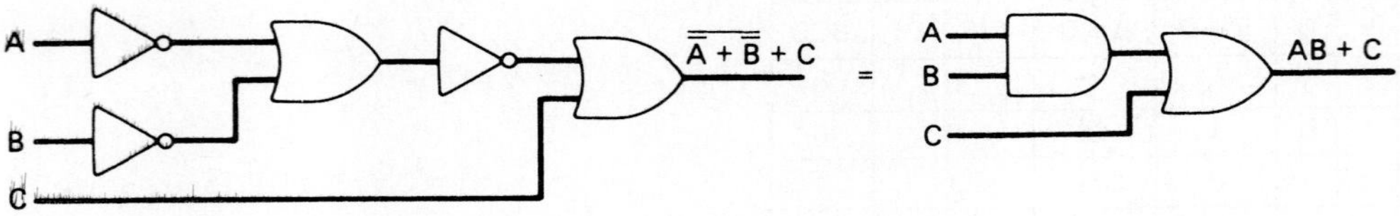

$\overline{\overline{A}+\overline{B}}+C$	Original expression
$\overline{\overline{A}}\cdot\overline{\overline{B}}+C$	14a
AB + C	13a

A	B	C	$\overline{A}$	$\overline{B}$	$\overline{A}+\overline{B}$	$\overline{\overline{A}+\overline{B}}$	$\overline{\overline{A}+\overline{B}}+C$	AB	AB + C
0	0	0	1	1	1	0	0	0	0
0	0	1	1	1	1	0	1	0	1
0	1	0	1	0	1	0	0	0	0
0	1	1	1	0	1	0	1	0	1
1	0	0	0	1	1	0	0	0	0
1	0	1	0	1	1	0	1	0	1
1	1	0	0	0	0	1	1	1	1
1	1	1	0	0	0	1	1	1	1

Equal

FIGURE 26-69 Answer to problem 8, Self-test 26-2.

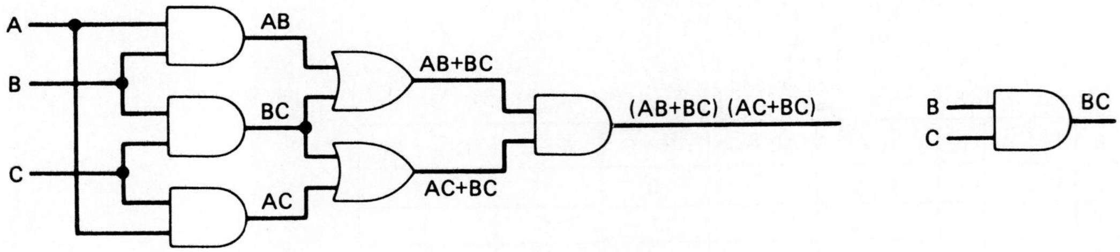

(AB+BC) (AC+BC)	Original expression
(BC+AB) (BC+AC)	6b
BC + (AB) (AC)	8b
BC + AABC	6a
BC + ABC	11a
BC (1+A)	8a
BC(1)	11b
BC	10a

FIGURE 26-70 Answer to problem 9, Self-test 26-2.

A	B	C	AB	BC	AC	AB+BC	AC+BC	(AB+BC)(AC+BC)
0	0	0	0	0	0	0	0	0
0	0	1	0	0	0	0	0	0
0	1	0	0	0	0	0	0	0
0	1	1	0	1	0	1	1	1
1	0	0	0	0	0	0	0	0
1	0	1	0	0	1	0	1	0
1	1	0	1	0	0	1	0	0
1	1	1	1	1	1	1	1	1

FIGURE 26-70 ***(cont.)***

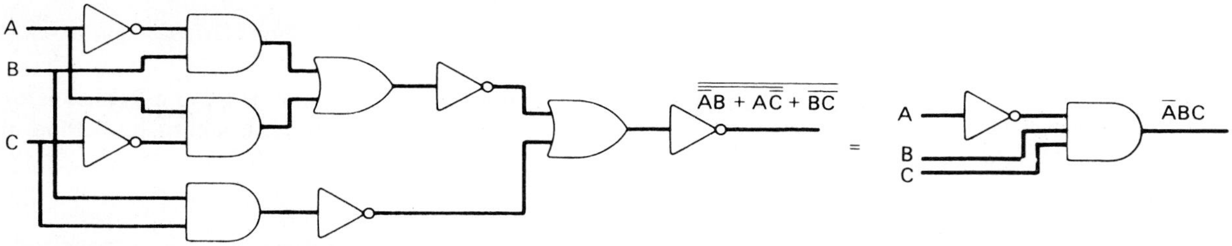

$\overline{\overline{\overline{A}B + A\overline{C}} + \overline{BC}}$	Original expression
$(\overline{A}B + A\overline{C})BC$	14a
$\overline{A}BBC + ABC\overline{C}$	8a
$\overline{A}BC + 0$	11a, 12a
$\overline{A}BC$	9b

A	B	C	$\overline{A}$	$\overline{C}$	$\overline{A}B$	$A\overline{C}$	$\overline{A}B + A\overline{C}$	$\overline{\overline{A}B + A\overline{C}}$	BC	$\overline{BC}$	$\overline{\overline{A}B + A\overline{C}} + \overline{BC}$	$\overline{\overline{\overline{A}B + A\overline{C}} + \overline{BC}}$	$\overline{A}BC$
0	0	0	1	1	0	0	0	1	0	1	1	0	0
0	0	1	1	0	0	0	0	1	0	1	1	0	0
0	1	0	1	1	1	0	1	0	0	1	1	0	0
0	1	1	1	0	1	0	1	0	1	0	0	1	1
1	0	0	0	1	0	1	1	0	0	1	1	0	0
1	0	1	0	0	0	0	0	1	0	1	1	0	0
1	1	0	0	1	0	1	1	0	0	1	1	0	0
1	1	1	0	0	0	0	0	1	1	0	1	0	0

Equal

FIGURE 26-71 Answer to problem 10, Self-test 26-2.

CHAPTER

27 KARNAUGH MAPS

INTRODUCTION

In Chapter 26, you learned how to simplify logic diagrams or logic expressions by using various laws, postulates, and theorems. In this chapter an alternate method of simplification is offered.

CHAPTER OBJECTIVES

In this chapter you will learn how to:

1. Interpret Karnaugh maps.
2. Simplify logic diagrams using Karnaugh maps.

27-1 TWO-VARIABLE EXPRESSION

When digital circuits are first designed, they may be more complex than necessary. By using Boolean algebra, theorems, and laws previously discussed, it is often possible to simplify logic expressions without changing the original logic functions. Simplified expressions are very desirable because they will result in less complex circuits with increased reliability. In addition, the smaller circuits result in increased speeds and greater economy.

Karnaugh maps provide a method by which a Boolean expression can be visualized and, more important, they represent a technique for simplification. A Karnaugh map is made up of a group of squares. The number of squares required depends on the number of variables. A two-variable expression requires four squares; three variables, eight squares; four variables, 16 squares; and so on. The number of squares required is equal to 2 raised to the number of variables. Karnaugh maps are seldom useful for more than five variables.

To begin, consider the truth table in Figure 27-1. The standard truth table for a two-input AND gate is shown in Figure 27-1(a). Note that a true output (1) exists only when A is true AND B is true. The other three outputs are indicated as not true (0).

In Figure 27-1(b), a similar table is produced. In this table, 1 is replaced by the variable, and 0 is replaced by the complement of the variable. The variables are then ANDed. Again, there exist four possible combinations. These are $\overline{A}\,\overline{B} + \overline{A}B + A\overline{B} + AB$. When both tables of Figure 27-1 are combined, the result is the Karnaugh map illustrated in Figure 27-2.

Note that the squares represent each of the four possible conditions for the two variables. The top-left square corresponds to the condition where both A and B are logical 0, and so it is expressed as $\overline{A}\,\overline{B}$. The bottom-right square corresponds to the condition where both A and B equal a logical 1 and therefore can be expressed as AB. In the top-right square, A is equal to a logical 1, but B is equal to a logical 0, and it is expressed as $A\overline{B}$. In the bottom-left square, A is equal to 0 and B is equal to 1, and it is expressed as $\overline{A}B$.

Two Karnaugh maps, each having two variables, appear in Figure 27-3. Note that a 1 appears in the top-right and bottom-right squares of Figure 27-3(a). The 1's indicate the presence of the term that corresponds to that square. The Karnaugh map expression for Figure 27-3(a) is

A	B	AB
0	0	0
0	1	0
1	0	0
1	1	1

(a)

A	B	AB
$\overline{A}$	$\overline{B}$	$\overline{A}\,\overline{B}$
$\overline{A}$	B	$\overline{A}B$
A	$\overline{B}$	$A\overline{B}$
A	B	AB

(b)

FIGURE 27-1

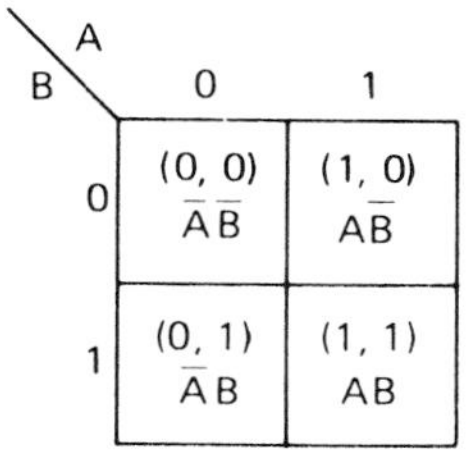

FIGURE 27-2

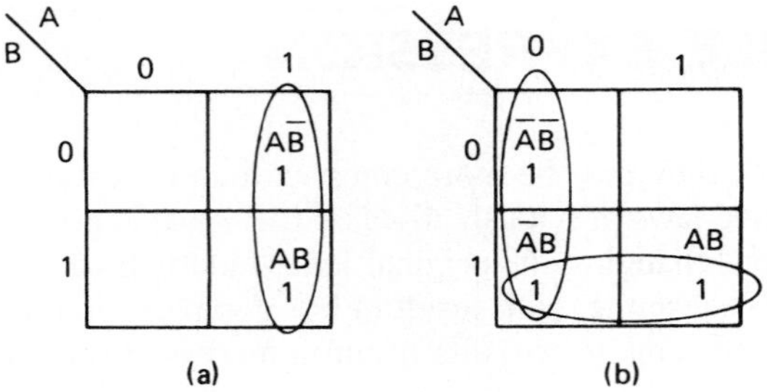

FIGURE 27-3

$$\text{map} = A\overline{B} + AB \qquad \text{(refer to Figure 27-2)}$$

The logical expression for Figure 27-3(b) is

$$\text{map} = \overline{A}\,\overline{B} + \overline{A}B + AB \qquad \text{(refer to Figure 27-2)}$$

The empty squares represent the absence of a term and are omitted to make the map easier to read.

The rules for a two-variable Karnaugh map are as follows:

1. Two adjacent squares are combined to represent a single variable.
2. The resulting variable is common to both terms.

Returning to Figure 27-3(a), note that there are two adjacent 1's that may be combined to represent one variable. This is shown in Figure 27-4.

In the expression $A\overline{B} + AB$, A is the common variable for both terms. Therefore, $A\overline{B} + AB = A$, which is much simpler than the original expression. That this is true may be easily verified through the use of some of the laws and theorems previously discussed.

$A\overline{B} + AB$	original expression
$A(\overline{B} + B)$	theorem 8a
$A(1)$	theorem 12b
A	theorem 10a

Therefore,

$$A\overline{B} + AB = A$$

For Figure 27-3(b), you will note that there are two sets of adjacent 1's, as shown in Figure 27-5. The two *adjacent vertical* 1's give $\overline{A}\,\overline{B} + \overline{A}B$,

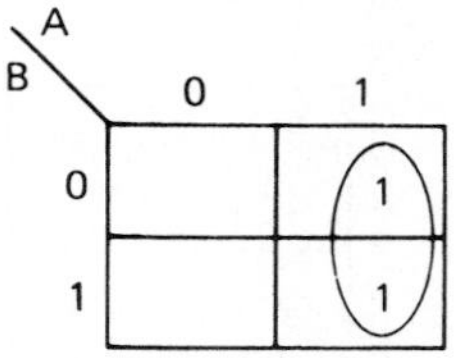

FIGURE 27-4

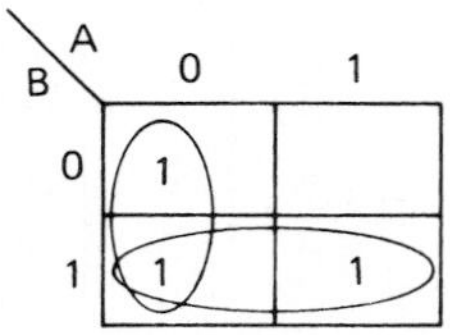

FIGURE 27-5 Karnaugh map showing two sets of adjacent squares. Map = $\overline{A}\,\overline{B} + \overline{A}B + AB$.

which simplifies to $\overline{A}$, because $\overline{A}$ is common to both terms. The two *adjacent horizontal* 1's give $\overline{A}B + AB$, which simplifies to B because B is common to both terms. Therefore,

$$\overline{A}\overline{B} + \overline{A}B + AB = \overline{A} + B$$

You may verify this expression by using previous laws and theorems, as follows:

$\overline{A}\overline{B} + \overline{A}B + AB$	original expression
$\overline{A}(\overline{B} + B) + AB$	theorem 8a
$\overline{A}(1) + AB$	theorem 12b
$\overline{A} + AB$	theorem 10a
$\overline{A} + B$	theorem 16b

Therefore,

$$\overline{A}\overline{B} + \overline{A}B + AB = \overline{A} + B$$

☐ EXAMPLE 27-1

Write the Boolean expression for the map in Figure 27-6.

Solution Map = $\overline{A}B + A\overline{B}$. All other boxes are omitted.

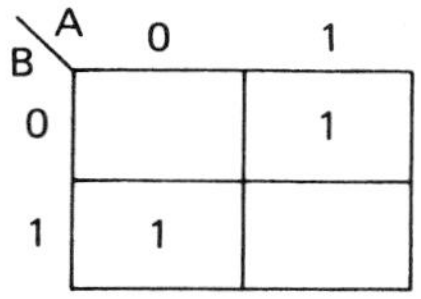

FIGURE 27-6 Karnaugh map for Example 27-1.

☐ EXAMPLE 27-2

Draw the map for the expression $AB + A\overline{B}$.

Solution Place a 1 in each box that represents a given expression. See Figure 27-7 and circle adjacent squares that contain terms. ☐

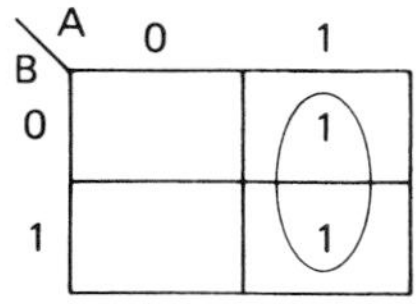

FIGURE 27-7 Karnaugh map of the expression $AB + A\overline{B}$.

27-2 THREE-VARIABLE EXPRESSION

Let's consider the truth table for a three-input AND gate, as shown in Figure 27-8. In the table for 27-8(b), 1's and 0's have been replaced by their respective variables. Note that for three variables there are eight possible combinations. The Karnaugh map for the three variables is shown in Figure 27-9. Each square represents the logical product of the three variables.

The rules for a three-variable Karnaugh map are as follows:

1. A single square represents a three-variable term.
2. Two adjacent squares are combined to represent a two-variable term.
3. Four adjacent squares are combined to represent one variable.

A	B	C	ABC
0	0	0	0
0	0	1	0
0	1	0	0
0	1	1	0
1	0	0	0
1	0	1	0
1	1	0	0
1	1	1	1

(a)

A	B	C	ABC
$\overline{A}$	$\overline{B}$	$\overline{C}$	$\overline{A}\,\overline{B}\,\overline{C}$
$\overline{A}$	$\overline{B}$	C	$\overline{A}\,\overline{B}C$
$\overline{A}$	B	$\overline{C}$	$\overline{A}B\overline{C}$
$\overline{A}$	B	C	$\overline{A}BC$
A	$\overline{B}$	$\overline{C}$	$A\overline{B}\,\overline{C}$
A	$\overline{B}$	C	$A\overline{B}C$
A	B	$\overline{C}$	$AB\overline{C}$
A	B	C	ABC

(b)

FIGURE 27-8 Truth table for three-input AND gate: (a) using 1's and 0's; (b) using variables.

C \ AB	0, 0	0, 1	1, 1	1, 0
0	(0, 0, 0) $\overline{A}\,\overline{B}\,\overline{C}$	(0, 1, 0) $\overline{A}B\overline{C}$	(1, 1, 0) $AB\overline{C}$	(1, 0, 0) $A\overline{B}\,\overline{C}$
1	(0, 0, 1) $\overline{A}\,\overline{B}C$	(0, 1, 1) $\overline{A}BC$	(1, 1, 1) ABC	(1, 0, 1) $A\overline{B}C$

FIGURE 27-9 General form of three-variable Karnaugh map.

Rule 1 for the three-variable Karnaugh map is exemplified in Figure 27-10. The terms in the squares cannot be combined. Therefore,

$$\text{map} = \overline{A}B\overline{C} + A\overline{B}\,\overline{C} + \overline{A}\,\overline{B}C$$

Figure 27-11 illustrates some possible combinations of the two adjacent squares (rule 2). Figure 27-11(a) shows how two end squares may be considered as adjacent to each other if we think of the map as a *wraparound cylinder*.

The map reading for Figure 27-11(a) is $\overline{A}\,\overline{B}\,\overline{C} + A\overline{B}\,\overline{C} = \overline{B}\,\overline{C}$, because $\overline{B}\,\overline{C}$ is common to both squares. Figure 27-11(b) shows two sets of adjacent squares. The two top-left squares read $\overline{A}\,\overline{B}\,\overline{C} + \overline{A}B\overline{C} = \overline{A}\,\overline{C}$; the two bottom-right squares read $ABC + A\overline{B}C = AC$. Therefore, the map reading for Figure 27-11(b) is

$$\text{map} = \overline{A}\,\overline{C} + AC$$

Rule 3 is illustrated by Figure 27-12. In Figure 27-12(a), the four outer boxes

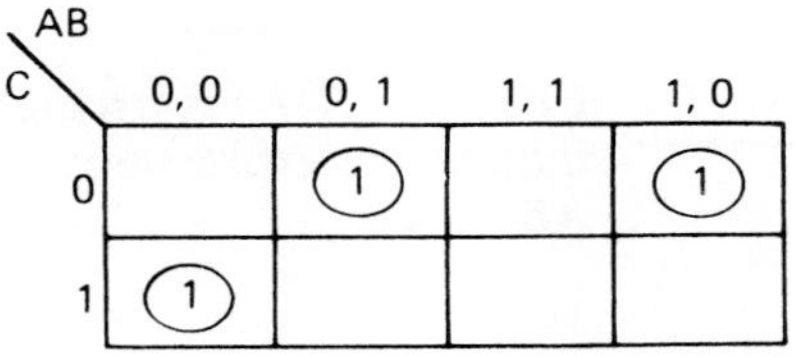

FIGURE 27-10 Single square represents a three-variable term.

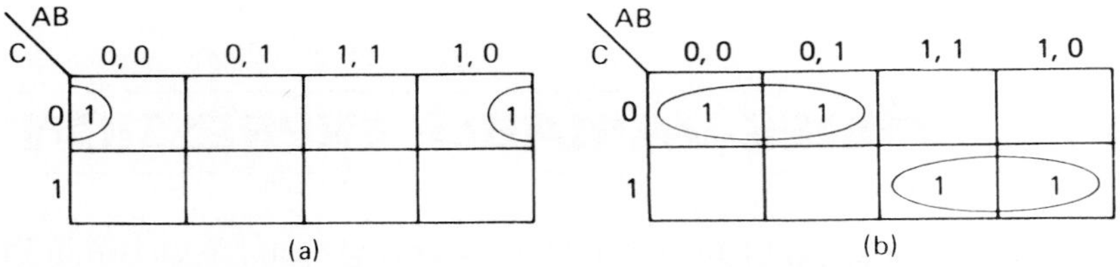

FIGURE 27-11 Two adjacent square combinations: (a) two adjacent squares rule (wraparound cylinder); (b) two adjacent squares rule.

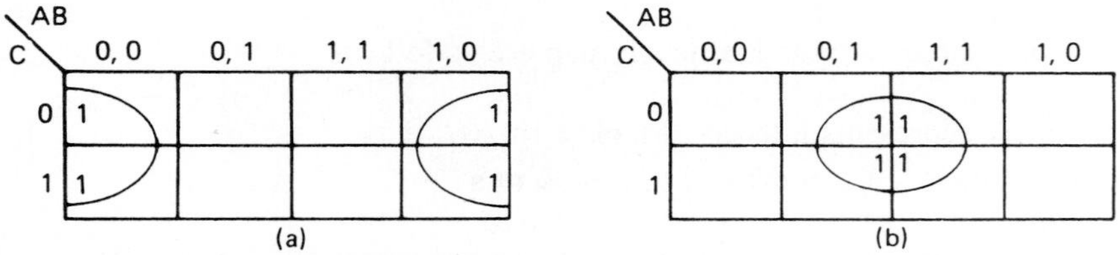

FIGURE 27-12 Four adjacent square combinations: (a) four adjacent squares (wraparound cylinder); (b) four adjacent squares.

may be combined because they are considered to be adjacent to each other (wraparound cylinder). The map reading for Figure 27-12(a) is $\overline{A}\overline{B}\overline{C} + A\overline{B}\overline{C} + \overline{A}\overline{B}C + A\overline{B}C = \overline{B}$. This is because $\overline{B}$ is the common variable for all the terms, which may be verified by the laws and theorems previously discussed in Chapter 26.

$\overline{A}\overline{B}\overline{C} + A\overline{B}\overline{C} + \overline{A}\overline{B}C + A\overline{B}C$	original expression
$(\overline{A}\overline{B}\overline{C} + \overline{A}\overline{B}C) + (A\overline{B}\overline{C} + A\overline{B}C)$	theorem 7b
$\overline{A}\overline{B}(\overline{C} + C) + A\overline{B}(\overline{C} + C)$	theorem 8a
$\overline{A}\overline{B}(1) + A\overline{B}(1)$	theorem 10a
$\overline{A}\overline{B} + A\overline{B}$	theorem 8a
$\overline{B}(\overline{A} + A)$	theorem 8a
$\overline{B}(1)$	theorem 12b
$\overline{B}$	theorem 10a

Figure 27-12(b) shows another possible combination involving four adjacent squares. The map reading for Figure 27-12(b) is

$$\overline{A}B\overline{C} + AB\overline{C} + \overline{A}BC + ABC = B$$

Note that B is the common variable in all the terms, which may be verified as follows:

$\overline{A}B\overline{C} + AB\overline{C} + \overline{A}BC + ABC$	original expression
$\overline{A}B\overline{C} + \overline{A}BC + AB\overline{C} + ABC$	theorem 7b
$\overline{A}B(\overline{C} + C) + AB(\overline{C} + C)$	theorem 8a
$\overline{A}B(1) + AB(1)$	theorem 10a
$\overline{A}B + AB$	theorem 8a
$B(\overline{A} + A)$	theorem 8a
$B(1)$	theorem 12b
B	theorem 10a

□ EXAMPLE 27-3

Write the simplified expression for the Karnaugh map in Figure 27-13.

Solution The four top squares (top row) may be combined to read $\overline{A}\overline{B}\overline{C} + \overline{A}B\overline{C} + AB\overline{C} + A\overline{B}\overline{C} = \overline{C}$, because $\overline{C}$ is common to all the terms. The middle four squares may be combined to read $\overline{A}B\overline{C} + AB\overline{C} + \overline{A}BC + ABC = B$, because B is common to all the terms. Therefore, the resulting expression is

$$\text{map} = B + \overline{C}$$

□

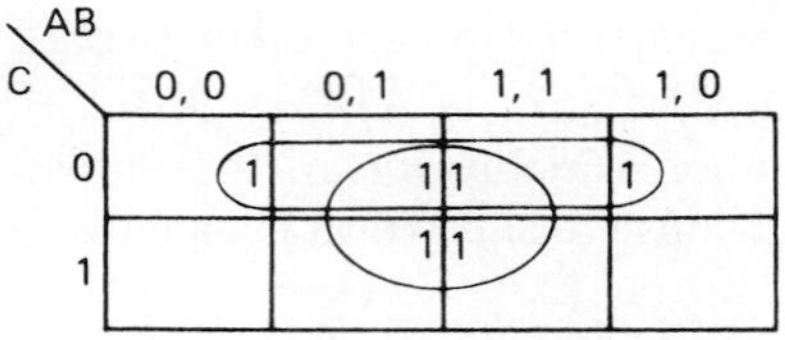

FIGURE 27-13 Karnaugh map solution. Map = B + $\bar{C}$.

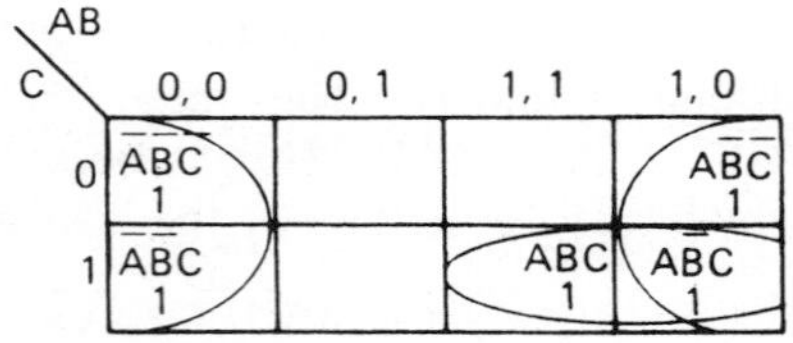

FIGURE 27-14 Karnaugh map solution. Map = $\bar{B}$ + AC.

Complex Boolean expressions may often be simplified by the construction of a Karnaugh map. Suppose we wish to simplify the following logic statement:

$$x = A\bar{B}\bar{C} + ABC + \bar{A}\bar{B}C + A\bar{B}C + \bar{A}\bar{B}\bar{C}$$

Since there are three variables, a Karnaugh map having eight squares is constructed, and 1's are placed in the square corresponding to each term of the expression, as shown in Figure 27-14. Note that the four outer squares may be combined because of the wraparound concept, resulting in

$$\bar{A}\bar{B}\bar{C} + A\bar{B}\bar{C} + \bar{A}\bar{B}C + A\bar{B}C = \bar{B}$$

The two lower-right squares may be combined to give

$$ABC + A\bar{B}C = AC$$

Therefore, the simplified expression is

$$x = \bar{B} + AC$$

☐ EXAMPLE 27-4

Write the Boolean expression that produced the Karnaugh map in Figure 27-15. Write the simplified expression.

Solution For the original expression,

$$\text{map} = \bar{A}\bar{B}\bar{C} + A\bar{B}\bar{C} + \bar{A}\bar{B}C + \bar{A}BC$$

The squares representing $\bar{A}\bar{B}\bar{C}$ and $A\bar{B}\bar{C}$ are adjacent squares. $\bar{B}$ and $\bar{C}$ are common to both, so the terms simplify to $\bar{B}\bar{C}$. The squares representing $\bar{A}\bar{B}C$ and $\bar{A}BC$ are adjacent squares. $\bar{A}$ and C are common to both, so the terms simplify to $\bar{A}C$. The squares representing $\bar{A}\bar{B}\bar{C}$ and $\bar{A}\bar{B}C$ are adjacent squares. $\bar{A}$ and $\bar{B}$ are common to both, so the terms simplify to $\bar{A}\bar{B}$. The simplified expression is

$$\text{map} = \bar{A}\bar{B} + \bar{A}C + \bar{B}\bar{C}$$

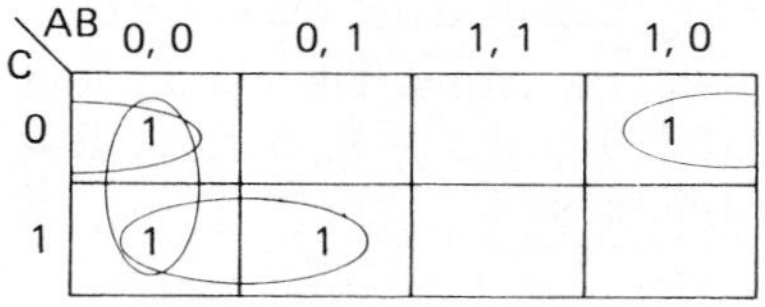

FIGURE 27-15 Karnaugh map for the expression $\bar{A}\bar{B}\bar{C} + A\bar{B}\bar{C} + \bar{A}\bar{B}C + \bar{A}BC$ in Example 27-4.

□ EXAMPLE 27-5

Construct a Karnaugh map and simplify the expression $\overline{A}B\overline{C} + \overline{A}BC + ABC$.

Solution Figure 27-16 is the Karnaugh map. An inspection of the map shows that the squares representing $\overline{A}B\overline{C}$ and $\overline{A}BC$ are adjacent, and $\overline{A}$ and B are common to both. Then $\overline{A}B\overline{C} + \overline{A}BC$ reduces to $\overline{A}B$. The squares representing $\overline{A}BC$ and ABC are adjacent, and B and C are common to both. Then $\overline{A}BC + ABC$ reduces to BC. Therefore,

$$\overline{A}B\overline{C} + \overline{A}BC + ABC = \overline{A}B + BC$$

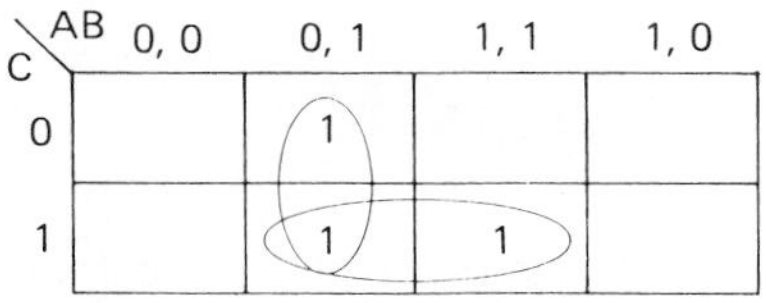

FIGURE 27-16 Karnaugh map for the expression $\overline{A}B\overline{C} + \overline{A}BC + ABC$ in Example 27-5.

□

PRACTICE PROBLEMS 27-1

1. Write the Boolean expression for the Karnaugh maps in Figure 27-17 and simplify the expression.

Draw a Karnaugh map and simplify the following Boolean expressions:

2. $A\overline{B}\overline{C} + \overline{A}\overline{B}C + A\overline{B}C$
3. $\overline{A}\overline{B}\overline{C} + \overline{A}\overline{B}C + A\overline{B}\overline{C} + A\overline{B}C$

Solutions

1. For Figure 27-17(a) the original expression is $\overline{A}\overline{B}\overline{C} + \overline{A}B\overline{C} + \overline{A}BC + ABC$, which reduces to $\overline{A}\overline{C} + \overline{A}B + BC$. There are three adjacent pairs and hence three terms in the answer. For Figure 27-17(b), the original expression is $AB\overline{C} + A\overline{B}\overline{C} + ABC + A\overline{B}C$, which reduces to A. There are four adjacent squares, which reduces to a single variable.
2. See Figure 27-18 for the Karnaugh map. There are two pairs of adjacent squares, so the expression reduces to $A\overline{B} + \overline{B}C$.
3. See Figure 27-19 for the Karnaugh map. There are four adjacent squares, so the expression reduces to a single variable, $\overline{B}$.

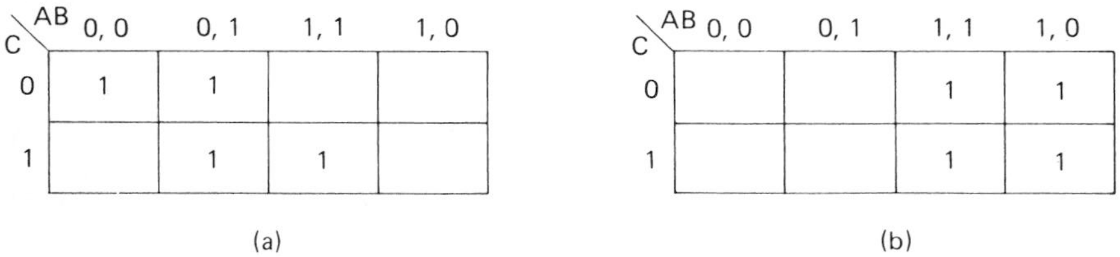

FIGURE 27-17 Karnaugh maps for problem 1, Practice Problems 27-1.

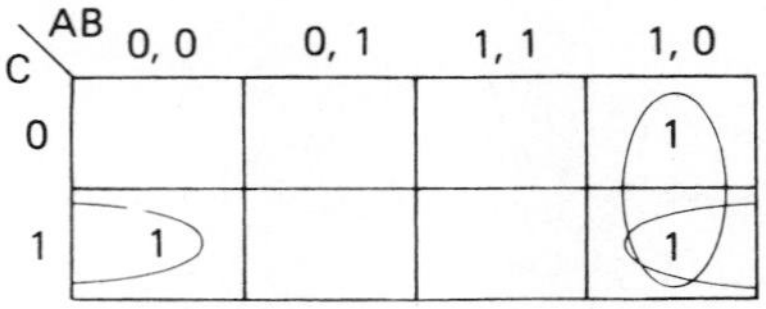

FIGURE 27-18 Karnaugh map for the expression $A\overline{B}\,\overline{C} + \overline{A}\,\overline{B}C + A\overline{B}C$ in problem 2, Practice Problems 27-1.

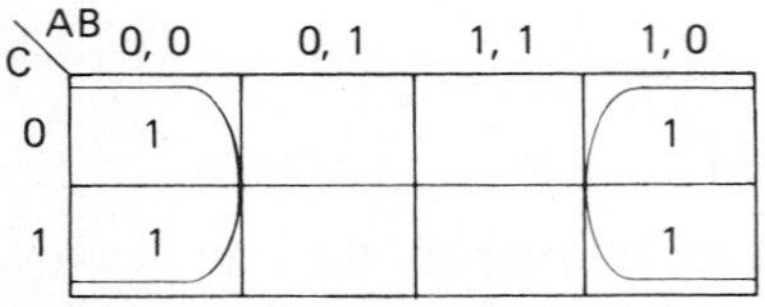

FIGURE 27-19 Karnaugh map for the expression $\overline{A}\,\overline{B}\,\overline{C} + \overline{A}\,\overline{B}C + A\overline{B}\,\overline{C} + A\overline{B}C$ in problem 3, Practice Problems 27-1.

Additional practice problems are at the end of the chapter.

27-3 FOUR-VARIABLE EXPRESSION

A four-variable Karnaugh map is shown in Figure 27-20. Because there are four variables, 16 combinations are possible.

The rules for a four-variable Karnaugh map are as follows:

1. A single square represents a four-variable term.
2. Two adjacent squares represent a three-variable term.
3. Four adjacent squares represent a two-variable term.
4. Eight adjacent squares represent one variable.

CD \ AB	0, 0	0, 1	1, 1	1, 0
0, 0	$\overline{A}\,\overline{B}\,\overline{C}\,\overline{D}$	$\overline{A}B\overline{C}\,\overline{D}$	$AB\overline{C}\,\overline{D}$	$A\overline{B}\,\overline{C}\,\overline{D}$
0, 1	$\overline{A}\,\overline{B}\,\overline{C}D$	$\overline{A}B\overline{C}D$	$AB\overline{C}D$	$A\overline{B}\,\overline{C}D$
1, 1	$\overline{A}\,\overline{B}CD$	$\overline{A}BCD$	$ABCD$	$A\overline{B}CD$
1, 0	$\overline{A}\,\overline{B}C\overline{D}$	$\overline{A}BC\overline{D}$	$ABC\overline{D}$	$A\overline{B}C\overline{D}$

FIGURE 27-20 General form of four-variable Karnaugh map.

□ EXAMPLE 27-6

Simplify the map in Figure 27-21.

Solution Part of the problem is solved by applying rule 4 to the eight adjacent squares that are full.

$$AB\overline{C}\,\overline{D} + A\overline{B}\,\overline{C}\,\overline{D} + AB\overline{C}D + A\overline{B}\,\overline{C}D + ABCD + A\overline{B}CD + ABC\overline{D} + A\overline{B}C\overline{D} = A$$

A is the only variable common to all eight terms. Use rule 3 for the four adjacent squares:

$$\overline{A}\,\overline{B}CD + \overline{A}BCD + ABCD + A\overline{B}CD = CD$$

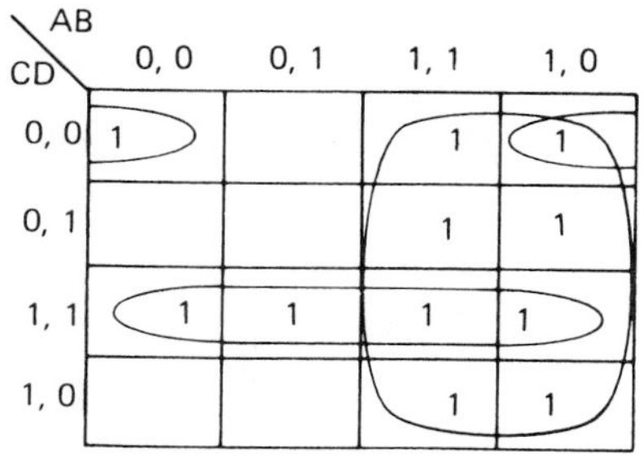

FIGURE 27-21 **Karnaugh map solution. Map = A + CD + $\overline{B}\overline{C}\overline{D}$.**

Variables C and D are common to all four terms. Use rule 2 for the two adjacent squares that are full:

$$\overline{A}\overline{B}\overline{C}\overline{D} + A\overline{B}\overline{C}\overline{D} = \overline{B}\overline{C}\overline{D}$$

Therefore, the map expression for Figure 27-21 is

$$\text{map} = A + CD + \overline{B}\overline{C}\overline{D}$$

□ EXAMPLE 27-7

Simplify the logic circuit in Figure 27-22 using Karnaugh maps.

Solution The terms are placed in the appropriate squares, as shown in Figure 27-23(a). Figure 27-23(b) shows how the two top and two bottom squares are combined by the four-adjacent-square rule. Note that $A\overline{D}$ is common to all terms. Therefore, map = $A\overline{D}$. The simplified logic circuit is shown in Figure 27-24.

□

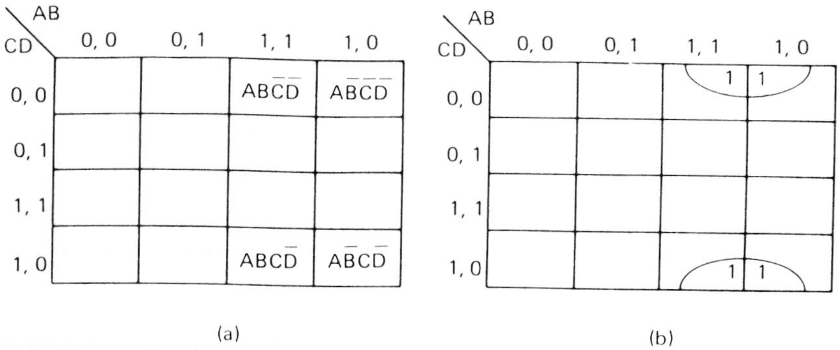

FIGURE 27-23 **Karnaugh map solution for the logic circuit in Figure 27-22: (a) variables placed in appropriate squares; (b) four adjacent square rule applied.**

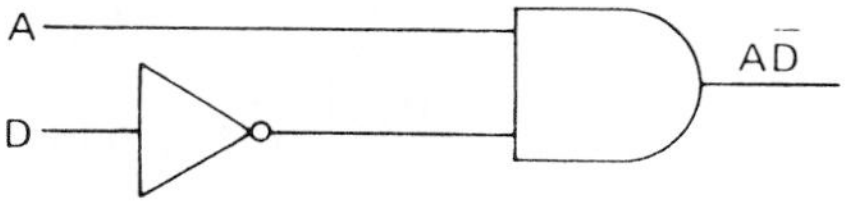

FIGURE 27-24 **Simplified circuit of Figure 27-22.**

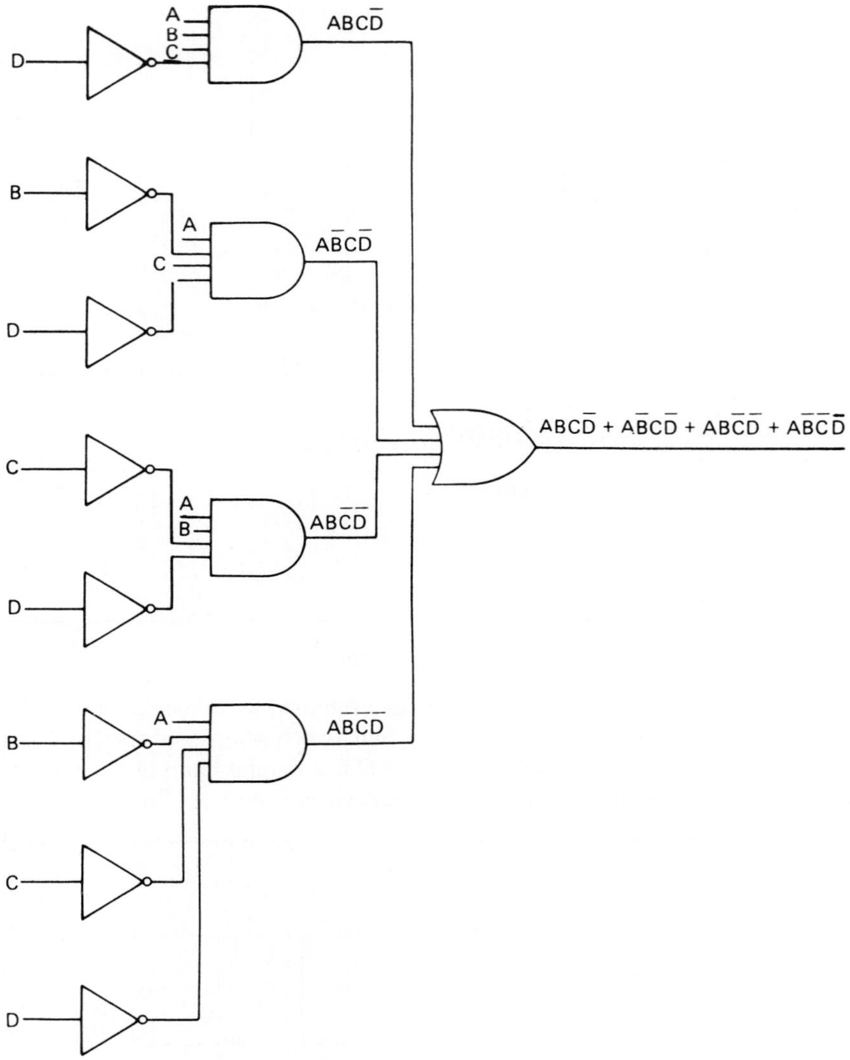

FIGURE 27-22 Logic circuit to be simplified.

PRACTICE PROBLEMS 27-2

Use Karnaugh maps to simplify the following expressions:

1. $AB\overline{C}\overline{D} + A\overline{B}\overline{C}\overline{D} + A\overline{B}CD + \overline{A}\overline{B}CD$
2. $\overline{A}\overline{B}\overline{C}\overline{D} + A\overline{B}\overline{C}D + \overline{A}\overline{B}\overline{C}D + A\overline{B}\overline{C}\overline{D} + A\overline{B}CD + A\overline{B}C\overline{D}$
3. Simplify the Boolean expression represented by the Karnaugh map in Figure 27-25.
4. Simplify the Boolean expression represented by the Karnaugh map in Figure 27-26.

CD \ AB	0, 0	0, 1	1, 1	1, 0
0, 0				
0, 1	1	1	1	1
1, 1		1		
1, 0				

FIGURE 27-25 Karnaugh map for problem 3, Practice Problems 27-2.

CD \ AB	0, 0	0, 1	1, 1	1, 0
0, 0		1		
0, 1				
1, 1				
1, 0	1	1	1	1

FIGURE 27-26 Karnaugh map for problem 4.

Solutions

1. See Figure 27-27. Map = $A\overline{C}\overline{D} + \overline{B}CD$.
2. See Figure 27-28. Map = $\overline{A}\overline{B}\overline{C} + A\overline{B}$.
3. $\overline{A}\overline{B}\overline{C}D + \overline{A}B\overline{C}D + AB\overline{C}D + A\overline{B}\overline{C}D$
   $= C\overline{D}$ (rule 3)
   $\overline{A}BC\overline{D} + \overline{A}BC\overline{D} = \overline{A}BD$ (rule 2)
   map = $\overline{A}BD + \overline{C}D$
4. $\overline{A}\overline{B}C\overline{D} + \overline{A}BC\overline{D} + ABC\overline{D} + A\overline{B}C\overline{D}$
   $= C\overline{D}$ (rule 3)
   $\overline{A}BC\overline{D} + \overline{A}B\overline{C}\overline{D} = \overline{A}B\overline{D}$ (rule 2)
   map = $\overline{A}B\overline{D} + C\overline{D}$

CD \ AB	0, 0	0, 1	1, 1	1, 0
0, 0			1	1
0, 1				
1, 1	1			1
1, 0				

FIGURE 27-27 Karnaugh map for problem 1.

CD \ AB	0, 0	0, 1	1, 1	1, 0
0, 0	1			1
0, 1	1			1
1, 1				1
1, 0				1

FIGURE 27-28 Karnaugh map for problem 2.

Additional practice problems are at the end of the chapter.

SELF-TEST 27-1

1. Simplify $AB\overline{C} + ABC + A\overline{B}C$ using Karnaugh maps.
2. Simplify the Boolean expression represented by the Karnaugh map in Figure 27-29.
3. Simplify $A\overline{B}\overline{C}\overline{D} + A\overline{B}\overline{C}D + A\overline{B}CD + A\overline{B}C\overline{D} + \overline{A}\overline{B}CD$ using Karnaugh maps.
4. Simplify the Boolean expression represented by the Karnaugh map in Figure 27-30.

C \ AB	0, 0	0, 1	1, 1	1, 0
0	1			1
1	1			

FIGURE 27-29 Karnaugh map for Self-test 27-1, problem 2.

CD \ AB	0, 0	0, 1	1, 1	1, 0
0, 0			1	
0, 1			1	
1, 1	1		1	
1, 0	1		1	

FIGURE 27-30 Karnaugh map for Self-test 27-1, problem 4.

Answers to Self-test 27-1 are at the end of the chapter.

CHAPTER 27 AT A GLANCE

Page	Rules	Example
672	The rules for a two-variable Karnaugh map are as follows: **1.** Two adjacent squares are combined to represent a single variable. **2.** The resulting variable is common to both terms.	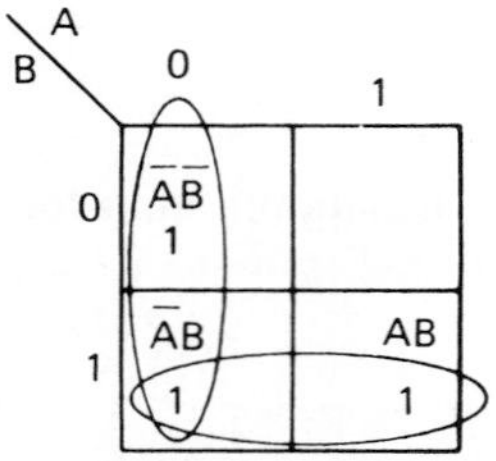 **FIGURE 27-31** $\overline{A}\overline{B} + \overline{A}B + AB = A + B$
673	The rules for a three-variable Karnaugh map are as follows: **1.** A single square represents a three-variable term. **2.** Two adjacent squares are combined to represent a two-variable term. **3.** Four adjacent squares are combined to represent one variable.	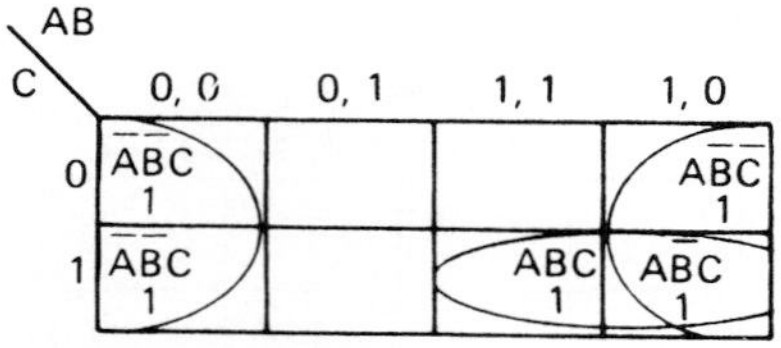 **FIGURE 27-32** $\overline{A}\overline{B}\overline{C} + A\overline{B}\overline{C} + \overline{A}\overline{B}C + A\overline{B}C$ $= \overline{B}$ $ABC + ABC = AC$ map $= \overline{B} + AC$

End of Chapter Problems 27-1

Simplify the following Boolean expressions using Karnaugh maps.

1. $\overline{A}\overline{B} + A\overline{B}$
2. $AB + A\overline{B}$
3. $\overline{A}\overline{B}\overline{C} + \overline{A}B\overline{C} + \overline{A}BC$
4. $AB\overline{C} + ABC + A\overline{B}C$
5. $\overline{A}\overline{B}\overline{C} + \overline{A}B\overline{C} + AB\overline{C} + A\overline{B}\overline{C}$
6. $\overline{A}\overline{B}C + \overline{A}BC + ABC + A\overline{B}C$
7. $\overline{A}B\overline{C} + AB\overline{C} + \overline{A}\overline{B}C + A\overline{B}C$
8. $\overline{A}\overline{B}\overline{C} + A\overline{B}\overline{C} + \overline{A}BC + ABC$
9. Simplify the Boolean expression represented by the Karnaugh map in Figure 27-33.
10. Simplify the Boolean expression represented by the Karnaugh map in Figure 27-34.

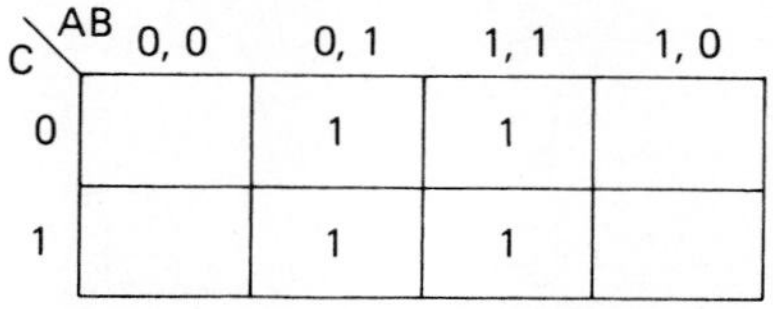

FIGURE 27-33 Karnaugh map for problem 9 of Problems 27-1.

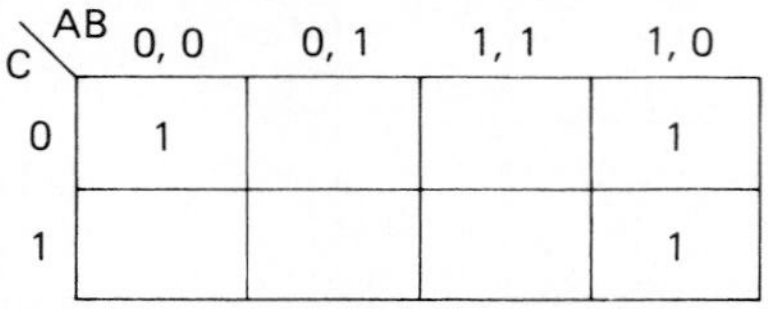

FIGURE 27-34 Karnaugh map for problem 10 of Problems 27-1.

11. Simplify the Boolean expression represented by the Karnaugh map in Figure 27-35.

C \ AB	0, 0	0, 1	1, 1	1, 0
0	1	1	1	1
1		1		

FIGURE 27-35 Karnaugh map for problem 11 of Problems 27-1.

12. Simplify the Boolean expression represented by the Karnaugh map in Figure 27-36.

C \ AB	0, 0	0, 1	1, 1	1, 0
0		1	1	
1	1			1

FIGURE 27-36 Karnaugh map for problem 12 of Problems 27-1.

End of Chapter Problems 27-2

Simplify the following Boolean expressions using Karnaugh maps.

1. $\bar{A}B\bar{C}\bar{D} + \bar{A}B\bar{C}D + AB\bar{C}D + ABCD$
2. $\bar{A}\bar{B}C\bar{D} + \bar{A}B\bar{C}\bar{D} + AB\bar{C}\bar{D} + A\bar{B}C\bar{D}$
3. $\bar{A}\bar{B}\bar{C}\bar{D} + \bar{A}\bar{B}\bar{C}D + \bar{A}B\bar{C}D + AB\bar{C}D +$ $A\bar{B}\bar{C}D$
4. $\bar{A}\bar{B}CD + \bar{A}BCD + ABCD + A\bar{B}CD + \bar{A}BC\bar{D}$
5. $\bar{A}\bar{B}\bar{C}\bar{D} + \bar{A}B\bar{C}\bar{D} + AB\bar{C}\bar{D} + A\bar{B}\bar{C}\bar{D} +$ $\bar{A}\bar{B}C\bar{D} + \bar{A}BC\bar{D} + ABC\bar{D} + A\bar{B}C\bar{D} + ABCD$
6. $\bar{A}\bar{B}\bar{C}\bar{D} + \bar{A}\bar{B}\bar{C}D + \bar{A}\bar{B}CD + \bar{A}\bar{B}C\bar{D} +$ $\bar{A}B\bar{C}\bar{D} + \bar{A}B\bar{C}D + \bar{A}BCD + \bar{A}BC\bar{D} +$ $AB\bar{C}D + A\bar{B}\bar{C}D$
7. Simplify the Boolean expression represented by the Karnaugh map in Figure 27-37.
8. Simplify the Boolean expression represented by the Karnaugh map in Figure 27-38.
9. Simplify the Boolean expression represented by the Karnaugh map in Figure 27-39.
10. Simplify the Boolean expression represented by the Karnaugh map in Figure 27-40.

CD \ AB	0, 0	0, 1	1, 1	1, 0
0, 0				
0, 1	1	1	1	1
1, 1	1	1	1	1
1, 0				

FIGURE 27-37 Karnaugh map for problem 7 of Problems 27-2.

CD \ AB	0, 0	0, 1	1, 1	1, 0
0, 0	1	1		
0, 1	1	1		
1, 1			1	1
1, 0			1	1

FIGURE 27-38 Karnaugh map for problem 8 of Problems 27-2.

CD \ AB	0, 0	0, 1	1, 1	1, 0
0, 0		1	1	
0, 1		1	1	
1, 1	1			1
1, 0				

FIGURE 27-39 Karnaugh map for problem 9 of Problems 27-2.

CD \ AB	0, 0	0, 1	1, 1	1, 0
0, 0				
0, 1	1			1
1, 1		1	1	
1, 0		1	1	

FIGURE 27-40 Karnaugh map for problem 10 of Problems 27-2.

Answers to Self-test

SELF-TEST 27-1

1. See Figure 27-41. Map = $AB + AC$.
2. Map = $\overline{A}\overline{B} + \overline{B}\overline{C}$.
3. See Figure 27-42. Map = $\overline{B}CD + A\overline{B}$.
4. Map = $\overline{A}\overline{B}C + AB$.

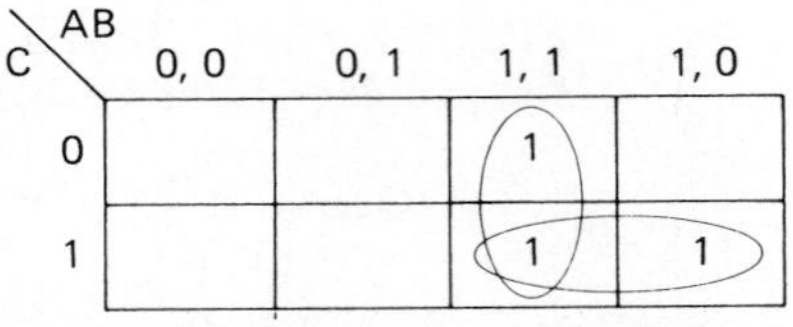

FIGURE 27-41 Karnaugh map for the expression $AB\overline{C} + ABC + A\overline{B}C$.

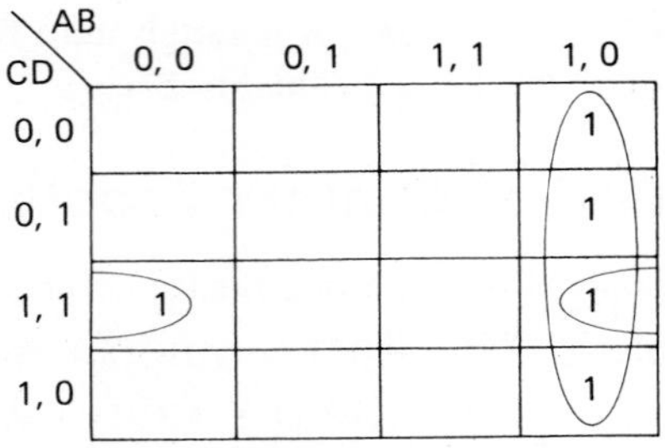

FIGURE 27-42 Karnaugh map for the expression $A\overline{B}\overline{C}\overline{D} + A\overline{B}\overline{C}D + A\overline{B}CD + A\overline{B}C\overline{D} + \overline{A}\overline{B}CD$.

APPENDIX A

THE CALCULATOR

There are many makes and models of calculator on the market today. Many full-function calculators are available for less than \$30. In addition to the basic functions that all calculators perform, you will need a calculator that includes trigonometric and logarithmic functions and displays the constant π (pi). It should also perform polar/rectangular conversions and be capable of performing basic functions in other number systems such as binary and octal. Such calculators are called *full-function* calculators. In addition to the functions mentioned above, full-function calculators will perform many other functions not mentioned here. Most full-function calculators will FIX the number of digits displayed to the right of the decimal point and will display answers in scientific (SCI) and engineering (ENG) notation.

All calculators handle basic functions the same way (except for Hewlett-Packard calculators) but may require different keystrokes to change from standard notation to scientific or engineering notation and to make other more advanced computations. All full-function calculators come with an instruction manual that provides step by step examples and sample problems. Many full-function calculators today are solar powered so there is never a need to replace batteries. Ask your instructor and advanced students for advice as to what calculators they like for the work you will be doing.

In this section of the text, we will look at the internal circuitry of a calculator to see how it works. An understanding of the internal operations of the calculator should help you better understand how to use it.

Internal to the calculator, there are two places where data are stored (there are probably others, but let's keep it simple). These places are digital circuits called *registers*. These registers are called the *x-register* and the *y-register*. The contents of the x-register are displayed. You cannot see the contents of the y-register.

Let's multiply 3×4 using the TI-36. Here is what you do: When first using the calculator, press [AC] to make sure that the calculator is cleared. Then:

1. Key in the 3.
2. Press the [×] key.
3. Key in the 4.
4. Press the [=] key. *Answer:* $3 \times 4 = 12$.

3 x-register

? y-register

FIGURE A-1

3 x-register

3 y-register

FIGURE A-2

4 x-register

3 y-register

FIGURE A-3

12 x-register

4 y-register

FIGURE A-4

When you went through these four steps, here is what happened internally in the calculator:

1. *Key in the 3*. The three is stored in the x-register and is displayed. See Figure A-1.
2. *Press the* [×] *key*. The 3 is copied into the y-register. Both the x- and y-registers now store the number 3. See Figure A-2. At the same time, the calculator is given the message "get ready to multiply the contents of the x-register by the contents of the y-register."
3. *Key in the 4*. The contents of the x-register are replaced with the number 4. The y-register is not affected. At this point, the x-register stores 4 and the y-register stores 3. See Figure A-3.
4. Press the [=] *key*. In step 2 we told the calculator to get ready to multiply the contents of the x-register by the contents of the y-register. Pressing the [=] key says "Do it!" At this point the calculator performs the multiply function and moves the product (12) into the x-register. The old contents of the x-register (4) go to the y-register and the original contents of the y-register (3) are lost. See Figure A-4. Remember, the calculator always displays the contents of the x-register.

Let's try a different one. Let's raise 4 to the third power. $4^3 = ?$. Here is the algorithm:

1. Key in the base, which is 4. 4 is stored in the x-register and is displayed.
2. Press [y^x]. 4 is moved into the y-register, and the calculator is instructed to get ready to raise the contents of the y-register (which is the base, 4) to some power. This exponent will be keyed into the x-register.
3. Key in the exponent, which is 3. The contents of the x-register (4) are replaced with the number 3. Now the y-register contains the base (4), and the x-register contains the exponent (3).
4. Press the [=] key. Pressing the equals key tells the calculator to raise the contents of the y-register (4) to the third power. As always, the result (64) is placed into the x-register and is displayed.

No matter what the function is, multiply, raise to a power, add, or whatever, the result is always moved to the x-register and displayed. There are various ways you can display the contents of the x-register; you can display from 1 to as many as 10 digits in floating decimal notation, or you can also display answers in scientific or engineering notation.

Algorithms for solving various problems exist throughout the text as problems occur for the first time.

All calculators that use *algebraic logic* follow the same rules when assigning priorities to the various mathematical operations. The algebraic hierarchy is as follows:

- Priority 1: Parentheses

- Priority 2: Reciprocals, squares and square roots, logs, trig functions
- Priority 3: Universal powers and roots
- Priority 4: Multiplication and division (same priority level)
- Priority 5: Addition and subtraction (same priority level)
- Priority 6: Equals

My colleague, Barbara Snyder, has simplified the above rules of priority and offers the PEMDAS rule, a variation on My Dear Aunt Sally (**M**ultiply, **D**ivide, **A**dd, and **S**ubtract), to help remember the priority order: *Please Excuse My Dear Aunt Sally*.

- **P**arentheses
- **E**xponents
- **M**ultiplication and **D**ivision
- **A**ddition and **S**ubtraction

Consult your users manual for more detailed information concerning your calculator.

APPENDIX B

TRIGONOMETRIC FUNCTIONS OF DEGREES

Angle°	*sin*	*tan*	*cot*	*cos*	*Angle°*
0.0	.00000	.00000	∞	1.00000	**90.0**
.1	.00175	.00175	572.96	1.00000	.9
.2	.00349	.00349	286.48	0.99999	.8
.3	.00524	.00524	190.98	.99999	.7
.4	.00698	.00698	143.23	.99998	.6
.5	.00873	.00873	114.59	.99996	.5
.6	.01047	.01047	95.489	.99995	.4
.7	.01222	.01222	81.847	.99993	.3
.8	.01396	.01396	71.615	.99990	.2
.9	.01571	.01571	63.657	.99988	.1
1.0	.01745	.01746	57.290	.99985	**89.0**
.1	.01920	.01920	52.081	.99982	.9
.2	.02094	.02095	47.740	.99978	.8
.3	.02269	.02269	44.066	.99974	.7
.4	.02443	.02444	40.917	.99970	.6
.5	.02618	.02619	38.188	.99966	.5
.6	.02792	.02793	35.801	.99961	.4
.7	.02967	.02968	33.694	.99956	.3
.8	.03141	.03143	31.821	.99951	.2
.9	.03316	.03317	30.145	.99945	.1
2.0	.03490	.03492	28.636	.99939	**88.0**
.1	.03664	.03667	27.271	.99933	.9
.2	.03839	.03842	26.031	.99926	.8
.3	.04013	.04016	24.898	.99919	.7
.4	.04188	.04191	23.859	.99912	.6
.5	.04362	.04366	22.904	.99905	.5
.6	.04536	.04541	22.022	.99897	.4
.7	.04711	.04716	21.205	.99889	.3
.8	.04885	.04891	20.446	.99881	.2
.9	.05059	.05066	19.740	.99872	.1
3.0	.05234	.05241	19.081	.99863	**87.0**
.1	.05408	.05416	18.464	.99854	.9
.2	.05582	.05591	17.886	.99844	.8
.3	.05756	.05766	17.343	.99834	.7
.4	.05931	.05941	16.832	.99824	.6
.5	.06105	.06116	16.350	.99813	.5
.6	.06279	.06291	15.895	.99803	.4
	cos	*cot*	*tan*	*sin*	

Angle°	*sin*	*tan*	*cot*	*cos*	*Angle°*
.7	.06453	.06467	15.464	.99792	.3
.8	.06627	.06642	15.056	.99780	.2
.9	.06802	.06817	14.669	.99767	.1
4.0	.06976	.06993	14.301	.99756	**86.0**
.1	.07150	.07168	13.951	.99744	.9
.2	.07324	.07344	13.617	.99731	.8
.3	.07498	.07519	13.300	.99719	.7
.4	.07672	.07695	12.996	.99705	.6
.5	.07846	.07870	12.706	.99692	.5
.6	.08020	.08046	12.429	.99678	.4
.7	.08194	.08221	12.163	.99664	.3
.8	.08368	.08397	11.909	.99649	.2
.9	.08542	.08573	11.664	.99635	.1
5.0	.08716	.08749	11.430	.99619	**85.0**
.1	.08889	.08925	11.205	.99604	.9
.2	.09063	.09101	10.988	.99588	.8
.3	.09237	.09277	10.780	.99572	.7
.4	.09411	.09453	10.579	.99556	.6
.5	.09585	.09629	10.385	.99540	.5
.6	.09758	.09805	10.199	.99523	.4
.7	.09932	.09981	10.019	.99506	.3
.8	.10106	.10158	9.8448	.99488	.2
.9	.10279	.10334	9.6768	.99470	.1
6.0	.10453	.10510	9.5144	.99452	**84.0**
.1	.10626	.10687	9.3572	.99434	.9
.2	.10800	.10863	9.2052	.99415	.8
.3	.10973	.11040	9.0579	.99396	.7
.4	.11147	.11217	8.9152	.99377	.6
.5	.11320	.11394	8.7769	.99357	.5
.6	.11494	.11570	8.6427	.99337	.4
.7	.11667	.11747	8.5126	.99317	.3
.8	.11840	.11924	8.3863	.99297	.2
.9	.12014	.12101	8.2636	.99276	.1
7.0	.12187	.12278	8.1443	.99255	**83.0**
.1	.12360	.12456	8.0285	.99233	.9
.2	.12533	.12633	7.9158	.99211	.8
.3	.12706	.12810	7.8062	.99189	.7
.4	.12880	.12988	7.6996	.99167	.6
.5	.13053	.13165	7.5958	.99144	.5
.6	.13226	.13343	7.4947	.99122	.4
.7	.13399	.13521	7.3962	.99098	.3
.8	.13572	.13698	7.3002	.99075	.2
.9	.13744	.13876	7.2066	.99051	.1
8.0	.13917	.14054	7.1154	.99027	**82.0**
.1	.14090	.14232	7.0264	.99002	.9
.2	.14263	.14410	6.9395	.98978	.8
.3	.14436	.14588	6.8548	.98953	.7
.4	.14608	.14767	6.7720	.98927	.6
.5	.14781	.14945	6.6912	.98902	.5
.6	.14954	.15124	6.6122	.98876	.4
.7	.15126	.15302	6.5350	.98849	.3
	cos	*cot*	*tan*	*sin*	

Angle°	*sin*	*tan*	*cot*	*cos*	*Angle°*
.8	.15299	.15481	6.4596	.98823	.2
.9	.15471	.15660	6.3859	.98796	.1
9.0	.15643	.15838	6.3138	.98769	**81.0**
.1	.15816	.16017	6.2432	.98741	.9
.2	.15988	.16196	6.1742	.98714	.8
.3	.16160	.16376	6.1066	.98686	.7
.4	.16333	.16555	6.0405	.98657	.6
.5	.16505	.16734	5.9758	.98629	.5
.6	.16677	.16914	5.9124	.98600	.4
.7	.16849	.17093	5.8502	.98570	.3
.8	.17021	.17273	5.7894	.98541	.2
.9	.17193	.17453	5.7297	.98511	.1
10.0	.17365	.17633	5.6713	.98481	**80.0**
.1	.17537	.17813	5.6140	.98450	.9
.2	.17708	.17993	5.5578	.98420	.8
.3	.17880	.18173	5.5026	.98389	.7
.4	.18052	.18353	5.4486	.98357	.6
.5	.18224	.18534	5.3955	.98325	.5
.6	.18395	.18714	5.3435	.98294	.4
.7	.18567	.18895	5.2924	.98261	.3
.8	.18738	.19076	5.2422	.98229	.2
.9	.18910	.19257	5.1929	.98196	.1
11.0	.19081	.19438	5.1446	.98163	**79.0**
.1	.19252	.19619	5.0970	.98129	.9
.2	.19423	.19801	5.0504	.98096	.8
.3	.19595	.19982	5.0045	.98061	.7
.4	.19766	.20164	4.9594	.98027	.6
.5	.19937	.20345	4.9152	.97992	.5
.6	.20108	.20527	4.8716	.97958	.4
.7	.20279	.20709	4.8288	.97922	.3
.8	.20450	.20891	4.7867	.97887	.2
.9	.20620	.21073	4.7453	.97851	.1
12.0	.20791	.21256	4.7046	.97815	**78.0**
.1	.20962	.21438	4.6646	.97778	.9
.2	.21132	.21621	4.6252	.97742	.8
.3	.21303	.21804	4.5864	.97705	.7
.4	.21474	.21986	4.5483	.97667	.6
.5	.21644	.22169	4.5107	.97630	.5
.6	.21814	.22353	4.4747	.97592	.4
.7	.21985	.22536	4.4373	.97553	.3
.8	.22155	.22719	4.4015	.97515	.2
.9	.22325	.22903	4.3662	.97476	.1
13.0	.22495	.23087	4.3315	.97437	**77.0**
.1	.22665	.23271	4.2972	.97398	.9
.2	.22835	.23455	4.2635	.97358	.8
.3	.23005	.23639	4.2303	.97318	.7
.4	.23175	.23823	4.1976	.97278	.6
.5	.23345	.24008	4.1653	.97237	.5
.6	.23514	.24193	4.1335	.97196	.4
.7	.23684	.24377	4.1022	.97155	.3
	cos	*cot*	*tan*	*sin*	

Angle°	*sin*	*tan*	*cot*	*cos*	*Angle°*
.8	.23853	.24562	4.0713	.97113	.2
.9	.24023	.24747	4.0408	.97072	.1
14.0	.24192	.24933	4.0108	.97030	**76.0**
.1	.24362	.25118	3.9812	.96987	.9
.2	.24531	.25304	3.9520	.96945	.8
.3	.24700	.25490	3.9232	.96902	.7
.4	.24869	.25676	3.8947	.96858	.6
.5	.25038	.25862	3.8667	.96815	.5
.6	.25207	.26048	3.8391	.96771	.4
.7	.25376	.26235	3.8118	.96727	.3
.8	.25545	.26421	3.7848	.96682	.2
.9	.25713	.26608	3.7583	.96638	.1
15.0	.25882	.26792	3.7321	.96593	**75.0**
.1	.26050	.26982	3.7062	.96547	.9
.2	.26219	.27169	3.6806	.96502	.8
.3	.26387	.27357	3.6554	.96456	.7
.4	.26556	.27545	3.6305	.96410	.6
.5	.26724	.27732	3.6059	.96363	.5
.6	.26892	.27921	3.5816	.96316	.4
.7	.27060	.28109	3.5576	.96269	.3
.8	.27228	.28297	3.5339	.96222	.2
.9	.27396	.28486	3.5105	.96174	.1
16.0	.27564	.28675	3.4874	.96126	**74.0**
.1	.27731	.28864	3.4646	.96078	.9
.2	.27899	.29053	3.4420	.96029	.8
.3	.28067	.29242	3.4197	.95981	.7
.4	.28234	.29432	3.3977	.95931	.6
.5	.28402	.29621	3.3759	.95882	.5
.6	.28569	.29811	3.3544	.95832	.4
.7	.28736	.30001	3.3332	.95782	.3
.8	.28903	.30192	3.3122	.95732	.2
.9	.29070	.30382	3.2914	.95681	.1
17.0	.29237	.30573	3.2709	.95630	**73.0**
.1	.29404	.30764	3.2506	.95579	.9
.2	.29571	.30955	3.2305	.95528	.8
.3	.29737	.31147	3.2106	.95476	.7
.4	.29904	.31338	3.1910	.95424	.6
.5	.30071	.31530	3.1716	.95372	.5
.6	.30237	.31722	3.1524	.95319	.4
.7	.30403	.31914	3.1334	.95266	.3
.8	.30570	.32106	3.1146	.95213	.2
.9	.30736	.32299	3.0961	.95159	.1
18.0	.30902	.32492	3.0777	.95106	**72.0**
.1	.31068	.32685	3.0595	.95052	.9
.2	.31233	.32878	3.0415	.94997	.8
.3	.31399	.33072	3.0237	.94943	.7
.4	.31565	.33266	3.0061	.94888	.6
.5	.31730	.33460	2.9887	.94832	.5
.6	.31896	.33654	2.9714	.94777	.4
.7	.32061	.33848	2.9544	.94721	.3
	cos	*cot*	*tan*	*sin*	

Angle°	*sin*	*tan*	*cot*	*cos*	*Angle°*
.8	.32227	.34043	2.9375	.94665	.2
.9	.32392	.34238	2.9208	.94609	.1
19.0	.32557	.34433	2.9042	.94552	**71.0**
.1	.32722	.34628	2.8878	.94495	.9
.2	.32887	.34824	2.8716	.94438	.8
.3	.33051	.35020	2.8556	.94380	.7
.4	.33216	.35216	2.8397	.94322	.6
.5	.33381	.35412	2.8239	.94264	.5
.6	.33545	.35608	2.8083	.94206	.4
.7	.33710	.35805	2.7929	.94147	.3
.8	.33874	.36002	2.7776	.94088	.2
.9	.34038	.36199	2.7625	.94029	.1
20.0	.34202	.36397	2.7475	.93969	**70.0**
.1	.34366	.36595	2.7326	.93909	.9
.2	.34530	.36793	2.7179	.93849	.8
.3	.34694	.36991	2.7034	.93789	.7
.4	.34857	.37190	2.6889	.93728	.6
.5	.35021	.37388	2.6746	.93667	.5
.6	.35184	.37588	2.6605	.93606	.4
.7	.35347	.37787	2.6464	.93544	.3
.8	.35511	.37986	2.6325	.93483	.2
.9	.35674	.38186	2.6187	.93420	.1
21.0	.35837	.38386	2.6051	.93358	**69.0**
.1	.36000	.38587	2.5916	.93295	.9
.2	.36162	.38787	2.5782	.93232	.8
.3	.36325	.38988	2.5649	.93169	.7
.4	.36488	.39190	2.5517	.93106	.6
.5	.36650	.39391	2.5386	.93042	.5
.6	.36812	.39593	2.5257	.92978	.4
.7	.36975	.39795	2.5129	.92913	.3
.8	.37137	.39997	2.5002	.92849	.2
.9	.37299	.40200	2.4876	.92784	.1
22.0	.37461	.40403	2.4751	.92718	**68.0**
.1	.37622	.40606	2.4627	.92653	.9
.2	.37784	.40809	2.4504	.92587	.8
.3	.37946	.41013	2.4383	.92521	.7
.4	.38107	.41217	2.4262	.92455	.6
.5	.38268	.41421	2.4142	.92388	.5
.6	.38430	.41626	2.4023	.92321	.4
.7	.38591	.41831	2.3906	.92254	.3
.8	.38752	.42036	2.3789	.92186	.2
.9	.38912	.42242	2.3673	.92119	.1
23.0	.39073	.42447	2.3559	.92050	**67.0**
.1	.39234	.42654	2.3445	.91982	.9
.2	.39394	.42860	2.3332	.91914	.8
.3	.39555	.43067	2.3220	.91845	.7
.4	.38715	.43274	2.3109	.91775	.6
.5	.39875	.43481	2.2998	.91706	.5
.6	.40035	.43689	2.2889	.91636	.4
.7	.40195	.43897	2.2781	.91566	.3
	cos	*cot*	*tan*	*sin*	

Angle°	*sin*	*tan*	*cot*	*cos*	*Angle°*
.8	.40355	.44105	2.2673	.91496	.2
.9	.40514	.44314	2.2566	.91425	.1
24.0	.40674	.44523	2.2460	.91355	**66.0**
.1	.40833	.44732	2.2355	.91283	.9
.2	.40992	.44942	2.2251	.91212	.8
.3	.41151	.45152	2.2148	.91140	.7
.4	.41310	.45362	2.2045	.91068	.6
.5	.41469	.45573	2.1943	.90996	.5
.6	.41628	.45784	2.1842	.90924	.4
.7	.41787	.45995	2.1742	.90851	.3
.8	.41945	.46206	2.1624	.90778	.2
.9	.42104	.46418	2.1543	.90704	.1
25.0	.42262	.46631	2.1445	.90631	**65.0**
.1	.42420	.46843	2.1348	.90557	.9
.2	.42578	.47056	2.1251	.90483	.8
.3	.42736	.47270	2.1155	.90408	.7
.4	.42894	.47483	2.1060	.90334	.6
.5	.43051	.47698	2.0965	.90259	.5
.6	.43209	.47912	2.0872	.90183	.4
.7	.43366	.48127	2.0778	.90108	.3
.8	.43523	.48342	2.0682	.90032	.2
.9	.43680	.48557	2.0594	.89956	.1
26.0	.43837	.48773	2.0503	.89879	**64.0**
.1	.43994	.48989	2.0413	.89803	.9
.2	.44151	.49206	2.0323	.89726	.8
.3	.44307	.49423	2.0233	.89649	.7
.4	.44464	.49640	2.0145	.89571	.6
.5	.44620	.49858	2.0057	.89493	.5
.6	.44776	.50076	1.9970	.89415	.4
.7	.44932	.50295	1.9883	.89337	.3
.8	.45088	.50514	1.9797	.89259	.2
.9	.45243	.50733	1.9711	.89180	.1
27.0	.45399	.50953	1.9626	.89101	**63.0**
.1	.45554	.51173	1.9542	.89021	.9
.2	.45710	.51393	1.9458	.88942	.8
.3	.45865	.51614	1.9375	.88862	.7
.4	.46020	.51835	1.9292	.88782	.6
.5	.46175	.52057	1.9210	.88701	.5
.6	.46330	.52279	1.9128	.88620	.4
.7	.46484	.52501	1.9047	.88539	.3
.8	.46639	.52724	1.8967	.88458	.2
.9	.46793	.52947	1.8887	.88377	.1
28.0	.46947	.53171	1.8807	.88295	**62.0**
.1	.47101	.53395	1.8728	.88213	.9
.2	.47255	.53620	1.8650	.88130	.8
.3	.47409	.53844	1.8572	.88048	.7
.4	.47562	.54070	1.8495	.87965	.6
.5	.47716	.54296	1.8418	.87882	.5
.6	.47869	.54522	1.8341	.87798	.4
.7	.48022	.54748	1.8265	.87715	.3
	cos	*cot*	*tan*	*sin*	

Angle°	*sin*	*tan*	*cot*	*cos*	*Angle°*
.8	.48175	.54975	1.8190	.87631	.2
.9	.48328	.55203	1.8115	.87546	.1
29.0	.48481	.55431	1.8040	.87462	**61.0**
.1	.48634	.55659	1.7966	.87377	.9
.2	.48786	.55888	1.7893	.87292	.8
.3	.48938	.56117	1.7820	.87207	.7
.4	.49090	.56347	1.7747	.87121	.6
.5	.49242	.56577	1.7675	.87036	.5
.6	.49394	.56808	1.7603	.86949	.4
.7	.49546	.57039	1.7532	.86863	.3
.8	.49697	.57271	1.7461	.86777	.2
.9	.49849	.57503	1.7391	.86690	.1
30.0	.50000	.57735	1.7321	.86603	**60.0**
.1	.50151	.57968	1.7251	.86515	.9
.2	.50302	.58201	1.7182	.86427	.8
.3	.50453	.58435	1.7113	.86340	.7
.4	.50603	.58670	1.7045	.86251	.6
.5	.50754	.58905	1.6977	.86163	.5
.6	.50904	.59140	1.6909	.86074	.4
.7	.51054	.59376	1.6842	.85985	.3
.8	.51204	.59612	1.6775	.85895	.2
.9	.51354	.59849	1.6709	.85806	.1
31.0	.51504	.60086	1.6643	.85717	**59.0**
.1	.51653	.60324	1.6577	.85627	.9
.2	.51803	.60562	1.6512	.85536	.8
.3	.51952	.60801	1.6447	.85446	.7
.4	.52101	.61040	1.6383	.85355	.6
.5	.52250	.61280	1.6319	.85264	.5
.6	.52399	.61520	1.6255	.85173	.4
.7	.52547	.61761	1.6191	.85081	.3
.8	.52697	.62003	1.6128	.84989	.2
.9	.52844	.62245	1.6066	.84897	.1
32.0	.52992	.62487	1.6003	.84805	**58.0**
.1	.53140	.62730	1.5941	.84712	.9
.2	.53288	.62973	1.5880	.84619	.8
.3	.53435	.63217	1.5818	.84526	.7
.4	.53583	.63462	1.5757	.84433	.6
.5	.53730	.63707	1.5697	.84339	.5
.6	.53877	.63953	1.5637	.84245	.4
.7	.54024	.64199	1.5577	.84151	.3
.8	.54171	.64446	1.5517	.84057	.2
.9	.54317	.64693	1.5458	.83962	.1
33.0	.54464	.64941	1.5399	.83867	**57.0**
.1	.54610	.65189	1.5340	.83772	.9
.2	.54756	.65438	1.5282	.83676	.8
.3	.54902	.65688	1.5224	.83581	.7
.4	.55048	.65938	1.5166	.83485	.6
.5	.55194	.66189	1.5108	.83389	.5
.6	.55339	.66440	1.5051	.83292	.4
.7	.55484	.66692	1.4994	.83195	.3
	cos	*cot*	*tan*	*sin*	

Angle°	*sin*	*tan*	*cot*	*cos*	*Angle°*
.8	.55630	.66944	1.4938	.83098	.2
.9	.55775	.67197	1.4882	.83001	.1
34.0	.55919	.67451	1.4826	.82904	**56.0**
.1	.56064	.67705	1.4770	.82806	.9
.2	.56208	.67960	1.4715	.82708	.8
.3	.56353	.68215	1.4659	.82610	.7
.4	.56497	.68471	1.4605	.82511	.6
.5	.56641	.68728	1.4550	.82413	.5
.6	.56784	.68985	1.4496	.82314	.4
.7	.56928	.69243	1.4442	.82214	.3
.8	.57071	.69502	1.4388	.82115	.2
.9	.57215	.69761	1.4335	.82015	.1
35.0	.57358	.70021	1.4281	.81915	**55.0**
.1	.57501	.70281	1.4229	.81815	.9
.2	.57643	.70542	1.4176	.81714	.8
.3	.57786	.70804	1.4124	.81614	.7
.4	.57928	.71066	1.4071	.81513	.6
.5	.58070	.71329	1.4019	.81412	.5
.6	.58212	.71593	1.3968	.81310	.4
.7	.58354	.71857	1.3916	.81208	.3
.8	.58496	.72122	1.3865	.81106	.2
.9	.58637	.72388	1.3814	.81004	.1
36.0	.58779	.72654	1.3764	.80902	**54.0**
.1	.58920	.72921	1.3713	.80799	.9
.2	.59061	.73189	1.3663	.80696	.8
.3	.59201	.73457	1.3613	.80593	.7
.4	.59342	.73726	1.3564	.80489	.6
.5	.59482	.73996	1.3514	.80386	.5
.6	.59622	.74267	1.3465	.80282	.4
.7	.59763	.74538	1.3416	.80178	.3
.8	.59902	.74810	1.3367	.80073	.2
.9	.60042	.75082	1.3319	.79968	.1
37.0	.60182	.75355	1.3270	.79864	**53.0**
.1	.60321	.75629	1.3222	.79758	.9
.2	.60460	.75904	1.3175	.79653	.8
.3	.60599	.76180	1.3127	.79547	.7
.4	.60738	.76456	1.3079	.79441	.6
.5	.60876	.76733	1.3032	.79335	.5
.6	.61015	.77010	1.2985	.79229	.4
.7	.61153	.77289	1.2938	.79122	.3
.8	.61291	.77568	1.2892	.79016	.2
.9	.61429	.77848	1.2846	.78908	.1
38.0	.61566	.78129	1.2799	.78801	**52.0**
.1	.61704	.78410	1.2753	.78694	.9
.2	.61841	.78692	1.2708	.78586	.8
.3	.61978	.78975	1.2662	.78478	.7
.4	.62115	.79259	1.2617	.78369	.6
.5	.62251	.79544	1.2572	.78261	.5
.6	.62388	.79829	1.2527	.78152	.4
.7	.62524	.80115	1.2482	.78043	.3
	cos	*cot*	*tan*	*sin*	

Angle°	*sin*	*tan*	*cot*	*cos*	*Angle°*
.8	.62660	.80402	1.2437	.77934	.2
.9	.62796	.80690	1.2393	.77824	.1
39.0	.62932	.80978	1.2349	.77715	**51.0**
.1	.63068	.81268	1.2305	.77605	.9
.2	.63203	.81558	1.2261	.77494	.8
.3	.63338	.81849	1.2218	.77384	.7
.4	.63473	.82141	1.2174	.77273	.6
.5	.63608	.82434	1.2131	.77162	.5
.6	.63742	.82727	1.2088	.77051	.4
.7	.63877	.83022	1.2045	.76940	.3
.8	.64011	.83317	1.2002	.76828	.2
.9	.64145	.83613	1.1960	.76717	.1
40.0	.64279	.83910	1.1918	.76604	**50.0**
.1	.64412	.84208	1.1875	.76492	.9
.2	.64546	.84507	1.1833	.76380	.8
.3	.64679	.84806	1.1792	.76267	.7
.4	.64812	.85107	1.1750	.76154	.6
.5	.64945	.85408	1.1708	.76041	.5
.6	.65077	.85710	1.1667	.75927	.4
.7	.65210	.86014	1.1626	.75813	.3
.8	.65342	.86318	1.1585	.75700	.2
.9	.65474	.86623	1.1544	.75585	.1
41.0	.65606	.86929	1.1504	.75471	**49.0**
.1	.65738	.87236	1.1463	.75356	.9
.2	.65869	.87543	1.1423	.75241	.8
.3	.66000	.87852	1.1383	.75126	.7
.4	.66131	.88162	1.1343	.75011	.6
.5	.66262	.88473	1.1303	.74896	.5
.6	.66393	.88784	1.1263	.74780	.4
.7	.66523	.89097	1.1224	.74664	.3
.8	.66653	.89410	1.1184	.74548	.2
.9	.66783	.89725	1.1145	.74431	.1
42.0	.66913	.90040	1.1106	.74314	**48.0**
.1	.67043	.90357	1.1067	.74198	.9
.2	.67172	.90674	1.1028	.74080	.8
.3	.67301	.90993	1.0990	.73963	.7
.4	.67430	.91313	1.0951	.73846	.6
.5	.67559	.91633	1.0913	.73728	.5
.6	.67688	.91955	1.0875	.73610	.4
.7	.67816	.92277	1.0837	.73491	.3
.8	.67944	.92601	1.0799	.73373	.2
.9	.68072	.92926	1.0761	.73254	.1
43.0	.68200	.93252	1.0724	.73135	**47.0**
.1	.68327	.93578	1.0686	.73016	.9
.2	.68455	.93906	1.0649	.72897	.8
.3	.68582	.94235	1.0612	.72777	.7
.4	.68709	.94565	1.0575	.72657	.6
.5	.68835	.94896	1.0538	.72537	.5
.6	.68962	.95229	1.0501	.72417	.4
.7	.69088	.95562	1.0464	.72294	.3
	cos	*cot*	*tan*	*sin*	

Angle°	*sin*	*tan*	*cot*	*cos*	*Angle°*
.8	.69214	.95897	1.0428	.72176	.2
.9	.69340	.96232	1.0392	.72055	.1
44.0	.69466	.96569	1.0355	.71934	**46.0**
.1	.69591	.96907	1.0319	.71813	.9
.2	.69717	.97246	1.0283	.71691	.8
.3	.69842	.97586	1.0247	.71569	.7
.4	.69966	.97927	1.0212	.71447	.6
.5	.70091	.98270	1.0176	.71325	.5
.6	.70215	.98613	1.0141	.71203	.4
.7	.70339	.98958	1.0105	.71080	.3
.8	.70463	.99304	1.0070	.70957	.2
.9	.70587	.99652	1.0035	.70834	.1
45.0	.70711	1.00000	1.0000	.70711	**45.0**
	cos	*cot*	*tan*	*sin*	

APPENDIX C ANSWERS TO SELECTED PROBLEMS

CHAPTER 1

Problems 1-1

1. (a) 8 (b) 2 (c) 0 (d) 2 (e) 8000 (f) 4
3. (a) 6 (b) 3 (c) 7 (d) 0 (e) 7000 (f) 4
5. (a) 8 (b) 7 (c) 3 (d) 9 (e) 800,000 (f) 4

Problems 1-2

1. (a) 0.3 (b) 0.016 (c) 0.00278 (d) 0.1763 (e) 0.435 (f) 0.2060

3. (a) $\frac{7}{1000}$ (b) $\frac{432}{10,000}$ (c) $\frac{174}{1000}$ (d) $\frac{65}{1,000,000}$ (e) $\frac{16}{100,000}$ (f) $\frac{1234}{100,000}$

5. (a) $\frac{17}{1000} = 0.017$ (b) $\frac{4}{100} = 0.04$ (c) $\frac{460}{10,000} = 0.0460$ (d) $\frac{27}{1,000,000} = 0.000027$
(e) $\frac{1780}{100,000} = 0.01780$ (f) $\frac{65}{1000} = 0.065$

7. (a) Thousandths (b) Millionths (c) Ten-thousandths
9. (a) Ten-thousandths (b) Millionths (c) Hundredths
11. (a) Six-thousandths (b) One hundred forty-seven thousandths (c) Ninety-two hundred-thousandths
(d) Seven-millionths (e) Four hundred thirteen ten-thousandths (f) One hundred one ten-thousandths

Problems 1-3

1. (a) $7\frac{14}{100}$ (b) $50\frac{2}{100}$ (c) $710\frac{143}{1000}$ (d) $9\frac{99}{1000}$ (e) $73\frac{653}{1000}$ (f) $207\frac{7834}{10,000}$ (g) $28\frac{736}{100,000}$ (h) $8\frac{706}{10,000}$

3. (a) 5.68 (b) 25.007 (c) 7.0165 (d) 70.4 (e) 473.025 (f) 80.00743 (g) 2475.000035 (h) 307.00008

5. (a) $93.7 = 93\frac{7}{10}$ (b) $30.04 = 30\frac{4}{100}$ (c) $11.0001 = 11\frac{1}{10,000}$ (d) $905.052 = 905\frac{52}{1000}$

7. (a) $273.00025 = 273\frac{25}{100,000}$ (b) $704.000704 = 704\frac{704}{1,000,000}$
(c) $2044.0504 = 2044\frac{504}{10,000}$ (d) $10,101.00089 = 10,101\frac{89}{100,000}$

Problems 1-4

1. 20 **3.** 50 **5.** 60 **7.** 130 **9.** 870

	Tens	*Hundreds*
11.	270	300
13.	360	400
15.	1380	1400
17.	1410	1400
19.	8710	8700

	Tens	*Hundreds*	*Thousands*
21.	4820	4800	5000
23.	85,470	85,500	85,000
25.	78,670	78,700	79,000
27.	2780	2800	3000
29.	35,490	35,500	35,000
31.	68,450	68,400	68,000
33.	73,650	73,700	74,000
35.	470,000		

Problems 1-5

	Hundredths	*Tenths*	*Units*	*Tens*
1.	163.78	163.8	164	160
3.	9.46	9.5	9	10
5.	88.89	88.9	89	90
7.	749.49	749.5	749	750
9.	39.28	39.3	39	40
11.	63.75	63.7	64	60
13.	478.67	478.7	479	480
15.	47.47	47.5	47	50
17.	16.55	16.5	17	20
19.	76.75	76.7	77	80

Problems 1-6

1. 5 **3.** 14 **5.** −6 **7.** 34 **9.** −28 **11.** 7 **13.** −14 **15.** 18 **17.** 21 **19.** −33

Problems 1-7

1. −42 **3.** 54 **5.** −84 **7.** 225 **9.** 930 **11.** −7 **13.** −24 **15.** 12 **17.** −32 **19.** 28

Problems 1-8

1. −17 **3.** 0 **5.** 17 **7.** 21 **9.** 18 **11.** 26 **13.** −21 **15.** 12 **17.** 15 **19.** −19

Problems 1-9

1. 16 **3.** 7 **5.** −36 **7.** −83 **9.** 4 **11.** 51 **13.** 130 **15.** 45 **17.** −144 **19.** 20

CHAPTER 2

Problems 2-1

1. 1000 **3.** 1 **5.** 1,000,000

Problems 2-2

1. 10^1 **3.** 10^3 **5.** 10^2

Problems 2-3

1. $\frac{1}{100}$ **3.** $\frac{1}{1000}$ **5.** $\frac{1}{10}$

Problems 2-4

1. 10^{-4} **3.** 10^{-6} **5.** 10^{-4} **7.** 10^{-3}

Problems 2-5

1. 10^{10} **3.** 10^6 **5.** 10^4 **7.** 10^{-4} **9.** 10^3 **11.** 10^{-10} **13.** 10^{-12} **15.** 10^{-9}

Problems 2-6

1. 10^4 **3.** 10^{-5} **5.** 10^8 **7.** 10^{12} **9.** 10^7 **11.** 10^{-3} **13.** 10^{-6} **15.** 10^{-7} **17.** 10^3 **19.** 10^{-4}

Problems 2-7

1. 10^{-7} **3.** 10^{-7} **5.** 10^{11} **7.** 10^{10} **9.** 10^{-12} **11.** 10^8 **13.** 10^{13} **15.** 10^{-4} **17.** 10^{-36}

Problems 2-8

1. (a) 4.75×10^2 (b) 0.0475×10^4 (c) 0.000475×10^6
3. (a) 0.659×10^2 (b) 0.00659×10^4 (c) 0.0000659×10^6
5. (a) 934×10^2 (b) 9.34×10^4 (c) 0.0934×10^6
7. (a) 47.8×10^2 (b) 0.478×10^4 (c) 0.00478×10^6
9. (a) 0.465×10^{-2} (b) 46.5×10^{-4} (c) 4650×10^{-6}
11. (a) 0.000555×10^{-2} (b) 0.0555×10^{-4} (c) 5.55×10^{-6}
13. (a) 0.0673×10^{-2} (b) 6.73×10^{-4} (c) 673×10^{-6}
15. (a) $0.00000108 \times 10^{-2}$ (b) 0.000108×10^{-4} (c) 0.0108×10^{-6}
17. (a) 0.325×10^3 (b) $325{,}000 \times 10^{-3}$
19. (a) 0.00475×10^3 (b) 4750×10^{-3}

Problems 2-9

1. 2.76×10^4 **3.** 4.78×10^1 **5.** 1.77×10^6 **7.** 2.73×10^2 **9.** 1.73×10^5
11. 7.89×10^1 **13.** 1.67×10^6 **15.** 5.79×10^4 **17.** 8.10×10^4 **19.** 4.71×10^5

Problems 2-10

1. 4.78×10^{-3} **3.** 7.47×10^{-1} **5.** 1.64×10^{-2} **7.** 2.78×10^{-6} **9.** 1.71×10^{-2}
11. 8.96×10^{-4} **13.** 4.54×10^{-3} **15.** 5.01×10^{-5} **17.** 2.55×10^{-1} **19.** 4.00×10^{-2}

Problems 2-11

1. 2.13×10^4 **3.** 1.28×10^{-1} **5.** 1.05×10^4 **7.** 6.13×10^0 **9.** 2.99×10^0 **11.** 1.31×10^{-3}
13. 1.23×10^{10} **15.** 2.18×10^{-6} **17.** 6.70×10^0 **19.** 1.82×10^{-2} **21.** 9.41×10^{-5}
23. 2.59×10^6

Problems 2-12

1. 1.10×10^4 **3.** 2.00×10^{-1} **5.** 3.86×10^3 **7.** 4.82×10^5 **9.** 2.88×10^{-1} **11.** 3.87×10^{-2}
13. 9.93×10^4 **15.** 4.26×10^5

Problems 2-13

1. 1.06×10^2 **3.** 1.27×10^{-4} **5.** 3.26×10^{-3} **7.** 1.67×10^{-3} **9.** 3.91×10^{-5} **11.** 5.16×10^{-6}
13. 1.10×10^{-4} **15.** 2.56×10^3 **17.** 8.46×10^1 **19.** 3.07×10^4 **21.** 2.44×10^3 **23.** 1.39×10^0
25. 1.74×10^0 **27.** 1.09×10^0 **29.** 2.53×10^0 **31.** 1.03×10^{-2} **33.** 6.75×10^{-3}

Problems 2-14

1. 10^6 **3.** 10^{10} **5.** 10^{-12} **7.** 10^{12} **9.** 10^{-8}

Problems 2-15

1. 10^3 **3.** 10^{-2} **5.** 10^{-2} **7.** 10^5 **9.** 10^3

Problems 2-16

1. 6.29×10^3 **3.** 3.17×10^6 **5.** 2.26×10^{-3} **7.** 7.62×10^{-5} **9.** 2.03×10^9 **11.** 9.00×10^{10} **13.** 8.76×10^{-8} **15.** 8.15×10^{-9} **17.** 6.75×10^0 **19.** 2.17×10^0

Problems 2-17

1. 9.64×10^0 **3.** 1.54×10^1 **5.** 8.40×10^{-2} **7.** 1.02×10^{-2} **9.** 5.20×10^3 **11.** 6.71×10^3 **13.** 3.04×10^{-3} **15.** 5.88×10^0 **17.** 1.07×10^{-2} **19.** 7.96×10^3 **21.** 1.58×10^4

CHAPTER 3

Problems 3-1

1. $56 \times 10^3 = 0.056 \times 10^6$ **3.** $220 \times 10^3 = 0.22 \times 10^6$ **5.** $390 \times 10^3 = 0.39 \times 10^6$ **7.** $180 \times 10^3 = 0.18 \times 10^6$ **9.** $43 \times 10^3 = 0.043 \times 10^6$ **11.** $0.22 \times 10^{-3} = 220 \times 10^{-6}$ **13.** $2.13 \times 10^{-3} = 2130 \times 10^{-6}$ **15.** $0.0556 \times 10^{-3} = 55.6 \times 10^{-6}$ **17.** $0.122 \times 10^{-3} = 122 \times 10^{-6}$ **19.** $0.256 \times 10^{-3} = 256 \times 10^{-6}$ **21.** $667 \times 10^{-9} = 667{,}000 \times 10^{-12}$ **23.** $17.9 \times 10^{-9} = 17{,}900 \times 10^{-12}$ **25.** $6.74 \times 10^{-9} = 6740 \times 10^{-12}$ **27.** $0.0177 \times 10^{-9} = 17.7 \times 10^{-12}$ **29.** $70 \times 10^{-9} = 70{,}000 \times 10^{-12}$ **31.** $7.3 = 7300 \times 10^{-3} = 0.0073 \times 10^3$ **33.** $5.67 = 5670 \times 10^{-3} = 0.00567 \times 10^3$ **35.** $0.0178 = 17.8 \times 10^{-3} = 0.0000178 \times 10^3$ **37.** $78.3 = 78{,}300 \times 10^{-3} = 0.0783 \times 10^3$ **39.** $845 = 845{,}000 \times 10^{-3} = 0.845 \times 10^3$

Problems 3-2

1. 0.26 mA = 260 μA **3.** 0.632 mS = 632 μS **5.** 7.63 kΩ = 0.00763 MΩ **7.** 17.3 mA = 17,300 μA **9.** 713 kΩ = 0.713 MΩ **11.** 5630 kHz = 5.63 MHz **13.** 2000 kΩ = 2 MΩ **15.** 0.237 mS = 237 μS **17.** 0.3 μF = 300 nF **19.** 0.062 mA = 62 μA

Problems 3-3

1. 0.8 A **3.** 0.0025 S **5.** 33,000 Ω **7.** 470 Ω **9.** 12,500 Hz **11.** 0.0001 F **13.** 0.00025 A **15.** 0.0009 S **17.** 750,000 Ω **19.** 0.03 A

Problems 3-4

1. 0.26 mA = 260 μA **3.** 0.00632 mS = 6.32 μS **5.** 56.2 mS = 56,200 μS **7.** 0.613 mS = 613 μS **9.** 7.63 kΩ = 0.00763 MΩ **11.** 470 kΩ = 0.470 MΩ **13.** 127 kHz = 0.127 MHz **15.** 713 kΩ = 0.173 MΩ **17.** 46,300 μA = 0.0463 A **19.** 0.00005 μF = 50 pF **21.** 3200 Ω = 0.0032 MΩ **23.** 0.270 MΩ = 270,000 Ω **25.** 0.000403 S = 0.403 mS **27.** 1,030,000 Hz = 1030 kHz **29.** 1430 pF = 0.00143 μF **31.** 5.5 mS = 5500 μS **33.** 0.00106 μF = 1060 pF **35.** 463,000 Ω = 0.463 MΩ **37.** 96.3 kHz = 0.0963 MHz **39.** 7.8 mA = 7800 μA **41.** 0.176 V = 176,000 μV **43.** 1730 mW = 0.00173 kW **45.** 25 mH = 25,000 μH **47.** 0.173 S = 173,000 μS **49.** 2.5 μF = 2500 nF

Problems 3-5

1. 2.04 mA = 2040 μA **3.** 0.0957 mA = 95.7 μA **5.** 0.167 mS = 167 μS **7.** 0.0582 mS = 58.2 μS **9.** 1.42 mS = 1420 μS **11.** 964 Ω = 0.964 kΩ **13.** 68,700 Ω = 68.7 kΩ = 0.0687 MΩ **15.** 6850 Ω = 6.85 kΩ = 0.00685 MΩ **17.** 115 Ω = 0.115 kΩ **19.** 4690 Ω = 4.69 kΩ **21.** 2.30 mS = 2300 μS **23.** 0.273 mA = 273 μA **25.** 70,000 Ω = 70 kΩ **27.** 0.0417 mA = 41.7 μA **29.** 9.52 V = 9520 mV **31.** 4.48 V = 4480 mV

33. 84,900 Ω = 84.9 kΩ **35.** 318 Ω = 0.318 kΩ **37.** 0.00491 μF = 4.91 nF **39.** 919 Hz = 0.919 kHz

Problems 3-6

1. 39.37 in **3.** 15.75 in **5.** 2.3 m = 90.55 in = 2.516 yd **7.** 10.94 yd **9.** 10,940 yd = 6.214 mi **11.** 25.4 cm **13.** 90.42 cm **15.** 804.7 m = 0.8047 km **17.** 88.50 km/h **19.** 160.9 km/h **21.** 1640 ft/s **23.** 328.1 ft/s **25.** 2.116 oz **27.** 5.291 oz **29.** 28.22 oz = 1.764 lb **31.** 176.4 oz = 11.03 lb **33.** 0.6124 kg **35.** 113.4 g **37.** 7.796 g

CHAPTER 4

Problems 4-1

1. 7 **3.** 10 **5.** 13 **7.** 23 **9.** 38 **11.** 56 **13.** 101 **15.** 119 **17.** 240 **19.** 195 **21.** 101_2 **23.** 1100_2 **25.** 10101_2 **27.** 101101_2 **29.** 111011_2 **31.** 1000100_2 **33.** 1100000_2 **35.** 10000111_2 **37.** 11010010_2

Problems 4-2

1. 12_{10} **3.** 63_{10} **5.** 190_{10} **7.** 2174_{10} **9.** 5068_{10} **11.** 24_8 **13.** 120_8 **15.** 550_8 **17.** 2611_8 **19.** 7315_8

Problems 4-3

1. 26_{10} **3.** 76_{10} **5.** 512_{10} **7.** 4522_{10} **9.** $41{,}003_{10}$ **11.** 16_{16} **13.** 61_{16} **15.** 200_{16} **17.** $A8C_{16}$ **19.** $17BB_{16}$

Problems 4-4

1. $312_8 = CA_{16}$ **3.** $233_8 = 9B_{16}$ **5.** $201_8 = 81_{16}$ **7.** $6311_8 = CC9_{16}$ **9.** $5303_8 = AC3_{16}$ **11.** 1111_2 **13.** 100100_2 **15.** 10000000_2 **17.** 101110000_2 **19.** 1000000000_2 **21.** 1100_2 **23.** 100100_2 **25.** 1001111_2 **27.** 11100011_2 **29.** 110011010_2

Problems 4-5

1. $75_8 = 111101_2 = 61_{10} = 3D_{16}$ **3.** $140_8 = 1100000_2 = 96_{10} = 60_{16}$
5. $2A_{16} = 101010_2 = 52_8 = 42_{10}$ **7.** $1A7_{16} = 110100111_2 = 647_8 = 423_{10}$
9. $100110_2 = 46_8 = 38_{10} = 26_{16}$ **11.** $111000101_2 = 705_8 = 453_{10} = 1C5_{16}$
13. $10_{10} = 1010_2 = 12_8 = A_{16}$ **15.** $290_{10} = 100100010_2 = 442_8 = 122_{16}$

Problems 4-6

1. 357_8 **3.** 775_8 **5.** 776_8 **7.** 1423_8 **9.** 1107_8

Problems 4-7

1. ED_{16} **3.** $CA5_{16}$ **5.** $D56_{16}$ **7.** $179D_{16}$ **9.** $1DA6C_{16}$

Problems 4-8

1. 10101_2 **3.** 110011_2 **5.** 101001_2 **7.** 1100010_2 **9.** 1101001_2

Problems 4-9

1. 45_8 **3.** 42_8 **5.** 245_8 **7.** 337_8 **9.** 1515_8 **11.** $1A_{16}$ **13.** $3E1_{16}$ **15.** $5F1_{16}$ **17.** AFD_{16} **19.** FDC_{16} **21.** 101_2 **23.** 10_2 **25.** 10_2 **27.** 11_2 **29.** 10110_2

Problems 4-10

1. 0110_2 **3.** 010011_2 **5.** 14_8 **7.** 143_8 **9.** $5D_{16}$ **11.** $3E2_{16}$ **13.** 01101_2 **15.** 100011_2 **17.** 01011_2 **19.** 1000011_2 **21.** 36_8 **23.** 227_8 **25.** 16_{16} **27.** $1A3_{16}$

CHAPTER 5

Problems 5-1

1. Two, binomial **3.** Three, trinomial **5.** Two, binomial **7.** Two, binomial **9.** Three, trinomial
11. The coefficients are 3, -4, and 2. The literal numbers are a, b, and c.
13. The coefficients are $\frac{2}{3}$ and 7. The literal numbers are x and y.
15. The coefficients are $\frac{5}{2}$ and -4. The literal numbers are a and b.

Problems 5-2

1. 5.29 **3.** 54.8 **5.** 256 **7.** 512 **9.** 46.7 **11.** 166 **13.** 16.0 **15.** 64.0 **17.** 256

Problems 5-3

1. 30.0 **3.** 150 **5.** 900 **7.** 450 **9.** 117 **11.** 0 **13.** 9.00 **15.** -6.89×10^4
17. -6.78×10^4 **19.** 5.86×10^{14}

Problems 5-4

1. 5.48 **3.** 6.32 **5.** 14.1 **7.** 32.2 **9.** 44.7

Problems 5-5

1. 4.47 **3.** 4.00 **5.** 21.2 **7.** 7.55 **9.** 19.6 **11.** 23.0 **13.** 15.2 **15.** 67.5
17. 84.4 **19.** 823 **21.** 3.66×10^3

Problems 5-6

1. 4.49 V **3.** 66.6 V **5.** 135 V **7.** 241 V **9.** 38.3 mW **11.** 4.40 W **13.** 7.20 W
15. 447 mW **17.** 3.42 mA **19.** 1.67 mA **21.** 678 mA **23.** 3.41 mA **25.** 15.0 V
27. 9.87 V **29.** 11.1 V **31.** 3.16 V **33.** 37.6 mW **35.** 5.36 mW **37.** 33.1 mW
39. 4.79 W

Problems 5-7

1. 318 Ω **3.** 3.98 kΩ **5.** 12.6 Ω **7.** 9.73 kΩ **9.** 1.63 kΩ **11.** 1.45 kHz **13.** 58.1 kHz
15. 1.42 kHz **17.** 3.56 kHz **19.** 63.7 Hz **21.** 144 kΩ **23.** 224 Ω **25.** 38.6 kΩ
27. 8.72 kΩ **29.** 7.10 MΩ

CHAPTER 6

Problems 6-1

1. $2 \cdot 3 \cdot 3$ **3.** $2 \cdot 2 \cdot 11$ **5.** $3 \cdot 3 \cdot 7$ **7.** $2 \cdot 2 \cdot 23$ **9.** $3 \cdot 7 \cdot 11$ **11.** $2 \cdot 2 \cdot 2 \cdot 7$ **13.** $3 \cdot 7 \cdot 7$
15. $2 \cdot 2 \cdot 3 \cdot 7$ **17.** $2 \cdot 3 \cdot 5 \cdot 7$ **19.** $5 \cdot 7 \cdot 13$

Problems 6-2

1. $\frac{1}{7}$ **3.** $\frac{5}{8}$ **5.** $\frac{11}{16}$ **7.** $\frac{2}{5}$ **9.** $\frac{21}{40}$ **11.** $\frac{3}{5}$ **13.** $\frac{14}{15}$ **15.** $\frac{13}{21}$ **17.** $\frac{49}{55}$ **19.** $\frac{1}{3}$

Problems 6-3

1. $\frac{1}{6}$ **3.** $\frac{1}{12}$ **5.** $\frac{3}{10}$ **7.** $\frac{5}{14}$ **9.** $\frac{21}{32}$ **11.** $\frac{1}{9}$ **13.** $\frac{1}{30}$ **15.** $\frac{2}{35}$ **17.** $\frac{1}{15}$ **19.** $\frac{11}{75}$

Problems 6-4

1. $1\frac{1}{3}$ **3.** $\frac{3}{14}$ **5.** $1\frac{1}{6}$ **7.** $\frac{6}{7}$ **9.** $\frac{7}{10}$ **11.** $\frac{1}{2}$ **13.** $1\frac{3}{11}$ **15.** $1\frac{1}{6}$ **17.** $3\frac{1}{9}$ **19.** $\frac{63}{64}$

Problems 6-5

1. $\frac{1}{2}$ **3.** $\frac{5}{7}$ **5.** $\frac{1}{2}$ **7.** $\frac{2}{7}$ **9.** $1\frac{11}{12}$ **11.** $\frac{5}{16}$ **13.** $\frac{13}{24}$ **15.** $-\frac{2}{9}$ **17.** $\frac{9}{16}$ **19.** $\frac{11}{18}$

Problems 6-6

1. 30 **3.** 715 **5.** 1260 **7.** 286 **9.** 72 **11.** 1575 **13.** 72 **15.** 260 **17.** 350 **19.** 225

Problems 6-7

1. $\frac{5}{12}$ **3.** $\frac{2}{3}$ **5.** $\frac{1}{2}$ **7.** $\frac{11}{16}$ **9.** $-\frac{1}{8}$ **11.** $-\frac{19}{48}$ **13.** $\frac{11}{12}$ **15.** $\frac{18}{25}$ **17.** $\frac{3}{4}$ **19.** $-\frac{11}{56}$ **21.** $\frac{152}{275}$
23. $\frac{1}{13}$

Problems 6-8

1. $8\frac{1}{2}$ **3.** $9\frac{1}{4}$ **5.** $3\frac{4}{5}$ **7.** $6\frac{1}{7}$ **9.** $6\frac{3}{4}$ **11.** $9\frac{4}{7}$ **13.** $5\frac{7}{16}$ **15.** $7\frac{5}{12}$ **17.** $6\frac{2}{3}$ **19.** $2\frac{7}{15}$
21. $\frac{21}{2}$ **23.** $\frac{67}{16}$ **25.** $\frac{29}{8}$ **27.** $\frac{28}{3}$ **29.** $\frac{31}{6}$ **31.** $\frac{71}{8}$ **33.** $\frac{37}{3}$ **35.** $\frac{22}{3}$
37. $\frac{53}{5}$ **39.** $\frac{67}{16}$

Problems 6-9

1. $1\frac{1}{4}$ **3.** $1\frac{1}{5}$ **5.** $\frac{5}{8}$ **7.** $4\frac{19}{20}$ **9.** $9\frac{3}{4}$ **11.** $7\frac{7}{12}$ **13.** 14 **15.** $21\frac{3}{8}$ **17.** $10\frac{7}{18}$ **19.** $25\frac{2}{3}$
21. $2\frac{5}{8}$ **23.** $1\frac{3}{4}$ **25.** $2\frac{1}{7}$ **27.** $9\frac{3}{5}$ **29.** $1\frac{7}{20}$ **31.** $\frac{5}{8}$ **33.** $\frac{21}{26}$ **35.** $\frac{14}{39}$ **37.** $1\frac{17}{25}$ **39.** $3\frac{1}{3}$

Problems 6-10

1. $6\frac{8}{15}$ **3.** $5\frac{23}{42}$ **5.** $4\frac{2}{15}$ **7.** $8\frac{3}{8}$ **9.** $\frac{21}{32}$ **11.** $3\frac{1}{12}$ **13.** $1\frac{7}{12}$ **15.** $1\frac{7}{8}$ **17.** $2\frac{1}{6}$ **19.** $3\frac{5}{12}$
21. $8\frac{5}{12}$ **23.** $3\frac{15}{16}$ **25.** $5\frac{11}{32}$ **27.** $6\frac{8}{15}$ **29.** $9\frac{23}{24}$

Problems 6-11

1. $\frac{3}{8}+\frac{3}{32}=0.469$
$\frac{3}{8}-\frac{3}{32}=0.281$
$\frac{3}{8}\times\frac{3}{32}=0.0352$
$\frac{3}{8}\div\frac{3}{32}=4.00$

3. $\frac{7}{16}+\frac{2}{7}=0.723$
$\frac{7}{16}-\frac{2}{7}=0.152$
$\frac{7}{16}\times\frac{2}{7}=0.125$
$\frac{7}{16}\div\frac{2}{7}=1.53$

5. $\frac{5}{8}+\frac{37}{64}=1.20$
$\frac{5}{8}-\frac{37}{64}=0.0469$
$\frac{5}{8}\times\frac{37}{64}=0.361$
$\frac{5}{8}\div\frac{37}{64}=1.08$

7. $\frac{3}{5}+\frac{3}{10}=0.900$
$\frac{3}{5}-\frac{3}{10}=0.300$
$\frac{3}{5}\times\frac{3}{10}=0.180$
$\frac{3}{5}\div\frac{3}{10}=2.00$

9. $3\frac{3}{8}+2\frac{1}{3}=5.71$
$3\frac{3}{8}-2\frac{1}{3}=1.04$
$3\frac{3}{8}\times 2\frac{1}{3}=7.88$
$3\frac{3}{8}\div 2\frac{1}{3}=1.45$

11. $4\frac{2}{5}+2\frac{15}{64}=6.63$
$4\frac{2}{5}-2\frac{15}{64}=2.17$
$4\frac{2}{5}\times 2\frac{15}{64}=9.83$
$4\frac{2}{5}\div 2\frac{15}{64}=1.97$

13. $7\frac{1}{2}+3\frac{5}{9}=11.1$
$7\frac{1}{2}-3\frac{5}{9}=3.94$
$7\frac{1}{2}\times 3\frac{5}{9}=26.7$
$7\frac{1}{2}\div 3\frac{5}{9}=2.11$

15. $4\frac{7}{8}+8\frac{7}{16}=13.3$
$4\frac{7}{8}-8\frac{7}{16}=-3.56$
$4\frac{7}{8}\times 8\frac{7}{16}=41.1$
$4\frac{7}{8}\div 8\frac{7}{16}=0.578$

17. $3\frac{2}{3} + 6\frac{2}{3} = 10.3$
$3\frac{2}{3} - 6\frac{2}{3} = -3.00$
$3\frac{2}{3} \times 6\frac{2}{3} = 24.4$
$3\frac{2}{3} \div 6\frac{2}{3} = 0.550$

19. $1\frac{7}{12} + 3\frac{2}{3} = 5.25$
$1\frac{7}{12} - 3\frac{2}{3} = -2.08$
$1\frac{7}{12} \times 3\frac{2}{3} = 5.81$
$1\frac{7}{12} \div 3\frac{2}{3} = 0.432$

CHAPTER 7

Problems 7-1

1. $2\cdot2\cdot2\cdot a\cdot a\cdot b\cdot b$ **3.** $2\cdot2\cdot2\cdot3\cdot x\cdot x\cdot y$ **5.** $2\cdot3\cdot7\cdot c\cdot c\cdot c$ **7.** $2\cdot2\cdot2\cdot11\cdot y\cdot y\cdot z$
9. $3\cdot3\cdot3\cdot5\cdot a\cdot a\cdot c\cdot c$

Problems 7-2

1. a^3b^3 **3.** $a^3b^3c^3$ **5.** $x^2y^3z^4$ **7.** a^3b^3 **9.** $a^4b^2c^3$ **11.** $a^4b^3c^2$ **13.** $78x^2y^3$
15. $32x^3y^2z^2$ **17.** $150a^2b^2c$ **19.** $270x^2y^4z^3$

Problems 7-3

1. $21a^4b^3$ **3.** $20a^{-4}b^{-5}$ **5.** $20x^6y^3z^4$ **7.** $6a^3b^{-2}c^{-1}$ **9.** $60x^5y^{-4}z^4$

Problems 7-4

1. $9a^2$ **3.** $8ab^2$ **5.** $5a^3bc^{-1}$ **7.** $\frac{xy}{5}$ **9.** $11x^{-5}y^{-1}z^{-4}$

Problems 7-5

1. $\frac{b^2}{10}$ **3.** $\frac{15}{14b^2}$ **5.** $\frac{10a^3}{21}$ **7.** $\frac{3x^3y^{-2}}{16}$ **9.** $\frac{2xy^{-1}}{3}$ **11.** $\frac{16xy^{-2}z}{3}$ **13.** $\frac{a^4b^{-2}c^{-1}}{24}$ **15.** $\frac{x^2y^{-4}}{18}$
17. $\frac{a^4b^{-4}}{3}$ **19.** $\frac{4x^{-1}y^{-1}z^2}{3}$

Problems 7-6

1. $\frac{9x^{-2}}{56}$ **3.** $\frac{2}{3x}$ **5.** $18b^{-3}$ **7.** $\frac{2abc^{-1}}{45}$ **9.** 1 **11.** $\frac{10a^{-1}b^{-1}c}{9}$ **13.** $\frac{a^3c^{-1}}{10}$ **15.** $\frac{2x^2y}{9}$
17. $\frac{9a^4b^{-4}c^{-6}}{8}$ **19.** $\frac{4x^{-3}y^{-1}z^{-1}}{5}$

Problems 7-7

1. $\frac{5x^{-1}}{7}$ **3.** $\frac{4x^{-1}}{7}$ **5.** $\frac{3a^{-1}b^{-1}}{4}$ **7.** $\frac{a^{-1}}{2}$ **9.** $\frac{x}{2}$ **11.** $\frac{2x}{5}$ **13.** $\frac{3ab}{5}$ **15.** $\frac{3ab}{5}$ **17.** $\frac{10x}{9}$
19. $-\frac{2x}{7}$ **21.** $-\frac{5x}{24}$ **23.** $\frac{2ab+3b}{8}$ **25.** $\frac{7abc-3ac}{4d}$ **27.** $-\frac{2abc+3ab}{4x}$ **29.** $\frac{6xy-4xz+3yz}{5}$

Problems 7-8

1. $\frac{11x}{15}$ **3.** $\frac{13a}{6}$ **5.** $-\frac{x}{6}$ **7.** $\frac{x}{2}$ **9.** $\frac{4y+5xy}{8x}$ **11.** $\frac{9x^2+8y^2}{12xy}$ **13.** $\frac{4by+3x}{10xy}$ **15.** $\frac{10ab^2c-15ab}{32c^2}$

17. $\frac{3b}{16}$ **19.** $\frac{44a}{105}$ **21.** $\frac{x}{30y^2}$ **23.** $\frac{40x^2 + 12y^2z - 5xz^2}{30xyz}$ **25.** $\frac{4a^2c + 3ab^2 - 9bc^2}{33abc}$
27. $\frac{75y^3 + 24x^3y - 20x^4y}{90x^2y^2}$ **29.** $\frac{9b^4 + 42ab^2c - 25a^3}{60a^2b^3}$

CHAPTER 8

Problems 8-1

1. 10 **3.** 2 **5.** 3 **7.** 9 **9.** 4 **11.** −4 **13.** 4 **15.** −9 **17.** 12 **19.** −7 **21.** 12
23. −10 **25.** 16 **27.** 12 **29.** $\frac{1}{10}$ **31.** 8 **33.** $\frac{1}{4}$ **35.** −6 **37.** $-\frac{2}{7}$ **39.** $-1\frac{1}{3}$ **41.** 6
43. $1\frac{1}{14}$ **45.** $\frac{3}{28}$ **47.** $\frac{2}{3}$ **49.** $1\frac{7}{20}$

Problems 8-2

1. 5 **3.** 7.35 **5.** 7.48 **7.** 1.22 **9.** 2.45 **11.** 49 **13.** 36 **15.** 190 **17.** $\frac{1}{16}$ **19.** $\frac{1}{144}$
21. $\frac{1}{64}$ **23.** 81 **25.** $a = \frac{6}{x}, x = \frac{6}{a}$ **27.** $b = \frac{7}{4c}, c = \frac{7}{4b}$ **29.** $R = \frac{E^2}{P}, E = \sqrt{PR}$
31. $f = \frac{B_C}{2\pi C}, C = \frac{B_C}{2\pi f}$ **33.** $a = \frac{1}{6\sqrt{b}}, b = \frac{1}{36a^2}$ **35.** $y = \frac{36}{25x}, x = \frac{36}{25y}$ **37.** $f = \frac{X_L}{2\pi L}, L = \frac{X_L}{2\pi f}$
39. $L = \frac{1}{(2\pi f_r)^2C}, C = \frac{1}{(2\pi f_r)^2L}$

Problems 8-3

1. $R_1 = 6.8$ kΩ **3.** $G_1 = 17.9\ \mu$S **5.** $V_o = 8.25$ V **7.** $R = 113$ Ω **9.** $R = 1.35$ Ω
11. $R = 34.1$ kΩ **13.** $E = 13.5$ V **15.** $E = 136$ V **17.** $E = 285$ V **19.** $I = 29.7$ mA
21. $I = 13.6$ mA **23.** $I = 383$ mA **25.** $R = 5.00$ kΩ **27.** $R = 268$ Ω **29.** $R = 7.32$ Ω
31. $N_P = 1770$ **33.** $N_P = 14.1$ **35.** $N_P = 79.1$ **37.** $Z_S = 6.94$ Ω **39.** $Z_S = 745$ Ω **41.** $Z_S = 1.83$ Ω
43. $N_S = 268$ **45.** $N_S = 141$ **47.** $Z_P = 3.20$ Ω **49.** $Z_P = 1.00$ kΩ **51.** $C = 84.2$ nF
53. $f = 3.18$ kHz **55.** $L = 128$ mH **57.** $f = 500$ Hz **59.** $L = 281$ mH **61.** $C = 25.3\ \mu$F

CHAPTER 9

Problems 9-1

1. $3a + 9$ **3.** $4a - 4$ **5.** $28 - 7b$ **7.** $-3a + 2$ **9.** $-6a - 6$ **11.** $6x - 12$ **13.** $6x + 6xy$
15. $6a^2 + 9ab$ **17.** $9x^2 + 18x - 36$ **19.** $8a^2 - 6a + 4$ **21.** $a^2 - 11ab$ **23.** $-a^2b + 8ab$
25. $5x^2y - 9xy$ **27.** $2x^2y - 20xy$ **29.** $3x^3 + 6x^2 + 3x$ **31.** $a^3 - 27a^2 + 42a$ **33.** $a^2 + 4a - 4ab$
35. $3x^2 + 17x - 18xy$

Problems 9-2

1. $\frac{44x^2}{3}$ **3.** $5y^2$ **5.** $8a^2 + 9a$ **7.** $5b^2 - 6b$ **9.** $6a^2 - 16a + 8$ **11.** $2a^2 - 6a + \frac{9}{2}$
13. $-2x^2 + 7x$ **15.** $8x^2 - 15x$ **17.** $5b^2 - 6b$ **19.** $\frac{6x^2}{5} + \frac{21b}{4}$

Problems 9-3

1. $a^2 + 6a + 9$ **3.** $a^2 - 12a + 36$ **5.** $4x^2 - 16x + 16$ **7.** $16x^2 + 24x + 9$ **9.** $4a^2 + 12ab + 9b^2$
11. $x^6 - 8x^3y + 16y^2$ **13.** $4x^2 - 12xy + 9y^2$ **15.** $a^2 + 4a + 3$ **17.** $y^2 - 3y - 18$ **19.** $a^2 + 4a - 12$

21. $x^2 - 5x + 4$ **23.** $a^2 - 16$ **25.** $4x^2 - 9$ **27.** $x^2 + 4xy + 3y^2$ **29.** $x^2 - xy - 20y^2$ **31.** $4a^2 + 5ab - 6b^2$ **33.** $2a^2 - 10ab + 8b^2$ **35.** $6a^2 + ab - 12b^2$ **37.** $6x^2 + 23xy + 20y^2$ **39.** $4x^2 - 16xy + 15y^2$

Problems 9-4

1. $x^2 + 3x + 2$ **3.** $3x^2 + x + 2$ **5.** $4x^2 - 3x + 2$ **7.** $x + 3y$ **9.** $x - 3$ **11.** $x - 7$ **13.** $3x - 2$ **15.** $a - 5$ **17.** $2x + 3y$ **19.** $3a - 4b$ **21.** $x - 9y$ **23.** $8 + 4x$ **25.** $4x - 4 + \dfrac{3}{3x + 2}$ **27.** $2x + 3y - \dfrac{2xy}{4x + 5y}$ **29.** $4a - 2 + \dfrac{4}{a + 1}$ **31.** $6a + 3 + \dfrac{1}{a - 1}$ **33.** $a + b + \dfrac{2ab}{a + b}$ **35.** $6x^2 - 3x + 7 - \dfrac{3}{x + 1}$

Problems 9-5

1. $3(a + 3)$ **3.** $7(b - 3)$ **5.** $2xy(2x + 3)$ **7.** $2a(8a + 1)$ **9.** $8abc^2(3ac - 2b)$ **11.** $3a^2b^2(4a + 3b^2 - 2a^2b^2)$ **13.** $5xy^2(2x^2 - 4x^3y^2 + 3y)$ **15.** $3b^2c(5a^3c^3 - a^2bc^3 - 4)$ **17.** $5x^2y^2z^2(7x^2z + 5y^2z^2 - 6)$ **19.** $6x^2yz^2(4xyz^2 - 7y^2 - 12)$ **21.** $\dfrac{7x^2z^2}{3}\left(\dfrac{z}{4} - \dfrac{2xy^2}{3}\right)$ **23.** $\dfrac{6x^2y^2z}{5}\left(\dfrac{2x^2}{7} + \dfrac{4yz^2}{11} - \dfrac{3xyz^3}{5}\right)$ **25.** $\dfrac{7a^2c^2}{11b^2}\left(\dfrac{3a}{2} + \dfrac{4}{3b} - \dfrac{6ac}{4}\right)$

Problems 9-6

1. $(x + 2)(x + 3)$ **3.** $(x + 8)(x - 4)$ **5.** $(a + 3)^2$ **7.** $(a + 4)(a - 4)$ **9.** $(x + 7)(x - 7)$ **11.** $(x - 5)(x - 3)$ **13.** $(a - 7)(a - 5)$ **15.** $(a - 4)(a - 7)$ **17.** $(y + 9)(y - 4)$ **19.** $(x - 7)(x + 5)$ **21.** $(x + 4)(x - 6)$ **23.** $(2x + 3)(3x - 2)$ **25.** $(4a - 2)(2a - 4)$ **27.** $(3x + 4)(3x - 2)$ **29.** $(3b - 6)(2b - 2)$ **31.** $(5a + 3)(2a + 5)$ **33.** $(2x + 3y)(x + 2y)$ **35.** $(4a - 3b)(5a + 2b)$ **37.** $(2x + 6y)(4x - 2y)$ **39.** $(3x + 3y)(4x + 2y)$

CHAPTER 10

Problems 10-1

1. $x = 1\frac{1}{5}$ **3.** $a = 5\frac{1}{4}$ **5.** $I = 3$ **7.** $x = -12$ **9.** $R = 1\frac{4}{5}$ **11.** $R = -4\frac{3}{5}$ **13.** $R = -4\frac{3}{7}$ **15.** $c = -1$ **17.** $I = 6$ **19.** $R = 6$ **21.** $x = -\dfrac{14}{23}$ **23.** $a = -\dfrac{15}{17}$ **25.** $x = -4$ **27.** $y = -7$ **29.** $G = 3\frac{4}{7}$ **31.** $R = -2\frac{7}{13}$ **33.** $a = \dfrac{b}{2b - 1}$, $b = \dfrac{a}{2a - 1}$ **35.** $a = \dfrac{2b}{1 + 3b}$, $b = \dfrac{a}{2 - 3a}$ **37.** $a = \dfrac{3bx}{12b - 2x}$, $b = \dfrac{2ax}{12a - 3x}$, $x = \dfrac{12ab}{3b + 2a}$ **39.** $a = \dfrac{9bx}{24b + 2x}$, $b = \dfrac{2ax}{9x - 24a}$, $x = \dfrac{24ab}{9b - 2a}$ **41.** $a = \dfrac{2b + 2}{b - 5}$, $b = \dfrac{5a + 2}{a - 2}$ **43.** $a = \dfrac{4b - 1}{12 - 6b}$, $b = \dfrac{12a + 1}{6a + 4}$ **45.** $b = 1\frac{1}{2}$

Problems 10-2

1. $V_{OC} = \dfrac{V_L(R_{TH} + R_L)}{R_L}$, $R_L = \dfrac{V_L R_{TH}}{V_{OC} - V_L}$, $R_{TH} = \dfrac{V_{OC}R_L - V_L R_L}{V_L}$

3. $I_{SC} = \dfrac{I_L(G_L + G_N)}{G_L}$, $G_L = \dfrac{I_L G_N}{I_{SC} - I_L}$, $G_N = \dfrac{I_{SC} G_L - I_L G_L}{I_L}$

5. $R_o = \dfrac{R_S R_i}{R_i - R_S}$, $R_i = \dfrac{R_S R_o}{R_o - R_S}$ **7.** $r_b = r_i - \beta r_e$ **9.** $R_1 = \dfrac{R_T R_2}{R_2 - R_T}$, $R_2 = \dfrac{R_T R_1}{R_1 - R_T}$ **11.** $\alpha = B - 1$

13. $R_1 = \dfrac{1}{2\pi fC} - R_2$ **15.** $V_B = V_{CC} - I_B(R_1 + R_2)$, $R_1 = \dfrac{V_{CC} - V_B - I_B R_2}{I_B}$

Problems 10-3

1. (a) $R_2 = 4.8\ \text{k}\Omega$ (b) $R_1 = 773\ \Omega$ **3.** (a) $I_T = 12$ mA (b) $G_2 = 1.79$ mS
5. (a) $h_{oe} = 74.3\ \mu$S (b) $R_L = 20.8\ \text{k}\Omega$ **7.** $R_1 = 3\ \text{k}\Omega$ **9.** $Z_2 = 220\ \text{k}\Omega$ **11.** $R_2 = 3.94\ \text{k}\Omega$
13. $R_B = 550\ \text{k}\Omega$ **15.** $R_1 = 9.34\ \text{k}\Omega$

Problems 10-4

1. $x = 4, x = -4$ **3.** $x = -4, x = -2$ **5.** $x = 3, x = -6$ **7.** $x = 2, x = 4$ **9.** $x = 8, x = -7$
11. $x = 8, x = -5$ **13.** $x = 4, x = 9$ **15.** $x = 6, x = -3$ **17.** $x = -7, x = -3$ **19.** $x = 1, x = 2$

Problems 10-5

1. $x = -6, x = -2$ **3.** $x = -4.73, x = -1.27$ **5.** $x = -4.87, x = 2.87$ **7.** $x = 15.9, x = -0.941$
9. $x = 7.53, x = -0.531$ **11.** $x = 0.333, x = -1$ **13.** $x = 3.27, x = -0.766$ **15.** $x = -2, x = -1$
17. $x = -1.77, x = -0.566$ **19.** $x = 5.81, x = -1.15$

CHAPTER 11

Problems 11-1

1. $I = 9$ A **3.** $I_3 = 3$ mA **5.** $I_2 = 45\ \mu$A **7.** $I_1 = 960$ mA **9.** $I_T = 25$ mA, $I_3 = 15$ mA
11. $I_1 = 800$ mA, $I_2 = 800$ mA **13.** $I_2 = 200\ \mu$A, $I_3 = 200\ \mu$A **15.** $I_T = 1.6$ A, $I_3 = 800$ mA, $I_4 = 800$ mA
17. $I_T = 3.2$ mA, $I_1 = 3.2$ mA, $I_2 = 500\ \mu$A **19.** $I_1 = 1.7$ mA, $I_3 = 3.8$ mA **21.** $I_T = 14.2$ mA, $I_2 = 4.8$ mA
23. $I_1 = 30$ mA, $I_3 = 10$ mA, $I_4 = 10$ mA, $I_6 = 70$ mA

Problems 11-2

1. $V_2 = 18$ V **3.** $E = 13$ V **5.** $V_3 = 11$ V **7.** $E = 18$ V **9.** $V_1 = 14$ V **11.** $V_1 = 12$ V, $V_2 = 8$ V
13. $V_2 = 13$ V, $V_3 = 13$ V **15.** $E = 17$ V, $V_2 = 5$ V **17.** $V_1 = 3$ V, $V_4 = 9$ V
19. $V_1 = 1.7$ V, $V_3 = 800$ mV **21.** $E = 60$ V, $V_3 = 10$ V **23.** $V_2 = 25$ V, $V_4 = 17$ V, $V_6 = 50$ V
25. $E = 25$ V, $V_2 = 12$ V, $V_3 = 5.5$ V

Problems 11-3

1. (a) 10 V (b) 5 V (c) 25 V (d) 30 V (e) −5 V
3. (a) 12 V (b) 43 V (c) −43 V (d) −35 V (e) −8 V **5.** (a) 11 V (b) 5 V (c) 19 V (d) −13 V
7. (a) $V_1 = 34$ V (b) $V_2 = 16$ V (c) 6 V (d) −50 V

CHAPTER 12

Problems 12-1

1. (a) $R_T = 45\ \text{k}\Omega$ (b) $R_T = 1.95\ \text{k}\Omega$ (c) $R_T = 16.8\ \text{k}\Omega$ (d) $R_T = 300\ \text{k}\Omega$
3. (a) $R_2 = 1.8\ \text{k}\Omega$ (b) $R_2 = 39\ \text{k}\Omega$
5. (a) $R_T = 2.29\ \text{k}\Omega$ (b) $R_T = 130\ \text{k}\Omega$ (c) $R_T = 297\ \Omega$ (d) $R_T = 1.36\ \text{M}\Omega$
7. (a) $R_3 = 5.6\ \text{k}\Omega$ (b) $R_3 = 82\ \text{k}\Omega$
9. (a) $G_T = 150\ \mu$S, $R_T = 6.67\ \text{k}\Omega$ (b) $G_T = 1.2$ mS, $R_T = 831\ \Omega$ (c) $G_T = 76.8\ \mu$S, $R_T = 13.0\ \text{k}\Omega$
(d) $G_T = 20\ \mu$S, $R_T = 50.0\ \text{k}\Omega$
11. (a) $G_T = 685\ \mu$S, $R_T = 1.46\ \text{k}\Omega$ (b) $G_T = 31.9\ \mu$S, $R_T = 31.3\ \text{k}\Omega$ (c) $G_T = 2.39$ mS, $R_T = 419\ \Omega$
(d) $G_T = 26.7\ \mu$S, $R_T = 37.4\ \text{k}\Omega$
13. (a) $R_2 = 4.7\ \text{k}\Omega$ (b) $R_2 = 199\ \Omega$

Problems 12-2

1. (a) $R_T = 1.01\ \text{k}\Omega$ (b) $R_T = 18.3\ \text{k}\Omega$ (c) $R_T = 36.6\ \text{k}\Omega$ (d) $R_T = 735\ \text{k}\Omega$
3. (a) $R_T = 5.08\ \text{k}\Omega$ (b) $R_T = 657\ \Omega$ (c) $R_T = 25\ \text{k}\Omega$ (d) $R_T = 755\ \Omega$
5. (a) $R_T = 23.3\ \text{k}\Omega$ (b) $R_T = 324\ \Omega$ (c) $R_T = 685\ \text{k}\Omega$ (d) $R_T = 7.98\ \text{k}\Omega$
7. 1.11 kΩ **9.** 95.1 kΩ

Problems 12-3

1. (a) $R_T = 950\ \Omega$, $I = 9.47$ mA, $V_1 = 2.56$ V, $V_2 = 6.44$ V
(b) $R_T = 4.5\ \text{k}\Omega$, $I = 2.67$ mA, $V_1 = 3.2$ V, $V_2 = 8.80$ V
(c) $R_T = 340\ \text{k}\Omega$, $I = 118\ \mu\text{A}$, $V_1 = 14.1$ V, $V_2 = 25.9$ V
(d) $R_T = 13.3\ \text{k}\Omega$, $I = 752\ \mu\text{A}$, $V_1 = 7.52$ V, $V_2 = 2.48$ V

3. (a) $R_T = 1.43\ \text{k}\Omega$, $I = 7$ mA, $V_1 = 7$ V, $R_2 = 429\ \Omega$
(b) $R_T = 16.5\ \text{k}\Omega$, $I = 2.43$ mA, $V_1 = 16.5$ V, $R_2 = 9.68\ \text{k}\Omega$
(c) $R_T = 41.0\ \text{k}\Omega$, $I = 366\ \mu\text{A}$, $V_1 = 3$ V, $R_2 = 32.8\ \text{k}\Omega$
(d) $R_T = 311\ \text{k}\Omega$, $I = 321\ \mu\text{A}$, $V_1 = 70.7$ V, $R_2 = 91.2\ \text{k}\Omega$

5. (a) $R_1 = 60.9\ \text{k}\Omega$, $R_T = 93.9\ \text{k}\Omega$, $V_1 = 13$ V, $V_2 = 7.03$ V
(b) $R_1 = 15\ \text{k}\Omega$, $R_T = 25\ \text{k}\Omega$, $V_1 = 15$ V, $V_2 = 10$ V
(c) $R_1 = 560\ \Omega$, $R_T = 1.56\ \text{k}\Omega$, $V_1 = 4.31$ V, $V_2 = 7.69$ V
(d) $R_1 = 270\ \text{k}\Omega$, $R_T = 450\ \text{k}\Omega$, $V_1 = 5.40$ V, $V_2 = 3.60$ V

7. (a) $V_2 = 225$ mV, $R_1 = 82.7\ \Omega$, $R_T = 3.08\ \text{k}\Omega$, $E = 231$ mV
(b) $V_2 = 24$ V, $R_1 = 7.15\ \text{k}\Omega$, $R_T = 19.2\ \text{k}\Omega$, $E = 38.3$ V
(c) $V_2 = 5.50$ V, $R_1 = 27\ \text{k}\Omega$, $R_T = 60\ \text{k}\Omega$, $E = 10$ V
(d) $V_2 = 1.22$ V, $R_1 = 120\ \Omega$, $R_T = 202\ \Omega$, $E = 3$ V

9. (a) $R_T = 61\ \text{k}\Omega$, $I = 148\ \mu\text{A}$, $V_1 = 4.87$ V, $V_2 = 1.48$ V, $V_3 = 2.66$ V
(b) $R_T = 388\ \text{k}\Omega$, $I = 103\ \mu\text{A}$, $V_1 = 22.7$ V, $V_2 = 10.3$ V, $V_3 = 7.01$ V
(c) $R_T = 2.58\ \text{k}\Omega$, $I = 9.69$ mA, $V_1 = 7.95$ V, $V_2 = 11.6$ V, $V_3 = 5.43$ V
(d) $R_T = 8.60\ \text{k}\Omega$, $I = 1.74$ mA, $V_1 = 3.14$ V, $V_2 = 7.50$ V, $V_3 = 4.36$ V

11. (a) $V_2 = 10$ V, $V_3 = 10$ V, $R_1 = 5\ \text{k}\Omega$, $R_3 = 10\ \text{k}\Omega$, $R_T = 25\ \text{k}\Omega$
(b) $V_2 = 7.05$ V, $V_3 = 6.25$ V, $R_1 = 44.7\ \text{k}\Omega$, $R_3 = 41.7\ \text{k}\Omega$, $R_T = 133\ \text{k}\Omega$
(c) $V_2 = 13.6$ V, $V_3 = 36.5$ V, $R_1 = 18\ \text{k}\Omega$, $R_3 = 22\ \text{k}\Omega$, $R_T = 48.2\ \text{k}\Omega$
(d) $V_2 = 1.41$ V, $V_3 = 1.16$ V, $R_1 = 27.9\ \Omega$, $R_3 = 74.6\ \Omega$, $R_T = 194\ \Omega$

13. (a) $E = 36$ V, $R_2 = 2.43\ \text{k}\Omega$, $R_3 = 4.87\ \text{k}\Omega$, $V_1 = 14.1$ V, $V_3 = 14.6$ V
(b) $E = 11.7$ V, $R_2 = 13.7\ \text{k}\Omega$, $R_3 = 14.6\ \text{k}\Omega$, $V_1 = 4.05$ V, $V_3 = 3.95$ V
(c) $E = 30$ V, $R_2 = 100\ \text{k}\Omega$, $R_3 = 150\ \text{k}\Omega$, $V_1 = 13.3$ V, $V_3 = 10$ V
(d) $E = 2$ V, $R_2 = 300\ \Omega$, $R_3 = 99.6\ \Omega$, $V_1 = 1.08$ V, $V_3 = 229$ mV

Problems 12-4

1. (a) $G_T = 5.49$ mS, $R_T = 182\ \Omega$, $V = 547$ mV, $I_1 = 2.02$ mA, $I_2 = 977\ \mu\text{A}$
(b) $G_T = 40.3\ \mu\text{S}$, $R_T = 24.8\ \text{k}\Omega$, $V = 17.3$ V, $I_1 = 255\ \mu\text{A}$, $I_2 = 444\ \mu\text{A}$
(c) $G_T = 8.25$ mS, $R_T = 121\ \Omega$, $V = 4.56$ V, $I_1 = 20.7$ mA, $I_2 = 16.9$ mA
(d) $G_T = 48.2\ \mu\text{S}$, $R_T = 20.8\ \text{k}\Omega$, $V = 2.08$ V, $I_1 = 37.1\ \mu\text{A}$, $I_2 = 62.9\ \mu\text{A}$

3. (a) $G_T = 3.68$ mS, $R_T = 272\ \Omega$, $V = 6.8$ V, $I_1 = 10$ mA, $R_2 = 453\ \Omega$
(b) $G_T = 694\ \mu\text{S}$, $R_T = 1.44\ \text{k}\Omega$, $V = 7.20$ V, $I_1 = 4$ mA, $R_2 = 7.20\ \text{k}\Omega$
(c) $G_T = 391\ \mu\text{S}$, $R_T = 2.56\ \text{k}\Omega$, $V = 5.88$ V, $I_1 = 1.25$ mA, $R_2 = 5.6\ \text{k}\Omega$
(d) $G_T = 360\ \mu\text{S}$, $R_T = 2.78\ \text{k}\Omega$, $V = 83.4$ V, $I_1 = 17.7$ mA, $R_2 = 6.8\ \text{k}\Omega$

5. (a) $G_T = 500\ \mu\text{S}$, $R_T = 2\ \text{k}\Omega$, $I_1 = 4.26$ mA, $I_2 = 5.74$ mA, $R_2 = 3.48\ \text{k}\Omega$
(b) $G_T = 14\ \mu\text{S}$, $R_T = 71.4\ \text{k}\Omega$, $I_1 = 139\ \mu\text{A}$, $I_2 = 211\ \mu\text{A}$, $R_2 = 118\ \text{k}\Omega$
(c) $G_T = 2.05$ mS, $R_T = 488\ \Omega$, $I_1 = 11.9$ mA, $I_2 = 8.12$ mA, $R_2 = 1.2\ \text{k}\Omega$
(d) $G_T = 41.2\ \mu\text{S}$, $R_T = 24.3\ \text{k}\Omega$, $I_1 = 1.13$ mA, $I_2 = 870\ \mu\text{A}$, $R_2 = 55.9\ \text{k}\Omega$

7. (a) $G_T = 94.2\ \mu\text{S}$, $R_T = 10.6\ \text{k}\Omega$, $I_T = 1.13$ mA, $I_2 = 600\ \mu\text{A}$, $R_1 = 22.6\ \text{k}\Omega$
(b) $G_T = 267\ \mu\text{S}$, $R_T = 3.75\ \text{k}\Omega$, $I_T = 1.33$ mA, $I_2 = 333\ \mu\text{A}$, $R_1 = 5\ \text{k}\Omega$
(c) $G_T = 155\ \mu\text{S}$, $R_T = 6.44\ \text{k}\Omega$, $I_T = 4.8$ mA, $I_2 = 3.09$ mA, $R_1 = 18.1\ \text{k}\Omega$
(d) $G_T = 133\ \mu\text{S}$, $R_T = 7.53\ \text{k}\Omega$, $I_T = 287\ \mu\text{A}$, $I_2 = 23.7\ \mu\text{A}$, $R_1 = 8.21\ \text{k}\Omega$

9. (a) $G_T = 2.2$ mS, $R_T = 454\ \Omega$, $V = 785$ mV, $I_1 = 291\ \mu\text{A}$, $I_2 = 654\ \mu\text{A}$, $I_3 = 785\ \mu\text{A}$
(b) $G_T = 469\ \mu\text{S}$, $R_T = 2.13\ \text{k}\Omega$, $V = 1.07$ V, $I_1 = 131\ \mu\text{A}$, $I_2 = 142\ \mu\text{A}$, $I_3 = 227\ \mu\text{A}$
(c) $G_T = 89.0\ \mu\text{S}$, $R_T = 11.2\ \text{k}\Omega$, $V = 5.62$ V, $I_1 = 255\ \mu\text{A}$, $I_2 = 144\ \mu\text{A}$, $I_3 = 101\ \mu\text{A}$
(d) $G_T = 1.18$ mS, $R_T = 846\ \Omega$, $V = 1.69$ V, $I_1 = 627\ \mu\text{A}$, $I_2 = 434\ \mu\text{A}$, $I_3 = 942\ \mu\text{A}$

11. (a) $G_T = 317\ \mu S$, $R_T = 3.15\ k\Omega$, $R_1 = 18\ k\Omega$, $R_3 = 5.6\ k\Omega$, $I_2 = 1.05\ mA$, $I_3 = 2.25\ mA$
(b) $G_T = 15\ \mu S$, $R_T = 66.7\ k\Omega$, $R_1 = 267\ k\Omega$, $R_3 = 133\ k\Omega$, $I_2 = 74.1\ \mu A$, $I_3 = 151\ \mu A$
(c) $G_T = 12.8\ \mu S$, $R_T = 78.0\ k\Omega$, $R_1 = 678\ k\Omega$, $R_3 = 99.9\ k\Omega$, $I_2 = 31.2\ \mu A$, $I_3 = 234\ \mu A$
(d) $G_T = 4.36\ mS$, $R_T = 230\ \Omega$, $R_1 = 910\ \Omega$, $R_3 = 560\ \Omega$, $I_2 = 20.3\ mA$, $I_3 = 24.6\ mA$

13. (a) $I_1 = 3.2\ mA$, $I_3 = 8.8\ mA$, $I_T = 15\ mA$, $R_1 = 4.7\ k\Omega$, $R_2 = 5\ k\Omega$, $R_3 = 1.7\ k\Omega$, $R_T = 1\ k\Omega$
(b) $I_1 = 60\ mA$, $I_3 = 120\ mA$, $I_T = 240\ mA$, $R_1 = 500\ \Omega$, $R_2 = 500\ \Omega$, $R_3 = 250\ \Omega$, $R_T = 125\ \Omega$
(c) $I_1 = 6.09\ mA$, $I_3 = 3.04\ mA$, $I_T = 20\ mA$, $R_1 = 1\ k\Omega$, $R_2 = 560\ \Omega$, $R_3 = 2\ k\Omega$, $R_T = 304\ \Omega$
(d) $I_1 = 70.6\ \mu A$, $I_3 = 26.5\ \mu A$, $I_T = 150\ \mu A$, $R_1 = 75\ k\Omega$, $R_2 = 100\ k\Omega$, $R_3 = 200\ k\Omega$, $R_T = 35.3\ k\Omega$

Problems 12-5

1. (a) $R_T = 29.4\ k\Omega$, $I_T = 510\ \mu A$, $I_1 = I_T = 510\ \mu A$, $I_2 = 300\ \mu A$, $I_3 = 211\ \mu A$, $V_1 = 5.1\ V$, $V_2 = V_3 = 9.9\ V$
(b) $R_T = 487\ \Omega$, $I_T = 18.5\ mA$, $I_1 = 18.5\ mA$, $I_2 = 10.1\ mA$, $I_3 = 8.34\ mA$, $V_1 = 3.33\ V$, $V_2 = V_3 = 5.67\ V$, $V_3 = 5.67\ V$

3. (a) $V_1 = 6\ V$, $V_2 = V_3 = 4\ V$, $I_2 = 85.1\ \mu A$, $I_3 = 415\ \mu A$, $R_3 = 9.64\ k\Omega$, $R_T = 20\ k\Omega$
(b) $R_1 = 4.05\ k\Omega$, $R_T = 20\ k\Omega$, $V_1 = 2.43\ V$, $V_2 = 9.57\ V$, $V_3 = 9.57\ V$, $I_2 = 245\ \mu A$, $I_3 = 355\ \mu A$

5. (a) $R_T = 11.8\ k\Omega$, $I_T = 5.10\ mA$, $I_1 = I_2 = 1.77\ mA$, $I_3 = 3.33\ mA$, $V_1 = 21.2\ V$, $V_2 = 38.9\ V$, $V_3 = 60\ V$
(b) $R_T = 19.9\ k\Omega$, $I_T = 151\ \mu A$, $I_1 = 87.2\ \mu A$, $I_2 = 87.2\ \mu A$, $I_3 = 63.8\ \mu A$, $V_1 = 652\ mV$, $V_2 = 2.35\ V$, $V_3 = 3\ V$

7. (a) $R_T = 1.22\ k\Omega$, $I_T = 32.8\ mA$, $I_4 = 8.51\ mA$, $I_1 = 24.3\ mA$, $I_2 = 15.6\ mA$, $I_3 = 8.69\ mA$, $V_1 = 16.5\ V$, $V_2 = V_3 = 23.5\ V$, $V_4 = 40\ V$
(b) $R_T = 263\ k\Omega$, $I_T = 114\ \mu A$, $V_1 = 10.5\ V$, $V_2 = 19.5\ V$, $V_3 = 19.5\ V$, $V_4 = 30\ V$, $I_1 = 69.9\ \mu A$, $I_2 = 41.5\ \mu A$, $I_3 = 28.7\ \mu A$, $I_4 = 44.1\ \mu A$

9. (a) $R_T = 8.96\ k\Omega$, $I_T = 2.79\ mA$, $I_1 = 2.79\ mA$, $I_2 = 1.11\ mA$, $I_3 = 756\ \mu A$, $I_4 = 924\ \mu A$, $V_1 = 8.37\ V$, $V_2 = V_3 = V_4 = 16.6\ V$
(b) $R_T = 100\ k\Omega$, $R_1 = 60\ k\Omega$, $R_2 = 75\ k\Omega$, $I_2 = 133\ \mu A$, $I_3 = 66.7\ \mu A$, $I_4 = 50\ \mu A$, $I_1 = 250\ \mu A$

11. (a) $R_T = 61.2\ k\Omega$, $I_T = 653\ \mu A$, $I_1 = I_4 = I_T = 653\ \mu A$, $I_2 = 441\ \mu A$, $I_3 = 213\ \mu A$, $V_1 = 6.53\ V$, $V_2 = V_3 = 11.9\ V$, $V_4 = 21.6\ V$
(b) $R_T = 578\ k\Omega$, $I_T = I_1 = I_4 = 26\ \mu A$, $I_2 = 15.3\ \mu A$, $I_3 = 10.6\ \mu A$, $V_1 = 3.89\ V$, $V_2 = V_3 = 7.21\ V$, $V_4 = 3.89\ V$

CHAPTER 13

Problems 13-1

1. $V_1 = 41.4\ V$, $V_2 = 16.4\ V$, $V_3 = 33.6\ V$, $I_1 = 18.8\ mA$, $I_2 = 9.13\ mA$, $I_3 = 28\ mA$
3. $V_1 = 36.8\ V$, $V_2 = 45\ V$, $V_3 = 18.2\ V$, $V_4 = 21.8\ V$, $I_1 = 13.6\ mA$, $I_2 = 13.6\ mA$, $I_3 = 18.2\ mA$, $I_4 = 4.64\ mA$
5. $V_1 = 1\ V$, $V_2 = 10\ V$, $V_3 = 8\ V$, $I_1 = 0.303\ mA$, $I_2 = 2.13\ mA$, $I_3 = 2.42\ mA$
7. $V_1 = 14.8\ V$, $V_2 = 10.2\ V$, $V_3 = 0.2\ V$, $I_1 = 21.8\ mA$, $I_2 = 21.6\ mA$, $I_3 = 0.2\ mA$
9. $V_1 = 14.2\ V$, $V_2 = 7.73\ V$, $V_3 = 18.1\ V$, $V_4 = 3.12\ V$, $I_1 = 6.44\ mA$, $I_2 = 6.44\ mA$, $I_3 = 5.49\ mA$, $I_4 = 945\ \mu A$

Problems 13-2

1. $V_{OC} = 20.4\ V$, $R_{TH} = 15\ k\Omega$, $I_L = 8.16\ \mu A$, $V_L = 8.16\ V$
3. $V_{OC} = 12\ V$, $R_{TH} = 1\ k\Omega$, $I_L = 6.86\ mA$, $V_L = 5.14\ V$
5. $V_{OC} = 25.4\ V$, $R_{TH} = 16.2\ k\Omega$, $I_L = 814\ \mu A$, $V_L = 12.2\ V$
7. $V_{OC} = 29.6\ V$, $R_{TH} = 2.56\ k\Omega$, $I_L = 8.3\ mA$, $V_L = 8.3\ V$
9. $V_{OC} = 1.5\ V$, $R_{TH} = 3.78\ k\Omega$, $I_L = 314\ \mu A$, $V_L = 0.314\ V$
11. $V_{OC} = 2.53\ V$, $R_{TH} = 5.94\ k\Omega$, $I_B = 43.9\ \mu A$, $I_C = 3.51\ mA$, $V_C = 13\ V$

Problems 13-3

1. (a) $I_{SC} = 1.36\ mA$, $G_N = 66.7\ \mu S$, $V_L = 8.16\ V$, $I_L = 816\ \mu A$
(b) $I_{SC} = 2.78\ mA$, $G_N = 85.9\ \mu S$, $V_L = 24.9\ V$, $I_L = 639\ \mu A$

3. (a) $I_{SC} = 12\ mA$, $G_N = 1\ mS$, $V_L = 5.14\ V$, $I_L = 6.86\ mA$
(b) $I_{SC} = 496\ \mu A$, $G_N = 49.6\ \mu S$, $V_L = 3.31\ V$, $I_L = 331\ \mu A$

5. (a) $I_{SC} = 1.56$ mA, $G_N = 61.6\ \mu S$, $V_L = 12.2$ V, $I_L = 812\ \mu A$
(b) $I_{SC} = 1.52$ mA, $G_N = 299\ \mu S$, $V_L = 1.90$ V, $I_L = 951\ \mu A$

CHAPTER 14

Problems 14-3

1. x-intercept = 3, 0; y-intercept = 0, 6; slope = −2
3. x-intercept = −1, 0; y-intercept = 0, $\frac{5}{2}$; slope = $\frac{5}{2}$
5. x-intercept = 5, 0; y-intercept = 0, 3; slope = $-\frac{3}{5}$
7. x-intercept = −7, 0; y-intercept = 0, 2; slope = $\frac{2}{7}$
9. x-intercept = 8, 0; y-intercept = 0, −6; slope = $\frac{3}{4}$

Problems 14-5

1. $R_{dc} \approx 1.67\ k\Omega$, $r_{ac} \approx 1.14\ k\Omega$
3. $\Delta I_f \approx 6.2$ mA, slope = 31 mS, $r_d = 32.3\ \Omega$
5. (a) $V_C \approx 40$ V, $V_R \approx 60$ V (b) $V_C \approx 95$ V, $V_R \approx 5$ V
7. (a) $V_C \approx 15.6$ V, $V_R \approx 4.4$ V (b) $V_C \approx 19.6$ V, $V_R \approx 0.4$ V

CHAPTER 15

Problems 15-2

1. $x = 3, y = 2$
3. $x = 2, y = -4$
5. $x = 2, y = 1$
7. $x = -2, y = -1$
9. $x = 4, y = 3$
11. $x = 4.57, y = 1.71$
13. $x = 1.03, y = 0.147$

Problems 15-3

1. $x = 1.60, y = 1.20$
3. $x = 1.14, y = 0.786$
5. $x = 1.31, y = -0.517$
7. $x = 5.00, y = 5.00$
9. $x = -2.83, y = -1.17$

Problems 15-4

1. $x = 1.60, y = 1.20$
3. $x = 1.14, y = 0.786$
5. $x = 1.31, y = -0.517$
7. $x = 5.00, y = 5.00$
9. $x = -2.83, y = -1.17$

Problems 15-5

1. $x = 1$, $y = 2$, $z = 3$
3. $x = -21\frac{1}{3}$, $y = 20$, $z = 16\frac{1}{3}$
5. $x = 3.8$, $y = 33.2$, $z = -25$

Problems 15-6

1. $I_1 = 1.76$ mA
$I_2 = 1.17$ mA
$I_3 = 587\ \mu A$
$V_1 = 8.27$ V
$V_2 = 11.7$ V
$V_3 = 11.7$ V

Problems 15-7

1. $I_1 = 12.7$ mA, $I_2 = 9.08$ mA, $I_3 = 21.8$ mA, $V_1 = 38.2$ V, $V_2 = 18.2$ V, $V_3 = 21.8$ V
3. $I_1 = 4.63$ mA, $I_2 = 220\ \mu A$, $I_3 = 4.44$ mA, $V_1 = 55.6$ V, $V_2 = 4.44$ V, $V_3 = 44.4$ V
5. $I_1 = 251$ mA, $I_2 = 45.3$ mA, $I_3 = 204$ mA, $V_1 = 25.1$ V, $V_2 = 14.9$ V, $V_3 = 44.9$ V
7. $I_1 = 4$ mA, $I_2 = 2.79$ mA, $I_3 = 1.2$ mA, $V_1 = 10.8$ V, $V_2 = 9.2$ V, $V_3 = 1.2$ V

CHAPTER 16

Problems 16-1

1. $j13$ **3.** j **5.** -25 **7.** -18 **9.** $j35$ **11.** $j6.60$ **13.** $-j18$ **15.** 12 **17.** $j3$
19. $-j2$ **21.** -7 **23.** $-j20$

Problems 16-2

1. $-4 - j4$ **3.** $-4 - j8$ **5.** $3.1 - j4.5$ **7.** $11 + j7$ **9.** $1.9 - j4.2$ **11.** $-1 - j7$
13. $10 - j6$ **15.** $0 + j0$ **17.** $6 + j0$ **19.** $0 - j6$

Problems 16-3

1. $15 + j27$ **3.** $25 + j25$ **5.** $27 - j8$ **7.** $30 + j10$ **9.** $20 + j2$ **11.** $20 - j8$ **13.** $10 - j6$
15. $17 - j10$ **17.** $-10 - j12$ **19.** $-21 - j20$ **21.** $5 + j3$ **23.** $-1 - j5$ **25.** $1 + j20$
27. $20 + j50$ **29.** $12 - j10$ **31.** $4 - j24$ **33.** $-2 + j0$ **35.** $3 - j2$ **37.** $2 + j2$
39. $-9 + j0$

Problems 16-4

1. $19 - j4$ **3.** $13 + j0$ **5.** $7 - j24$ **7.** $-12 - j18$ **9.** $14 + j38$ **11.** $33 - j13$ **13.** $1 - j1$
15. $0.8 + j1.6$ **17.** $0.310 + j0.276$ **19.** $-1.04 + j0.280$ **21.** $0.780 + j0.0244$

CHAPTER 17

Problems 17-1

1. (a) $\phi = 55°$ (b) Side R is greater (c) Side R (d) Side R **3.** (a) 4.83 kΩ (b) 487 Ω (c) 98.9 kΩ
5. (a) 2.99 mS (b) 800 μS (c) 425 μS

Problems 17-2

1. $c = 26.1$; angle ϕ is greater. **3.** $b = 245$ **5.** $a = 68.1$ **7.** $Z = 976$ Ω; angle θ is greater.
9. $Z = 731$ kΩ; angle θ is greater. **11.** $Z = 84.5$ kΩ; angle ϕ is greater. **13.** $X = 6.8$ kΩ; angle θ is greater.
15. $X = 150$ kΩ; angle θ is greater. **17.** $R = 27$ kΩ; angle θ is greater. **19.** $R = 90.5$ Ω; angle ϕ is greater.
21. $Y = 832$ μS; angle θ is greater. **23.** $Y = 4.37$ mS; angle ϕ is greater. **25.** $B = 21.3$ μS; angle ϕ is greater.
27. $B = 4.54$ mS; angle ϕ is greater. **29.** $G = 556$ μS; angle ϕ is greater. **31.** $G = 13.3$ mS; angle θ is greater.

Problems 17-3

1. (a) $\sin\theta = \frac{X}{Z}$, $\cos\theta = \frac{R}{Z}$, $\tan\theta = \frac{X}{R}$ (b) $\sin\phi = \frac{R}{Z}$, $\cos\phi = \frac{X}{Z}$, $\tan\phi = \frac{R}{X}$

Problems 17-4

1. 0.4586 **3.** 0.8957 **5.** 0.2028 **7.** 0.9157 **9.** 2.391 **11.** 0.8481

Problems 17-5

1. 23.4° **3.** 60.5° **5.** 56.2° **7.** 38.7° **9.** 55.8°

Problems 17-6

1. $\theta = 32°$, $\phi = 58°$, $c = 47.2$
3. $\theta = 55°$, $b = 140$, $c = 244$
5. $a = 11.1$, $b = 13.3$
7. $R = 42.9$ kΩ, $Z = 46.4$ kΩ
9. $R = 1.58$ kΩ, $Z = 1.71$ kΩ
11. $Z = 11.5$ kΩ, $\theta = 60.8°$
13. $Z = 33.4$ kΩ, $\theta = 36.1°$
15. $X = 11.5$ kΩ, $Z = 23.1$ kΩ
17. $X = 207$ kΩ, $Z = 340$ kΩ
19. $R = 15.3$ kΩ, $X = 37.0$ kΩ
21. $R = 599$ Ω, $X = 508$ Ω
23. $X = 18.4$ kΩ, $\theta = 56.9°$
25. $X = 1.81$ kΩ, $\theta = 45.1°$
27. $R = 33.7$ kΩ, $\theta = 27.4°$
29. $R = 22$ kΩ, $\theta = 34.8°$

31. $B = 28.0\ \mu S$, $Y = 66.2\ \mu S$
33. $B = 290\ \mu S$, $Y = 319\ \mu S$
35. $\theta = 34.9°$, $Y = 15.3$ mS
37. $\theta = 53.7°$, $Y = 21.5\ \mu S$
39. $G = 185$ mS, $Y = 189$ mS
41. $G = 6.40$ mS, $Y = 7.20$ mS
43. $\theta = 47.2°$, $B = 6.86$ mS
45. $\theta = 45.1°$, $B = 2.34$ mS
47. $G = 59.4\ \mu S$, $B = 37.1\ \mu S$
49. $G = 3.77$ mS, $B = 5.38$ mS
51. $\theta = 40.7°$, $G = 8.72$ mS
53. $\theta = 32.1°$, $G = 271\ \mu S$

CHAPTER 18

Problems 18-1

1. angle $B = 60°$, side $a = 28.53$, side $b = 32.26$
3. angle $A = 124.9°$, angle $B = 20.13°$, side $a = 143.0$
5. angle $B = 43.51°$ or angle $B = 136.5°$; angle $C = 108.5°$ or angle $C = 15.52°$; side $c = 30.30$ or side $c = 8.549$
7. angle $C = 80°$, side $b = 18.79$, side $c = 19.70$
9. angle $A = 121.2°$ or angle $A = 14.82°$; angle $B = 36.82°$ or angle $B = 143.2°$; side $a = 228.4$ or side $a = 68.28$
11. angle $B = 57.7°$ or angle $B = 122.3°$; angle $C = 97.3°$ or angle $C = 32.7°$; side $c = 23.47$ or side $c = 12.78$
13. angle $A = 25°$, side $a = 30.34$, side $c = 69.35$
15. angle $B = 32.66°$, angle $C = 52.34°$, side $c = 95.37$
17. angle $A = 100.4°$ or angle $A = 11.5°$; angle $B = 45.55°$ or angle $B = 134.5°$; side $a = 8.266$ or side $a = 1.68$
19. angle $B = 10°$, side $a = 16.22$, side $b = 3.438$
21. angle $B = 40°$, side $a = 186.2$, side $b = 239.4$
23. angle $A = 102.3°$ or angle $A = 21.7°$; angle $C = 49.72°$ or angle $C = 130°$; side $a = 83.25$ or side $a = 31.5$
25. angle $A = 29.31°$, angle $C = 50.70°$, side $a = 174.0$
27. angle $A = 63°$, side $a = 15.21$, side $c = 6.393$
29. angle $A = 140°$, side $a = 178.5$, side $c = 95$
31. angle $A = 61.26°$, angle $C = 48.74°$, side $a = 233.3$
33. angle $A = 24.74°$, angle $C = 35.26°$, side $a = 72.47$
35. angle $A = 83.02°$ or angle $A = 30.98°$; angle $C = 63.98°$ or angle $C = 116.0°$; side $a = 36.45$ or side $a = 18.9$
37. angle $B = 38°$, side $a = 80.76$, side $b = 67.98$
39. angle $B = 60°$, side $a = 5.87$, side $b = 5.266$

Problems 18-2

1. angle $A = 30.80°$, angle $B = 108.0°$, angle $C = 41.17°$
3. angle $A = 108.2°$, angle $B = 22.33°$, angle $C = 49.47°$
5. angle $B = 35.1°$, angle $C = 49.9°$, side $a = 104$
7. angle $B = 19.21°$, angle $C = 80.79°$, side $a = 35.92$
9. angle $A = 45.33°$, angle $C = 34.67°$, side $b = 692.4$
11. angle $A = 121.9°$, angle $B = 39.57°$, angle $C = 18.57°$
13. angle $A = 60°$, angle $B = 81.79°$, angle $C = 38.21°$
15. angle $A = 57.02°$, angle $C = 93.0°$, side $b = 5.007$
17. angle $A = 28.10°$, angle $C = 46.90°$, side $b = 82.02$
19. angle $B = 52.88°$, angle $C = 32.12°$, side $a = 749.6$
21. angle $B = 23.41°$, angle $C = 36.59°$, side $a = 21.79$
23. angle $A = 48.19°$, angle $B = 48.19°$, angle $C = 83.62°$
25. angle $A = 59.75°$, angle $B = 45.35°$, angle $C = 74.90°$
27. angle $A = 54.03°$, angle $B = 91.12°$, angle $C = 34.84°$
29. angle $A = 36.69°$, angle B = 71.66°, angle $C = 71.66°$

CHAPTER 19

Problems 19-1

1. 1.67 ms **3.** 833 μs **5.** 568 ns **7.** 2.50 kHz **9.** 752 Hz **11.** 40.0 kHz **13.** 0.436 rad

15. 1.57 rad **17.** 3.14 rad **19.** 1.31 rad **21.** 10.5 rad **23.** 8.37° **25.** 28.6° **27.** 90° **29.** 100° **31.** 573°

Problems 19-2

1. (a) 377 rad/s, 21,600°/s (b) 942 rad/s, 54,000°/s (c) 1260 rad/s, 7.20×10^4 degrees/s (d) 1880 rad/s, 1.08×10^5 degrees/s (e) 3770 rad/s, 2.16×10^5 degrees/s (f) 5030 rad/s, 2.88×10^5 degrees/s (g) 7850 rad/s, 4.50×10^5 degrees/s
3. (a) 1.91 Hz (b) 7.96 Hz (c) 63.7 Hz (d) 143 Hz (e) 398 Hz (f) 796 Hz (g) 1.11 kHz
5. (a) 27.8 Hz (b) 41.7 Hz (c) 139 Hz (d) 333 Hz (e) 556 Hz (f) 2.04 kHz (g) 4.58 kHz
7. (a) 0.314 rad, 18° (b) 0.785 rad, 45° (c) 3.14 rad, 180° (d) 8.8 rad, 504°
9. (a) 1.6 rad, 91.8° (b) 4.01 rad, 230° (c) 16 rad, 918° (d) 44.9 rad, 2570°
11. (a) 14.3 kHz (b) 1.03 kHz (c) 9.26 kHz (d) 62.2 kHz
13. (a) 175 Hz (b) 1.72 kHz (c) 61.1 kHz (d) 1.67 MHz
15. (a) 31.8 μs (b) 3.98 μs (c) 151 μs (d) 11.1 μs (e) 50 μs (f) 167 μs
17. (a) 10.6 μs (b) 1.33 μs (c) 50.2 μs (d) 3.70 μs (e) 16.7 μs (f) 55.6 μs

Problems 19-3

1. (a) 3.91 V (b) 9.57 V (c) 17.7 V (d) −9.57 V (e) −25 V (f) −9.57 V
3. (a) 8.59 V (b) 10.9 V (c) −6.75 V (d) −9.37 V (e) −11.2 V (f) −12.0 V
5. (a) 700 μA (b) 0 (c) −700 μA (d) 0 (e) 700 μA (f) 0
7. (a) 86.7 μA (b) 173 μA (c) 260 μA (d) 346 μA (e) 432 μA (f) 518 μA

Problems 19-4

1. (a) 65 V = 91.9 V_{pk} = 184 V_{p-p} (b) 110 V = 156 V_{pk} = 311 V_{p-p}
3. (a) 220 V = 311 V_{pk} = 622 V_{p-p} (b) 440 V = 622 V_{pk} = 1.24 kV_{p-p}
5. (a) 600 μA = 848 μA_{pk} = 1.70 mA_{p-p} (b) 1.2 mA = 1.70 mA_{pk} = 3.39 mA_{p-p}
7. (a) 37.8 mA = 53.5 mA_{pk} = 107 mA_{p-p} (b) 3 A = 4.24 A_{pk} = 8.48 A_{p-p}
9. (a) 350 μV_{pk} = 700 μV_{p-p} = 247 μV (b) 400 mV_{pk} = 800 mV_{p-p} = 283 mV
11. (a) 80 V_{pk} = 160 V_{p-p} = 56.6 V (b) 200 V_{pk} = 400 V_{p-p} = 141V
13. (a) 200 μA_{pk} = 400 μA_{p-p} = 141 μA (b) 9.45 mA_{pk} = 18.9 mA_{p-p} = 6.68 mA
15. (a) 500 mA_{pk} = 1.00 A_{p-p} = 354 mA (b) 1.75 A_{pk} = 3.50 A_{p-p} = 1.24 A
17. (a) 320 μV_{p-p} = 160 μV_{pk} = 113 μV (b) 70 mV_{p-p} = 35 mV_{pk} = 24.7 mV
19. (a) 17 V_{p-p} = 8.50 V_{pk} = 6.01 V (b) 283 V_{p-p} = 142 V_{pk} = 100 V
21. (a) 50 μA_{p-p} = 25 μA_{pk} = 17.7 μA (b) 700 μA_{p-p} = 350 μA_{pk} = 247 μA
23. (a) 9.38 mA_{p-p} = 4.69 mA_{pk} = 3.32 mA (b) 650 mA_{p-p} = 325 mA_{pk} = 230 mA

Problems 19-5

1. (a) i = 2.35 mA, e = 8.75 V (b) i = 3.99 mA, e = 9.31 V (c) i = 0, e = −4.23 V (d) i = −3.8 mA, e = −7.31 V
3. (a) i = 23.6 mA, e = 16.6 V (b) i = 20.5 mA, e = 28.3 V (c) i = 12.4 mA, e = 25.6 V (d) i = −11.0 mA, e = −24.8 V
5. (a) i = 140 μA, e = −53 V (b) i = 485 μA, e = −1.40 V (c) i = 185 μA, e = 76.5 V (d) i = −600 μA, e = −45.9 V
7. (a) i = 21.9 mA, e = 23.5 V (b) i = 24.7 mA, e = 55.8 V (c) i = −12.2 mA, e = 12.2 V (d) i = −3.31 mA, e = 34.8 V

CHAPTER 20

Problems 20-1

1. E = 28.3 V, θ = −45° **3.** E = 14.4 V, θ = −33.7° **5.** V_C = −7.73 V, θ = −50.6°
7. V_C = −34 V, θ = −58.3° **9.** E = 19.6 V, V_C = −12.6 V **11.** E = 6.07 V, V_C = −2.56 V

13. $V_R = 14.6$ V, $V_C = -10.5$ V **15.** $V_R = 62.4$ V, $V_C = -31.8$ V **17.** $V_R = 22.4$ V, $\theta = -41.8°$
19. $V_R = 30$ V, $\theta = -53.1°$ **21.** $E = 10.4$ V, $V_R = 8.82$ V **23.** $E = 31.9$ V, $V_R = 10.9$ V

Problems 20-2

1. $X_C = 6.37$ kΩ, $Z = 11.9$ kΩ, $\theta = -32.5°$ **3.** $X_C = 796$ Ω, $Z = 1.06$ kΩ, $\theta = -48.7°$
5. $X_C = 3.18$ kΩ, $Z = 3.76$ kΩ, $\theta = -57.9°$ **7.** $X_C = 20.5$ kΩ, $Z = 33.9$ kΩ, $\theta = -37.3°$
9. $X_C = 692$ Ω, $Z = 3.37$ kΩ, $\theta = -11.8°$
11. $X_C = 1.33$ kΩ, $Z = 2.4$ kΩ, $\theta = -33.6°$, $V_R = 25$ V, $V_C = -16.6$ V, $I = 12.5$ mA
13. $X_C = 15.9$ kΩ, $Z = 18.8$ kΩ, $\theta = -57.9°$, $V_R = 5.32$ V, $V_C = -8.47$ V, $I = 532$ μA
15. $X_C = 5.31$ kΩ, $Z = 6.25$ kΩ, $\theta = -58.1°$, $V_R = 15.8$ V, $V_C = -25.5$ V, $I = 4.8$ mA
17. $X_C = 49$ kΩ, $Z = 50$ kΩ, $\theta = -78.5°$, $V_R = 4$ V, $V_C = -19.6$ V, $I = 400$ μA
19. $X_C = 398$ Ω, $Z = 616$ Ω, $\theta = -40.3°$, $V_R = 76.3$ V, $V_C = -64.6$ V, $I = 162$ mA
21. $\theta = -46.9°$, $X_C = 8.76$ kΩ, $f = 90.8$ kHz, $V_R = 27.3$ V, $V_C = -29.2$ V, $I = 3.33$ mA
23. $\theta = -51.3°$, $X_C = 937$ Ω, $f = 17$ kHz, $V_R = 9.38$ V, $V_C = -11.7$ V, $I = 12.5$ mA
25. $\theta = -39.5°$, $X_C = 2.23$ kΩ, $f = 2.38$ MHz, $V_R = 23.1$ V, $V_C = -19.1$ V, $I = 8.57$ mA
27. $\theta = -29.6°$, $X_C = 31.8$ kΩ, $f = 200$ Hz, $V_R = 6.52$ V, $V_C = -3.70$ V, $I = 116$ μA
29. $\theta = -44.7°$, $X_C = 802$ Ω, $f = 198$ Hz, $V_R = 8.53$ V, $V_C = -8.44$ V, $I = 10.5$ mA

Problems 20-3

1. $E = 12.8$ V, $\theta = 51.3°$ **3.** $E = 24.4$ V, $\theta = 55°$ **5.** $E = 7.52$ V, $V_L = 3.41$ V
7. $E = 71.3$ V, $V_L = 65.8$ V **9.** $V_L = 8.62$ V, $\theta = 45.9°$ **11.** $V_L = 41.1$ V, $\theta = 43.3°$
13. $V_R = 26$ V, $\theta = 30°$ **15.** $V_R = 13.8$ V, $\theta = 36°$ **17.** $E = 3.4$ V, $V_R = 1.6$ V
19. $E = 31$ V, $V_R = 26.8$ V **21.** $V_R = 4.77$ V, $V_L = 7.63$ V **23.** $V_R = 35.4$ V, $V_L = 35.4$ V

Problems 20-4

1. $X_L = 3.77$ kΩ, $Z = 5.43$ kΩ, $\theta = 44°$ **3.** $X_L = 377$ Ω, $Z = 626$ Ω, $\theta = 37°$
5. $X_L = 13.4$ kΩ, $Z = 30$ kΩ, $\theta = 26.3°$ **7.** $X_L = 1.26$ kΩ, $Z = 2.2$ kΩ, $\theta = 34.9°$
9. $X_L = 785$ Ω, $Z = 1.27$ kΩ, $\theta = 38.1°$, $V_R = 11.8$ V, $V_L = 9.27$ V, $I = 11.8$ mA
11. $X_L = 471$ Ω, $Z = 559$ Ω, $\theta = 57.5°$, $V_R = 26.9$ V, $V_L = 42.2$ V, $I = 89.4$ mA
13. $X_L = 19.9$ kΩ, $Z = 33.6$ kΩ, $\theta = 36.4°$, $V_R = 16.1$ V, $V_L = 11.9$ V, $I = 595$ mA
15. $X_L = 78.5$ Ω, $Z = 127$ Ω, $\theta = 38.1°$, $V_R = 9.44$ V, $V_L = 7.41$ V, $I = 94.4$ mA
17. $X_L = 3.73$ kΩ, $\theta = 38.4°$, $f = 913$ Hz, $E = 12.1$ V, $V_R = 9.46$ V, $I = 2.02$ mA
19. $X_L = 26.5$ kΩ, $\theta = 55.8°$, $f = 56.2$ kHz, $E = 4.47$ V, $V_R = 2.51$ V, $I = 140$ μA
21. $X_L = 28.1$ kΩ, $\theta = 52°$, $f = 5.59$ kHz, $E = 21.6$ V, $V_R = 13.3$ V, $I = 605$ μA
23. $X_L = 1.6$ kΩ, $\theta = 64.9°$, $f = 2.55$ kHz, $E = 29.8$ V, $V_R = 12.6$ V, $I = 16.9$ mA
25. $X_L = 8.65$ kΩ, $L = 320$ mH, $\theta = 51.8°$, $V_R = 4.72$ V, $E = 7.63$ V, $I = 694$ μA
27. $X_L = 4.27$ kΩ, $L = 90.5$ mH, $\theta = 62.7°$, $V_R = 2.32$ V, $E = 5.06$ V, $I = 1.05$ mA
29. $X_L = 68.4$ kΩ, $L = 1.98$ H, $\theta = 55.5°$, $V_R = 13.8$ V, $E = 24.3$ V, $I = 293$ μA
31. $X_L = 5.51$ kΩ, $L = 73.1$ mH, $\theta = 54.7°$, $V_R = 6.01$ V, $E = 10.4$ V, $I = 1.54$ mA

Problems 20-5

1. $Z_{polar} = 30.9\angle 48°$ kΩ
$Z_{rect} = 20.7$ kΩ $+ j23$ kΩ

3. $Z_{polar} = 952\angle -51°$ Ω
$Z_{rect} = 599$ Ω $- j740$ Ω

5. $Z_{polar} = 857\angle -17°$ Ω
$Z_{rect} = 820$ Ω $- j251$ Ω

7. $Z_{polar} = 6.35\angle -45°$ kΩ
$Z_{rect} = 4.49$ kΩ $- j4.49$ kΩ

9. $Z_{polar} = 50\angle -36.9°$ kΩ
$Z_{rect} = 40$ kΩ $- j30$ kΩ

11. $Z_{polar} = 216\angle 33.7°$ kΩ
$Z_{rect} = 180$ kΩ $+ j120$ kΩ

13. $Z_{polar} = 4.6\angle -40.7°$ kΩ
$Z_{rect} = 3.49$ kΩ $- j3$ kΩ

15. $Z_{polar} = 5.35\angle 44.5°$ kΩ
$Z_{rect} = 3.82$ kΩ $+ j3.75$ kΩ

17. $Z_{polar} = 20.6\angle 61°$ kΩ
$Z_{rect} = 10$ kΩ $+ j18$ kΩ

19. $Z_{polar} = 2.31\angle -38.9°$ kΩ
$Z_{rect} = 1.8$ kΩ $- j1.45$ kΩ

21. $E_{polar} = 16.2\angle 51.8°$ V
$E_{rect} = 10$ V $+ j12.7$ V

23. $E_{polar} = 7.60\angle -55.4°$ V
$E_{rect} = 4.32$ V $- j6.25$ V

25. $E_{polar} = 12\angle -28.4°$ V
$E_{rect} = 10.6$ V $- j5.7$ V

27. $E_{polar} = 15\angle 34.5°$ V
$E_{rect} = 12.4$ V $+ j8.5$ V

29. $E_{polar} = 15\angle -40°$ V
$E_{rect} = 11.5$ V $- j9.64$ V

31. $E_{polar} = 14 \angle 34.6^\circ$ V
$E_{rect} = 11.5$ V $+ j7.93$ V

33. $E_{polar} = 136 \angle -28^\circ$ V
$E_{rect} = 120$ V $- j64$ V

35. $E_{polar} = 5 \angle 58.7^\circ$ V
$E_{rect} = 2.6$ V $+ j4.27$ V

37. $E_{polar} = 30 \angle 32.2^\circ$ V
$E_{rect} = 25.4$ V $+ j16$ V

39. $E_{polar} = 10 \angle -36.9^\circ$ V
$E_{rect} = 8$ V $- j6$ V

Problems 20-6

1. $4.64 \angle 47.2^\circ$ kΩ
3. $25.9 \angle -25.8^\circ$ kΩ
5. $9.00 \angle -7.41^\circ$ kΩ
7. $82.6 \angle 1.87^\circ$ kΩ
9. $420 \angle 30^\circ$ kΩ
11. $12.8 \angle -11^\circ$ kΩ

Problems 20-7

1. $Z_{rect} = 1.76$ kΩ $+ j3$ kΩ
$Z_{polar} = 3.48 \angle 59.6^\circ$ kΩ

3. $Z_{rect} = 1.15$ kΩ $+ j1.3$ kΩ
$Z_{polar} = 1.74 \angle 48.5^\circ$ kΩ

5. $R = 3.4$ kΩ and $L = 623$ mH

7. $Z_{rect} = 200$ Ω $- j281$ Ω
$Z_{polar} = 345 \angle -54.5^\circ$ Ω
$V_R = 14.5$ V
$V_C = -65.9$ V; $I = 72.5$ mA
$V_L = 45.5$ V; in the ESC, $R = 200$ Ω and $C = 1.13$ μF

Problems 20-8

1. $f_r = 2.6$ kHz (a) $X_L = 12.2$ kΩ, $X_C = 12.2$ kΩ, $Z_{polar} = 470 \angle 0^\circ$ Ω, $I = 21.3$ mA, $V_R = 10$ V, $V_L = 260$ V, $V_C = -260$ V. The ESC consists of 470 Ω of resistance. (b) $X_L = 10.8$ kΩ, $X_C = 13.8$ kΩ, $Z_{polar} = 3.04 \angle -81.1^\circ$ kΩ, $I = 3.29$ mA, $V_R = 1.55$ V, $V_L = 35.5$ V, $V_C = -45.4$ V. In the ESC $R = 470$ Ω and $C = 23.1$ nF. (c) $X_L = 12.7$ kΩ, $X_C = 11.8$ kΩ, $Z_{polar} = 1.10 \angle 62.4^\circ$ kΩ, $I = 9.56$ mA, $V_R = 4.49$ V, $V_L = 122$ V, $V_C = -113$ V. In the ESC $R = 470$ Ω and $L = 55.1$ mH.
3. $f_r = 225$ Hz (a) $X_L = 1.41$ kΩ, $X_C = 1.41$ kΩ, $Z_{polar} = 300 \angle 0^\circ$ Ω, $I = 83.3$ mA, $V_R = 25$ V, $V_L = 118$ V, $V_C = -118$ V. The ESC consists of 300 Ω of resistance. (b) $X_L = 1.26$ kΩ, $X_C = 1.59$ kΩ, $Z_{polar} = 450 \angle -48.1^\circ$ Ω, $I = 55.6$ mA, $V_R = 16.8$ V, $V_L = 69.9$ V, $V_C = 88.5$ V. The ESC consists of 300 Ω of resistance and 2.41 μF of capacitance. (c) $X_L = 1.88$ kΩ, $X_C = 1.06$ kΩ, $Z_{polar} = 872 \angle 69.9^\circ$ Ω, $I = 28.7$ mA, $V_R = 8.60$ V, $V_L = 53.9$ V, $V_C = -30.4$ V. The ESC consists of 300 Ω of resistance and 437 mH of inductance.

CHAPTER 21

Problems 21-1

1. $I_{polar} = 40 \angle 54^\circ$ mA
$I_{rect} = 23.5$ mA $+ j32.4$ mA

3. $I_{polar} = 2.75 \angle 67.4^\circ$ mA
$I_{rect} = 1.05$ mA $+ j2.54$ mA

5. $I_{polar} = 175 \angle 22^\circ$ μA
$I_{rect} = 162$ μA $+ j65.6$ μA

7. $I_{polar} = 442 \angle 37.6^\circ$ μA
$I_{rect} = 350$ μA $+ j270$ μA

9. $I_{polar} = 8.51 \angle 55.9^\circ$ mA
$I_{rect} = 4.77$ mA $+ j7.05$ mA

11. $I_{polar} = 141 \angle 45^\circ$ μA
$I_{rect} = 100$ μA $+ j100$ μA

13. $I_{polar} = 1.19 \angle 33^\circ$ mA
$I_{rect} = 1$ mA $+ j649$ μA

15. $I_{polar} = 683 \angle 20.5^\circ$ μA
$I_{rect} = 640$ μA $+ j239$ μA

17. $I_{polar} = 53.9 \angle 62^\circ$ mA
$I_{rect} = 25.3$ mA $+ j47.6$ mA

19. $I_{polar} = 47.6 \angle 71^\circ$ mA
$I_{rect} = 15.5$ mA $+ j45$ mA

21. $I_{polar} = 6.31 \angle 59.5^\circ$ mA
$I_{rect} = 3.2$ mA $+ j5.44$ mA

23. $I_{polar} = 333 \angle 55.7^\circ$ μA
$I_{rect} = 188$ μA $+ j275$ μA

25. $I_{polar} = 12 \angle 51.3^\circ$ mA
$I_{rect} = 7.5$ mA $+ j9.37$ mA

27. $I_{polar} = 1 \angle 49.5^\circ$ mA
$I_{rect} = 650$ μA $+ j760$ μA

29. $I_{polar} = 3.15 \angle 36.0^\circ$ mA
$I_{rect} = 2.55$ mA $+ j1.85$ mA

31. $I_{polar} = 43 \angle 35.5^\circ$ mA
$I_{rect} = 35$ mA $+ j25$ mA

33. $I_{polar} = 400 \angle 30.0^\circ$ μA
$I_{rect} = 346$ μA $+ 200$ μA

35. $I_{polar} = 6.08 \angle 35.7^\circ$ mA
$I_{rect} = 4.94$ mA $+ 3.55$ mA

Problems 21-2

1. $Y_{rect} = 303$ μS $+ j377$ μS
$Y_{polar} = 484 \angle 51.2^\circ$ μS
$I_{rect} = 2.51$ mA $+ j3.12$ mA
$V = 8.26$ V

3. $Y_{rect} = 1.33$ mS $+ j1.88$ mS
$Y_{polar} = 2.31 \angle 54.7^\circ$ mS
$I_{rect} = 577$ μA $+ j816$ μA
$V = 433$ mV

5. $Y_{rect} = 147$ μS $+ j94.2$ μS
$Y_{polar} = 175 \angle 32.7^\circ$ μS
$I_{rect} = 16.8$ mA $+ j10.8$ mA
$V = 115$ V

7. $Y_{rect} = 100\ \mu S + j188\ \mu S$
$Y_{polar} = 213\ \angle 62.1^\circ\ \mu S$
$I_{rect} = 4.69\text{ mA} + j8.83\text{ mA}$
$V = 46.9\text{ V}$

9. $Y_{rect} = 66.7\ \mu S + j66\ \mu S$
$Y_{polar} = 93.8\ \angle 44.7^\circ\ \mu S$
$I_{rect} = 1.42\text{ mA} + j1.41\text{ mA}$
$V = 21.3\text{ V}$

11. $R = 5.96\text{ k}\Omega$
$f = 1.27\text{ kHz}$
$I = 4.67\text{ mA}$
$V = 17.9\text{ V}$

Problems 21-3

1. $I_{polar} = 3.8\ \angle -30^\circ\text{ mA}$
$I_{rect} = 3.29\text{ mA} - j1.9\text{ mA}$

3. $I_{polar} = 210\ \angle -39^\circ\ \mu A$
$I_{rect} = 163\ \mu A - j132\ \mu A$

5. $I_{polar} = 1.34\ \angle -23.4^\circ\text{ mA}$
$I_{rect} = 1.23\text{ mA} - j532\ \mu A$

7. $I_{polar} = 41.8\ \angle -73.3^\circ\text{ mA}$
$I_{rect} = 12\text{ mA} - j40\text{ mA}$

9. $I_{polar} = 50\ \angle -53.1^\circ\text{ mA}$
$I_{rect} = 30\text{ mA} - j40\text{ mA}$

11. $I_{polar} = 234\ \angle -50.2^\circ\ \mu A$
$I_{rect} = 150\ \mu A - j180\ \mu A$

13. $I_{polar} = 2.28\ \angle -28.5^\circ\text{ mA}$
$I_{rect} = 2\text{ mA} - j1.09\text{ mA}$

15. $I_{polar} = 4.45\ \angle -73^\circ\text{ mA}$
$I_{rect} = 1.3\text{ mA} - j4.25\text{ mA}$

17. $I_{polar} = 42.4\ \angle -27.4^\circ\ \mu A$
$I_{rect} = 37.6\ \mu A - j19.5\ \mu A$

19. $I_{polar} = 1.03\ \angle -45^\circ\text{ mA}$
$I_{rect} = 730\ \mu A - j730\ \mu A$

21. $I_{polar} = 14\ \angle -40^\circ\text{ mA}$
$I_{rect} = 10.7\text{ mA} - j9\text{ mA}$

23. $I_{polar} = 690\ \angle -43.5^\circ\ \mu A$
$I_{rect} = 501\ \mu A - j475\ \mu A$

25. $I_{polar} = 27\ \angle -47.8^\circ\text{ mA}$
$I_{rect} = 18.1\text{ mA} - j20\text{ mA}$

27. $I_{polar} = 300\ \angle -30^\circ\ \mu A$
$I_{rect} = 260\ \mu A - j150\ \mu A$

29. $I_{polar} = 1.83\ \angle -30.4^\circ\text{ mA}$
$I_{rect} = 1.58\text{ mA} - j925\ \mu A$

31. $I_{polar} = 6.7\ \angle -45.5^\circ\text{ mA}$
$I_{rect} = 4.7\text{ mA} - j4.77\text{ mA}$

33. $I_{polar} = 1.35\ \angle -52.3^\circ\text{ mA}$
$I_{rect} = 825\ \mu A - j1.07\text{ mA}$

35. $I_{polar} = 100\ \angle -42.4^\circ\text{ mA}$
$I_{rect} = 73.8\text{ mA} - j67.5\text{ mA}$

Problems 21-4

1. $Y_{polar} = 63.9\ \angle -38.5^\circ\ \mu S$
$Y_{rect} = 50\ \mu S - j39.8\ \mu S$
$I_{polar} = 40\ \angle -38.5^\circ\text{ mA}$
$I_{rect} = 31.3\text{ mA} - j24.9\text{ mA}$

3. $Y_{polar} = 166\ \angle -53^\circ\ \mu S$
$Y_{rect} = 100\ \mu S - j133\ \mu S$
$I_{polar} = 200\ \angle -53^\circ\text{ mA}$
$I_{rect} = 120\text{ mA} - j160\text{ mA}$

5. $Y_{polar} = 266\ \angle -36.8^\circ\ \mu S$
$Y_{rect} = 213\ \mu S - j159\ \mu S$
$I_{polar} = 100\ \angle -36.8^\circ\ \mu A$
$I_{rect} = 80.1\ \mu A - j59.9\ \mu A$

7. $Y_{polar} = 48.8\ \angle -40.8^\circ\ \mu S$
$Y_{rect} = 37\ \mu S - j31.9\ \mu S$
$I_{polar} = 400\ \angle -40.8^\circ\ \mu A$
$I_{rect} = 303\ \mu A - j261\ \mu A$

9. $Y_{polar} = 1.49\ \angle -26.4^\circ\text{ mS}$
$Y_{rect} = 1.33\text{ mS} - j663\ \mu S$
$I_{polar} = 30\ \angle -26.4^\circ\text{ mA}$
$I_{rect} = 26.9\text{ mA} - j13.4\text{ mA}$

Problems 21-5

1. $269\ \angle -27^\circ\ \Omega$ 3. $158\ \angle 40^\circ\ \mu S$ 5. $657\ \angle 44.1^\circ\ \Omega$ 7. $8.55\ \angle -27.8^\circ\text{ k}\Omega$ 9. $2.59\ \angle -38.2^\circ\text{ k}\Omega$

Problems 21-6

1. $I_R = 1.21$ mA, $I_C = 1.59$ mA, $Y_{polar} = 110\ \angle 52.8^\circ\ \mu S$, $Z_{polar} = 9.06\ \angle -52.8^\circ\text{ k}\Omega$, $V = 18.1$ V. In the ESC $R = 5.47\text{ k}\Omega$ and $C = 630$ pF.
3. $I_R = 24.1$ mA, $I_C = 32.0$ mA, $Y_{polar} = 3.54\ \angle 53^\circ$ mS, $Z_{polar} = 283\ \angle -53^\circ\ \Omega$, $V = 11.3$ V. In the ESC $R = 170\ \Omega$ and $C = 236$ nF.
5. $I_R = 192\ \mu A$, $I_C = 160\ \mu A$, $Y_{polar} = 27.7\ \angle -39.7^\circ\ \mu S$, $Z_{polar} = 36.1\ \angle 39.7^\circ\text{ k}\Omega$, $V = 9.04$ V. In the ESC $R = 27.8\text{ k}\Omega$ and $L = 24.5$ mH.
7. $I_R = 4.06$ mA, $I_C = 2.92$ mA, $Y_{polar} = 1.81\ \angle -35.8^\circ$ mS, $Z_{polar} = 552\ \angle 35.8^\circ\ \Omega$, $V = 2.76$ V. In the ESC $R = 447\ \Omega$ and $L = 10.3$ mH.
9. $I_R = 10.4$ mA, $I_C = 4.88$ mA, $I_L = -22.0$ mA, $Y_{polar} = 193\ \angle -58.8^\circ\ \mu S$, $Z_{polar} = 5.18\ \angle 58.8^\circ\text{ k}\Omega$, V = 104 V. In the ESC $R = 2.68\text{ k}\Omega$ and $L = 28.2$ mH
11. $I_R = 257\ \mu A$, $I_C = 4.95$ mA, $I_L = -4.64$ mA, $Y_{polar} = 228\ \angle 49.9^\circ\ \mu S$, $Z_{polar} = 4.38\ \angle -49.9^\circ\text{ k}\Omega$, $V = 1.75$ V. In the ESC $R = 2.82\text{ k}\Omega$ and $C = 3.17$ nF

Problems 21-7

1. $Y_{rect} = 621\ \mu S + j726\ \mu S$, $C = 28.9$ nF
3. $Y_{rect} = 37.4\ \mu S + j33.1\ \mu S$, $C = 43.9$ pF
5. $Y_{rect} = 289\ \mu S - j219\ \mu S$, $L = 274$ mH
7. $Y_{rect} = 39.2\ \mu S - j32.8\ \mu S$, $L = 485$ mH
9. $Z_{rect} = 3.15\text{ k}\Omega - j1.55\text{ k}\Omega$, $C = 10.3$ nF
11. $Z_{rect} = 216\ \Omega - j361\ \Omega$, $C = 136$ nF
13. $Z_{rect} = 52.6\ \Omega + j28.5\ \Omega$, $L = 45.3$ mH
15. $Z_{rect} = 249\ \Omega + j1.48\text{ k}\Omega$, $L = 29.2$ mH

Problems 21-8

1. $Z = 2.00\ k\Omega + j19.9\ k\Omega$ 3. $Z = 7\ k\Omega - j1\ k\Omega$ 5. $Z = 16.9\ k\Omega - j26.3\ k\Omega$
7. $Z = 8.61\ k\Omega - j9.21\ k\Omega$
9. $R = 1.13\ k\Omega$ and $L = 528$ mH. The phase angle is 71.2°
11. $R = 8.45\ k\Omega$, $L = 895$ mH, $\theta = 73.2°$ 13. $R = 23.9\ k\Omega$, $C = 8.42$ nF, $\theta = -38.3°$
15. $R = 6.18\ k\Omega$, $C = 6.75$ nF, $\theta = -29.7°$

CHAPTER 22

Problems 22-1

1. $I_{max} = 31.9$ mA, $P_{max} = 479$ mW, $V_{o(max)} = 15$ V. At f_{co}, $X_C = 470\ \Omega$, $I = 22.6$ mA, $P = 239$ mW, $V_o = 10.6$ V.
3. $I_{max} = 53.6\ \mu A$, $P_{max} = 16.1\ \mu W$, $V_{o(max)} = 300$ mV. At f_{co}, $X_C = 5.6\ k\Omega$, $I = 37.9\ \mu A$, $P = 8.03\ \mu W$, $V_o = 212$ mV.
5. $f_{co} = 265$ Hz. At f_{co}, $I = 11.8$ mA, $V_o = 14.1$ V, $P = 166$ mW.
7. $f_{co} = 362$ Hz. At f_{co}, $I = 12.9$ mA, $V_o = 28.3$ V, $P = 364$ mW.

Problems 22-2

1. $V_{OC} = 6.47$ V, $R_{TH} = 2.16\ k\Omega$, $f_{co} = 147$ Hz, $V_o = 2.11$ V
3. $V_{OC} = 20$ V, $R_{TH} = 2\ k\Omega$, $f_{co} = 39.8$ Hz, $V_o = 14.1$ V
5. $V_{OC} = 10$ V, $R_{TH} = 5\ k\Omega$, $f_{co} = 31.8$ Hz, V_o at $f_{co} = 3.25$ V
7. $V_{OC} = 5.66$ V, $R_{TH} = 4.08\ k\Omega$, $f_{co} = 39$ Hz
9. $I_{SC} = 31.9$ mA, $G_N = 2.96$ mS, $f_{co} = 9.43$ kHz, V_o at $f_{co} = 7.62$ V
11. $I_{SC} = 5$ mA, $G_N = 891\ \mu S$, $f_{co} = 142$ kHz, V_o at $f_{co} = 3.97$ V

Problems 22-3

1. $f_1 = 169$ Hz, $f_2 = 183$ kHz, V_o at $f_m = 5.74$ V, V_o at $f_{co} = 4.06$ V
3. $f_1 = 15$ Hz, $f_2 = 128$ kHz, V_o at $f_m = 5.66$ V, V_o at $f_{co} = 4$ V
5. $f_1 = 21.7$ Hz, $f_2 = 5.66$ kHz, V_o at $f_m = 1.28$ V, V_o at $f_{co} = 904$ mV
7. $f_1 = 692$ Hz, $f_2 = 1.35$ MHz, V_o at $f_m = 2.14$ V, V_o at $f_{co} = 1.51$ V 9. $f_1 = 43.9$ Hz, $f_2 = 87.9$ kHz
11. $f_1 = 10.3$ Hz, $f_2 = 241$ kHz
13. BW = 80 Hz, $f_1 = 1960$ Hz, $f_2 = 2.04$ kHz. At f_r, $V_R = 20$ V, $I = 20$ mA, $P = 400$ mW. At f_1 and f_2, $V_R = 14.1$ V, $I = 14.1$ mA, $P = 200$ mW.
15. BW = 393 Hz, $f_1 = 5.30$ kHz, $f_2 = 5.70$ kHz. At f_r, $V_R = 5$ V, $I = 41.7$ mA, $P = 208$ mW. At f_1 and f_2, $V_R = 3.54$ V, $I = 29.5$ mA, and $P = 104$ mW.
17. $f_r = 1.01$ kHz, $Q = 14.1$, BW = 71.9 Hz, $f_1 = 974$ Hz, $f_2 = 1.05$ kHz
19. $f_r = 3.56$ kHz, $Q = 5.96$, BW = 597 Hz, $f_1 = 3.26$ kHz, $f_2 = 3.86$ kHz

CHAPTER 23

Problems 23-1

1. log 1000 = 3, $10^3 = 1000$ 3. log 1 = 0, $10^0 = 1$ 5. log 10,000,000 = 7, $10^7 = 10{,}000{,}000$
7. log 0.1 = −1, $10^{-1} = 0.1$ 9. log 0.000001 = −6, $10^{-6} = 0.000001$

Problems 23-2

1. log 3.42 = 0.5340, $10^{0.5340} = 3.42$ 3. log 4.05 = 0.6075, $10^{0.6075} = 4.05$
5. log 6.44 = 0.8089, $10^{0.8089} = 6.44$ 7. log 9.45 = 0.9754, $10^{0.9754} = 9.45$
9. log 2.50 = 0.3979, $10^{0.3979} = 2.50$

Problems 23-3

1. log 14 = 1.146, $10^{1.146} = 14$ 3. log 743 = 2.871, $10^{2.871} = 743$ 5. log 1070 = 3.029, $10^{3.029} = 1070$
7. log 43.4 = 1.638, $10^{1.638} = 43.4$ 9. log 10,000 = 4, $10^4 = 10{,}000$

11. $\log 23{,}400 = 4.369$, $10^{4.369} = 23{,}400$ **13.** $\log 47{,}700 = 4.679$, $10^{4.679} = 47{,}700$
15. $\log 14.6 = 1.164$, $10^{1.164} = 14.6$ **17.** $\log 37{,}000 = 4.568$, $10^{4.568} = 37{,}000$
19. $\log 2{,}000{,}000 = 6.301$, $10^{6.301} = 2{,}000{,}000$

Problems 23-4

1. $\log 0.00377 = -2.424$, $10^{-2.424} = 0.00377$ **3.** $\log 0.037 = -1.432$, $10^{-1.432} = 0.037$
5. $\log 0.746 = -0.1273$, $10^{-0.1273} = 0.746$ **7.** $\log 0.407 = -0.39$, $10^{-0.39} = 0.407$

Problems 23-5

1. $\log (27 \times 56) = \log 27 + \log 56 = 1.431 + 1.748 = 3.18$, $10^{3.18} = 1.51 \times 10^3$
3. $\log (107 \times 243{,}000) = \log 107 + \log 243{,}000 = 2.029 + 5.386 = 7.415$, $10^{7.415} = 2.6 \times 10^7$
5. $\log (43 \times 0.143) = \log 43 + \log 0.143 = 1.633 + (-0.8447) = 0.7888$, $10^{0.7888} = 6.15$
7. $\log (0.043 \times 0.0106) = \log 0.043 + \log 0.0106 = -1.367 + (-1.975) = -3.341$, $10^{-3.341} = 4.56 \times 10^{-4}$
9. $\log (0.0000176 \times 0.000357) = \log 0.0000176 + \log 0.000357 = -4.754 + (-3.447) = -8.202$, $10^{-8.202} = 6.28 \times 10^{-9}$
11. $\log (256 \times 0.00246) = \log 256 + \log 0.00246 = 2.408 + (-2.609) = -0.2008$, $10^{-0.2008} = 0.630$
13. $\log (78 \times 0.473) = \log 78 + \log 0.473 = 1.892 + (-0.3251) = 1.567$, $10^{1.567} = 36.90$
15. $\log (4730 \times 0.000506) = \log 4730 + \log 0.000506 = 3.675 + (-3.296) = 0.3790$, $10^{0.3790} = 2.393$
17. $\log (760 \times 7360) = \log 760 + \log 7360 = 2.881 + 3.867 = 6.748$, $10^{6.748} = 5.594 \times 10^6$
19. $\log (2560 \times 0.00457) = \log 2560 + \log 0.00457 = 3.408 + (-2.340) = 1.068$, $10^{1.068} = 11.70$

Problems 23-6

1. $\log \dfrac{1760}{43} = \log 1760 - \log 43 = 3.246 - 1.633 = 1.612$, $10^{1.612} = 40.9$

3. $\log \dfrac{65}{467} = \log 65 - \log 467 = 1.813 - 2.669 = -0.856$, $10^{-0.856} = 0.139$

5. $\log \dfrac{0.00273}{172} = \log 0.00273 - \log 172 = -2.564 - 2.236 = -4.799$, $10^{-4.799} = 1.59 \times 10^{-5}$

7. $\log \dfrac{4300}{0.15} = \log 4300 - \log 0.15 = 3.633 - (-0.824) = 4.457$, $10^{4.457} = 2.87 \times 10^4$

9. $\log \dfrac{0.0675}{0.00103} = \log 0.0675 - \log 0.00103 = -1.171 - (-2.987) = 1.816$, $10^{1.816} = 65.5$

11. $\log \dfrac{0.00467}{73.2} = \log 0.00467 - \log 73.2 = -2.331 - 1.865 = -4.196$, $10^{-4.196} = 6.375 \times 10^{-5}$

13. $\log \dfrac{0.147}{0.0432} = \log 0.147 - \log 0.0432 = -0.8327 - (-1.365) = 0.5318$, $10^{0.5318} = 3.406$

15. $\log \dfrac{4730}{0.443} = \log 4730 - \log 0.443 = 3.675 - (-0.3536) = 4.029$, $10^{4.029} = 1.068 \times 10^4$

17. $\log \dfrac{0.445}{635} = \log 0.445 - \log 635 = -0.3516 - 2.803 = -3.154$, $10^{-3.154} = 7.015 \times 10^{-4}$

19. $\log \dfrac{0.00275}{4.35} = \log 0.00275 - \log 4.35 = -2.561 - 0.6385 = -3.199$, $10^{-3.199} = 6.322 \times 10^{-4}$

Problems 23-7

1. $\log 3.4^3 = 3 \log 3.4 = 3 \times 0.5315 = 1.594$, $10^{1.594} = 39.3$
3. $\log 6^{-1.3} = -1.3 \log 6 = -1.3 \times 0.7782 = -1.012$, $10^{-1.012} = 9.74 \times 10^{-2}$

5. $\log 0.432^{2.5} = 2.5 \log 0.43 = 2.5 \times -0.3665 = -0.9163,\ 10^{-0.9163} = 0.121$
7. $\log 6^{4.3} = 4.3 \log 6 = 4.3 \times 0.7782 = 3.346,\ 10^{3.346} = 2.22 \times 10^3$
9. $\log 0.27^{-4.4} = -4.4 \log 0.27 = -4.4 \times -0.5686 = 2.502,\ 10^{2.502} = 3.18 \times 10^2$
11. $\log 8.7^{6.3} = 6.3 \log 8.7 = 6.3 \times 0.9395 = 5.919,\ 10^{5.919} = 8.3 \times 10^5$
13. $\log 3.4^{3.2} = 3.2 \log 3.4 = 3.2 \times 0.5315 = 1.701,\ 10^{1.701} = 50.20$
15. $\log 6.35^{-2.3} = -2.3 \log 6.35 = -2.3 \times 0.8028 = -1.846,\ 10^{-1.846} = 1.424 \times 10^{-2}$
17. $\log 0.045^{3.6} = 3.6 \log 0.045 = 3.6 \times (-1.347) = -4.848,\ 10^{-4.848} = 1.42 \times 10^{-5}$
19. $\log 0.00735^{-2.5} = -2.5 \log 0.00735 = -2.5 \times (-2.134) = 5.334,\ 10^{5.334} = 2.16 \times 10^5$

Problems 23-8

	Logarithmic Form	*Exponential Form*		*Logarithmic Form*	*Exponential Form*
1.	$\ln 2 = 0.6931$	$e^{0.6931} = 2$	**17.**	$\ln 8.43 = 2.132$	$e^{2.132} = 8.43$
3.	$\ln 0.932 = -0.07042$	$e^{-0.07042} = 0.932$	**19.**	$\ln 0.346 = -1.061$	$e^{-1.061} = 0.346$
5.	$\ln 60 = 4.094$	$e^{4.094} = 60$	**21.**	$\ln 3.67 = 1.30$	$e^{1.30} = 3.67$
7.	$\ln 0.073 = -2.617$	$e^{-2.617} = 0.073$	**23.**	$\ln 0.107 = -2.235$	$e^{-2.235} = 0.107$
9.	$\ln 247 = 5.509$	$e^{5.509} = 247$	**25.**	$\ln 176 = 5.17$	$e^{5.17} = 176$
11.	$\ln 0.00146 = -6.529$	$e^{-6.529} = 0.00146$	**27.**	$\ln 0.293 = -1.228$	$e^{-1.228} = 0.293$
13.	$\ln 1100 = 7.003$	$e^{7.003} = 1100$	**29.**	$\ln 873 = 6.772$	$e^{6.772} = 873$
15.	$\ln 0.00027 = -8.217$	$e^{-8.217} = 0.00027$			

Problems 23-9

1. 10.38 **3.** 43.38 **5.** 70.11 **7.** 200.3 **9.** 4.711 **11.** 1.537 **13.** 0.9955 **15.** 1.162
17. 0.7298 **19.** 0.8676

CHAPTER 24

Problems 24-1

1. $10^x = 340$ **3.** $10^x = 14$ **5.** $10^x = 7.4$ **7.** $10^y = 0.014$ **9.** $10^y = 0.734$ **11.** $10^y = -0.00178$
13. $x = 10^{-2.43}$ **15.** $x = 10^{-3.643}$ **17.** $x = 10^{-0.143}$ **19.** $y = 10^{1.44}$ **21.** $y = 10^{2.834}$ **23.** $y = 10^{4.146}$
25. $\log y = 3.6$ **27.** $\log y = 4.73$ **29.** $\log y = 2.73$ **31.** $a = \log 732$ **33.** $a = \log 47.3$
35. $a = \log 276$ **37.** $y = \log 0.043$ **39.** $y = \log 0.1778$ **41.** $y = \log 0.00178$ **43.** $\log x = -3.4$
45. $\log x = -2.76$ **47.** $\log x = -1.17$

Problems 24-2

1. 2.291×10^7 **3.** 1.002 **5.** 3.776×10^{-1} **7.** 0.9917 **9.** 2.665 **11.** −1.762 **13.** −2.602
15. 4.668 **17.** 1.445×10^4 **19.** 1.079 **21.** 2.512×10^{-4} **23.** 6.714×10^{-1} **25.** 2.873
27. −2.145 **29.** 2.778×10^3 **31.** -5.332×10^{-1} **33.** 7.915 **35.** 1×10^3 **37.** 2.291×10^7
39. 1.458 **41.** 5.849×10^{-1}

Problems 24-3

1. −0.3550 **3.** 1.171 **5.** 2.105×10^3 **7.** 1×10^{-4} **9.** 5.561 **11.** 1.570 **13.** 1×10^8
15. 2×10^5 **17.** 2.850 **19.** 1.269 **21.** 5×10^{-6} **23.** $y = 1.796 \times 10^5$ **25.** 2.347
27. 0.3162 **29.** $y = 0.4225$

Problems 24-4

1. 2.398 **3.** 2.708 **5.** 5.298 **7.** −7.824 **9.** −2.708 **11.** 14.88 **13.** 3.829 **15.** 1.832×10^{-2}

Problems 24-5

1. 0.9819 **3.** 0.47 **5.** 13.19 **7.** 31.57 **9.** 2.749 **11.** 3.641 **13.** 2.606 **15.** 2.885
17. 0.9502 **19.** 0.4716 **21.** 0.5489

CHAPTER 25

Problems 25-1

1. 60 dB **3.** $A_p = 43$ dB **5.** $A_p = 44$ dB **7.** 3.79 mW **9.** $P_{in} = 1$ W **11.** $P_{in} = 5$ mW
13. 1.58 W **15.** $P_{out} = 3.15$ W **17.** $P_o = 316$ mW **19.** 100 mW **21.** $P_o = 1.58$ W **23.** $P_o = 100$ W
25. 88.6 dB **27.** $A_v = 49.5$ dB **29.** $A_v = 106$ dB **31.** 500 mV **33.** $V_{out} = 27.3$ mV
35. $V_{out} = 103$ V **37.** 126 μV **39.** $V_{in} = 2.67$ mV **41.** $V_{in} = 60\ \mu$V **43.** 706 mV **45.** $V_{in} = 2.00$ V
47. $V_{in} = 7.06$ V **49.** $V_{in} = 3.16$ V

Problems 25-3

11. A_v at 30 Hz $= -18.1$ dB, A_v at 50 kHz $= -10.6$ dB

Problems 25-4

1. $V_C = 34.7$ V, $V_R = 65.3$ V **3.** 2.47 ms **5.** 145 μs **7.** 5.41 ms **9.** 9.76 kΩ **11.** 9.53 nF

CHAPTER 26

Problems 26-2

1. $A\overline{B} + \overline{B}C$ **3.** $\overline{\overline{X}Y}Z$ **5.** $\overline{\overline{A}\,\overline{B}} + C$ **7.** $(A + B)(\overline{A + C})D$ **9.** $Z(X + Y)$ **11.** $\overline{\overline{AB}(A + B)}$
13. $\overline{X + \overline{Y}} \cdot \overline{\overline{Y} + Z}$ **15.** $\overline{\overline{A + B}(B + C)(C + D)}$ **17.** $\overline{XY + \overline{YZ}}$ **19.** $\overline{A\overline{B} + \overline{B}C}$

INDEX

A

B

C

D

E

F

G

H

I

J

K

L

M

N

O

P

Q

R

S

T

U

V

W

X

Y